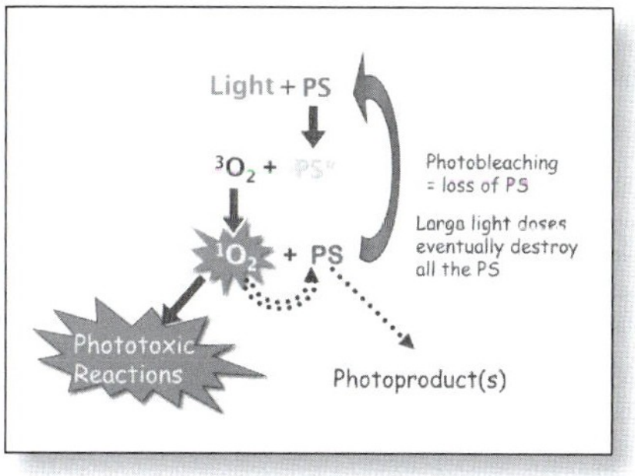

HANDBOOK OF PHOTODYNAMIC THERAPY

Updates on Recent Applications of Porphyrin-Based Compounds

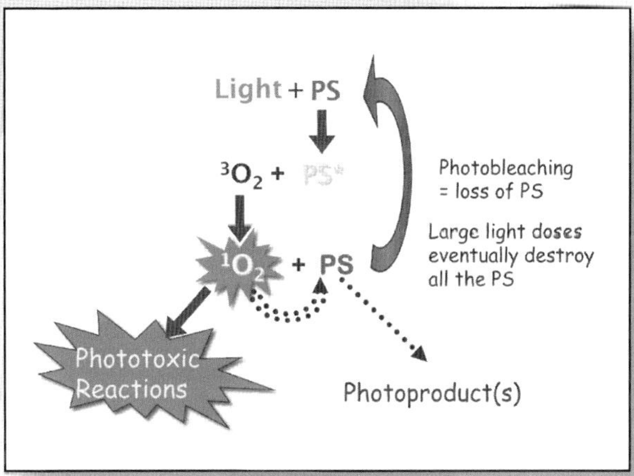

HANDBOOK OF PHOTODYNAMIC THERAPY

Updates on Recent Applications of Porphyrin-Based Compounds

edited by

Ravindra K. Pandey
Roswell Park Cancer Institute, USA

David Kessel
Wayne State University, USA

Thomas J. Dougherty
Roswell Park Cancer Institute, USA

World Scientific

NEW JERSEY • LONDON • SINGAPORE • BEIJING • SHANGHAI • HONG KONG • TAIPEI • CHENNAI • TOKYO

Published by

World Scientific Publishing Co. Pte. Ltd.
5 Toh Tuck Link, Singapore 596224
USA office: 27 Warren Street, Suite 401-402, Hackensack, NJ 07601
UK office: 57 Shelton Street, Covent Garden, London WC2H 9HE

Library of Congress Cataloging-in-Publication Data
Names: Pandey, Ravindra K., editor. | Kessel, David, editor. | Dougherty, Thomas J. (Thomas John), 1933–
Title: Handbook of photodynamic therapy : updates on recent applications of porphyrin-based compounds / [edited by] Ravindra K. Pandey, Roswell Park Cancer Institute, USA, David Kessel, Wayne State University, USA, Thomas J. Dougherty, Roswell Park Cancer Institute, USA.
Description: New Jersey : World Scientific, 2016.
Identifiers: LCCN 2016003721 | ISBN 9789814719643 (hardcover : alk. paper)
Subjects: LCSH: Cancer--Photochemotherapy.
Classification: LCC RC271.P43 H36 2016 | DDC 616.99/40631--dc23
LC record available at http://lccn.loc.gov/2016003721

British Library Cataloguing-in-Publication Data
A catalogue record for this book is available from the British Library.

Typeset by Stallion Press
Email: enquiries@stallionpress.com

Printed in Singapore

Contents

Preface

Photodynamic therapy (PDT), a relatively new modality for the treatment of variety of oncological diseases, is based on the preferential accumulation of photosensitizing molecules (photosensitizers) by target tissues. Upon activation with light at a wavelength corresponding to an absorbance band, the photosensitizer produces reactive oxygen species which damage both malignant cells, their neovasculature and also initiate antitumor inflammatory and immune responses. Most of the photosensitizers either approved for commercialization or undergoing various stages of clinical trials are related to tetrapyrrolic system (porphyrins, chlorins, bacteriochlorins, corroles and phthalocyanines) or the extended ring systems (texaphyrin). This is due to their desired photophysical properties, ability to generate reactive oxygen species and their tumor-avidity. The latter property also depends on overall lipophilicity and the nature of the substituents present at peripheral position(s) of the macrocyclic rings. However, several non-porphyrin based compounds have also been reported to be effective photosensitizers. The clinical use of PDT has been extended to various non-oncological applications such as cardiovascular, dermatological, microbial and ophthalmic diseases. A large number of porphyrin-based compounds are also under investigation for cancer imaging (magnetic resonance, nuclear and optical) and PDT. The use of certain nanoparticles to develop target-specific multifunctional platforms (Theranostics) for cancer-imaging and/or image-guided therapy is also being explored, and promising findings are reported.

This book is intended to summarize recent advances in various aspects of photodynamic therapy and is divided in three parts: Part 1 is mainly focused of the design of improved PDT agents related to chlorophyll-a and bacteriochlorophyll-a based analogs, peptide conjugates with enhanced target specificity, improved phthalocyanine-based agents for cancer-imaging and therapy and porphyrins containing heteroatoms. It also illustrates a process for the design of nanoparticles as efficient drug delivery vehicles for cancer-imaging and therapeutic agents. This chapter also indicates the importance of light dosimetry in optimizing the PDT response. Part 2 involves mechanistic aspects of PDT, e.g., mode of cell death, vascular effect of PDT, possibilities for combination therapy, the PDT mechanism of non-porphyrin based compounds related to the photosensitizer ETNBS, and combination therapy for ophthalmology applications. Part 3 covers the clinical use of photosensitizers in oncological and non-oncological applications. Only those indications were selected where PDT has shown significant advantages over other treatment modalities. This chapter also provides a fascinating example of Photofrin development from bench to bedside, and great insights for developing new agents.

The editors are highly grateful to the authors, for their excellent contributions and their willingness to make compromises in view of space limitations. The editors take the responsibility for the mistakes, whereas the credit for the hopeful success of the book goes to chapter authors. We sincerely thank Sandhya Devi and the other members of the editorial staff of World Scientific Publishers for their help in preparing this monograph.

Ravindra K Pandey, David Kessel and Thomas J Dougherty

Part 1

Chapter 1

Multifunctional Agents for Cancer-Imaging and Photodynamic Therapy: Impact of Polyacrylamide-Based Nanoplatforms

Aimee J. Marko[†], *Nayan J. Patel*[†], *Penny Joshi*[†,‡], *Joseph R. Missert*[†] *and Ravindra K. Pandey*[†,*]

[†]*Photodynamic Therapy Center, Cell Stress Biology, Roswell Park Cancer Institute, Buffalo, NY14263*
[‡]*Department of Chemistry, Kumaun University, Nainital, India*

Abstract

The work summarized in this chapter is focused on the chemical constituents of Photofrin, its limitations and the strategy to take the advantage of porphyrins and reduced porphyrin-based compounds in developing improved imaging and image-guided dual function agents. Advantages of polyacrylamide (PAA)-based nanoparticles (NPs) for enhancing the tumor-avidity are also discussed. Such an approach under proper strategies could help in developing other targeted multifunctional, biodegradable, biocompatible and non-toxic NPs for cancer-imaging and therapy, which is otherwise difficult to achieve.

1. Introduction

A. Brief History of Photodynamic Therapy

Photodynamic therapy (PDT), sometimes called photochemotherapy, is a form of therapy that uses non-toxic light-sensitive compounds which upon exposure to a specific wavelength of light become toxic to targeted malignant and other diseased cells.[1] PDT has also shown the ability to kill viruses, microbial cells and bacteria.[2] It is used clinically to treat a wide range of medical conditions,

*Corresponding author

including treatment of acne, wet age-related macular degeneration (AMD)[3] and malignant cancers,[4] and is recognized as a treatment strategy which is both minimally invasive and minimally toxic.

While the applicability and potential of PDT has been known for over a hundred years, the development of modern PDT has been a gradual one, involving scientific progress in the fields of photobiology and cancer biology, as well as the development of modern photonic devices, such as lasers and LEDs.[5] It was John Toth[6] who acknowledged the importance of "photodynamic chemical effect" and supported in setting up several clinical sites in Japan where the term "radiation" had negative connotations.[7] PDT received even greater interest as result of clinical PDT data reported by Thomas Dougherty in early 1980.[8]

The earliest recorded treatments that exploited the photosensitizer and a light source, in this case sunlight, for medical effect can be found in ancient Egyptian and Indian sources.[9] The use of topically applied vegetable and plant substances to produce photoreactions in skin and cause a re-pigmentation of skin lesions, as seen with vitiligo and leukoderma[10] has been known for a long time. The photosensitizing agents used in these ancient therapies have been characterized with modern science as belonging to the psoralen family of chemicals.[11] Psoralens are still in use today in PDT regimes to treat a variety of skin conditions, including vitiligo, psoriasis, neurodermatitis, eczema, cutaneous T-cell lymphoma and lichen ruber planus.

The first detailed scientific evidence that photosensitive synthetic dyes, in combination with a light source and oxygen could have potential therapeutic effect was recognized at the turn of the 20th century in the laboratory of von Tappeiner in Munich, Germany.[12] Historically, this was a time when Germany was leading the world in the industrial synthesis of dyes.

While studying the effects of acridine on paramecia cultures, Oscar Raab,[13] a student of von Tappeiner observed a toxic effect. Raab also observed that light was necessary for the killing of paramecia cultures to take place. Subsequent work in the laboratory of von Tappeiner[14] showed that oxygen was essential for the 'photodynamic action' — a term proposed by von Tappeiner.

With the discovery of photodynamic effects, von Tappeiner and colleagues went on to perform the first PDT trial in patients with skin carcinoma using the photosensitizer eosin.[15] Out of six patients with a facial basal cell carcinoma, treated with a 1% eosin solution and a long-term exposure either to sunlight or to arc-lamp light, four patients showed total tumor resolution and a relapse-free period of 12 months.

It was only much later, when Thomas Dougherty and co-workers at Roswell Park Cancer Institute, Buffalo NY, clinically tested PDT again. In 1978, they published striking results in which they treated 113 cutaneous or subcutaneous malignant tumors and observed a total or partial resolution of 111 tumors.[8]

The active photosensitizer used in the clinical PDT trial by Dougherty was Hematoporphyrin Derivative (HpD),[16] which was first characterized in 1960 by Lipson.[17] In his research, Lipson's objective was to find a diagnostic agent suitable for the detection of tumors in patients. With the discovery of HpD, Lipson went onto pioneer the use of endoscopes and HpD fluorescence to detect tumors.[18] HpD derived from hematoporphyrin is a complex mixture of large number of porphyrin monomers and higher oligomers. As a mixture, it has long been considered suitable agent for tumor photodiagnosis and PDT because cancerous cells exhibit a significantly greater uptake and affinity for porphyrins compared to surrounding tissues. This important observation, which underlies the success of PDT to treat cancers, had been established by a number of scientific researchers prior to the discoveries made by Lipson. In 1924, Policard[19] first revealed the diagnostic capabilities of hematoporphyrin fluorescence when he observed that ultraviolet radiation elicited red fluorescence in the sarcomas of laboratory rats. Policard hypothesized at the time that the fluorescence was associated with endogenous hematoporphyrin accumulation. In 1948, Figge with co-workers[20] showed in laboratory animals that porphyrins exhibit a preferential affinity for rapidly dividing cells, including malignant, embryonic, and regenerative cells, and because of this, they proposed that porphyrins should be used in the treatment of cancer. Subsequently, many scientific authors have repeated the observation that cancerous cells naturally accumulate porphyrins and have proposed a number of mechanisms to explain the reason of porphyrin-based photosensitizers' tumor-avidity.[21]

Purified HpD, under the pharmaceutical name Photofrin, was the first PDT agent approved for clinical use in 1993 to treat a form of bladder cancer in Canada. Over the next decade, PDT received significant international attention and its clinical use was explored for various cancer indications, which lead to its approval by a number of health organizations all over the world (Table 1).

Chemically, Photofrin is a complex mixture, comprising porphyrin monomers, dimers, trimers and higher oligomers joined with ether, ester and/or carbon–carbon linkages.[22–34] The reports from several research groups, e.g., Bonnett *et al.* (UK), Ward *et al.* (Australia), Chen & Kessel *et al.* (USA) helped to characterize Photofrin chemically. However, the synthesis of individual dimeric, trimetic and oligomeric structures in which the hematoporphyrin

Table 1. Historical development of PDT and the utility of porphyrin-based compounds in cancer-imaging and/or therapy.

Year	Development of PDT
3000 BC — 1900	Origin of phototherapy in Egypt, India and Greece.
1900	Acridine orange and Light kills paramecia (Raab).
1903	Topical eosin as a photosensitizer (von Tappeiner).
1907	"Photodynamic" term introduced (von Tappenier).
1913	Hematoporphyrin (HP) sensitizer used on himself (Meyer Betz).
1924	Localization and fluorescence of Porphyrins seen in tumors (Policard).
1940–1960	Porphyrins for tumor detection by fluorescence (Figge).
1955–1956	HpD isolated (Schwartz) and used for the PDT of cancer (Lipson).
1970–1980	HPD and its purified version (Photofrin) helped to revive clinical PDT (Dougherty).
1986–1990	Oncologic Foundation of Buffalo was established by Dougherty (Funded by J & J and Lederle) and the components of Photofrin (porphyrin dimers and higher oligomers joined with ether, ester and carbon-carbon bonds were synthesized and evaluated for photodynamic activity, which helped in approval of Photofrin by the US FDA (Pandey, Smith and Dougherty). Australian group headed by Ward in collaboration with Kessel and Chang (USA) also helped in identification of Photofrin components.
1987–1995	Photofrin development/commercialization by QLT.
2001	Visudyne (benzoporphyrin derivative-BPD) approved for AMD (QLT).
2001	Foscan approved in Europe for head/neck cancer (Scotia/Biolitec).
2010	ALA for skin cancer (DUSA).
Approved PDT agents	Photofrin, Foscan, Visudyne, ALA and its analog.
PSs at advanced clinical stages	HPPH (RPCI/Photolitec), Tookad (Biolitec), Phthalocyanine (PC4, Cleveland).
PDT Agents under advanced stages (under IND submission)	Photobac (787 nm): Fluorescence guided PDT (Photolitec/HISUN). HPPH-CD: Fluorescence-guided surgery with an option of PDT (HISUN/Photolitec). PET-ONCO: Multimodality agent for PET-Imaging with an option of fluorescence-guided PDT (Photolitec/HISUN). Certain chlorin and bacteriochlorin analogs (Europe, Korea, China and Russia).
Current basic research	Targeted photosensitizers, photocleavable multimodality agents, liposomal, polymer and nanoparticle PS-delivery systems and multifunctional nanoparticles.

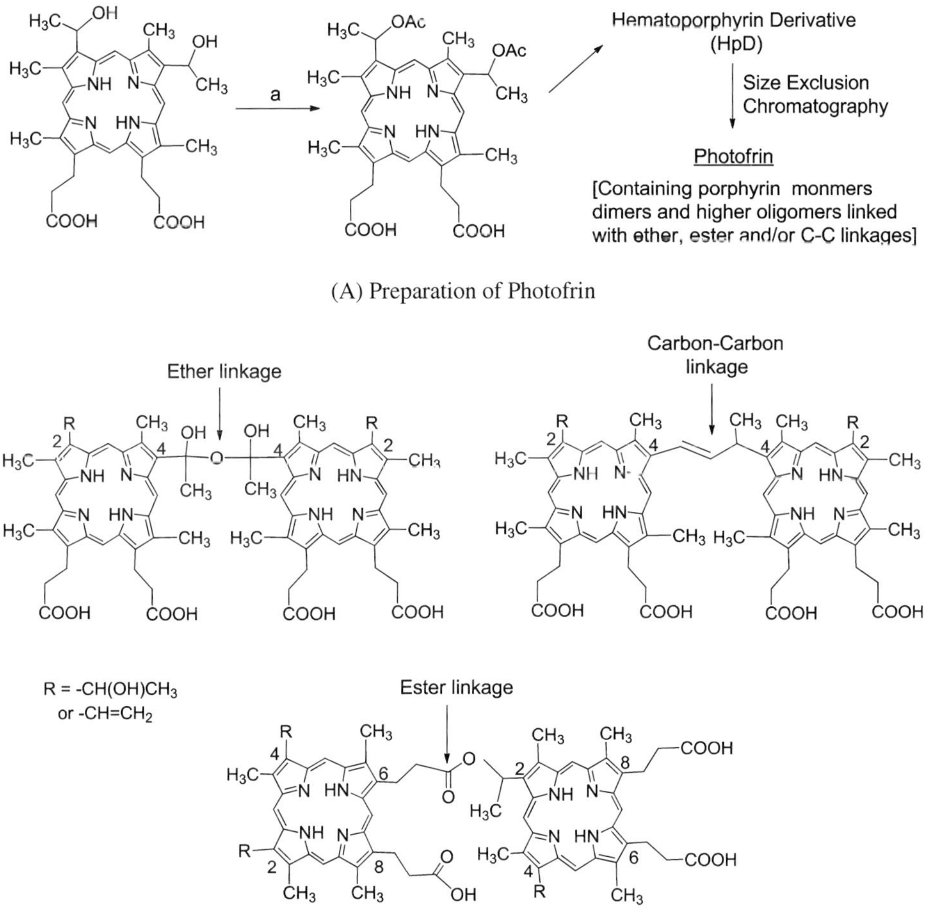

(A) Preparation of Photofrin

(B) Structures of some of the active components of Photofrin

Fig. 1. (A) Typical procedure for the preparation of Photofrin from HpD. (B) Structures of some of the components of Photofrin linked with ether, carbon–carbon and ester linkages. The structures of related components and higher oligomers joined with these linkers at other peripheral positions are not shown. Reagents: a: 5% H_2SO_4/acetic acid, b: 0.1N NaOH.

or its dehydrated monomers were linked with ether, ester or C–C linkages by Pandey *et al.*[32] at Roswell Park Cancer Institute, Buffalo in collaboration with Kevin Smith, University of California, Davis and Lederle group, New Jersey led to the identification of the active components of Photofrin (Fig. 1). Biological studies in mice further revealed that the dimeric and trimeric structures were quite effective with limited toxicity, whereas the higher oligomers present in significant amount in Photofrin tend to stay in the skin for a longer period and are responsible for its skin phototoxicity, which is the major limitation of Photofrin, and is why the patients are advised to be away from direct for sunlight at least 6–8 weeks after the PDT treatment.

B. Basic Concept of PDT and Mechanism of Tumor Destruction

Most modern PDT applications involve three key components:[35] (a) photosensitizer, (b) source of light and (c) tissue oxygen. The combination of these three components leads to the chemical destruction of any tissues which have either selectively taken up the photosensitizer or have been locally exposed to light. The wavelength of the light source needs to be appropriate for exciting the photosensitizer to produce reactive oxygen species. These reactive oxygen species generated through PDT are free radicals (Type I reaction) or a highly reactive state of oxygen known as singlet oxygen (Type II reaction). The mechanisms of Type I and Type II reactions are depicted in a modified Jablonski diagram (Fig. 2).

The photosensitizer should be able to generate a triplet state of appropriate energy (approximately 0.95 eV) which is the minimum energy required to excite the triplet ground state of molecular oxygen ($^3O_2$) to its excited singlet state ($^1O_2$). A photosensitizer could be an efficient PDT agent if the quantum yield of singlet oxygen and/or other reactive oxygen species is high enough (>0.4). That is, at least 40% of excited photosensitizer molecules will create singlet oxygen or reactive oxygen species instead of disbursing energy through

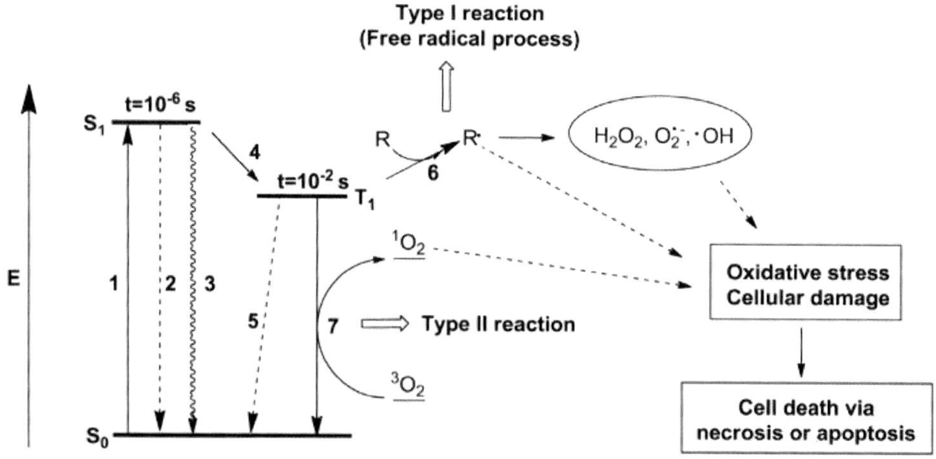

Fig. 2. A modified Jablonski diagram *adapted from* Sternberg *et al.*, Tetrahedron, 54, 4151 (1998). 1 — absorption, 2 — fluorescence, 3 — internal conversion, 4 — intersystem crossing, 5 — phosphorescence, 6 — Formation of free radicals by energy transfer from T_1 of the photosensitizer to biological substrates, 7 — Formation of singlet oxygen by energy transfer from T_1 photosensitizers to triplet oxygen.

fluorescence, phosphorescence or other means. In addition, the longer the life time of the excited photosensitizer's triplet state (>1 ms), the better the interaction with surrounding molecules, resulting in the generation of more cytotoxic species. To understand the mechanism of PDT[36] it is important to distinguish it from other light-based and laser therapies such as laser wound healing and rejuvenation or intense pulsed light hair removal, which do not require a photosensitizer.

There are three major inter-related mechanisms by which PDT mediates tumor destructions: direct cytotoxic effects on tumor cells,[35] damage to tumor vasculature,[36] and induction of an inflammatory immune reaction that can lead to the development of systemic immunity.[37] It is believed that the intracellular localization of the photosensitizer coincides with the primary site of photodamage (singlet oxygen half-life: 0.04 ms, radius of action: 0.02 mm). Most of the effective photosensitizers reported so far localize mainly either in mitochondria, lysosomes and/or endoplasmic reticulum.[38–41] The mitochondrial localizing photosensitizers were initially thought to result in the apoptotic cell death mechanism.[42] However, most photosensitizers show both apoptotic and necrotic paths for tumor destruction, and this can be a function of the PDT dose. During the treatment process, the enzymes that are needed for apoptosis are destroyed and there will be enough cell damage to cause a necrotic result (plasma membrane damaged). It is also not clear if it is the primary, secondary, or both sites of photosensitizer localization that are responsible for tumor destruction. Necrosis is an un-programmed accidental cell death, initiated by cellular damage and characterized by cytoplasm swelling, fragmentation of cell organelles and disruption of the plasma membrane, leading to the release of intracellular contents and inflammation *in vivo*.[43] Recent studies suggest that autophagy may also be initiated by PDT when organelles such as mitochondria and ER are damaged by PDT and survival mechanisms are initiated to remove damaged organelles from cells.[44] Another of the main causes of tumor destruction is reported to be through vascular shutdown limiting the supply of oxygen and nutrients to tumor, leading to tissue hypoxia and tumor cell death. Finally, it has also been shown that PDT activates the immune response, which causes infiltration of immune cells such as lymphocytes, leukocytes and macrophages into treated tissue.[45] The inflammatory cytokines interleukin 6 (IL-6) and interleukin 1 (IL-1), but not tumor necrosis factor-α (TNF-α) have been shown to be up-regulated in response to PDT, and is nicely presented in a review article published by Hamblin *et al.*[46]

Recent work from Roswell Park Cancer Institute with a series of photosensitizers related to HPPH in a co-culture system of primary tumor and normal lung cells[47] has shown that (i) the structure of the photosensitizers determines the type of cellular uptake and retention, (ii) cell-type-specific photosensitizer level, but not subcellular site of photosensitizer accumulation, correlates with the magnitude of photoreaction, cellular signaling and cell death, (iii) epithelial cells show transformation-dependent changes in photosensitizer uptake and retention, (iv) cell type-specific retention of the photosensitizers permits selective killing of cells and finally (v) the cell-type retention of HPPH involves passive and active components. This study also revealed that the level of HPPH retention in certain transformed epithelial cells is strongly influenced by active export processes, correlating with the elevated expression of ABCG2. The ABCG2 transporter effects were confirmed by Morgan *et al.*[48] On investigating the phototoxic response against a series of chlorophyll-a analogs in side population stem cell-like cancer cells treated with PDT. These findings revealed that it is important to verify ABCG2 expression in tumors when choosing the most appropriate photosensitizer for treatment, especially if the photosensitizer is a substrate and ABCG2 positive stem cells are part of tumor initiating population. If the non-stem cells or ABCG2 negative population are treated *in vitro* with a tyrosine kinase inhibitor (TIC), then the substrate specificity and affinity of the photosensitizer may be less important, since it will not be pumped out.

An exciting finding from the Roswell Park group also suggests that photo-induced STAT3 dimerization directly correlates to photosensitizing efficacy, and may be a useful prognostic indicator of treatment efficacy.[49] In this study, murine and human cancer cell lines PDT caused an immediate covalent cross-linking of STAT3 to homodimeric and other complexes.[50–52] The magnitude of this effect was strictly a function of the PDT reaction that is determined by the photosensitizer concentration and light dose. The cross-link reaction of STAT3 was proportional to the subsequent cytotoxic outcome of PDT. An equivalent photoreaction as detected *in vitro* occurred in tumors treated *in situ* with PDT. The light dose-dependent STAT3 cross-linking indicated the relative effectiveness of PDT as a function of the distance of the tissue to the treating laser light source. Absence of cross-links correlated with treatment failure. The data suggest that the relative amount of cross-linked STAT3 predicts the probability for beneficial outcome, whereas absence of cross-links predicts treatment failure. Determination of STAT3 cross-links after PDT might be clinically useful for early assessment of PDT response.

C. Recent Advances in Developing Improved Photosensitizers:

At present several photosensitizers are commercially available for clinical use [e.g., Photofrin, Foscan and aminolevulinic acid (ALA)]. Both Photofrin and Foscan are associated with limited tumor-selectivity and high skin phototoxicity. A large number of 2nd generation PDT agents [e.g., Tookad, Visudyne, HPPH, phthalocuanine (Pc 4)] are at various stages of clinical trials.[53–60] These 2nd generation compounds do not show any significant skin phototoxicity, but have not proven to be "ideal" candidates. For example, Tookad, a near infrared (NIR) photosensitizer has a very short treatment window (drug infusion and treatments are done at the same time) which could create problems in treatment reproducibility. Visudyne also clears from the tumor quickly and shows specificity only for certain tumor types, however, it has shown excellent use in treating macular degeneration. HPPH, a chlorophyll-a analog,[61–66] is undergoing global clinical trials for treatment of head and neck cancer. The Phase I clinical results from Roswell Park Cancer Institute are highly promising. It has also shown great promise in treating esophageal and early-stage lung cancers (Phase I/II clinical trials). But its maximal absorbance in the far red window is at 665 nm *in vivo*. Because this wavelength does not travel as deeply into tissue as light of 800–900 nm, HPPH may not be useful for treating large and deeply seated tumors. However, clinical trials of HPPH-iPDT (interstitial) will be undergoing soon to show its capability of treating deeply seated tumors.

In recent years the development of improved PDT agents has been focused on active and passive approaches, and to translate these approaches to photosensitizers with long wavelength absorption in the range of 700–800 nm.[67] The passive approach has been focused on structure-activity relationship (SAR) studies, a classical approach used in drug development to understand the impact of substitution at various positions in the molecule and overall lipophilicity on PDT efficacy. This approach is suitable and has been useful in selecting an effective agents from a particular series of analogs. So far, most of the PDT agents either approved by the various health organizations or are at various stages of clinical trials have been developed on the basis of SAR/QSAR studies. This approach has been extremely useful in developing improvement PDT agents related to chlorins,[68–73] bacteriochlorins[74,75] and phthalocyanines[76,77] with limited skin phototoxicity.

The active approach of drug development is based on conjugating photosensitizers to certain monoclonal antibodies[78] or small molecules,[79] e.g., peptides,[80–82] carbohydrates,[83–85] etc., known to bind the receptors with overexpression in tumor cells. The monoclonal antibody approach showed excellent *in vitro* targeting in respective receptor positive cells, but *in vivo* limited efficacy was observed. This could be due to the presence of the small amount of the photosensitizer compared to high molecular weight antibodies, and/or due to poor reproducibility of photosensitizer amount in preparations. Conjugating cRGD with a proper linker to HPPH[86] showed drug localization on the cell surface, and higher cell uptake in integrin positive cells, but did not show specificity in an inhibition assay. In other words, incubating the integrin positive cells with cRGD first (100–fold) and then adding HPPH-cRGD conjugate did not inhibit the uptake of the photosensitizer. However, presence of the cRGD moiety in HPPH-cRGD conjugate changes the pharmacokinetic profile of the molecule, and compared to HPPH the HPPH-cRGD showed higher uptake and PDT efficacy at 4 h compared to 24 h shown for HPPH alone.

To explore a different targeting option, a series of carbohydrate-photosensitizer conjugates have been reported in the literature.[87] In an effort to enhance the efficacy and selectivity of photosensitizers currently under development by the Roswell Park group, HPPH was conjugated with mono-, di- or tetrasaccharides with the hope that these conjugates would target carbohydrate binding molecules known to be differentially expressed on the surface of many tumor cells such as carbohydrate binding lectins galectin-1 and -3.[88] Several of these carbohydrate conjugates exhibited superior antitumor activity. This could not be explained by enhanced photosensitizer binding to cellular galectin receptors, therefore this study suggests that perhaps it is not the target-specificity, but the overall all lipophilicity and lack of conjugate export by the ABCG2 transporter that may be the contributing factors responsible for the difference in pharmacokinetic profile and PDT efficacy of the conjugates.

Several mitochondria-localizing photosensitizers are known to target peripheral benzodiazepine receptors (PBR),[89] which has recently been identified as translocator protein (TSPO 18 kDa) by Papadopoulos *et al.*[90] based on its structure and molecular function. TSPO is involved in numerous functions, including a role in steroidogenesis mitochondrial respiration and apoptosis regulation.[91] A number of findings argue in favor of the development of TSPO targeting approaches in the treatment of human cancers, as it is a component of the central regulatory complex of apoptosis.[92] TSPO binding by high affinity ligands enhances apoptosis induction of numerous inducers, moreover TSPO ligands are able to reverse the Bcl-2 cytoprotective effect.[93] In order to find TSPO targeted photosensitizers, PK11195, which binds to TSPO was conjugated with HPPH and the preliminary *in vitro/in vivo* results indicate that the conjugate enhances its target-specificity and the PDT efficacy.[94] However, this study needs further confirmation by repeating this study using larger groups of mice.

Besides overall lipophilicity, the nature of the metal present at the center of a photosensitizer also seems to play an important role in PDT efficacy.[95–97] The impact of metallation of porphyrin-based compounds on PDT efficacy has been nicely reviewed in one of our previous review article.[98] This approach was extended in pyropheophorbide analogs containing In(III) as a central metal. The indium complex was more photostable than the corresponding free-base analog, with higher singlet oxygen producing ability.[99] Interestingly, insertion of indium into the photosensitizer did not alter its intracellular localization properties; both analogs were found to localize in mitochondria. However, the presence of indium made a significant improvement (8–10-fold) in PDT efficacy in tumored mice.

D. Multifunctional agents for cancer-imaging and PDT:

In recent years, medical imaging has become a vital part in the early detection, diagnosis and treatment of cancer.[100–105] In some cases medical imaging is the first step in preventing the spread of cancer through early detection, and in many cases makes it possible to cure or eliminate the cancer altogether. Computed tomography (CT) imaging, Magnetic Resonance (MR) imaging,[106] Mammography, Nuclear Medicine (NM) imaging,[107] ultrasound (US)[108] imaging and X-ray imaging[109] are all very important tools in the fight against cancer. After cancer has been diagnosed, imaging is often used to follow the course of cancer treatment and to monitor growth or remission. Medical imaging is being used more frequently to build precise computer models which allow doctors to guide the radiation treatment of cancer (Fig. 3).

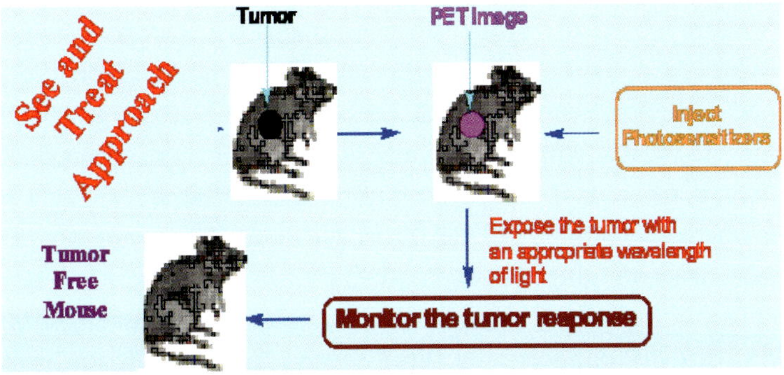

Fig. 3. Concept of a “See and Treat” approach.

All the imaging techniques currently being used have certain advantages and limitations. However this article will focus mainly on the use of porphyrin-based compounds in designing nuclear [positron emission tomography (PET)], MR and fluorescence imaging with an option of PDT.[110]

1. Fluorescence/PET/MRI agents with an option of therapy

In the case of cancer, the detection/diagnosis of the lesion at its early stage is generally attended by long term complete response and cure of the disease. It is well known that detection of early stage (superficial) cancer using standard imaging techniques and/or endoscopic devices with the use of white light is difficult and often unrewarding in the presence of occult cancer.[111] This is because of the low discrimination power of these techniques to distinguish between normal and slightly abnormal tissue. In contrast, spectroscopic and fluorescence imaging provides a better visual expression of early lesions.[112] Fluorescence refers to the reemission of optical radiation, following the absorption of light energy. In molecular systems, the fluorescence consists of an emission band at lower energies (longer wavelengths) than that of the excited wavelength (the absorbed light). This method for cancer detection is based on displaying differential autofluorescence emission between normal and abnormal diseased tissue, on exposure to blue light. However, in comparison with autofluorescence,[113] induced fluorescence of a photosensitizer can exhibit a stronger contrast and, therefore, should have a higher sensitivity with fewer false positive results.

The advantages of fluorescence imaging compared with other modalities are superior sensitivity, extremely low energy radiation, and the ability to monitor multiple independent optical biomarker reporters simultaneously, with relatively simple imaging hardware. Optical techniques provide an accurate and rapid means to detect regions of cancer, but a challenge in optical cancer screening is to locate tumors lying deep within organs. The absorption and emission spectra of most native fluorophores (e.g., tryptophan, collagen, elastin, NADH and flavin etc.) within tissue run from ultraviolet (UV) to visible range. A spectral window in the far-red to the NIR range (650–900 nm) corresponds to low absorption and scattering of light by tissue and allows light to penetrate up to several centimeters into tissue. However, this method is dependent on the selectivity of the fluorophore at the target site. Despite its utility in real time imaging/short acquisition time (sec–min), relatively high spatial resolution, high sensitivity, and multiplexing, fluorescence imaging can be applied externally or internally (endoscopy), it suffers from limited depth penetration and whole-body imaging is not yet possible[114] in humans.

On the other hand, PET provides (i) unlimited depth of penetration, (ii) whole body imaging, (iii) quantitative molecular imaging, and (iv) can be combined with CT or MRI for anatomical information. The disadvantages are (i) radiation exposure, (ii) cost, (iii) low spatial resolution (1–2 mm) and (iv) long acquisition time. PET provides better resolution than fluorescence.

Similar to PET, MRI[115] also provides (i) unlimited depth penetration, (ii) whole body imaging, (iii) quantitative molecular imaging, (iv) can be combined with CT for anatomical information, (v) whole body imaging is possible, (v) no ionizing radiation, but (i) it is expensive, (ii) with a long acquisition time (minute–hours), and (iii) limited sensitivity for detection at lower concentration of the contrast agents. Therefore, PET/optical imaging and/or MRI/optical imaging could be complimentary to each other, and in order not have to control the pharmacokinetic/pharmacodynamics profiles of two different imaging agents, it would be beneficial to develop a single tumor-avid agent, which incorporates dual imaging capability.[116]

The tumor-selectivity of porphyrin derivatives and their photodynamic action subsequently lead to their clinical use as sensitizers for PDT of various cancers. Wrenn *et al.* developed a simple method to prepare ^{64}Cu-tetrasulphophthalocyanine which showed good selectivity for brain injury and experimental brain tumors in animals.[117] A few years later, a series of naturally occurring porphyrins labeled with ^{64}Cu ($t_{1/2}$–12.8 h) were administered to cancer patients in an attempt to visualize neoplasms. Compared to the results obtained with animals, localization in humans was found to be disappointing. However, recently ^{64}Cu-tetracarboxytetraphenyl porphyrin[118] was reported as a potential agent for inflamed lymph node detection. Certain ^{111}In(III) porphyrins in which Indium is sitting at the center of porphyrin core or ^{111}In(DTPA) conjugated with porphyrins and HPPH (a tumor-avid chlorophyll-a analog) have shown potential for imaging cancer by SPECT (single photon emission computed tomography).[119,120] However, these compounds also have high liver-uptake (Fig. 4).

Similar to porphyrins, selected phthalocyanine derivatives exhibit tumor retention in animal models, which led to investigate their utility as carriers of clinically accepted radioisotopes such as ^{99m}Tc or ^{111}In chelated with DPTA and DOTA moieties. Such conjugates showed significant liver and spleen retention in animal models, with minimal tumor uptake.[95]

Certain metalloporphyrins have also been reported as contrast agents for magnetic resonance imaging (MRI).[121] Mn (III) is an interesting metal ion for medical application due to its good MRI contrast enhancing properties; Mn(III) ions do form strong chelates with porphyrins which are not readily

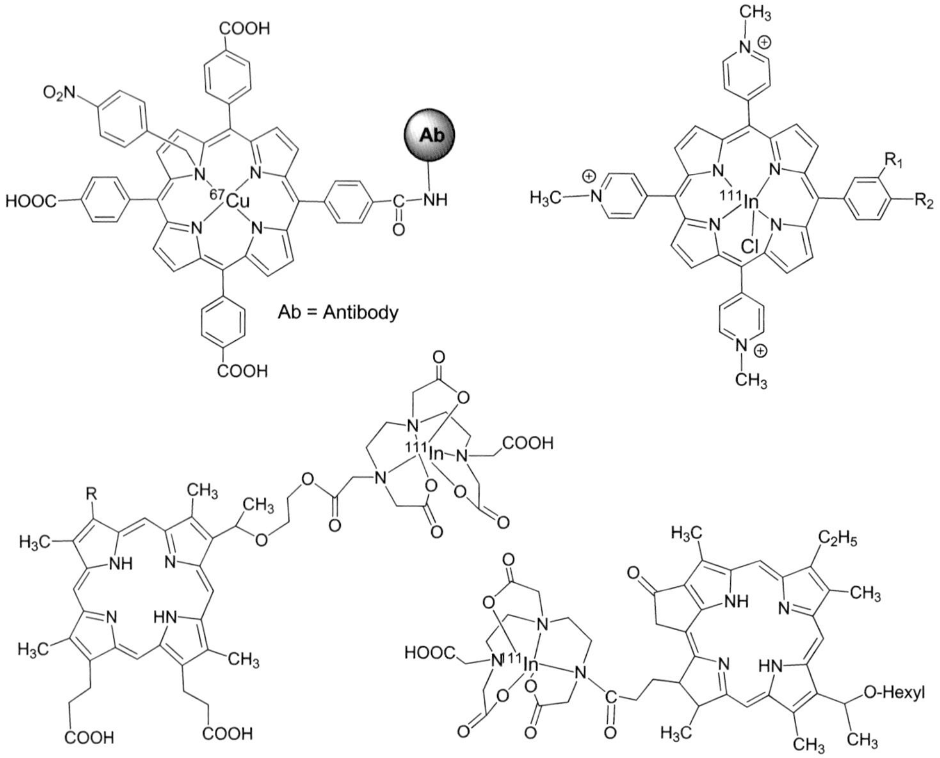

Fig. 4. Porphyrin and chlorin — based nuclear imaging agents. R, R_1 and R_2 = Various substituents.

demetalated *in vivo*. This approach has been used for the preparation of several porphyrin-based MRI agents (Fig. 5). However, their detailed *in vivo* stability is not known, and should be investigated in depth, because, high concentrations of free manganese is severely toxic *in vivo*.

NIR PDT Agents:

After the discovery of Photofrin and realizing its advantages and limitations in treating cancer, one of the objectives of many laboratories, including that of the Roswell Park group has been to develop long-wavelength (660–800 nm) multi-functional agents for cancer imaging (fluorescence/PET or MRI/fluorescence) with an option of PDT.[121–123] and with limited skin phototoxicity. In this regard, a series of tumor-avid photosensitizers (chlorins & bacteriochlorins) with long wavelength absorption in the range of 660–800 nm were synthesized and evaluated for imaging and PDT efficacy. Among the photosensitizers developed was HPPH, one of the chlorine analogs derived from chlorophyll-a, which has successfully completed Phase I clinical trials of oral, lung, esophageal

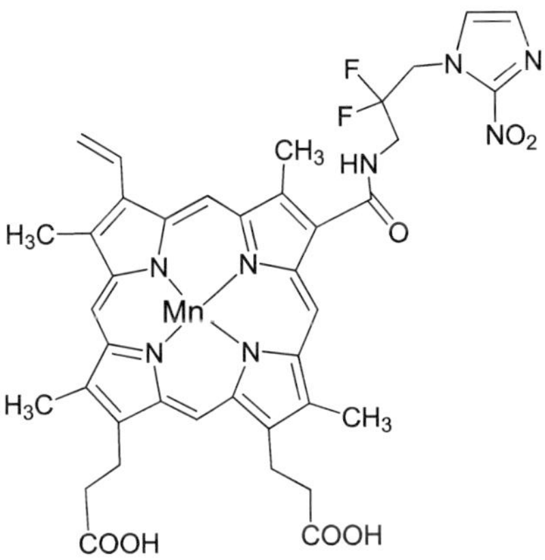

Fig. 5. Mn(III) porphyrin for tumor-imaging by MR.

cancers, and will be undergoing Phase II clinical trials for oral cancer in the United State, India and China. Another chlorophyll-a derivative, an iodobenxyloxy ethyl analog of pyropheophorbide a, shows excellent PDT efficacy in various tumor time at 24 h post-injection.[122] The corresponding ^{124}I-analog (PET-ONCO) is being developed by Photolitec, a spin-off of Roswell in collaboration with a Chinese company (Hisun Pharma) is at an advanced stage of preclinical development for PET imaging. The bacteriopurpurinimide analog Photobac[123] with long wavelength absorption near 800 nm, is also being developed by Photolitec for fluorescence guided therapy of cancer. Finally, the HPPH-3Gd(III)DTPA conjugate[124] (Fig. 6) is also in advanced stages of preclinical studies, and the *in vivo* imaging (fluorescence/MRI) and phototherapy results in small animals are quite promising.

2. Nanotechnology and Medicine: Advantages and Limitations

Despite rapid advances in diagnostic procedures and treatments, the overall survival rate from cancer has not improved substantially. Therefore efforts are currently underway in many laboratories all over the world to use various approaches for the development of novel pharmaceuticals for use in cancer-imaging and therapy. One area of research that holds potential of providing great benefits for society in the future is in medicinal uses of nanotechnology.[125,126] Nanotechnology is already being used as the basis for new, more effective drug delivery systems (Fig. 7). The National Cancer Institute has created the Alliance for Nanotechnology in Cancer in the hope that investment in nanomedicine could lead to new and more effective agents for detecting,

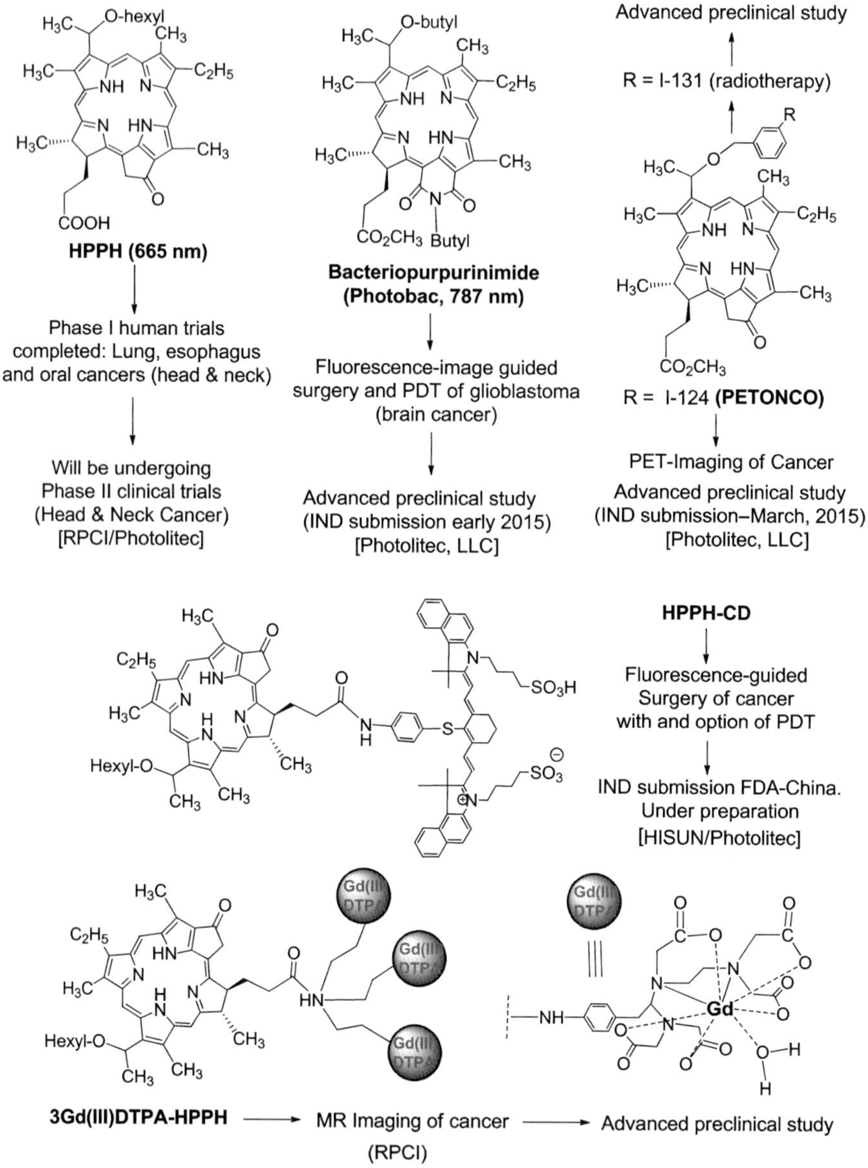

Fig. 6. Cancer-imaging and therapeutic agents under human clinical and advanced preclinical trials, developed at the PDT Center, Roswell Park Cancer Institute, Buffalo.

diagnosing and treating various forms of cancer. The passive targeting approach of nanoparticles, however, suffers from several limitations. Targeting cancer cells using the enhanced permeability and retention (EPR) effect is not feasible in all tumors because the degree of vascularization and porosity of tumor vessels vary with the tumor type. One approach in overcoming these problems

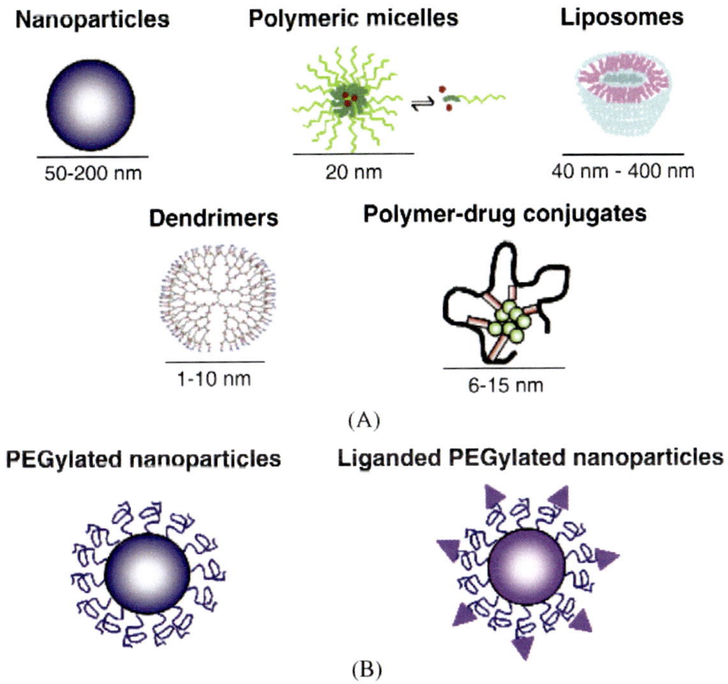

Fig. 7. Nanomedicine in drug delivery. (A) Types of nanocarriers currently described in preclinical and clinical studies. (B) Schematic representation of pegylation and ligand grafting (*J. Control. Release* **148**, 135–146 (2010)).

to achieve tumor-specificity is to introduce desired targeting moieties at the surface of the nanoparticles with *in vivo* cleavable linkers. The multifunctional properties of nanoparticles present several advantages for the cancer-specific delivery of imaging and therapeutic agents. A variety of ligands with therapeutic or diagnostic capability can be introduced at the nanoparticles' large surface area in a single nanoparticle system.

Nanocarriers are also of sufficient size that they may be taken up by the reticuloendothelial system (RES) instead of accumulating in tumors. The common method to protect nanocarriers from the RES consists of coating the surface of the particles with polyethylene glycol (PEG), a procedure called pegylation.[127] There are certain factors which determine the "stealth" characteristics of PEGylated nanoparticles, which are as follows: (a) the molecular weight of the PEG chain, (b) the surface chain density and (c) the conformation. The coating of PEG chains at the surface of the nanoparticles results in an increase in the blood circulation half-life by several orders of magnitude. By creating a hydrophilic layer around the nanoparticles, steric repulsive forces repel the attachment of opsonin proteins, thereby blocking the opsonization process.

One of the major concerns of using nanoparticles in medicine has been their unknown health impact in humans.[128] As nanotechnology is an emerging field, it becomes extremely important to determine that at what extent the material being used for developing a particular nanoconstruct will benefit or pose risks to human health. The extremely small size of nanomaterials means that they are much more readily disbursed in the human body than larger sized particles. How these nanoparticles behave inside the organism is one of the significant issues, which needs to be resolved. Therefore before further modification of any nanoparticle, the long term *in vivo* toxicity of each type of nanoparticle should be determined in small animal models. Another factor that is equally important is the reproducibility of the nanoparticle preparations should be confirmed on a regular basis by simple analytical methods.

A. PAA-based Nanoparticles:

The PAA-based nanotechnology approach has been investigated to improve the delivery and tumor-specificity of hydrophobic photosensitizers and imaging agents.[129] These nanoparticles are promising in that (i) their hydrophilicity and charge can be altered, (ii) they possess enormous surface area that can be modified with functional groups possessing a diverse array of chemical and biochemical properties, including tumor selective ligands, (iii) owing to their subcellular and sub-micron size, they can penetrate deep into tissues and are generally taken up efficiently by cells, and (iv) photosensitizer-loaded nanoparticles can be made by numerous methods, such as encapsulation,[130] conjugation[131] and post-loading[132] depending on the desired characteristics. For example, a fluorescent tag may be covalently conjugated while a porphyrin analog could be postloaded in order to trace both nanoparticle location and drug delivery disbursement area. In recent years a variety of nanoparticle and liposomal formulations have been developed by several research groups, and have been the subject of several recent articles[133] and reviews.[134–136] Therapeutic examples include nanoparticles containing PDT agents, folate receptor targeted nanoparticles, boron containing dendrimers for neutron capture therapy, and NP-directed thermal therapy. When you say “NP” do you mean nanoparticle? Because this is the first time I’m seeing the short form of the word in the paper. Diagnostically, nanoparticle formulations have been used for imaging with multiple modalities, notably MR, PET and fluorescence.[137] Kopelman’s group has developed nanoplatforms called PEBBLEs consisting of a PAA matrix surrounding a fluorophore and specific chromoionophores for subcellular, real-time measurements of primary biochemical and physiological processes, including the level of oxygenation.[138] The development of sensors for non-invasive determination of oxygen levels in cells and tissue is critical for understanding cellular functions, as well as for monitoring the status of disease, and for predicting the efficacy of therapy. The Kopelman group[139] has reported such a non-toxic targeted oxygen nanosensor made of PAA hydrogel, NIR luminescent dyes, and surface conjugated tumor specific peptides, which enabled non-invasive real-time monitoring of oxygen levels in living cancer cells under normal and hypoxic conditions. The required sensitivity, brightness, selectivity and stability were achieved by tailoring the interaction between the nanomatrix and indicator dyes.

Kopelman's group[140] has also shown that PAA-based hydrogel nanoparticles with covalently linked coomassie blue could help in delineation of tumor margins during brain surgery. In this approach, a tumor-targeting deep-blue nanoparticle-based visible contrast agent showed tumor-specific visible color staining. This technology thus enables color-guided tumor resection in real time, with no need for extra equipment or special lighting conditions.

Boyle and co-workers[141] have also reported the utility of certain porphyrin–nanosensor conjugates as new tools for the measurement of intracellular response to reactive oxygen species. The sensing component was synthesized by the formation of reverse micelles, with the inner micelle core containing the polymeric components.

Besides porphyrins, methylene blue-conjugated hydrogel nanoparticles were prepared through a microemulsion polymerization, after conjugation of the dye with a monomer.[142] The nanoparticles have a 50–60 nm diameter in solution. This conjugation method enables a large increase in loading of methylene blue per nanoparticle and also minimizes dye leaching out of the nanoparticle. Furthermore, it has been shown that the dye content can be controlled by variation of the dye amount, enabling a more refined control of the singlet oxygen producing ability. The nanoparticles are coated with F3 peptides[143] which gave specific targeting in selected tumor cells (9L, MDA-MB-435 and F98) *in vitro*. MTT cytotoxicity assay revealed that PAA nanoparticles have no dark toxicity but excellent PDT efficacy increasing with the nanoparticle dose and irradiation time. Tissue distribution and phramacokinetic profiles of both pegylated and non-pegylated PAA nanoparticles following intravenous injection in rats have shown that these nanoparticles are not associated with acute toxicity, as measured by typical serum markers of inflammation and cellular damage.[144]

For the last ten years, Roswell Park group in collaboration with Prof. Kopelman, University of Michigan, Prof. Prasad, University of Buffalo, Prof. Kim, Pohang University of Science and Technology in the Republic of Korea and Prof. Chen, University of Arkansas has been investigating the use of various types of nanoparticles (e.g., ORMOSIL,[145–147] gold,[148] and PAA) and nanocrystals.[149] Among the nanoparticles investigated so far, the PAA have the advantages of (i) a relatively large knowledge base for cancer imaging, PDT, chemical sensing, stability and biodegradation, (ii) no known *in vivo* toxicity, (iii) long plasma circulation time without surface modification, but with biodegradation and bioelimination rates controllable via the type and amount of selective cross-linking (introduced during polymerization inside reverse micelles), (iv) the ability to scale-up to 400 g material has been demonstrated, as well as (v) storage stability over extended periods. Limitations include the leaching of small hydrophilic components unless they are anchored and unknown limitation on bulk tumor permeability because of hydrogel swelling.

1. Synthetic Approaches

The NP preparations were characterized by NMR, the size was determined by transmission/scanning electron microscopy (TEM/SEM). Dynamic light scattering (DLS) was used to determine hydrodynamic diameter and surface charge. Repeated washes with ethanol helped in removing the undesired monomers and surfactants from the NPs. To incorporate photosensitizers into nanoparticles for cancer treatment by photodynamic therapy, in an initial study at RPCI and Univ. of Michigan HPPH was incorporated into non-toxic biodegradable amine-functionalized polyacrylamide (AFPAA) using three different methods[150] (i) encapsulation, (ii) conjugation and (iii) post-loading. HPPH or HPPH-linked acrylamide was added into a monomer mixture and polymerized in a microemulsion for encapsulation or conjugation. For post-loading, HPPH was added to an aqueous suspension of pre-formed nanoparticles (Fig. 8). The free HPPH was removed by spin filtration, before characterization, stability, dye leaching, particle size, singlet oxygen production, dark toxicity, whole-body fluorescence imaging and *in vivo* PDT efficacy were examined. The tumor uptake and *in vivo* PDT data were compared with the photosensitizer HPPH in 1.0% Tween80/D5W formulation.

Further studies indicated that HPPH was successfully encapsulated, conjugated or post-loaded into AFPAA nanoparticles. The resultant nanoparticles had a mean diameter of 27–32 nm as determined by DLS and SEM.

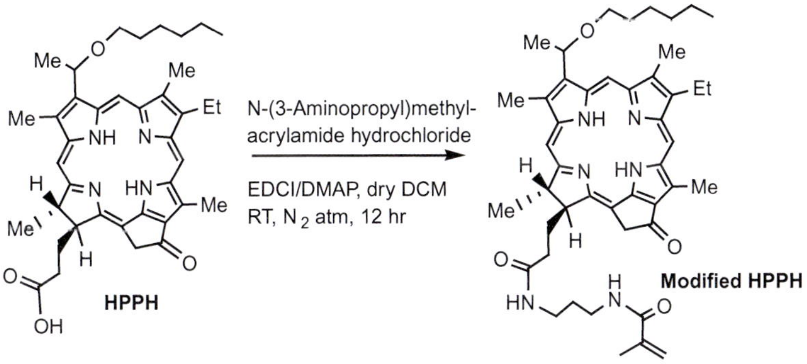

(A) Preparation of modified HPPH

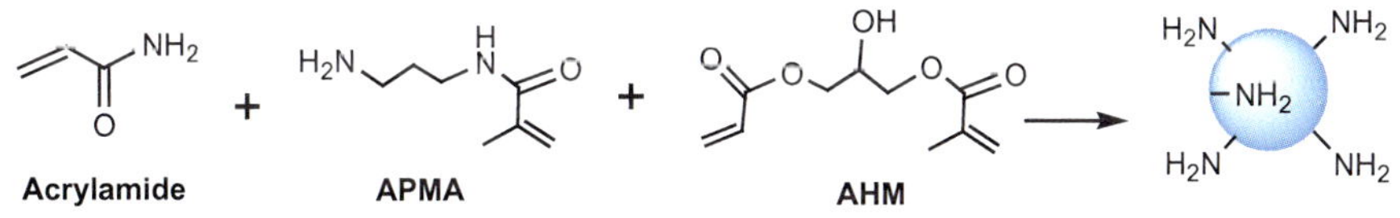

(B) Preparation of PAA nanoparticles

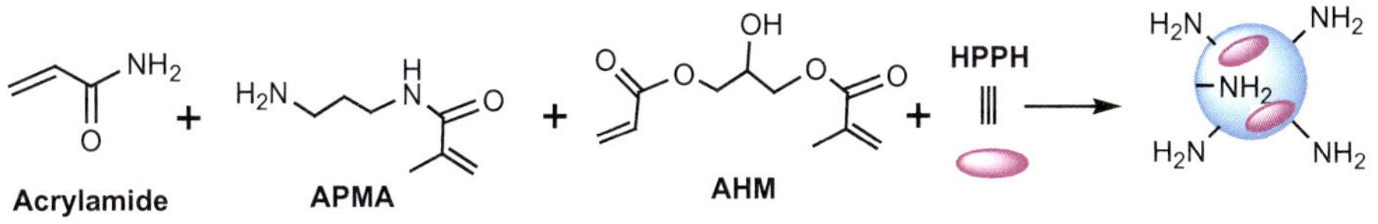

(C) Preparation of HPPH encapsulated PAA Nanoparticles

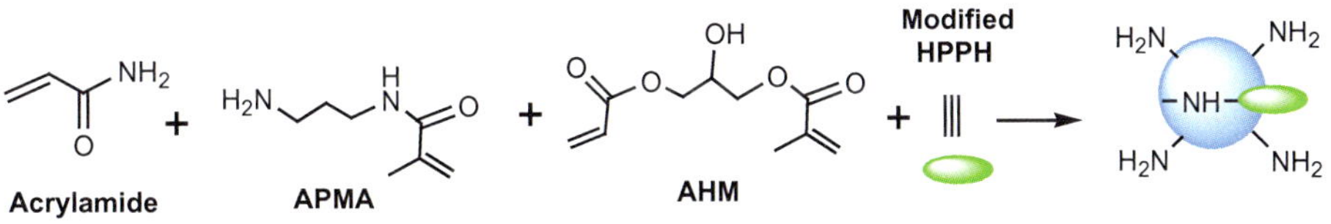

(D) Preparation of HPPH - PAA NP conjugate:

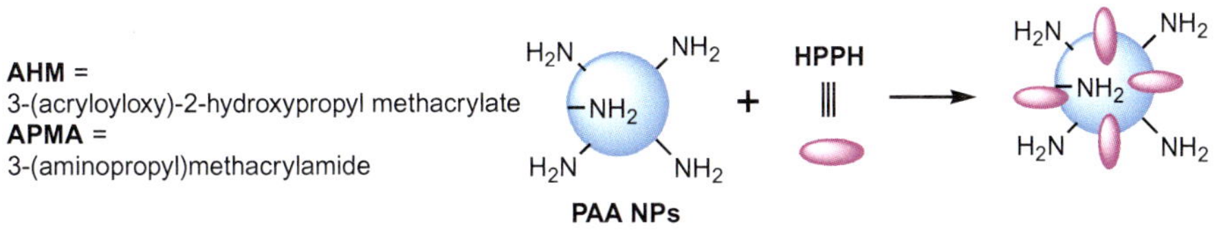

(E) Preparation of HPPH post-loaded PAA Nanoparticles

Fig. 8. Preparation of modified HPPH, blank NPs and the corresponding nanoformulations by using encapsulation, conjugation and post loading of the desired photosensitizers.

The HPPH remained intact after entrapment, and the HPPH leaching out of nanoparticles was negligible in all three formulations. The highest singlet oxygen was achieved by the post-loaded formulation, which also showed the highest phototoxicity. The lower photosensitizing activity (lower cell kill) of HPPH encapsulated and conjugated in nanoparticles could be due to aggregation and/or the close proximity of the photosensitizer molecules allowing energy transfer that caused reduction of singlet oxygen production and reduced phototoxicity.

On the basis of advantages of PAA nanoparticles, a large variety of corresponding nanoconstructs were developed by the Roswell Park and University of Michigan groups on incorporating effective photosensitizers and imaging agents in a single platform. Their comparative *in vitro/in vivo* cancer imaging and therapy were investigated in cell culture and small animals. The methodologies used for preparing such nanoconstructs and their comparative biological data with respective photosensitizers and imaging agents are briefly discussed.

B. Multifunctional Nanoparticles for Cancer Imaging and Therapy

1. NIR Fluorescence Imaging and PDT

In an ideal situation, the surgeon would prefer a tumor resection with continuous guidance from high-contrast fluorescence observed directly in the resection cavity emanating from the tumor cells themselves. The penetration of light through tissue increases as its wavelength increases from 630 to 800 nm. The intrinsic absorption of most tissues is dominated by hemoglobin, with strong absorption band peak at wavelengths shorter than 630 nm. The tails of these bands extend above 630 nm but grow weaker with increasing wavelength. Thus, the probability of a photon being absorbed by endogenous chromophores decreases with increasing wavelength from 630 to 800 nm with a concomitant decrease in scattering,[151,152] which also increases tissue penetration. The other characteristic which is important is a significant Stokes shift, or red shift, in wavelength wider than 20 nm between the longest wavelength absorption and emission bands of the fluorophore. Most of the porphyrin-based compounds (e.g., porphyrins, chlorins and bacteriochlorins) generally show intense fluorescence, but have a small Stokes shift in the long-wavelength region.[153] For example, HPPH is a highly effective PDT agent with significant fluorescence quantum yield. Excitation of HPPH at 665 nm *in vivo* exhibits two emission bands at 670 nm (strong) and 720 nm (weak). Due to a strong overlap between the broad absorption (665 nm) and fluorescence (670 nm), there are two options for using fluorescence image guided PDT with HPPH: (i) either excite the molecule at a lower wavelength and take the advantage of strong emission at 670 nm or (ii) excite the molecule at 665 nm and image the tumor at the weak emission (720 nm) of HPPH, which is not very efficient.

In contrast to porphyrin or reduced porphyrin systems, the heterocyclic polymethine dyes (cyanine dyes or CDs) have created enormous interest due to their desirable photophysical properties (long wavelength absorption and fluorescence, large Stokes shift, and high extinction coefficient.[154] Only a few organic dye molecules have been approved by the FDA for use in humans and one of these compounds belongs to a class of cyanine dyes (CDs). Unfortunately, most of the CDs related to indocyanine green (ICG),[155] an FDA approved agent, do not retain in tumors for a significant period of time, which diminishes their utility in imaging and monitoring the tumor response. There are several approaches reported in the literature to enhance the tumor-specificity of this class of compounds. The Roswell Park group showed for the first time the utility of photosensitizer-CD conjugate as a bifunctional agent by covalently linking an amine functionalized tumor-avid photosensitizer (e.g., HPPH and purpurinimide) to a CD bearing benzoic acid functionality.[156,157] Such compounds showed excellent fluorescence imaging ability *in vivo* in various tumor models at a very low dose.[158] These conjugates showed significant PDT efficacy also. However, the PDT drug dose was higher than the imaging dose, because a portion of the singlet oxygen produced by exciting the photosensitizer was quenched by the CD. Thus, less singlet oxygen was available for the destruction of tumor, causing reduced long-term tumor response at low drug doses.

In order to solve this problem, the Roswell Park group used both encapsulating and post-loading approaches to add therapeutic and imaging agents to PAA nanoparticles, and a recent report reveals that compared with encapsulation, the post-loading approach was more effective when hydrophobic compounds were used in conjunction with PAA nanoparticles. Due to its porous nature, it provides a great platform for post-loading, wherein the hydrophobic part of the molecule may interact preferentially. This approach provides an opportunity to post-load the therapeutic and imaging agents at the desired ratio of each component. It was shown that the concentration of HPPH and the amino-functionalized CD in a ratio of 2 to 1 minimized[159] the undesirable quenching of the dye, which afforded excellent tumor-imaging (NIR fluorescence) and good phototherapeutic efficacy of the nanoconstruct formulation without diminishing the PDT efficacy of HPPH alone (Fig. 9).

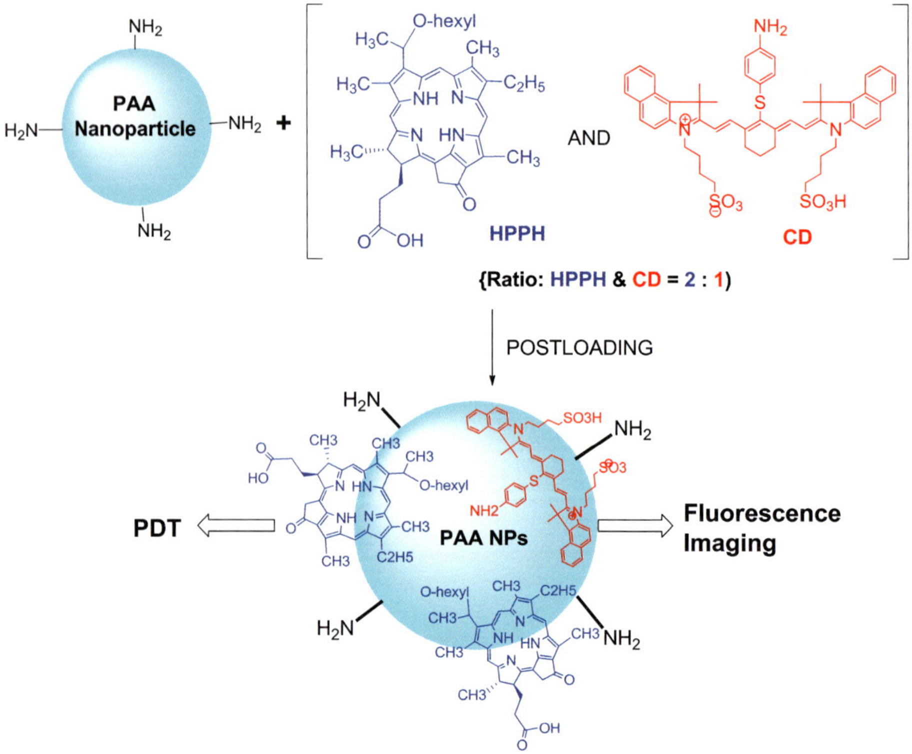

Fig. 9. Post-loading approach for the preparation of nanoconstructs in fluorescence-imaging and PDT of cancer, a "See and Treat" approach.

This approach certainly illustrates the advantage of nanoparticles in developing multifunctional agents over the classical synthetic approach of drug design.

2. PET/fluorescence imaging with an option of PDT

Multifunctional nano-platforms that combine both imaging and therapy, while improving selective uptake in tumors, have been of interest for the treatment of cancer, especially when the platforms are non-toxic, biodegradable and bio-eliminable.[160–162] Such a safe multimodal and multifunctional nanoformulation, designed by Gupta *et al.* combines optical and PET imaging, with an option of PDT. It shows favorable pharmacokinetics and selective tumor uptake in mice bearing Colon 26 tumors.

To investigate the utility of porphyrin-based compounds in PET/fluore scence imaging and photodynamic therapy, the Roswell Park group synthesized and evaluated a series of ^{124}I-labeled compounds related to pyropheophorbide-*a* and purpurinimides derived from chlorophyll-*a* (Fig. 10). Among the

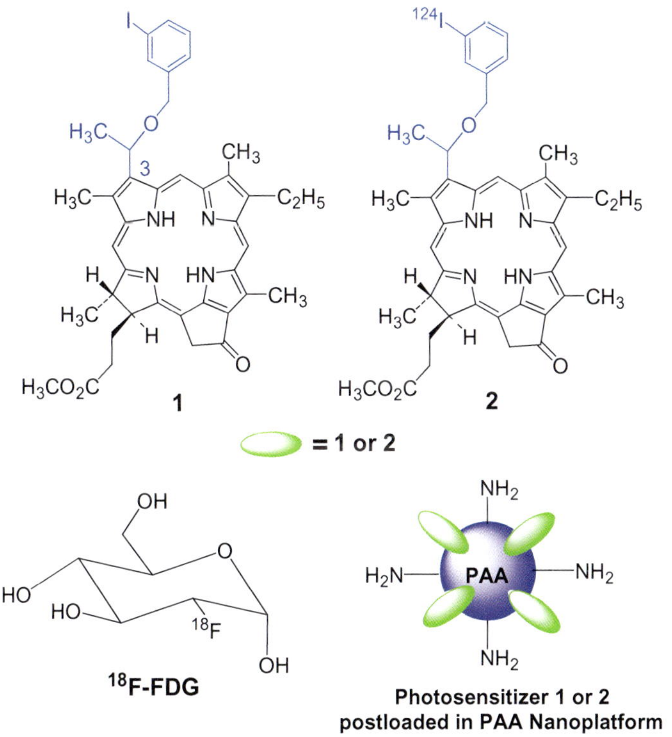

Fig. 10. Structures of photosensitizer 1 (PS 1) the corresponding ^{124}I- labeled analog 2, ^{18}F-fluoro-dexoyglucose (FDG) and a schematic representation of photosensitizer (PS1 or PS2) post-loaded in polyacrylamide nanoparticles (PAANPs).

compounds evaluated, methyl 3-(*m*-iodobenzyloxy)ethyl-3-devinyl-pyropheo-pheophorbide-*a*^{127}I-(non-radioactive) and its (^{124}I, radioactive analog) showed the highest potential for both PET/fluorescence imaging with an option of PDT in mice bearing Colon26 and other tumors (U87, RIF, lung, 4T1 and Panc-1). A comparative study indicated that in some of the indications (e.g., brain, colon, pancreatic, prostate), where^{18}F-FDG showed limited potential, the ^{124}I- agent showed excellent PET imaging ability and PDT efficacy. However, it suffers from two main drawbacks: (i) though the compound showed significant tumor-avidity and 4–5% of the injected dose accumulated in the tumor 24 h post injection, a significant uptake was also observed in the spleen and liver and (ii) the Stokes shift between the absorption (665 nm) and fluorescence (670 nm) wavelengths is small. However, an additional band (weak) is present at 710 nm, which can be used for fluorescence imaging. To investigate the utility of this compound in a nanoplatform, biodegradable amine-fun ctionalized PAA NPs (25–30 nm) were post-loaded with ^{124}I-labeled PS2 in high concentration. The loading efficiency of the PS was determined by measuring the radioactivity present in the NP formulation after centrifuge filtering the formulation. This approach was repeated several times and an average >99% post-loading efficiency was observed, based on the radioactivity present pre and post-centrifuge filtering. Changes in the absorbance and fluorescence spectra proved to be very informative in assessing the post-loading of PS into the PAA NPs. Measuring the absorption and fluorescence spectra of the filtrate throughout the post-loading process established the parameters to optimize the post-loading of the ^{124}I- labeled and/or non-labeled agents in nanoparticles. Upon completion of the postloading process (120 min), the optical absorption across the whole electronic absorption spectra of PS1 post-loaded to NPs (NP1) at variable concentrations were obtained. In the nanoparticle formulation, absorbance peaks were found to be sharper with a higher intensity than the photosensitizer alone indicating less aggregation, which was further confirmed by measuring the fluorescence intensity in 17% BSA (bovine serum albumin); the more disaggregated the formulation, the better it interacts with light, and the more fluorescence is produced. The aggregation/disaggregation properties of the PS can be observed in the electronic absorption and fluorescence spectra and were found to be extremely useful in monitoring the progress of post-loading of the PS in PAA NPs. The measurement of absorption and fluorescence spectra of the PS post-loaded in the NP and in filtrates at various time intervals provide great advantage in monitoring the amount of the PS released and retained in NPs.[162] Thus, this approach along with DLS and TEM images were extremely helpful in establishing the reproducibility of the NPs. The PET images in Fig. 11, clearly illustrate that the post-loading formulation of ^{124}I- chlorophyll-a analog, demarcated tumor at 24, 48 and 72 h post-injection.

A comparative biodistribution study of PS2, NP2 and ^{18}F-FDG in BALB/c mice bearing Colon26 tumors (Fig. 12) showed that by post-loading PS2 to PAA NPs, the percent injected dose/gram (%ID/gram) in the tumor at 24 h post-injection significantly increased from an average value of 4.61% to 10.28% ($P = 0.008$). The nanoparticles (NP2) showed a remarkable decrease in accumulation of PS2 in the spleen and liver and significantly less in the heart and muscle as compared to ^{18}F-FDG alone. The %ID/gram of PS2 present in the spleen and liver at 24 h post-injection was 15.51 and 9.32%, respectively. With the nanoparticle formulation NP2, the amount decreased to 2.02 and 3.99% and 1.61 and 2.91% for the 1.4 and 0.5 mg formulation respectively. The %ID/g for ^{18}F-FDG in the same tumor model was also compared against NP2 and a significant difference (4.31% versus 10.28%, $P = 0.019$) was observed in the tumor. This reflects a 123% increase in PS2 (NP formulation) present in

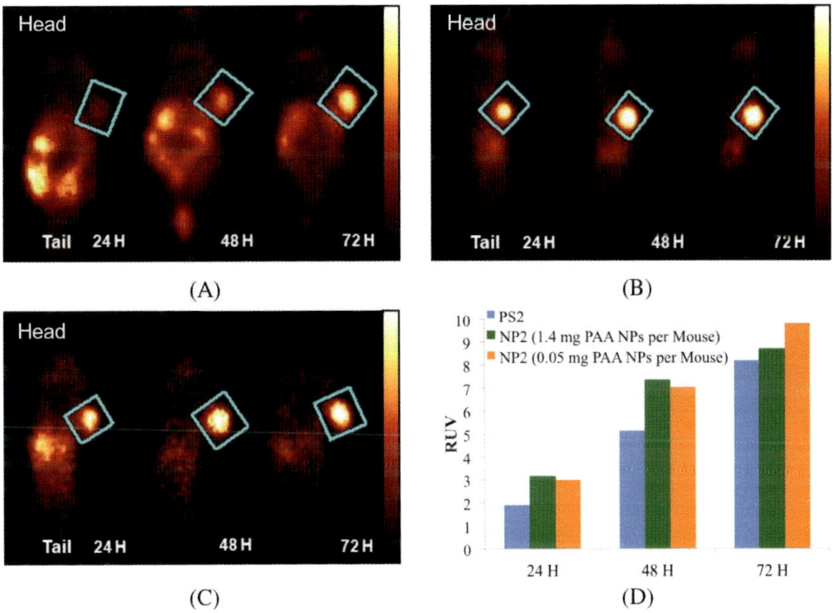

Fig. 11. Whole body PET Images of BALB/c mice bearing subcutaneous Colon26 rumors on the right shoulder with (A) ^{124}I-PS at 24, 48, and 72 h post-injection (*i.v.*), (B) ^{124}I-PS post-loaded in PAA NPs (NP2 – 1.4 mg PAA NPs per mouse) at 24, 48, and 72 h post-injection (*i.v.*) (C)^{124}I-**PS** post-loaded in PAA NPs (NP2–0.05 mg PAA NPs per mouse) at 24, 48, and 72 h post-injection (*i.v.*) and (D) the relative uptake values (RUV) of PS and NP2 (1.4 mg and 0.05 mg PAA NPs containing same radioactivity (20 μCi/mouse) of the^{124}I-PS, suggesting that the amount of NPs used in formulation do not make any significant diffence in tumor uptake of the PET agent. For OSEM 2D rendered digital PET imaging movie of NP2, see Fig. 6, Supporting Material.

the tumor as compared to free PS2 (without NP formulation) and interestingly an increase of 139% if compared against^{18}F-FDG.

The RPCI group also explored the possibility of developing a platform for PET/NIR fluorescence imaging in which PS2 (radioactive PS) was post loaded to PAA NPS conjugated with NIR CD.[163,164] Such a construct (Fig. 13) provided an opportunity to vary the quantities of both the agents for optimal imaging. Whole-body fluorescence and PET imaging was performed in BALB/c mice bearing subcutaneous Colon26 tumors. As shown in Fig. 14, the tumors are clearly identifiable by PET and reduced signals were observed from the non-target organs, e.g., liver and gut, with a substantial increase in relative tumor visibility observed from 4 to 48 h post-injection. This approach allowed for a higher degree of differentiation of the tumor versus the non-target organs. With NP5, PET imaging was accomplished with 16.6 μCi or 0.6142 MBq, which was ~10 times lower than radioactivity that was initially

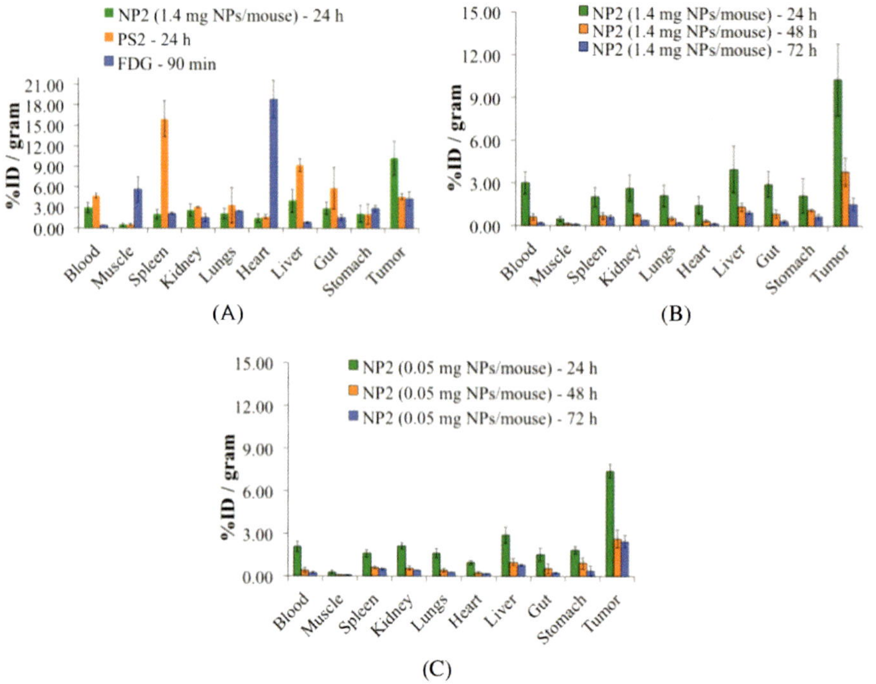

Fig. 12. Comparative *in vivo* biodistribution in BALB/c mice (3 mice/group) bearing Colon26 tumors on the right shoulder (A): PS2 (^{124}I-PS1), NP2 (^{124}I-analog post loaded in PAA NPs) and ^{18}F-FDG (B) & (C): Optimization of the imaging at variable amount of PAA NPS (1.4 mg and .05 mg/mouse respectively) and time intervals. The amount of radioactivity injected: **PS2** (^{124}I-PS1 free) and PS2 post loaded in PAA NPs (NP2): 20 μCi/mouse; ^{18}F-FDG: 70 μCi/mouse (not optimized).

used. Thus, with this approach the RPCI group was able to reduce the dose of the NPs PET (0.05 mg of PAA NP/mouse), while maintaining its ability to perform real-time imaging to validate the presence and extent of tumors for fluorescence image-guided surgery (Fig. 14).

The RPCI group also explored the utility of the post-loading approach in evaluating the PDT efficacy of the corresponding non-labeled analog. The PS was post-loaded by following the methodology used for the labeled agent, as discussed above. The biological efficacy of PS1 and the corresponding nanoformulation NP1 (10 mg/ml stock) were investigated in BALB/c mice bearing Colon26 tumors (10 mice/group). The mice were injected with PS or the corresponding NP formulation at a dose of 1.0 μmole/kg. The tumors were exposed to light at 665 nm (dose: 135 J/cm^2, 75 mW/cm^2) at 24 h post-injection and the tumor response was recorded daily following the animal protocol approved by the institute's IACUC committee. The survival curves

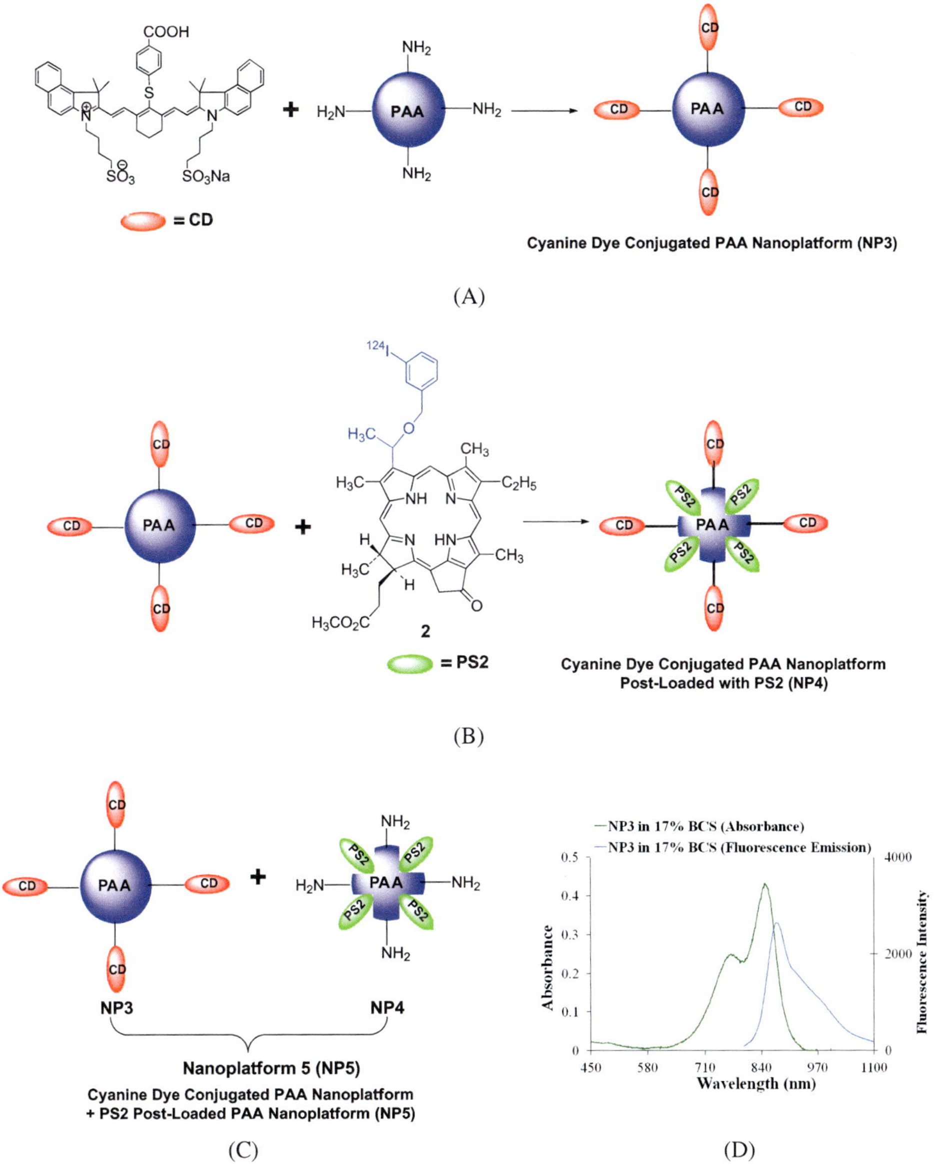

Fig. 13. Schematic for creating the multifunctional PAA NPs for NIR and PET imaging (A) NIR CD conjugated PAA nanoplatform (NP3). (B) NIR CD conjugated PAA nanoplatform post-loaded with PS2 (NP4) (C) NIR CD conjugated PAA nanoplatform (NP3) combined with PS2 post-loaded nanoplatform to formulate nanoplatform 5 (NP5). (D) The absorbance and fluorescence spectra for NP3 in 17% bovine calf serum (BCS) in phosphate buffered saline (PBS). The excitation wavelength was 785 nm and the fluorescence was collected beyond 800 nm.

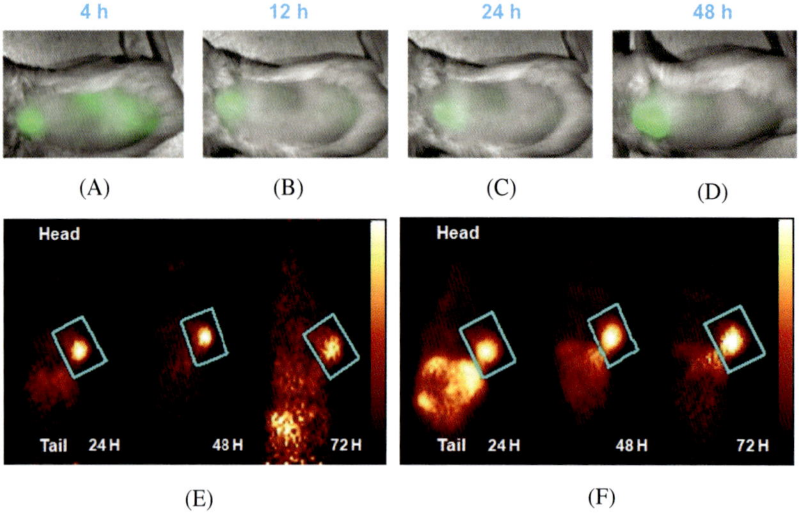

Fig. 14. Multimodal imaging of NP3, NP4, and NP5 in BALB/c mice bearing Colon26 tumors. (A–D): Whole-body NIR fluorescence imaging of NP3, 4, 12, 24, and 48 h post-injection. (E) PET imaging of NP4. (F) PET imaging of NP5.

are shown in Fig. 15(A) which highlights the remarkable enhancement of long-term tumor cure with the NP formulation, from 20% (PS alone: 2/10 mice were tumor free) to 80% (8/10 mice were tumor free) with the corresponding nanoparticles (NP1) formulation.

One of the main concerns in developing nanoparticles in the area of health care has been their unknown long-term toxicity to organs.[164] Therefore, the organ toxicity of the PAA nanoparticles was investigated by the RPCI group at four doses: 100 mg/kg (40-fold higher); 200 mg/kg (80-fold higher), 300 mg/kg (120-fold higher); and 400 mg/kg (160-fold higher), which were significantly higher than the required doses of the NPs for PET imaging (2.5 mg/kg or 0.05 mg per mouse). At all doses, the mice (3 BALB/cmice/group) were monitored for weight loss/gain and other signs of distress for 30 days post-injection of the nanoparticles. None of the mice showed any weight loss or gain in weight profile as reported by Taconic Farms for BALB/c mice. On day 30, the mice were necropsied and the organs were analyzed for toxicity by H&E staining. No acute toxicity was observed even at the 160-fold higher dose. The H&E staining for the liver, spleen, heart, kidney and lung (dose: 400 mg/kg/group) along with the corresponding controls (BALB/c mice void of nanoparticles), are summarized in Fig. 15(B).

This study illustrates the utility of a biodegradable/biocompatible, non-toxic PAA-based nanoparticles-photosensitizer formulation for developing a highly efficient "Multimodality Platform" for tumor-imaging by

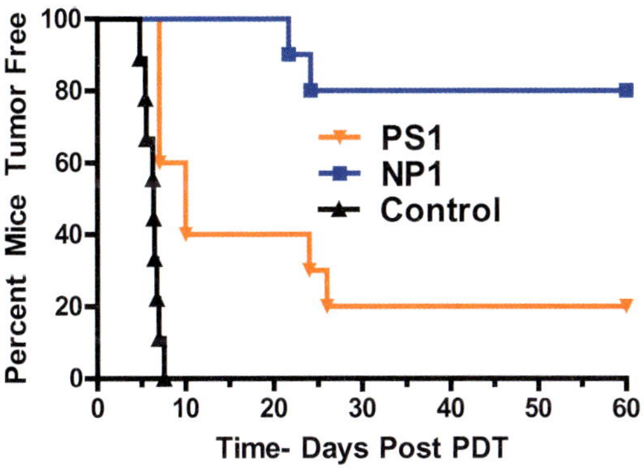

(A)

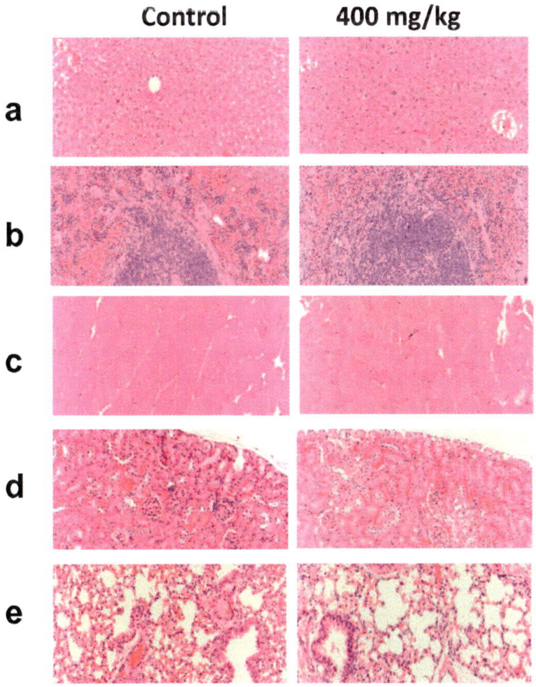

(B)

Fig. 15. (A) *In vivo* PDT data (% mice cured) by Kaplan-Meier survival curve show a significant difference in PDT efficacy of PS1 with and without NPs at a dose of 1.0 μmole/kg. The tumors were exposed to light (135 J/cm^2 and 75 mW/cm^2) 24 h post-injection. The P value for the two survival curves is <0.0001 as determined by the Mantel–Cox test. (B): Formalin-fixed, paraffin embedded hematoxylin-eosin (H&E.) stained tissue sections (representative sample for 400 mg/kg): (a) Liver, (b) spleen, (c) heart, (d) kidney, and (e) lung [Magnification: 200x].

PET/Fluorescence and photodynamic therapy. Comparative tumor-imaging, biodistribution and PDT efficacy data demonstrate the advantage of our approach with significant translational value. The PAA NPs formulation makes a significant difference in the pharmacokinetic profile of the PET and PDT agents. Thus it could provide an efficient approach to improve the drug delivery of various hydrophobic cancer therapeutic agents, not limited to PDT. So far, this happens to be the first example demonstrating the use of a non-toxic nanoconstruct for the use in PET and fluorescence image-guided PDT.

3. Targeted PAA nanoparticles for a "See and Treat" approach

Kopelman, Pandey and co-workers[165] have recently described the development of multifunctional nanocarriers, based on amine-functionalized biodegradable PAA nanoparticles, for cancer theranostics (Therapy + Diagnostics) including active tumor targeting, fluorescence imaging, and PDT (Fig. 16). The structural design involves adding primary amino groups and biodegradable

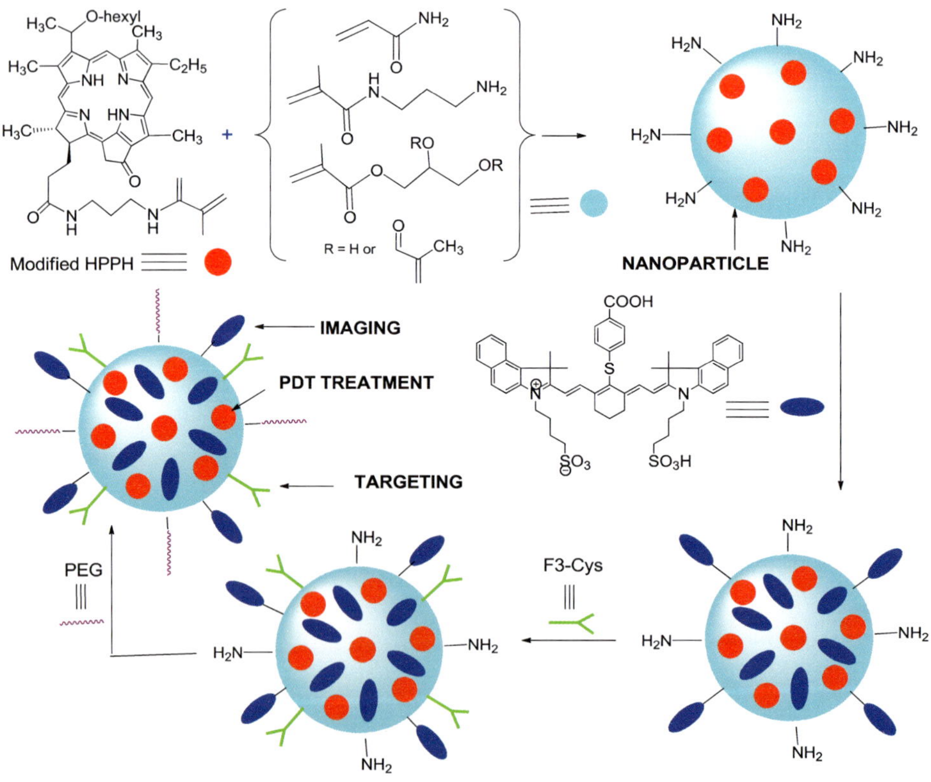

Fig. 16. A synthetic approach for the multifunctional PAA-based nanoconstruct.

cross linkers during the nanoparticle preparation. An accelerated biodegradation study indicated that that the polymer matrix chain collapsed after degradation. In vitro targeting studies on human breast cancer cells indicate that the targeted nanoparticles can be transported efficiently into tumor cells. Incubating the multifunctionalized nanocarriers with cancer cells *in vitro* enabled strong fluorescence imaging. Irradiation of the photosensitizing drug incorporated within the nanoparticles, with light of the suitable wavelength, causes significant and selective damage to tumors. The incorporation of photodynamic and fluorescent imaging agents into the nanoparticle matrix, and conjugating of PEG and tumor targeting moieties on the surface of the nanoparticles provides an easy way to create targeted NPs for imaging and therapy of cancer. The targeting moieties that have been incorporated so far in PAA-based nanoparticles are cyclic RGD (cRGD) and F3 peptide.

3. Conclusions and Outlook

The main objective of this particular chapter is to discuss (i) the inherent problems associated with Photofrin, (ii) a brief report on the strategies developed at Roswell Park Cancer Institute for developing PDT agents with fluorescence and MR imaging capabilities, (iii) to discuss the development of multifunctional agents for PET imaging with an option of fluorescence-guided surgery and PDT (iv) to demonstrate the advantages of biodegradable, biocompatible and non-toxic PAA-based nanoparticles as vehicles to deliver the imaging and therapy agents in desired ratio to tumors, and finally (v) to outline a versatile method for developing a targeted multimodality nanoplatform for a "See and Treat" approach for cancer imaging and therapy.

Recent exciting reports from various laboratories on nanotechnology indicate that this approach is having a significant impact in delivering the hydrophobic compounds to disease sites, with enhanced imaging and therapy capabilities. The target-specific nanoparticles show slower delivery of chemotherapeutic drugs to tumor retention for a longer period of time, and this may improve long-term tumor cure with low toxicity, reduced metastasis and improved quality-of-life. Such a combination therapy approach is already showing great promise in developing dual therapeutic agents with significant synergic effect with and without imaging ability. It has also been shown that multifunctional e.g., PET/fluorescence and MR/fluorescence, which are complimentary to each other, can be achieved by careful designing of the nanoparticles. This technology also provides an opportunity for both active and passive targeting. In passive targeting, the selective accumulation of nanocarriers and drug occurs by the EPR effect, the exploitation of which is becoming the gold standard in cancer-targeting drug designs. Nevertheless, to reach the tumor passively, some limitations exist: passive targeting depends on the degree of tumor vascularization and angiogenesis. Thus extravasation of nanocarriers will vary with tumor types and anatomical sites. Besides, the high interstitial fluid pressure of tumors associated with the poor lymphatic drainage explains the size relationship with the EPR effect - larger and long-circulating nanocarriers (100 nm) are better retained in the tumor than smaller molecules which easily diffuse. In active targeting ligands are attached at the surface of the nanoparticles with a goal of specific delivery of the therapeutic and imaging agents to cancer cells. In the active targeting strategy, the cancer targeting can be achieved by targeting surface molecules on the the cancer cell membrane, or the targeting of tumor endothelium.

PDT is a localized treatment, and the major limitation has been its inability to treat metastases. Various synthetic approaches in which tumor-avid photosensitizers were conjugated with a chemotherapeutic agent by *in vivo* cleavable or non-cleavable linkers did not show desired specificity. However, the nanoparticle approach may help overcome this problem by incorporating the desired amount of PDT and chemotherapeutic agents in nanoparticles which have higher tumor specificity.

In recent years, a large number of nanoparticles for use in medicine have been reported in the literature with several advantages and limitations and it may take some time before this technology gets approval from various health organizations.This article is mainly focused on PAA nanoplatforms due to their synthetic reproducibility, biocompatibility, biodegradability, non-toxicity, and ability to be surface-modified and targeted. In animal models, the PAA nanoparticles certainly show advantages in drug delivery of hydrophobic, water-insoluble photosensitizers and imaging agents with improved tumor-specificity. In the future, it is expected that further studies with other classes of nanoparticles with limited toxicity and improved cancer-imaging and therapy may help in developing desired theranostic agents.

Acknowledgments

We wish to acknowledge the significant contributions from the authors who graciously gave permission for use of their work and illustrations in this article. The contributions from the RPCI group reported herein have been supported generously by the financial support from the NIH (CA119358, CA55791,

CA127369, CA176154, CA109914), Roswell Park Alliance, Photolitec, LLC, and the NCI Cancer Support Grant CA016156.

References

1. M. Ethirajan, Y. Chen, P. Joshi and R. K. Pandey, *Chem. Soc. Rev.,* **40**, 340 (2011).
2. M. Magaraggia, and G. Josi in *Photodynamic Therapy* Basic, Translational and Clinic, Artech House, Boston 337, (2008).
3. P. K. Kaiser, *Curr. Med Res. Opin,* **23**, 477 (2007).
4. T. J. Dougherty and J. G. Levy, Clinical Applications of Photodynmic Therapy in *Organic Photochemistry and Photobiology* Eds., W. Horspool, and F. Lenci, (CRC Handbook, Boca Raton, 2004).
5. B. L. Aronoff, *J. Surg. Oncol,* **64**, 84 (1997).
6. J. Toth, Private Communication.
7. K. Harubumi, J. *Photochem. Photobiol.*, **42**, 96 (1998).
8. T. J. Dougherty and J. G. Levy, *Photodynamic Therapy (PDT) and Clinical Applications in Biomedical Photonics Handbook* (Ed. T.Vo-Dinh), CRC Press, Boca Raton, 2003).
9. M. F. Edelson, *Sci. Am.* **68**, 259 (1988).
10. J. D. Spikes, *The historical development of ideas on applications of photosensitized reactions in the health sciences.* NATO ASI Series, Series A: Life Sciences, **85**, 209 1985; (Primary Photo-Processes Biol. Med.).
11. T. M. Schmitt, S. Chi-Menti and F. P. Gasparro, *J. Photochem. Photobiol. B,* **27**, 101 (1995).
12. H. Von Tappeiner, A, Jodlbauer, and H. Lehmann, *Dtsch. Arch. Klin. Med.*, **82**, 217 (1903).
13. O. Raab, *Zeitechrift fuer Biologie (Munich)*, **39**, 324 (1900).
14. H. von Tappeiner and H. Jodlbauer, *Dtsch. Arch. Klin. Med.* **80**, 427 (1904).
15. H. H. Jesionek, and H. von Tappiner, *Dtsch. Arch. Klin. Med.*, **82** 223 (1905).
16. T. J. Dougherty, W. R. Potter, K. R. Weishaupt, Eds. D. Doiron, and C. J Gomer, in *Porphyrin Localization and Treatment of Tumors* 301 (Alan R. Liss., New York, 301, 1984,).
17. T. J. Dougherty, K. Weishaupt and W. Potter, *U. S. Patent,* 4, 649, 151 (1987)
18. R. L. Lipson and E. J. Baldes, *Archives of Dermatology,* **82**, 508 (1960).
19. A. Policard, *CR Soc. Biol.*, **91**, 1423 (1924).
20. F. H. Figge, G. S. Weiland and L.O. Manganeiello, *Anat. Res.* **101**, 657(1948).
21. B. W. Henderson and T. J. Dougherty *Photochem. Photobiol.*, **55**, 145 (1992).
22. R. Bonnett and M. C. Berenbaum, *Adv. Exp. Med. Biol.*, **160**, 241 (1983).
23. T. J. Dougherty, W. R. Potter and K. R. Weishaupt *Ad. Med. Biol.*, **301** (1984).
24. D. Kessel, *Photochem. Photobiol.*, **44**, 193 (1986).
25. D. Kessel, P. Thompson, B. Musselman and C. K. Chang, *Cancer Res.*, **47**, 4542 (1987).
26. I. K. Morris, and A. D. Ward, *Tetrahedron Lett.*, **2** (1988).
27. R. K. Pandey and T.J. Dougherty T. J. *Cancer Res.*, **49**, 2501 (1989).
28. R. K. Pandey and. K. M. Smith, *Tetrahedron Lett.*, **29**, 4557 (1988).
29. R. K. Pandey and T. J. Dougherty *Photochem. Photobiol.*, **47**, 769 (1988).
30. R. K. Pandey, K. M. Smith and T. J. Dougherty, *J. Med. Chem.* **33**, 2032 (1990).
31. R. K. Pandey, F-Y, Shiau, T. J Dougherty, and K. M. Smith, *Tetrahedron*, **47**, 9571 (1991).

32. R. K. Pandey, M. M., Siegel, R, Tsao, J. H. McReynolds, and T. J. Dougherty, *Biomed and Environ. Mass Spectrometry*, **19** 405 (1990).
33. R. K. Pandey, M. G. H. Vicenta, F. Y. Dhiau, T. J. Dougherty and K. M. Smith, Proc. SPIE, 1426 (1991).
34. R. K. Pandey, F-Y, Shiau, C. J. Medforth, T. J. Dougherty and K. M. Smith Tetrehedron Lett., **31**, 7399 (1990).
35. I. J. MacDonald and T. J. Dougherty, *J. Porphyrins and Phthalocyanines*, **5**, 105 (2001).
36. B.W. Henderson, and V.H. Fingar *Photochem. Photobiol.*, 49, 299 (1989).
37. D. E. Dolmans, A. Kadambi, J. S. Hill, C. A. Waters, B. C. Robinson, J. P. Walker, D. Fukumura and Jain, R. K. *Cancer Res.*, **62**, 2151 (2002).
38. A. Ferrario, K.F. von Tiehl, N. Rucker, M. A. Schwarz, P. S. Gill and C. Gomer, *J. Cancer Res.*, **60**, 4066, (2000).
39. D. Kessel, Y. Luo, P. Mathiau, and J.R. Reiners *Photochem. Photobiol.*,**71**, 196 (2000).
40. M. Lam, N. L. Oleinick, A. L. Nieminen, *J. Biol. Chem.*, **276**, 47379 (2001).
41. D. Kessel, and N.L. Oleinick, N. L. *Methods Mol. Biol.*, **635**, 419 (2010).
42. J .Usuda, K. Azizudin, S. M. Chiu, and N.L. Oleinick, N. *Photochem. Photobiol.*, **78**,1 (2003).
43. S.O. Gollnick, S. S. Evans, H. Baumann, B. Owczarczak, P. Maier, L. Vaughan, W. C. Wang, E. Unger and B.W. Henderson, *Br. J. Cancer*, **88**, 1772 (2003).
44. D. W. Hunt and J. G. Levy, *Expert Opin Investing Drugs* 7, 57 (1998).
45. E. Reginato, P. Wolf, and M.R. Hamblin, *World J. Immunol*, **27**,1 (2014).
46. A. P Castano, P. Mroz and M. R. Hamblin, *Nat. Cancer* **6**, 535 (2006).
47. E. C. Tracy, M. J Bowman, R. K., Pandey, B. W. Henderson, and H. Baumann, *Photochem. Photobiol.*, **87**, 1405 (2011).
48. J. Morgan, J. D. Jackson, X. Zheng, S. K. Pandey and R. K. Pandey, *Mol. Pharmaceutics*, 7, 1789 (2010).
49. E.C. Tracy, M.J. Bowman, R. K. Pandey, B. W. Henderson and H. Baumann, *Photochem. Photobiol.*, 1405, (2011).
50. B. W. Henderson, C. Daroqui, E. C. Tracy, L. A., Vaughan, G. M. Loewen, M. T.Cooper, and Baumann, H. *Clinical Cancer Res.* **13**, 3156 (2007).
51. R. K. Pandey, A. B. Sumlin, S. Constantine, M. Aoudia, W. R. Potter, D. A. Bellnier, B. W. Henderson, M. A. Rodgers, J. K. M. Smith, and T. Dougherty, *Photochem. Photobiol.*, **64**,194 (1996).
52. Y. Chen, K. Ohkubo, M, Zhang , E. Wenbo, W. Liu, , S. K. Pandey, M. Ciesielski, H., Baumann, E. Tracy, S. Fukuzumi, K. M. Kadish, R. Fenstermaker, A. Oseroff, and R. K. Pandey, *Photochem. Photobiol. Sci.*, **6**, 1257 (2007).
53. M. Ethirajan, N. Patel, and R. K. Pandey, Eds. K. M. Kadish, K. M. Smith, and R. Guilard, Porphyrin-based multifunctional agents for tumor-imaging and photodynamic therapy (PDT) in Handbook of Porphyrin Science, (World Scientific, New Jersey, 2010).
54. C Moore, D. Huebler, T. Zimmerman, L. A. J. Heinemann, F. Saad, and D. M., Thai, *Eur. Urol.*, **46**, 80. (2004).
55. M. Kramer, J. W. Miller, N. Michaud, R. S. Moulton, T. Hasan, T. J. Flotte and E. S. Gragoudas, *Ophthalmology*, **103**, 427 (1996).
56. T. J. Kinsella, E. D. Baron, V. C. Colussi, K. D. Cooper, C. L. Hoppel, S. T. Ingalls, M. E. Kenney, X. Li, N. L. Oleinick, S. R. Stevens, and S. C. Remick, *Front. Oncol.*, **1** (2011).

57. R. Biswas, J. H. Mooh and J. C. Ahn, *Photochem. Photobiol.*, **90**, 1108 (2014).
58. N. S. Rigual, G. Shafirstein, M .T. Cooper, H. Baumann, D. A. Bellnier, U. Sunar, E. C. Tracy, D. J. Rohrbach, D. J. Eilding, W. Tan, N. A. Sullivan, M. Maureen, Merzianu, and B. W. Henderson. Clinical Cancer Res., **19**, 6605 (2013).
59. H. Ali, and E. van Lier, Porphyrins and phthalocyanines as photosensitizers anradiosensitizers in Handbook of Porphyrin Science (Eds. K. M. Kadish, Smith, and Guilard, R.), World Scientific, New Jersey, 2010, 1.
60. R. K. Pandey, L. N. Goswami, Y. Chen, A. Gryshuk, J. H. Missert, A. Oseroff, and T. Dougherty, *J. Lasers in Surgery and Medicine* **38**, 445 (2006).
61. B. W. Henderson, D. A. Bellnier, W. R. Greco, A. Sharma, R. K. Pandey, L. A. Vaughan, K. R. Weishaupt, and Dougherty, T. *J. Cancer Research*, **64**, 194 (1997).
62. D. A. Bellnier, B. W. Henderson, R. K. Pandey, W. R. Potter and Dougherty, T. *J. Photochem. Photobiol.*, **20**, 55 (1993).
63. D. A., Bellnier, W. R. Greco, H. Nava, G. M. Loewen, A. Oseroff, R. K. Pandey and T. Dougherty, J. *Cancer Research*, **63**, 1806 (2003).
64. G. M. Loewen, R. K. Pandey, D. A. Bellnier, B. W. Henderson and T. J. Dougherty., *Lasers in Surgery and Medicine*, **38**, 364 (2006).
65. D. A. Bellnier, W. R. Greco, H. Nava, G. M. Loewen, A. R. Oseroff, and T. J. Dougherty, *Cancer Chemother. Pharmacol.*, **57**, 40 (2005).
66. T. J. Dougherty, R. K. Pandey, H. Nava, J. A. Smith, H. O. Douglass, S. B. Edge, D. A. Bellnier, L. O'Malley, and M. Cooper, *Proc. SPIE*, **25**, 3909 (2000).
67. R. K. Pandey and G. Zheng, Porphyrins as photosensitizers in photodynamic therapy, in *The Porphyrin Handbook* Eds. , K. M. Kadish, K. M. Smith, and R. Guilard, **6**, 157 (2000).
68. M. R. Hamblin, Covalent photosensitizer conjugates, Part I: Antibodies and other proteins for targeted photodynamic therapy in *Advances in Photodynamic Therapy. Basic, translational and clinical* Eds. M. R. Hamblin, and P. Mroz, (Atrech House, Boston, 2008) 193.
69. W. R. Potter, B. W. Henderson, D. A. Bellnier, R. K. Pandey, L. A. Vaughan, W. R. Kenneth Dougherty, T. J. *Photochem. Photobiol.*, **70**, 781 (1999).
70. G. Zheng, W. R. Potter, A. Sumlin, T. J. Dougherty and R. K. Pandey, *Bioorg. Med. Chem. Lett.*, **10**, 123 (2000).
71. A. Rungta, G. Zheng, J. R. Missert, W. R. Potter, and T. J. Dougherty, and R. K. Pandey, *Biorg. Med. Chem. Lett.* **10**, 1463 (2000).
72. G. Zheng, W. R. Potter, S. H. Camacho, J. R. Missert, G. Wang, D. A. Bellnier, B. W. Henderson, M. A. Rodgers, T. J. Dougherty, K. M. Smith, and R. K. Pandey, *J. Med. Chem.*, **44**, 1540 (2000).
73. S. Mettath, G. Li, T. Srikrishnan, R. Mehta, Z. D. Grossman, T. J. Dougherty, and R. K. Pandey, *Organic Lett.*, **1**, 1961 (1999).
74. A. Kozyrev, V. Suresh, S. Das, M. O. Senge, M. Shibata, T. J. Dougherty, and Pandey, R. K. *Tetrahedron*, **56**, 3353 (2000).
75. A. L. Gryshuk, A. Graham, S. K. Pandey, W. R. Potter, J. R. Missert, A. Oseroff, T. J. Dougherty and R. K. Pandey, *Photochem. Photobiol.*, **76**, 555 (2002).
76. C. Liu, M. P. Dobhal, M. Ethirajan, J. R. Missert, R. K. Pandey, S. Balasubramanian, D. Sukumaran, M. Zhang, K. M. Kadish, K. Ohkubo and S. Fukuzumi, *J. Am. Chem. Soc.*, **130**, 14311 (2008).

77. A. Gryshuk, Y. Chen, L. N. Goswami, S. Pandey, J. R. Missert, T. Phulchanskyy, W. Potter, P. S. Prasad, A. Oseroff, and R. K. Pandey, *J. Med. Chem.*, **30**, 1754, (2007).
78. M. R Hamblin, Covalent photosensitizer conjugates, Part 2: Peptides, polymers and small molecules for targeted photodynamic therapy Basic, translational and clinical Eds. M R., Hamblin, and P. Mroz, (Atrech House, Boston, 2008) 217.
79. X. Jiang, R. K. Pandey and K. M Smith, *Tetrahedron Lett.*, **36**, 365 (1995).
80. Y. Chen, M. Sajjad, Y. Wang, C. Batt, H. A. Nabi, and R. K. Pandey, *J. Med. Chem.*, **2**, 136 (2011).
81. C. Dubuc, R. Langlois, F. Benard, N. Vauchon, K, Klarskov, P. Tone and J. van Lier, *Biorg. Med. Chem. Lett.*, **18**, 2424 (2008).
82. B. G. Ongarora, S. D. Satyanarayana-Jois, I. Seghal, K. R. Fontenot, X. Hu and M. G. H, Vicente, J. Med. Chem. **55** 3725 (2012).
83. X. Zheng and R. K. Pandey *Anti-Cancer Agents in Medicinal Chemistry* **8**, 241 (2008).
84. G. Zheng, A. S. Graham, S. Shibata, J. R. Missert, , A. Oseroff, T. J. Dougherty, and Pandey, R. K. *J. Org. Chem.*, **66**, 8709 (2001).
85. X. Chen, and C. M. Drain, *Drug Design Reviews* **1**, 215, (2004).
86. A. Srivatsan, M. Ethirajan, S. K. Pandey, S. Dubey, X. Zheng, T-H, Liu, M. Shibata, J. Missert, J. Morgan and R. K Pandey, *Molecular Pharmaceutics*, **8**, 1186 (2011).
87. D. V. Titov, M. L. Gening, Y. E. Tsvetkov and N. E. Nifantiev, *Russ. Chem. Rev.*, **83**, (2014).
88. S. K., Pandey, X. Zheng, J. Morgan, S. Camacho, W. Potter, D. A. Belliner, and R. K. Pandey, *Molecular Pharmaceuitics*, **4**, 448 (2007).
89. L. Veenman, M. Gavish, *Pharmacol. Ther.* **51**, 629 (2006).
90. V. Papadopoupos, M. Baraldi, T. R. Guilarte, T. B. Kundsen, J. J. Lacapere, P. Lindermann, M. D. Novenberg, D. Nutt, A. Weizman, M. R. Zhang and M. Gavish, *Trends Pharmacol. Sci.*,**27**, 402 (2006).
91. A. P. Sutter, K. Master, M. Hopfner, B. Barthel, P. Grabowski, S. Paiss, P. Carayon, M. H. Zeitz, H. Hand Sherubi, *Radiat. Oncol. Invest.* **4**, 318 (2002).
92. D. Kessel, M. Antolovich, and K. M Smithy, *Photochem. Photobiol.* **74**, 346, 2001.
93. M. Srivastava, N. Ahmed, S. Gupta, and H. Mukhtar, *Journal of Biol. Chem.*, **276** 15481, (2001).
94. Y. Chen, M. Sajjad, Y. Wang, C. Batt, H. A. Nabi and R. K. Pandey, *ACS Medical Chemistry Letters*, **2**, 136 (2011).
95. H. Ali and J. E. van Lier, *Chem. Rev.*, **99**, 2379 (1999).
96. A. Scherz, A. Y. Salomon, A. Brandis, and H. Scheer, US Patent No. US 6, **569**, 846 B1, (2003).
97. N. V. Koudinova, J. H. Pinthus, A. Brandis, O. Brenner, P. Bendel, J. Ramon, Z, Eshhar A. Scherz, Y. Salomon, *Int. J. Cancer*,**104**, 782, (2003).
98. M. Ethirajan, C. Saenz, A. Gupta, M. P. Dobhal, and R. K. Pandey, Photosensitizers for photodynamic therapy and imaging in Advances in Photodynamic Therapy Eds. M. R. Hamblin, and P. Mroz, (Artech House, Boston, 2008).
99. Y. Chen, X, Zheng, M. P. Dobhal, A. Gryshuk, J. Morgan, T. J. Dougherty, A. Oseroff, and R. K. Pandey, *J. Med. Chem.*, **48**, 3692 (2005).
100. R. J. Hargreaves, *Clinical Pharmacology and Therapeutics*, **83**, 349, (2008).
101. J. K. Willmamm, H. van Bruggen, L. M. Dinkelborgand and S. S Gambhir, N*ature Reviews Drug Discovery*, 7, 591 (2008).

102. A. M. Master, Y. Qi, N. L. Oleinick, and A. S Gupta, *Nanomedicine: Nanotechnology, Biology and Medicine* **8**, 655, (2012).
103. P. M. Matthews, I. Rabiner and R. Gumm, *Current Opinion in Pharmacology*, **11**, 501 (2011).
104. R. Wissleder, *Science*, **312**, 1168 (2006).
105. M. L. James and S. S. Gambhir, *Physiol. Rev.*, **92**, 897 (2012).
106. Z. Liang and P. Lauterbur, Principles of magnetic resonance imaging. A signal processing perspective. IEEE Press Series in Biomedical Engineering, (New York: IEEE Press, 2000).
107. H. D. Burns, R. E. Gibson, R. F. Dannals, and P. K. S. Siegel, (Eds), Nuclear Imaging in Drug Discovery, Development and Approval, (Birkhauser, Boston, 1993).
108. S. V. Bachawal, K. C. Jensen, A. M. Lutz, S. Gambhir, S. Tranquart, F. Tian, L. Willmann, *Cancer Research*, **73**, 1689, (2013).
109. N. Hollings, and P. Shaw. *European Respiratory Journal (ERJ)*, **19**, 722 (2003).
110. C. M. Carpenter and B. W. Pogue Diffuse optical spectroscopy and magnetic resonance imaging in Translational Multimodality Optical Imaging Eds. Azar, F. S. and Intes, X. M. B (Artech House, Boston, 2008.)
111. M. B Bestari, T. L Ang and S. A. Abdurachman, *Endoscopy Ultrasound*, **41**, 144 (2009).
112. Y. Pu, W. B. Wang, M. Xu, G. C. Tang, Y. Budansky, M. Sharanov, S. Achilefu, J. A. Eastham and R. R. Alfano, *Technol. Cancer Res. Treat.* **10**, 507, 2011.
113. V. Masilamani, K. Zl-Zhrani, M. Al-Salhi, A. Al-Diab, and M. Al-Ageily, *Journal of Luminescence*, **109**, 143, (2004).
114. S. K., Piper, C. Habermehl, C. H. Schmitz, Kuebler, W. M., Obrig, H., Steinbrink, J. and Mahnert, J. *J. PloS ONE*, **8**, e8749 (2013).
115. J. F. Dunn, *Adv. Exper. Med. Biol.* **454**, 103, (1998).
116. K. Haedicke, S. Grafe, F. Lehmann and I. Hilger *Biomaterials*, **34**, 10075 (2013).
117. F. R. Wrenn, Jr., M. L. Good, P. Handler,*Science* **18**, 525 (1951).
118. D. A. Cole, J. A. Mercer-Smith, S. A. Schreyer, J. K. Norman, and D. K. Lavallee, *Nucl. Med. Biol.*, **17**, 457 (1990).
119. B. Ma, G. Li, P. Kanter, D. Lamonica, Z. Grossman, and R. K. Pandey, *J. Porphyrins and Phthalocyanines*, **6**, 500. 2003.
120. K. Iwai, S. Krimura, T. Ido, R. Iwata, *Int. J. Radiat. Appl. Instrum., Part B.* 775 (1990).
121. M. Ethirajan, N. J. Patel, and R. K. Pandey, Porphyrin-based multifunctional agents for tumor-imaging and photodynamic therapy (PDT) in Handbook of Porphyrin Science (Eds. Kadish, K. M., Smith, K. M. and Guilard R.), (World Scientific, New Jersey, 2010) 249.
122. S. K. Pandey, A. L. Gryshuk, M. Sajjad, X. Zheng, Y. Chen, M. M. Abouzeid, J. Morgan, I Charamisinau, H. Nabi, A. Oseroff, and R. K. Pandey, *J. Med. Chem.*, **48**, 6286 (2005).
123. N. J. Patel, Ph. D. Thesis, Roswell Park Graduate Division, SUNY, Buffalo, **2014**.
124. J. A. Spernyak, W. H. White, L. N. Goswami, M. Ethirajan, Y. Chen, J. R. Missert, J. Morgan, R. Mazurchuk, and R. K. Pandey, *Bioconjugate Chem.*, **21**, 828, (2010).
125. A. Gupta, and R. K. Pandey, in Supramolecular Systems in Biomedical Fields (Ed. H-J Schneider,), Chapter 15, (*The Royal Society of Chemistry*, 2013).
126. D. A. LaVan, and R. Langer, Medical Nanotechnology in Introduction to Nanoscience & Nanotechnology (Eds. G. L. Hornyal, H. F. Tibbals, J. Dutta, and J. J. Moore,), (CRC Press, Boca Raton, 2009) 1405.
127. D. Fabienne, O. Feron and P. Veronique, *Journal of Controlled Release*, **148**, 135 (2010).

128. I. Bergin, and F. Witzmann, *Int. J. Biomed Nanosci Nanotechnol.*, **3**, 163 (2013).
129. L. Karamchand, G. Kim, S. Wang, J. J. Hah, A. Ray, R. Jiddou, Y-E Koo Lee, M. A. Philbert, and R. Kopelman, *Nanoscale*, **5**, 10327 (2013).
130. S. M. Buck, H. Xu, M. Brasuel, M. A. Philbert, and R. Kopelman, *Talenta*, 63, 41, (2004).
131. M. Kuruppurachchi, H. Savoie, A. Lowry, C. Alonso, and R. W. Boyle, *Mol. Pharmaceutics*, **8**, 920 (2011).
132. S. Wang, W. Fan, G. Kim, H-J Hah, Y-E Koo Lee, R. Kopelman, M. Ethirajan, A. Gupta, L. N. Goswami, P. Pera, J. Morgan, and R. K Pandey,*Lasers in Surgery and Medicine*, **43**, 686, (2011).
133. J. Funkhouser, *Curr. Drug Discovery*, **2**, 17, (2002).
134. H. Xu, S. M. Buck, R. Kopelman, M. A. Philbert, M. Brasuel, B. D. Ross, and A. Rehemtulla *Isr. J. Chem.*, **44**, 317 (2004).
135. V. Torchilin, *Advanced Drug Delivery Reviews*, **58**, 1532 (2006).
136. A. D. Bangaham, and R. W. Horne, *J. Mol. Bio.*, **8**, 660 (1964).
137. A. Gupta, S. Wang, A. Marko, M. Ethirajan, Y. Chen, R. Yao, M. Sajjad, R. Kopelman, and R. K. Pandey, *Theranostics*, **4**, 614 (2014).
138. D. Si, T. Epstein, Y-E. Koo Lee, and R. Kopemnan, *Anal. Chem.* **64**, 978 (2012).
139. H. K. Yoon, X. Lou, Y-C. Chen, Y-E, Koo Lee and R. Kopelman, *Chem. Mater.*, **26**, 1592 (2014).
140. G. Nie, H. J. Hah, G. Kim, Y-E Koo Lee, M. Qin, T.S. Ratani, P. Fatiads, A. Miller, A. Kochi, D. Gao, T. Chen, D. A. Orringer, O. Sagher, M. A. Philbert, and R. Kopelman, *Small*, **8**, 884, (2012).
141. L. B. Josefsen, J. W. Aylott, A. Beeby, P. Warburton, J. P. Boyle, C .Leers, and R. W. Boyle, *Mol. Pharmaceutics*, **9**, 801, (2010).
142. W. Tang, H. Xu, R. Kopelman, and M. A. Philbert, *Photochem. Photobio.*, **81**, 242 (2005).
143. H. J. Hah, G. Kim, Y-E. Koo Lee, D. A. Orringer, O. Sagher, M. A. Philbert, and R. Kopelman, *Macromol. Biosci.*, **11**, 90 (2011).
144. Wenger, Y., Schneider, R. J., Reddy, G. R. Kolpelman, R., Jolliet, O. and Philbert, M. A. Toxicol. Appl. Pharmacol. **25**, 181, 2011.
145. A. Gupta, L. N. Goswami, M. Ethirajan, J. Missert, K. V. R. Rap, T. Ohulchanskyy, I. Roy, J. Morgan, P. N. Prasad and R. K. Pandey, *J. Porphyrins and Phthalocyanines*, **15**, 401 (2011).
146. S. Kim, T. Y. Ohulchanskyy, D. Bharali, Y. Chen, R. K. Pandey, and P.N. Prasad, *J. Phy. Chem. C.* **113**, 12641 (2009).
147. T. Y. Ohulchanskyy, I. Roy, L. N. Goswami, Y. Chen, E. J. Burgy, R. K. Pandey, A. O. Oseroff, and P. N. Prasad, *Nano Lett.*, 7, 2835 (2007).
148. A. Srivatsan, Jenkins S.V., Jeon M., Wu Z., Kim C., Chen J. and Pandey R.K. *Theranostics*, **4**, 163 (2014).
149. K. Baba, H. E. Pudavar, I. Roy, T. Y. Ohulchanskyy, Y. Chen, R. K. Pandey, and P. N. Prasad, *Mol. Pharmaceutics*, **4**, 289, (2007).
150. S. Wang, W. Fan, G. Kim, H-J. Hah, Y-E Koo Lee, R. Kopelman, M. Ethirajan, A. Gupta, L. N. Goswami, P. Pera, J. Morgan, and R. K. Pandey, *Lasers in Surgery and Medicine*, **43**, 686, (2011).
151. P. C. Wilson, and M. S. Patterson, *Phys. Med. Biol.*, **5**, R61, 2008.
152. F. S. Azar, and X. Intes, (Eds.), Translational Multimodality Optical Imaging, (Artech House, Boston, 2008).

153. R. K. Pandey, G. Zheng, in The Porphyrin Handbook (Eds. K. M. Kadish, K. M. Smith, and R. Guilard), Vol. 6, Academic Press, 2010.
154. R. K., Pandey, N., James, Y. Chen, and , M. P. Dobhal in Heterocyclic Polymethine Dyes Ed L Strekowski (Springer, 2008).
155. Indocyanine Green (ICG) was approved by the United States Food and Drug Administration (FDA).
156. Y. Chen, A. Gryshuk, S. Achilefu, T. Ohulchanskyy, W. Potter, T. Zhong, J. Morgan, B. Chance, P. N. Prasad, B. W. Henderson, A. Oseroff, R. K. Pandey, *Bioconjugate Chem.*, **16**, 1264 (2005).
157. M. P. A., Williams, M. Ethirajan, K. Ohkubo, Y. Chen, J. Morgan, W. H. White, M. Shibata, S. Fukuzumi, and K. M. Kadish, *Bioconjugate Chem.*, **22**, 2283 (2011).
158. D. Muffoletto, A. Gupta, Z. Xu, R. K. Pandey, and U. Sunar, in Optical Methods for Tumor Treatment and Detection: Mechanisms and Techniques in PDT XVII Ed. Kessel, D. H., (Proceedings of SPIE, **7164**, 716409 2009).
159. A. Gupta, S. Wang, P. Pera, , K. V. R. Rao, N. Patel, T. Y. Ohulchanskyy, J. Missert, J. Morgan, Y-E Koo Lee, R. Kopelman, and R. K Pandey, *Nanomedicine: Nanotechnology, Biology and Medicine*, **8**, 941 (2012).
160. G.R. Reddy, M. Bhojani, P. McConville *Clin Can Res.* **12**: 6677 (2006).
161. Y-EL R. Koo, G. R. Reddy, M. Bhojani, R. Schneider, M. A. Philbert, A Rehemtulla, B. D. Ross, R. Kopelman; 58, 1556, (2006).
162. M.S Bhojani., *et al.*, Cancer Nanotechnology, *American Scientific Publishers*, (2006).
163. A. Gupta, S. Wang, A. Marko, P. Joshi, M. Ethirajan, Y. Chen, R. Yao, M. Sajjad, R. Kopelman, and R. K. Pandey, *Theranostics*, **4**, 614 (2014).
164. G. L, Hornyak H. F. Tibbals, J. Dutta J. J. Moore Introduction to nanoscience & nanotechnology. Boca Raton, Florida, USA: (*CRC Press;* 2009).
165. W. Wang, G. Kim, Y-E Koo, Lee, H. J. Hah, M. Ethirajan, R. K. Pandey, and R. Kopelman *ACS Nano*,**6**, 6843 (2012).

Chapter 2

Design and Synthesis of Photosensitizer-Peptide Conjugates for PDT

Tyrslai M. Williams,[†] *Martha Sibrian-Vazquez*[‡] *and M. Graça H. Vicente*[†,*]

[†] *Department of Chemistry, Louisiana State University Baton Rouge, LA 70803, USA*
[‡] *Department of Chemistry, Portland State University Portland, OR 97201, USA*

Abstract

The design and synthesis of covalently linked photosensitizer-peptide conjugates are reviewed in this chapter. Peptides of relatively small size (<50 amino acids) are most commonly used for enhancing the photosensitizer targeting specificity, due to their straightforward synthesis and ease of conjugation to photosensitizers. Porphyrin, chlorin and phthalocyanine macrocycles are the most common PDT photosensitizers used in the conjugation reactions to peptides, generally via a low molecular weight linker, following either solution or solid phase protocols. These so-called third-generation photosensitizers have shown increased selectivity for targeting specific cell enzymes or receptors, and enhanced PDT efficacy compared with unconjugated photosensitizers.

1. Introduction

Photodynamic therapy (PDT) is a tri-component treatment that combines a photosensitizer, light and oxygen to produce reactive oxygen species (ROS), which cause oxidative damage to targeted cells and tissues.[1–3] As a minimally invasive treatment for cancer and other diseases, PDT constitutes an

*Corresponding author.

alternative to radiation and chemotherapy, with the potential for becoming highly selective via the use of cell-targeted photosensitizers, combined with precise light delivery. Porphyrin and porphyrin-derived compounds are the most common PDT photosensitizers due to their unique photophysical and pharmacokinetic properties, and their usually low dark toxicity. In addition, photothermal therapy (PTT) is a related cancer treatment that involves light activation of a photosensitizer to generate electronic excitation energy, which causes local temperature increase and subsequent tumor cell destruction, both in the presence and absence of oxygen.[4,5] Photofrin is a FDA-approved porphyrin for the PDT treatment of various cancers, including lung, skin, cervical, gastric and bladder.[6,7] Another porphyrin derivative, Visudyne, is FDA-approved for the treatment of the wet form of age-related macular degeneration (AMD). However, both of these porphyrin-based drugs are mixtures of compounds with limited selectivity for tumor tissue. In the last few decades a considerable amount of research has centered on the development of new porphyrin-based sensitizers with improved physicochemical properties, in particular with increased absorption within the therapeutic window (600–800 nm), and enhanced ROS production, and clearance from normal tissues. New second-generation photosensitizers emerged from these studies, and several are currently in clinical trials, including Foscan (temoporfin, tetra(meta-hydroxyphenyl)chlorin, mTHPC), HPPH (2-[1-hexyloxyethyl]-2-devinyl-pyropheophorbide a, Photochlor), NPe_6 (mono-L-aspartyl chlorin e_6, talaporfin sodium, LS11), $SnEt_2$ (tin (IV) ethyl etiopurpurin, rostaporfin, Photrex), mTHBC (tetra(meta-hydroxyphenyl)bacteriochlorin), and phthalocyanine 4 (Pc4). Some of these photosensitisers and their formulations have shown preferential tumor accumulations relative to the surrounding healthy tissues, due to the so-called enhanced permeability and retention (EPR) effect.[8,9] Most recently, research has centered on the development of tumor-targeted photosensitizers, containing biomolecule(s) directed against antigens or ligands over-expressed on tumor cells, commonly referred to as third-generation photosensitizers.[10] Among these, photosensitizer-biomolecule conjugates bearing covalent linkage(s) to molecules such as monoclonal antibodies, hormones, vitamins, oligonucleotides, carbohydrates and peptides, have received particular attention. The choice of biomolecule is crucial for the targeting ability of the conjugates; in addition, the degree of receptor expression in the targeted cells, the binding affinity for the ligand, and the structural and photochemical properties of the conjugates are determinants of biological efficacy.[11,12] In most studies reported to date the photophysical properties of the photosensitizer remain unchanged upon bioconjugation, and the tumor-targeted conjugates show enhanced PDT activity in comparison with the non-conjugated photosensitizers. The use of small bioactive peptides conjugated to photosensitizers is a particularly convenient strategy due to their readily availability, low immunogenicity, low cost (relative to proteins and antibodies), and the easy tuning of their specific recognition properties.[13,14] In addition, the use of cyclic peptides, N-alkylated derivatives, and D-amino acids can be used to increase the stability of the peptides.[15,16] This chapter gives a general overview of the design strategies and the most common conjugation methodologies leading to the synthesis of third-generation PDT photosensitizers bearing covalently attached peptides, and reviews the available biological activity of the conjugates.

2. Common Conjugation Methodologies

2.1. *General Considerations*

The general design of a photosensitizer-peptide conjugate for PDT involves three components, as shown in Fig. 1. Each component plays an important role in modulating the physicochemical properties and biological efficacy of the conjugate. The vast majority of PDT photosensitizers are based on a tetrapyrrolic macrocycle, such as a porphyrin, chlorin, bacteriochlorin or phthalocyanine. Recently, boron dipyrromethene (BODIPY) derivatives have also been investigated as PDT photosensitizers.[17,18] A linker is generally used

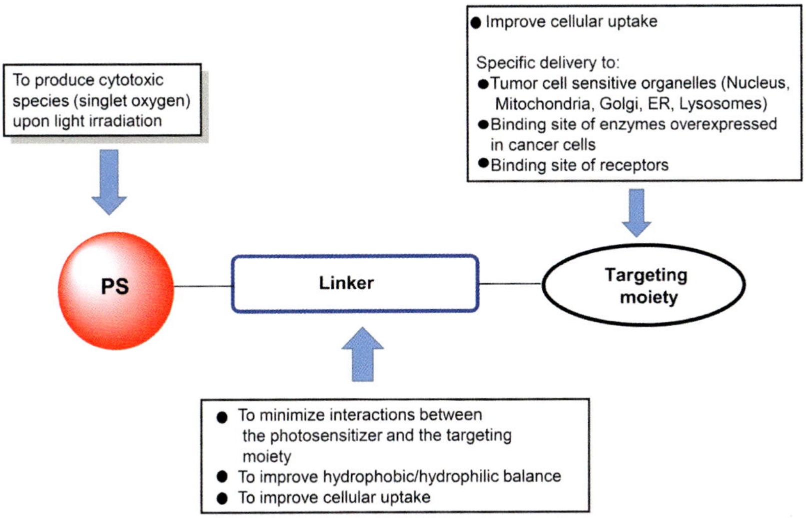

Fig. 1. General design of third-generation photosensitizers.

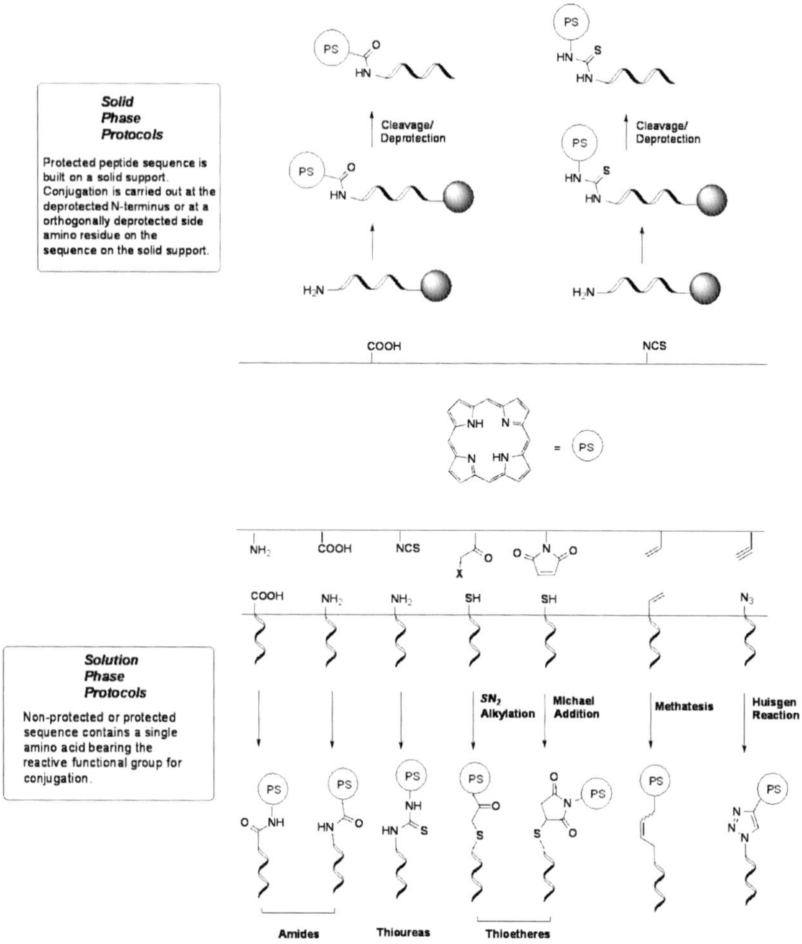

Fig. 2. Conjugation protocols normally used for preparing photosensitizer-peptide conjugates.

to minimize intramolecular interactions between the photosensitizer and the peptide.

Photosensitizer-peptide conjugates can be synthesized using either solution or solid phase protocols (Fig. 2). The method of choice depends on the structural characteristics of both the photosensitizer and the peptide sequence. Solution phase-based protocols for conjugation in aqueous or organic media are limited to peptides bearing a single amino acid that contains the proper functionality for conjugation. The conjugation reactions typically used for the preparation of photosensitizer-peptide conjugates involve: (a) amide bond formation via activated carboxylic acids and amine reactive species; (b) Huisgen cycloaddition reactions; (c) cross-metathesis, and (d) SN_2 alkylations and

Michael additions for thiol containing peptides. These are discussed in detail below.

Protocols for conjugation of photosensitizers on solid phase support are in general based on standard solid phase peptide synthesis (SPPS) methodology. SPPS is an attractive approach for the synthesis of conjugates, particularly those containing peptides with 10 or more amino acid residues, due to the shorter time required for peptide synthesis compared with solution phase. SPPS uses excess of reagents to drive the reaction to completion, and avoids the separation and purification of intermediates since these are simply washed out. In this approach, the peptide sequence is first fully assembled on the solid support. Then, either the alpha amino group of the last amino acid on the sequence, or an orthogonally protected side chain amino group are deprotected, thus providing a free amino group available for conjugation to an amine reactive photosensitizer. Finally, a one step deprotection and cleavage from the resin affords the photosensitizer-peptide conjugate. Although SPPS offers several advantages over solution phase protocols, the major disadvantages of solid phase support-based conjugations are the low yields often obtained, associated with poor solvation of the conjugate that leads to photosensitizer- or peptide-resin adsorption, and/or to degradation of the photosensitizer during the cleavage/deprotection step.

2.2. *Carboxylic Acid Activation*

Photosensitizer peptide conjugates prepared *via* amide bond formation require the use of coupling reagents to produce an activated carboxylic acid-containing photosensitizer which reacts with the *N*-terminus of the peptide sequence or with an orthogonally deprotected amino side residue, to afford the desired conjugate. Several coupling reagents have been used to produce the conjugates, including mixed anhydrides, carbonic anhydrides or active esters. The choice of the proper coupling reagent depends on the chemical properties (reactivity, steric hindrance, stability) of both the photosensitizer and the terminal amino acid on the peptide sequence to be conjugated.[19]

Activation of the carboxylic acid group at the photosensitizer periphery can be carried out either by synthesis, isolation and purification of the corresponding activated ester, or by generating it *in situ* using the appropriate activating reagents. *N*-hydroxysuccinimyl activated photosensitizer esters have been used in solution phase protocols for conjugate synthesis (Table 1, entries 1–3). Advantages of *N*-hydroxysuccinimyl esters rely on their readily reaction with amines under mild conditions, and minimum racemization of the amino acid

Table 1. Photosensitizer-peptide conjugates prepared *via* amide bond formation using solution phase protocols.

Entry	Photosensitizer	Coupling reagent	Peptide sequence	Reaction conditions (yield)	Ref
1	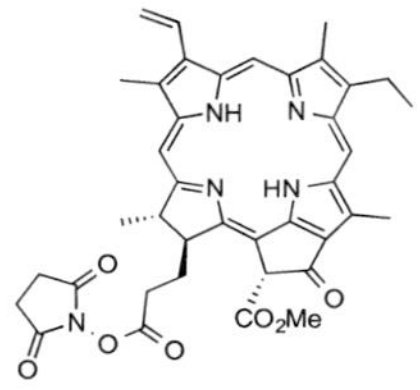	—	GFTRSAGGGG	DMSO	23, 24, 25
2	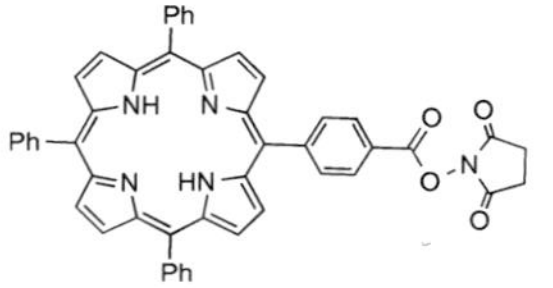	—	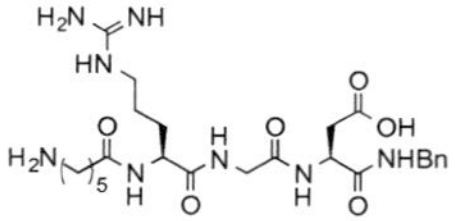	DIEA	26
3	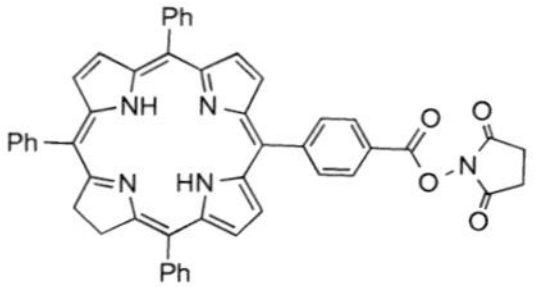	—	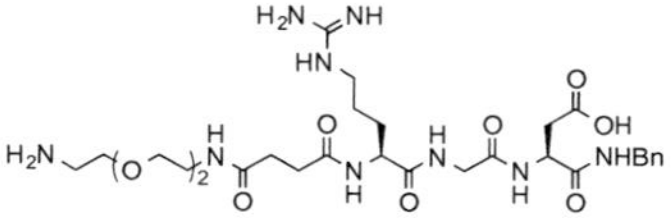	DIEA	26
4	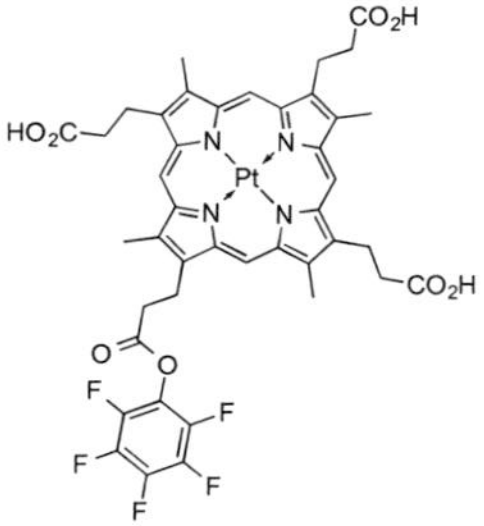	—	Bacteneсin 7 (15-24)	DMSO, rt, 24–72 h	27

5	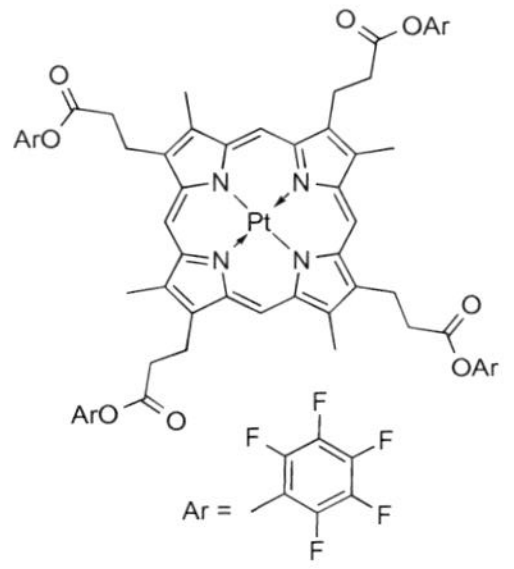	—	GRRRRRRRRR RR RRA RRS RRR RGR	DMSO, rt, 24–72 h	28
6	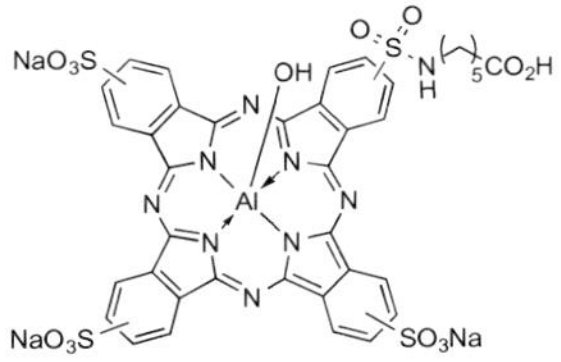	—	Bombesin	Carbonate buffer, pH 8.5	29
7	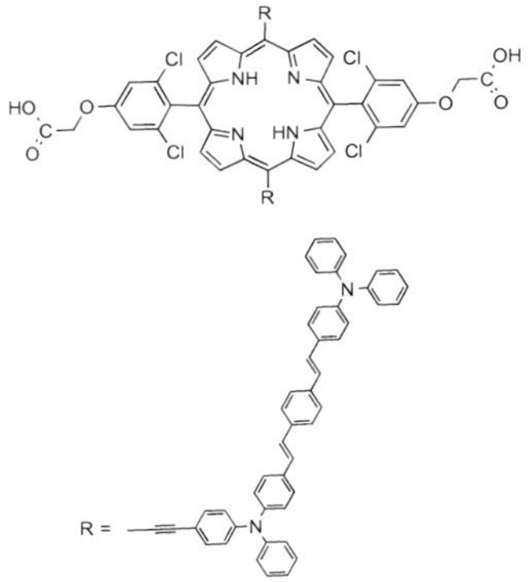	DCC DMAP	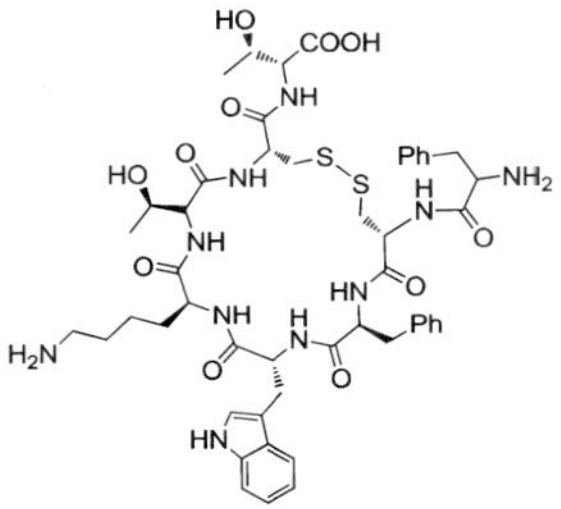	DMF	30

(*Continued*)

Table 1. (*Continued*)

Entry	Photosensitizer	Coupling reagent	Peptide sequence	Reaction conditions (yield)	Ref
8	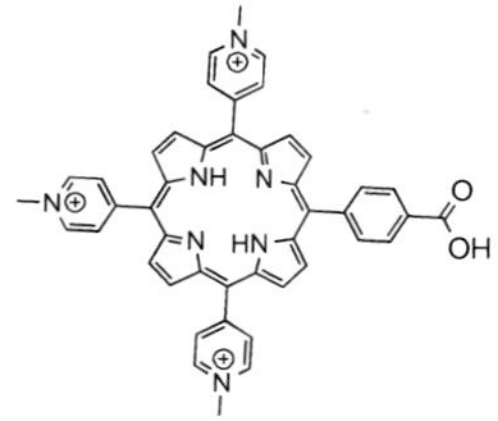	EDC	AcK(AAA)	DMF, DIEA	31
9	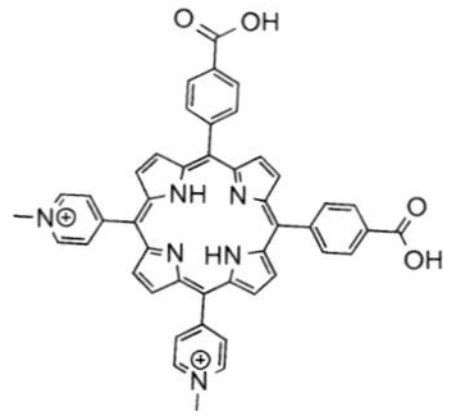	BOP	AcK(AAA)	DMF, DIEA, (47%)	31
10	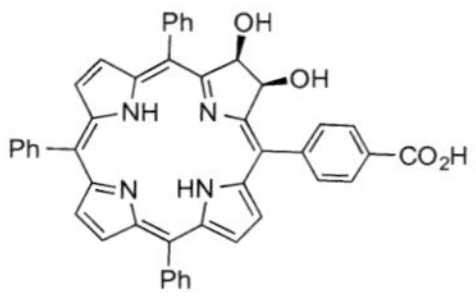	HBTU HOBt	RRRRRRR	DIEA Solid phase, (28%)	32

11	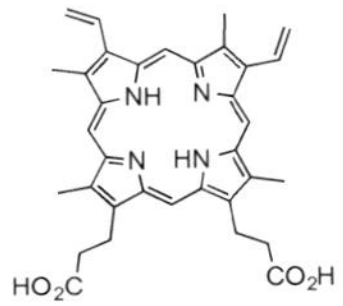	PyBOP	AA_1-AA_2-AA_3-SY-GnRH agonist: AA_1 = pGlu, AA_2 = His, AA_3 = Trp or AA_1 = (*D*)-pGlu, AA_2 = (*D*)-Phe, AA_3 = (*D*)-Trp	NMM, DMF (65%)	33
12	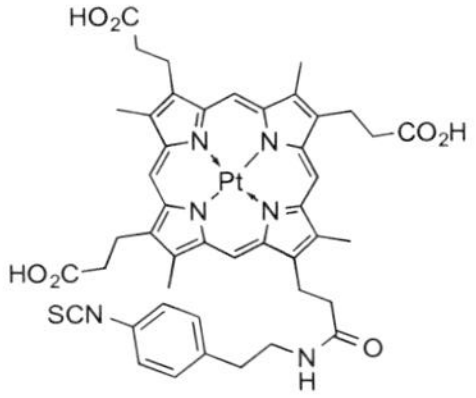	—	FRRRRRRRR Bactenectin GPRPLPFPRPG	DMSO	28

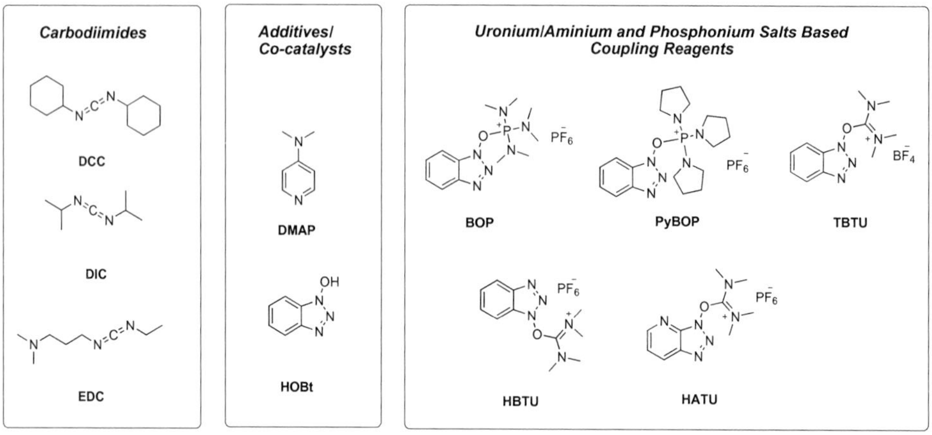

Fig. 3. Coupling reagents normally used for the preparation of photosensitizer-peptide conjugates, in solution and solid phase protocols.

on the peptide sequence. Pentafluorophenyl-activated photosensitizer esters have also been used for the synthesis of several photosensitizer-cell penetrating peptide conjugates (Table 1, entries 4–5). Carbodiimides, uronium/aminium and phosphonium salts-based coupling reagents (Fig. 3) are commonly used in the synthesis of photosensitizer-peptide conjugates, using both solution and solid phase protocols. DMAP and HOBt are normally used as additives that act as acyl transfer reagents in conjugation reactions involving carbodiimides as the coupling reagents. DMAP and HOBt react with the *O*-acylurea intermediate to form a more reactive acylpyridinium carboxylate ion pair and the benzotriazole ester, respectively.[19,20] In addition to improving the conjugation reaction yield, the conditions decrease epimerization levels; a problem found when carbodiimides are used as the coupling reagents, is usually obtained when HOBt is used as an additive.[21,22] Tables 1 and 2 provide a summary of the reaction conditions used for the synthesis of photosensitizer-peptide conjugates in solution and solid phase, respectively.

2.3. *Huisgen [3+2] Cycloadditions*

The copper(I)-catalyzed addition of organic azides to acetylenes (both internal and external) to form triazoles, known as the Huisgen [3 + 2] cycloaddition reaction, has been widely exploited for conjugation reactions. Since the azide and alkyne groups are relatively easy to incorporate into organic compounds via nucleophilic or electrophilic reactions, the scope of this reaction is broad.[64–66] For example, Ikawa *et al.*[67] reported the synthesis of a hydrophilic

Table 2. Photosensitizer-peptide conjugates prepared *via* amide bond formation using solid phase protocols.

Entry	Photosensitizer	Coupling reagent	Peptide sequence	Reaction conditions (yield)	Ref
1	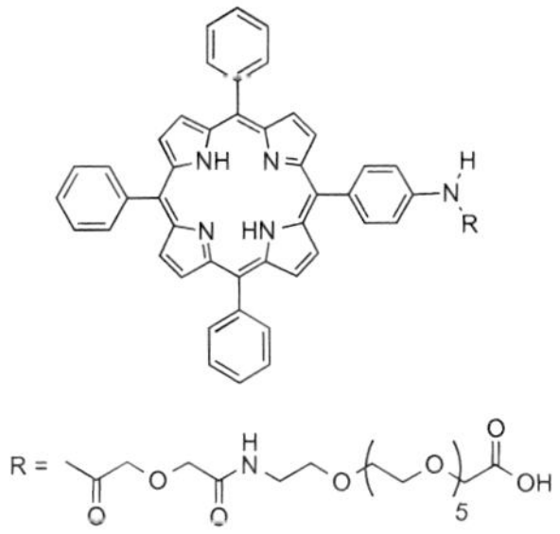	TBTU, HOBt	(1) MSVLTPLLLR GLTGSARRLP VPRAKIHSL (2) GPKKKRKV (3) GRRRRRRRR (4) GRKKRRQ RRRPPQ	DMF, rt, 24 h (13–52%)	34, 35, 36, 37
2	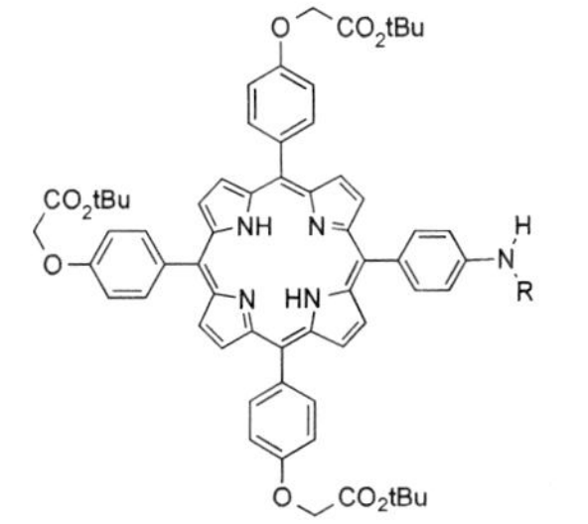 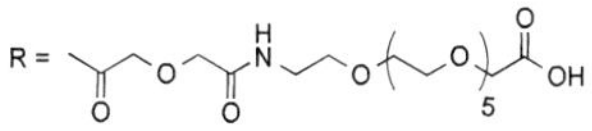	TBTU, HOBt	(1) GPKKKRKV (2) GRKKRRQR RRPPQ	DMF, rt, 24 h (13–52%)	35
3	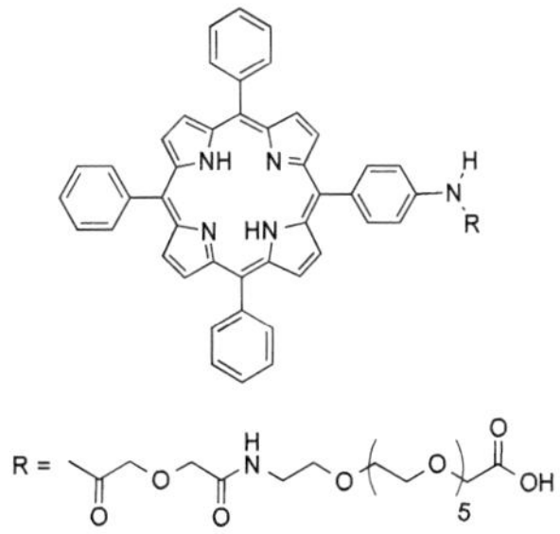	TBTU, HOBt	(1) APPKKKRKVE DPRKKRRQR RRPPQGG (2) MGLGLHLLV LAAALQGAW SQAPPKKKRKVG (3) RQIKIWFQ NRRMKWKKA PPKKKRKVGG (4) KRPAATKKAGQ AKKKLEDPRKK RRQRRRPPQG	DMF, DIEA rt. 24 h, (18–72%)	38
4	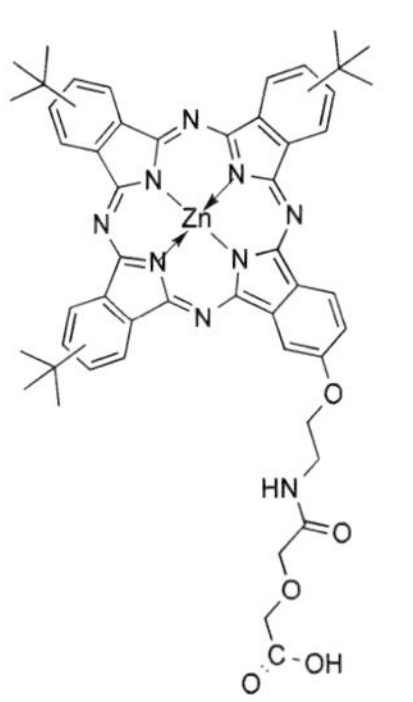	TBTU, HOBt	(1) KRPAATKKAGQ AKKKLEDPRKK RRQRRRPPQG	DMF, DIEA, rt, 24 h, (57%)	39

(*Continued*)

Table 2. (*Continued*)

Entry	Photosensitizer	Coupling reagent	Peptide sequence	Reaction conditions (yield)	Ref
4	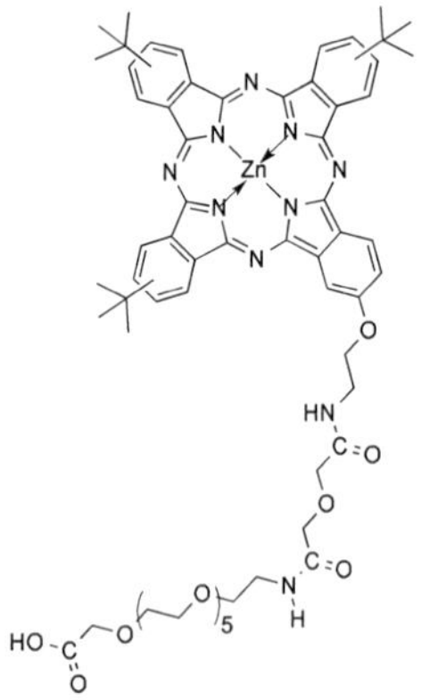	TBTU, HOBt	KRPAATKKA GQAKKKLED PRKKRRQR RRPPQG	DMF, DIEA, rt, 24 h, (42%)	39
5	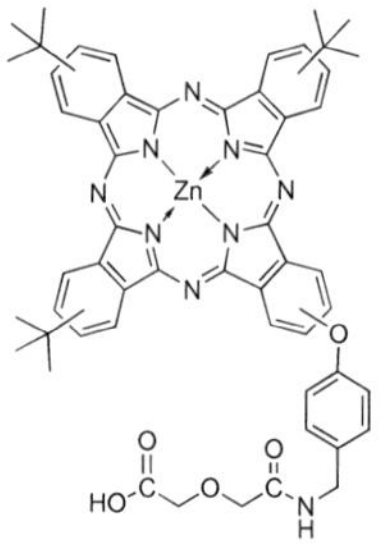	TBTU, HOBt	(1) LARLLT (2) GYHWY GYTPQNVI	DMF, DIEA	40
6	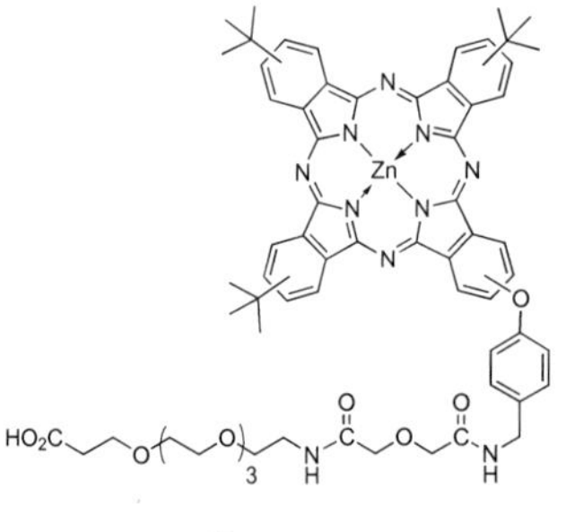	TBTU, HOBt	(1) LARLLT (2) GYHWY GYTPQNVI	DMF, DIEA	40
7	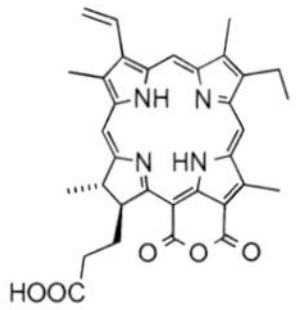	HATU	GGVKRK KKPGYG	DCM, DMF, DMS, DIEA, (85%)	41

(*Continued*)

Table 2. (*Continued*)

Entry	Photosensitizer	Coupling reagent	Peptide sequence	Reaction conditions (yield)	Ref
8	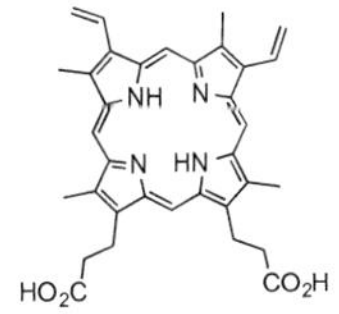	HATU	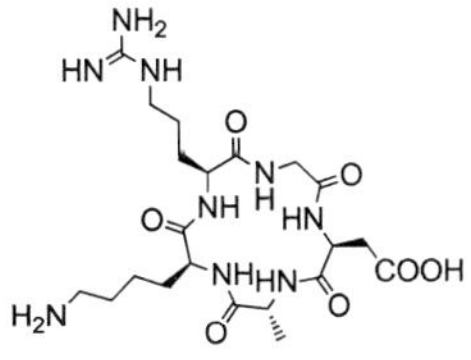	DCM, DMF, DMS, DIEA	42
9	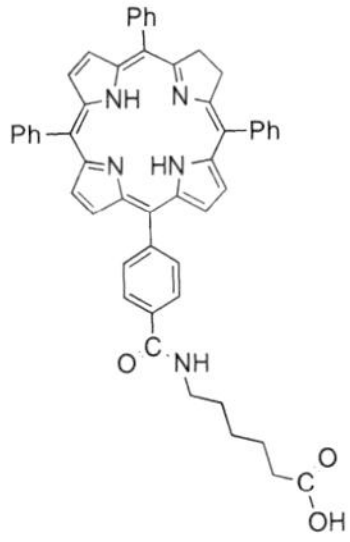	HBTU, HOBt	ATWLPPR	DMF, DIEA	43
10	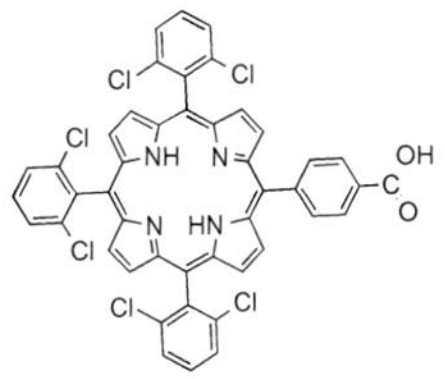		KDYMGWMDF		44
11	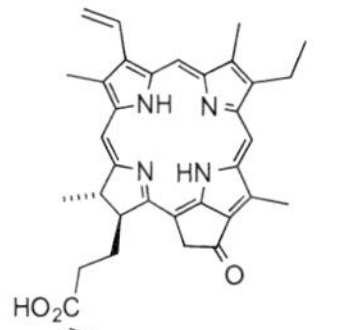	HBTU, HOBt	(1) GDEVDGSGK (2) GDEVDGSGK (3) TSGPNQEQK (4) GPLGLARK	NMP, (95%)	45, 46, 47
12	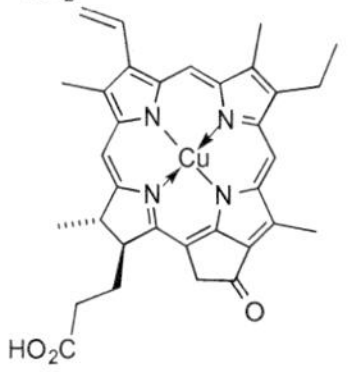		GDEVDGSGK-Folic acid		48
13	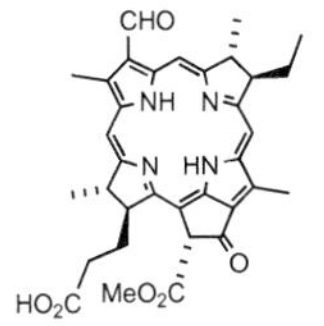		GDEVDGSGK-Folic acid		49

(*Continued*)

Table 2. (*Continued*)

Entry	Photosensitizer	Coupling reagent	Peptide sequence	Reaction conditions (yield)	Ref
14	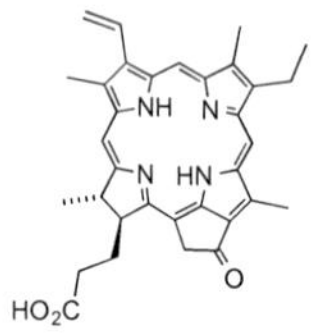		(1) RRRRRRRR-GPLGLARK-EEEEEEEE (2) RRRRRRRR-RPLALWRSK-EEEEEEEE		50
15	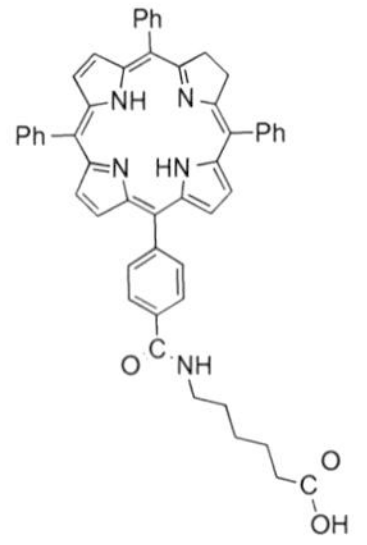		(1) ATWLPPR (2) D-Ala-TWLPPR (3) [CH_2NH_2]Ala-TWLPPR		51
16	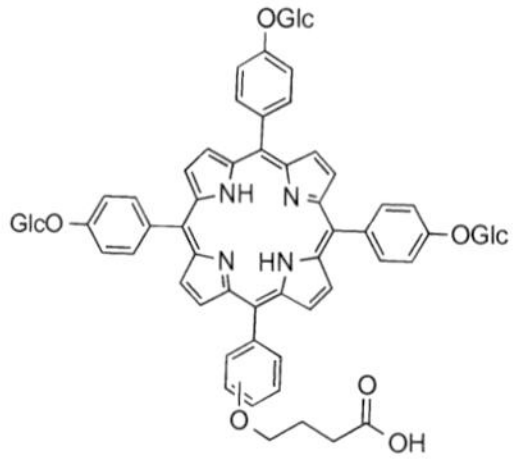	DIC, HOBt	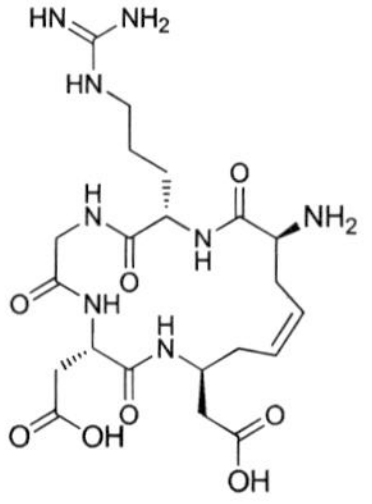		52
17	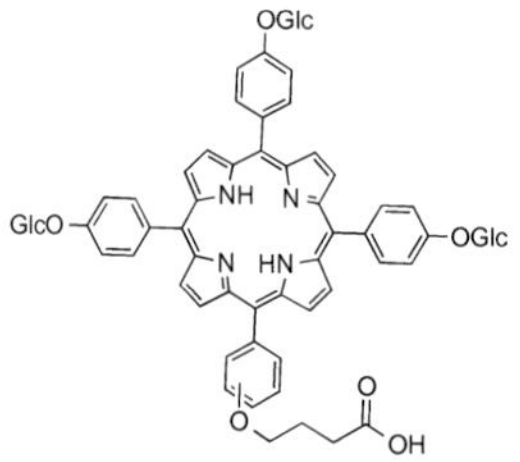	DIC, HOBt	RGD		42
18	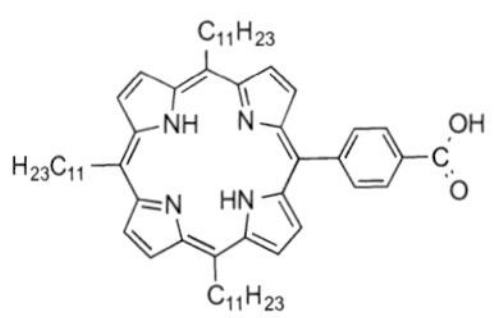	PyBOP	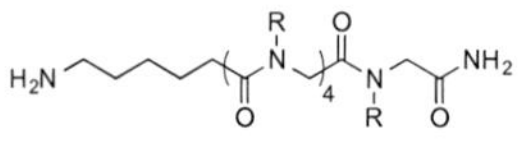 R = $(CH_2)_4NH_2$ (oligo-Ψ-Lys) $(CH_2)_6NH_2$ $(CH_2)_8NH_2$	DIEA	53

(*Continued*)

Table 2. (*Continued*)

Entry	Photosensitizer	Coupling reagent	Peptide sequence	Reaction conditions (yield)	Ref
19	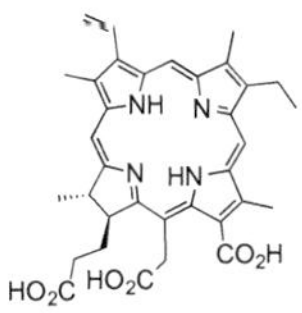	DCC	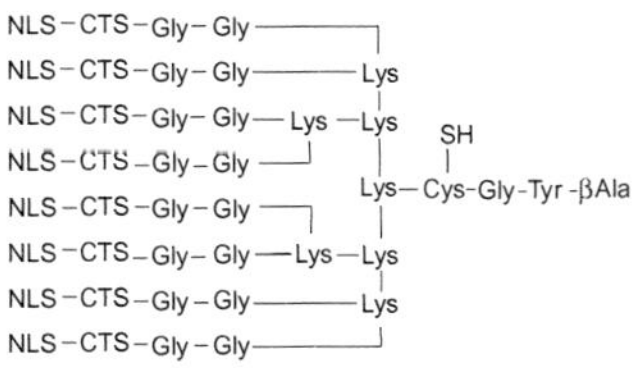 NLS= T124A simian virus SV40 large Ta ntigen (124-135) CTS= KKKKK	rt, 40 min	54
20	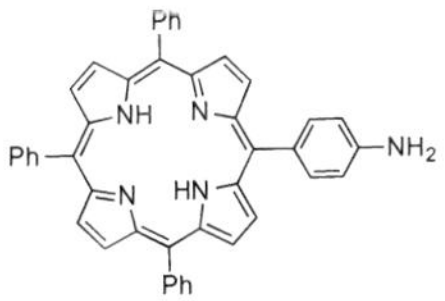	EDCI, HOBt	K KK KKK RRR KRRR RRKR	DIEA, DMF, rt, 24 h (45–95%)	55
21	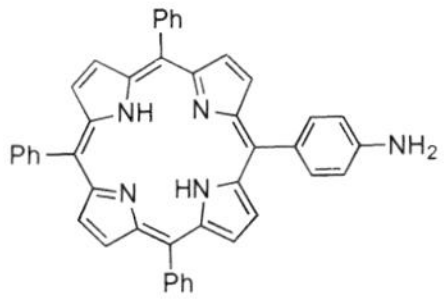	DCC, DMAP	K KK KKK RRR KRRR RRKR	DIEA, DMF, rt, 24 h (18–65%)	55
22	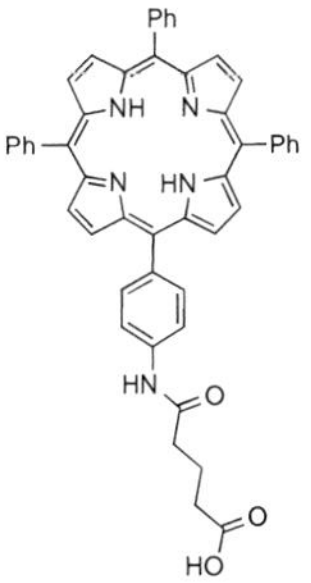	TBTU, HOBt	K KK KKK RRKR	DIEA, DMF rt, 24 h	55
23	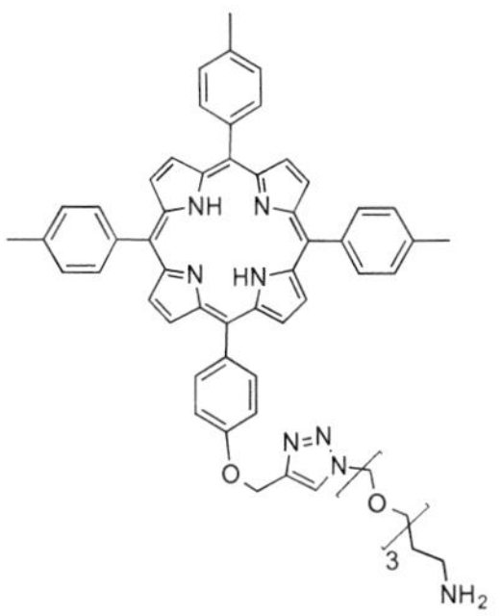	EDC, NHS	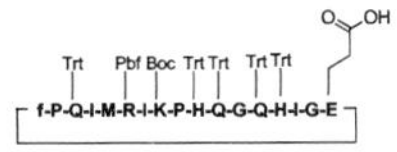	DMF	56

(*Continued*)

Table 2. (*Continued*)

Entry	Photosensitizer	Coupling reagent	Peptide sequence	Reaction conditions (yield)	Ref
24	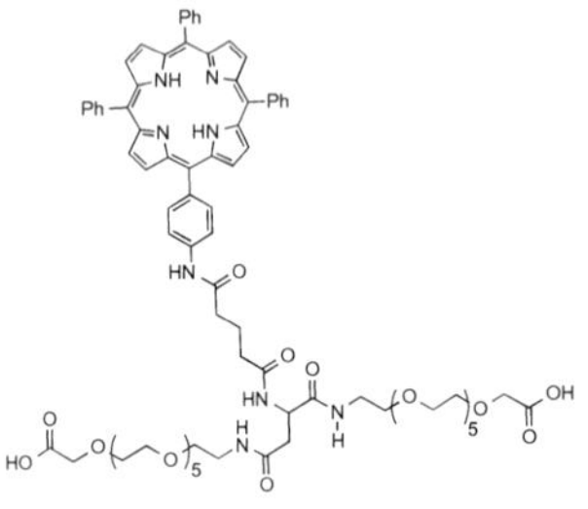	TBTU, HOBt	GPKKKRKV	DMF	57
25	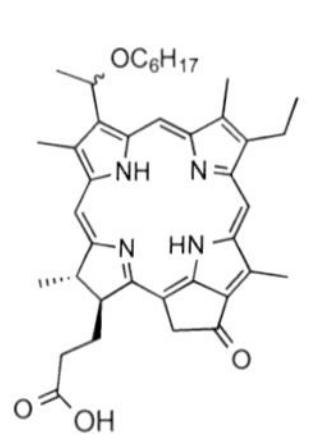	DMAP DCC	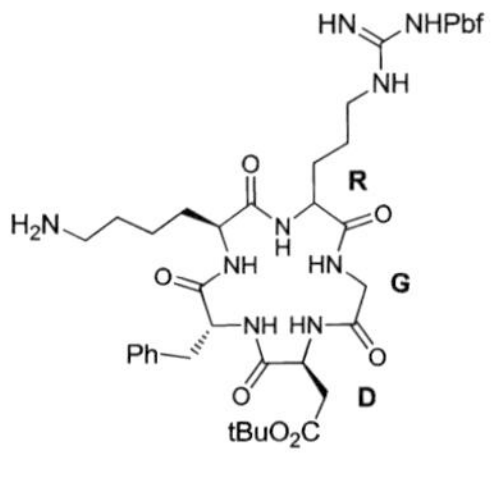	DMF	58
			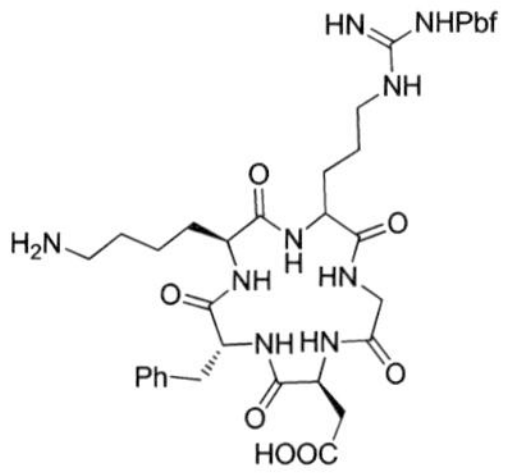		
			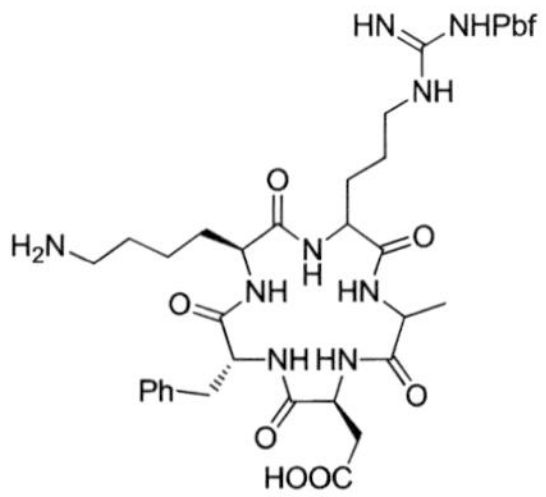		

(*Continued*)

Table 2. (*Continued*)

Entry	Photosensitizer	Coupling reagent	Peptide sequence	Reaction conditions (yield)	Ref
26	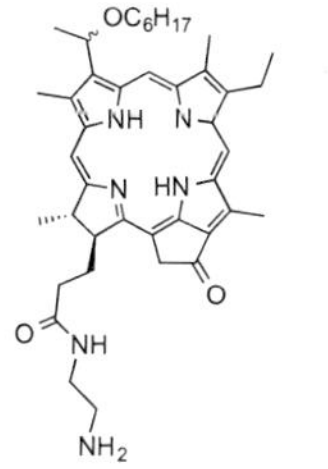	DCC, DMAP	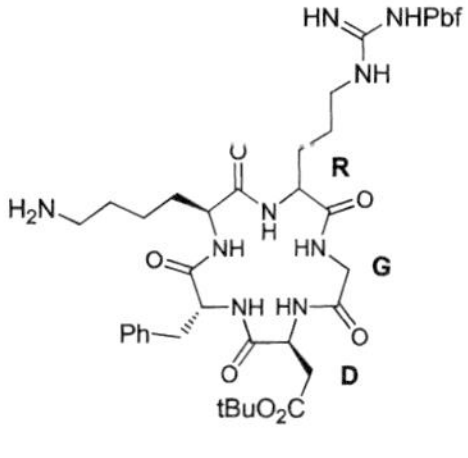	DMF	58
			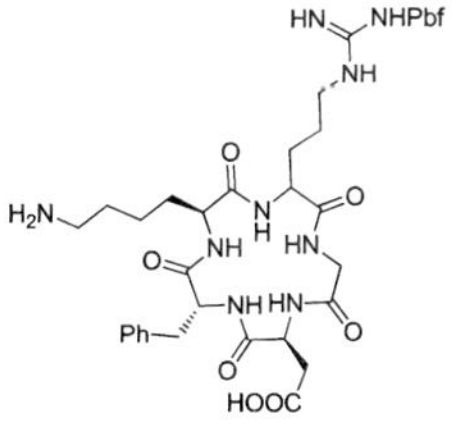		
			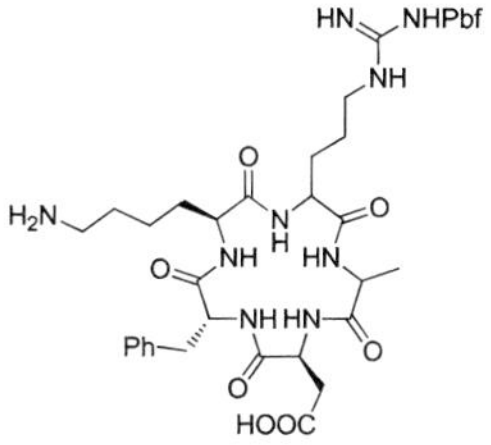		
27	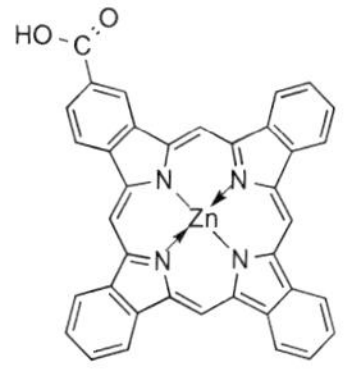	HOBT, HBTU	GHWSYGLRPG	DIEA, DMF	59
28	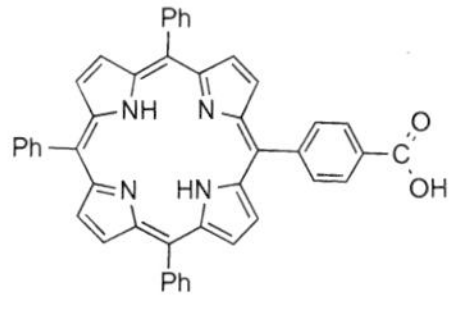	TBTU, HOBT	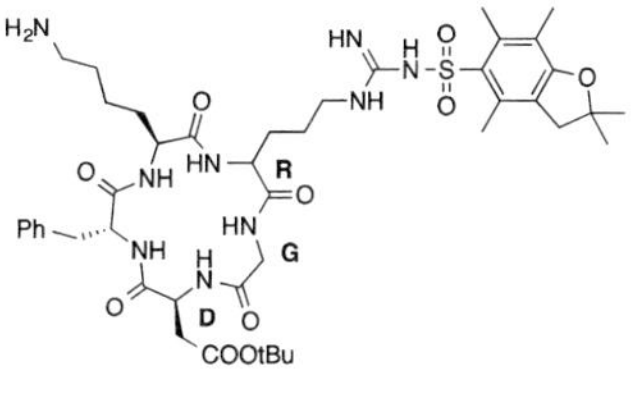 H-(BuOt)DFKR(Pbf)G	DIEA, DMF	60

(*Continued*)

Table 2. (*Continued*)

Entry	Photosensitizer	Coupling reagent	Peptide sequence	Reaction conditions (yield)	Ref
29	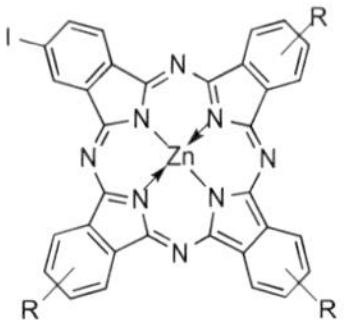	Pd(OAc)$_2$, (o-tolyl)$_3$P	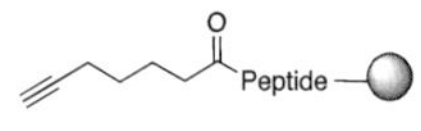 YVG= TyrValGlyNH- RGD= ArgGlyAspNH- BBN= NH(CH$_2$CH$_2$O)$_2$ CH$_2$(O)D- TyrGlnTrpAlaValb- AlaHis3-AlaNleNH-	DIPEA, DMF or THF	61
30	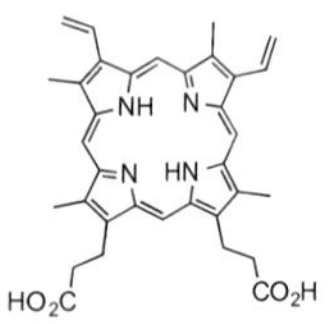	PyBOP	AAhexPQRRSARLSA ATWLPPR 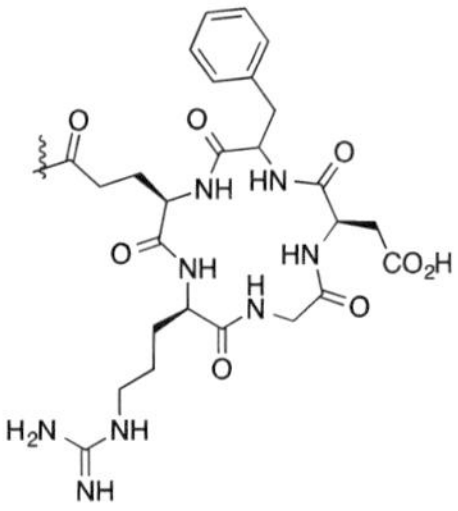 (cERGDPhe)	DMF, DIEA	62
31	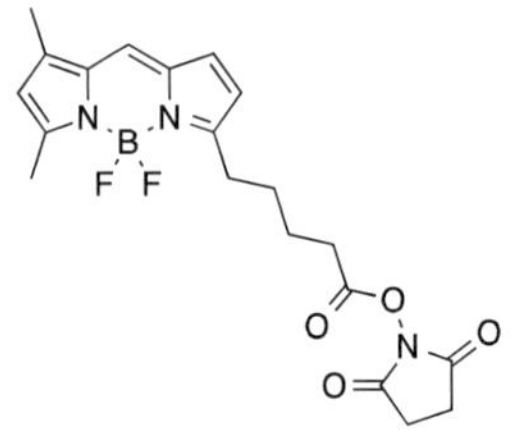		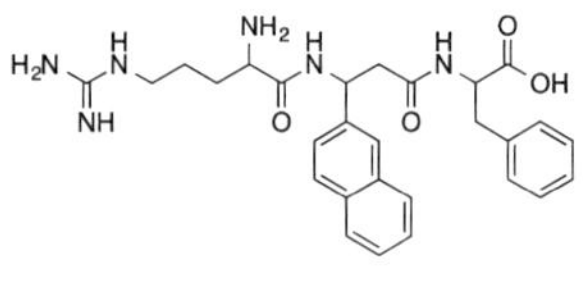	1. Et$_3$N, DMF, 4h 2. DMF, 72h	63

nona-arginine *N*-fused tetraphenylporhyrin conjugate (Scheme 1) on solid phase support using a copper-catalyzed cycloaddition reaction.

2.4. *Cross-metathesis*

Ring opening cross-metathesis (RCM) reactions are often used for crosslinking and labeling of proteins and peptides.[68] Carbon–carbon double bond forming reactions offer an alternative approach for orthogonal labeling of biological molecules. This approach was exploited by Sol *et al.*[69] who prepared in solid

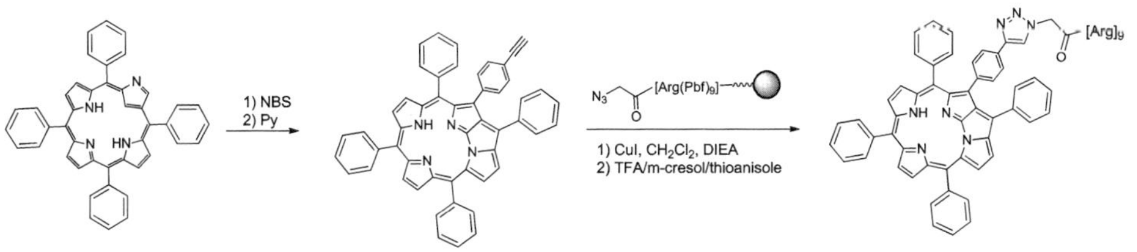

Scheme 1. Synthesis of a nona-arginine *N*-fused tetraphenylporhyrin conjugate *via* a copper-catalyzed cycloaddition reaction.[67]

phase support a glucosylated porphyrin dimer containing, as a spacer, the pentapeptide allylglycyl-RGD-allylglycine sequence. As shown in Scheme 2, the desired conjugate was prepared *via* a cross-metathesis reaction using Grubb's catalyst, in 57–63% yield after cleavage and deprotection from the solid support.

2.5. *Isothiocyanate Reactions*

Isothiocyanate-containing compounds are considered amine reactive compounds because they selectively react with primary amines (α-amino group on amino acids or the ε-amino group of lysine). The isothiocyanate functionality can be installed in organic molecules by reaction of an aromatic amine with thiophosgene, or by using 1,1'-thiocarbonyldi-2(1*H*)-pyridone.[55,70] Synthesis of conjugates using isothiocyanate-derived photosensitizers, involves the nucleophilic attack of the primary amine on the electrophilic carbon of the isothiocyanate group to form a thiourea (Fig. 2).[71] Dmitriev *et al.*[72] prepared a series of Pd(II)-coproporphyrin-I isothiocyanate conjugates containing the HIV-derived Tat (48–56) peptide sequence and the bactenecin 7-derived (15–24) peptide sequence (Table 3, entry 1).

The conjugation of a mono-functional isothiocyanate tetraphenylporhyrin derivative (Table 3, entry 2) with protected amino acids and positively charged peptide sequences was evaluated by us, in both solution and solid phase.[55]

2.6. *Thiol SN_2 Alkylation and Michael Addition*

Cysteine residues in proteins and peptides, either native or specifically incorporated into a particular peptide sequence, provide a reactive functionality for selective conjugations *via* SN_2-type alkylation to halide derivatives, or by Michael addition to maleimide moieties properly installed in the photosensitizer framework (Fig. 4). The higher nucleophilicty of the thiol group (cysteine pKa $\sim$ 8.0) compared with those of the hydroxyl group in serine, threonine, tyrosine, and the amine group on the side chain of lysine (pKa $\sim$10),

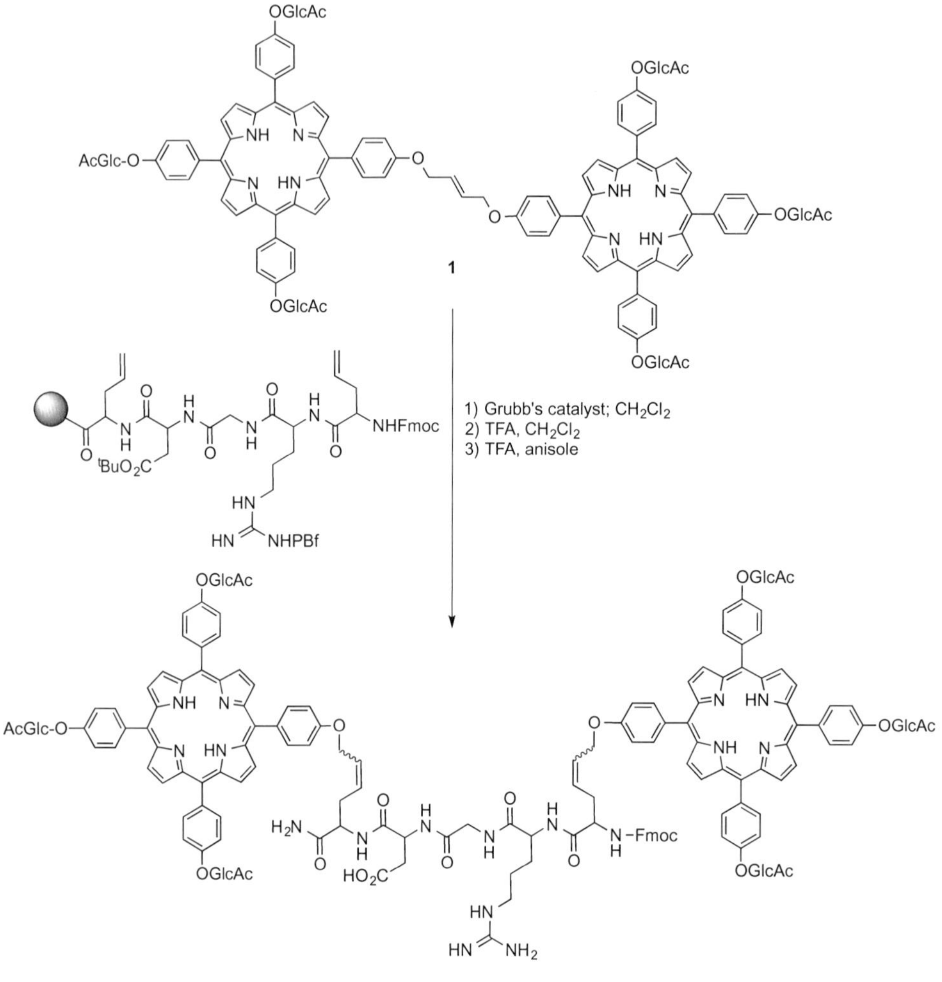

Scheme 2. Synthesis of an RGD-porphyrin dimer conjugate *via* cross-metathesis as reported by Sol *et al.*[69]

ornithine or the *N*-terminus of a peptide sequence (pKa 9–11 depending on the terminal amino acid), allow these reactions to occur under controlled pH conditions. This strategy was used successfully by Chaloin *et al.*[73] for the conjugation of a Mn-porphyrin to the chimeric peptide (AcMGLGH-LLVLAAALQGAKKKRKVC) incorporating a nuclear localization peptide sequence (NLS) in aqueous conditions at pH 8.2 (Table 4, entry 1). To achieve this, the corresponding photosensitizer was initially derivatized by introducing the required iodoacetamide functionality at the porhyrin periphery.

Table 3. Conjugation of photosensitizers to peptides *via* isothiocyanate groups.

Entry	Photosensitizer	Peptide sequence	Reaction conditions (yield)	Ref
1	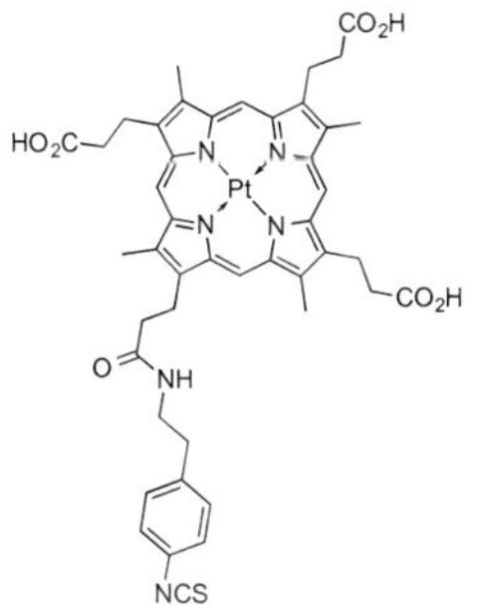	FRRRRRRRR GPRPLPFPRPG	DMSO, TEA, rt, 24 h (100% based on Abs)	72
2	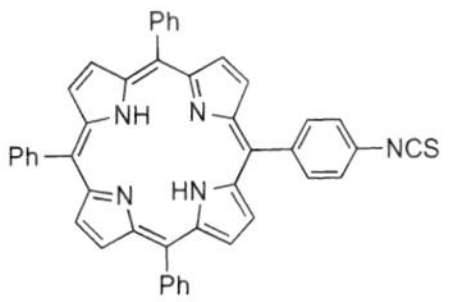	K KK KKK RRKR	DMF, rt, 24 h (35–93%)	55

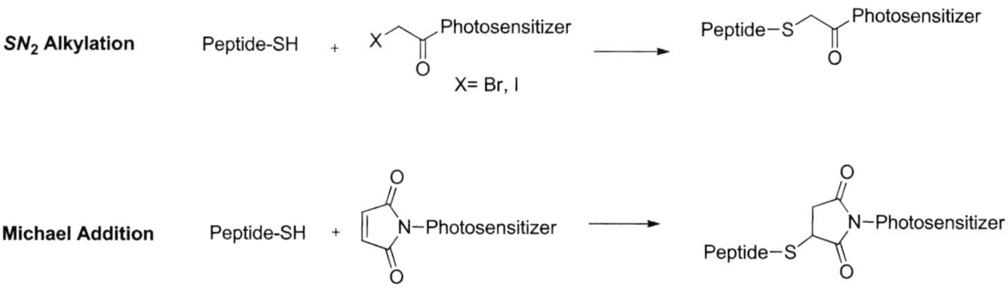

Fig. 4. General conjugation reactions of peptide sequences containing thiol reactive groups with activated photosensitizers.

Michael addition of the thiol group to a double bond can occur via a free radical or ionic mechanism. This type of reaction is efficient and usually gives high yield of the desired conjugate. Peptide conjugation via thiol double bound addition to maleimide-photosensitizers has been reported by Asayama *et al.*[74] Wang *et al.*[75] and Bourre *et al.*[76] The maleimide moiety can be introduced by reaction of the amino-functionalized photosensitizer with the bifunctional activated *N*-hydroxysuccinimyl SMCC reagent (succinimidyl-4-*N*-maleimido)-cyclohexane-1-carboxylate) (Table 4, entry 2), or with *N*-hydroxysuccinimyl *N*-propionic acid maleimide (Table 4, entries 3–7).

3. Common Targets

3.1. *Hormone Receptors*

Dubuc *et al.*[29] designed sulfonated aluminum phthalocyanine ($AlPcS_4$) conjugates that target gastrin-releasing peptide receptors (GRPR). The GRPR are 7-transmembrane G-protein coupled receptors that are overexpressed in several cancers, including colon, lung and prostate. For example, in prostate cancer, GRPR have been shown to be expressed early in tumor development and their expression correlates with the aggressive behavior of the tumors.[79] The $AlPcS_4$ was conjugated to 8-Aco-bombesin[80] using a C-14 linker (Table 1, entry 6). The biological activity of the $AlPcS_4$ conjugate was evaluated in PC-3 human prostate tumor cells, and its relative binding affinity (RBA) to GRPR was found to be 2.3. The conjugate was shown to have 2.5 times greater photodynamic efficacy compared with unconjugated $AlPcS_4$.

Starkey *et al.*[30] designed a two-photon activatable photosensitizer with therapeutic window 780–950 nm, to target the somatostatin receptor type 2 (SST2r). The authors used two cyclic peptides (SST2r and SST5r) to target the receptors (Table 1, entry 7). The conjugates were investigated *in vitro* using human small cell lung cancer (NCI-H69) and breast cancer (MDA-MB-231) cell lines. The conjugates were found to have significant uptake in SST2r-positive MDA-MB-231 cells compared with SST2r-negative human pancreatic carcinoma cells (CAPAN-1). The conjugates were also tested in xenografts of MCID mice bearing NCI-H69 or MDA-MB-231 tumors. The authors were able to find notable regression of tumors at depths as low as 2 cm.

Rahimipour *et al.*[33] chose to target the GnRH receptor by way of a protoporphyrin IX (PpIX) conjugated to a gonadotropin-releasing hormone (GnRH) peptide agonist (pGlu-His-Trp-Ser-Tyr-D-Lys-Leu-Arg-Pro-Gly-NH_2) and a GnRH peptide antagonist (D-pGLU-D-Phe-D-Trp-Ser-Tyr-D-Lys-Leu-Arg-Pro-Gly-NH2) (Table 1, entry 11). The phototoxicity of the PpIX conjugates was evaluated in both mouse pituitary gonadotrope (αT3-1) and in human breast cancer (MCF-7) cells. The selectivity of the conjugates was evaluated via co-incubation studies. When the PpIX-peptide agonist conjugate was co-incubated with increasing concentrations of the agonist peptide, a dose-dependent increase in cell survival was observed. On the other hand, when the agonist peptide was co-incubated with unconjugated PpIX the survival rate was

Table 4. Conjugation of photosensitizers to peptide containing cysteine residues.

Entry	Photosensitizer	Peptide sequence	Reaction conditions, yield	Ref
1	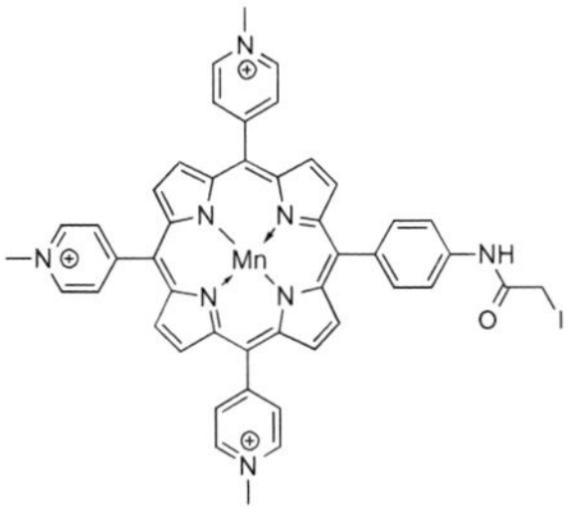	AcMGLGHLLVLAA ALQGAKKKRKVC	50 mM HEPES buffer pH 8.2	73
2	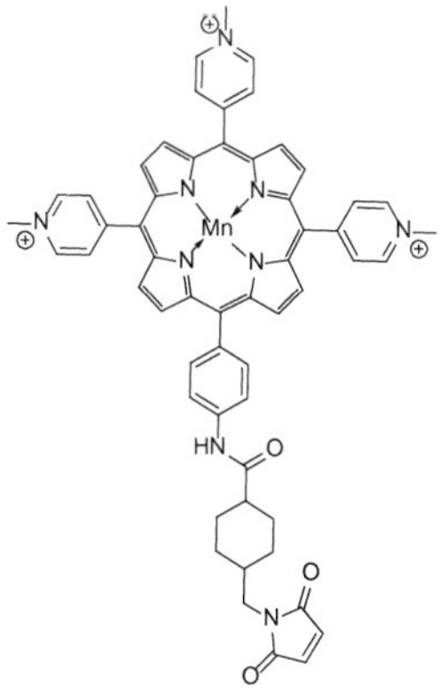	MLSLRQSIRFFKGC	DMF, 37°C, 3 d	74, 77, 78
3	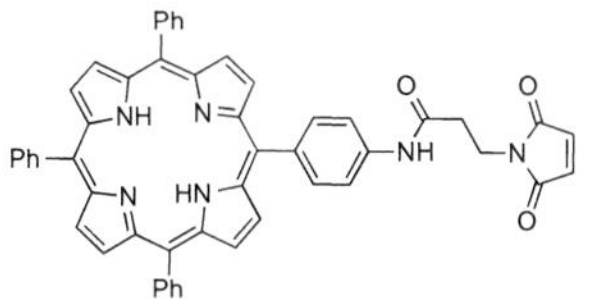	GEKKEEQRRR-GYKC	DMSO, pyridine, 7.5%	76
4	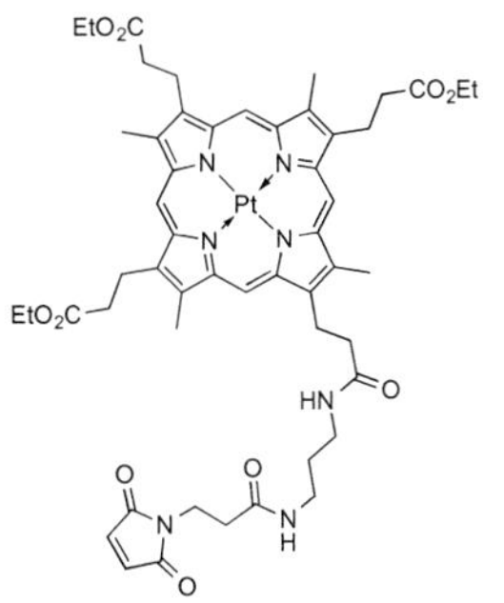	CFRRRRRRRRR MGRTVVVLGGGIS GLAAGCRRRRR RRRR	DMSO, rt, 24–72 h	28

(*Continued*)

Table 4. (*Continued*)

Entry	Photosensitizer	Peptide sequence	Reaction conditions, yield	Ref
5	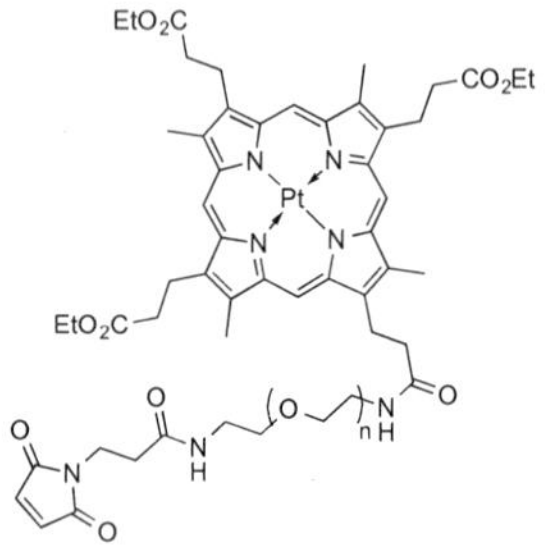	CFRRRRRRRR MGRTVVVLGGGIS GLAAGCRRRRR RRRR	DMSO, rt, 24–72 h	28
6	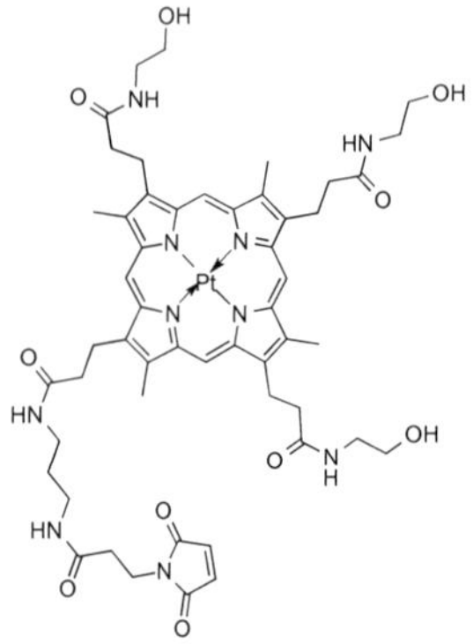	GRRRRRRRR RR RRA RRS RRR RGR	DMSO, rt, 24–72 h	27, 28
7	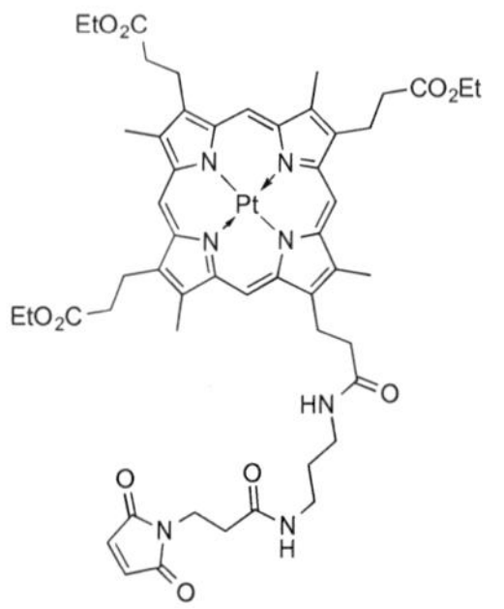	CysF-Tat (48-56) Cys-FRRRRRRRR	DMSO, rt, 24–72 h	28

not affected. The PpIX-peptide conjugate was shown to have 10 times higher selectivity for cells overexpressing GnRH receptors.

Xu *et al.*[59] designed a phthalocyanine conjugate that also targets the GnRH receptor by using a GnRH or luteinizing releasing hormone (LHRH) peptide as its receptor ligand. GnRH is a decapeptide with sequence GHWSYGLRPG (Table 2, entry 28). The GnRH receptor is most prominent in breast, ovarian, endometrial, and prostate tumors.[81,82] The ZnPc-GnRH conjugate was tested on breast cancer (MCF-7) cells, and compared with antagonist GnRH cetrorelix. The ZnPc-GnRH conjugate showed increased uptake in the tumor versus normal cells, and was able to compete with cetrorelix for GnRH receptor binding, showing specificity for GnRH receptors.

3.2. $\alpha_v\beta_3$ *Integrin Receptors*

The $\alpha_v\beta_3$ integrin is a heterodimeric transmembrane glycoprotein receptor that has been identified on several tumor cell types, including osteosarcomas, neuroblastomas and lung carcinomas.[83,84] There are 25 known $\alpha\beta$ heterodimers, and the peptide sequence RGD has been shown to bind to most of these.

Pandey *et al.*[58] designed a series of chlorophyll *a*-based photosensitizers bearing different cyclic RGD (cRGD) derivatives and evaluated their biological effects in several cell lines. The cRGD derivatives included analogs where the aspartic acid or arginine binding sites were free, and an analogue where the glycine amino acid was replaced with alanine. The three cRGD derivatives were conjugated to 2-(1'-hexyloxyethyl)-2-devinylpyropheophorbide *a* (HPPH) (Table 2, entry 26). The compounds showed different cellular uptake in $\alpha_v\beta_3$-positive brain tumor (U87) and breast tumor (4T1) cells vs. $\alpha_v\beta_3$-negative A431 cells. In particular, the conjugate possessing both binding sites free showed exceptional cellular uptake and PDT activity. The uptake was altered when one binding site was blocked, and the replacement of glycine with alanine blocked binding of the cRGD peptide due to the additional methyl group. Molecular modeling was performed to gain more insight about the properties of the conjugates.

Frochot *et al.*[60] reported the synthesis of porphyrin- and chlorin-based photosensitizers conjugated to a linear or cyclic RGD-containing peptide (Table 2, entry 29). The biological properties of the conjugates were tested on HUVEC cells that positively express $\alpha_v\beta_3$ integrins and on murine mammary carcinoma EMT-6 cells that negatively express $\alpha_v\beta_3$ integrins. The integrin expression was evaluated using flow cytometry. The conjugates were found to have 98-fold higher cellular uptake for linear RGD and 8-fold higher uptake for cyclic RGD.

Johan van Lier *et al.*[61] synthesized a series of ZnPc-peptide conjugates to target the GRPR and integrin receptors. Peptides of different lengths and structures were chosen, including YVG, RGD, and BBN (6–14), for conjugation to ZnPc (Table 2, entry 30). The conjugates were tested against cell lines that expressed different levels of GRPR and integrin receptors: human

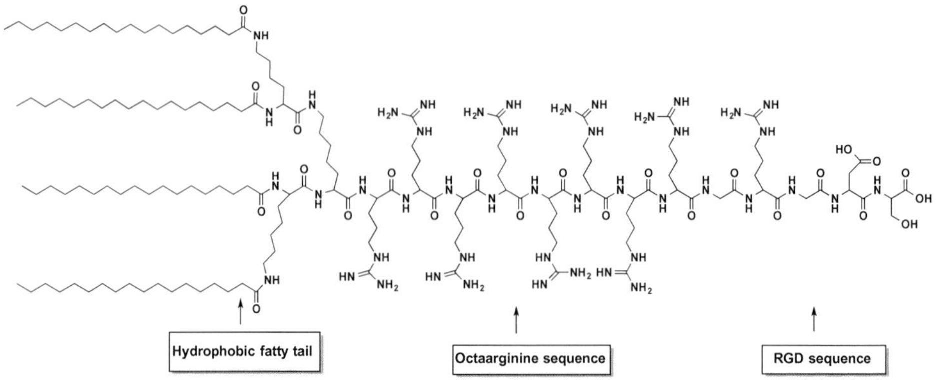

Fig. 5. Surfactant-like tetra-tail amphiphilic peptide synthesized by Chen *et al.*[85]

breast cancer MDA-MB-231, human lung adenocarcinoma A549, human prostate cancer PC-3, and murine mammary carcinoma EMT-6 cells. The ZnPc-RGD conjugate showed receptor binding activity in *in vitro* blocking studies, however its PDT efficacy was considerably low. This was thought to be due to conjugation of the ZnPc to a single monomeric RGD peptide, and could possibly be alleviated by conjugation to multiple RGD peptides. The subcellular distribution of conjugates ZnPc-RGD and ZnPc-BBN was investigated in PC-3 and MDA-MB-231 cells. Both conjugates were found to localize preferentially within mitochondria, lysosomes, and Golgi apparatus.

Chen *et al.*[85] designed a targeted porphyrin-loaded peptide micelle containing a surfactant-like tetra-tail amphiphilic peptide $[(C_{18})_2K]KR_8GRGDS$ (Fig. 5) composed of four hydrophobic fatty tails to promote self-assembly. The porphyrin-loaded micelle was tested using the drugs Ibuprofen and DOX to determine the loading/unloading ability of the micelles. The size of the porphyrin was shown to disturb the loading of the micelles. The porphyrin-loaded micelles were evaluated in HeLa cells and shown to have low dark toxicity and significantly higher phototoxicity. The cellular uptake of the conjugates was determined in HeLa and human embryonic kidney transformed (293T) cells. The uptake was found to be greater in HeLa cells vs. 293T cells due to the overexpression of the integrin receptors $\alpha_v\beta_3$ and $\alpha_v\beta_5$.

We have recently reported five amphiphilic protoporphyrin IX-peptide conjugates containing the sequences ATWLPPR, AAhexPQRRSSARLSA, and a cyclic peptide cERGDPhe (Table 2, entry 31).[62] These conjugates were investigated in both human squamous carcinoma (HEp2) and human myeloid leukemia (HL60) cell lines *in vitro*. Conjugates were found to have low dark and phototoxicities in both cell lines with $IC_{50} > 100\,\mu M$, however one conjugate containing the sequence AAhexPQRSARLSA showed higher cytotoxicity ($IC_{50} = 74\,\mu M$). The conjugates also showed 3–10 fold higher cellular uptake in HL-60 cells when compared with HEp2 cells. The 10-fold increase in cellular uptake was observed for the cyclic peptide cERGDPhe conjugate. All conjugates mainly localized in endosomal vesicles within HEp2 cells and in mitochondria in HL60 cells.

3.3. *EGF Receptors*

EGF receptors (EGFR) are glycoproteins that are overexpressed in several cancer cells, including breast, gastric colorectal, esophageal, prostate, bladder, non-small cell lung cancer (NSCLC), pancreatic, ovarian, renal, head and neck (SCCHN) cancers.[86] The overexpression is a result of mutations occurring during the binding of the endogenous ligand EGF to EGFR. These mutations have prompted studies on replacement peptide ligands that bind with the same affinity as the natural ligand EGF. Vicente *et al.*[40] reported the conjugation of ZcPc to two EGFR-targeting peptides with the sequences LARLLT and GYHWYGYTPQNVI (Table 2, entries 5, 6). The conjugates were evaluated in human carcinoma A431 and HEp2 cells, human colorectal HT-29, and kidney Vero (negative control) cells. The LARLLT conjugates were efficiently taken up by cells overexpressing EGFR, whereas the conjugates bearing the longer hydrophobic peptide sequence were poorly taken up, probably due to their low solubility. Intravenous administration of a ZcPc-LARLLT conjugate into nude mice bearing A341 and HT-29 human tumor xenografts resulted in strong tumor fluorescence, 24 h after administration.

Gupta *et al.*[87] synthesized a silicon phthalocyanine (Pc4)-loaded micelle for EGFR targeting. One such peptide, YHWYGYTPQNVI, was chosen for conjugation to a Pc4 micelle and the micelles were tested in human epidermoid carcinoma A431 cells that overexpress EGFR, and human breast cancer MCG-7c3 cells that do not overexpress EGFR. The micelle conjugates showed higher cellular uptake compared with unconjugated micelles in A431 cells, and no cellular uptake in MCG-7c3 cells demonstrating receptor selectivity.

Satyanarayanajois *et al.*[63] reported the conjugation of a BODIPY dye to a peptidomimetic (HERP5) (Table 2, entry 32). The conjugates were synthesized via SPPS techniques and were evaluated for cellular targeting and uptake in breast cancer cell lines BT-474 and SKBR-3, due to their overexpression of HER2 proteins. The authors were able to find that the BODIPY-HERP5 conjugate was able to bind to the HER2 extracellular domain and remain on the membrane for 24 h. They also noticed internalization of the conjugate.

3.4. *VEGF Receptors*

These receptors consist of powerful signal proteins that are created by cells to assist in angiogenesis. There are two key receptor sites of this family, VEGFR-1 (Flt-1) and VEGFR-2 (Flk-1/KDR).[88,89] Several studies have probed peptide phage libraries to discover ligands for VEGF. From these studies one peptide emerged, ATWLPPR, due to its high binding affinity to an anti-VEGF neutralizing monoclonal antibody.[90] Barberi-Heyob *et al.*[43] evaluated the binding affinity of the ATWLPPR peptide against the natural $VEGF_{165}$ ligand on neuropilin-1 receptors in the presence of KDR receptors. The chlorin derivative TPC (5-(4-carboxyphenyl)-10,15.20-triphenylchlorin) was conjugated to ATWLPPR via a spacer (6-aminohexanoic acid, Ahx) (Table 2, entry 9). The conjugate was evaluated in HUVEC cells for phototoxicity, dark cytotoxicity, and cellular uptake. It was found that TPC-Ahx-ATWLPPR accumulated in HUVEC cells rapidly, 25-fold more than TPC alone. In addition, the conjugate was found to bind to NRP-1 but not to KDR.

Miller *et al.*[51] targeted VEGFR2 using the homing peptide ATWLPPR conjugated to verteporfin-polyvinyl alcohol (PVA) (Table 2, entry 15). The authors evaluated the biological efficacy of the verteporfin-PVA-peptide conjugate to choroidal neovascularization (CNV) in rat eyes, and the conjugate was compared to the control unconjugated verteporvin-PVA. The verteporfin-PVA-peptide conjugate showed retention of the photosensitizer photophysical properties, and higher selectivity compared with the standard liposomal verteporfin and unconjugated control. The targeted verteporfin-PVA-peptide conjugate showed localization 1 hour after injection and enhanced treatment in inducing CNV closure (100%) at low drug doses ($3\,mg/m^2$). The authors showed similar CNV closure with the control, however at higher light dose ($25\,J/cm^2$) compared with the targeted verteporfin-PVA-peptide conjugate, which needed a lower light dose ($10\,J/cm^2$).

Bakleh *et al.*[56] chose to target VEGF-R2 in their studies on vascular endothelial growth factor-like peptides. The authors focused their studies on improving the *in vivo* half-life of porphyrin conjugates with an emphasis on oligopeptides in a cyclic structure (Table 2, entry 24). The peptide used, CBO-P11, is a cyclopeptide that has been under investigation as an anti-cancer agent in preclinical trials.

3.5. *Folate Receptors*

Folate receptors (FR) are anchored cell surface receptors. These receptors are generally overexpressed on the surface of many cancer cells, but distinctively very limited expression is found in healthy tissues and organs.[91] This selective function of FR makes them especially interesting when developing anti-cancer agents. Folic acid/folate is normally used as the targeting moiety for FR, however incorporation of a peptide sequence as a linker between the folate and the fluorescent molecule can promote cellular uptake and determine the biodistribution of the photosensitizer.[47] Zheng *et al.*[48,49] synthesized a variety of porphyrin-based probes (PPF) with targeting moieties for FR. One PPF consisted of a radio-metalloporphyrin, a PKM peptide linker (GDEVDGSGK), and folate (Table 2, Entry 12). The resulting conjugate and imaging probe was tested *in vivo* in mice bearing KB (FR-positive human epidermal cancer) or MT-1 (FR-positive human breast carcinoma) tumors. The authors found that utilizing ^{64}Cu-PPF maintained the biodistrubition and selective tumor uptake of the PPF.[48]

Zheng *et al.*[49] showed that bacteriochlorophyll (Bchl) conjugated to GDEVDGSGK and folate also provided multimodal capabilities for imaging as well as PDT, functioning as a theranostic probe (BPF) (Table 2, Entry 13). The authors investigated the properties of the conjugate in epithelial carcinoma KB and HT1080 cells. The BPF probe was able to significantly decrease cell viability in the presence of light. The tumor specificity was shown *in vivo* using mice bearing KB (high FR expression) and HT1080 (low FR expression) tumors. The authors observed, at all times, that the fluorescent signal was significantly higher in KB vs. HT1080 tumors, demonstrating the specificity of BPF for FR. Folate has also been recently conjugated via a cycloaddition reaction to a BODIPY-based photosensitizer.[92]

3.6. *Nuclear Localizing Sequences (NLS)*

Incorporation of a NLS peptide sequence has been shown to increase the cellular uptake and to enhance the phototoxicity of photosensitizer-NLS conjugates, compared with the corresponding non-conjugated photosensitizers. We have reported[34,35,37,38,57] (Table 5, entries 2-11) a series of porphyrin-peptide conjugates bearing a NLS, or an NLS-embedded within a cell penetrating peptide sequence. The NLS of the simian virus (SV40) large T antigen,[93] PKKKRKV, and the minimal sequence KRPAATKKAGQAKKKL from nucleoplasmin,[94,95] were used on the design of these conjugates. Although no subcellular localization within the nucleus of HEp2 or PC-3M cells was observed by microscopy, enhancement of cellular uptake and phototoxicity compared with the non-conjugated porphyrins were generally observed in the *in vitro* studies of these conjugates.

Meunier *et al.*[73] have reported the synthesis of a Mn(II)porphyrin-peptide conjugate bearing an amphipihilic peptide sequence with the NLS peptide KKKRKV embedded (Table 5, entry 1). *In vitro* subcellular localization of this conjugate in HS68, NIH 3T3 fibroblasts and L1210 murine lymphocytic cells showed that the conjugate was efficiently taken up by cells and localized around the nuclei. This conjugate exhibited 100-fold efficacy on inducing tumor cell death, compared with the non-conjugated porphyrin.

Lo *et al.*[96] synthesized a ZnPc-peptide conjugate using a 1,4-bis(triethylene glycol)-substituted ZnPc and the NLS peptide GGPKKKRKV (Table 5, entry 12). The conjugates were evaluated in human colorectal carcinoma HT29 cells, and found to be non-cytotoxic in the dark but to display higher phototoxicity (IC_{50} = 0.21 μM) than the unconjugated ZnPc (IC_{50} = 0.39 μM). The ZnPc-NLS conjugate was also tested *in vivo* using nude mice bearing HT29 tumors. The NLS-conjugate and unconjugated ZnPc were injected intravenously (1 nmol/g) and monitored continuously. After 24–30 h post-injection bright fluorescence was observed only in the tumors of mice injected with the ZnPc-NLS conjugate, indicating higher tumor retention for the conjugate.

Verwilst *et al.*[97] synthesized a NLS-containing dihalogenated BODIPY conjugate bearing a peptide with the sequence GPKKKRKV (Table 5, entry 13). The BODIPY-NLS conjugate was evaluated in human bladder carcinoma T24 cells. The NLS-BODIPY conjugate showed low dark and photocytotoxicities, and preferential localization in the cell ER and lysosomes. No nuclear localization was observed with the conjugate or the BODIPY alone.

3.7. *Mitochondria Localizing Sequences (MLS)*

The targeting of cancer cell mitochondria by porphyrin conjugates bearing a MLS has been reported.[28,37,74] MLS are specific peptide sequences present in mitochondrial proteins that play an important role on their translocation within the mitochondrial matrix.[98–100] Dmitriev *et al.*[28] reported the synthesis and *in vitro* studies of Pt(II) coproporphyrin-peptide conjugates bearing the sequence MGRTVVVLGGGISGLAAGC (Table 6, entries 1,2). These conjugates were designed to contain a short or a PEG-based linker as well as nona-arginine to tune the hydrophilic/hydrophobic properties of the resulting conjugates. Although these conjugates localized mainly in the lysosomes and no mitochondrial localization was observed, they accumulated to a higher extent within MEF cells in comparison to the non-conjugated porphyrin.

Table 5. Photosensitizer conjugates containing nuclear localizing sequences (NLS).

Entry	Photosensitizer	Peptide sequence	Cell line	IC_{50} dark	IC_{50} photo	Subcell localization	Ref
1	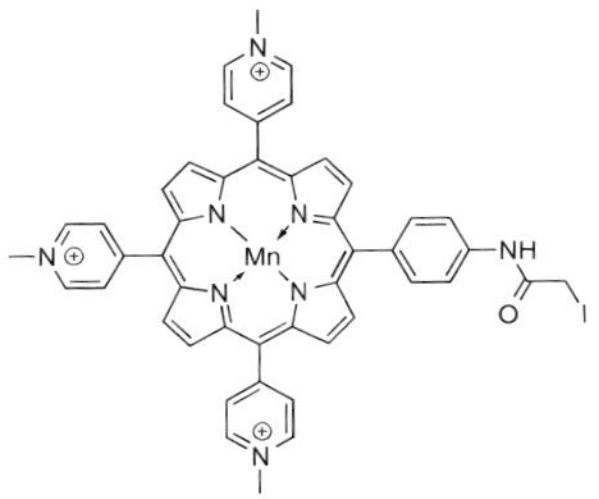	AcMGLGHL LVLAAA LQGAKK KRKVC	HS68 NIH-3T3 L1210	— — — —	— — — 0.6	Nuclei	73
2	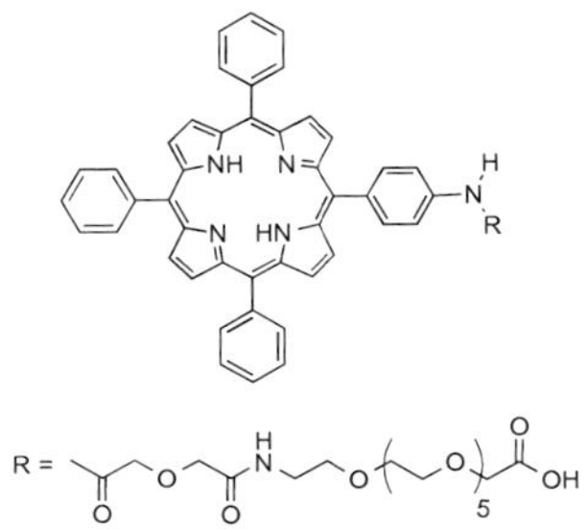	(1) APPKK KRKVEDPR KKRRQR RRPPQGG	HEp2	34	7.5	Vesicles	38
		(2) MGLGLH LLVLAAALQ GAWSQAPP KKKRKVG		100	10	Vesicles,	
		(3) RQIKIW FQNRRMKWK KAPPKKK RKVGG		53	9.3	Lysosomes	
		(4) KRPAAT KKAGQAKKKL EDPRKKRR QRRRPPQG		27	6.9	Lysosomes, ER	
3	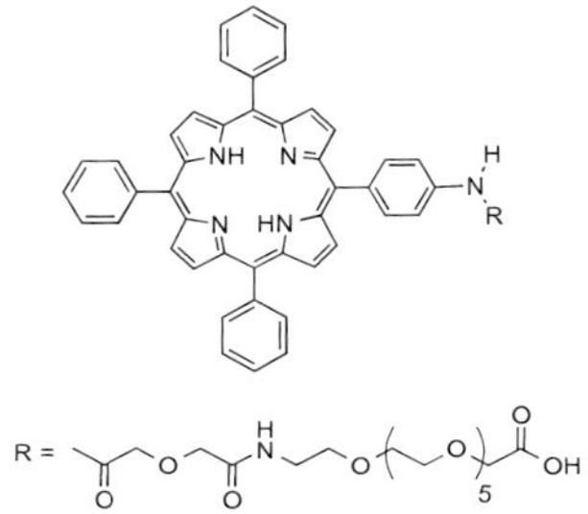	GPKKKRKV	HEp2	90	1.7	ER	35

Table 5. (*Continued*)

Entry	Photosensitizer	Peptide sequence	Cell line	IC_{50} dark	IC_{50} photo	Subcell localization	Ref
4	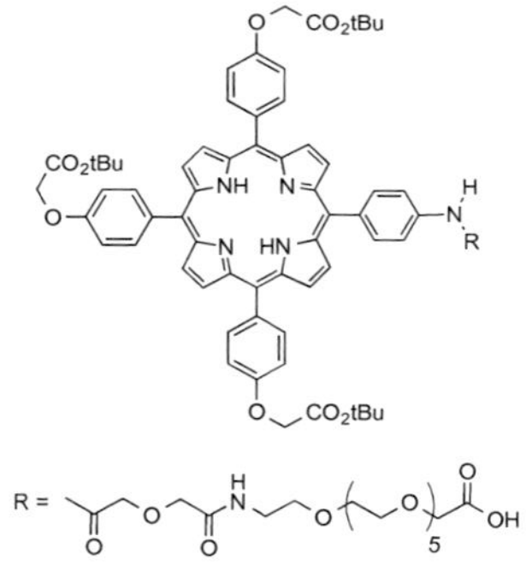	GPKKKRKV	HEp2	100	67	Lysosomes	35
5	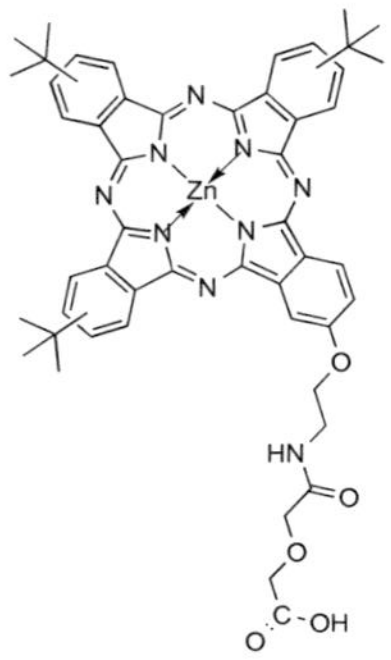	KRPAATK KAGQAKKK LEDPRKKRR QRRRPPQG	HEp2	77	1.44	Lysosomes	39

6	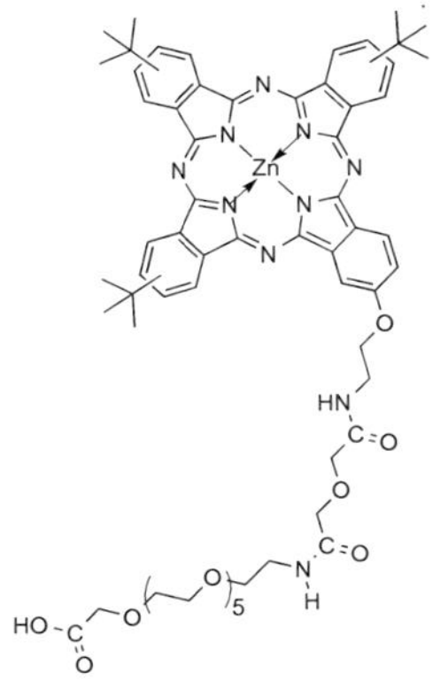	KRPAATK KAGQAKK KLEDPRKK RRQRRRPPQG	HEp2	87	1.87	Lysosomes	39
7	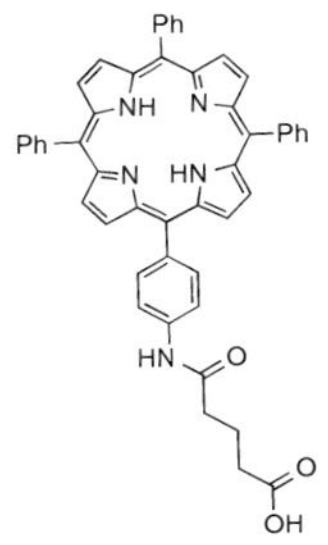	RRKR	PC3M	32.9	1.48	—	57
8	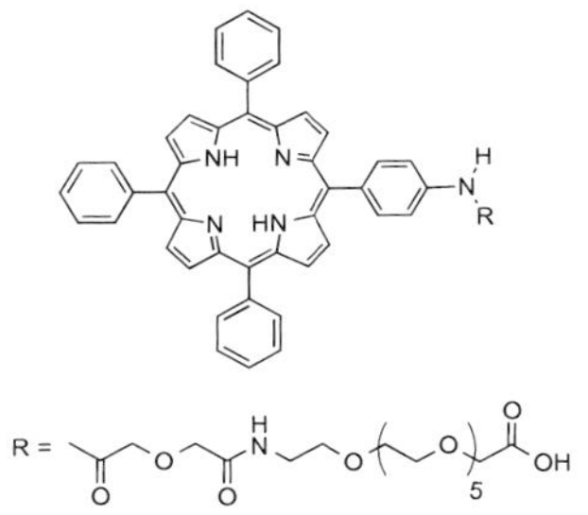	GPKKKRKV	PC3M	38.3	0.40	—	57

(*Continued*)

Table 5. (*Continued*)

Entry	Photosensitizer	Peptide sequence	Cell line	IC_{50} dark	IC_{50} photo	Subcell localization	Ref
9	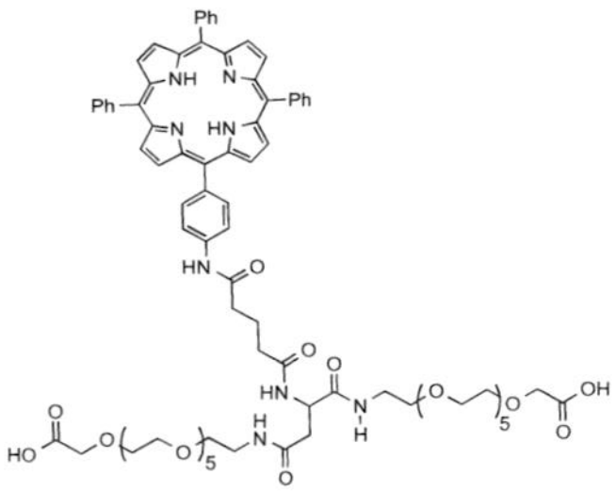	GPKKKRKV	PC3M	11.4	2.95	—	57
10	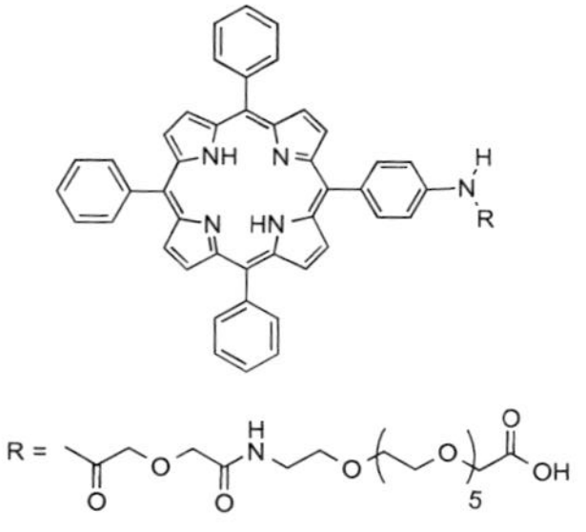	RQIKIWFQ NRRMKWKKA PPKKKRKVG	PC3M	10.5	1.34	—	57

11	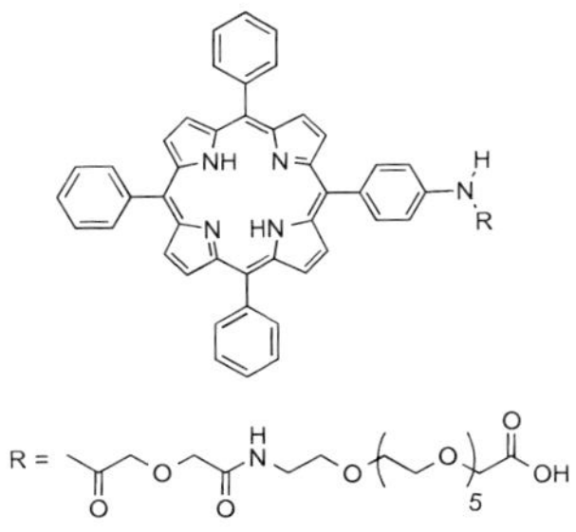	KRPAAT KKAGQAK KKLEDPR KKRRQRR RPPQG	PC3M	20.8	1.21	—	57
12	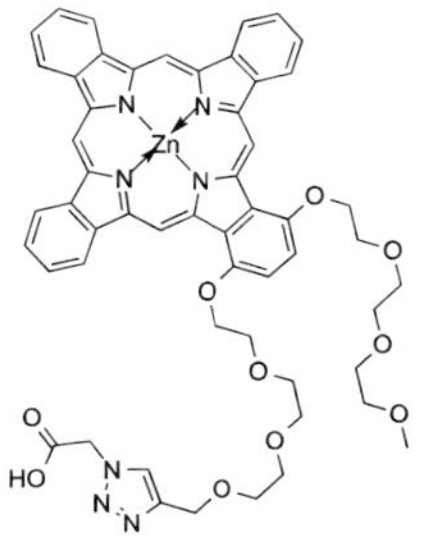	GGPKKKRKV	HT29	—	.21	Plasma mem- brane	96
13	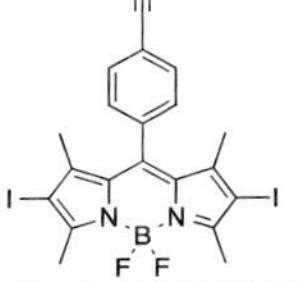	azidoGPKKKRKV	T24	—	.016	Lysosomes, ER	97

Asayama *et al.*[74] used the sequence MLSLRQSIRFFKGC to prepare a cationic Mn(II)-porphyrin conjugate (Table 6, entry 3) that exhibited enhanced mitochondrial localization *via* a mitochondrial swelling assay on isolated rat mitochondria. We used the sequence MSVLTPLLLRGLTGSAR-RLPVPRAKIHSL derived from human cytochrome *c* oxidase to prepare a tetraphenylporphyrin conjugate (Table 6, entry 4).[37] *In vitro* studies on HEp2 cells showed that this conjugate localized mainly within the cell lysosomes.

3.8. *Protease-Activated Conjugates*

Peptides used as homing devices are normally required to remain intact to ensure the targeting of specific sites, however there is a branch of photosensitizers that require the encoded peptide sequence to be cleaved upon interaction with certain proteases. Protease-activatable photosensitizers (PAPs) are structurally different in that they are non-toxic until enzymatic cleavage occurs, removing the quencher.

Activatable photosensitizers in PDT have explored the activity of the peptide sequences involved in the enzymatic cleavage. These studies have progressed to the development of a class of photosensitizers called molecular beacons (MBs).[101] Molecular beacons are based on the interaction of the fluorescent and the quencher molecules of the probe. This interaction uses Förster resonance energy transfer (FRET) to control targeted activation.[102] The FRET effect is facilitated by an electrostatic interaction between the oligopeptide polyanionic and polycationic chains. MBs usually consist of three components: a photosensitizer, a disease-specific linker and a quencher (Fig. 6). These components need to remain close to allow the photosensitizer to remain quenched through FRET.[103]

The first PDT activatable photosensitizer reported by Zheng *et al.*[104] combined a pyropheophorbide a caspase-3 specific peptide sequence (Gly-Asp-Glu-Val-Asp-Gly-Ser-Gly-Lys) and a carotene FRET acceptor (Fig. 7). The authors were able to confirm by HPLC that the addition of caspase-3 induced cleavage of the linker providing photo-activation. The authors confirmed a 4-fold increase of oxygen production after incubation with capase-3 (Table 7, entry 1).

Zheng *et al.*[105] developed a MB containing a lysine linker that was used to determine if FRET can be used effectively to deactivate the photosensitizer by coupling to various quenchers. The authors were able to find that FRET-based quenchers could be used to quench singlet oxygen, and a Förster radius of 5 nm was required for linker separations of ~2 nm. Zheng *et al.*[45] investigated

Table 6. Photosensitizer conjugates containing mitochondria localization peptide sequences (MLS).

Entry	Photosensitizer	Peptide sequence	Cell line	Subcellular localization	Ref
1	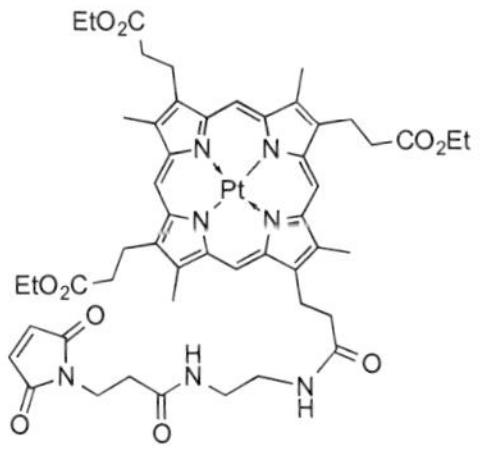	MGRTVVVLG GGISGLAAGC RRRRR RRRR	MEF HeLa HTC116 dPC12	Lysosomes Cytoplasm	28
2	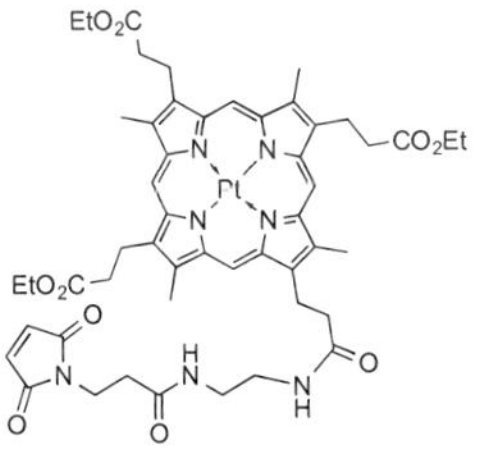	MGRTVVVLG GGISGLAAG CRRRRR RRRR	MEF HeLa HTC116 dPC12	Lysosomes Cytoplasm	28
3	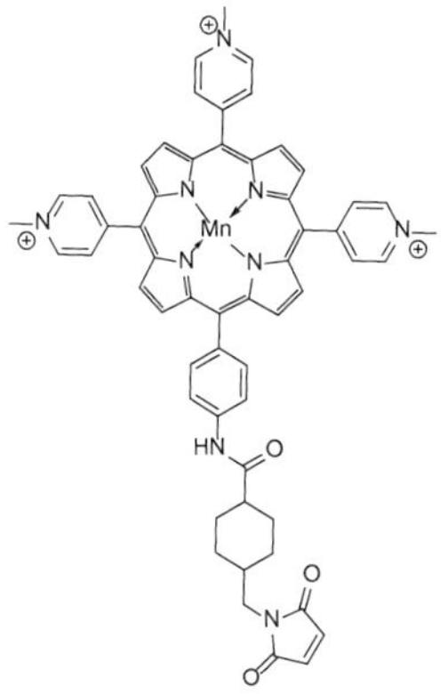	MLSLRQSI RFFKGC	RAW 264.7	Mitochondria	74
4	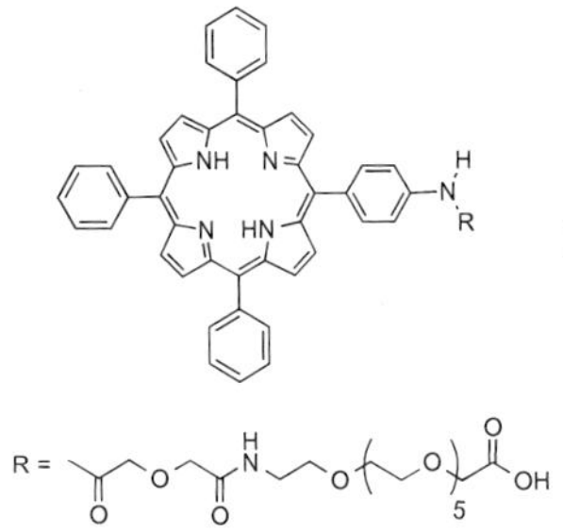	MSVLTPLL LRGLTGSARRL PVPRAKIHSL	HEp2	Lysosomes	37

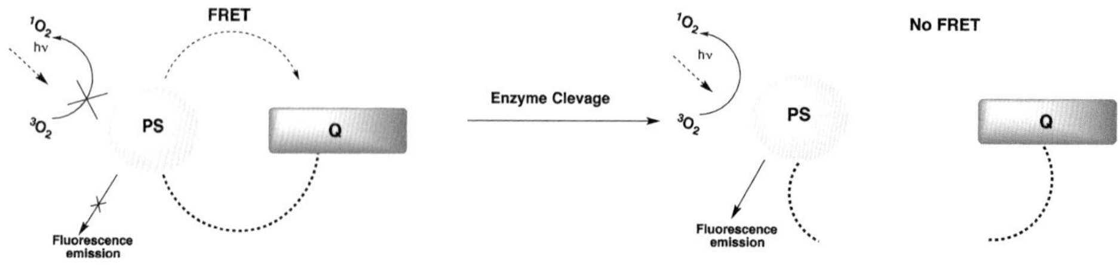

Fig. 6. Protease-activatable photosensitizer.

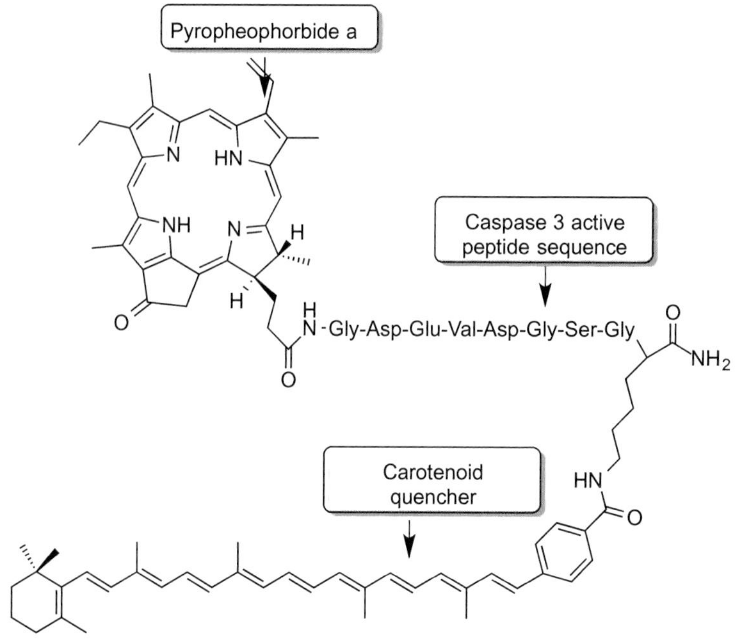

Fig. 7. Caspase 3 activatable photosensitizer by Zheng *et al.*[104]

pyropheophorbide *a* conjugated to a short peptide sequence (GPLGLARK) that is recognized by metalloproteinase 7 (MMP-7) (Table 7, entry 2), and quenched by the carotenoid BHQ-3 as a FRET acceptor (Fig. 8). The PDT efficacy of the conjugate was determined in human breast cancer cells (BT20), and in mice bearing human nasopharyngeal epidermoid carcinoma (KB) tumors. The conjugate increased the singlet oxygen production by 19-fold, after incubation in MMP-7-positive cells.

Zheng *et al.*[46] investigated the fibroblast activating protein (FAP) protease. FAP is a cell surface protein as well as a serine protease that is overexpressed on the surface of tumor-associated fibroblasts (Table 7, entry 3). FAP is known to cleave α-antiplasmin and target-peptide sequences between proline and

Table 7. Protease activatable photosensitizers (PAP)

Entry	Photosensitizer	Protease cleavable linker	Enzyme	Quencher	Ref
1	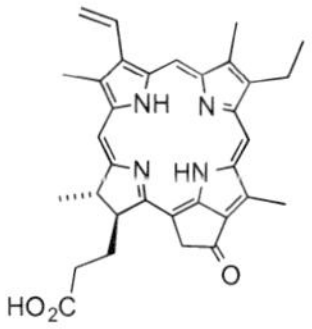	GDEVDGSGK	Caspase-3	Carotene (A)	104
2	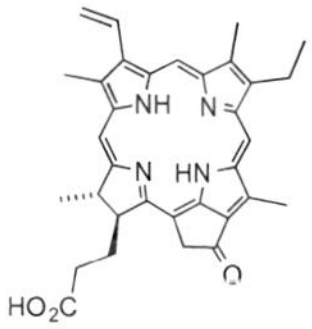	GPLGLARK	MMP7	BHQ-3	45
3	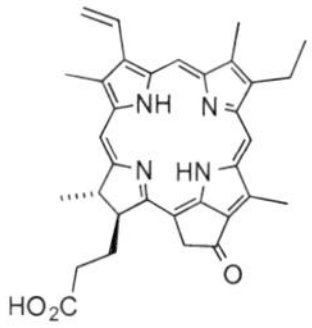	TSGPNQEQK	FAP	BHQ-3	46
4	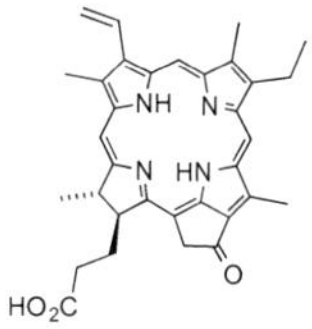	GDEVDGDGK	Caspase	Rox	107
5	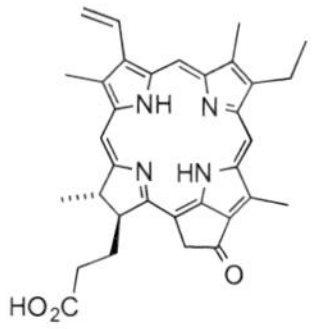	GPLGLARK or RPLALWRSK	MMP7	BHQ-3	50
6	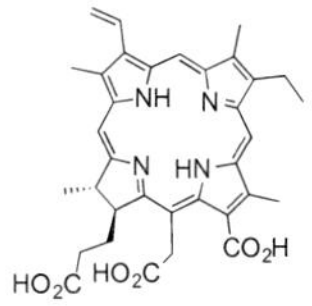	PEGylated Polylysine	Cathepsin B	Self-quenching	108
7	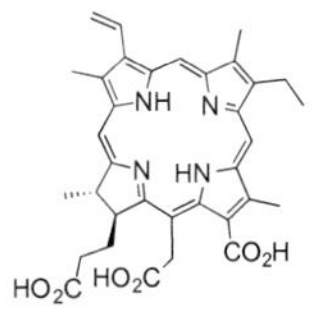	PEGylated Polylysine	Trypsin	Self-quenching	32

(*Continued*)

Table 7. (*Continued*)

Entry	Photosensitizer	Protease cleavable linker	Enzyme	Quencher	Ref
8	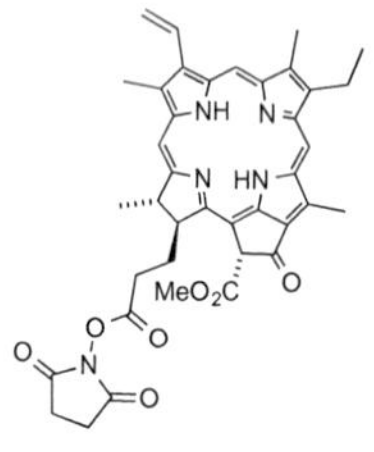	Polylysine	Trypsin	Self-quenching	23
9	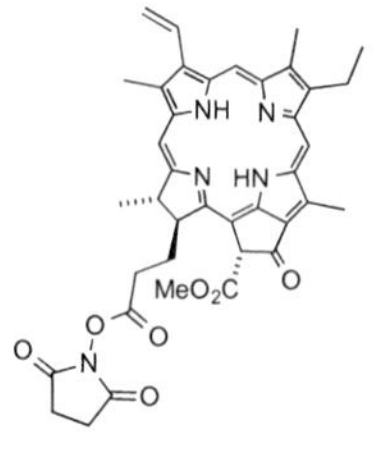	GFTRSAG, modified polylysine backbone	Trypsin	Self-quenching	24
10	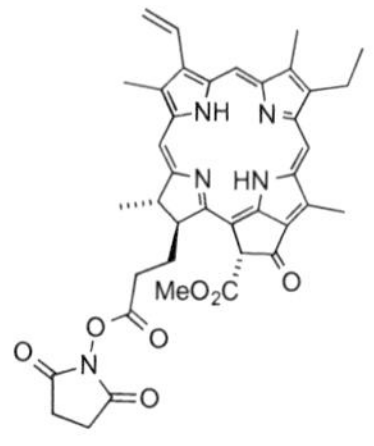	GFPIPRSGGGGG, Modified polylysine backbone	Thrombin	Self-quenching	25

asparagine residues.[106] The authors synthesized the MB with pyropheophorbide *a* as the photosensitizer, the peptide TSGPNQEQK, and BHQ-3 as FRET acceptor. The authors were able to monitor cleavage of the MB in solution by murine and human FAP. *In vitro* and *in vivo* studies suggested that the FAP enzyme could specifically activate the MB and provide a 200-fold increase in fluorescence.

Zheng *et al.*[107] designed a PAP that used pyropheophorbide *a* conjugated to the N-terminus of the capase-3 specific peptide GDEVDGDGK (Table 7, entry 4). After deprotection and cleavage, the resulting conjugate was then quenched with 5-carboxy-xrhodamine (Rox) at the epsilon amine on the C-terminus lysine residue. This PAP was found to have a 17-fold fluorescence dynamic range of Rox to Pyro. Investigations in HT-29 cells showed a robust FRET pair between the Rox donor and pyropheophorbide acceptor.

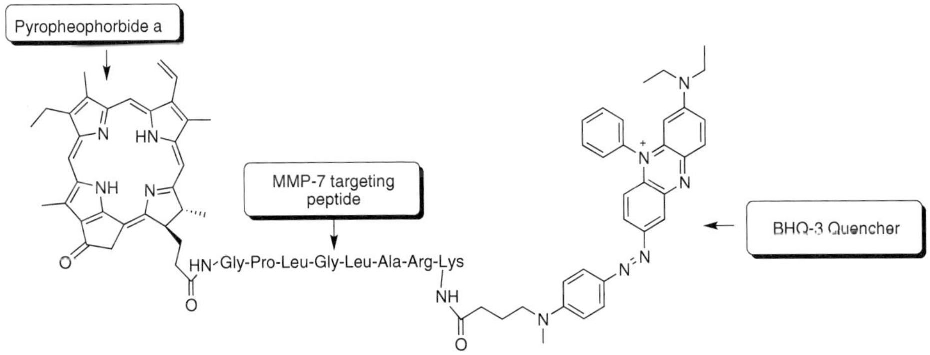

Fig. 8. MMP-7 activatable photosensitizer by Zheng *et al.*[45]

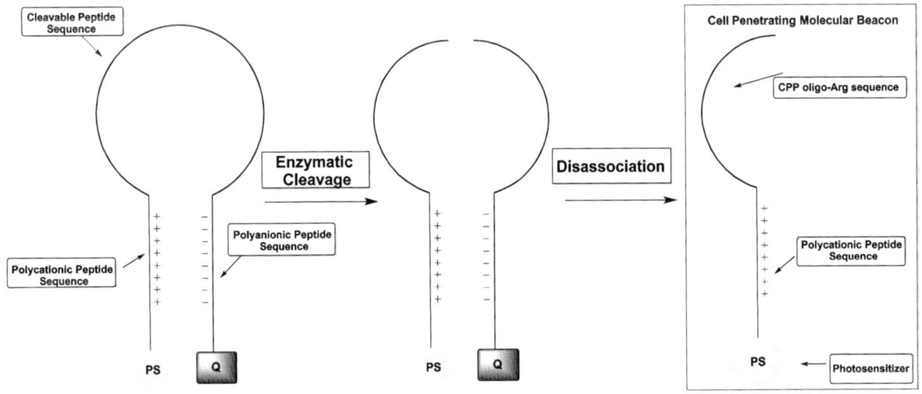

Fig. 9. General zipper molecular beacon (ZMB) mechanism.

Studies suggest that tuning MBs at the biolinker can result in high quenching. The approach uses a polyanion and polycation peptide zipper mechanism (Fig. 9) that provides several advantages: (1) electrostatic interaction between the polyanionic and polycationic chains, causing the quencher and photosensitizer to be close in proximity, (2) the polyanionic portion prevents cell penetration before cleavage, (3) the hairpin conformation minimizes blockage of the enzyme active site by bulky chromophores, and (4) causes the release of a conjugate bearing a photosensitizer and a cell-penetrating peptide, the polycationic portion.

Zheng *et al.*[50] reported a series of zipper molecular beacons (ZMBs) that contained pyropheophorbide *a*, a BHQ-3 quencher (Table 7, entry 5), and two oligopeptide strings, a polyanionic sequence (D-Asp_n) and a polycationic sequence (D-Arg_n). The series of ZMBs was tested for quenching efficiency,

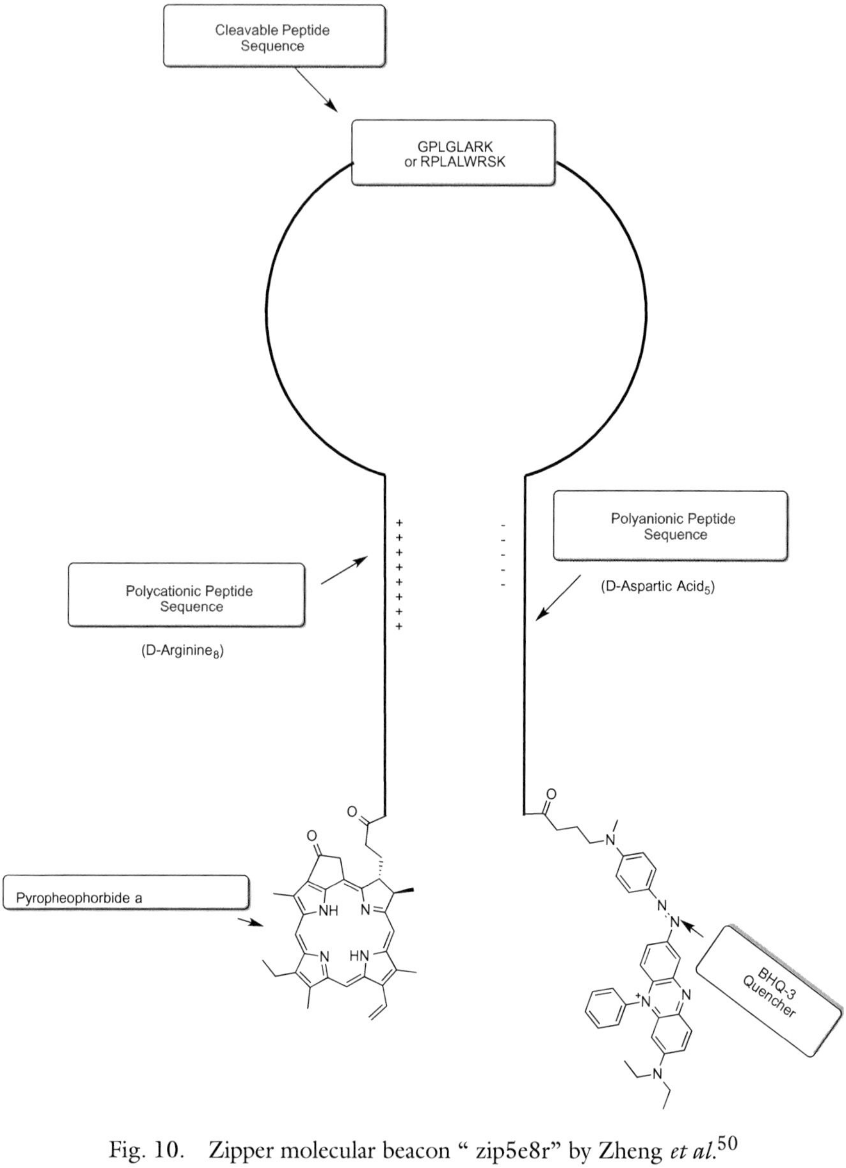

Fig. 10. Zipper molecular beacon " zip5e8r" by Zheng *et al.*[50]

optimal balance of high quenching, and rapid activation. These studies showed that the unsymmetric zip5e8r (Fig. 10) can enable rapid cell internalization, and the probe is not dependent on the specific peptide substrate sequence.

Polymeric photosensitizer prodrugs (PPPs) include active polymeric lysine backbones with short peptide sequences with at least one amino acid with a reactive amino group for conjugation to the backbone structure, to provide a cleavage location for several different proteases. This peptide sequence also

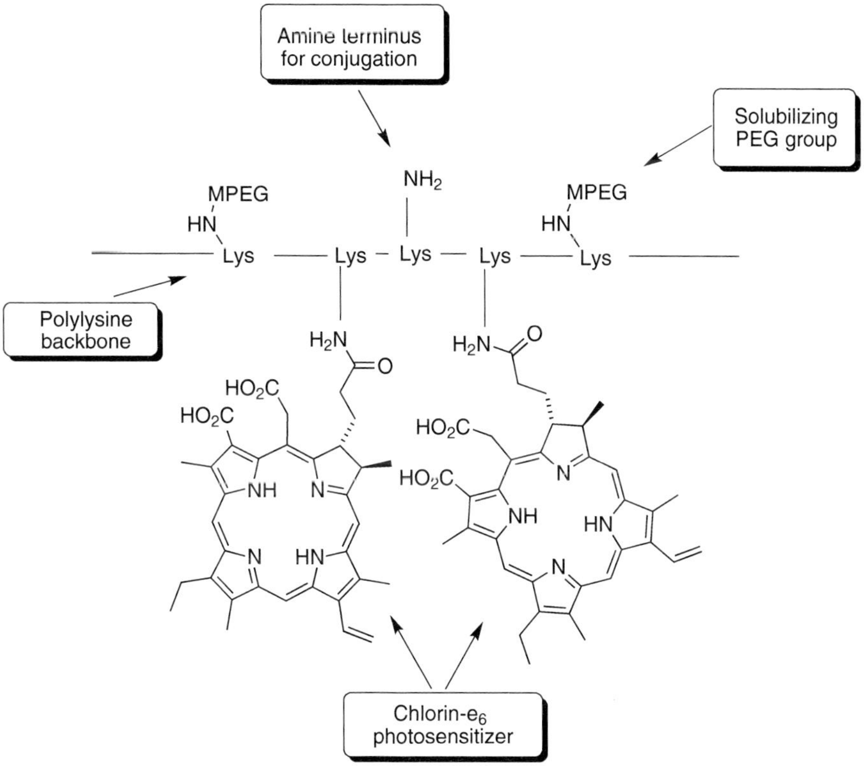

Fig. 11. Polymeric photosensitizer designed by Choi *et al.*[32,108]

provides arms for conjugation to additional photosensitizers for increased PDT activity. Choi *et al.*[32,108] made use of the self-quenching effects of aggregated porphyrin derivatives (Table 7, entries 6, 7). Multiple chlorin e_6 were conjugated to a polylysine backbone with monomethoxy-polyethylene glycol solubilizing linkers (Fig. 11). The PPPs were tested with several cathepsin proteases, in particular cathepsin B. *In vivo* PDT treatments using the PPPs resulted in attenuated tumor growth in mice.

Lange *et al.*[23] designed PPPs with a lysine backbone but coupled it to short peptide sequences to enhance the specificity of the PPP (Table 7, entry 8). The PPPs housed a lysine backbone consisting of 1–30 molecules per poly(L-lysine) chain. The use of pegylation is unfavorable to self-quenching molecules, and to counter these issues the use of 18 molecules of pheorphorbide per poly(L-lysine) chain was the most efficient in quenching. The authors tested the PPPs *in vitro* using bladder carcinoma T24 cells, and in mouse colon carcinoma *ex vivo*.

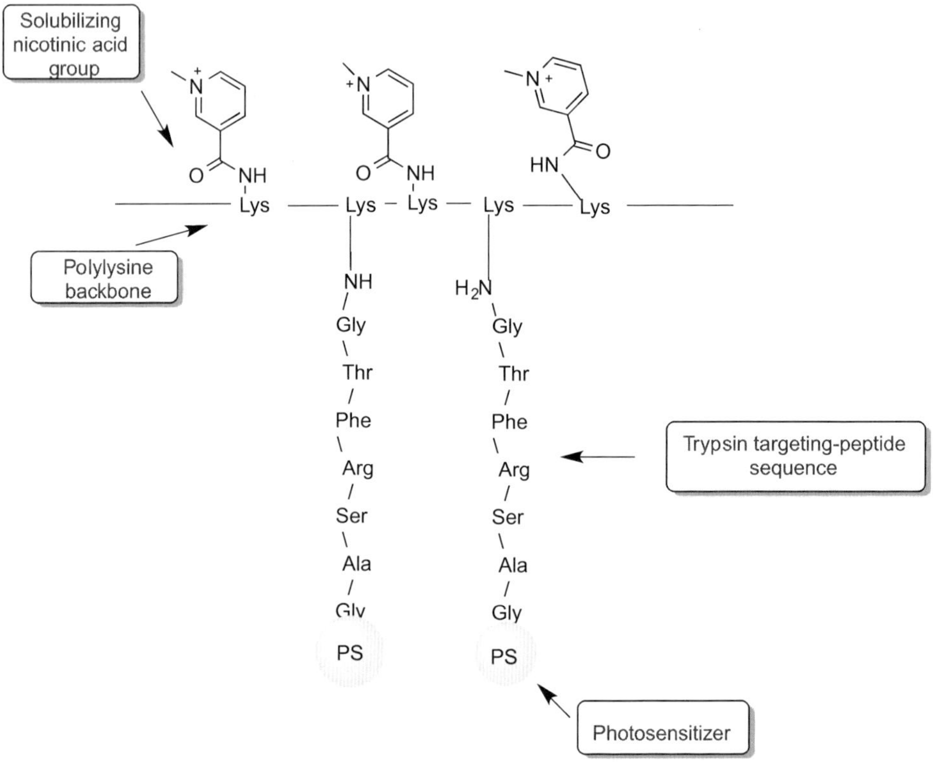

Fig. 12. Polymeric photosensitizer designed by Lange *et al.*[24]

To improve the solubility of the PPP Lange *et al.*[24,25] capped lysine groups with derivatized nicotinic acid containing a cationic quaternary ammonium group, which also prevented non-specific PPP activation. The PPP contained the peptide sequence GFTRSAG targeted to trypsin (Fig. 12) and its phototoxicity was evaluated in T24 cells (Table 7, entry 9). In another study thrombin was targeted using a different peptide sequence (GfPIPRSGGGGG). The conjugate was designed to target thrombin in synovial tissue of rheumatoid arthritis and was shown to be phototoxic in rheumatoid arthritis synovial fibroblasts (Table 7, entry 10).

4. Concluding Remarks

Third-generation PDT photosensitizers consisting of a peptide sequence linked to a porphyrin-based photosensitizer (usually a porphyrin, chlorin, bacteriochlorin or phthalocyanine macrocycle) are designed to increase the target selectivity and specificity of the photosensitizers. These conjugates have been synthesized using a variety of conjugation methodologies, including cycloaddition reactions, amide bond formation and cross-metathesis, and are usually readily internalized by cells and non-immunogenic. The peptide sequences are chosen to target specific cell surface receptors, enzymes, subcellular compartments and other biological targets, with the goal to enhance PDT efficacy. Biological studies have been conducted for some of the conjugates, both *in vitro* and *in vivo*, although additional pre-clinical studies are necessary before the potential of third-generation photosensitizers is fully realized. Additional studies on conjugate optimization could lead to highly selective and effective PDT drugs.

Acknowledgments

The authors thank the National Institutes of Health, grant number R01 CA179902 for financial support.

References

1. T. J. Dougherty, C. J. Gomer, B. W. Henderson, G. Jori, D. Kessel, M. Korbelik, J. Moan and Q. Peng, *J. Natl. Cancer Inst.*, **90**, 889 (1998).
2. R. K. Pandey and G. Zheng, in *The Porphyrin Handbook*, Vol 6, Eds. K. M. Kadish, K. M. Smith and R. Guilard (Academic Press, Boston, 2000), p. 157.
3. L. B. Josefsen and R. W. Boyle, *Theranostics* **2**, 916 (2012).
4. M. Camerin, S. Rello, A. Villanueva, X. Ping, M. E. Kenney, M. A. J. Rodgers and G. Jori, *Eur. J. Cancer* **41**, 1203 (2005).
5. A. Mazzaglia, M. Trapani, N. Micali, L. M. Scolaro, T. Parisi and M. T. Sciortino, *J. Biotechnol.* **150**, S192 (2010).
6. A. P. Castano, P. Mroz and M. R. Hamblin, *Nat. Rev. Cancer* **6**, 535 (2006).
7. S. B. Brown, E. A. Brown and I. Walker, *Lancet Oncol.* **5**, 497 (2004).
8. A. K. Iyer, G. Khaled, J. Fang and H. Maeda, *Drug Disc. Today* **11**, 812 (2006).
9. L. B. Josefsen and R. W. Boyle, *Br. J. Pharmacol.* **154**, 4 (2008).
10. F. Giuntini, R. W. Boyle, M. Sibrian-Vazquez and M. G. H. Vicente, in *The Handbook of Porphyrin Science*, Vol. 27, Ed. G. Ferreira (World Scientific Publishers, Singapore, 2013), p. 303.
11. T. M. Allen, *Nature Rev.* **2**, 750 (2002).
12. T. M. Allen and P. R. Cullis, *Science*, 1818 Vol: 303 (2004).
13. R. Schneider, L. Tirand, C. Frochot, R. Vanderesse, N. Thomas, J. Gravier, F. Guillemin and M. Barberi-Heyob, *Anti-Cancer Agents in Med. Chem.* **6**, 469 (2006).
14. F. Giuntini, C. M. A. Alonso and R. W. Boyle, *Photochem. Photobiol. Sci.* **10**, 759 (2011).
15. R. Tugyi, K. Uray, D. Ivan, E. B. Fellinger, A. Perkins and F. Hudecz, *Proc. Natl. Acad. Sci.* USA **102**, 413 (2005).
16. S. Hess, O. Ovadia, D. E. Shalev, H. Senderovich, B. Qadri, T. Yehezkel, Y. Salitra, T. Sheynis, R. Jelinek, C. Gilon and A. Hoffman, *J. Med. Chem.* **50**, 6201 (2007).

17. S. G. Awuah and Y. You, *RSC Adv.* **2**, 11169 (2012).
18. A. Kamkaew, S. H. Lim, H. B. Lee, L. V. Kiew, L. Y. Chung and K. Burgess, *Chem. Soc. Rev.* **42**, 77 (2013).
19. E. Valeur and M. Bradley, *Chem. Soc. Rev.* **38**, 606 (2009).
20. B. Neises and W. Steglich, *Angew. Chem. Int. Ed.* **17**, 522 (1978).
21. W. König and R. Geiger, *Chem. Ber.* **103**, 788 (1970).
22. W. König and R. Geiger, *Chem. Ber.* **103**, 2024 (1970).
23. M. A. Campo, D. Gabriel, P. Kucera, R. Gurny and N. Lange, *Photochem. Photobiol.* **83**, 958 (2007).
24. D. Gabriel, M. A. Campo, R. Gurny and N. Lange, *Bioconjugate Chem.* **18**, 1070 (2007).
25. D. Gabriel, N. Busso, A. So, H. van den Bergh, R. Gurny and N. Lange, *J. Control. Release* **138**, 225 (2009).
26. M. Boisbrun, R. Vanderesse, P. Engrand, A. Olie, S. Hupont, J. B. Regnouf-De-Vains and C. Frochot, *Tetrahedron* **64**, 3494 (2008).
27. R. I. Dmitriev, H. M. Ropiak, D. V. Yashunsky, G. V. Ponomarev, A. V. Zhdanov and D. B. Papkovsky, *FEBS J.* **277**, 4651 (2010).
28. R. I. Dmitriev, H. M. Ropiak, G. V. Ponomarev, D. V. Yashunsky and D. B. Papkovsky, *Bioconjugate Chem.* **22**, 2507 (2011).
29. C. Dubuc, R. Langlois, F. Bénard, N. Cauchon, K. Klarskov, P. Tone and J. E. van Lier, *Bioorg. Med. Chem. Lett.* **18**, 2424 (2008).
30. J. R. Starkey, A. K. Rebane, M. A. Drobizhev, F. Q. Meng, A. J. Gong, A. Elliott, K. McInnerney and C. W. Spangler, *Clin. Cancer Res.* **14**, 6564 (2008).
31. G. Mező, L. Herényi, J. Habdas, Z. Majer, B. Myśliwa-Kurdziel, K. Tóth and G. Csík, *Biophys. Chem.* **155**, 36 (2011).
32. Y. Choi, J. R. McCarthy, R. Weissleder and C. H. Tung, *Chem. Med. Chem.* **1**, 458 (2006).
33. S. Rahimipour, N. Ben-Aroya, K. Ziv, A. Chen, M. Fridkin and Y. Koch, *J. Med. Chem.* **46**, 3965 (2003).
34. M. Sibrian-Vazquez, E. Hao, T. J. Jensen and M. G. H. Vicente, *Bioconjugate Chem.* **17**, 928 (2006).
35. M. Sibrian-Vazquez, T. J. Jensen, R. P. Hammer and M. G. H. Vicente, *J. Med. Chem.* **49**, 1364 (2006).
36. M. Sibrian-Vazquez, T. J. Jensen and M. G. H. Vicente, *Org. Biomol. Chem.* **8**, 1160 (2010).
37. M. Sibrian-Vazquez, I. V. Nesterova, T. J. Jensen and M. G. H. Vicente, *Bioconjugate Chem.* **19**, 705 (2008).
38. M. Sibrian-Vazquez, T. J. Jensen and M. G. H. Vicente, *J. Med. Chem.* **51**, 2915 (2008).
39. M. Sibrian-Vazquez, J. Ortiz, I. V. Nesterova, F. Fernandez-Lazaro, A. Sastre-Santos, S. A. Soper and M. G. H. Vicente, *Bioconjugate Chem.* **18**, 410 (2007).
40. B. G. Ongarora, K. R. Fontenot, X. Hu, I. Sehgal, S. D. Satyanarayana-Jois and M. G. H. Vicente, *J. Med. Chem.* **55**, 3725 (2012).
41. I. Walker, D. I. Vernon and S. B. Brown, *Bioorg. Med. Chem. Lett.* **14**, 441 (2004).
42. C. L. Conway, I. Walker, A. Bell, D. J. H. Roberts, S. B. Brown and D. I. Vernon, *Photochem. Photobiol. Sci.* **7**, 290 (2008).
43. L. Tirand, C. Frochot, R. Vanderesse, N. Thornas, E. Trinquet, S. Pinel, M. L. Viriot, F. Guillemin and M. Barberi-Heyob, *J. Control. Release* **111**, 153 (2006).

44. S. De Luca, D. Tesauro, P. Di Lello, R. Fattorusso, M. Saviano, C. Pedone and G. Morelli, *J. Pept. Sci.* 7, 386 (2001).
45. G. Zheng, J. Chen, K. Stefflova, M. Jarvi, H. Li and B. C. Wilson, *Proc. Natl. Acad. Sci. USA* **104**, 8989 (2007).
46. P. C. Lo, J. Chen, K. Stefflova, M. S. Warren, R. Navab, B. Bandarchi, S. Mullins, M. Tsao, J. D. Cheng and G. Zheng, *J. Med. Chem.* **52**, 358 (2009).
47. K. Stefflova, H. Li, J. Chen and G. Zheng, *Bioconjugate Chem.* **18**, 379 (2007).
48. J. Shi, T. W. Liu, J. Chen, D. Green, D. Jaffray, B. C. Wilson, F. Wang and G. Zheng, *Theranostics* **1**, 363 (2011).
49. T. W. Liu, J. Chen, L. Burgess, W. Cao, J. Shi, B. C. Wilson and G. Zheng, *Theranostics* **1**, 354 (2011).
50. J. Chen, T. W. Liu, P. C. Lo, B. C. Wilson and G. Zheng, *Bioconjugate Chem.* **20**, 1836 (2009).
51. R. Z. Renno, Y. Terada, M. J. Haddadin, N. A. Michaud, E. S. Gragoudas and J. W. Miller, *Arch. Ophthalmol.* **122**, 1002 (2004).
52. V. Chaleix, V. Sol, M. Guilloton, R. Granet and P. Krausz, *Tetrahedron Lett.* **45**, 5295 (2004).
53. T. Schroder, K. Schmitz, N. Niemeier, T. S. Balaban, H. F Krug, U. Schepers and S. Brase, *Bioconjugate Chem.* **18**, 342 (2007).
54. S. K. Bisland, D. Singh and J. Gariepy, *Bioconjugate Chem.* **10**, 982 (1999).
55. M. Sibrian-Vazquez, T. J. Jensen, F. R. Fronczek, R. P. Hammer and M. G. H. Vicente, *Bioconjugate Chem.* **16**, 852 (2005).
56. M. E. Bakleh, V. Sol, K. Estieu-Gionnet, R. Granet, G. Deleris and P. Krausz, *Tetrahedron* **65**, 7385 (2009).
57. I. Sehgal, M. Sibrian-Vazquez and M. G. H. Vicente, *J. Med. Chem.* **51**, 6014 (2008).
58. A. Srivatsan, M. Ethirajan, S. K. Pandey, S. Dubey, X. Zheng, T. H. Liu, M. Shibata, J. Missert, J. Morgan and R. K. Pandey, *Mol. Pharm.* **8**, 1186 (2011).
59. P. Xu, J. Chen, Z. Chen, S. Zhou, P. Hu, X. Chen and M. Huang, *PLoS ONE* 7, 1 (2012).
60. C. Frochot, B. Di Stasio, R. Vanderesse, M. J. Belgy, M. Dodeller, F. Guillemin, M. L. Viriot and M. Barberi-Heyob, *Bioorg. Chem.* **35**, 205 (2007).
61. E. Ranyuk, N. Cauchon, K. Klarskov, B. Guérin and J. E. van Lier, *J. Med. Chem.* **56**, 1520 (2013).
62. M. Sibrian-Vazquez, X. Hu, T. J. Jensen and M. G. H. Vicente, *J. Porphyrins Phthalocyanines.* **16**, 609 (2012).
63. S. Banappagari, A. McCall, K. Fontenot, M. G. H. Vicente, A. Gujar and S. D. Satyanarayanajois, *Eur. J. Med. Chem.* **65**, 60 (2013).
64. V. V. Rostovtsev, L. G. Green, V. V. Fokin and K. B. Sharpless, *Angew. Chem.* **114**, 2708 (2002).
65. C. W. Tornøe, C. Christensen and M. Meldal, *J. Org. Chem.* **67**, 3057 (2002).
66. Q. Wang, T. R. Chan, R. Hilgraf, V. V. Fokin, K. B. Sharpless and M. Finn, *J. Am. Chem. Soc.* **125**, 3192 (2003).
67. Y. Ikawa, H. Harada, M. Toganoh and H. Furuta, *Bioorg. Med. Chem. Lett.* **19**, 2448 (2009).
68. Y. A. Lin, J. M. Chalker and B. G. Davis, *Chembiochem* **10**, 959 (2009).
69. V. Sol, V. Chaleix, R. Granet and P. Krausz, *Tetrahedron* **64**, 364 (2008).

70. J. M. Sutton, O. J. Clarke, N. Fernandez and R. W. Boyle, *Bioconjugate Chem.* **13**, 249 (2002).
71. G. T. Hermanson, in *Bioconjugate Techniques*, 3rd Edn., Eds. J. Audet and M. Preap (Academic Press, San Diego, 2013), p. 229.
72. R. I. Dmitriev, A. V. Zhdanov, G. V. Ponomarev, D. V. Yashunski and D. B. Papkovsky, *Anal. Biochem.* **398**, 24 (2010).
73. L. Chaloin, P. Bigey, C. Loup, M. Marin, N. Gaelotti, M. Piechaczyk, F. Heitz and B. Meunier, *Bioconjugate Chem.* **12**, 691 (2001).
74. S. Asayama, E. Kawamura, S. Nagaoka and H. Kawakami, *Mol. Pharm.* **3**, 468 (2006).
75. J. T. Wang, F. Giuntini, I. M. Eggleston, S. G. Bown and A. J. MacRobert, *J. Control. Release* **157**, 305 (2012).
76. L. Bourre, F. Giuntini, I. M. Eggleston, C. A. Mosse, A. J. MacRobert and M. Wilson, *Photochem. Photobiol. Sci.* **9**, 1613 (2010).
77. J. Bernadou, G. Pratviel, F. Bennis, M. Girardet and B. Meunier, *Biochemistry* **28**, 7268 (1989).
78. G. Pratviel, M. Pitie, J. Bernadou and B. Meunier, *Nucleic Acids Res.* **19**, 6283 (1991).
79. D. Xiao, J. Wang, L. L. Hampton and H. C. Weber, *Gene* **264**, 95 (2001).
80. T. J. Hoffman, H. Gali, C. J. Smith, L. G. Sieckman, D. L. Hayes, N. K. Owen and W. A. Volkert, *J. Nucl. Med.* **44**, 823 (2003).
81. A. Nagy and A. V. Schally, *Biol. Reprod.* **73**, 851 (2005).
82. G. Keller, A. V. Schally, T. Gaiser, A. Nagy, B. Baker, G. Westphal, G. Halmos and J. B. Engel, *Cancer Res.* **65**, 5857 (2005).
83. F. G. Giancotti and E. Ruoslahti, *Science* **285**, 1028 (1999).
84. S. I. Moriwaki, J. Misawa, Y. Yoshinari, I. Yamada, M. Takigawa and Y. Tokura, *Photodermatol. Photoimmunol. Photomed.* **17**, 241 (2001).
85. J.-X. Chen, H.-Y. Wang, C. Li, K. Han, X.-Z. Zhang and R.-X. Zhuo, *Biomaterials* **32**, 1678 (2011).
86. D. S. Salomon, R. Brandt, F. Ciardiello and N. Normano, *Crit. Rev. Oncol. Hematol.* **19**, 183 (1995).
87. A. M. Master, Y. Q. Qi, N. L. Oleinick and A. S. Gupta, *Nanomedicine NBM* **8**, 655 (2012).
88. S. Soker, S. Takashima, H. Q. Miao, G. Neufeld and M. Klagsbrun, *Cell* **92**, 735 (1998).
89. G. Neufeld, A. T. Cohen, S. Gengrinovitch and Z. Poltorak, *FASEB J.* **12**, 9 (1999).
90. R. Binetruy-Tournaire, C. Demangel, B. Malavaud, R. Vassy, S. Rouyre, M. Kraemer, J. Plouet, C. Derbin, G. Perret and J. C. Mazie, *EMBO J.* **19**, 1525 (2000).
91. H. Yoo and T. Park *J. Control Release* **100**, 247 (2004).
92. M.-R. Ke, S.-L. Yeung, D. K. P. Ng, W.-P. Fong and P.-C. Lo, *J. Med. Chem.* **56**, 8475 (2013).
93. D. Kalderon, B. L. Roberts, W. D. Richardson and A. E. Smith, *Cell* **39**, 499 (1984).
94. C. Dingwall and R. Laskey, *Semin. Cell Biol.* **1**, 11 (1990).
95. A. Prado, I. Ramos, L. J. Frehlick, A. Muga and J. Ausió, *Biochem. Cell Biol.* **82**, 437 (2004).
96. M.-R. Ke, S.-L. Y, W.-P. Fong, D. K. P. Ng and P.-C. Lo, *Chem. Eur. J.* **18**, 4225 (2012).
97. P. Verwilst, C. C. David, V. Leen, J. Hofkens, P. A. M. de Witte and W. M. De Borggraeve, *Bioorg. Med. Chem. Lett.* **23**, 3204 (2013).
98. C. M. Koehler, *FEBS Lett.* **476**, 27 (2000).

99. T. Lithgow, *FEBS Lett.* **476**, 22 (2000).
100. A. Muratovska, R. N. Lightowlers, R. W. Taylor, J. A. Wilce and M. P. Murphy, *Adv. Drug Delivery Rev.* **49**, 189 (2001).
101. J. F. Lovell, T. W. B. Liu, J. Chen and G. Zheng, *Chem. Rev.* **110**, 2839 (2010).
102. Y. Gao, G. Qiao, L. Zhuo, N. L. Ying and B. Tang, *Chem. Commun.* **47**, 5316 (2011).
103. A. Bugaj, *Photochem. Photobiol. Sci.* **10**, 1097 (2011).
104. J. Chen, K. Sefflova, M. J. Niedre, B. C. Wilson, B. Chance, J. D. Glickson and G. Zheng, *J. Am. Chem. Soc.* **126**, 11450 (2004).
105. J. F. Lovell, J. Chen, M. T. Jarvi, W.-G. Cao, A. D. Allen, Y. Liu, T. T. Tidwell, B. C. Wilson and G. Zheng, *J. Phys. Chem. B* **113**, 3203 (2009).
106. C. Edosada, C. Quan, T. Tran, V. Pham, C. Wiesmann, W. Fairbrother and B. Wolf, *FEBS Lett.* **580**, 1581 (2006).
107. J. F. Lovell, M. W. Chan, Q. Qi, J. Chen and G. Zheng, *J. Am. Chem. Soc.* **133**, 18580 (2011).
108. Y. Choi, R. Weissleder and C. Tung, *Cancer Res.* **66**, 7225 (2006).

Chapter 3

Current Developments in Using *MESO*-(TETRA) Substituted Porphyrins for PDT

Yasser M. Shaker[†,‡], *Ayman M. K. Sweed*[†,‡], *Claire Moylan*[†]
Luke Rogers[†], *Aoife A. Ryan*[†], *Rosine Petitdemange*[†]
and Mathias O. Senge[†,§,*]

[†] *School of Chemistry, SFI Tetrapyrrole Laboratory, Trinity Biomedical Science Institute, 152–160 Pearse Street, Trinity College Dublin, The University of Dublin, Dublin 2, Ireland*
[‡] *Department of Chemistry of Natural and Microbial Products, Division of Pharmaceutical and Drug Industries, National Research Centre, Dokki, 12622 Cairo, Egypt*
[§] *Medicinal Chemistry, Institute of Molecular Medicine, Trinity Centre for Health Sciences, Trinity College Dublin, St James's Hospital, Dublin 8, Ireland*
sengem@tcd.ie

Abstract

meso-Substituted porphyrins present a readily accessible class of porphyrins which has found wide applications in biomedical sciences. Contemporary syntheses allow the generation of large libraries of compounds with variable *meso* substituents or the preparation of bioconjugates. This chapter summarizes current synthetic strategies and medicinal chemistry applications of *meso*-substituted porphyrins in photodynamic therapy (PDT).

1. Introduction

The goal of this chapter is to describe recent advances in the development and use of *meso*-tetrasubstituted porphyrins in photodynamic therapy (PDT).

*Corresponding author

The focus is on analogues of 5,10,15,20-tetrakis(3-hydroxyphenyl) chlorin (*m*-THPC) and related second and third generation photosensitizers (PSs) currently under development. There has been a rapid growth in the literature on this topic. The chapter summarizes the conjugation of *meso*-tetrasubstituted porphyrin photosensitizers with biologically active elements (amino acids, bile acids, peptides, folic acid, sugars, oligonucleotides, monoclonal antibody), where most of the contemporary work also focuses. Porphyrin isomers and core modified porphyrins will also be discussed, followed by a brief discussion of nanoparticle derived systems.

The story began in 1912, with the first 'clinical trial' of porphyrins in humans when Friedrich Meyer-Betz injected himself intravenously with 200 mg of the crude naturally occurring material isolated from haemoglobin, haematoporphyrin (Hp), and investigated its effect on his skin. Erythema (reddening), edema (swelling) and pain developed in the area exposed to light.[1] Over 40 years later Schwartz[2] treated crude Hp with acetic acid and sulfuric acid, followed by treatment with strong alkali base. The resulting compound was given the name haematoporphyrin derivative (HpD) and it demonstrated increased tumor localization and fluorescence properties, which could be implemented in diagnostic tools.[3] HpD consists of a mixture of Hp monomer, dimer and oligomers. Elimination of the Hp monomer (the less active porphyrin) by partial purification produced porfimer sodium which was clinically approved for use in PDT in 1993 under the trade name Photofrin.[4] It is less heterogeneous than HpD but still contains a large number (approx 60) of porphyrin compounds.[5] Besides its poor water solubility, Photofrin has a low molar absorption coefficient (1170 $M^{-1}cm^{-1}$ at 630 nm) and thus requires high drug concentrations and light doses to achieve tumor destruction. Moreover, Photofrin possesses a low tumor/normal tissue selectivity of less than one. Another limitation is the long lasting cutaneous photosensitivity due to its localization in the skin during which the patient should avoid sunlight for 4–6 weeks. Thus, while Photofrin was perceived as a breakthrough "compound", in terms of opening a new treatment modality for cancer, it has significant drawbacks.

As a result, the search for improved photosensitizers with enhanced photophysical properties is of important requisite for future advances in PDT. The active phase of the synthesis of second generation photosensitizers focused on pure compounds with defined chemical composition and strong absorption at wavelengths between 650 nm and 800 nm for more effective PDT. Here, porphyrins, chlorins, and bacteriochlorins were the main targets as they have absorption maximum in the red region and have high singlet oxygen

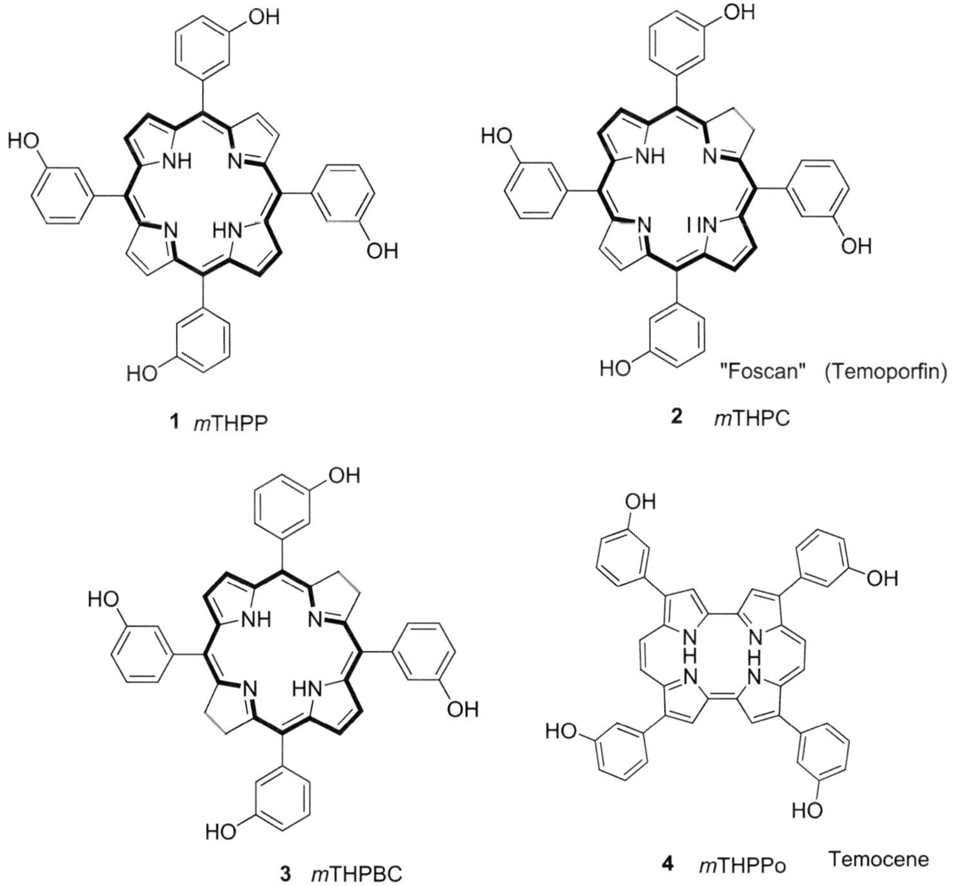

1 *m*THPP

2 *m*THPC

3 *m*THPBC

4 *m*THPPo

Fig. 1. Chemical formulae of *m*THPC related compounds.

quantum yields.[6] The classic example for a second generation porphyrin-based photosensitizer is 5,10,15,20-tetrakis (*m*-hydroxyphenyl) chlorin (*m*THPC, **2**, Fig. 1).[6,7]

This compound received regulatory approval in Europe in 2001 from the EMA under the name Foscan (Temoporfin). Currently it is used for the palliative treatment of head and neck cancers. For example, in its early stage in a confirmation study carried out with 29 patients, it gave complete local reduction of squamous cell carcinoma of the head and neck as an alternative modality for surgery and radiotherapy.[8] The parent porphyrin compound (5,10,15,20 tetrakis(*m*-hydroxyphenyl)porphyrin, *m*THPP, **1**) is 25–30 times as potent than HpD in PDT; however, smaller drug doses and less side effects are possible with *m*THPC and *m*THPBC.[9,10]

In a comparative study, *m*THPC showed superior uptake and tumor localization compared to *m*THPP.[11] The bacteriochlorin **3** was less photostable and thus **2** has received the most widespread use of these compounds.[12] It has a strong absorption band at 652 nm and a singlet oxygen quantum yield of 0.43, which is higher than *m*THPP. In addition, temocene (*m*THPPo) **4** is an interesting example of a porphycene analogue of *m*THPC. It has excellent photophysical properties which make it a promising photosensitizer for PDT applications.[13]

One possible pitfall of porphyrin-based photosensitizers is their tendency to form aggregates, which affects their biodistribution.[14] Here, introduction of a polar functional group can increase the solubility and enhance the pharmacokinetics of the drug. Thus, one aspect of contemporary research is a focus on amphiphilic porphyrins and related bioconjugates, e.g., glycoporphyrins.[15] The hydrophilic part increases the solubility of the drug in the blood stream and prevents aggregation while the hydrophobic part aids uptake, membrane passage and localization. For example, with *m*THPC, the introduction of electron withdrawing groups or alterations of the hydroxy groups resulted in a decrease in the partition coefficient and a significant increase in solubility.[16]

The main focus of this review is on such FGI and bioconjugate strategies and how controlling the type of the substituent at the *meso* position and conjugation with biologically active elements can increase the efficiency of the porphyrin structure unit for PDT. Figure 2 is a representative summary of the various strategies possible on tetrapyrrole **5** in this context. Note, this chapter focuses exclusively on *meso* substituted porphyrins of the A_x- and ABCD-type.[17]

2. *m*THPC (Temoporfin)

*m*THPC (Temoporfin) can be synthesized in acceptable yields as outlined in Fig. 3. Its synthesis involves the condensation of pyrrole **7** with *m*-aceto xybenzaldehyde **6** in propionic acid under heating at reflux for 30 minutes followed by deprotection of the acetyl group to yield 5,10,15,20-tetrakis (*m*-hydroxyphenyl)porphyrin **1**. Reduction of **1** without protection of the phenolic groups can be achieved using *p*-toluenesulphonylhydrazide to form 5,10,15,20-tetra(*m*-hydroxyphenyl)chlorin (*m*THPC, **2**).[9] A second reduction yields 5,10,15,20-tetra(*m*-hydroxyphenyl)bacteriochlorin (*m*THPBC, **3**).[18]

The pharmacological properties of the basic photosensitizer can easily be modulated through simple transformations or formulation strategies. For example, encapsulating *m*THPC into nontoxic liposomes (Foslip®) or modified pegylated liposomes (Fospeg®) yields carriers of high capacity for the otherwise

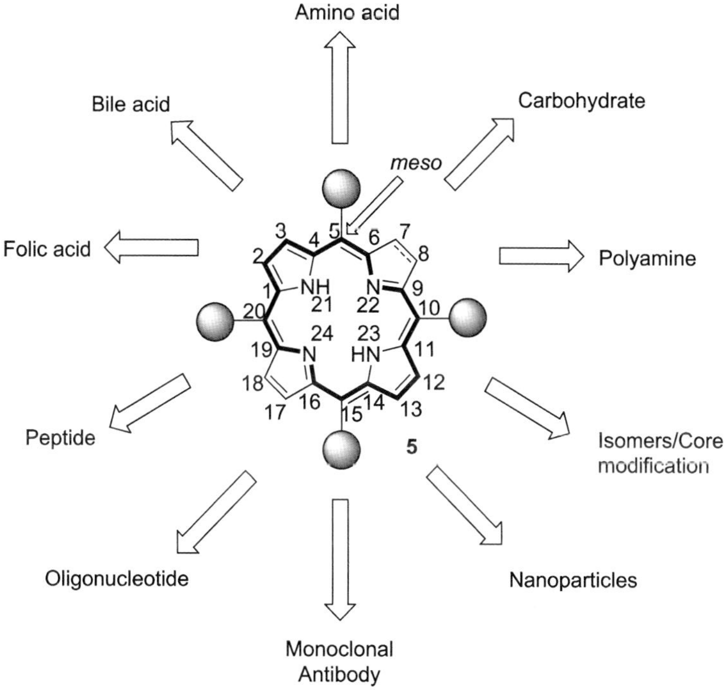

Fig. 2. Illustration of various bioconjugate strategies for porphyrin-based photosensitizers.

poorly water-soluble drugs.[19,20] Additional benefits are different drug release times,[21] different selectivities for tissue and cell types and more rapid tissue clearance,[22] improved pharmacokinetics and lower drug and light doses.[23–25]

We have recently reviewed the state-of-the-art for *m*THPC and its development and the reader is referred to these articles for more details on chemical and biological studies of Temoporfin.[7,10] This article focuses on recent advances with related compounds and aims to update these more comprehensive works.

3. New *meso*-Substituted Porphyrin Systems

3.1. *mTHPP-Related Systems*

An ideal photosensitizer must have a long lived triplet state ($\tau_T > 5\,\mu s$), as it is considered the most important precursor of the $O_2(^1\Delta_g)$ state. One such frequented strategy implemented to maximize the Φ_T of a given photosensitizer is through the heavy atom effect. Through the introduction of heavy atoms, e.g., the halogenation of porphyrins as described below; one can enhance the rate of the transition of the PS from its singlet to triplet state (intersystem crossing), which is a spin-forbidden process.

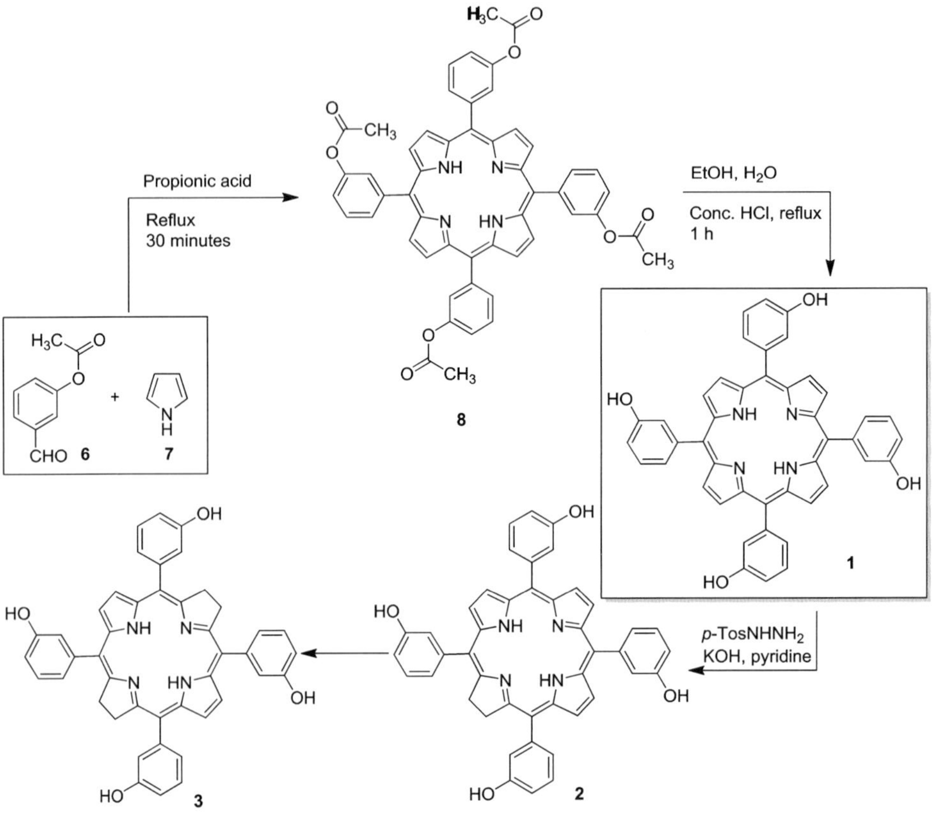

Fig. 3. Synthesis of *m*THPC.

As such, bromination of porphyrins favors reactive oxygen species (ROS) production by increasing the formation of triplet states and singlet oxygen generation. Laranjo *et al.* aimed to exploit this heavy atom effect by functionalizing hydroxylporphyrin derivatives with bromine *via* formation of 5,10,15,20-tetrakis(2-bromo-3-hydroxyphenyl)porphyrin **9** and 5,15-bis(2-bromo-3-hydroxyphenyl)porphyrin **10** as represented in Fig. 4.[26] *In vitro* studies with A375 and WiDr cells showed that for colorectal adenocarcinoma cells, IC_{50} values of 464 nM for **9** and 180 nM for **10** could be determined. Melanoma cells, with an IC_{50} of 100 nM for **10** and 50 nM for **9**, were more susceptible to both sensitizers. For *in vivo* studies Balb/c nu/nu mice were used and gave satisfactory results with **10**.

Despite the relative success of these brominated systems, the bulk of studies conducted tend to focus more on the creation of amphiphilic systems. A typical example is a study by Ben-Dror *et al.* using a series of *meso* substituted porphyrins **11** with various degrees of hydrophobicity, closely related to

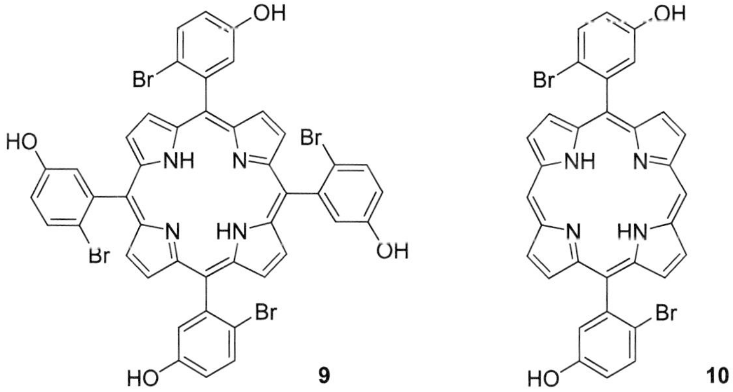

Fig. 4. Brominated hydroxyphenylporphyrins.

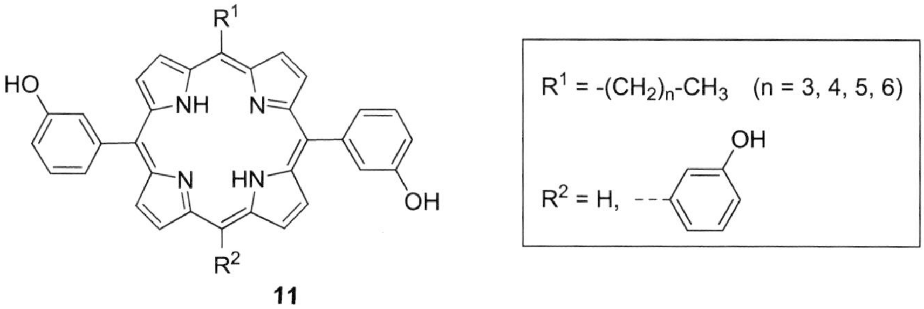

Fig. 5. *meso*-Substituted hydroxyphenylporphyrins with different degrees of hydrophobicity.

*m*THPP **1** and *m*THPC **2** (Fig. 5).[27] The porphyrins used had alkyl chains with different lengths at the 5 position and 3-hydroxyphenyl substituents at the 10, 15, and 20 positions. These porphyrins were found to be very hydrophobic, with log P values in the range of 8.9–11.8. The correlation between log P and K_b (the binding constant of these porphyrins into liposomes) was found to be negatively linear. This indicates that an increase in hydrophobicity results in a decreased binding affinity to liposomes. However, all the synthesized porphyrins were taken up and penetrated into cells.

More recently, Rojkiewicz *et al.* [2013] developed a series of tetrakis (hydroxyphenyl)porphyrin derivatives containing long alkyl chains (*via* ether, ester and amide linkages) as potential photosensitizers in PDT. They determined the absorption, fluorescence spectra, triplet state lifetime and singlet oxygen generation capacity of the synthesized compounds.[28]

More a step into the realm of compounds for electron transfer studies are the porphyrin-C_{60} (P-C_{60}) **12** and its metal complex (ZnP-C_{60}) **13** dyads.[29,30]

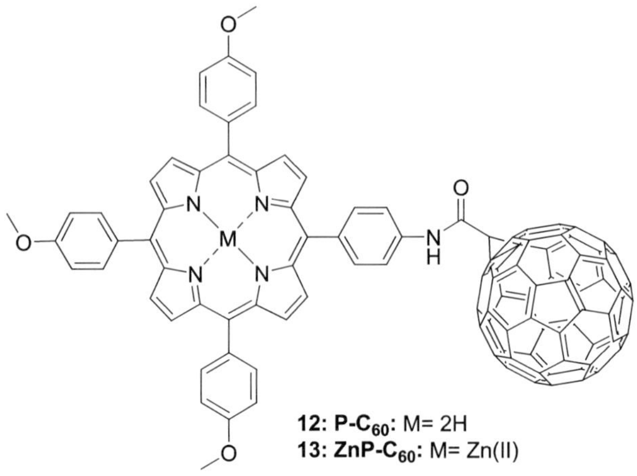

Fig. 6. Porphyrin-fullerene dyads investigated in PDT.

These studies represented the first evaluation of dyads with a high capacity to form a photo-induced charge-separated state for PDT. The fluorescence quantum yield for both dyads were obtained in toluene and DMF and in both solvents these dyads show only very weak emission from the porphyrin moiety, indicating strong quenching of the porphyrin excited singlet state by the attached fullerene ($\sim 3.4 \times 10^{-3}$ for **12** and $\sim 1.5 \times 10^{-3}$ for **13**). Compound **12** in toluene exhibited a singlet oxygen quantum yield of 0.80, whereas in DMF the value changed to 0.18. A similar result was observed for **13** which displayed a singlet oxygen quantum yield of 0.62, whereas in DMF the value diminished to 0.04. Porphyrin **12** showed an inactivation of 80% of Hep-2 human larynx-carcinoma cells with 54 J/cm^2 of irradiation; predominantly *via* apoptosis (Fig. 6).

3.2. *Cationic Porphyrins*

Cationic porphyrins have long captivated the attention of PDT researchers as they are endowed with a number of desirable traits, the most interesting of which being their increased water solubility. This family of tetrapyrroles is currently undergoing a renaissance of sorts in the field of anti-microbial PDT due to their rare ability to destroy Gram-negative bacteria. For example, four 5,10,15,20-tetraphenylporphyrin (TPP) derivatives bearing either triphenylphosphonium (**14** and **15**) or triethylammonium (**16** and **17**) and terminated alkoxy groups at either the *para*-(**14** and **16**) or *meta*-(**15** and **17**) position of a *meso* phenyl residue were synthesized by Lei *et al.* (Fig. 7).[31]

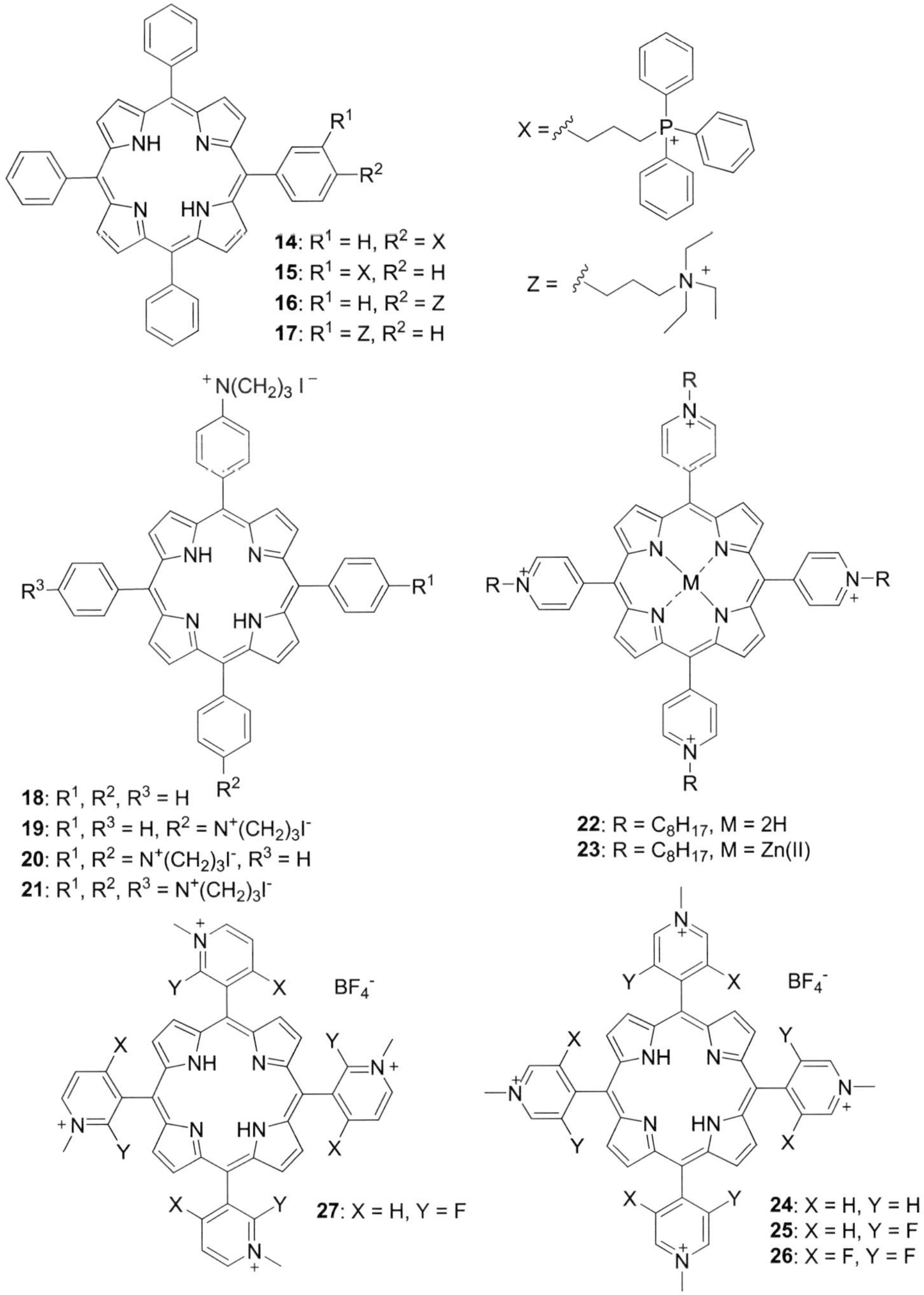

Fig. 7. Cationic porphyrins used in PDT studies.

All compounds exhibited similar absorption and fluorescence spectra along with similar $^{1}O_2$ quantum yields. The more lipophilic nature of triphenylphosphonium ion over triethylammonium ions results in higher octanol/water partition coefficients for **14** and **15** compared to **16** and **17**. Confocal fluorescence microscopy proved that **14–17** accumulate in the mitochondria. MTT assays showed that these derivatives have significant phototoxicity and that their dark toxicity is negligible at concentrations of 6–10 μM towards the human breast cancer cell line MCF-7. The $^{1}O_2$ quantum yields of **14–17** were measured in acetonitrile and found to be 0.53, 0.50, 0.57, and 0.58, respectively. Porphyrin **14** exhibited the least dark cytotoxicity and **16** possessed the highest phototoxicity among the four porphyrins.

In a similar context Jensen *et al.* [2010] synthesized a series of cationic porphyrins **18–21** bearing one to four -$N(CH_3)_3^+$ groups linked to the *p*-phenyl positions.[32] They investigated the effect of overall charge and its distribution on the cellular uptake, phototoxicity and intracellular localization using human carcinoma HEp2 cells. The dicationic porphyrin **19** accumulated the most within cells and preferentially localized within vesicular compartments and in mitochondria. The monocationic porphyrin **18** proved to be the most phototoxic of the series, localizing primarily in lipid membranes, including the plasma membrane, ER, mitochondria, and Golgi. Both the tricationic porphyrin **20** and the tetracationic porphyrin **21** localized subcellularly, primarily in the mitochondria, but of the two only **21** showed moderate phototoxicity ($IC_{50} = 8\ \mu$M at 1 J$\cdot$cm^{-2}).

Further studies looked at the role of amphiphilicity and zinc insertion on PDT efficacy. Pavani *et al.* evaluated the PSs **22** and **23** which exhibited similar photophysical properties ($\Phi_f \leq 0.02$; $\Phi_\Delta \sim 0.8$).[33] Both an increase in lipophilicity and the presence of zinc in the porphyrin resulted in higher vesicle and HeLa cell uptake. Porphyrin **22** was more enriched in the mitochondria and the zinc derivative **23** had the highest global uptake. Zinc insertion decreased the interaction with isolated mitochondria and with the mitochondria of HeLa cells. Phototoxicity increased proportionally with membrane binding efficiency. Cytotoxicity tests showed that these porphyrins have no dark toxicity, whilst **23** exhibited good cytotoxicity with ~85% tumor eradication with a light dose of 175 mJ$\cdot$cm^{-2}.

Water-soluble fluoro-substituted porphyrins were designed for similar studies by Ko *et al.*[34] of the porphyrins **24–27** the latter two exhibited significantly higher photocytotoxicity on HeLa cells. IC_{50} values were 0.87 μM for porphyrin **26** and 1.1 μM for porphyrin **27**, while IC_{50} values of 3 μM and 40 μM were obtained for **24** and **25** respectively. Regarding overall PDT efficacy, it was determined that compound **27** was the most effective PDT photosensitizer based on the studies conducted with HeLa cells.

Related studies focused on water-soluble tricationic porphyrins,[35] brominated cationic porphyrins,[36] and polymethine conjugated porphyrins as dual chromophores. These exhibited promising photodynamic activity either *in vitro* or *in vivo*.[37] This class of compounds still merits further investigations for both PDT and anti-microbial PDT.

3.3. *Porphyrin-Polyamine Conjugates*

Polyamines, i.e. molecules which contain two or more primary amino groups, are essential for normal cell growth. Rapidly proliferating cancer cells require more polyamines than their biosynthetic abilities allow, thus they tend to use a polyamine transport system (PTS) to fulfill their needs. The increased uptake of polyamines by cancer cells is an attractive chemotherapeutic approach. Previous studies have shown that the PTS can readily take up heterocyclic derivatives if conjugated to natural polyamines.[38]

Hence, recent studies in PDT have focused on comparing the efficacy of porphyin-polyamine conjugates in relation to Photofrin. These studies indicate a better efficiency of polyamine conjugates over the clinically approved PS. For example, Sarrazy *et al.*[39] synthesized a number of porphyrin-polyamine conjugates containing two or four spermidine or spermine units (Fig. 8).[40] Their phototoxicity was evaluated using human keratinocyte (HaCat) and human breast cancer (MCF7) cell lines and all conjugates were more efficient than Photofrin at inducing cell death *via* apoptosis. The *in vitro* cytotoxicity of the compounds was evaluated and their phototoxicity followed the order **31** ($65.4 \pm 22.5\ \mu M$)>**28** > **29**, **32** > **30** and **32** ($49.4 \pm 25.7\ \mu M$) > **28** > **29** > **31** > **30** according to IC_{50} values for HaCat and MCF7, respectively.

Polyamine conjugates have also been employed in conjunction with Boron Neutron Capture Therapy (BNCT). Vicente *et al.* provided promising results on the synthesis of carboranyl-containing porphyrins and chlorins aimed at developing dual modality photosensitizers suitable for both PDT and BNCT cancer treatment.[41,42] The results suggest that carboranylchlorins are good candidates for this dual treatment modality and warrant further investigations. Recently, seven polyamine conjugates of a tri(*p*-carboranylmethylthio)-tetrafluorophenylporphyrin (**33–39**) were prepared in high yields by sequential substitution of the *p*-phenyl fluoride of tetrakis(pentafluorophenyl)porphyrin (TPPF) (Fig. 9).[43] The polyamines used were derivatives of the natural-occurring spermine with different lengths of the carbon chains, terminal

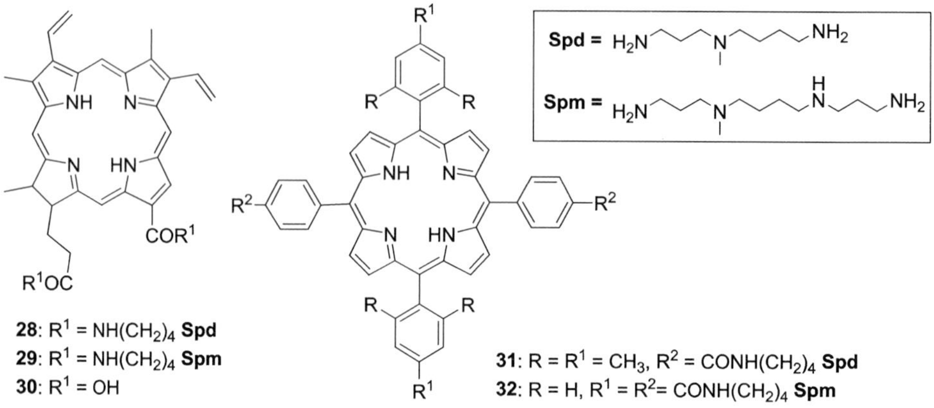

Fig. 8. Spermidine and spermine appended porphyrins.

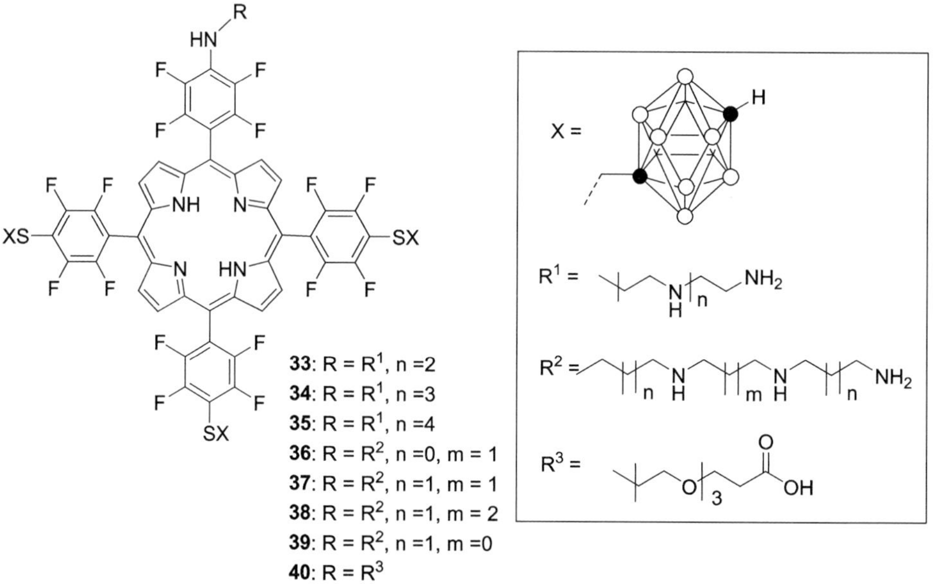

Fig. 9. Carboranyl- and polyamine-appended porphyrins.

primary amine groups and, in two of the conjugates, additional aminoethyl moieties. The tri(polyethylene glycol)conjugate **40** was also synthesized for comparison. The polyamine conjugates showed low dark cytotoxicity ($IC_{50} >$ 400 μM) and low phototoxicity ($IC_{50} > 40$ μM at 1.5 J·cm^{-2}). All polyamine conjugates, with one exception, showed higher uptake into human glioma T98G cells (up to 12-fold) than the PEG conjugate, and localized preferentially in the ER, Golgi and lysosomes.

3.4. *meso-Tetrasubstituted Chlorins*

Although *m*THPC is an effective PS for PDT it still suffers from drawbacks; amongst others is the ease of re-oxidation to the parent porphyrin **1**. Thus, attempts have been made to prepare related compounds with substituents at the reduced pyrrole ring. A classical approach is the use of dihydroxy derivatives which have been used by Rancan *et al.* to verify the *in vitro* effect of different substituents in dihydroxychlorins and the influence of amphiphilicity on intracellular uptake, subcellular localization, and photosensitizing activity (Fig. 10).[44] Variations in the substituents significantly influenced the intracellular uptake of these chlorins. For example, chlorin **42** which was the most polar sensitizer in this study, depending on incubation time, yielded an intracellular concentration in Jurkat cells 2–8 times higher than the unsubstituted chlorin **41**. These systems localize predominantly in lysosomes and increasing the amphiphilicity of the sensitizer molecules correlated with increased sensitizer uptake and increased number of necrotic cells after irradiation.

Recently, the synthesis of novel *β*-functionalized derivatives of *m*THPC was achieved *via* nucleophilic addition reactions to diketochlorin derivatives.[45] The intriguing *β*-substituted dihydroxychlorin products **43–45** exhibited strong absorption in the red spectral region, high singlet oxygen quantum yields (0.97, 1.56 and 1.10 respectively), and were found to be highly effective in *in vitro*

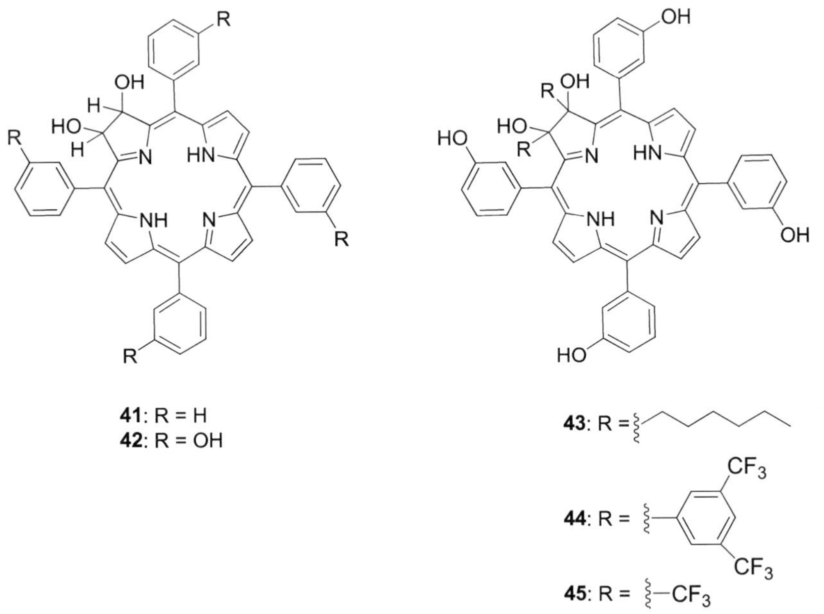

Fig. 10. Chlorins related to *m*THPC.

assays against HT-29 tumor cells. Chlorins **43** and **45** showed phototoxicity at concentrations of 2 and 10 μM respectively, and were similar in activity to *m*THPC **2**. Chlorin **44** displayed a lower activity and was only moderately effective at a concentration of 10 μM.

4. Bioconjugates of *meso*-Tetrasubstituted Porphyrins

4.1. *Glycoconjugated Porphyrins*

The synthesis of porphyrin-carbohydrate conjugates is based around the fact that on the surface of cancer cells an overexpression of specific extracellular proteins is observed. These proteins are known as lectins and they bind carbohydrates and therefore, porphyrins possessing carbohydrate groups could, in theory, enhance the uptake of the photosensitizer into, for example, retinoblastoma cells (Y79) bearing α-mannose and β-galactose receptors. Melanoma B16 cells and adenocarcinoma colon cells HT29 have been found to overexpress β-galactose and β-glucose receptors respectively, and these cell lines could provide additional targets for such PS conjugates.[46] As a result, a significant body of information on sugar-appended porphyrins is available.[47]

4.1.1. *Carbohydrates directly-attached to the meso position*

Two recent examples can serve to illustrate the synthetic strategies used for directly linked porphyrin-carbohydrate conjugates. Casiraghi *et al.* constructed porphyrins bearing two and four glycosidic units directly linked at the *meso* positions.[48] Notably, the two palladium derivatives **46** (*C*-*meso*-linked aryl-uridinyl porphyrin) and **47** (tetra-*C*-glycosylated porphyrin) were proven to be effective for the selective cleavage of double stranded DNA into form II nicked circular DNA on exposure to visible light at RT in aqueous media (Fig. 11). Thus, some of the compounds illustrated here may have additional uses next to PDT.

More directly related to PDT are photosensitizers prepared *via* fusion of the C6 of the pyranose sugar to the *meso* position of the porphyrin.[49] Examples are the C_2-symmetric conjugates **48–55** and D_2-symmetric derivative **56**. The tetraglyco-fused porphyrin **56** gave the highest cellular uptake and photocytotoxicity both in HeLa and HCT116 cells. Compounds **49**, **50** and **54** bearing electron donating groups showed high phototoxic effects in both human cancer cell lines compared to **51–53**.

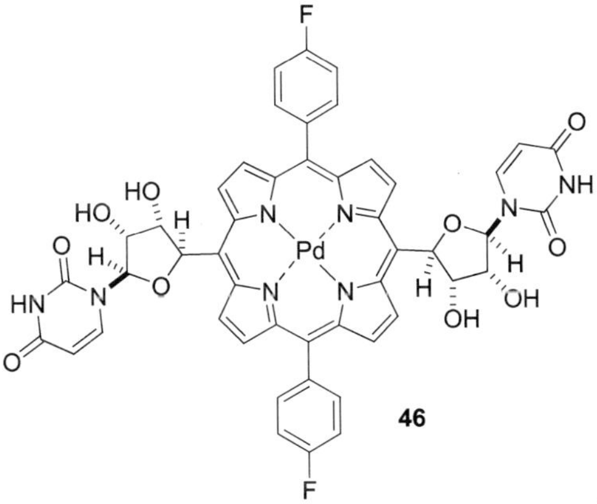

46

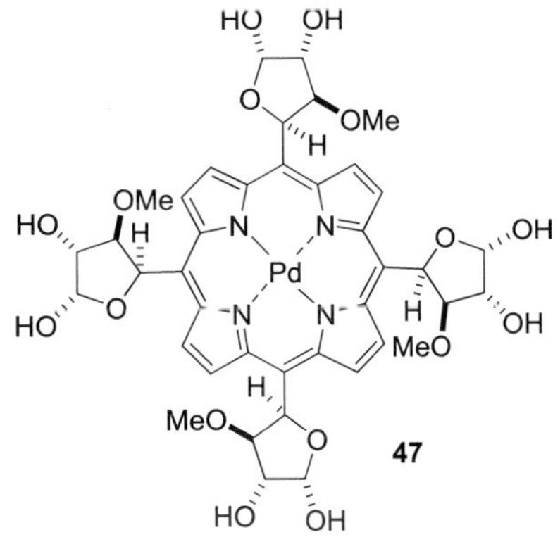

47

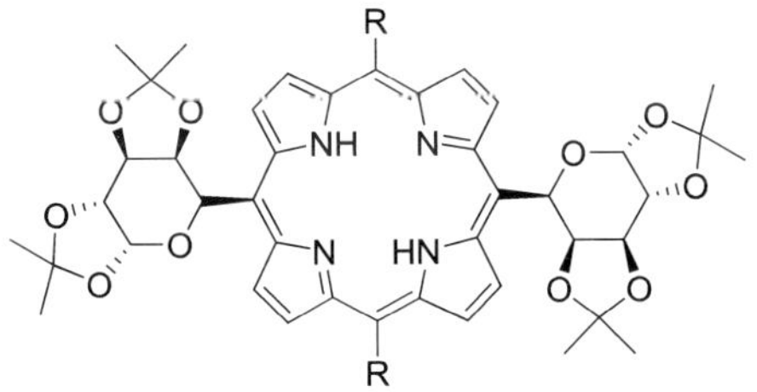

48: R = Ph
49: R = 4-OMePh
50: R = 4-SMePh
51: R = C_6F_5
52: R = 4-ClPh
53: R = 4-NO_2Ph
54: R = Thiophen-3-yl
55: R = 2,6-OMePh
56: R = Sugar

Fig. 11. Directly *meso*-linked glycoporphyrins.

4.1.2. *Porphyrin-Linker-Carbohydrates systems*

(a) ***O-Glycosylated Porphyrins and Chlorins***

From the outset of their pioneering studies in this area Maillard and coworkers identified the impact of porphyrin glycosylation on photosensitizers related to systems such as *m*THPC.[50] For example, they synthesized glucosyl conjugated tri- and tetra(*meta*-hydroxyphenyl)chlorins and investigated their PDT effect in HT29 human colorectal adenocarcinoma cells (Fig. 12).[51] Typically the symmetrical tetraglucosylated chlorins displayed poor internalization and photoactivity compared to the asymmetric, more amphiphilic derivatives (**57**, **58** {1 : 1}). The latter showed better photoactivity compared to *m*THPC with light doses of 5–25 J·cm^{-2}. A comparison of the log P values of *m*THPC (>3), **57**, **58** (1.45) and the tetraglycosylated derivative (−1.31) again shows the importance of amphiphilicity in such structures. The fluorescence emission spectra of all compounds were similar; however, the glycosylated derivatives exhibited slightly better fluorescence quantum yields (φ_f = ~0.12 versus *m*THPC, $\varphi_f = 0.095$).

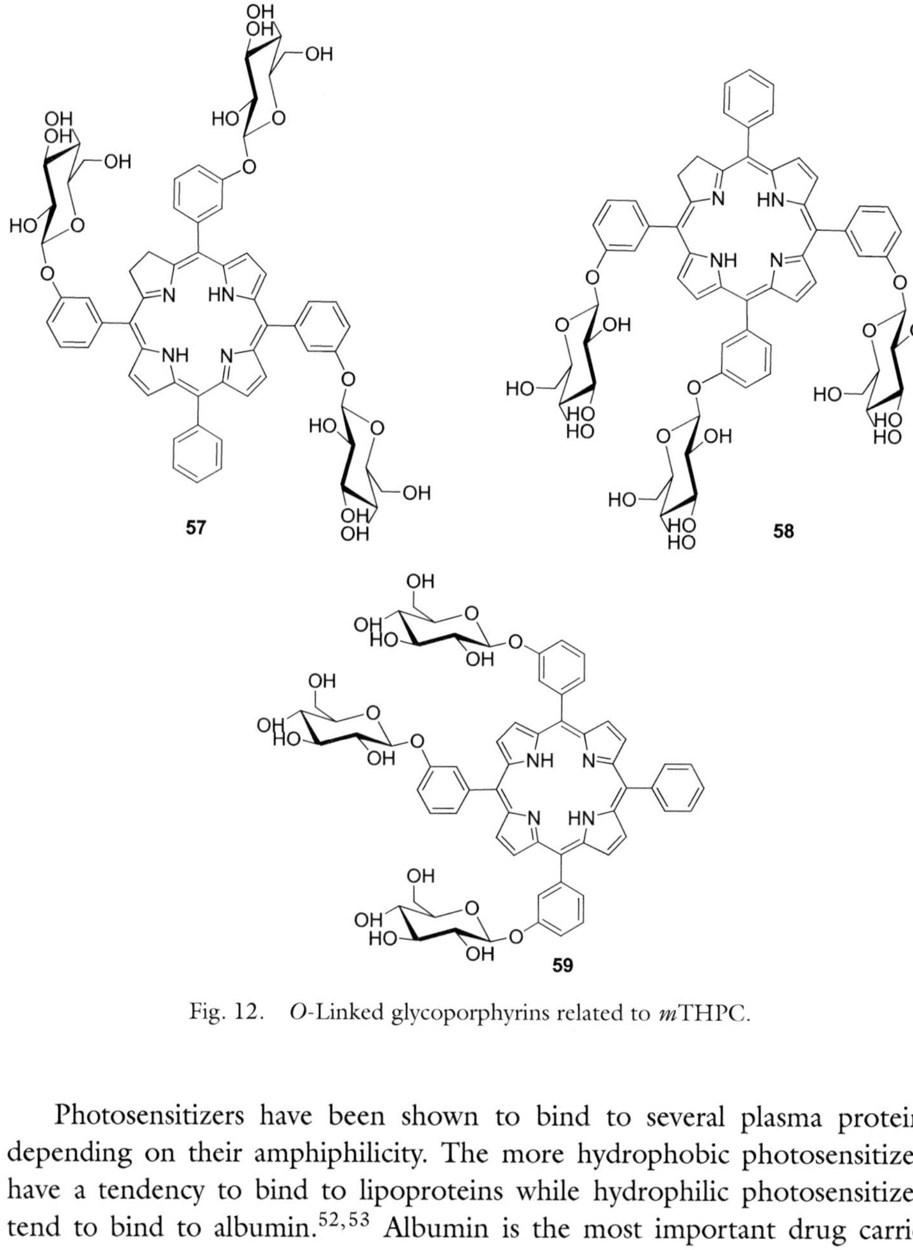

Fig. 12. *O*-Linked glycoporphyrins related to *m*THPC.

Photosensitizers have been shown to bind to several plasma proteins depending on their amphiphilicity. The more hydrophobic photosensitizers have a tendency to bind to lipoproteins while hydrophilic photosensitizers tend to bind to albumin.[52,53] Albumin is the most important drug carrier protein. During cellular uptake and phototoxicity tests, it was shown that **57** and **58** associated more to the albumin protein compared to *m*THPC. Both compounds were partly internalized by an active receptor-mediated endocytosis mechanism compared to *m*THPC which is mediated by an active uptake process *via* a low density lipoprotein (LDL).[54,55] Passive diffusion must also play a role in internalization. Cellular uptake of **57** and **58** was 30% lower compared to *m*THPC and exhibited saturation. Maximum phototoxicity in HT29 cells was observed with concentrations as low as 2 μM at light doses of 5–25 Jcm^{-2}. The roughly 4-fold better efficacy compared to *m*THPC was attributed to a higher accumulation of the former in the mitochondria.

The related porphyrin compound **59**, 5,10,15-tris(*meta-O-β*-D-glucosyloxyphenyl)-20-phenylporphyrin, was used for an *in vivo* study in healthy rats.[56] In order to establish the pharmacokinetics of this class of compounds, three different drug doses of **59** were administered (0.25, 0.5 and 1 $mg{\cdot}kg^{-1}$), and compared to *m*THPC at 0.3 $mg{\cdot}kg^{-1}$. The conjugate **59** was cleared more rapidly from the organism compared to *m*THPC, and its mean residence time was 5 h compared to *m*THPC, which had a residence time of 20 h with no metabolite of either PS identified. Looking at the biodistribution, a large amount of the glycoporphyrin was concentrated in organs such as the lungs, liver and spleen which are rich in reticulo-endothelial cells. The maximum PS concentration in the organs was observed at 14 h post-infection and within 48 h the PS was eliminated from all organs.

Many related studies have focused on chlorins and the impact of different types and arrangements of the carbohydrate units. Initially progress was hampered by the lower stability of chlorins but significant improvements have been made through the preparation of glycochlorins.[57–60] A typical *in vitro* study examined the cellular uptake and photocytotoxcity of glycoconjugated chlorin derivatives with *β*-D-glucopyranosyl, *β*-D-galactopyranosyl, *β*-D-xylopyranosyl and *β*-D-arabinopyranosyl groups using HeLa cells and compared them to tetraphenylporphyrin tetrasulfonic acid (TPPS) (Fig. 13).[61] The 5,10,15,20-tetrakis[3-(*β*-D-xylopyranosyloxy)phenyl]chlorin (**61*mc***) showed a 50 times higher uptake than TPPS. All glycochlorins exhibited higher uptake than TPPS, and either equal or higher uptake than the porphyrin derivatives. The order of cellular uptake was as follows: **61*mc*** > **61*pd*** > **61*md*** > **61*pa*** > **61*pb*** > **61*pc*** > **61*ma*** > **61*mb*** > TPPS in line with the higher lipophilicity of the glycochlorins.

Studies on the photocytotoxicity of these compounds used 5,10,15,20-tetrakis[3-(*β*-D-xylopyranosyloxy)phenyl]chlorin **61*mc*** and 5,10,15,20-tetrakis[3-(*β*-D-glucopyranosyloxy)phenyl]chlorin **61*ma*** in HeLa cells. At a drug dose concentration of 2×10^{-7} M, their toxicity decreased in the order of **61*ma*** > **61*mc*** > TPPS, indicating that the photocytotoxicity was not related to cellular uptake. The LD_{50} for **61*mc*** was $\sim 2 \times 10^{-7}$M and 5×10^{-6} M for **60*pa***.[62]

Tetraphenylporphyrin-type systems have been adopted by several groups to investigate the effect of glycosyl groups at the periphery of the porphyrin

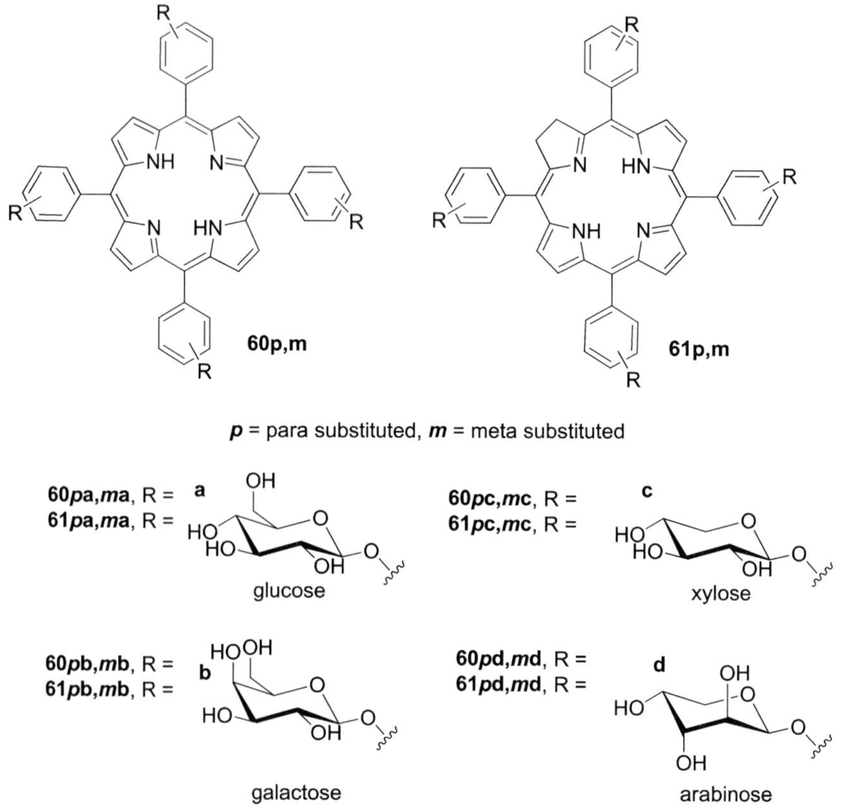

Fig. 13. Glycoporphyrins and –chlorins employed for uptake and PDT studies in HeLa cells.

macrocycle. Overall, the triglycosylated derivatives are more phototoxic then the parent tetrapyrrole or the symmetrical tetraglycosylated derivatives.[50] The photobiological properties of glycoporphyrins are dependent on the linker, the type of glycoside residue and the anomeric configuration.[47]

Returning to the fundamental contributions by Maillard, his group synthesized diethylene glycol (Deg) linked glycoporphyrins and studied their PDT effect in human retinoblastoma cells (Y79).[60] Compounds **63–67** (Fig. 14) showed maximum fluorescence emission at 632 nm and a weak band at 720 nm with relatively low fluorescence quantum yields of ~0.03. Compound **62** was internalized poorly by Y79 cell lines compared to the glycosylated derivatives. The most significant result was the high uptake of **65** and its good photoactivity ($LD_{50} = 0.35\,\mu M$) compared to the parent compounds **62** and **67**. The singlet oxygen production was similar for all glycosylated compounds ~0.47. The compounds were tested for uptake by murine melanocyte cells (B16) and

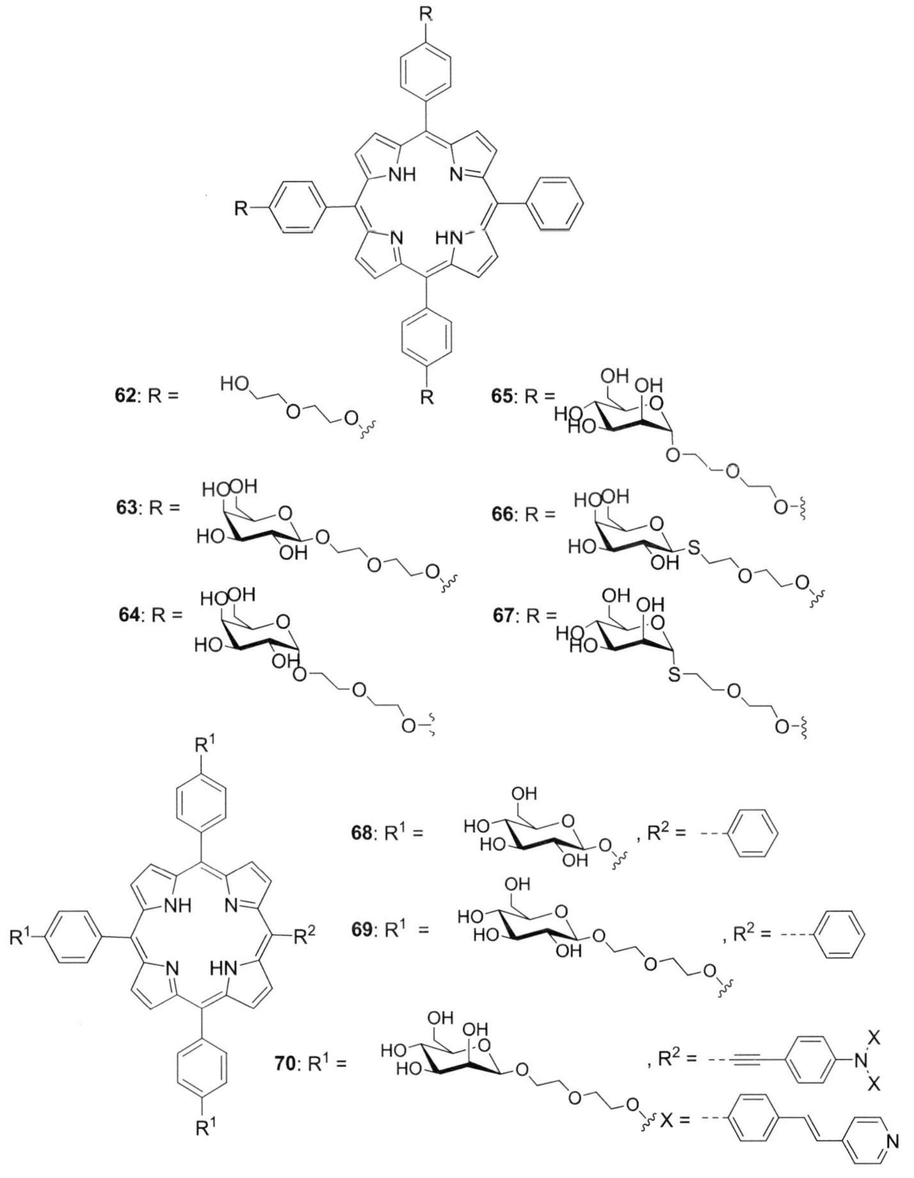

Fig. 14. Diethylene glycol (Deg) linked glycoporphyrins.

human adenocarcinoma cells (HT29) and behaved similarly at 4 h incubation at 4 μM concentration. Initial work carried out on the pre-incubation of cells in glycosylated albumin showed a 40–45% inhibition of uptake of TPP(*p*-Deg-*O*-β-GalOH)$_3$ and TPP(*p*-Deg-*O*-α-ManOH)$_3$ indicating a cell-sugar-receptor saturation.

Expanding on earlier work using **57**, **58** and **65** in HT29 and Y79 cell lines, a number of porphyrin and chlorin derivatives were explored for photocytotoxicity using weaker light doses of 1.8 $J.cm^{-2}$ and excitation at > 540 nm after 24 h incubation.[63] Even though the glycoconjugated derivatives had lower PDT activity compared to *m*THPC they displayed IC_{50} values of 2.4–0.05 μM against Y79 cells. The *para*-*O*-β-glucose porphyrin derivative **68** was found to have the best phototoxicity compared to the parent porphyrin system with an IC_{50} of 0.9 μM and no cytotoxicity in the absence of light with an $IC_{50} > 15\ \mu$M. However, this compound is less PDT active compared to the Deg-porphyrin conjugates described previously (**63–65**). The extension of the linker between the glycoconjugate and the chromophore appears to significantly increase the PDT activity. Also highlighted was the difference in the anomeric configuration with **64** (IC_{50} = 0.05 μM) showing a 10-fold greater phototoxicity than **69** (IC_{50} = 0.6 μM) in Y79 cells.

Hammerer *et al.* synthesized the carbohydrate-porphyrin-triphenylamine hybrid **70** containing a porphyrin linked at the *meso* position with triphenyl amine *via* an alkyne bond while the other three *meso* groups was conjugated with diethylene glycol mannose units (DEG-Man).[64] Interestingly, the hybrid **70** displayed high singlet oxygen quantum yields as well as good two photon absorption (2PA) cross section values as a result of the conjugating alkyne unit. However, compound **70** was inactive against HT-29 and retinoblastoma Y79 cells.

The number of available glycoporphyrins has increased significantly through the use of 'click'-reactions. The 1,3-dipolar cycloaddition between an azide and an alkyne, that results in the formation of a 1,2,3-triazole moiety has recently been implemented by a number of research groups to obtain a wide range of *O*-glycoporphyrins. This synthetic strategy is considered an effective and highly efficient method for the conjugation of sugar units to the porphyrin periphery due to the high synthetic yields and the relatively simple and accessible experimental conditions.[65–69]

Together with Scanlan we reported on mono-, di-, and trisubstituted glycoconjugated porphyrin compounds using 'click chemistry'.[70,71] This involved the Cu(I)-catalyzed reaction between various azido porphyrins and propargylic carbohydrates to form a library of glycoporphyrins using microwave mediated reactions. Figure 15 gives a representative example for the synthesis of the difunctional glycoporphyrin **73** from azido porphyrins **71** and propargyl mannose **72** *via* a "double-click" process. The compounds were evaluated against esophageal squamous-cell carcinoma cell lines.

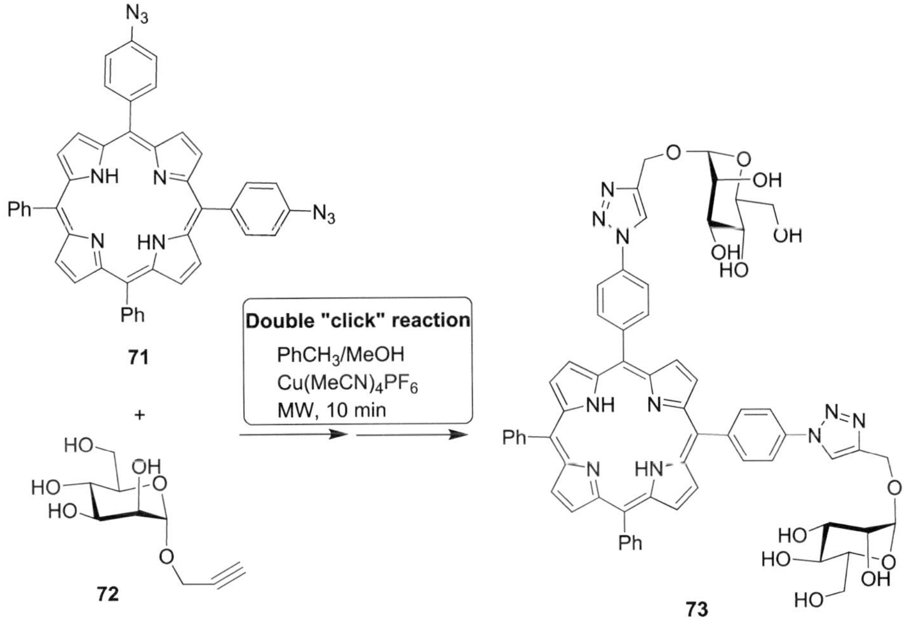

Fig. 15. Synthesis of a bisglycoporphyrin *via* 'click' reaction.

(b) *S-Glycosylated Porphyrins and Chlorins*

S-linked glycoconjugates are considered to be more stable than their *O*-linked counterparts and thus several porphyrin derivatives of the former have been reported. In several cases this has been combined with the use of fluorinated porphyrins which have high triplet quantum yields due to their enhanced intersystem crossing.[72] Examples are the *S*-glycoside conjugated porphyrins **74** and **75** (Fig. 16).[73] The tetraglucose-porphyrin conjugate **74** showed better uptake compared to the galactose derivative **75** in the human breast cancer MDA-MB-231 cell line and also displayed better PDT activity.[74]

A 'complete set' (mono-, *cis*-bis-, *trans*-bis-, tris- and tetrakis) of tetrakis(perfluorophenyl)porphyrins substituted with β-D-glucopyranosylthio groups was investigated by Hirohara *et al.*[75] The *S*-glucosylated porphyrin *trans*-**78** showed excellent photocytotoxicity ($EC_{50} < 5$ nM) in HeLa cells. The log P affecting the uptake of the porphyrin derivatives was determined and decreased from 5.1 to 0.8 in the order **76** > *cis*-**77** > *trans*-**78** > **79** > **74**. The photocytotoxicity was examined at a drug dose concentration of 100 nM in HeLa cells with photoirradiation at 16 J$\cdot$ cm^{-2} and 24 h incubation post treatment. All drug molecules displayed no dark toxicity and the photocytotoxicities of **76**, *cis*-**77**, *trans*-**78**, **79**, and **74** were much higher than TPPS.

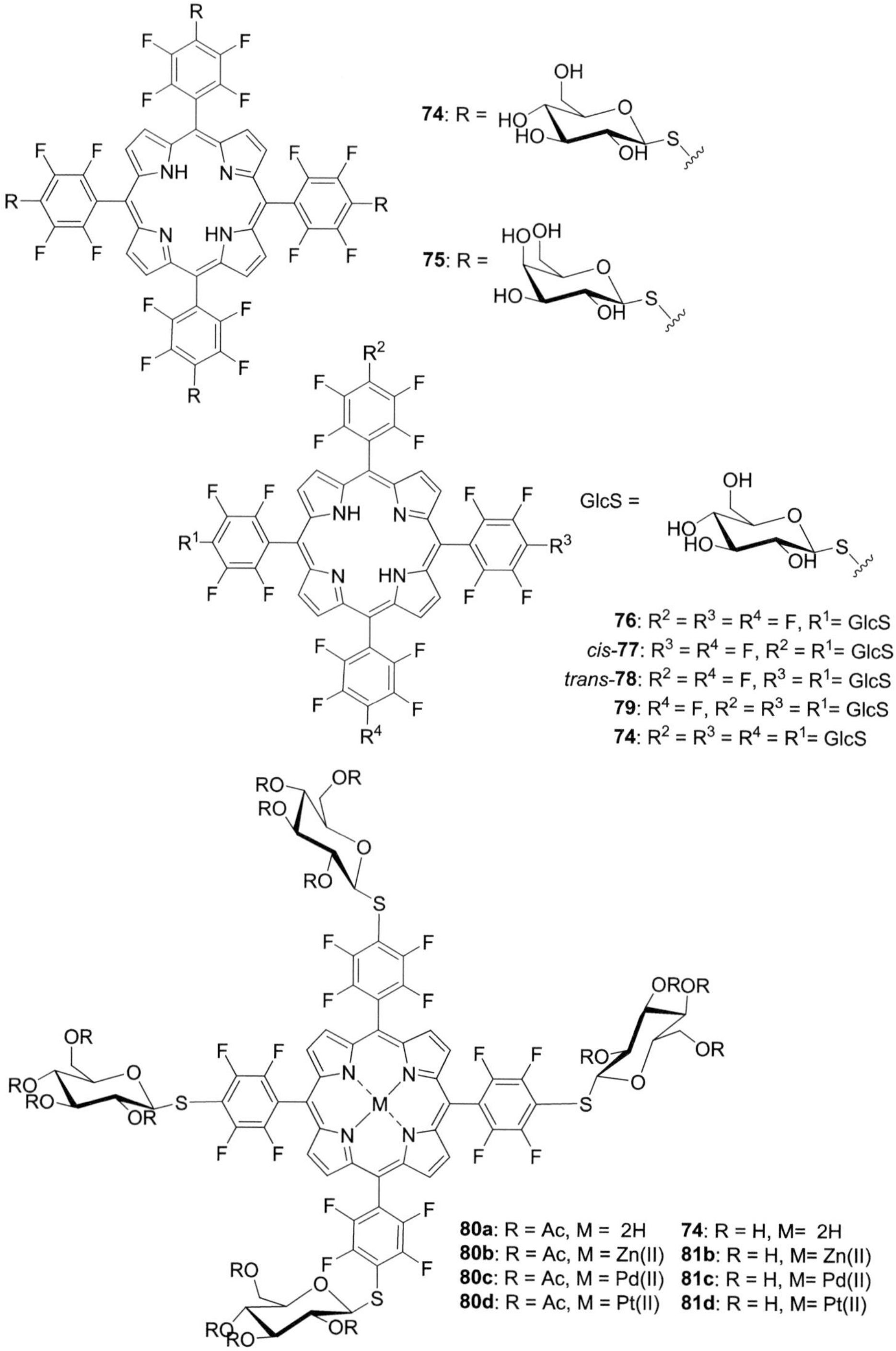

Fig. 16. *S*-Linked glycoporphyrins.

EC_{50} values were <5 nM (***trans*-78**), ca. 80 nM (**79**) and ca. 90 nM (**76**, ***cis*-77**, **74**). Porphyrin ***trans*-78** had a 20 times higher efficacy than the other *S*-glucosylated porphyrins.

In order to utilize the heavy atom effect to increase the $^1O_2$ quantum yield, several metalloporphyrins were investigated. For example, 5,10,15,20-tetrakis(4-(2,3,4,6-tetra-*O*-acetyl-β-D-glucopyranosylthio)-2,3,5,6-tetrafluorophenyl)porphyrin **80a** and its Zn(II), Pd(II), and Pt(II) complexes **80b**, **80c**, and **80d** were synthesized and deprotected to give the *S*-glucosylated derivative **74** and its metal complexes **81b**, **81c**, and **81d**.[76] The singlet oxygen production increased in the order of free-base fluoroporphyrins (**80a**, 0.08 and **74**, 0.06) < Zn(II) complexes (**80b**, 0.055 and **81b**, 0.056) <Pd(II) complexes (**80c**, 0.003 and **81c**, 0.002). However, the $^1O_2$-production of the Pt(II) complexes (**80d**, 0.001 and **81d**, 0.001) was very low. *In vitro* photocytotoxicity was measured with HeLa cells incubated for 24 h at a drug dose of 0.5 μM (16 J$\cdot$cm^{-2}). The protected derivatives **80a**, **80b**, **80c**, and **80d** showed no photocytotoxicity while **74**, **81b**, and **81c** were active. Compounds **74**, **81b**, **81c**, and **81d** induced 90% cell death (EC_{90}) at ~0.2 μM in each case. This is about 25 times lower than observed for the O-glucosylated derivatives (EC_{90} > 5 μM).[77]

Hirohara *et al.* performed comparative analyses of a range of glycosylated TPPs and chlorins.[78] The *O*-glycosylated chlorins had promising photocytotoxicity in HeLa cells but were difficult to purify.[79,80] Photosensitizers **80a**, **82**, **83**, **85**, **87** and **88** showed no significant photocytotoxicity at a concentration of 0.5 μM (24 h, 16 J$\cdot$cm^{-2}), while the deprotected photosensitizers **74**, **75**, **84** and **86** were photocytotoxic (Fig. 17). The *S*-glucosylated photosensitizers **74** and **84** showed higher photoactivity than *S*-galactosylated **75** and **86** respectively. The drug concentrations inducing 50% cell death (EC_{50}) of **74**, **75**, **84** and **86** were estimated to be ~0.15, 1.5, 0.07 and 0.3 μM respectively.[78]

Drain's group extended such studies to the bacteriochlorins and isobacteriochlorins.[81] Compared to the porphyrin **74**, **84** exhibited a 6-fold increase in Φ_f, **89** a 12-fold increase, **90** had a Φ_f similar to **74**, however, it also possessed a considerably red-shifted absorbance to 730 nm with a nearly 50-fold greater absorbance. Previous studies with **74** indicated good uptake and photodynamic effects on several cancer cell lines, e.g., K:Molv NIH 3T3 mouse fibroblasts and the MDA-MB-231 breast cancer cell line. Here, **84**, **89** and **90** were shown to be taken up at nanomolar concentrations.

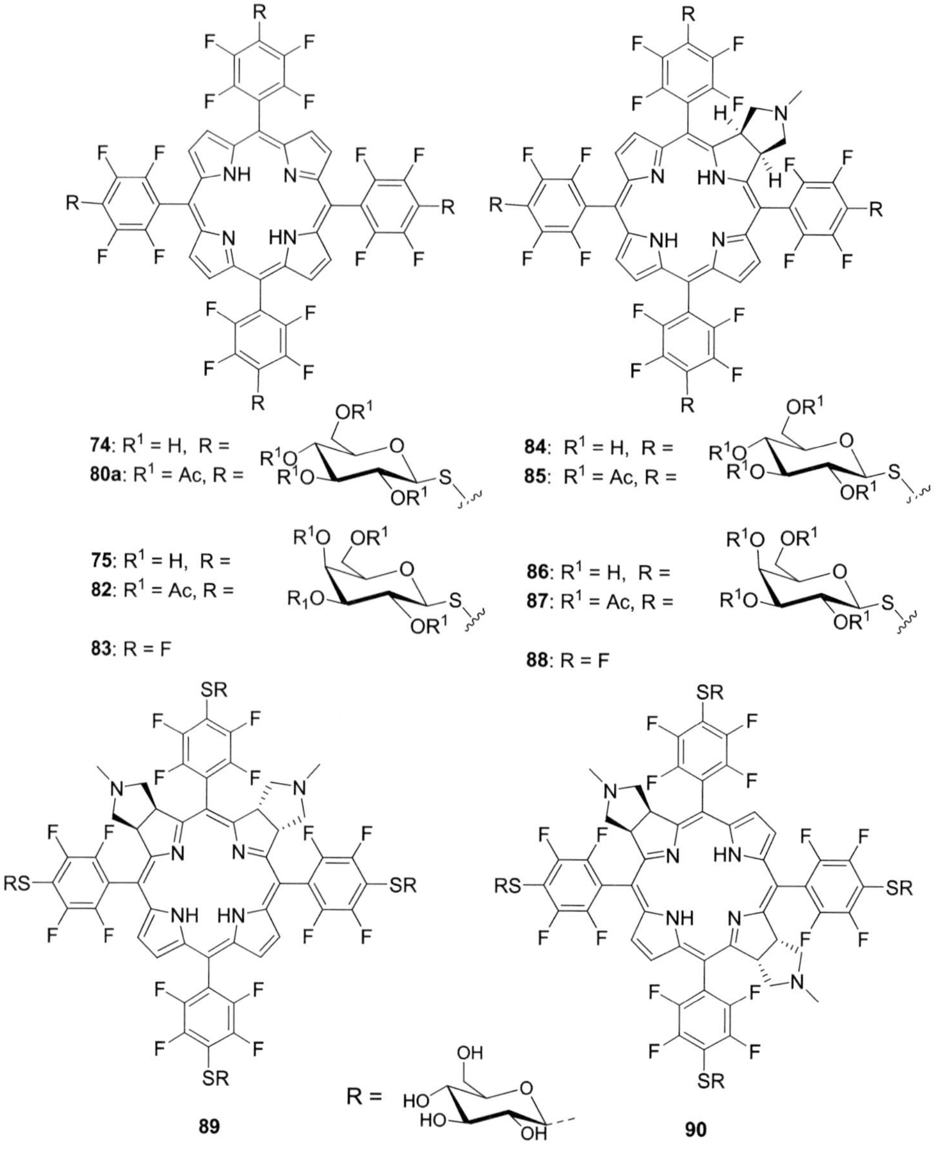

Fig. 17. *S*-Linked glycoporphyrins and -chlorins.

(c) *Glycosylated Porphyrin Dimers and Oligomers*

Additionally, dimeric and oligomeric porphyrin glycoconjugates have been studied. Typical examples for neutral and cationic porphyrin dimers linked at the *meso* position *via* a flexible hydrocarbon chain are shown in Figs. 18 and 19.[82] The porphyrins **91–95** exhibited a $^1O_2$ production efficiency similar to that of Hp. Photocytotoxicities of these compounds were evaluated with

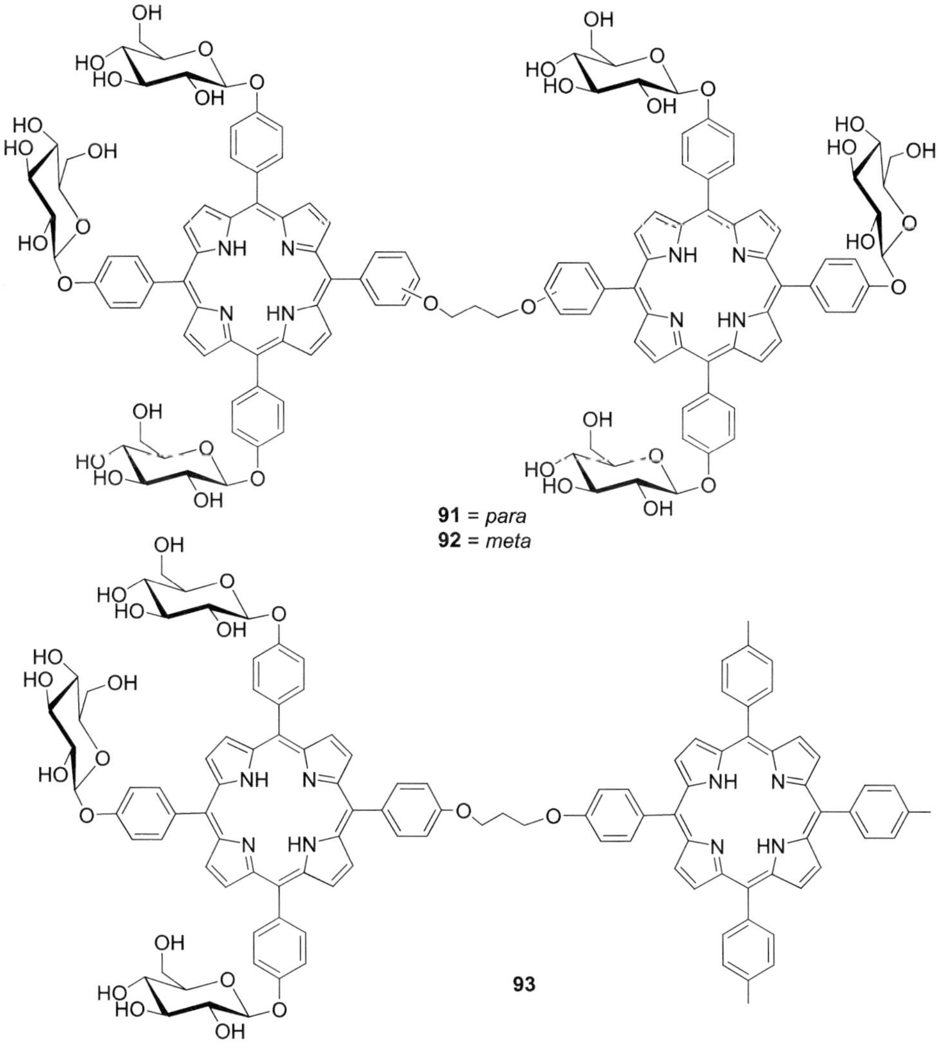

Fig. 18. Bisporphyrin glycoconjugates.

K562 human chronic leukaemia cell lines. The more hydrophilic bisporphyrin dimers **91**, **92** and **94** showed very low activity while compound **93** was the most active. It was however, only equivalent to Photofrin II after 120 minutes of irradiation and further incubation in the dark.

A different approach was taken by Maillard's group, who prepared a series of conjugated porphyrin dimers as new PDT agents (Fig. 20).[83] Apart from compound **96** all dimers produced twice as much $^1O_2$ than **65**. Determination of the cyto- and phototoxicity (HT-29 and Y79 cells) showed **96** and **98** to have only slight toxicity at high concentrations toward the Y79 cell line

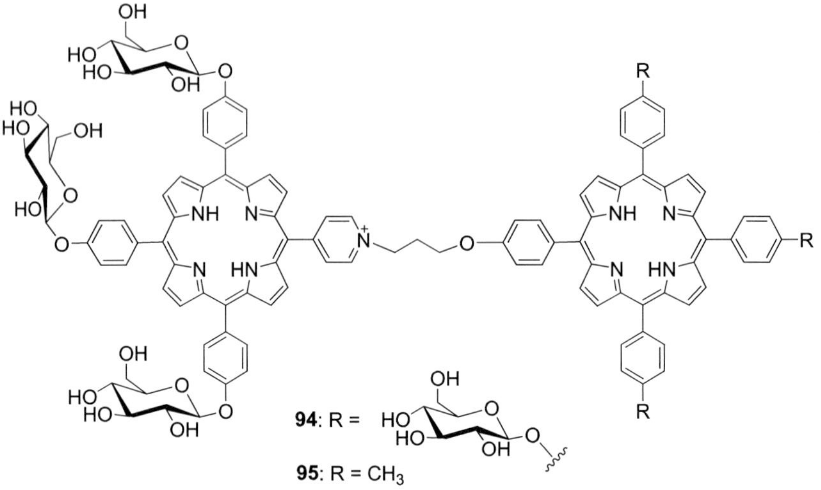

Fig. 19. Cationic bisporphyrin glycoconjugates.

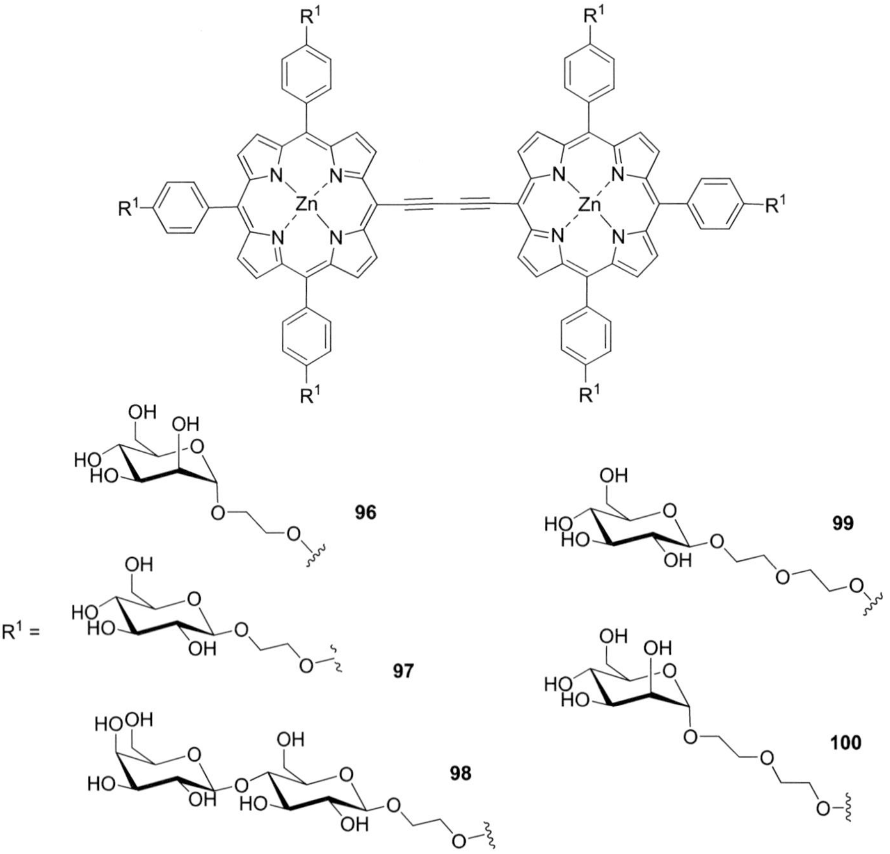

Fig. 20. Conjugated bisporphyrins with carbohydrate units.

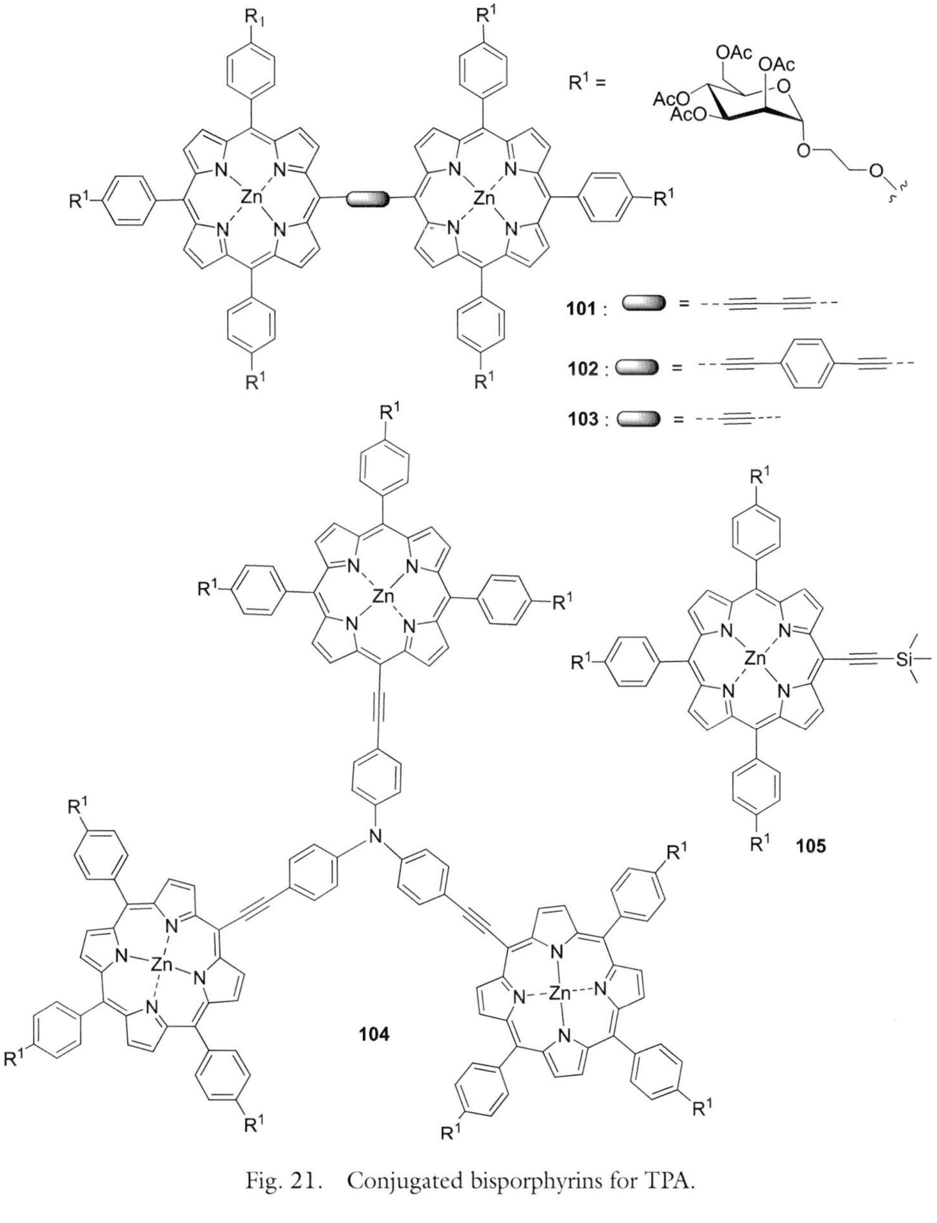

Fig. 21. Conjugated bisporphyrins for TPA.

but also exhibited dark toxicity. Overall, the low solubility of these compounds in the culture medium impeded toxicity tests.

More promising are compounds **101–105** shown in Fig. 21.[84] They exhibit high two-photon absorption (TPA) properties at wavelengths suitable for biological tissues, which, combined with their high singlet oxygen quantum yields (from 0.43 up to 0.75), makes them promising PS.

(d) *Aminoglycosamide Porphyrins and Chlorins*

In a further effort to improve the stability of glycoporphyrins to endogenous hydrolysis, aminoglycosamide derivatives were investigated (Fig. 22).[85] Similar to earlier results the HT29 cell line was more affected by the asymmetric and less hydrophobic glucosylated PSs **106** and **107**. The chlorin derivative **106** gave the better PDT effect ($\Phi^1O_2 = 0.73$, $\Phi_f = 0.16$) compared to the porphyrins**107** ($\Phi^1O_2 = 0.55$, $\Phi_f = 0.09$), **108** ($\Phi^1O_2 = 0.33$, $\Phi_f = 0.1$) and TPP ($\Phi^1O_2 = 0.55$, $\Phi_f = 0.11$). The cellular uptake of **107** (10 μM) after 24 h was ~11.5-fold higher than TPP, whereas the accumulation of **106**

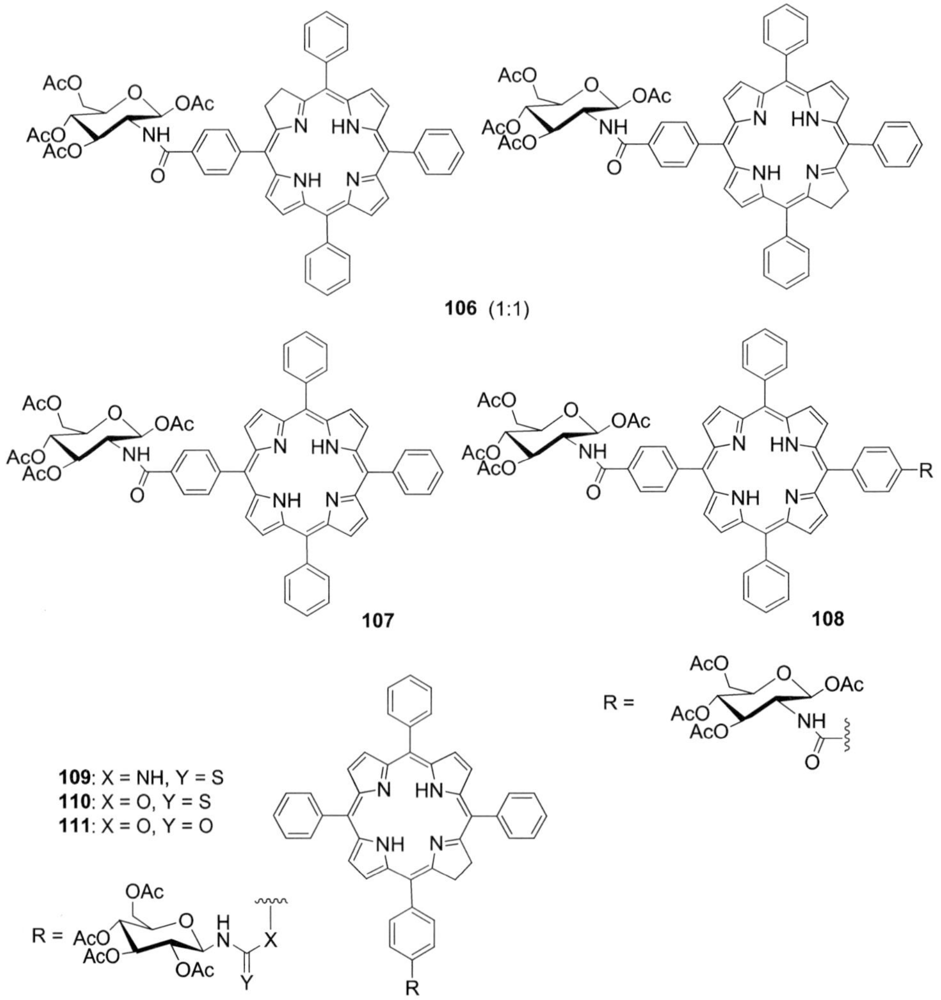

Fig. 22. Selected aminoglucosamide porphyrins.

and **108** was 6.4- and 5.0-fold respectively. *In vitro* studies indicated that **106** and **107** were more efficient than TPP and that these compounds accumulate in the endoplasmic reticulum.[86]

Related systems were described by Dallagnol *et al.* who synthesized porphyrin conjugates bearing thiourea, thiocarbamate and carbamate linkers (**109–111**).[87] They observed a strong correlation between the linker type and the photophysical properties of the porphyrin conjugates. The porphyrin derivatives with the carbamate linker gave the highest $^1O_2$ production ($\Phi_\Delta = 0.46$).

4.2. *Porphyrin-Amino Acid Conjugates*

Next to carbohydrates a wide range of amino acid conjugates have been prepared in the past. A typical example are contemporary studies by Weimin *et al.* on the synthesis of porphyrin derivatives **112–116** (Fig. 23).[88] This involved double nitration of TPP followed by reduction to two amino groups and subsequent coupling with the amino acid. *In vitro* PDT studies with the Ec9706 cell line indicated a notable effect of the amino acid on the photocytotoxicity. The porphyrin complexes with Mn(II) showed higher photocytotoxicity than those with Co(II). Inhibition of the Ec9706 cell line increased with Co(II) insertion from 44% to 74%, while Mn(II) insertion resulted in an increase from 55% to 100%. It is interesting that insertion of

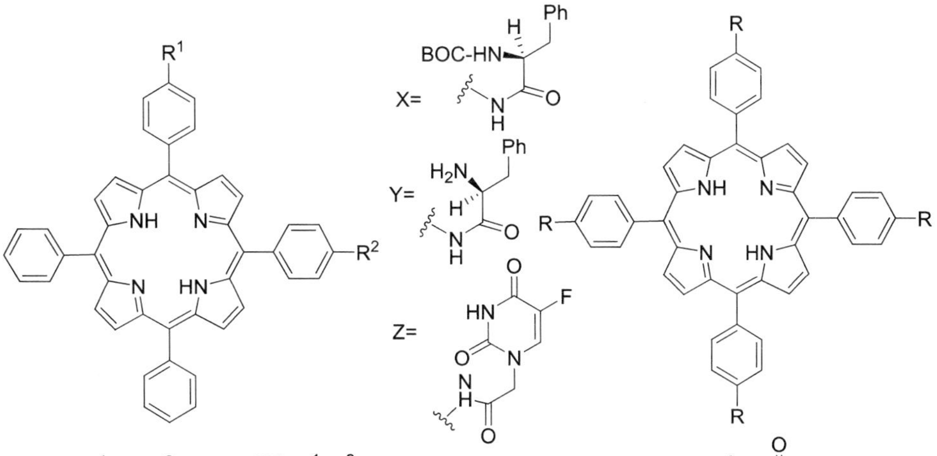

112: R¹= X, R² = NH₂ **113**: R¹, R² = X
115: R¹, R² = Y **114**: R¹ = Y, R² = NH₂
117: R¹, R² = Z **116**: R¹ = Z, R² = NH₂

Fig. 23. Amino acid appended porphyrins.

metal ion enhanced the activity of mono-substituted porphyrins **112, 115,** and **117** and, in contrast, decreased that of disubstituted porphyrins **113, 114** and **116**.

A valine-porphyrin conjugate **118** was prepared by Wang *et al.* [2008] using a modified Adler method.[89] The compound showed low cytotoxicity and good phototoxicity. In addition, porphyrin amino acid conjugates were also investigated in HeLa cells by Serra *et al.*[90] Such conjugates were stable in liposomes and penetrated efficiently into the cytoplasm of cultured cells and exhibited no dark cytotoxicity.

4.3. *Porphyrin-Bile Acid Conjugates*

In recent times, bile acids have become an attractive group for conjugation onto the porphyrin periphery due to their potential in substrate selectivity for saccharide cancer markers and also their affinity for apical sodium dependent bile acid transporters (ASBT), which are present in Barrett's esophagus. One of the first examples was the synthesis of the porphyrin-bile acid conjugates **119–122** by Koivukorpi *et al.* (Fig. 24).[91] They showed good PDT activity in *in vitro* studies with A431NS, HeLaS3, 4T1 and FHC cell lines and ablated tumors in BALB/c mice which were subcutaneously transplanted with 4T1 mammary carcinoma cells.[92] Binding studies showed that these compounds have the potential to recognize various oligosaccharides under physiological conditions and can be used for selective fluorescence detection of transformed cells expressing glycosylated markers.

We reported recently on a series of porphyrin and chlorin based bile acid conjugates (**123–126**) to impart greater selectivity and specificity on the photosensitizer for the treatment of esophageal cancer (Fig. 24).[93] Chronic esophageal exposure to bile acids in patients with gastro-esophageal reflux disease is associated with the development of Barrett's metaplasia and associated molecular markers of inflammation which have been shown to support transformation, initiation and progression of tumor development. Another feature of Barrett's is the development of bile acid transporters such as the apical sodium dependent bile acid transporter (ASBT).[94] All four compounds were found to be highly soluble and successfully entered the cell, however, no phototoxicity was observed in SKGT-4 and OE22 cells and no $^{1}O_2$ was produced by the conjugates upon illumination. Intracellular aggregation or the presence of the triazole ring may have caused the lack of toxicity observed. A change in the linker group or site of conjugation may be necessary for further improvements.

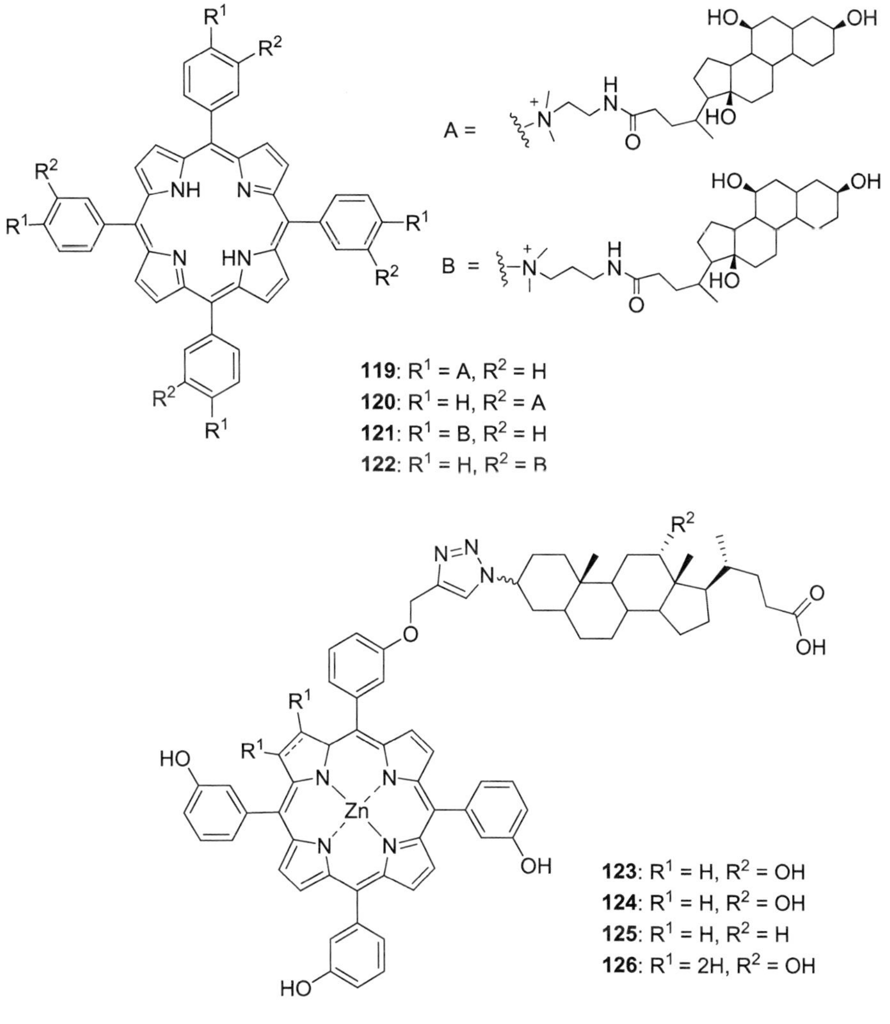

119: R^1 = A, R^2 = H
120: R^1 = H, R^2 = A
121: R^1 = B, R^2 = H
122: R^1 = H, R^2 = B

123: R^1 = H, R^2 = OH
124: R^1 = H, R^2 = OH
125: R^1 = H, R^2 = H
126: R^1 = 2H, R^2 = OH

Fig. 24. Bile acid appended porphyrins.

4.4. *Porphyrin-Folic Acid Conjugates*

Folic acid is an essential vitamin required for cell growth. It has a high affinity for specific binding to a folate receptor (FR). The combination of the photophysical properties of porphyrin with the targeting ability of folic acid successfully produced highly efficient conjugate systems for photodynamic therapy. Schneider *et al.* reported the design and synthesis of two folic acid conjugated TPPs **127** and **128** (Fig. 25).[95] The cellular uptake of both compounds by KB nasopharyngeal cells after 24 h incubation was 7-fold higher than the reference TPP, indicating an active transport *via* receptor-mediated endocytosis.

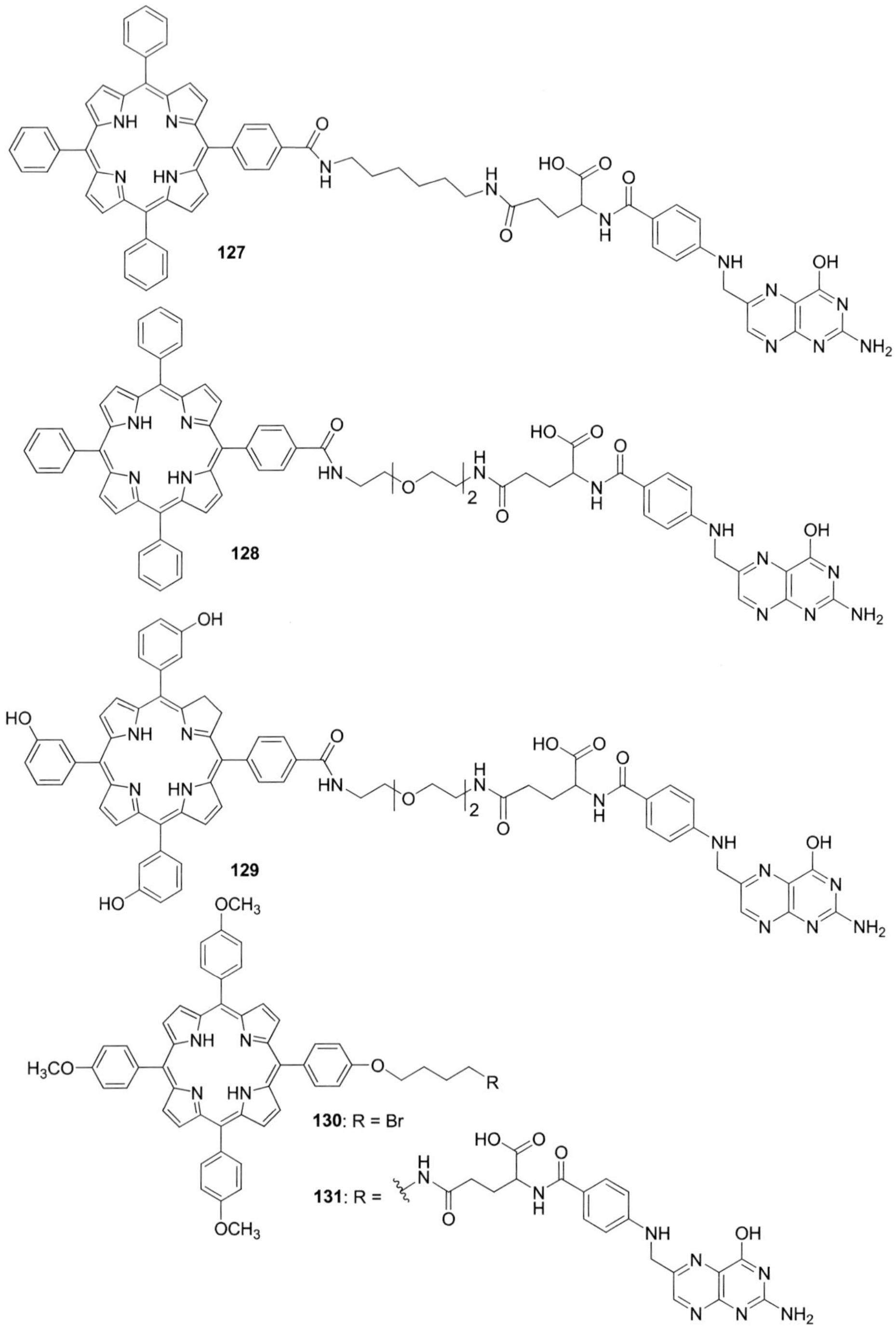

Fig. 25. Folic acid appended porphyrins.

Subsequently the same group reported the synthesis and photophysical properties of *m*THPC-like PS conjugated to folic acid *via* 2,2'-(ethylenedioxy)-bis-ethylamine units (**129**).[96] The conjugate **129** exhibited higher accumulation in KB tumors compared to tris(3-hydroxyphenyl)-4-carboxyphenylchlorin after 4 h post-injection.

A similar conjugate **131** showed 35 times higher cellular uptake in HeLa cells compared to the precursor porphyrin **130** after 24 h incubation.[97] Compound **131** also exhibited a very high photocytotoxicity (87% cell growth inhibition) after irradiation and low dark cytotoxicity. Additionally, a lower photocytotoxicity for normal cells and folate-receptor negative cells was observed.

4.5. *Porphyrin-Peptide Conjugates*

Each bioconjugate provides its own unique features to the porphyrin scaffold with peptides being no exception. Such derivatives have the potential as drug delivery agents, as they are relatively small in size and offer good tissue penetration. The excellent work carried out in proteomics makes them the most researched of all the bioconjugates and subsequently offers significant advanced targeting capabilities for medicinal applications including PDT.

One such example of porphyrin-peptide conjugates is the chlorin-ATWLPPR peptide conjugate **132**, which was prepared to specifically target tumor vasculature (Fig. 26).[98] The conjugate binds to recombinant NRP-1 protein and exhibits higher potency compared to the non-conjugated PS in human umbilical vein endothelial cells (HUVEC) expressing NRP-1. Further studies utilized MDA-MB-231 breast cancer cells, which strongly over-express the NRP-1 receptor.[99] A significant decrease of the conjugated PS after uptake was observed after RNA interference-mediated silencing of NRP-1. Use of a mouse model with U87 human malignant glioma cells demonstrated that only conjugated PS allowed a selective accumulation in endothelial cells lining the tumor vessels. The vascular effect induced by the conjugated PS was evident by a reduction in tumor blood flow of around 50% during PDT. Nonetheless, affinity for NRP-1 remains low and further improvements are possible.

Cationic peptide-porphyrin derivatives have also been reported, including compounds **133–135**.[100] These compounds showed a moderate inhibitory activity toward aminopeptidase N, an enzyme responsible for tumor cell growth. Compound **135** showed the greatest potential for PDT with an $IC_{50} = 9.48\ \mu M$.

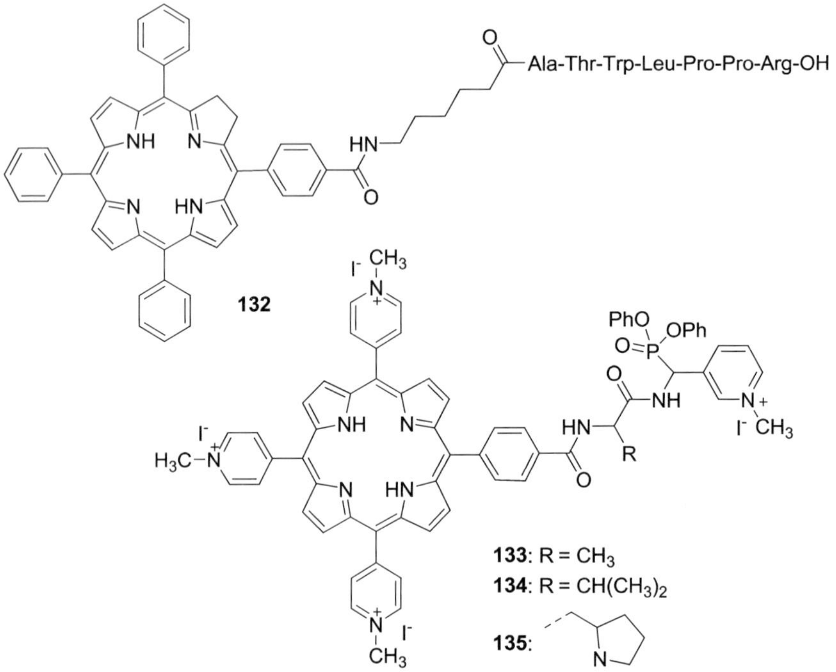

Fig. 26. Porphyrin-peptide conjugates.

Peptide PSs have also been developed for specific targeting in the newly emerging field of anti-microbial PDT. Doselli *et al.* conjugated neutral and cationic PSs to antimicrobial peptides (CAMPs), magainin and buforin (**136**, **137**) in order to increase the efficiency of bacterial inactivation (Fig. 27).[101] The porphyrin-magainin conjugate increased the binding of the cationic and hydrophilic porphyrins with bacterial cells as indicated by the high photoinactivation which was retained after washing the bacterial suspension.

Solid-phase synthesis and more complex reaction conditions have also found applications in this field. For example, the RGD tripeptide has been investigated as a bioconjugate group in PDT by Chaleix *et al.*[102] Here, the cyclic peptide-porphyrin conjugate **138** (Fig. 28) contained a constrained RGD moiety formed by ring-closing metathesis and displayed similar singlet oxygen production to that of HP.

4.6. *Porphyrin-Oligonucleotide Conjugates*

Porphyrin-oligonucleotide conjugates are of considerable interest in photomedical applications, specifically for the targeting of DNA and ultimately its photomodification.[103] A typical example is the synthesis of water-soluble

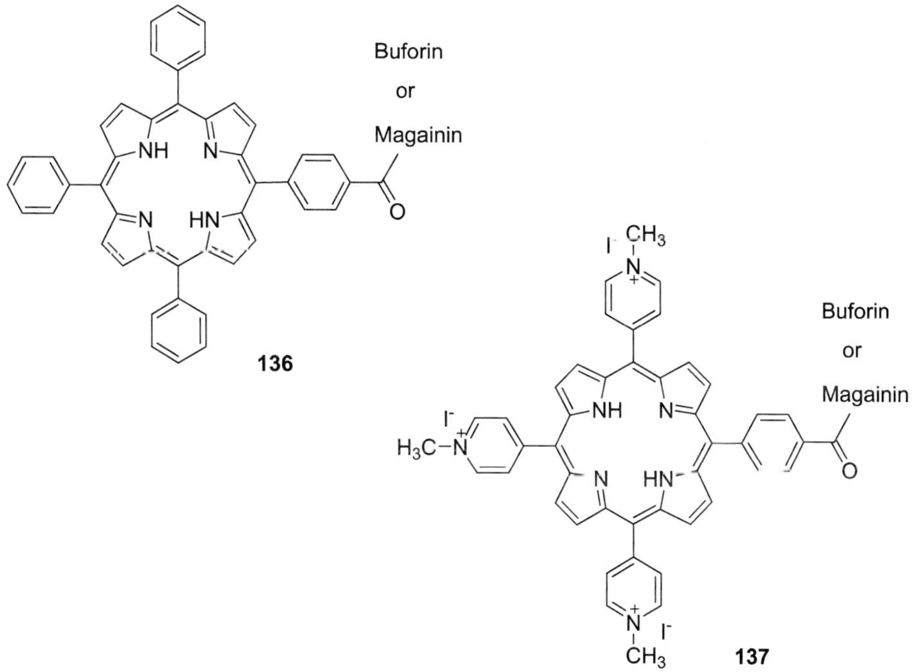

Fig. 27. Anti-bacterial porphyrin-peptide conjugates.

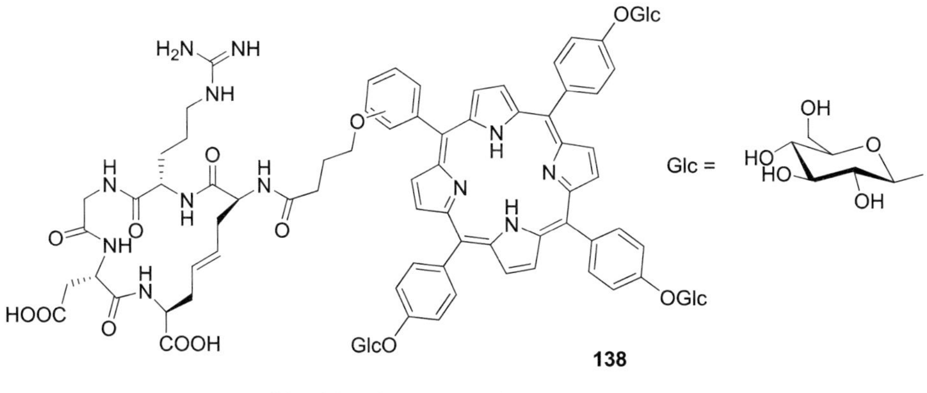

Fig. 28. A RGD-porphyrin conjugate.

porphyrin-oligonucleotide (ODN) conjugates **140a–c** by Mammana *et al.* (Fig. 29).[104] Similar to the situation with glycoporphyrins different linker units have been employed to modify the properties of such conjugates. For example, the zinc porphyrin-oligonucleotide conjugate **141** with a flexible linker was also prepared *via* solid support synthesis.[105]

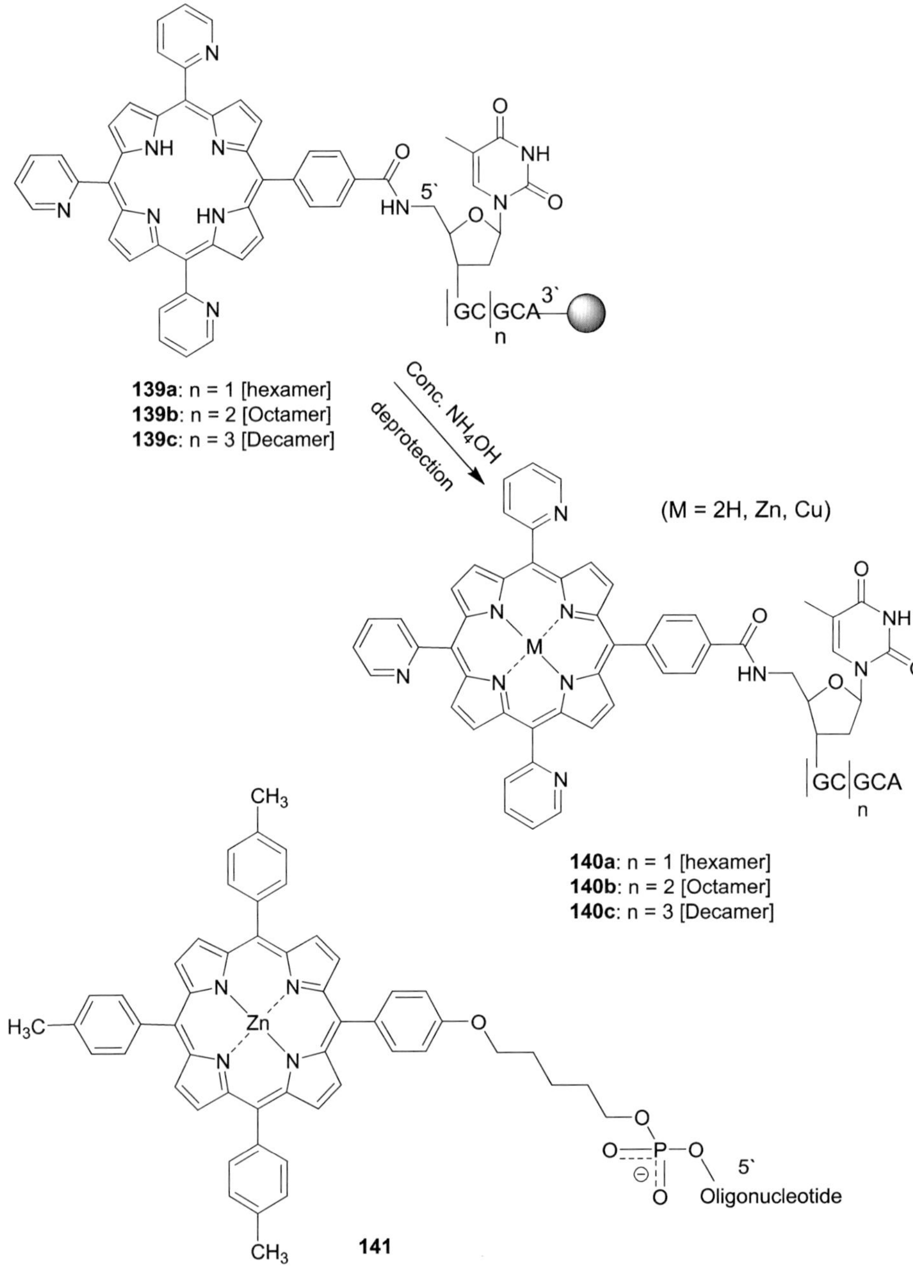

141

Fig. 29. Solid-phase synthesis of porphyrin-oligonucleotide conjugates.

4.7. *Porphyrin-Antibody Conjugates (Photoimmuno-conjugates)*

Some antigens are over-expressed in tumor tissues and thus targeting such antigens with their complimentary target-specific antibodies conjugated to porphyrins provides a logical targeting approach for PDT.[106] The utility of this approach is illustrated with the following examples.

Vrouenraets *et al.* conjugated typically unsuitable hydrophilic sensitizers with cMAb U36 and mMAb 425 by using a labile ester and a core PS (Fig. 30).[107] The conjugates showed no impairment of integrity on SDS-PAGE, were stable in serum and exhibited optimal immunoreactivity when the sensitizer: MAb ratio was <3. *In vitro* internalization experiments showed that **142** and ^{125}I-mMAb 425 conjugates were internalized by A431 cells, in contrast to ^{125}I-mMAbE48 conjugates. *In vitro* studies showed the internalized cMAb U36 and mMAb 425 conjugates were PDT active in A431 cells, unlike the non-internalized E48 conjugate and the unconjugated sensitizer. Biodistribution data of conjugates with sensitizer^{125}I-cMAb U36 ratios varying from 1:1 to 3:1 in tumor-bearing nude mice revealed selective accumulation within the tumor.

Boyle's group investigated various porphyrin systems for their utility as immuno-conjugates. For example, of the two isothiocyanato porphyrins **143** and **144**, only the latter could be linked to single chain Fv fragments (scFv) (Fig. 31).[108] Clearly, hydrophobic porphyrins are not useful in this context as they tend to form non-covalent complexes with the antibodies. However, the conjugate was found to be active against colorectal cancer Caco-2 cells.

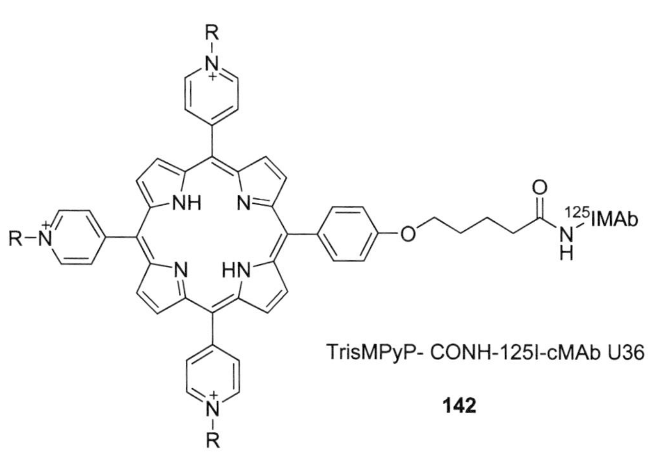

142

Fig. 30. A Porphyrin-immuno conjugate.

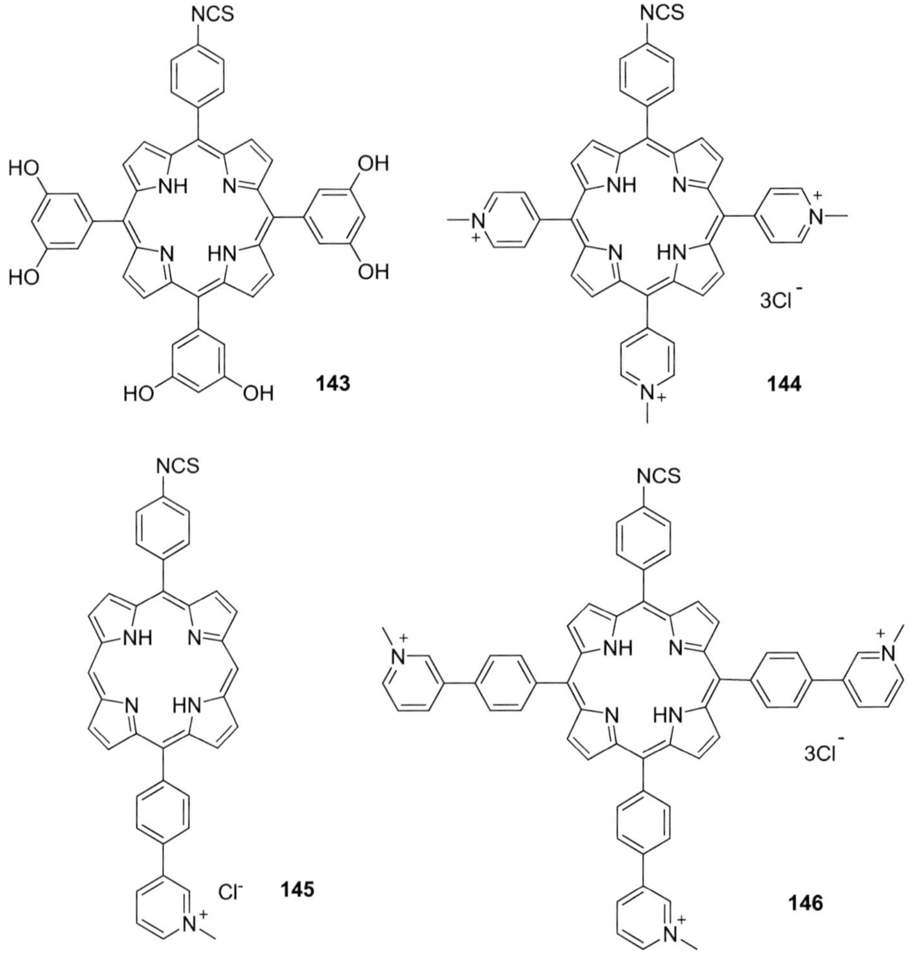

Fig. 31. Porphyrin building blocks utilized by Boyle for immune-conjugate studies.

This work was continued by taking two cationic porphyrins **145** and **146** and conjugating them with three different MAb's: CD 326, CD 146 and CD 104.[109] Their PDT efficacy was tested against the human colon adenocarcinoma (LoVo) and CORL23 human large cell lung carcinoma (ECACC) cell lines. While the immune-conjugates had approximately 10-fold lower LD_{50} values (ranging from 0.25–1.05 μM), a non-specific PDT effect was observed in LoVO cells, which do not express the CD146 antigen. The CD104 antigen is over-expressed in colorectal and bladder cancers,[110,111] and thus, the porphyrin-anti-CD104 conjugate allows for targeting of these cancer tissues.

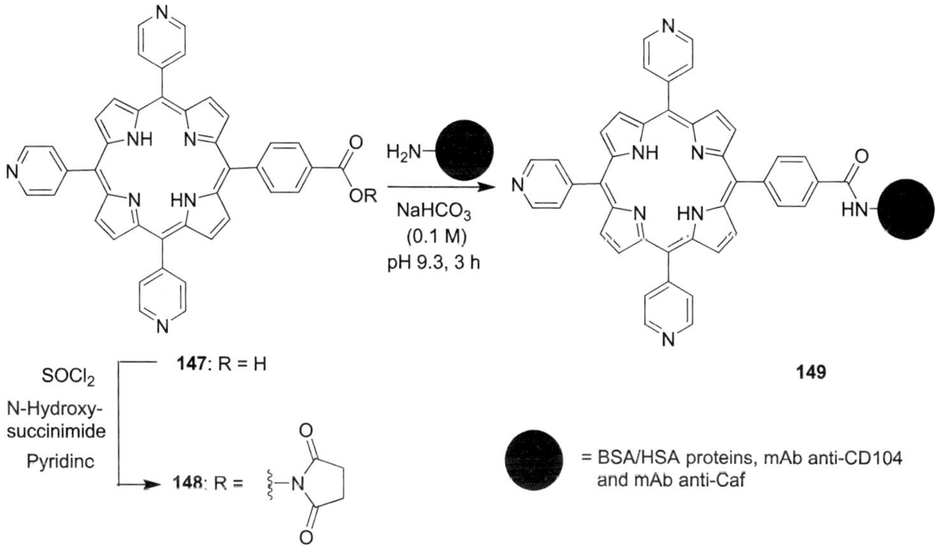

Fig. 32. Synthesis of porphyrin albumin/immune conjugates.

Very recently, Pereira *et al.* [2014] found that conjugation of a *meso* tetra-substituted porphyrin **148** with serum albumin and mAb anti-CD104 significantly increased the efficacy of the conjugate in targeted cancer destruction.[112] Compound **148** (derived from **147**) was conjugated with bovine and human serum albumin (BSA and HSA) and monoclonal antibodies (mAb anti-CD104 and mAb anti-Caf) to form the novel porphyrin-albumin and mAb immune-conjugates **149** for studies in the human bladder cancer cell line UM-UC-3 (Fig. 32).

5. Related Tetrapyrroles

5.1. *Tetrabenzoporphyrins*

Benzannulated porphyrins are of interest due to the increased stability of the macrocycle. Tetrabenzoporphyrin (TBP) is often considered an intermediate class between porphyrins and phthalocyanines.[113] The unsubstituted tetrabenzoporphyrin can be obtained starting from phthalic anhydride **150** which can be converted to **151**, followed by heating in aqueous ammonia to form the phthalimide derivative **152**. Condensation of **152** in the presence of zinc acetate forms the symmetric TBP **153**.[114] However, condensation of tetrahydroisoindoles **154** with 4-substituted benzaldehydes **155** in the presence of $BF_3 \cdot Et_2O$ in methylene chloride followed by oxidation with DDQ yields the *meso*-tetraaryltetrabenzoporphyrins **157** (Fig. 33).[115]

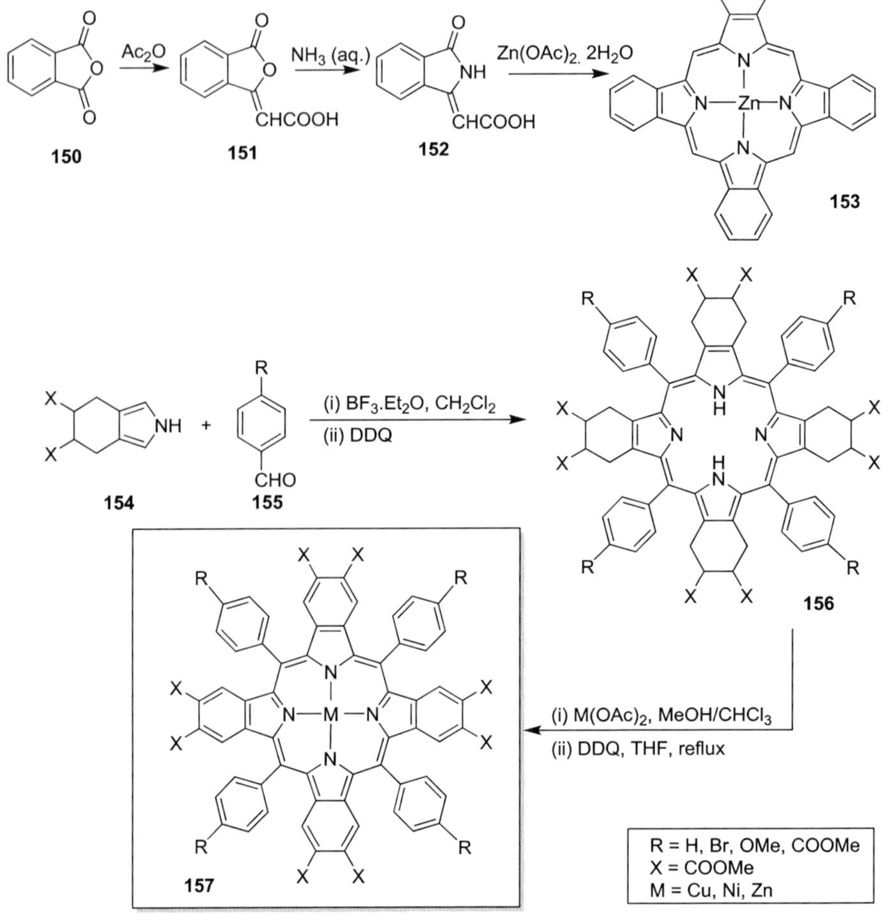

Fig. 33. Synthesis of tetrabenzoporphyrins.

These synthetic advances have now been used for the preparation of bioconjugates for PDT studies. For example, Ménard *et al.* designed the conjugated tetraglucosyl-tetrabenzoporphyrins **158** and **159** and the tetrapolyamine-tetrabenzoporphyrins **160** and **161** (Fig. 34).[116] All derivatives were tested against HaCaT and MCF-7 cells and proved less active than Photofrin II, possibly due to the altered photophysical properties of these highly nonplanar porphyrins.

5.2. *Corroles*

Corroles are aromatic tetrapyrroles containing one directly linked pyrrole-pyrrole unit which results in a smaller core cavity. By now their synthesis

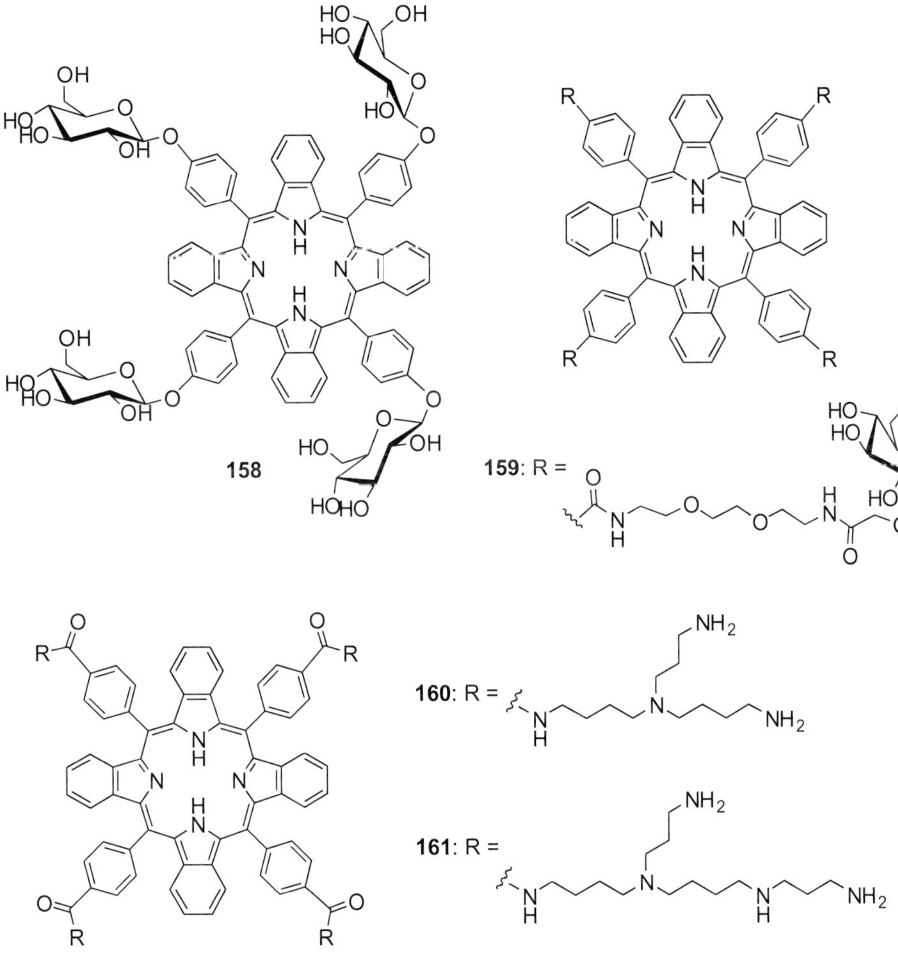

Fig. 34. Tetrabenzoporphyrin bioconjugates.

has advanced almost on par to that of porphyrins and Fig. 35 illustrates two approaches for the synthesis of *meso*-trisubstituted corroles (A_3-corroles). Paolesse *et al.* modified the Adler–Longo method for the condensation of pyrrole and aromatic aldehydes (3:1 equivalents instead of 1:1 equivalents).[117] Changing propionic acid to acetic acid yields corrole **162** as well as porphyrin **163** (Approach A). Alternatively, a condensation of excess pyrrole with aldehyde in a H_2O/MeOH/HCl mixture yields the bilane **164** which can be oxidatively cyclized with DDQ to the corrole **162** (Approach B).[118]

In terms of PDT applications, similar possibilities exist for corroles as has been shown with porphyrins. For example, a photophysical study of a series of halogenated mono-hydroxy corroles (**165–170**, Fig. 36) indicated

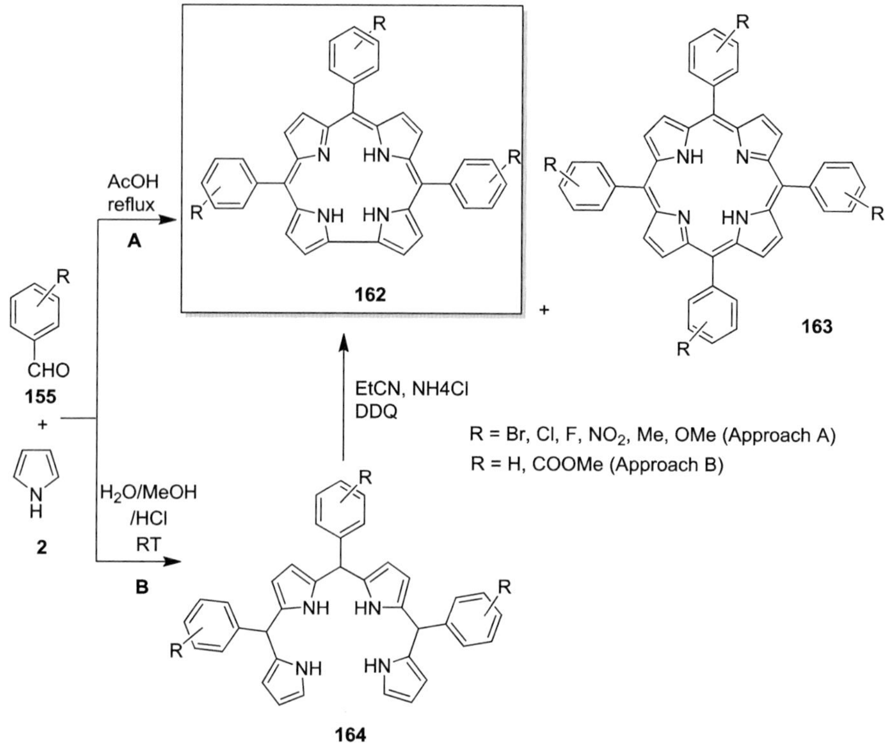

Fig. 35. Synthesis of *meso* substituted corroles.

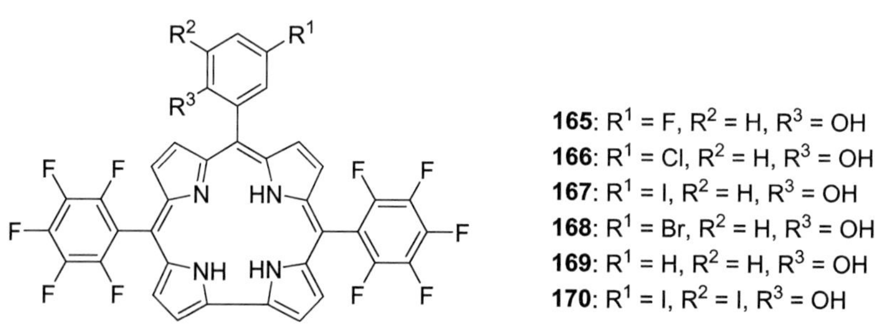

Fig. 36. Selected *meso*-hydroxyphenyl-corroles.

the usual heavy atom effect on the intersystem crossing (ISC).[119] Note, the long wavelength absorption of these corroles is much stronger than that of TPP, which can aid PDT.

5.3. N-*Confused Porphyrins*

N-Confused porphyrin (NCP) is a porphyrin isomer in which one inverted pyrrole ring is connected to the two adjacent *meso* positions at the α- and β-positions (instead of two α-positions in case of porphyrin). Typically, these systems were identified first during variations of the condensation of pyrrole and aromatic aldehydes (Fig. 37).[120,121]

Recently, Thomas *et al.* synthesized a water-soluble derivative of these N-confused porphyrins.[122] Compound **173** has long-wavelength absorption bands >700 nm while its singlet oxygen quantum yield (0.70 in methanol and 0.55 in water) is comparable to structurally similar porphyrin derivatives. The compound exhibited no cytotoxicity in the absence of light but was highly phototoxic. *In vitro* studies with a series of human cancer cell lines [colon (HCT-116), breast (MCF7-ER, PR positive, and MDA-MB-231-ER, PR negative), pancreatic (MIA-PaCa-2), cervical (HeLa and SiHa), and oral (SCC-172 and SCC-131] show promise for these kind of compounds. The

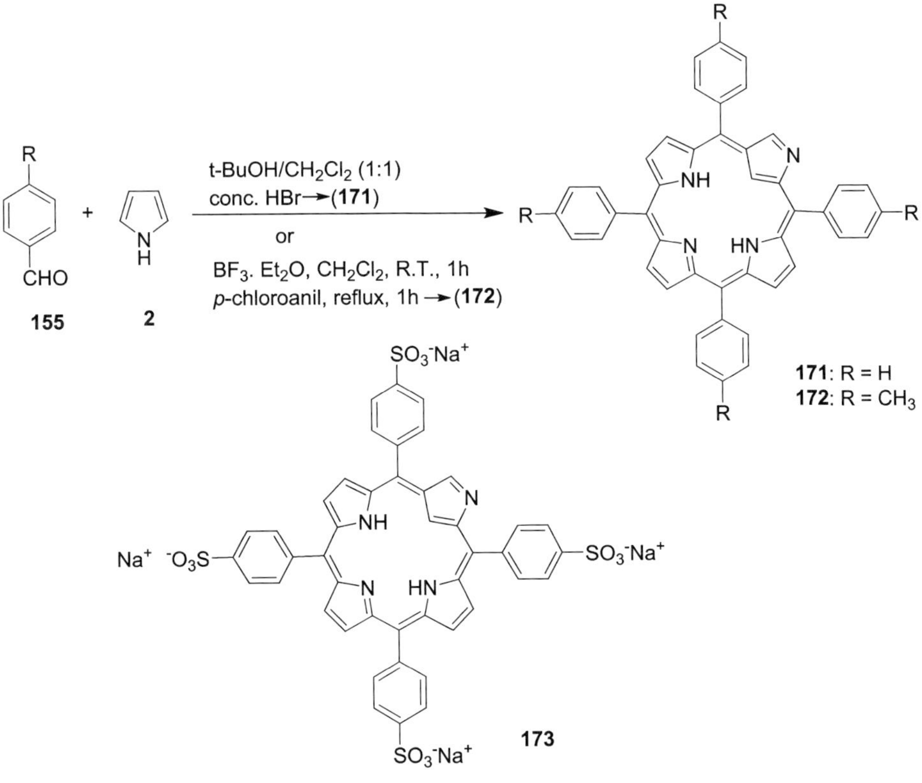

Fig. 37. Synthesis of *N*-confused porphyrins and a water-soluble derivative.

best IC_{50} value was found with breast adenocarcinoma cells ($IC_{50} = 6\,\mu M$), which is comparable to other second generation PSs.

5.4. *Thiaporphyrins*

Thiaporphyrin macrocycles are core-modified porphyrins which can be obtained *via* replacement of one or more pyrrole rings with thiophene. Such a modification significantly alters the chemical properties compared to the

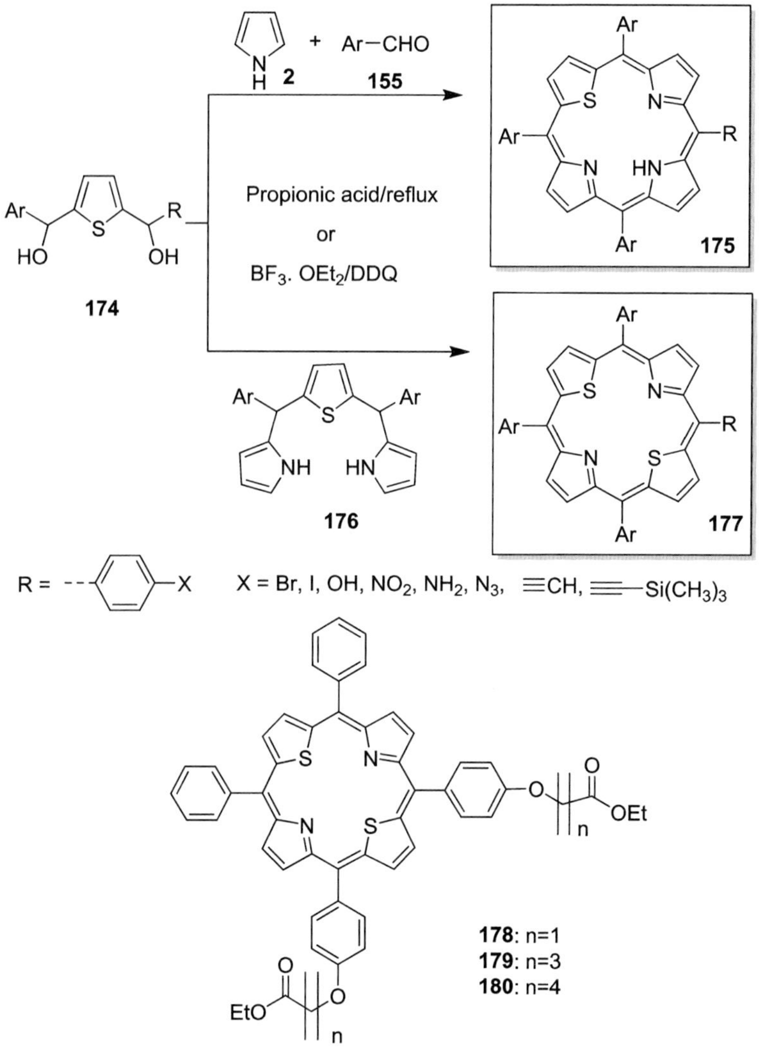

Fig. 38. Synthesis of thiaporphyrins.

N_4-porphyrins. Typical synthetic examples are shown in Fig. 38 where the functionalized *meso* tetrasubstituted 21-thiaporphyrins (N_3S porphyrins) and 21,23-dithiaporphyrins (N_2S_2 porphyrins) were prepared *via* condensation of the unsymmetrical thiophene diols **174** with either pyrrole and aryl aldehyde or 16-thiatripyrrane **176** under standard porphyrin synthesis conditions to yield **175** and **177** respectively.[123]

A series of analogues of 5,20-diphenyl-10,15-bis(4-carboxylatomethoxy) phenyl-21,23-dithiaporphyrin **178–180** were investigated as potential PSs by Minnes *et al.*[124] Changing the chemical structure by varying the length of the side chain which links the carboxylic group with phenol residues at the *meso* position gave no significant alteration in the absorption and emission spectra. All singlet oxygen production values in methanol were close to unity. However, singlet oxygen quenching was observed with the elongated side chain linker in liposomes while the opposite was observed with the shorter side chains in the cell lines, U936 and R3230Ac.

5.5. *Porphycenes*

Porphycene is a constitutional isomer of porphyrin which is synthesized easily *via* a McMurry reaction of 2,2'-bipyrrole-5,5'-dicarbaldehyde. Here, Temocene (*m*THPPo) is considered as the porphycene analogue of *m*THPC. Temocene **4** has excellent photophysical properties which makes it a promising PS for PDT applications. The synthesis of temocene **4** is illustrated in Fig. 39.[13,125]

García-Díaz *et al.* investigated the photodynamic activity of **4** and compared to Temoporfin, Temocene **4** shows lower activity but also lower dark toxicity and superior photostability. Indeed, its singlet oxygen quantum yield (Φ_Δ = 0.10) is lower than that of Temoporfin (Φ_Δ = 0.43), but the kinetics of temocene and temoporfin of photobleaching under the same irradiation conditions showed that Temocene is substantially more photostable than Temoporfin. Temocene was less cytotoxic in the dark in HeLa cells and complete cell inactivation was achieved with 5 mM *m*THPPo at light doses of 3.5 $J{\cdot}cm^{-2}$.[13]

The position and intensity of the Q-bands might give porphycenes an advantage for PDT applications relative to porphyrins as they require lower doses of drug and light to produce the same quantities of excited states. Other porphycene-type PS such as **187–189** with different substituents at position 9 of the tetrapyrrole macrocycle, were investigated *in vitro* by Scherer *et al.* using

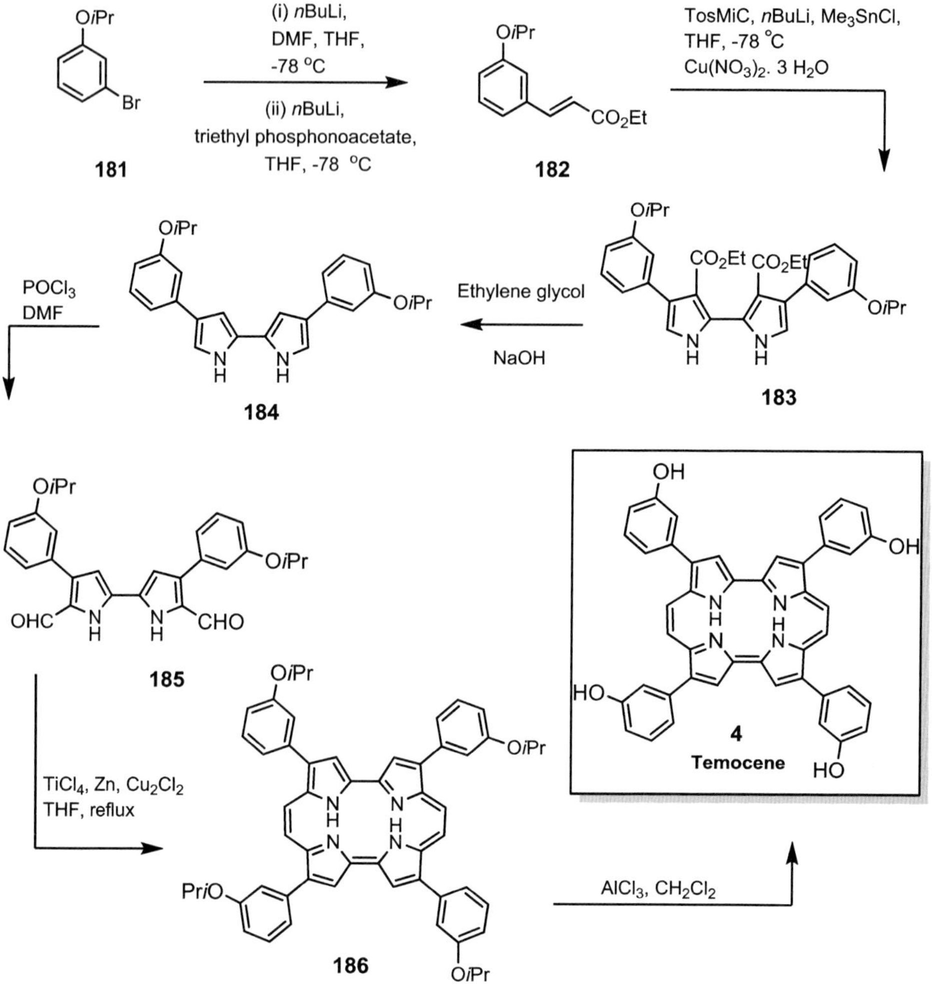

Fig. 39. Synthesis of Temocene.

different human skin-derived cell lines (HaCaT, SCL I, SCL II) (Fig. 40).[126] HexoTMPn **187** was the most efficient in all three cell lines, especially in HaCaT cells with an EC_{50} of 14 nmol·L^{-1}, compared to 62 nmol·L^{-1} for **189** and 89 nmol·L^{-1} for **188**. Singlet quantum yields ranged from 0.27–0.36.

Similar photophysical data was reported by Arad *et al.* for 2,7,12,17-tetraphenylporphycene (TPPo) **190**, its Pd(II) **191** and Cu(II) **192** complexes.[127] The triplet ($\Phi_T = 0.33$) and singlet oxygen ($\Phi_\Delta = 0.23$) quantum yields of the former are comparable to that of other porphycenes, but much lower in comparison to the porphyrin isomer TPP ($\Phi_\Delta = 0.68$). In the case

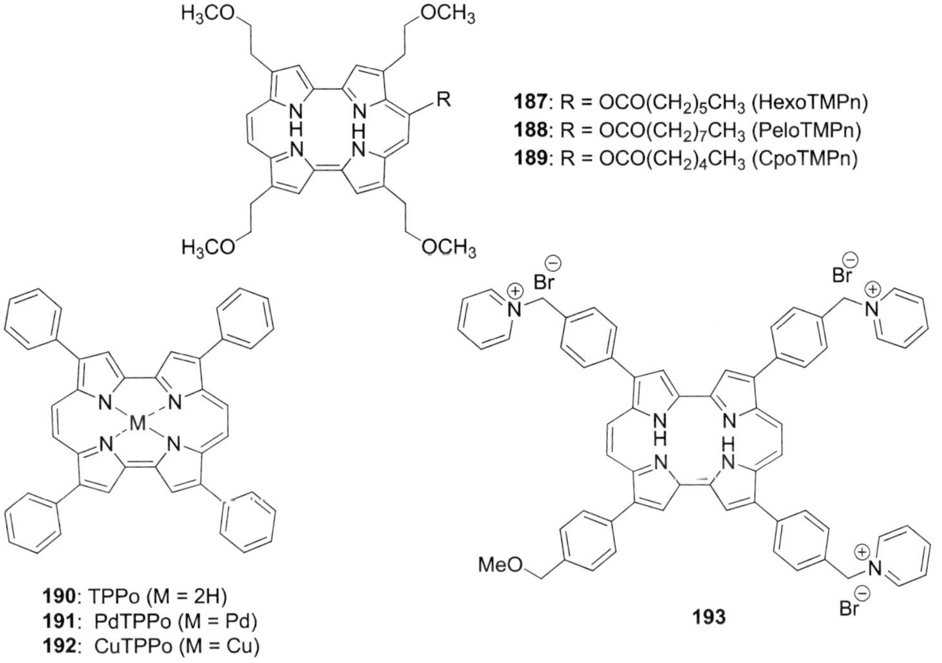

Fig. 40. Porphycenes used in PDT studies.

of **191**, fluorescence is inhibited by the internal heavy-atom effect and the triplet and singlet oxygen quantum yields are enhanced ($\Phi_T = \Phi_\Delta = 0.78$). Ragas *et al.* reported the synthesis of a tricationic water-soluble porphycene (Py_3MeO-TBPo) **193**.[128] While its photophysical parameters were similar to the other porphycenes, aggregation in aqueous media might limit its use in PDT. However, *in vitro* PDT cell killing *via* apoptosis was shown by Ruiz-González *et al.* in HeLa cells.[129]

6. Nano-Formulation Strategies

Nanoparticles (NP) offer an attractive route for the administration of PSs and a relatively recent review details current strategies.[20] NPs allow for passive and active targeting of cancer tissues and choroidal neovascularization. Depending on the NP size, passage of the NP through endothelial barriers can be controlled, as has been shown by encapsulation of *m*THPP into biodegradable NP.[130] Even a very hydrophobic compound such as chloro(5,10,15,20-tetraphenylporphyrinato)indium(III), when encapsulated into NPs of diameters <200 nm, can be used as a PS to kill LNCaP prostate cancer tumor cells (conditions: 2 h, 1.8 μmol·L^{-1}, 45 J·cm^{-2}) *via* apoptosis.[131]

Another recent example involves the use of an *m*THPP-type compound with four brucine quaternary ammonium salts bound to the *meta*-phenyl positions. This pigment was bound to gold nanoparticles and its PDT utility investigated both *in vitro* and *in vivo* using the basaloid squamous cell carcinoma PE/CA-PJ34.[132] *In vitro* PDT experiments showed that the Au-NP-bound conjugates were less phototoxic than unbound conjugates, as they are more susceptible to aggregation in the culture media. In contrast, the *in vivo* experiments showed that the NP-bound PS were more effective than their free counterparts, possibly due to interaction of the conjugated-NPs with plasma proteins to yield supramolecular complexes subject to the EPR-effect and complete regression of the PE/CA-PJ34 carcinoma after photodynamic treatment was observed.

NPs can also result in pH-controllable photosensitization, as shown through the interaction of the cationic 5,10,15,20-tetrakis(*N*-methyl-4-pyridyl)porphyrin (TMPyP) with bare SiO_2 nanoparticles.[133] At alkaline pH the functionalized SiO_2 nanoparticles act as a quencher of triplet and singlet oxygen with the PSs photoactivity "turned off". In acidic solutions the PS is released from the SiO_2 surface and thus can generate $^1O_2$ efficiently. This was indicated by singlet oxygen quantum yields of 0.45 in acidic solutions and 0.08 in basic solutions. The biological relevance of this approach was confirmed in initial studies with SK-BR-3 breast cancer cells.

Another intriguing example is the free base porphyrin-cyclodextrin conjugate **194** developed by Král and coworkers.[134,135] This nanosystem for multifunctional drug delivery and multimodal cancer therapy utilizes a combination of the binding of selected drugs in the cavity of the cyclodextrin with the photosensitizing properties of the porphyrin macrocycle (Fig. 41) and gave excellent accumulation of the conjugate in cancer tissue. The multimodality of this approach is illustrated by the four parts of the system: drug (chemotherapy); porphyrin (targeted photodynamic therapy); metal (coordination with protein); protein (immunotherapy). This nanosystem can be used for efficiently targeting various therapeutic proteins.

7. Outlook

Looking forward and trying to predict the next paradigm shift in PDT is a difficult undertaking as most of the rational enhancements made to 2nd generation PS have not worked, as of yet. Nanoparticles designed for PDT have reached their zenith in terms of research and publications, with concerns still, over their bio-toxicity due to clearance rates. Bio-conjugates have become an ever increasing field with the belief that these biologically pertinent molecules

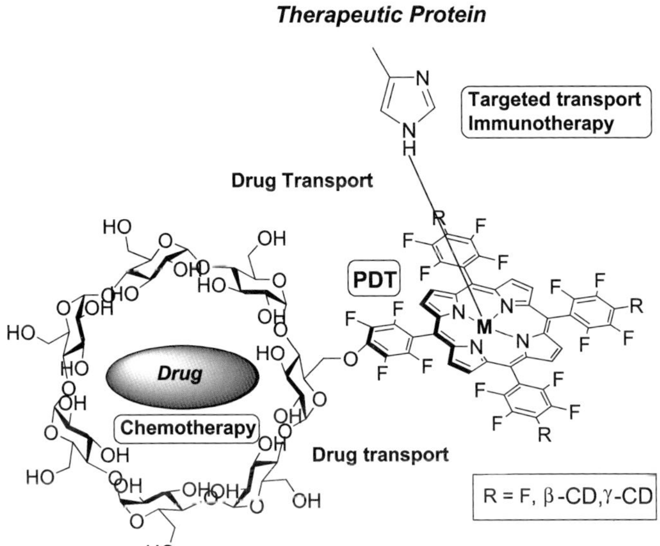

194

Fig. 41. Král's cyclodextrin-based multimodality drug delivery system.

attached to a PS will impart greater selectivity and improve the PS's targeting for the cancer cells. Sugars have received the biggest draw from researchers and, although some mild success has been achieved, none have managed to pose a threat to replace Photofrin®, Foscan® or any of the other clinically approved PSs. Likewise, no matter the biological fragment; choice and location of the linker can be pivotal to the success of the PS. For example, while the ever popular "click" reaction is easy to perform, the biologically stable heterocycle formed has often resulted in no *in vitro* PDT activity due to $^1O_2$ scavenging.

New emerging biological compounds of interest are folic acid, antibodies and bile acids. Since the discovery that folate receptors are over-expressed on the surface of numerous cancer cell lines, folic acid conjugates have been a logical route for improved targeting. Studies have shown folic acid conjugates with impressive $^1O_2\,\Delta_\Phi$, whether these results can transfer into *in vitro* experiments, it is yet to be seen. In addition, issues regarding solubility and ease of purification arise as specialized equipment is generally needed to purify these conjugates due to the poor solubility of folic acid in organic solvent.

Acknowledgments

This work was supported by Science Foundation Ireland (IvP 13/IA/1894) and the Egyptian Cultural and Educational Bureau (Channel Project AMKS).

References

1. R. Bonnett, *Chemical Aspects of Photodynamic Therapy* (Gordon and Breach Publ., Amsterdam, The Netherlands, 2000).
2. S. K. Schwartz, K. Absolon and H. Vermund *Univ. Minn. Med. Bull.* 7, 27 (1955).
3. R. L. Lipson, E. J. Baldes and A. M. Olsen, *J. Thorac. Cardiovasc. Surg.* **42**, 623 (1961).
4. J. Moan, *Photochem. Photobiol.* **43**, 681 (1986).
5. R. K. Pandey, F. Y. Shiau, T. J. Dougherty and K. M. Smith, *Tetrahedron* **47**, 9571 (1991).
6. R. Bonnett, *Chem. Soc. Rev.* **24**, 19 (1995).
7. M. O. Senge, *Photodiagn. Photodyn. Ther.* **9**, 170 (2012).
8. M. P. Copper, I. B. Tan, H. Oppelaar, M. C. Ruevekamp and F. A. Stewart, *Arch. Otolaryngol. Head. Neck. Surg.* **129**, 709 (2003).
9. R. Bonnett, R. D. White, U. J. Winfield and M. C. Berenbaum, *Biochem. J.* **261**, 277 (1989).
10. M. O. Senge and J. C. Brandt, *Photochem. Photobiol.* **87**, 1240 (2011).
11. Q. Peng, J. Moan, L. W. Ma and J. M. Nesland, *Cancer Res.* **55**, 2620 (1995).
12. R. Bonnett and G. Martinez, *Org. Lett.* **4**, 2013 (2002).
13. M. García-Díaz, D. Sánchez-García, J. Soriano, M. L. Sagristà, M. Mora, A. Villanueva, J. C. Stockert, M. Cañete and S. Nonell, S., *Med. Chem. Commun.* **2**, 616 (2011).
14. S. Sasnouski, E. Pic, D. Dumas, V. Zorin, M. A. D'Hallewin, F. Guillemin and L. Bezdetnaya, *Radiat Res.* **168**, 209 (2007).
15. D. V. Titov, M. L. Gening, Y. E. Tsvetkov and N. E. Nifantiev, *Russ. Chem. Rev.* **83**, 523 (2014).
16. S. P. Songca and B. Mbatha, *J. Pharm. Pharmacol.* **52**, 1361 (2000).
17. M. O. Senge, *Chem. Commun.* **47**, 1943 (2011).
18. H. W. Whitlock, R. Hanauer, M. Y. Oester and B. K. Bower, *J. Am. Chem. Soc.* **91**, 7485 (1969).
19. J. W. Hofman, M. G. Carstens, F. von Zeeland, C. Helwig, F. M. Flesch, W. E. Hennink and C. F. V. and Nostrum, *Pharm. Res.* **25**, 2065 (2008).
20. E. Paszko, C. Ehrhardt, M. O. Senge, D. P. Kelleher and J. V. Reynolds, *Photodiagn. Photodyn. Ther.* **8**, 14 (2011).
21. V. Reshetov, D. Kachatkou, T. Shmigol, V. Zorin, M. A. D'Hallewin, F. Guillemin and L. Bezdetnaya, *Photochem. Photobiol. Sci.* **10**, 911 (2011).
22. H. P. Lassalle, D. Dumas, S. Gräfe, M. A. D'Hallewin, F. Guillemin and L. Bezdetnaya, *J. Contr. Rel.* **134**, 118 (2009).
23. J. Buchholz, B. Kaser-Hotz, T. Khan, C. R. Bley, K. Melzer, R. A. Schwendener, M. Roos and H. Walt, *Clin. Cancer Res.* **11**, 7538 (2005).
24. W. J. Syu, H. P. Yu, C. Y. Hsu, Y. C. Rajan, Y. H. Hsu, Y. C. Chang, W. Y. Hsieh, C. H. Wang and P. S. Lai, *Small* **9**, 2060 (2012).
25. F. Moret, D. Scheglmann and E. Reddi, *Photochem. Photobiol. Sci.* **12**, 823 (2013).
26. M. Laranjo, A. C. Serra, M. Abrantes, M. Piñeiro, A. C. Gonçalves, J. Casalta-Lopes, L. Carvalho, A. B. Sarmento-Ribeiro, A. Rocha-Gonsalves and F. Botelho, *Photodiagn. Photodyn. Ther.* **10**, 51 (2013).
27. S. Ben-Dror, I. Bronshtein, A. Wiehe, B. Röder, M. O. Senge and B. Ehrenberg, *Photochem. Photobiol.* **82**, 695 (2006).
28. M. Rojkiewicz, P. Kuś, P. Kozub and M. Kempa, *Dyes Pigments* **99**, 627 (2013).

29. M. E. Milanesio, M. G. Alvarez, V. Rivarola, J. J. Silber and E. N. Durantini, *Photochem. Photobiol.* **81**, 891 (2005).
30. M. G. Alvarez, C. Prucca, M. E. Milanesio, E. N. Durantini and V. Rivarola, *Int. J. Biochem. Cell Biol.* **38**, 2092 (2006).
31. W. Lei, J. Xie, Y. Hou, G. Jiang, H. Zhang, P. Wang, X. Wang and B. Zhang, *J. Photochem. Photobiol. B: Biol.* **98**, 167 (2010).
32. T. J. Jensen, M. G. H. Vicente, R. Luguya, J. Norton, F. R. Fronczek and K. M. Smith, *J. Photochem. Photobiol. B: Biol.* **100**, 100 (2010).
33. C. Pavani, A. F. Uchoa, C. S. Oliveira, Y. Iamamoto and M. S. Baptista, *Photochem. Photobiol. Sci.* **8**, 233 (2009).
34. Y.-J. Ko, K.-J. Yun, M.-S. Kang, J. Park, K.-T. Lee, S. B. Park and J.-H. Shin, *Bioorg. Med. Chem. Lett.* **17**, 2789 (2007).
35. J. N. Silva, A. Galmiche, J. P. C. Tome, A. Boullier, M. G. P. M. S. Neves, E. M. P. Silva, J.-C. Capiod, J. A. S. Cavaleiro, R. Santus, J.-C. Maziere, P. Filipe and P. Morliere, *Biochem. Pharmacol.* **80**, 1373 (2010).
36. N. Picard, H. Ali, J. E. van Lier, K. Klarskov and B. Paquette, *Photochem. Photobiol. Sci.* **8**, 224 (2009).
37. T. Bříza, J. Králová, P. Cígler, Z. Kejík, P. Poučková, P. Vašek, I. Moserová, P. Martásek and V. Král, *Bioorg. Med. Chem. Lett.* **22**, 82 (2012).
38. J. G. Delcros, S. Tomasi, S. Duhieu, M. Foucault, B. Martin, M. Le Roch, V. Eifler-Lima, J. Renault and P. Uriac, *J. Med. Chem.* **49**, 232 (2006).
39. V. Sarrazy, G. Garcia, J. P. Mbakidi, C. Le Morvan, G. Begaud-Grimaud, R. Granet, V. Sol and P. Krausz, *J. Photochem. Photobiol. B: Biol.* **103**, 201 (2011).
40. G. Garcia, V. Sarrazy, V. Sol, C. Le Morvan, R. Granet, S. Alves and P. Krausz, *Bioorg. Med. Chem.* **17**, 767 (2009).
41. R. Luguya, T. J. Jensen, K. M. Smith and M. G. H. Vicente, *Bioorg. Med. Chem.* **14**, 5890 (2006).
42. E. Hao, E. Friso, G. Miotto, G. Jori, M. Soncin, C. Fabris, M. Sibrian-Vazquez and M. G. H. Vicente, *Org. Biomol. Chem.* **6**, 3732 (2008).
43. N. V. S. D. K. Bhupathiraju and M. G. H. Vicente, *Bioorg. Med. Chem.* **21**, 485 (2013).
44. F. Rancan, A. Wiehe, M. Nöbel, M. O. Senge, S. Al Omari, F. Böhm, M. John and B. Röder, *J. Photochem. Photobiol. B: Biol.* **78**, 17 (2005).
45. D. Aicher, S. Gräfe, C. B. W. Stark and A. Wiehe, *Bioorg. Med. Chem. Lett.* **21**, 5808 (2011).
46. S. Griegel, M. F. Rajewsky, T. Ciesiolka and H. J. Gabius, *Anticancer Res.* **9**, 723 (1989).
47. C. Moylan, E. M. Scanlan and M. O. Senge, *Curr. Med. Chem.*, in press (2015).
48. M. Cornia, M. Menozzi, E. Ragg, S. Mazzini, A. Scarafoni, F. Zanardi and G. Casiraghi, *Tetrahedron* **56**, 3977 (2000).
49. S. Vedachalam, B.-H. Choi, K. K. Pasunooti, K. M. Ching, K. Lee, H. S. Yoon and X.-W. Liu, *Med. Chem. Commun.* **2**, 371 (2011).
50. M. Momenteau, P. Maillard, M.-A. De Be'linay, D. L. Carrez and A. Croisy, *J. Biomed. Opt.* **4**, 298 (1999).
51. I. Laville, T. Figueiredo, B. Loock, S. Pigaglio, P. Maillard, D. S. Grierson, D. Carrez, A. Croisy and J. Blais, *Bioorg. Med. Chem.* **11**, 1643 (2003).
52. J. C. Maziere, P. Morliere and R. Santus, *J. Photochem.Photobiol. B: Biol.* **8**, 351 (1991).

53. M. Mori, T. Kuroda, A. Obana, I. Sakata, T. Hirano, S. Nakajima, M. Hikida and T. Kumagai, *Jpn. J. Cancer Res.* **91**, 845 (2000).
54. D. Kessel, *Int. J. Clin. Pract.* **53**, 263 (1999).
55. A. T. Michael-Titus, R. Whelpon and Z. Yaqub, *Br. J. Clin. Pharmacol.* **40**, 594 (1995).
56. M.-C. Desroches, A. Bautista-Sanchez, C. Lamotte, B. Labeque, D. Auchère, R. Farinotti, P. Maillard, D. S. Grierson, P. Prognon and A. Kasselouri, *J. Photochem. Photobiol. B: Biol.* **85**, 56 (2006).
57. Y. Mikata, Y. Onchi, K. Shibata, T. Kakuchi, H. Ono, S. I. Ogura, I. Okura and S. Yano, *Bioorg. Med. Chem. Lett.* **8**, 3543 (1998).
58. G. Zheng, A. Graham, M. Shibata, J. R. Missert, A. R. Oserroff, T. J. Dougherty and R. K. Pandey, *J. Org. Chem.* **66**, 8709 (2001).
59. A. M. G. Silva, A. C. Tomé, M. G. P. M. S. Neves, A. M. S. Silva, J. A. S. Cavaleiro, D. Perrone and A. Dondoni, *Tetrahedron Lett.* **43**, 603 (2002).
60. I. Laville, S. Pigaglio, J. C. Blais, F. Doz, B. Loock, P. Maillard, D. S. Grierson and J. Blais, *J. Med. Chem.* **49**, 2558 (2006).
61. S. Hirohara, M. Obata, S. Ogata, C. Ohtsuki, S. Higashida, S. Ogura, I. Okura, M. Takenaka, H. Ono, Y. Sugai, Y. Mikata, M. Tanihara and S. Yano, *J. Photochem. Photobiol. B: Biol.* **78**, 7 (2005).
62. S. Hirohara, M. Obata, A. Saito, S. Ogata, C. Ohtsuki, S. Higashida, S. Ogura, I. Okura, Y. Sugai, Y. Mikata, M. Tanihara and S. Yano, *Photochem. Photobiol.* **80**, 301 (2004).
63. P. Maillard, B. Loock, D. S. Grierson, I. Laville, J. Blais, F. Doz, L. Desjardins, D. Carrez, J. L. Guerquin-Kern and A. Croisy, *Photodiagn. Photodyn. Ther.* **4**, 261 (2007).
64. F. Hammerer, G. Garcia, S. Chen, F. Poyer, S. Achelle, C. Fiorini-Debuisschert, M. P. Teulade-Fichou and P. Maillard, *J. Org. Chem.* **79**, 1406 (2014).
65. E. Hao, T. J. Jensen and M. G. H. Vicente, *J. Porphyrins Phthalocyanines* **13**, 51 (2009).
66. M. Okada, Y. Kishibe, K. Ide, T. Takahashi and T. Hasegawa, *Int. J. Carbohydr. Chem.*, ID 305276 (2009).
67. G. Garcia, D. Naud-Martin, D. Carrez, A. Croisy and P. Maillard, *Tetrahedron* **67**, 4924 (2011).
68. S. Cecioni, S. E. Matthews, H. Blanchard, J.-P. Praly, A. Imberty and S. Vidal, *Carbohydr. Res.* **356**, 132 (2012).
69. M. L. Gening, D. V. Titov, S. Cecioni, A. Audfray, A. G. Gerbst, Y. E. Tsvetkov, V. B. Krylov, A. Imberty, N. E. Nifantiev and S. Vidal, *Chem. Eur. J.* **19**, 9272 (2013).
70. O. B. Locos, C. C. Heindl, A. Corral, M. O. Senge and E. M. Scanlan, *Eur. J. Org. Chem.*, 1026 (2010).
71. R. Daly, G. Vaz, A. M. Davies, M. O. Senge and E. M. Scanlan, *Chem. Eur. J.* **18**, 14671 (2012).
72. S. I. Yang, J. Seth, J.-P. Strachan, S. Gentemann, D. Kim, D. Holten, J. S. Lindsey and D. F. Bocian, *J. Porphyrins Phthalocyanines* **3**, 117 (1999).
73. P. Pasetto, X. Chen, C. M. Drain and R. W. Franck, *Chem. Commun.* 81 (2001).
74. X. Chen, L. Hui, D. A. Foster and C. M. Drain, *Biochemistry* **43**, 10918 (2004).
75. S. Hirohara, M. Nishida, K. Sharyo, M. Obata, T. Ando and M. Tanihara, *Bioorg. Med. Chem.* **18**, 1526 (2010).
76. S. Hirohara, M. Obata, H. Alitomo, K. Sharyo, T. Ando, S. Yano and M. Tanihara, *Bioconjugate Chem.* **20**, 944 (2009).

77. M. Obata, S. Hirohara, K. Sharyo, H. Alitomo, K. Kajiwara, S. Ogata, M. Tanihara, C. Ohtsuki and S. Yano, *Biochim. Biophys. Acta* **1770**, 1204 (2007).
78. S. Hirohara, M. Obata, H. Alitomo, K. Sharyo, T. Ando, M. Tanihara and S. Yano, *J. Photochem. Photobiol. B: Biol.* **97**, 22 (2009).
79. S. Hirohara, M. Obata, S. Ogata, K. Kajiwara, C. Ohtsuki, M. Tanihara and S. Yano, *J. Photochem. Photobiol. B: Biol.* **84**, 56 (2006).
80. S. Hirohara, M. Obata, H. Alitomo, K. Sharyo, S.-L. Ogata, C. Ohtsuki, S. Yano, T. Ando and M. Tanihara, *Biol. Pharm. Bull.* **31**, 2265 (2008).
81. S. Singh, A. Aggarwal, S. Thompson, J. P. C. Tome, X. Zhu, D. Samaroo, M. Vinodu, R. Gao and C. M. Drain, *Bioconjugate Chem.* **21**, 2136 (2010).
82. V. Sol, V. Chaleix, Y. Champavier, R. Granet, Y.-M. Huang and P. Krausz, *Bioorg. Med. Chem.* **14**, 7745 (2006).
83. G. Garcia, F. Hammerer, F. Poyer, S. Achelle, M. P. Teulade-Fichou and P. Maillard, *Bioorg. Med. Chem.* **21**, 153 (2013).
84. S. Achelle, P. Couleaud, P. Baldeck, M.-P. Teulade-Fichou and P. Maillard, *Eur. J. Org. Chem.*, 1271 (2011).
85. B. Di Stasio, C Frochot, D. Dumas, P. Even, J. Zwier, A. Müller, J. Didelon. F Guillemin, M. L. Viriot and M. Barberi-Heyob, *Eur. J. Med. Chem.* **40**, 1111 (2005).
86. M. L. Barberi-Heyob, P. O. Védrine, J. L. Merlin, R. Millon, J. Abecassis, M. F. Poupon and F. Guillemin, *Int. J. Oncol.* **24**, 951 (2004).
87. J. C. C. Dallagnol, D. R. B. Ducatti, S. M. W. Barreira, M. D. Noseda, M. E. R. Duarte and A. G. Gonçalves, *Dyes Pigments* **107**, 69 (2014).
88. S. Weimin, Z. Gen, D. Guifu, Z. Yunxiao, Z. Jin and T. Jingchao, *Bioorg. Med. Chem.* **16**, 5665 (2008).
89. H. M. Wang, J. Q. Jiang, J. H. Xiao, R. L. Gao, F. Y. Lin and X. X. Liu, *Chem. Biol. Interact.* **2**, 154 (2008).
90. V. V. Serra, A. Zamarrón, M. A. Faustino, M. C. Cruz, A. Blázquez, J. M. Rodrigues, M. G. Neves, J. A. Cavaleiro, A. Juarranz and F. Sanz-Rodríguez, *Bioorg. Med. Chem.* **18**, 6170 (2010).
91. J. Koivukorpi, E. Sievänen, E. Kolehmainen and V. Král, *Molecules* **12**, 13 (2007).
92. J. Králová, J. Koivukorpi, Z. Kejík, P. Poučková, E. Sievänen, E. Kolehmainen and V. Král, *Org. Biomol. Chem.* **6**, 1548 (2008).
93. L. Rogers, F. Majer, N. N. Sergeeva, E. Paszko, J. F. Gilmer and M. O. Senge, *Bioorg. Med. Chem. Lett.* **23**, 2495 (2013).
94. R. Sharma, A. Long and J. F. Gilmer, *Curr. Med. Chem.* **18**, 4029 (2011).
95. R. Schneider, F. Schmitt, C. Frochot, Y. Fort, N. Lourette, F. Guillemin, J. F. Müller and M. Barberi-Heyob, *Bioorg. Med. Chem.* **13**, 2799 (2005).
96. J. Gravier, R. Schneider, C. Frochot, T. Bastogne, F. Schmitt, J. Didelon, F. Guillemin and M. Barberi-Heyob, *J. Med. Chem.* **51**, 3867 (2008).
97. D. Li, J. Diao, D. Wang, J. Liu and J. Zhang, *J. Porphyrins Phthalocyanines* **14**, 547 (2010).
98. L. Tirand, C. Frochot, R. Vanderesse, N. Thomas, E. Trinquet, S. Pinel, M.-L. Viriot, F. Guillemin and M. Barberi-Heyob, *J. Control. Release* **111**, 153 (2006).
99. N. Thomas, D. Bechet, P. Becuwe, L. Tirand, R. Vanderesse, C. Frochot, F. Guillemin and M. Barberi-Heyob, *J. Photochem. Photobiol. B: Biol.* **96**, 101 (2009).
100. J. Habdas and B. Boduszek, *J. Pept. Sci.* **15**, 305 (2009).

101. R. Dosselli, R. Ruiz-González, F. Moret, V. Agnolon, C. Compagnin, M. Mognato, V. Sella, M. Agut, S. Nonell, M. Gobbo and E. Reddi, *J. Med. Chem.* **57**, 1403 (2014).
102. V. Chaleix, V. Sol, Y.-M. Huang, M. Guilloton, R. Granet, J. C. Blais and P. Krausz, *Eur. J. Org. Chem.*, 1486 (2003).
103. H. Li, O. S. Fedorova, W. R. Trumble, T. R. Fletcher and L. Czuchajowski, *Bioconjugate Chem.* **8**, 49 (1997).
104. A. Mammana, T. Asakawa, K. Bitsch-Jensen, A. Wolfe, S. Chaturantabut, Y. Otani, X. Li, Z. Li, K. Nakanishi, M. Balaz, G. A. Ellestad and N. Berova, *Bioorg. Med. Chem.* **16**, 6544 (2008).
105. A. Onoda, M. Igarashi, S. Naganawa, K. Sasaki, S. Ariyasu and T. Yamamura, *Bull. Chem. Soc. Jpn.* **82**, 1280 (2009).
106. R. Hudson, M. Carcenac, K. Smith, L. Madden, O. J. Clarke, A. Pèlegrin, J. Greenman and R. W. Boyle, *Br. J. Cancer* **92**, 1442 (2005).
107. M. B. Vrouenraets, G. W. M. Visser, C. Loup, B. Meunier, M. Stigter, H. Oppelaar, F. A. Stewart, G. B. Snow and G. A. M. S. Van Dongen, *Int. J. Cancer* **88**, 108 (2000).
108. C. Staneloudi, K. Smith, R. Hudson, N. Malatesti, H. Savoie, R. W. Boyle and J. Greenman, *J. Immunol.* **120**, 512 (2007).
109. K. Smith, N. Malatesti, N. Cauchon, D. Hunting, R. Lecomte, J. E. Van Lier, J. Greenman and R. W. Boyle, *J. Immunol.* **132**, 256 (2011).
110. P. Simon-Assmann, C. Leberquier, N. Molto, T. Uezato, F. Bouziges and M. Kedinger, *J. Cell Sci.* **107**, 577 (1994).
111. R. Falcioni, A. Sacchi, J. Resau and S. J. Kennel, *Cancer Res.* **48**, 816 (1988).
112. P. M. Pereira, J. J. Carvalho, S. Silva, J. A. S. Cavaleiro, R. J. Schneider, R. Fernandes and J. P. Tomé, *Org. Biomol. Chem.* **12**, 1804 (2014).
113. R. B. M. Koehorst, J. F. Kleibeuker, T. J. Schaafsma, D. A. de Bie, B. Geursten, R. N. Henrie and H. van der Plas, *J. Chem. Soc., Perkin Trans.* **2**, 1005 (1981).
114. V. N. Kopranenkov, E. A. Makarova and E. A. Luk'yanets, *Zh. Obshch. Khim.* **51**, 2727 (1981).
115. M. A. Filatov, A. V. Cheprakov and I. P. Beletskaya, *Eur. J. Org. Chem.*, 3468 (2007).
116. F. Ménard, V. Sol, C. Ringot, R. Granet, S. Alves, C. L. Morvan, Y. Queneau, N. Ono and P. Krausz, *Bioorg. Med. Chem.* **17**, 7647 (2009).
117. R. Paolesse, S. Mini, F. Sagone, T. Boschi, L. Jaquinod, D. J. Nurco and K. M. Smith, *Chem.Commun.* 1307 (1999).
118. J.-W. Ka, W.-S. Cho and C.-H. Lee, *Tetrahedron Lett.* **41**, 8121 (2000).
119. L. Shi, H.-Y. Liu, H. Shen, J. Hu, G.-L. Zhang, H. Wang, L.-N. Ji, C.-K. Chang and H.-F. Jiang, *J. Porphyrins Phthalocyanines* **13**, 1221 (2009).
120. H. Furuta, T. Asano and T. Ogawa, *J. Am. Chem. Soc.* **116**, 767 (1994).
121. P. J. Chmielewski, L. Latos-Grazynski, K. Rachlewicz and T. Glowiak, *Angew. Chem. Int. Ed. Engl.* **33**, 779 (1994).
122. A. P. Thomas, P. S. S. Babu, S. A. Nair, S. Ramakrishnan, D. Ramaiah, T. K. Chandrashekar, A. Srinivasan and M. R. Pillai, *J. Med. Chem.* **55**, 5110 (2012).
123. Y. Pareek and M. Ravikanth, *RSC Adv.* **4**, 7851 (2014).
124. R. Minnes, H. Weitman, Y. You, M. R. Detty and B. Ehrenberg, *J. Phys. Chem. B* **112**, 3268 (2008).
125. D. Sanchez-Garcia, J. I. Borrell and S. Nonell, *Org. Lett.* **11**, 77 (2009).

126. K. Scherer, C. Abels, W. Bäumler, G. Ackermann and R. M. Szeimies, *Arch. Dermatol. Res.* **295**, 535 (2004).
127. O. Arad, N. Rubio, D. Sánchez-García, J. I. Borrell and S. Nonell, *J. Porphyrins Phthalocyanines* **13**, 376 (2009).
128. X. Ragàs, D. Sánchez-García, R. Ruiz-González, T. Dai, M. Agut, M. R. Hamblin and S. Nonell, *J. Med. Chem.* **53**, 7796 (2010).
129. R. Ruiz-González, P. Acedo, D. Sánchez-García, S. Nonell, M. Cañeteb, J. C. Stockert and A. Villanueva, *Eur. J. Med. Chem.* **63**, 401 (2013).
130. A. Vargas, M. Eid, M. Fanchaouy, R. Gurny and F. Delie, *Eur. J. Pharm. Biopharm.* **69**, 43 (2008).
131. A. R. da Silva, A. C. Pelegrino, A. C. Tedesco and R. A. Jorge, *J. Braz. Chem. Soc.* **19**, 491 (2008).
132. K. Zaruba, J. Kralova, P. Rezanka, P. Pouckova, L. Veverkova and V. Král, *Org. Biomol. Chem.* **8**, 3202 (2010).
133. W. Li, W. Lu, Z. Fan, X. Zhu, A. Reed, B. Newton, Y. Zhang, S. Courtney, P. T. Tiyyagura, R. R. Ratcliff, S. Li, E. Butler, H. Yu, P. C. Ray and R. Gao, *J. Mater. Chem.* **22**, 12701 (2012).
134. J. Králová, Z. Kejík, T. Bříza, P. Poučková, A. Král, P. Martásek and V. Král, *J. Med. Chem.* **53**, 128 (2010).
135. Z. Kejík, T. Bříza, J. Králová, P. Poučková, A. Král, P. Martásek and V. Král, *Bioorg. Med. Chem. Lett.* **21**, 5514 (2011).

Chapter 4

Core-Modified Porphyrins as Photosensitizers in Photodynamic Therapy

Michael R. Detty

Department of Chemistry, University at Buffalo,
The State University of New York Buffalo, NY USA
mdetty@buffalo.edu

Abstract

The synthesis, photophysical properties, biological properties, and phototoxicity *in vitro* and *in vivo* are summarized for core-modified porphyrins in which pyrrolic NH groups are replaced with O, S, Se, or Te atoms at the 21- or 21, 23-positions.

1. Introduction

One of the first porphyrin sensitizers approved for use in photodynamic therapy (PDT) was hematoporphyrin derivative, which was a mixture of monomeric, dimeric, and oligomeric porphyrin derivatives. Purified porfimer sodium[1] or Photofrin® is the commercial analogue of hematoporphyrin derivative and PDT with Photofrin® as a photosensitizer has regulatory approval in the USA, Canada, The Netherlands, France, Germany, and Japan for cervical cancer, endobronchial cancer, esophageal cancer, bladder cancer, and gastric cancers.[2] While Photofrin® is effective as a photosensitizer, it is far from an ideal photosensitizer. Photofrin and other porphyrin-related sensitizers absorb weakly in the red region of the spectrum ($\geq$600 nm where penetration of light in tissue is optimal)[3] and induce long-lasting skin photosensitivity (4–6 weeks) through retention in cutaneous tissue.[4] Porphyrins do produce reactive oxygen species (ROS) efficiently and modification of the porphyrin structure to give longer wavelengths of absorption and reduced skin photosensitivity while retaining the efficient generation of ROS was a worthwhile research endeavor.

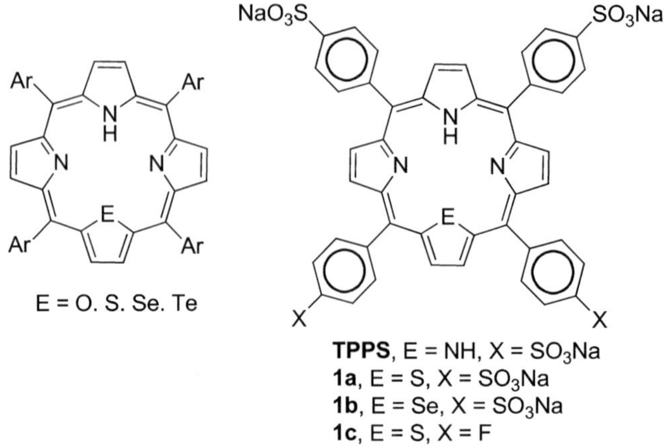

Chart 1. Structures of 21-chalcogeno core-modified porphyrins **1** and tetraphenylporphyrin tetrasulfonatesulfonate (**TPPS$_4$**).

It is difficult to increase the wavelengths of absorption of the porphyrin chromophore through changes at the *meso* positions of the porphyrins or through modifications of groups attached to the individual pyrrole rings. While these types of changes allow tailoring of biological properties, it limits the absorption maxima of porphyrin molecules to the shorter wavelength region of the biological window for PDT. Longer-wavelength-absorbing porphyrins (band I absorption maxima of $\geq$650 nm) have been prepared by substituting a chalcogen atom (O, S, Se or Te) for an NH at the 21-position of the porphyrin ring (Chart 1).[5–9] Such molecules are called core-modified porphyrins, and sulfonated analogues of the 21-thia- and 21-selenaporphyrins (**1a** and **1b**, respectively, Chart 1), which are analogues of tetraphenylporphyrin tetrasulfonate (**TPPS$_4$**) have been evaluated as photosensitizers for PDT.[10–12] The primary focus of this chapter will be the thio-, seleno-, and tellurophene-containing derivatives with chalcogen atoms at either just the 21-position or at the 21,23-positions.

2. Synthetic Routes to the 21-Chalcogenoporphyrins

In 1967, Adler and coworkers published a simplified synthesis of the meso-tetraphenylporphyrin using benzaldehyde and pyrrole, which optimized several variables including concentration, solvent, acidity, and a stoichiometric oxidant.[13] The basic tenets of the Adler synthesis have been applied to the synthesis of 21-chalcogeno core-modified porphyrins. A chalcogenophene disubstituted at the 2- and 5-positions with phenylhydroxymethyl substituents [compounds **3a-6a**, path 1) in Scheme 1] reacts with benzaldehyde and

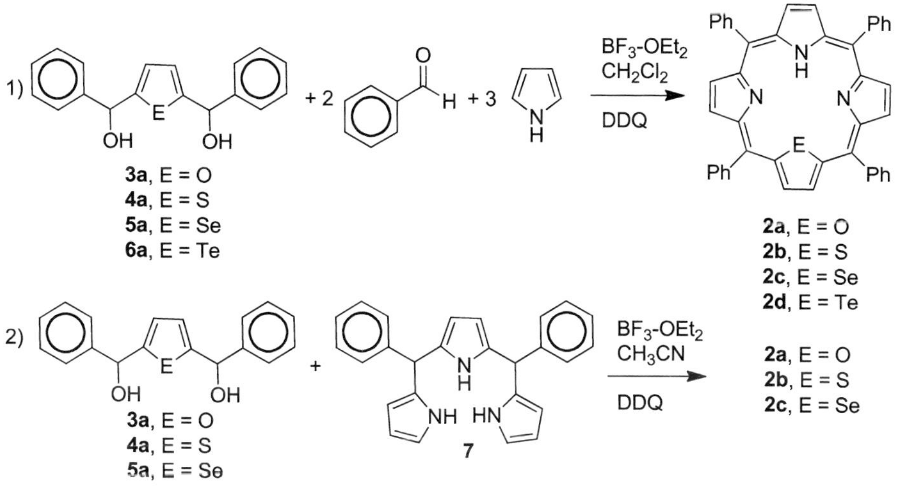

Scheme 1. Major synthetic routes to 21-chalcogeno core-modified porphyrins.

pyrrole in the presence of borontrifluoride etherate and DDQ as an oxidant in dichloromethane to give the tetraphenyl-substituted derivatives **2**.[7,9,14] The heavier chalcogen analogues **2b**, **2c**, and **2d** were isolated in 12%, 12%, and 6% isolated yields, respectively.[14]

A second route that is a modification of the Adler procedure is shown is path 2) of Scheme 1. Condensation of the chalcogenphene diols **3a**, **4a**, and **5a** with the tripyrrane **7** in the presence of borontrifluoride etherate and DDQ in acetonitrile gave the 21-chalcogeno core-modified porphyrins **2a–c**.[15–17]

The chalcogenophene diol derivatives are the key intermediates in these syntheses. The synthesis of the thiophene diol derivatives **4** and the selenophene diol derivatives **5** is illustrated in Scheme 2. Thiophene can be lithiated at both the 2- and 5-positions with *n*-butyllithium to give 2,5-dilithiothiphene.[18] The addition of an aromatic aldehyde (such as benzaldehyde) gives the corresponding di(arylhydroxymethyl)thiophene **4** in excellent yield (89% as shown for the preparation of **4a**). Similarly, selenophene is dilithiated at the 2- and 5-positions with *n*-butyllithium to give 2,5-dilithioselenophene, which can then react with aromatic aldehydes to give the corresponding selenophene diols **5** (path 2), Scheme 2).[18] The addition of benzaldehyde to 2,5-dilithioselenophene gave diol **5a** in 42% isolated yield.

Paths 1) and 2) in Scheme 2 lead only to symmetrical 2,5-disubstituted thiophenes or selenophenes. An approach to unsymmetrical 2,5-disubstituted diols is illustrated in path 3) of Scheme 2.[19] Mono-lithiation of thiophene with 1.0 equivalent of *n*-butyllithium gives 2-lithiothiophene, which then reacts with benzaldehyde to give 2-(phenylhydroxymethyl)thiophene. The hydroxyl

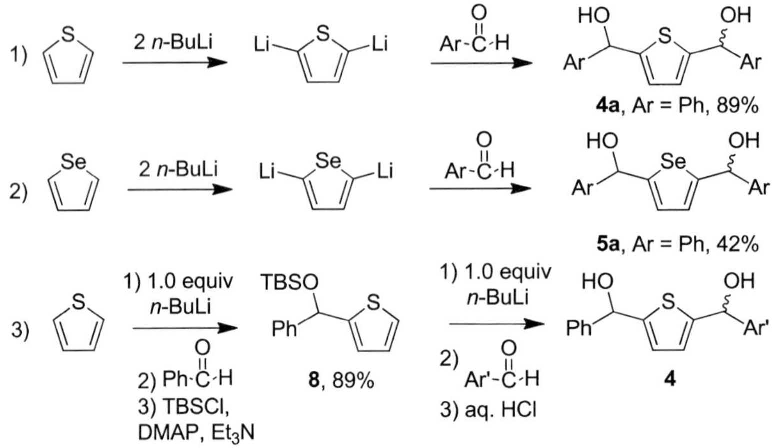

Scheme 2. Synthesis of chalcogenophene diols **4** and **5**.

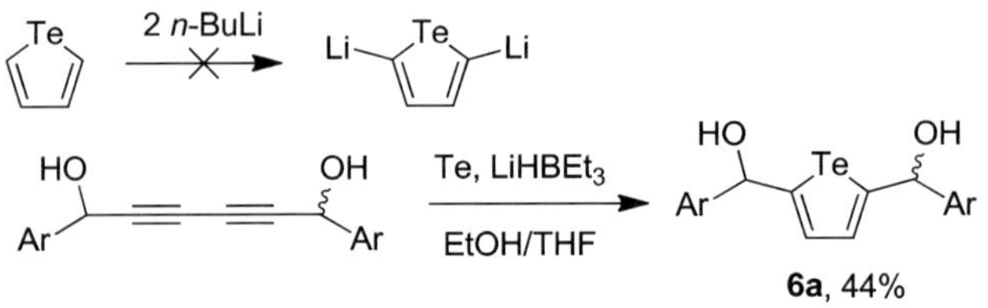

Scheme 3. Synthesis of tellurophene diols **6**.

group is protected as a *tert*-butyldimethylsilyl ether to give the 2-substituted derivative **8** in 89% isolated yield. Thiophene **8** can then be treated with a second equivalent of *n*-butyllithium to give unsymmetrical thiophene diol derivatives **4**.

The lithiation approach does not work for the synthesis of tellurophene diols **6** as illustrated in Scheme 3. Attempts to dilithiate tellurophene wth *n*-butyllithium were uniformly unsuccessful. Loss of tellurium was observed and ring-opened products were also generated. Alternatively, the addition of disodium[20] or dilithium telluride[21] to 1,6-diaryl-1,6-dihydroxyhexa-2,4-diynes **9** gave the tellurophene diols **6** in modest yield (path 2), Scheme 3). The diphenyl derivative **6a** was isolated in 44% yield by this method.[21]

The synthetic routes in Scheme 2 and 3 have allowed a variety of different chalcogenophene diol derivatives **4**–**6** to be prepared as shown in Chart 2. These compounds can be converted to the 21-chalcogeno core-modified porphyrins using the chemistry shown in Scheme 1.

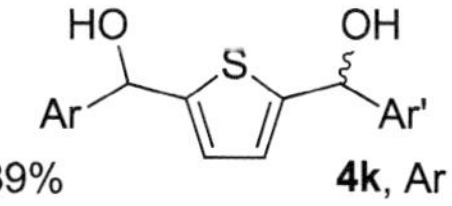

4a, Ar = Ar' = Ph, 89%
4b, Ar = Ar' = 4-FC_6H_4, 43%
4c, Ar = Ph, Ar' = 4-FC_6H_4, 56%
4d, Ar = Ar' = 4-$CF_3C_6H_4$, 52%
4e, Ar = Ph, Ar' = 4-$CF_3C_6H_4$, 57%
4f, Ar = Ar' = 4-$Me_2NC_6H_4$, 65%
4g, Ar = Ar' = 4-MeC_6H_4, 82%
4h, Ar = Ar' = 4-$MeOC_6H_4$, 82%
4i, Ar = Ph, Ar' = 4-$MeOC_6H_4$, 68%
4j, Ar = Ar' = 2-thienyl, 80%
4k, Ar = Ph, Ar' = 2-thienyl, 73%
4l, Ar = Ar' = 4-EtC_6H_4, 60%
4m, Ar = Ar' = 4-*t*-BuC_6H_4, 76%
4n, Ar = Ph, Ar' = 4-*t*-BuC_6H_4, 43%
4o, Ar = Ar' = 4-*n*-BuC_6H_4, 63%
4p, Ar = Ar' = 2,4,6-$Me_3C_6H_2$, 53%
4q, Ar = Ph, Ar' = 2,4,6-$Me_3C_6H_2$, 68%
4r, Ar = Ar' = 4-*i*-PrC_6H_4, 73%
4s, Ar = Ar' = 4-PhC_6H_4, 70%

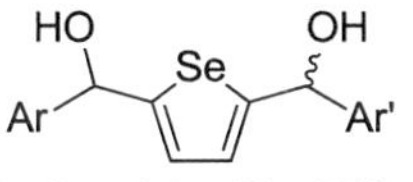

5a, Ar = Ar' = Ph, 42%
5b, Ar = Ar' = 4-FC_6H_4, 37%
5c, Ar = Ar' = 4-$MeOC_6H_4$, 74%

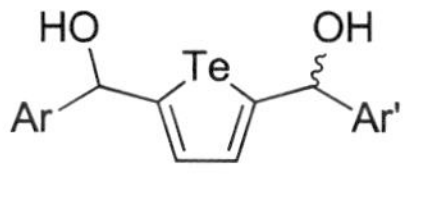

6a, Ar = Ar' = Ph, 44%
6b, Ar = Ar' = 4-$MeOC_6H_4$, 71%
6c, Ar = Ar' = 3,4,5-$(MeO)_3C_6H_2$, 72%

Chart 2. 2,5-Di(arylhydroxymethyl)chalcogenophenes prepared by the synthetic methods of Schemes 2 and 3.

Table 1. Band I absorption maxima (λ_{max}), extinction coefficients (ε), and quantum yields for the generation of singlet oxygen [$\Phi(^1O_2)$] for tetraphenylporphyrin tetrasulfonate (**TPPS$_4$**), tetraphenylporphyrin (**TPP**), and 21-chalcogeno core-modified porphyrins **1a**, **1c**, and **2a–d**.

Compd	λ_{max}, nm	ε, M^{-1} cm^{-1}	$\Phi(^1O_2)$	Reference
TPPS$_4$	630	3900	0.71	12
1a	666	4200	0.80	12
1c	691	3200	0.78	12
TPP	649	3800	0.85	12
2a	650	7200	—	6
2b	675	7200	—	14
2c	677	7300	—	14
2d	657	6400	—	14

3. Photophysical Properties of the 21-Chalcogenoporphyrins

The substitution of an oxygen atom for an NH of a porphyrin has very little impact on the absorption maximum (λ_{max}) of band I in a comparison of tetraphenyl porphyrin (**TPP**) and tetraphenyl-21-oxoporphyrin **2a** (λ_{max} of 649 and 650 nm, respectively, Table 1).[6] Replacing an NH with S or Se, however, gives a significant bathochromic shift of band I to 675 nm for **2b** and 677 nm for **2c** relative to **TPP** (Table 1).[14] Replacing an NH in **TPP** with a Te atom gives a very small bathochromic shift for band I (657 nm, Table 1).

Replacing an NH with S in the 21-thio core-modified porphyrins has minimal impact on values of the quantum yield for the generation of singlet oxygen [$\Phi(^1O_2)$] as shown in Table 1. Values of $\Phi(^1O_2)$ for **TPPS$_4$** and **TPP** are 0.71 and 0.85,[22] respectively. 21-Thio core-modified porphyrins **1a** and **1c** (structures in Chart 1) have values of $\Phi(^1O_2)$ of 0.80 and 0.78, respectively, which are intermediate between those for **TPPS$_4$ and** TPP.

4. Phototoxicity of the 21-Chalcogenoporphyrins

The core-modified porphyrins **1a–c** have been evaluated as photosensitizers *in vitro* in Colo-26 cells. Colo-26 cells are a murine colon carcinoma cell line that express P-glycoprotein (P-gp), but are not deemed truly drug resistant.[23] Photosensitizers effective against this cell line may have promise in treating truly drug-resistant cell lines and tumors. The tetra-sulfonated derivatives **1a** and **1b** were not particularly effective photo-sensitizers toward Colo-26 cells with values of EC_{50} (the concentration of photosensitizer to give 50% cell kill with, in these examples, $4\,J\,cm^{-2}$ of filtered 590–800-nm light) of $>100\,\mu M$.[12] However, the unsymmetrical thiaporphyrin **1c** — with two 4-sulfonatophenyl and 2-4-fluorophenyl substituents in the *meso*-positions — was far more effective with an EC_{50} of $2.1\,\mu M$ with $4\,J\,cm^{-2}$ of filtered 590–800-nm light.[12] 21-Oxoporphyrins and 21-thioporphyrins have also shown some tumor necrosis following PDT treatment.[24]

5. The 21,23-Dichalcogenoporphyrins

Core-modified porphyrins bearing two sulfur and/or selenium atoms at the 21- and 23-positions have been found to be more effective photosensitizers for PDT in both *in vitro* and *in vivo* studies than the 21-chalcogeno core modified porphyrins with one chalcogen atom. The 21,23-dithio core-modified porphyrins **9** shown in Chart 3 have been prepared and evaluated as photosensitizers *in vitro*.[12,25–28]

Similarly, the 21-seleno-23-thio core-modified porphyrins **9** and the 21,23-diseleno core-modified porphyrins **10** (Chart 4) have been prepared and evaluated *in vitro*.[12,25,26] The synthetic methods to prepare the 21,23-dithio, 21-seleno-23-thio, and 21,23-diseleno core-modified porphyrins are similar.

6. Synthetic Routes to the 21,23-Dichalcogenoporphyrins

The 21,23-dichalcogeno core-modified porphyrins have been prepared analogously to path 2) of Scheme 1. The key intermediates are the dipyrranechalcogenophenes **12–14**, which are prepared as shown in

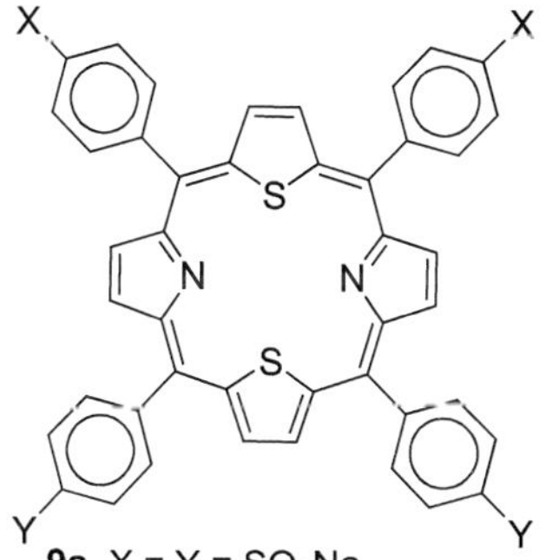

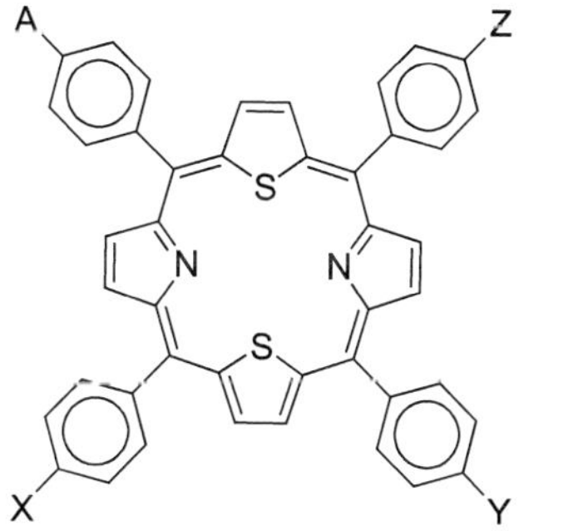

9a, X = Y = SO_3Na
9b, X = F, Y = SO_3Na
9c, X = CF_3, Y = SO_3Na
9d, X = NMe_2, Y = SO_3Na
9e, X = 4-Me, 2-SO_3Na; Y = SO_3Na
9f, X = H, Y = OCH_2CO_2H
9g, X = F, Y = OCH_2CO_2H

9h, A = OCH_2CO_2H, X = Y = Z = H
9i, A = X = Y = OCH_2CO_2H, Z = H
9j, A = X = Z = Y = OCH_2CO_2H

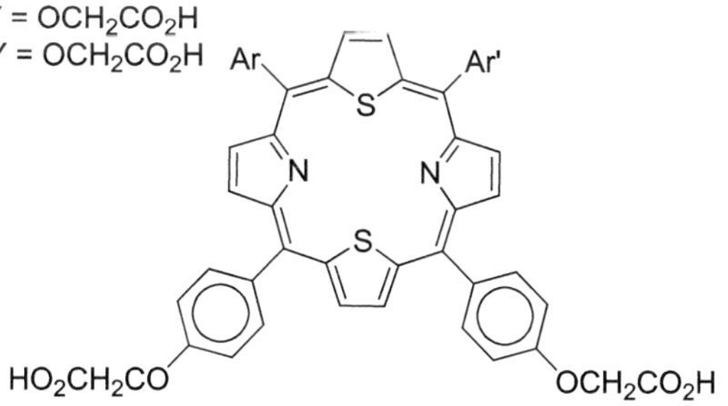

9k, Ar = Ar' = 2-thienyl
9l, Ar = C_6H_5, Ar' = 2-thienyl
9m, Ar = Ar' = 2,4,6-$Me_3C_6H_2$
9n, Ar = C_6H_5, Ar' = 2,4,6-$Me_3C_6H_2$
9o, Ar = Ar' = 4-*tert*-BuC_6H_4
9p, Ar = C_6H_5, Ar' = 4-*tert*-BuC_6H_4
9q, Ar = Ar' = Me
9r, Ar = Ar' = Et
9s, Ar = Ar' = *n*-Bu
9t, Ar = 4-FC_6H_4, Ar' = C_6H_5
9u, Ar = Ar' = 4-ClC_6H_4
9v, Ar = Ar' = 4-$CF_3C_6H_4$
9w, Ar = C_6H_5, Ar' = 4-$CF_3C_6H_4$
9x, Ar = Ar' = 4-$MeOC_6H_4$
9y, Ar = Ar' = $Me_2NC_6H_4$

Chart 3. 21,23-Dithio core-modified porphyrins **9** evaluated as photosensitizers *in vitro*.

(Scheme 4). The chalcogenophene diols **4–6** react with pyrrole in the presence of boron trifluoride etherate to give the corresponding dipyrranechalcogenophenes **12–14**, respectively in 51–99% isolated yields.[12,25,26] The dipyrrane chalcogenophenes **12–14** then react with chalcogenophene diols **4–6** to give the 21,23-dithio core-modified porphyrins **9**, the 21-seleno-23-thio core-modified porphyrins **10**, the 21,23-diseleno core-modified porphyrins **11**, and the 21-chalcogeno-23-telluro core-modified porphyrins **15** (Scheme 5). The 21-telluro-23-thio and 21,23-ditelluro core-modified porphyrins, which were prepared via the method of Scheme 5, are shown in Chart 5.

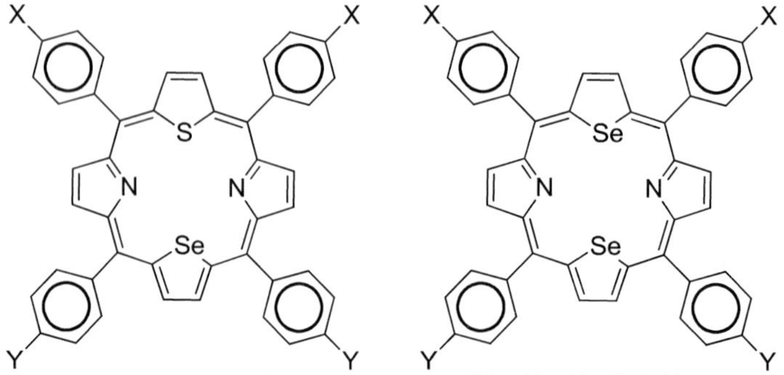

10a, X = F, Y = SO_3Na
10b, X = SO_3Na, Y = F

11a, X = Y = SO_3Na
11b, X = F, Y = SO_3Na
11c, X = H, Y = OCH_2CO_2H

Chart 4. 21-Seleno-23-thio core-modified porphyrins **10** and 21,23-diseleno core-modified porphyrins **11** evaluated as photosensitizers *in vitro*.

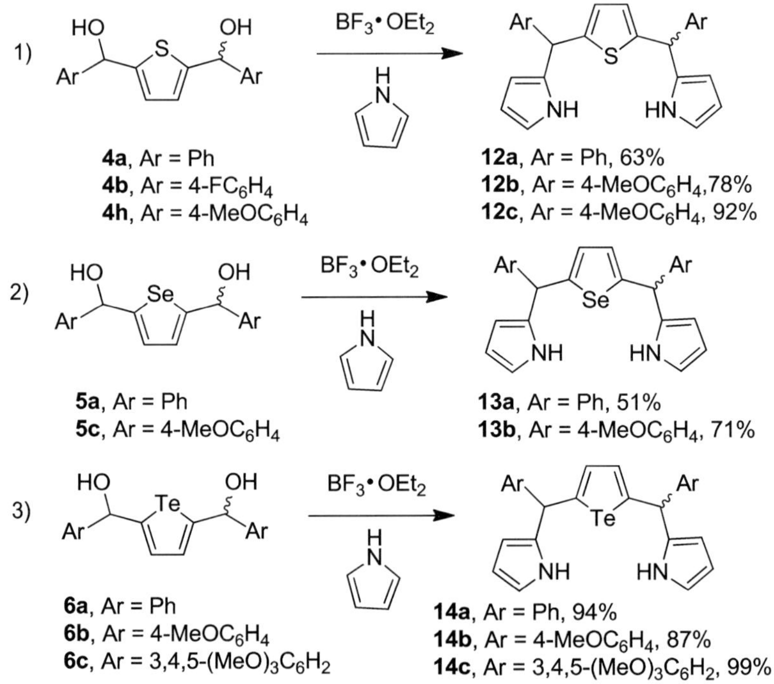

Scheme 4. Preparation of dipyrranechalcogenophenes **12–14**.

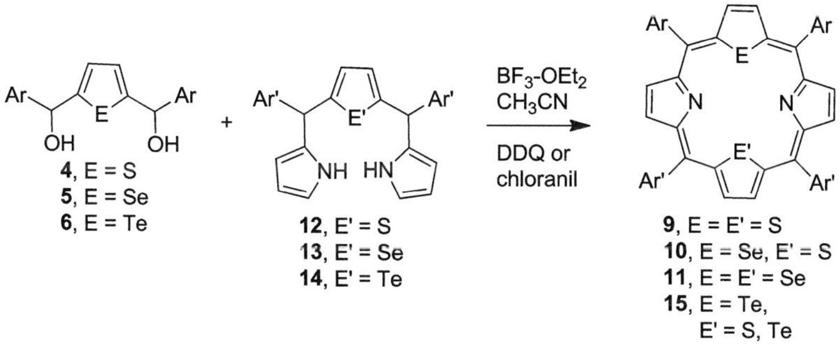

Scheme 5. Synthesis of 21,23-dichalcogeno core-modified porphyrins **9–11** and **15**.

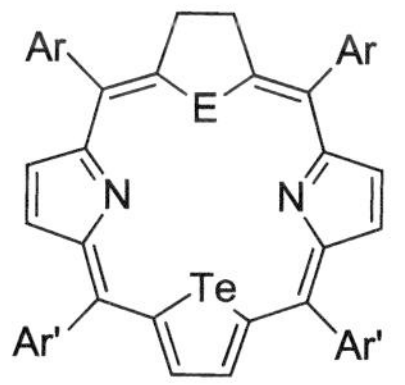

16a, E = S, Ar = Ar' = Ph (17%)
16b, E = Te, Ar = Ar' = Ph (34%)
16c, E = Te, Ar = Ar' = 2,4,6-$Me_3C_6H_2$ (12%)
16d, E = Te, Ar = Ph, Ar' = 4-$MeOC_6H_4$ (28%)
16e, E = Te, Ar = Ar' = 4-$MeOC_6H_4$ (19%)
16f, E = Te, Ar = 4-$MeOC_6H_4$,
Ar' = 2,4,6-$Me_3C_6H_2$ (13%)

Chart 5. 21-Telluro-23-chalcogeno core-modified porphyrins prepared according to Scheme 5.

7. Physical and Photophysical Properties of the 21-Telluro-23-thioporphyrins and 21,23-Ditelluroporphyrins

Structurally, the tellurium-containing core-modified porphyrins are a complicated class of molecules. The tellurium atom is much larger than the pyrrolic nitrogen atoms in the core-modified porphyrins and can create a "competition" for the intramolecular space in the plane of the greater porphyrin ring. Telluroporphyrin **2d** (Scheme 1) is nearly planar with Te…N contacts (3.13 Å) that are less than the sum of van der Waals radii (3.55 Å).[8] In 21-telluro-23-thioporphyrin **16a**, the chalcogen atoms are above and below the plane of

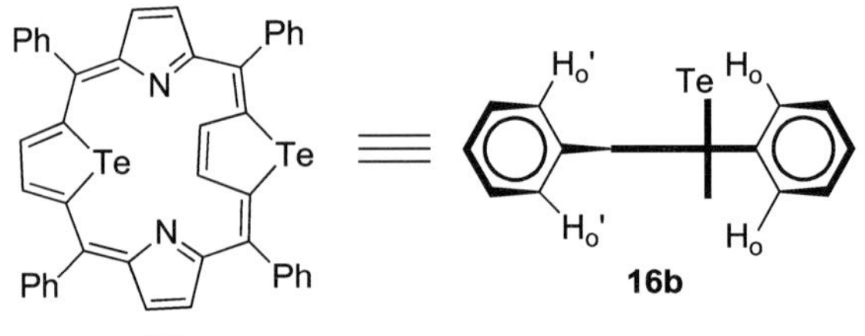

Chart 6. Ring "flipped" tellurophene in 21,23-ditelluroporphyrins.

Table 2. Band I absorption maxima (λ_{max}), extinction coefficients (ε), and quantum yields for the generation of singlet oxygen [$\Phi(^1O_2)$] for 21-telluro-23-thioporphyrin **16a** 21,23-ditelluroporphyrins **16b–f**.

Compd	λ_{max}, nm	ε, M^{-1} cm^{-1}	$\Phi(^1O_2)$	Reference
16a	686	3,700	0.05	20
16b	668	16,000	<0.05	28
16c	692	18,600	<0.05	21
16d	690	18,200	<0.05	21
16e	698	16,300	<0.05	21
16f	694	18,100	<0.05	21

the porphyrin periphery, but are still nearly planar.[20] Again, the intramolecular Te…S contacts (2.65 Å) are much shorter than the sum of van der Waals radii for the two atoms (3.81 Å). In the 21,23-ditelluraporphyrins **16b–f**, the dominant structure both in solution and in the crystalline state is one in which the two Te atoms are so large that they cannot both occupy the core and one tellurophene is "flipped" with the Te atom pointed away from the porphyrin core as illustrated in Chart 6.[21,29] This arrangement disrupts the porphyrin aromaticity and leads to thermal interconversion of conformations. As a consequence, the 21-telluro-23-thio- and 21,23-ditelluroporphyrins generate singlet oxygen inefficiently and are poor photosensitizers for PDT *in vitro* with Colo-26 cells ($EC_{50} > 100\,\mu M$ with $5\,J\,cm^{-2}$ of 695-nm light from a tunable dye laser).[30] Photophysical properties for the 21-telluroporphyrins are summarized in Table 2.

8. Physical and Photophysical Properties of 21,23-Dithioporphyrins, 21-Seleno-23-thioporphyrins, and 21,23-Diselenoporphyrins

Among the core-modified porphyrins, the 21,23-dithioporphyrins **9** provide the most potent photosensitizers both *in vitro* and *in vivo* in animal models.

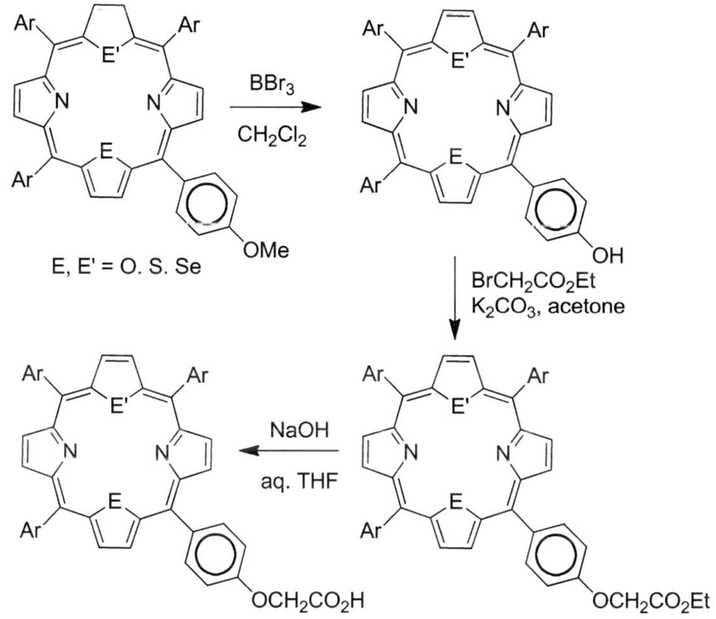

Scheme 6. Conversion of methoxyphenyl substituents to carboxylate derivatives.

Water solubility within this series of compounds has been provided by exhaustive sulphonation of *meso*-phenyl substituents to provide derivatives with two and four sulfonatophenyl substituents (examples are compounds **9a–e** in Chart 3) as one approach.[12,25] Alternatively, as shown in Scheme 6, 4-methoxyphenyl substituents have been demethylated with boron tribromide to the corresponding phenols, which can then react with ethyl bromoacetate in the presence of sodium carbonate and the resulting esters saponified to give the carboxylic acid derivatives **9f–y** of Chart 3 with 1 to 4 carboxylic acid groups.[12,26] The sulfonates **9a–e** are soluble at concentrations of 10 mg per mL of saline (approximately 10^{-2} M in photosensitizer) while the carboxylate derivatives **9f–y** are an order of magnitude less soluble at 1 mg per mL of saline (approximately 10^{-3} M in photosensitizer).[12]

The band I values of λ_{max} are in the range 689–714 nm as compiled in Table 3 and are ≈30–50 nm longer than those observed for the 21-chalcogenoporphyrins with values of ε of 4200–8900 $M^{-1}\ cm^{-1}$. The quantum yields for the generation of singlet oxygen [$\Phi(^1O_2)$] vary from 0.32 for **9y** (with two 4-dimethylaminophenyl substituents) to 0.90 for **9m** (with two 2,4,6-trimethylphenyl substituents). The tetrasulfonato derivative **9a** has $\Phi(^1O_2)$ of 0.50, but all of the other 21,23-dithioporphyrins **9** have values of $\Phi(^1O_2)$ in the 0.64–0.90 range (Table 2), which makes them interesting as potential

Table 3. Band I absorption maxima (λ_{max}), extinction coefficients (ε), quantum yields for the generation of singlet oxygen [$\Phi(^1O_2)$], and effective concentrations to give 50% cell kill in Colo-26 or R3230AC cells (EC_{50}) with 5 J cm^{-2} of filtered 350–800-nm light for 21,23-dithio core-modified porphyrins **9**.

Compd	λ_{max}, nm	ε, M^{-1} cm^{-1}	$\Phi(^1O_2)$	EC_{50}, μM	Reference
9a	695	4000	0.50	30	25
9b	701	4600	0.74	1.6	12
9c	701	4200	—	>100	12
9d	716	7500	0.64	0.64	12
9e	689	4300	—	12	12
9f	699	5600	0.80	0.15	12, 26
9g	717	4800	0.71	0.43	12
9h	696	3700	0.78	10	14
9i	700	5700	0.74	>100	26
9j	701	4200	0.78	>100	26
9k	707	5300	0.67	0.10	27
9l	702	6300	0.86	0.08	27, 31
9m	697	4800	0.90	>10	27
9n	697	6900	0.80	0.10	27
9o	702	4800	0.80	>10	27
9p	700	5100	0.80	>10	27
9r	701	4000	—	>10	27
9s	700	3200	—	>10	27
9t	698	5300	0.83	0.1	28
9u	699	4600	0.76	0.5	28
9v	697	4600	0.75	—	28
9w	698	5600	0.87	—	28
9x	704	4100	0.81	0.1	28
9y	714	8900	0.32	0.5	28

photosensitizers for PDT. For **9m**, the mesityl substituents sterically limit rotation of these *meso*-substituents, which reduces rates of internal conversion and increases the relative contribution of intersystem crossing from the singlet excited state. The dimethylaminophenyl groups of **9y** are good donor substituents for internal donor–acceptor interactions, which would decrease the relative contribution of intersystem crossing in the singlet excited state.

Replacing one or both sulfur atoms with selenium to give 21-selena-23-thia- (**10**) or 21,23-diselenporphyrins (**11**) has little impact on wavelengths for values of λ_{max} for band I (Table 4). Porphyrins **10a** and **10b** interchange the two chalcogen atoms in the 21-selena-23-thiaporphyrins with essentially no change in λ_{max} for band I (4 nm difference) or $\Phi(^1O_2)$ (0.64 and 0.65, respectively). As observed with tetrasulfonato dithiaporpyrin **9a** where the observed $\Phi(^1O_2)$ of 0.50 was lower than derivatives with at least two different *meso* substituents, tetrasulfonato 21,23-diselenoporphyrin **11a** has $\Phi(^1O_2)$ of

Table 4. Band I absorption maxima (λ_{max}), extinction coefficients (ε), quantum yields for the generation of singlet oxygen [$\Phi(^1O_2)$], and effective concentrations to give 50% cell kill in Colo-26 or R3230AC cells (EC_{50}) with 5 J cm^{-2} of filtered 350–800-nm light for 21-seleno-23-thio core-modified porphyrins **10** and 21,23-diseleno core-modified porphyrins **11**.

Compd	λ_{max}, nm	ε, M^{-1} cm^{-1}	$\Phi(^1O_2)$	EC_{50}, μM	Reference
10a	698	7100	0.65	2.1	12
10b	702	6300	0.64	2.2	12
11a	695	4500	0.17	>100	12, 25
11b	703	3800	0.55	7.9	12
11c	695	2800	0.30	10	25

only 0.17. Values of $\Phi(^1O_2)$ for the 21,23-diselenoporphyrin **11b** and **11c** are lower than corresponding 21,23-dithio- or 21-seleno-23-thioporphyrins.

9. Structure and Its Impact on Phototoxicity of the 21,23-Dichalcogeno-porphyrins and Other Biological Properties *in vitro*

The 21,23-dichalcogenoporphyrins offer two different approaches to structural diversity: The identity of the chalcogen atoms at the 21- and 23-positions and the identity of the *meso*-substituents. The tellurium-containing core-modified porphyrins are poor photosensitizers for PDT *in vitro* with EC_{50}'s >100 μM and poor quantum yields for the generation of singlet oxygen [$\Phi(^1O_2) \leq 0.05$, Table 2]. Table 3 contains EC_{50} values for 21,23-dithioporphyrins and Table 4, EC_{50} values for 21-seleno-23- and 21,23-diselenoporphyrins. Within the latter two classes, photosensitizers with $EC_{50} < 10\,\mu$M with 5 J cm^{-2} of 350–800-nm light toward Colo-26 and/or R3230AC rat tmammary adenocarcinoma cells.

Core-modified porphyrins with three or four water-solubilizing groups appear to be poor photosensitizers *in vitro.* 21,23-Dithioporphyrin **9a** and 21,23-diselenoporphyrin **11a** with four phenylsulfonato groups have values of EC_{50} of 30 μM (Table 3) and >100 μM (Table 4), respectively. 21,23-Dithioporphyrins **9i** with three carboxylates and **9j** with four carboxylates each have values of EC_{50} > 100 μM (Table 3). In contrast, within the carboxylate series, 21,23-dithioporphyrin **9f** and 21,23-diselenoporphyrin **11c** with two carboxylates have values of EC_{50} of 0.15 μM and 10μ M, respectively (Table 3). 21,23-Dithioporphyrin **9h** with one carboxylate substituent has an intermediate value of EC_{50} of 10 μM.

9.1. *Impact of the Number of Water-solubilizing Groups*

Within the series of 21,23-dithioporphyrin derivatives **9f** and **9h–j**, the efficacy as a photosensitizer is strictly a function of the number of water solubilizing groups and not values of $\Phi(^1O_2)$. For these four compounds, values of $\Phi(^1O_2)$ are quite similar with a range of 0.74–0.80. Values of $\Phi(^1O_2)$ in the phenylsulfonato-substituted derivatives have a much wider range. For tetrasulfonates **9a** and **11a**, values of $\Phi(^1O_2)$ are 0.50 (Table 3) and 0.17 (Table 4), respectively. Disulfonato derivative **11c** (Table 4) has a value of $\Phi(^1O_2)$ of 0.30, which is intermediate between the other two.

The cellular uptake of **9f** and **9h–j** was evaluated in R3230AC cells and was found to be very dependent on the number of water-solubilizing groups.[26] 21,23-Dithioporphyrin **9j** with four carboxylate groups gave <2 femtomol per cell of **9j** after a 3-h incubation with 10 μM **9j**. 21,23-Dithioporphyrin **9i** with three carboxylic acid residues gave 6 ± 4 femtomol per cell under the same conditions while **9h** gave 20 ± 4 femtomol per cell. 21,23-Dithioporphyrin **9f** with two carboxylic acid residues gave 30 ± 4 femtomol per cell after a 3-h incubation with 10 μM photosensitizer. These values were different by statistically significant amounts with the **9f–9h** comparison giving $p < 0.038$.

Within the 21,23-dithioporphyrins **9** with two carboxylate substituents, the most potent photosensitizers *in vitro* are **9d**, **9h**, **9f**, **9g**, **9k**, **9l**, **9t**, **9u**, **9x**, and **9y**, which all have values of $EC_{50} < 1.0$ μM (Table 3). With the exception of **9y** with $\Phi(^1O_2)$ of 0.32, the remaining members of this series have values of $\Phi(^1O_2)$ in the range 0.64–0.86 (Table 3), which suggests that the molecules should efficiently produce singlet oxygen as photosensitizers. With the exception of **9h** with two *meso* phenyl substituents, the remaining compounds in this series have phenyl substituents substituted at the 4-position with *ortho*, *para*-directing substituents (NMe_2, F, Cl, MeO) or π-excessive heterocycles (2-thienyl). It is likely that this "push" (electron-donating substituents)/"pull" from the substituents (i.e. electron-withdrawing water-solubilizing groups) contributes to the effectiveness of these molecules as photosensitizers. In contrast, 21,23-dithioporphyrin **9c** with two 4-trifluoromethylphenyl substituents and 21,23-dithioporphyrins **9o–9s** with alkyl substituents all have values of $EC_{50} > 10$ μM.

In summary, the 21,23-dithioporphyrins appear to be more effective as photosensitizers *in vitro* than selenium-containing analogues. Within the 21,23-dithioporphyrins, derivatives with two water-solubilizing groups on adjacent *meso*-positions are more effective photosensitizers than derivatives with one, three, or four water-solubilizing groups. Substitution patterns that give a push/pull of electrons appear to be more effective as photosensitizers. Steric effects within alkyl series do not appear to be important. The carboxylate water-solubilizing groups are preferred relative to the sulfonates.

9.2. *The Impact of Spacer Length*

The length of the alkyl spacer between the aryl ring and the carboxylic acid residue in structures such as **9f** was examined.[31] The series of 21,23-dithioporphyrins **9a**, **9z**, **9aa**, **9ab**, **9ac**, and **9ad** was prepared as shown in Scheme 7. The number of methylene groups between the phenolic oxygen and the carboxylic acid was varied from 1 to 10. For this series of compounds, the absorption maxima for all of the porphyrin bands were essentially identical and values of $\Phi(^1O_2)$ in MeOH were essentially identical [$\Phi(^1O_2) \approx 1.00$, Table 5]. However, when the photosensitizers of Scheme 7 were buried in the membranes of liposomes, the chain length had a pronounced impact on values of $\Phi(^1O_2)$. With values of n of 1 and 3 (compounds **9f** and **9z**, respectively), values of $\Phi(^1O_2)$ in the liposome were on the order of 0.70. Beyond $n = 3$, further increases in the chain length gave systematic decreases in $\Phi(^1O_2)$: from 0.54 for $n = 4$ to 0.14 for $n = 10$ (Table 5). These results suggest that phototoxicity may be a function of spacer length between water-solubilizing groups and the porphyrin core and that longer spacers may give photosensitizers that remain membrane bound. These differences should be apparent in the cellular phototoxicity of the **9a**, **9z**, **9aa**, **9ab**, **9ac**, and **9ad** series of core-modified porphyrins.

The dark toxicity of the **9a**, **9z**, **9aa**, **9ab**, **9ac**, and **9ad** series was evaluated toward U297 cells — a human lymphoma.[31] The 21,23-dithioporphyrins **9f** and **9z** with $n = 1$ and 3, respectively, were dark toxic at lower concentrations

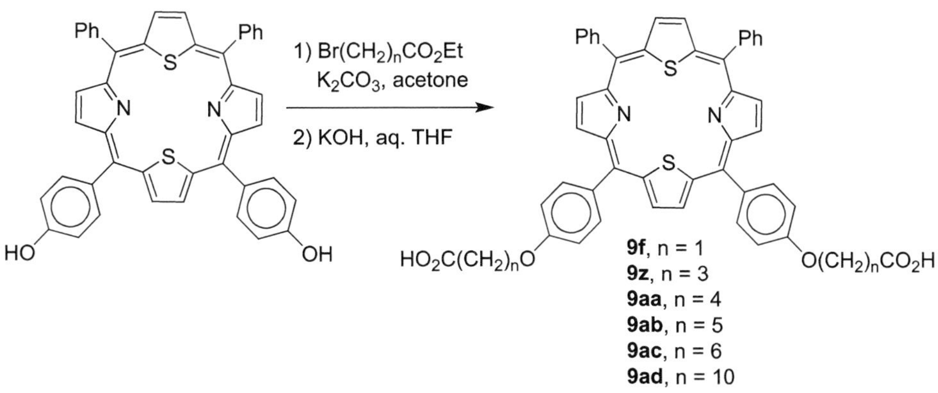

Scheme 7. Synthesis of 21,23-dithioporphyrins with variable linkage length between carboxylic acid residue and phenolic oxygen.

Table 5. The effect of spacer length on photophysical properties and phototoxicity with 21,23-dithioporphyrins **9a**, **9z**, **9aa**, **9ab**, **9ac**, and **9ad**.

Compd	*n*	$\Phi(^1O_2)$ MeOH	$\Phi(^1O_2)$ Liposome	EC_{50}, μM R3230AC cells 3.8 J cm^{-2} of 350–750-nm light	EC_{50}, J cm^{-2} U937 cells 0.5 μM porphyrin, 380–480-nm light
9f	1	1.00	70	0.06	1.5
9z	3	1.00	71	0.2	7.5
9aa	4	1.00	54	1.0	30
9ab	5	1.00	42	2.0	>50
9ac	6	1.00	19	>10	>50
9ad	10	1.00	14	>10	>50

than their longer chain counterparts. However, these two compounds were the most phototoxic members of the series toward both U937 cells and R3230AC rat mammary adenocarcinoma cells as summarized in Table 5. Compounds **9ac** and **9ad** with $n = 6$ and 10, respectively, were essentially non-phototoxic.

From the *in vitro* studies, 21,23-dithioporphyrins **9f** and **9l** appear to be the optimal core-modified porphyrin photosensitizers among those examined. The phototoxicity of **9f** *in vitro* in Colo-26 cells was compared to the phototoxicity of HPPH (a pyropheophorbide derivative),[32] Photofrin®, 21,23-dithioporphyrin **9b** with two 4-fluorophenyl and two 4-sulfonatophenyl *meso*-substituents (Chart 3), and 21,23-diselenoporphyrin **11b** with two 4-fluorophenyl and two 4-sulfonatophenyl *meso*-substituents (Chart 4). With 4 J cm^{-2} of 590–800-nm light, EC_{50} for **9f** was 0.15 μM,[26] EC_{50} for HPPH was 1.0 μM,[26] EC_{50} for Photofrin® was 10 μM,[26] EC_{50} for **9b** was 1.1 μM,[26] and EC_{50} for **11b** was 7.9 μM.[12] 21,23-dithioporphyrin **9l** with a *meso*-phenyl and a *meso*-2-thienyl substituent displayed an EC_{50} of 0.080 μM with 5 J cm^{-2} of broad-band, 350–800-nm light toward R3230AC cells.[33]

10. *in vivo* Studies

10.1. *21,23-Dithioporphyrin 9b and 21,23-Diselenoporphyrin 11b*

Dithioporphyrin **9b** and diselenoporphyrin **11b** were examined in several *in vivo* studies to evaluate systemic toxicity, distribution, photodynamic efficacy, and long-term skin photosensitization.[12] With porphyrin-related photosensitizers, a typical therapeutic dose is typically 1–5 mg/kg of body mass.[34,35] BALB/c mice were given a single dose of 10 mg/kg ($\approx$10 μmol/kg) of either dithioporphyrin **9b** or diselenoporphyrin **11b** as a saline solution via tail-vein injection. The animals tolerated this treatment quite well with no mortality and no signs of morbidity after 90 days.[12]

Distribution studies in BALB/c mice bearing Colo-26 tumors were conducted following a single tail-vein injection of 5 mg/kg (5 μmol/kg) of 21,23-dithioporphyrin **9b**. Relative concentrations in selected tissues were examined by fluorescence spectroscopy 1, 4, and 24 h post injection. There were no statistically significant differences in concentrations in the Colo-26 tumors at 1, 4, and 24 h post injection. After 24 h, concentrations in liver, kidney, lung, and heart had decreased to levels below those found in the tumors. Only the spleen maintained comparable concentrations to those found in the tumors.[12] These results suggested that treatment 24 h post injection would minimize the presence of photosensitizer in tissues other than in the tumor while maintaining PDT-useful levels in the tumor.

As a point of comparison, the uptake of dithiaporphyrin **9b** with two water-solubilizing sulfornato groups in tumor was much greater than the uptake of dithioporphyrin **9a** with four water-solubilizing sulfonate groups. At a dose of 5 mg (5 μmol)/kg via tail-vein injection, a concentration of 19 ± 7 ng of **9a**/g of protein was measured in tumors of BALB/c mice bearing Colo-26 tumors 24 h post injection. In contrast, a concentration of 494 ± 90 ng of **9b**/g of protein was measured in the tumors of animals 24 h post injection of 5 mg (5 μmol) of **9b**/kg, which represents a greater than 25-fold difference in photosensitizer uptake and is consistent with two water-solubilizing groups being better than four.[12]

PDT with core-modified porphyrins **9a**, **9b**, and **11b** was evaluated in BALB/c mice bearing implanted Colo-26 tumors. PDT with Photofrin® was used as a control for comparison purposes. Based on values of EC_{50} *in vitro*, 21,23-diselenoporphyrin **11b** and Photofrin® should have similar treatment doses. Values of EC_{50} with 21,23-dithioporphyrin **9b** were an order of magnitude smaller than thse for **11b** and Photofrin® while 21,23-dithioporphyrin **9a** with four sulfonate groups had a much higher value of EC_{50}.

BALB/c mice bearing Colo-26 tumors were given 10 mg (10 μmole)/kg of 21,23-dithioporphyrin **9a**, 0.125 mg (0.13 μmole)/kg of 21,23-dithioporphyrin **9b** or 2.5 mg (2.5 μmole)/kg of 21,23-diselenoporphyrin **11b** as a 5% dextrose in water solution via tail vein injection. Four hours post injection, the animals were irradiated with 135 J cm^{-2} of 694-nm red light from a dye laser (75 mW cm^{-2} for 30 min). An additional group of animals was treated with Photofrin® at 2.5 mg (4 μmole)/kg of body mass and were given 135 J cm^{-2} of 630-nm light 24 h post injection. The time in days to

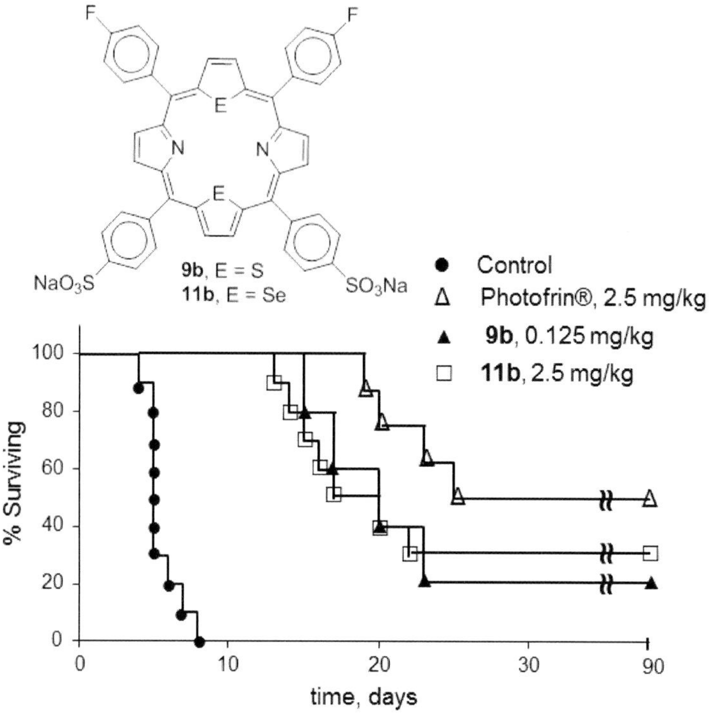

Fig. 1. PDT with 21,23-dithioporphyrin **9b** (filled triangles), and 21,23-diselenoporphyrin **11b** (open squares) against implanted Colo-26 tumors in BALB/c mice. Animals received 0.125 mg/kg of **9b** or 2.5 mg/kg of **11b** and 135 J cm^{-2} of 694-nm light 4 h post injection. Times were measured post treatment until tumor volumes reached 400 mm^3. These data can be compared to a control group receiving neither drug nor light (filled circles) and to a group treated with 2.5 mg/kg of Photofrin® and 135 J cm^{-2} of 630-nm light (open triangles) 24 h post injection.

400 mm^3 tumor volume was noted for each animal in each group, and the results are presented as a Kaplan–Meyer plot in Fig. 1.[12]

The untreated control animals gave a mean time to 400 mm^3 of 6 ± 2 days. The four animals treated with 21,23-dithioporphyrin **9a** gave a mean time to 400 mm^3 of 13 ± 6 days, which is significantly greater ($p < 0.014$) than untreated controls, but no cures were obtained. Of the five animals treated with dithiaporphyrin **9b** and light, one cure was obtained, and the other four animals gave a mean time to 400 mm^3 of 18 ± 4 days, which is a 300% increase relative to untreated controls ($p < 0.001$). Of the 10 animals treated with diseleno-porphyrin **11b** and light, three cures were obtained and the other seven animals gave a mean time to 400 mm^3 of 17 ± 3 days, which is significantly greater than untreated controls ($p < 0.001$). The treatment protocols with **9b** and **11b** showed no significant difference with respect to the time to reach 400 mm^3.[12]

The results with **9b** and **11b** can be compared with the animals treated with Photofrin®. The intereval between administration of photosensitizer and irradiation, the fluence, and the fluence rate with Photofrin® have been optimized with this model.[3,4,35] In this particular group of animals, four cures were obtained and the time to 400 mm^3 with the remaining four animals was 21 ± 4 days, which is not significantly different than the time to 400 mm^3 with either **9b** or **11b**.[12]

10.2. *21,23-Dithioporphyrin 91*

The 21,23-dithioporphyrin **91** (Chart 7) with two carboxylic acid water-solubilizing groups was not as water soluble as core-modified porphyrins **9b** and **11b** with two sulfonato water-solubilizing groups. Formulation of **91** for *in vivo* studies was successful with a 1% Tween-80/5% dextrose aqueous formulation.[33] BALB/c mice given tail-vein injections of 25 μmole/kg of **91** showed no statistically significant differences in concentration in tumor tissue at 24, 48, and 72 h post injection. At the 48 h and 72 h time points, significantly greater amounts of **91** were found in tumor relative to lungs, kidneys, skin, or muscle. Higher concentrations were found in the liver relative to tumor.[32]

BALB/c mice with implanted Colo-26 tumors were treated with **91** (2.1 μmole/kg or 3.7 μmole/kg) and 350 J cm^{-2} of 690-nm light delivered from a diode laser at 200 mW cm^{-2}. Irradiation was 24 h post injection. In the untreated control group and in the group receiving **91** alone and no light, tumor volume increased by 15-fold and 10-fold, respectively, on day 7 following treatment. The treated groups showed a significant delay in tumor growth. The group receiving 2.1 μmol/kg exhibited a growth inhibition of 72% compared to the control group on day 7 following PDT treatment while the group receiving 3.7 μmol/kg showed reduced tumor growth to nonmeasurable levels from days 5–11 after PDT. Unfortunately, increased tumor volumes were observed two weeks following treatment.[33]

11. Long-term Skin Photosensitization

With Photofrin® and related porphyrin sensitizers, long-term skin photosensitization is a complication of treatment. Patients must avoid sunlight for significant periods of time.[4,36] Long-term skin photosensitization was evaluated in a comparison of 21,23-dithioporphyrin **9b**, 21,23-diselenoporphyrin **11b**, and Photofrin® using the well-established murine ear-swelling response.[37,38]

In BALB/c mice treated with **9b** at 0.125 mg/kg, the ear-swelling response disappeared between days 3 and 4. In BALB/c mice treated with **11b** at

5.0 mg/kg, the ear-swelling response disappeared between days 5 and 7. Even at 31 days, there remained a positive ear-swelling response in Photofrin®-treated animals. Importantly, concentrations of core-modified porphyrins **9b** or **11b** that gave comparable results to Photofrin® for PDT efficacy gave significantly reduced duration of cutaneous photosensitivity.

Animals treated with 21,23-dithioporphyrin **9l** showed a time-dependent response to foot-pad irradiation following treatment. The physical damage and healing time were much shorter with irradiation 96 h post injection relative to foot-pad irradiation 48 h post injection. These results suggest that photosensitizer **9l** is clearing the animal 96 h post injection.[33]

12. Conclusions

The core-modified porphyrins offer several advantages over conventional porphyrin derivatives. The core-modified porphyrins generate singlet oxygen efficiently, absorb at longer wavelengths than conventional porphyrins, and have sub-micromolar values of EC_{50} in cellular studies. The 21,23-dithioporphyrin **9b** was effective *in vivo* in BALB/c mice bearing implanted Colo-26 cells at a concentration of 0.125 mg/kg of body mass. At this concentration, skin photosensitization disappeared 3–4 days post injection, which is a tremendous advantage of **9b** relative to Photofrin®, while efficacy was comparable between **9b** and Photofrin®.

The next generation of core-modified porphyrins may feature conjugation to other functionality, which will aid in targeting the photosensitizer to the desired site of action.[39] The delocalized cations rhodamine 123 and triphenylphosphonium were tethered to 21,23-dithioporphyrin **9h** with one carboxylic acid residue to give conjugates **17** and **18**, respectively (Chart 7).

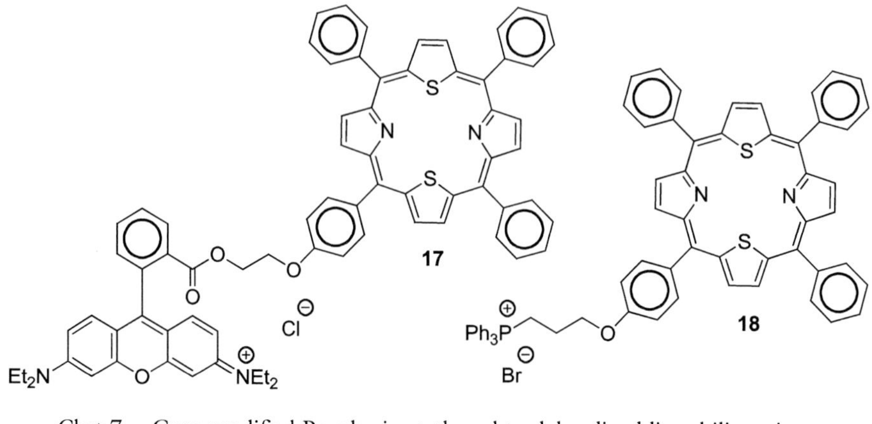

Chat 7. Core-modified Porphyrins tethered to delocalized lipophilic cations.

The uptake of core-modified porphyrin **9h** (without a tethered group) was 0.4 ± 0.1 femtomole/cell, when delivered at $10\,\mu M$ to Colo-26 cells.[39] Rhodamine-tethered derivative **17** gave a cellular uptake of 3.2 ± 0.4 femtomole/cell in Colo-26 cells, which was 8-fold greater than the uptake of **9h**. Triphenylphosphonium derivative **18** gave a cellular uptake of 6.2 ± 1.1 femtomole/cell in Colo-26 cells, which represents a 15-fold increase relative to **9h**. The conjugates remained phototoxic *in vitro*. Future studies may examine *in vivo* efficacy of the bifunctional materials in PDT.

References

1. T. J. Dougherty, W. R. Potter and K. R. Weishaupt, *Prog. Clin. Biol. Res.* **170**, 301 (1984).
2. E. J. Dennis, G. C. Dolmans, D. Fukumura and R. K. Jain, *Nat. Rev. Cancer* **3**, 380 (2003).
3. W. M. Sharman, C. M. Allen and J. E. van Lier, *Drug Discov. Today* **4**, 507 (1999).
4. D. A. Bellnier and T. J. Dougherty, *J. Clin. Laser Med. Surg.* **14**, 311 (1996).
5. M. J. Broadhurst, R. Grigg and A. W. Johnson, *J. Chem. Soc. D*, 1480 (1969).
6. M. J. Broadhurst and R. Grigg, *J. Chem. Soc. C*, 3681 (1971).
7. L. Latos-Grazynski, J. Lisowski, M. M. Olmstead and A. L. Balch, *Inorg. Chem.* **28**, 1183 (1989).
8. L. Latos-Grazynski, E. Pacholska, P. J. Chmielewski, M. M. Olmstead and A. L. Balch, *Angew. Chem. Int. Ed. Engl.* **34**, 2252 (1995).
9. L. Latos-Grazynski, E. Pacholska, P. J. Chmielewski, M. M. Olmstead and A. L. Balch, *Inorg. Chem.* **35**, 566 (1996).
10. P. Ziolkowski, J. Milach, K. Symonowicz, P. Chmielewski, L. Latos-Grazynski and E. Marcinkowska, *Tumori* **81**, 364 (1995).
11. E. Marcinkowska, P. Ziolkowski, E. Pacholska, L. Latos-Grazynski, P. Chmielewski and C. Z. Radzikowski, *Anticancer Res.* **17**, 3313 (1997).
12. D. G. Hilmey, M. Abe, M. I. Nelen, C. E. Stilts, G. A. Baker, S. N. Baker, F. V. Bright, S. R. Davies, S. O. Gollnick, A. R. Oseroff, S. L. Gibson, R. Hilf and M. R. Detty, *J. Med. Chem.* **45**, 449 (2002).
13. A. D. Adler, F. R. Longo, J. D. Finarelli, J. Goldmacher, J. Assoub and L. Korsakoff, *J. Org. Chem.* **32**, 476 (1967).
14. M. Abe, D. G. Hilmey, C. E. Stilts, D. K. Sukumaran, M. R. Detty, *Organometallics* **21**, 2986 (2002).
15. P.-Y. Heo, K. Shin and C.-H. Lee, *Tetrahedron Lett.* **37**, 197 (1996).
16. P.-Y. Heo, K. Shin and C.-H. Lee, *Bull. Korean Chem. Soc.* **17**, 515 (1996).
17. C.-H. Lee, J.-Y. Park and H.-J. Kim, *Bull. Korean Chem. Soc.* **21**, 97 (2000).
18. A. Ulman and J. Manassen, *J. Chem. Soc. Perkin Trans.* **4**, 1066 (1979).
19. Y. You, S. L. Gibson, R. Hilf, S. R. Davies, A. R. Oseroff, I. Roy, T. Y. Ohulchanskyy, E. J. Bergey and M. R. Detty, *J. Med. Chem.* **46**, 3734 (2003).
20. A. Ulman, J. Manassen, F. Frolow and D. Rabinovich, *Tetrahedron Lett.* **1978**, 1885 (1978).
21. B. Sathyamoorthy, A. Axelrod, V. Farwell, S. M. Bennett, B. D. Calitree, J. B. Benedict, D. K. Sukumaran and M. R. Detty, *Organometallics* **29**, 3431 (2010).
22. F. Wilkinson, W. P. Helman and A. B. Ross, *J. Phys. Chem. Ref. Data* **22**, 113 (1993).

23. E. C. Spoelstra, H. Dekker, G. J. Schuurhuis, H. J. Broxterman and J. Lankelma, *Biochem. Pharmacol.* **41**, 349 (1991).
24. P. Ziolkowski, K. Symonowicz, P. Chmielewski, L. Latos-Grazynski, G. Streckyte, R. Rotomskis and J. Rabczynski, *J. Cancer Res. Clin. Oncol.* **125**, 563 (1999).
25. C. E. Stilts, M. I. Nelen, D. G. Hilmey, S. R. Davies, S. O. Gollnick, A. R. Oseroff, S. L. Gibson, R. Hilf and M. R. Detty, *J. Med. Chem.* **43**, 2403 (2000).
26. Y. You, S. L. Gibson, R. Hilf, S. R. Davies, A. R. Oseroff, I. Roy, T. Y. Ohulchanskyy, E. J. Bergey and M. R. Detty, *J. Med. Chem.* **46**, 3734 (2003).
27. Y. You, S. L. Gibson, R. Hilf and M. R. Detty, *Biorg. Med. Chem.* **13**, 2235 (2005).
28. Y. You, S. L. Gibson and M. R. Detty, *Biorg. Med. Chem.* **13**, 5968 (2005).
29. E. Pacholska, L. Latos-Grażyński and Z. Ciunik, *Angew. Chem. Int. Ed.* **40**, 4466 (2001).
30. Detty, M. R. unpublished data.
31. R. Minnes, H. Weitman, Y. You, M. R. Detty and B. Ehrenberg, *J. Phys. Chem. B* **112**, 3268 (2008).
32. W. R. Potter, B. W. Henderson, D. A. Bellnier, R. K. Pandey, L. A. Vaughan, K. R. Weishaupt and T. J. Dougherty, *Photochem. Photobiol.* **70**, 781 (1999).
33. B. Paul, P. Rajaputra and Y. You, *Photochem. Photobiol.* **87**, 1468 (2011).
34. T. J. Dougherty, C. J. Gomer, B. W. Henderson, G. Jori, D. Kessel, M. Korbelik, J. Moan and Q. Peng, *J. Natl. Cancer Inst.* **90**, 889 (1998).
35. B. W. Henderson and T. J. Dougherty, *Photochem. Photobiol.* **55**, 145 (1992).
36. M. R. Detty, S. L. Gibson and S. Wagner, *J. Med. Chem.* **47**, 3897 (2004).
37. C. W. Hawkins, D. R. Bickers, H. Mukhtar and C. A. Elmets, *J. Invest. Dermatol.* **86**, 638 (1986).
38. D. A. Bellnier and T. J. Dougherty, *Photochem. Photobiol.* **49**, 369 (1989).
39. R. Rajaputra, G. Nkepang, R. Watley and Y. You, *Bioorg. Med. Chem.* **21**, 379 (2013).

Chapter 5

Phthalocyanines in Cancer-Imaging and Therapy

Hasrat Ali and Johan E. van Lier

Department of Nuclear Medicine and Radiobiology,
Faculty of Medicine and Health Sciences, Université de Sherbrooke,
3001, 12th Avenue North, Sherbrooke (QC) J1H 5N4, Canada
johan.e.vanlier@usherbrooke.ca

Abstract

Photodynamic therapy (PDT) is a form of phototherapy using nontoxic, light-sensitive compounds that upon exposure to light generate toxic, reactive oxygen species (ROS) targeting malignant and other diseased cells (phototoxicity). Most modern PDT applications involve three key components: A photosensitizer (PS), a light source and molecular oxygen. PDT is used clinically to treat a wide range of medical conditions, including wet age-related macular degeneration and various cancers, and it is recognised as a treatment strategy that is both minimally invasive and minimally toxic. Phthalocyanines (Ps) and its derivatives are structurally related to porphyrins. They have been advanced as PS for PDT due to their strong absorption in the visible region (strong *Q* band bands around 670 nm) and high quantum yield for the generation of ROS. To optimize PDT efficacy for selected applications their lipophilic and hydrophilic properties can be modulated by modifying peripheral substituants and the central metal ion. Furthermore their target specificity can be improved through conjugation to biomolecules. In this chapter we address potential structure-activity relationships of phthalocyanine-based PS and review their clinical potential.

1. Introduction

Cancer (malignant tumors) is a disease involving abnormal cell growth with the potential to invade or spread to other parts of the body.[1] There are over 100 different known cancers that affect humans.[2] Accordingly to WHO report 2012 about 14.1 million new cancer cases per year occur globally,[3] being responsible for about 8.2 million, or 14.6%, of all human deaths.[3,4] Most cancers are initially diagnosed because of their appearance, secondary symptoms or through screening. Neither of these observations are conclusive, a definitive diagnosis requires examination of a tissue biopsy sample by a pathologist. Additional medical tests commonly include blood tests, x-rays, CT scans, MRI, PET and endoscopy. The principal treatment options include surgery, chemotherapy, radiation therapy, and while in advanced disease cases palliative care is often the only alternative. Selective treatments depend on the type, location and stage of the cancer as well as the person's health and personal wishes. Complementary and alternative cancer treatments comprise a diverse group of health care approaches, practices, and products that are not part of conventional medicine.[5] Most alternative treatments and medications for cancer have not been rigorously studied or tested. Experimental cancer treatment protocols are studied in clinical trials and compared to the best existing treatment. They may include entirely new treatments, or they may concern treatments that have shown promise in one type of cancer, and are now being tested for their effectiveness in another type of cancer.[6] Increasingly such treatments are developed alongside diagnostic tests to identify the right drugs for the patient, based on his/hers individual biology.[7] The improved understanding of molecular biology and cellular biology of cancer has led to a number of new treatments. In this chapter we focus on photodynamic therapy (PDT) as treatment modality, emphasizing phthalocyanines (Pc) as photosensitizers (PS), as well as their applications in imaging.

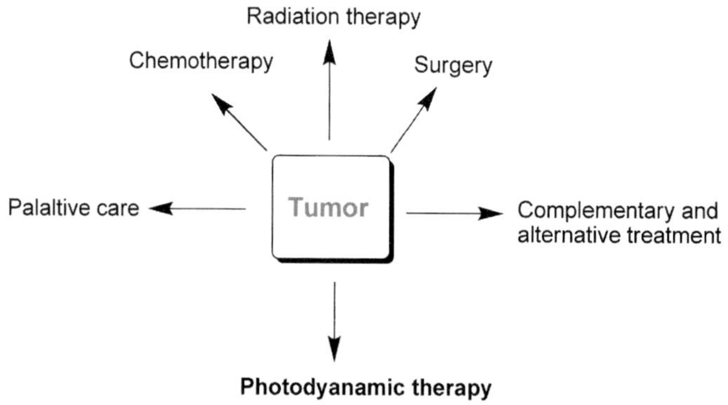

2. PDT

PDT requires a PS (a molecule that absorbs the photon), oxygen and light. Photosensitization can occur within living cells or tissues (photobiology) or in pure chemical system (photochemistry). In photobiology, the excited PS may react with adjacent biomolecules or with triplet ground state molecular oxygen, to initiate biological damage via a Type I or Type II reaction.

In the Type I reaction, the excited PS reacts directly with the substrate, in a one-electron or hydrogen transfer reaction, to produce radical ions or free radicals. Type I processes can be sub-divided into two pathways. (A) The transfer of an electron (oxidation) from a substrate to the excited PS*, generating radical anion $PS^{\bullet -}$ and a substrate radical cation. The majority of the radicals react instantaneously with oxygen, generating a superoxide radical anion ($O_2^{\bullet -}$) and a complex mixture of reactive oxygen species (ROS) including the highly reactive hydroxyl radical ($OH^{\bullet}$). The ROS initiate oxidative chain reactions producing a cascade of cytotoxic free radical species, a process common in the oxidative damage of fatty acids and other lipids. (B) The transfer of a hydrogen atom (reduction) to the excited PS* generating free radicals that rapidly react with molecular oxygen, creating likewise ROS initiating further reactions with biomolecules. In the Type II reaction, the excited sensitizer transfers its excess energy to ground-state molecular oxygen ($^3O_2$), producing excited singlet oxygen ($^1O_2$), while regenerating the ground-state sensitizer. Singlet oxygen is a highly ROS that has an excited state lifetime of a few microseconds in most biological environments, e.g., 3–4.5 μs in water. The singlet oxygen produced in the Type-II process rapidly reacts with a variety of electron-rich substrates including cholesterol,[8,9] unsaturated fatty acids in lipid layers of membrane structures, amino acid residues such as cysteine, histidine, and tryptophan of protein structures as well as nucleic acid bases of DNA, particularly guanine and thymine. Characteristic differences in products obtained in Type I or Type II reactions from cholesterol and 2′-deoxyguanosine allow for the detection of reaction pathways in biological systems.[10,11]

2.1. *Spectral Properties*

The electronic spectra of porphyrins can be separated in two parts, (A) visible and (B) UV (near UV). The electronic heart of the structure is the inner 16-membered conjugated carbon ring with 18-π electrons. Generally, one very intense band is observed in the near-UV region of the spectrum, the Soret band. Metallation results, in most cases, in the collapse of the four-band spectrum to yield two absorption bands in the visible region between 500 and 600 nm, while an intense Soret band between 389 and 420 nm usually remains unaltered.[12] Replacement of four meso-carbon with nitrogen atom generates a tetra-aza-porphyrin (TAP). Benzo-, naphtha- and anthraco-ring fused TAP generate Pc, naphthalocyanine (NPc) and anthracyanine (Ac), respectively.[13] Metal free Pc exhibit a clear split Q-band, while monomeric MPc show a sharp, intense Q-band at 670 nm, with a molar extinction coefficients in the range of 10^5 M^{-1} cm^{-1}. A weak absorption Soret band at ~350 nm arises from a strong electronic transition from the ground state to the second excited singlet state ($S_0 \rightarrow S_2$), whereas the *Q* bands result from a weak transition to the first excited singlet state ($S_0 \rightarrow S_1$). All absorption bands are interpreted as resulting from π-π^* interactions. Further expansion of the Pc macrocycle with benzo rings on each of the *iso*-indole subunits gives the NPc macrocycle, with strong, red shifted *Q*-band absorption maxima around 740–780 nm. The position of the *Q* band also depends on the nature of the substituents, their position and the solvent.

3. Ps

One of the earliest reports[14] on Pcs dates back to 1907 and reports on an unidentified, dark insoluble material, which was subsequently identified as metal-free Pc. Similarly, in 1927, Swiss researchers[15] accidentally obtained an exceptionally stable blue material during the reaction of 1, 2-dibromobenzene with CuCN. The product was later characterized as a mixture of metal free and CuPc. The authors remarked on the high stability of these complexes but did not provide further characterization. The name for this class of compounds was cornered by Sir Reginald Linstead in the early 1930's during his pioneering work on the subject. Based on their intensely blue-green colour he proposed the name phthalocyanine (Pc) derived from the Greek words "naphtha" for rock oil and "cyanine" for dark blue. In 1933, Linstead *et al.*[16] were the first to determine their structures, and subsequently reported on the formation of several MPc complexes. Pc are closely related to other tetrapyrrole macrocycles, including the naturally occurring porphyrins and the synthetic porphyrazines (Pz) (Fig. 1).

A general method for the synthesis of MPc involves a straightforward, one step condensation reaction from *ortho*-dicarboxylic acid or their various derivatives,[17] which remains the most widely used methodology (Scheme 1). However, using these reaction conditions the isolation of asymmetrically substituted Pc is tedious and difficult. An alternative route for their synthesis involves the use of sub-Pc as intermediates. The serendipitous discovery of sub-Pcs was made in 1972 by Meller and Ossko.[18] The most important application of sub-Pc involves their use as intermediates in the "Kobayashi" ring-opening reaction.[19] Treatment of a tri-*t*-butyl sub-Pc with succinimide

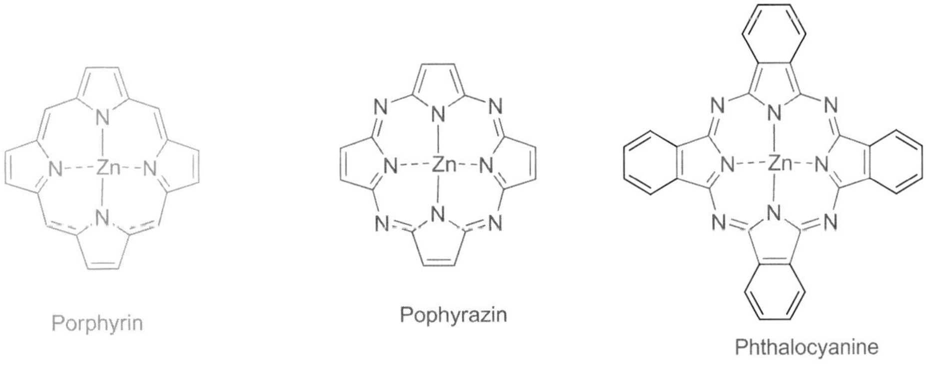

Fig. 1. Tetrapyrrole Macrocycle Derviatives.

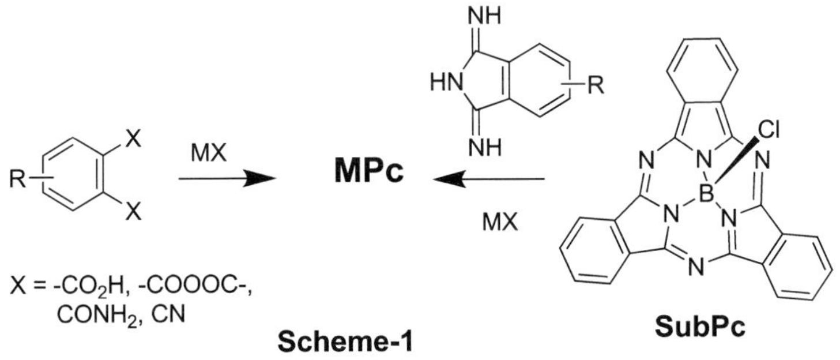

Scheme 1. General scheme for the preparation of Metallo phthalocyanines.

or diiminoisoindoline gave substituted Pc and this reaction has widely been adopted to prepare a variety of asymmetric Pc (Scheme 1).[20]

The Pc macrocycle is a hydrophobic ring system and the intrinsic structure of non-substituted Pc render them insoluble in nearly all solvents. Appropriate substitutions can remediate this problem. Water-solubility is a central theme for the development of novel Pc for biological applications. To this end various substitution pattern at the macrocyclic ring, nature of the central metal ion as well as the introduction of axial ligands have been advanced. Bio-compatible modifications often require attachment of hydrophilic moieties, preferably resulting in an amphiphilic structure. In addition to the overall amphiphilic balance, other parameters such as tendency to aggregate, nature of the aqueous medium in terms of pH,[21] ionic strength and presence of a co-solvent, further contribute to the solubility of the Pc.[22] PD activities of Pc and their analogs correlate well with their tissue retention and cellular distribution pattern, which in turn are guided by overall charge and solubility of the Pcs. We will review the biomedical application of Pc accordingly to charge (anionic, cationic and non-charged) and conjugation to other molecules.

3.1. *Anionic Ps*

To overcome the Pc solubility problem, one approach involves introduction of anionic substituents at the periphery of the ring structure. Commonly used anionic substituents are sulfonate, phosphorus or carboxylate-based functions that are attached directly, or through various spacers onto the macrocycle. The degree of water-solubility depends on the number of anionic substituents, the nature of the peripheral ring substituents, the central metal ion, as well as the pH of the medium. Most of the anionic Pc considered for PDT constitute substitutions with sulfonate or sulfonic acid groups.

Two different pathways were reported for the synthesis of sulfonated Pc (PcS): (A) Cyclotetramerization of the monosodium salt of 4-sulfopthalic acid, alone or mixed with other phthalonitrile derivatives, in the presence of metal salt, and (B) direct sulfonation of the MPc macrocycle in oleum.[23] Method A gives complex mixtures of isomers and differently sulfonated products. The known yield enhancing effect of microwave[24] was successfully applied for the synthesis of Pc.[25] To improve the yield of the conventional condensation procedure, indole-protected sulfophthalonitrile was also used.[26] Method B furnishes similar reaction mixture as method A, with the addition of sulfonation products at either 3- and 4-positions of the peripheral phenyl ring, which further contributes to the high complexity of the reaction mixtures. Controlling product composition and ratios between the various sulfonation products remains tedious, being affected by temperature, reaction time, as well as the nature of the central metal ion. The products can be separated by reversed phase HPLC and resolved isomers have been studied in various bio-assays to establish structure-PD activity relationships. Both *in vitro* and *in vivo* evaluation of HPLC-purified $MPcS_n$ (**1**) revealed that fractions enriched in amphiphilic, adjacently disulfonated $MPcS_{2adj}$ showed the highest PDT activity.[27] Depending on the degree of sulfonation, $MPcS_n$ exhibit varying hydrophobic and hydrophilic properties, affecting individual determinants for PDT efficacy, including tendency to aggregate, capacity to generate $^1O_2$, pharmacokinetics, cell uptake and intracellular distribution pattern. Hydrophobic PS tend to aggregate in aqueous solution, resulting in decreased capacity for $^1O_2$ production, while hydrophilic PS display decreased cell uptake, localizing at less photosensitive subcellular sites, and rapid clearance from the blood pool, and excretion.

It is generally assumed that the presence of a central metal ion is essential for the PDT efficacy of the Pc. However Feofanov *et al.*[28] demonstrated that the metal-free H_2PcS_n possess considerable PD activity, including the metal-free disulfonated fraction. A series of amphiphilic derivatives of $AlPcS_4$ was prepared by substituting one of the four sulfonate groups with aliphatic side chains of 4–16 carbon atoms.[29] Substitution with long aliphatic chains greatly enhances PD efficacy both *in vitro* and *in vivo*. Since $^1O_2$ yields were similar for the four substituted $AlPcS_4$ the increased cytotoxic activity likely reflects enhanced subcellular localization facilitated by the long aliphatic side chains.

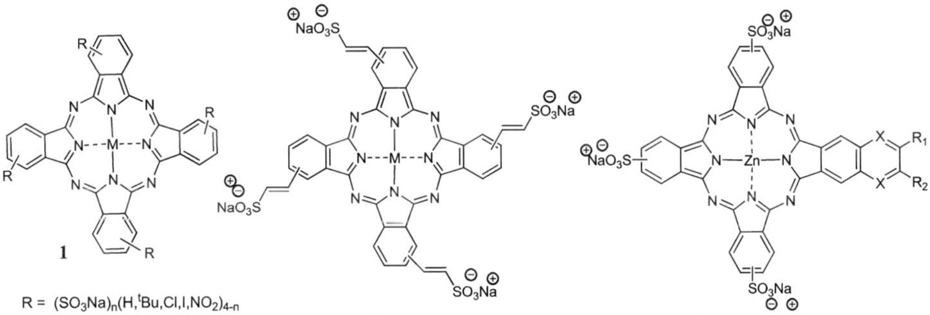

1

R = $(SO_3Na)_n(H,{}^tBu,Cl,I,NO_2)_{4-n}$

M = Zn, Ga^{3+}, Al^{3+}

2a: M = Zn
2b: M = AlOH
2c: M = GaOH

3a: X = CH; R_1 = ^{t}Bu; R_2 = H
3b: X = CH; R_1 = R_2 = ^{t}Bu
3c: X = N; R_1 = R_2 = Ph-4-^{t}Bu

The enhanced $AlPcS_4$-PDT effect on HEP-2 cells induced by quercetin (QCT), used as a supplement, has also been studied.[30] At concentrations of QCT >50 mM the augmented PDT-induced cytotoxicity was attributed to changes in the cell membrane potential, which significantly increased apoptosis and/or necrosis and cell mortality.

Two-photon excitation, originally predicted by the Nobel Laureate Maria Goeppert-Mayer in the 1930s,[31] is a non-linear optical process in which a molecule is promoted to an excited state by simultaneous absorption of two photons. Two-photon activated PDT, denoted here as 2-γ PDT, has also been explored since 1990, and has several potential advantages over conventional one-photon (1-γ) PDT using (quasi) continuous-wave (CW) illumination.[32] The feasibility to induce oxygen-independent tumour cell kill by sequential 2-γ excitation of $CuPcS_4$ was studied in Jurkat cells *in vitro*.[33] Cell viability decreased 10–20% following 532 nm two-photon excitation while 680 nm one-photon illumination did not affect cell survival. These data suggested that two-photon excitation of $CuPcS_4$ to the upper excited triplet state results in the formation of toxic species. Another study that indicated that sequential 2-γ activated $CuPcS_4$ is capable of inactivating biological active molecules involved studies with acetylcholinesterase (ACE).[34] Simultaneous irradiation of ACE in the presence of $CuPcS_4$ with 514 nm and 670 nm continuous wave (CW) light induced a 20–50% increase in enzyme inhibition as compared to 1-γ irradiation. The enzyme activity was not affected by $CuPcS_4$ or light alone, decreased linearly with the irradiation time, but was shown to be oxygen-dependent. The authors concluded that the photoinactivation of ACE involved ROS produced by the interaction of the upper excited T_n state of $CuPcS_4$ with molecular oxygen. The efficiency of Pc dyes for simultaneous 2-γ absorption was estimated by measurement of 2-γ absorption cross sections of a number of metallated Pc, using a femtosecond laser.[35] Although the MPc dyes showed up to 10-fold higher 2-γ absorption as compared to that obtained for Photofrin, the values indicate the requirement of unacceptable high light doses for their excitation in a clinical setting.

To address the effect of the site of periphery substitution on the PDT potency of MPc, a series of *ortho*- and *meta*-tetravinylsulfonated derivatives MPc-*o*/*m*-$(VS)_4$ (**2**), were prepared.[36] Vinyl sulfonate groups were introduced via coupling of $MPcI_4$ (M= Zn, AlOH or GaOH) with sodium vinyl sulfonate, using the Pd-catalyzed Heck cross-coupling reaction in aqueous medium. Compared to the analogous $MPcS_4$, the tetravinyl-MPc$(VS)_4$ derivatives are more hydrophobic and their *Q*-band absorption maxima are red-shifted. *Ortho*-substituted MPc-*o*-$(VS)_4$ are more photocytotoxic as compared to their corresponding *meta* analogs. Intracellular localization at both mitochondrial and lysosomal membranes appeared to be the major factor contributing to the augmented PDT activity. Using modified sulfonation conditions to introduce at the same time phthalimidomethyl and sulfonate groups lead to the formation of comparable amphiphilic Pc.[37] In this approach non-substituted Pc was stirred in oleum in the presents of phthalimide *para*-formaldehyde. The resulting di-sulfonato-di-phthalimidomethyl AlPc exhibited enhanced hydrophobicity.[38] PDT in a mouse tumor model in conjunction with vaccination suggested its potential for the treatment of lymphoma.[39] *In vitro*, $ZnPcS_{mix}$ mediated PDT was shown to induce cell death via an apoptotic pathway.[40]

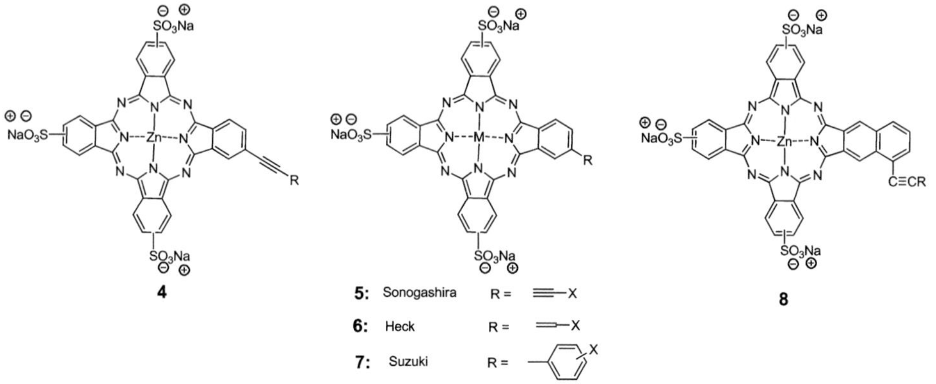

The promising properties of sulfonated Pc for clinical PDT prompted a search for alternative synthetic approaches to yield well-characterized, single isomeric products. To this end the ring expansion of boron sub-Pc using the pyridinium salt of tris(4-chlorosulfonyl) sub-PcB(Br) was evaluated.[41] The ring expansion with diiminoisoindoline derivatives yielded the corresponding H_2PcS_3, which was subsequently metaled to yield the corresponding $ZnPcS_3$ (**3**). PD inactivation of EMT-6 tumor cells showed that adding a (*t*-Bu)benzo or (*t*-Bu)naphtho group to $ZnPcS_3$ increased the PDT efficacy, whereas addition of a fourth sulfonatobenzo (i.e. $ZnPcS_4$) or bulky diphenylpyrazino group decreased the activity of the parent $ZnPcS_3$.[42] The use of quantum dots (QDs) to improve PDT efficacy was also evaluated. QDs are semiconductor nanocrystals and photoluminescent QDs were envisaged to be capable of indirect activation of the PS through energy transfer. Thiol stabilized CdTe QDs synthesized in aqueous phase were used as energy donors to $AlPcS_4$ through fluorescence resonance energy transfer (FRET). The PS in the presence of the QD revealed a high triplet state yield, hence confirming the possibility to use QDs in combination with Pcs for PDT. Likewise, 3-mercaptopropionic acid capped QD mixed with $AlPcS_4$ resulted in long triplet lifetime and high triplet yield of the PS, and high energy transfer efficiency, hence was proposed for further studies on its potential for PDT of cancer.[43] Water-soluble $ZnPcS_4$ encapsulated into virus-like particles (VLPs) gave a stable Pc-loaded protein nanocontainer with potential use as a PS/vehicle system for PDT.[44]

An alternative method for the preparation of amphiphilic $MPcS_3$ derivatives involves cyclotetramerisation of indole-protected sulfophthalonitrile with iodophthalonitrile in the presence of Zn salts to yield asymmetric Pc bearing three indolylsulfonates and one iodo substituent.[45] The products are easily purified over silica gel providing a versatile precursor for the synthesis of a variety of PcS_3 derivatives via Pd-catalysed coupling reactions. Protected indolyl groups can readily be removed by lithium methylate. In this manner a series of alkynyl-substituted PcS_3 were prepared (**4**). Correlating amphiphilic properties to different biological PD parameters allowed for the selection of $MPcS_3$ with optimal structural features for PDT. Varying the length of the substituted alkynyl side-chain modulates the hydrophobic/hydrophilic properties of the dyes and a linear relationship between *n*-octanol/water partition coefficients and retention times on reversed-phase HPLC were established. Aggregate formation of the dyes in aqueous solution increased hydrophobicity while monomer formation was favored by the addition of serum proteins or organic solvent. $ZnPcS_3$ bearing hexynyl and nonynyl substituents exhibited high cellular uptake with strong localization at the mitochondrial membranes, which coincided with effective photocytotoxicity toward tumor cells in culture. Further increase in the length of the alkynyl chains (dodecynyl, hexadecynyl) did not improve phototoxicity, most likely due to extensive aggregation and reduced cell uptake in aqueous medium. Aggregation is evident from blue-shifts in the electronic spectra and reduced capacity to generate singlet oxygen. These studies indicated that amphiphilic properties facilitate cell uptake and intracellular localization at photosensitive sites, particularly mitochondria.[46] Without protection of the sulfo group, the sodium salt of 4-sulfophthalic acid and 4-iodophthalonitrile in presence metal salt gave a water-soluble asymmetric $MPcS_3I$ that served as a precursor for Pd-catalyzed cross-coupling reactions (Heck, Suzuki and Sonogashira) to produce amphiphilic water-soluble Pc (**5**-**7**).[47] AB_3 type asymmetric Pc bearing three sulfonate groups and one *tert*-butyl were tested for their anti-viral activity. Relative activities increased with the nature of the central metal ion in the following order: Ga(III) < Al(III) < Zn(II), and varied inversely with the degree of sulfonation. $AlPcS_3$(*t*-Bu) exhibited the most favorable anti-viral properties combined with a low toxicity index.[48]

A series of water-soluble, mononaphtho-trisulfobenzo-Pc (**8**) bearing an alkynyl side chain of varying lengths on the naphthalo ring, were prepared and tested for their efficacy to inhibit plasma extravasation of the retina in the rat.[49] As observed with the analogous $ZnPcS_3$ series, the hexynyl substituted PS was the most potent, capable inducing complete inhibition of extravasation at low PS and light doses, rendering this analog of particular interest as a PS for PDT of retinal and sub-retinal edema.

An alternative route for the synthesis of monosulfonated Pc involves substitution of a monoamino Pc, using the Meerwein reaction.[50] In this procedure the monodiazonium salt of 11,18,25,tri(*tert*-butyl)-4-diaza-Pc is treated with SO_2 in AcOH to yield the monosulfonated analog as a major product, along with a chlorinated derivative. Introduction of fluorine atoms to biologically active compounds can modulate their physical–chemical and pharmacological properties, and the potential applications of fluorinated PS in biology and medicine has been reviewed.[51] An attempt to prepare the 4-aminododecafluoro ZnPc in a similar manner failed. However protecting the amino group as an acetamido derivative gave the pure, monosulfo $ZnPcF_{12}S_1$ (**9**) in good yield. Compared to the $ZnPcF_{16}$, the monosulfonated analog **9** showed improved pharmacokinetics in a mouse tumor model with lower liver and spleen retention and higher tumor-to-non-target ratios.[52]

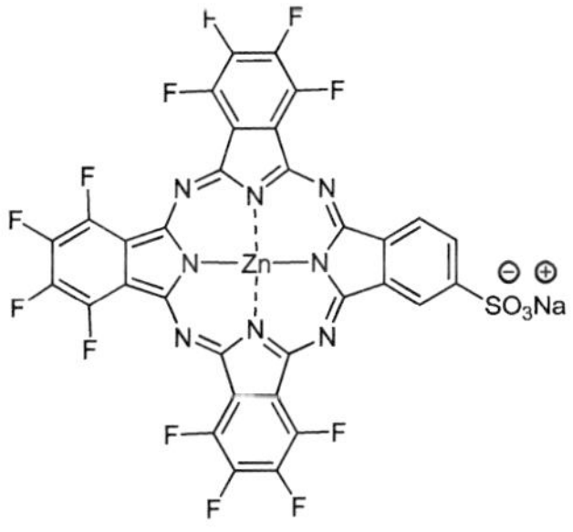

9

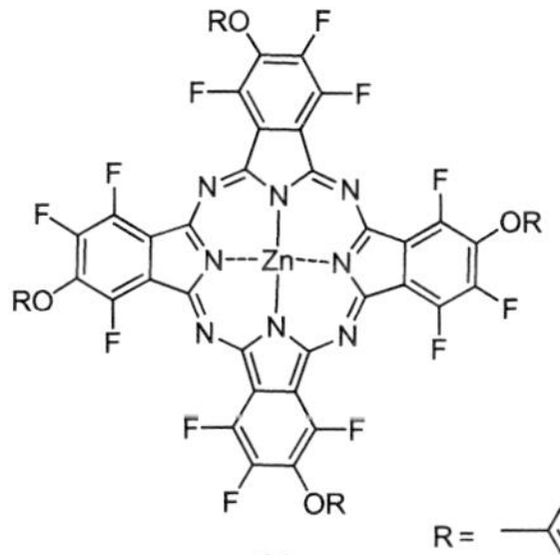

10

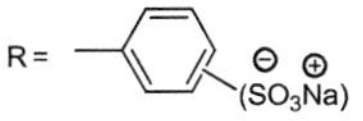

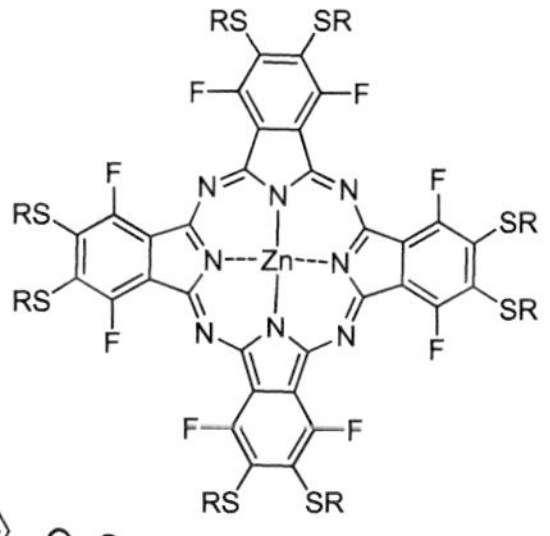

11

Other analogues of $ZnPcF_{16}$ that have been advanced include a water-soluble fluorinated ZnPc bearing different numbers of additional sulfophenyl substituents (**10** and **11**) resulting in improved water-solubility and increased *in vitro* PD activity as compared to $ZnPcS_4$.[53] A new hexadeca-substituted ZnPc (**12**) bearing hexyloxy groups in the 3,6-positions and *p*-sulfonylphenoxy and chloro groups in the 4 and 5-positions showed good phototoxicity against HepG2 cells.[54] Comparison of ZnPc bearing either four trifluoromethylbenzyloxy or methoxy substituents showed that fluorinated derivatives exhibited higher photocytotoxicity than the methoxy analogs.[55]

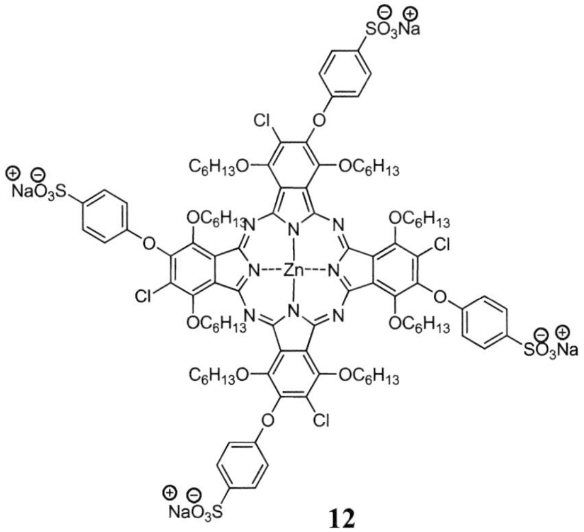

12

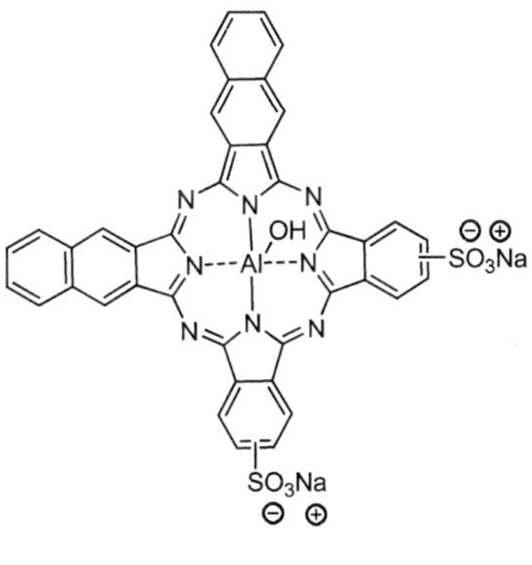

13

Pc analogs incorporating both phthalo- and naphthalo-moieties offer a wide choice of red-shifted excitation wavelengths.[56] Such water soluble naphtho-sulfobenzo-porphyrazines (M-NSBP) were formed by mixed condensation of the monosodium salt of 4-sulfophthalic acid, naphthalene-2,3-dicarboxylic acid together with a metal salt.[57] The presence of both sulfobenzo- and naphthalo-groups on the same macrocycle enhanced PDT efficacy, with the amphiphilic disulfonated AlOH-NSBP (**13**) showing the best tumor response.[58]

Pc functionalized with phosphorus include tetrabenzotriazacorrole (TBC) derivatives that comprise a novel species to this family of macrocyclic compounds, featuring a dihydroxy phosphorus(V) with a non-planar structure.[59] Sulfonated dihydroxyphosphorus(V) (**14**) was shown to act at the DNA level inducing photocleavage.[60] Phosphorus-substituted Pc bearing either phosphate, phosphonate or phosphinate substituents have also been reported as potential PDT agents (**15–17**). Their structures are particularly useful for comparison of basic properties with those of their homologous derivatives.[61] The synthesis of a Pc phosphate derivative featuring butyl spacer chains gave a tetra[1-(*O*-ethylphosphatobutyl)]tetrasodium ZnPc salt (**17**) as a major product, along with products resulting from hydrolysis of the butoxy-phosphorus bonds (with one/three *O*-ethylphosphato groups).[62] The *in vivo* PDT activity of **17** is comparable to that of the tetra(hydroxypropyl)ZnPc, but the former has the advantage of being water soluble, facilitating its formulation.[63]

A series of ZnPc tetrasubstituted with carboxy groups at α- or β-positions were synthesized and the 4-carboxylphenoxy derivative showed the highest PD activity.[64] Two hexadeca-carboxyl ZnPc derivatives showed photocytotoxicity against cells of macrophage origin, suggesting an action mechanism involving receptor-mediated endocytosis without the need of bioconjugation to a carrier.[65]

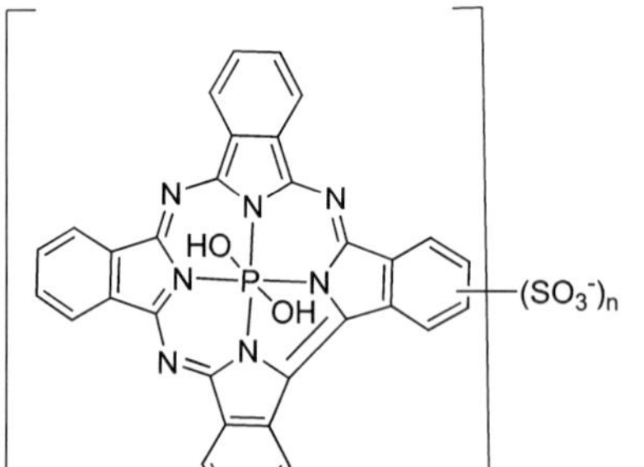

14

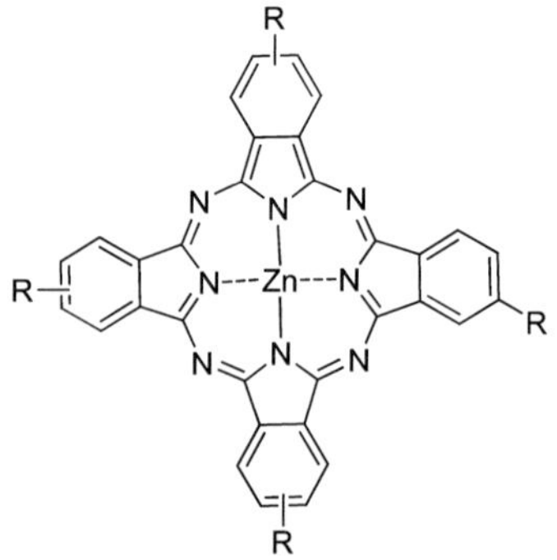

15: R = PO_3H
16: R = $(CH_2)P(O)(Et)(ONa)$
17: R = $(CH_2)_4OP(O)(OEt)(ONa)$

An amphiphilic tetracarboxyoctafluoro ZnPc (**18**) was tested for PD efficiency against HeLa cells and its activity was compared to that of the hydrophilic octacarboxy ZnPc and the hydrophobic hexadecafluoro ZnPc.[66]

Compound **18** showed the highest PD effect, which appeared to correlate to accumulation and monomerization of the PS in the hydrophobic lipid membrane of the HeLa cells.

Conjugation of $ZnPc(COOH)_4$ with QDs gave CdTe@ZnS–GSH and resulted in decrease of both fluorescence quantum yields and excited state lifetimes of the Pc.[67] The interaction of $MPc(COOH)_8$ [M = Al(III) or Co(II)] with BSA was studied and shown to involve binding to specific domains of the protein.[68] An octacarboxyFePc complex mixed with polyurethane (PU) (FePc(OCAP)/polyurethane) showed good singlet oxygen generating and cell inactivating properties.[69] Hydroxy aluminium octacarboxy Pc (Al(OH)OCPc) was conjugated with a diamine Pt complex to give water soluble conjugates containing two or three Pt complexes $Al(OH)OCPc(Pt)_2$ and $Al(OH)OCPc(Pt)_3$ with improved singlet oxygen quantum yields.[70] Covalent conjugation of amine and diamine Pt(II) and $ZnPc(CO_2)_8$ bearing one-four amine Pt and ethylenediamine-N,N′-diacetic acid platinum moieties (**19**) showed increased capacity to generate singlet oxygen and higher photocytotocicity against cancer cells *in vitro*.[71]

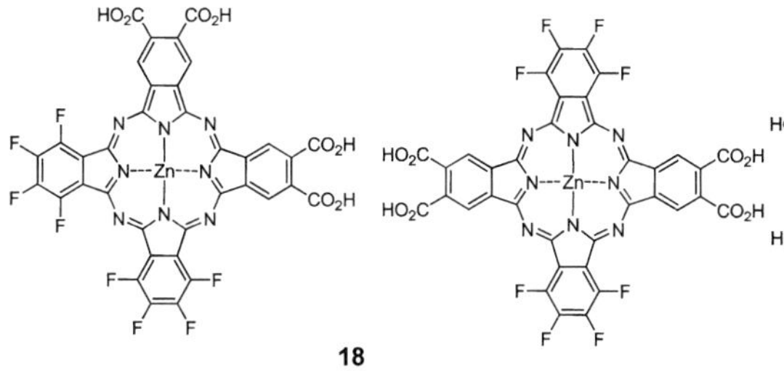

18

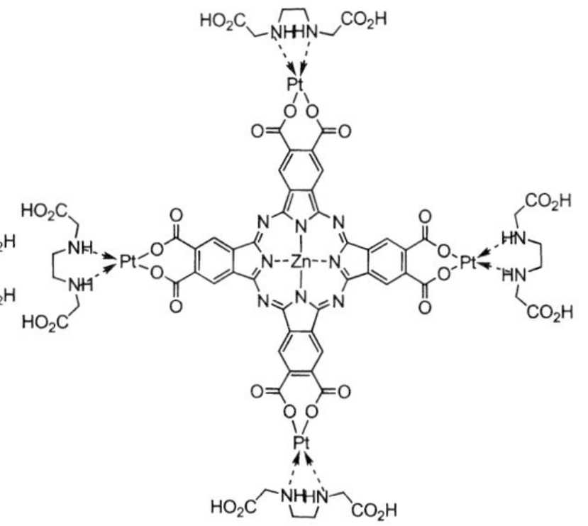

19

3.2. *Cationic Ps*

Quaternization of aliphatic or aromatic nitrogen atoms produces cationic dyes that have been shown to localize in mitochondria and to bind to DNA.[72] On such account both alkyl and aromatic types of cationic quaternized Pc have been prepared and evaluated for their biological activities. The quaternized Pc nitrogen atoms can be part of amines or pyridine derivatives, or can be incorporated in the macrocycle. The quaternization of the Pc is implemented at the end of the synthetic pathway.

The first cationic Pc, bearing quaternized amines attached through an alkyl chain, were reported by Leznoff *et al.*[73] The *in vitro* PD activity of the cationic ZnPc **20** was less than that of the neutral analog. Subsequently, Wohrle *et al.*[74] reported that Pc containing alkyl chains of different lengths attached to the Pc through an oxygen atom (**21**) influenced the hydrophilic versus lipophilic character of the compounds. Positively charged Pc showed higher *in vitro* PD activity than Photofrin II. Compounds with long aliphatic side chains versus methyl groups at the quarternary nitrogen of the pyridine or amino groups exhibited higher *in vitro* PD activity. Peripherally and non-peripherally tetra-2-[2-(dimethylamino)ethoxy]ethoxy substituted ZnPc complexes and their quaternized amphiphilic derivatives (**22**) show excellent solubility in both organic and aqueous solutions, which makes them potential PS for use in PDT.[75] The water-soluble complexes bind strongly to serum albumin; hence they can easily be transported in the blood. Similarly peripherally and non-peripherally, tetra-substituted quaternized 2-diethylaminoethanethiol Pc **23** and peripherally octa-substituted derivatives **24** were also reported and shown to give high triplet quantum yields and long lifetimes in DMSO.[76] In aqueous media strong binding to BSA occurred resulting in low triplet yields and lifetimes, which improved upon addition of Cremophore EL as surfactant. Strong binding to BSA occurred in the order **24** > **23b** > **23a** due to steric effect at the non-peripheral position.

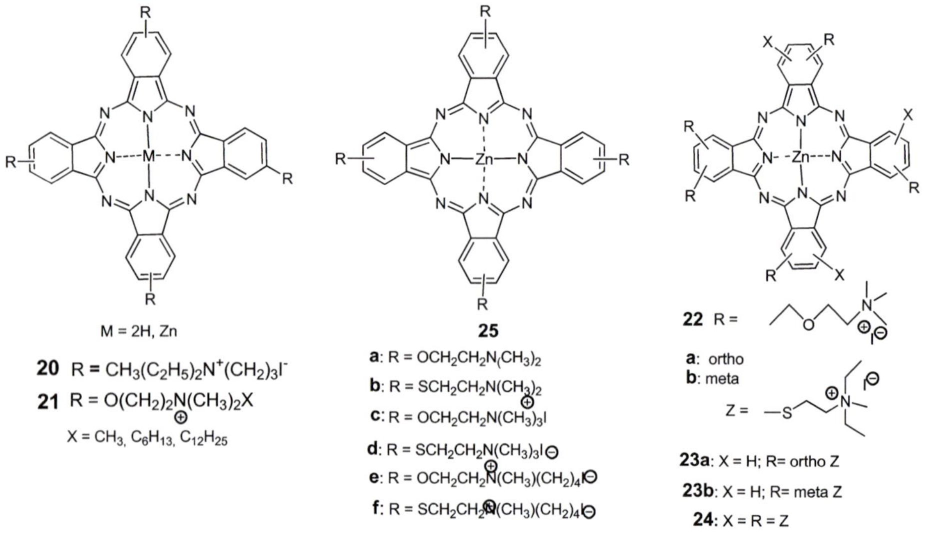

Among four sulfur-linked cationic ZnPc (**25 b,d,f**) compound **25d** was the best PS with a high singlet oxygen quantum yield and bathochromic shift of the Q-band.[77] Increasing the lipophilicity of Pc **25d**, to yield **25e** and **25f** resulted in improved PD activity.[78] These cationic Pc were shown to interact with CT-DNA.[79] Four tetra alkyl-substituted cationic ZnPc featuring different hydrocarbon chains lengths attached to the pyridyloxy group were capable of photoinactivating A. hydrophila. This bacterial strain is characterized as a multidrug-resistant, ubiquitous Gram-negative bacterium that causes disease in mammals and humans. The PD inactivation paralleled the uptake behavior of the PS, such that substitution from methyl- to dodecyl improved the PDT efficacy.[80]

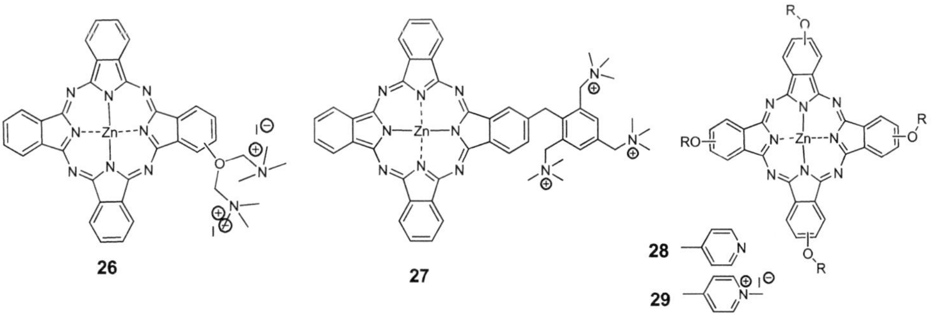

Mono-substituted, amphiphilic ZnPc bearing a 1,3-bis(dimethylamino)-2-propoxy group at the α- or β-position and the corresponding *N*,*N*-dimethylated derivatives (**26**), were shown effective generators of singlet oxygen and good PS for *in vitro* PDT. Affinity for cell membranes was higher for the dicationic derivatives as compared to the non-ionic analogs.[81] A ZnPc substituted with a triamino moiety and its tri-N-methylated analog **27** remained non-aggregated in water containing 0.05% Cremophor EL and effectively generated singlet oxygen.[82] Both PS are highly cytotoxic, particularly against Gram-positive bacterial strains and various viruses such as the enveloped influenza A virus (H1N1) and herpes simplex type 1 (HSV1), while lacking activity towards the non-enveloped viruses such as adenovirus type 3 (Ad3) and coxsackie virus (Cox B1).

The pyridyloxy ZnPc (**28**) and its *N*-methyl cationic derivative (**29**) were compared *in vitro* for their capacity to photoinactivate Gram-negative *Escherichia coli*. Only the cationic dye was shown to be an efficient for the photoinactivation of these bacteria.[83] The quaternized tetra- (non-peripheral) and octa-(peripheral) derivatives of ZnPc substituted with 2-mercaptopyridine are good singlet oxygen producers and strongly bind to blood plasma proteins suggesting their potential as PS for PDT.[84] Peripherally substituted tetra(2-mercaptopyridino)-Pc containing Si, Ga, Sn and In as central metal ions were also reported and their photophysical and photochemical properties suggest their potential for PDT.[85] Two water soluble Pcs, tetramethyl tetrakis-2,(3)-[(4-methyl-2-pyridyloxy)ZnPc, and the negatively charged analog tetrakis-2,(3)-(3-carboxylicacid-6-sulfanylpyridine)-ZnPc, shown good photocytotoxicity against pancreatic carcinoid cells.[86] The binding of cationic Pc to DNA has received ample attention.[87,88] The interaction between calf thymus (CT)-DNA and an octacationic ZnPc showed strong binding to CT-DNA in a non-intercalative mode.[89] Titrimetric investigations have indicated that electrostatic binding between tetra substituted, cationic Pc and anionic DNA phosphate groups occurs to a large extent and that PDT stacking between DNA and Pc moieties occurs subsequently. A water-soluble, quaternized tetra-substituted 4-[20,30,50,60-tetrafluoro-40-(2-dimethylaminoetanethio) benzyloxy] ZnPc (**30**) was shown to interact with calf thymus DNA resulting in a change in the thermal denaturation profile of the DNA.[90] Among tetra- and octa-fluoro thio-pyridinium ZnPc derivatives compound **31a** was shown the most photoactive against E. coli. The dye is non-aggregated in PBS, highly fluorescent and a promising candidate as a fluorescence probe.[91]

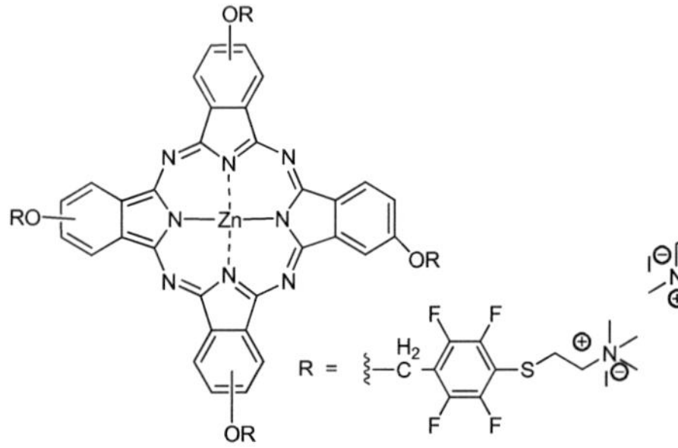

30

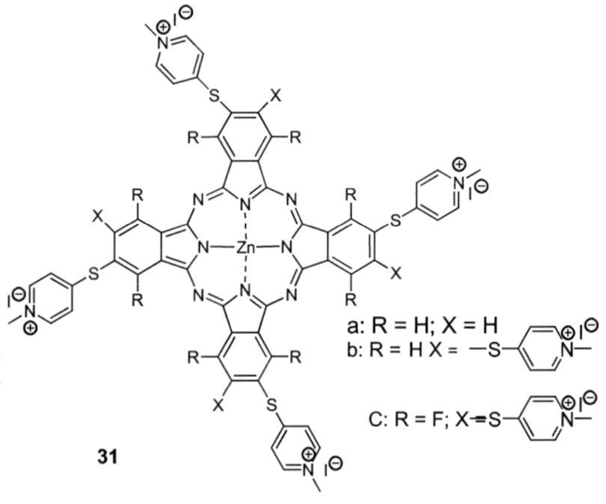

31

The photophysical and photosensitizing properties of two octacationic oxotitanium Pc (TiOPcs), bearing pyridiniomethyl (**32**) or cholinyl (**33**) substituents, showed that both compounds were monomeric in water with high quantum yields for fluorescence and singlet oxygen formation, and photobactericidal potential.[92]

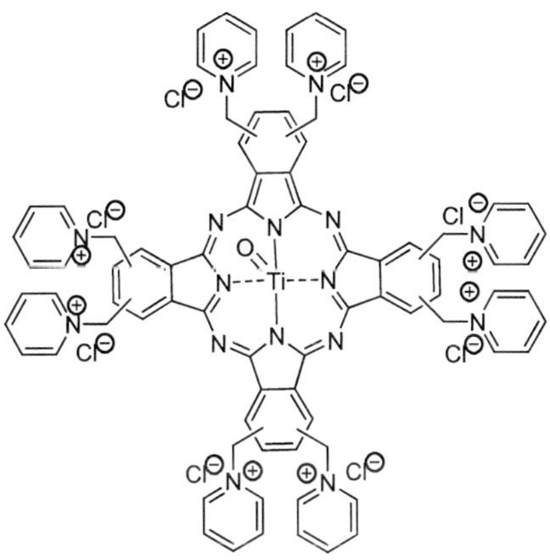

32

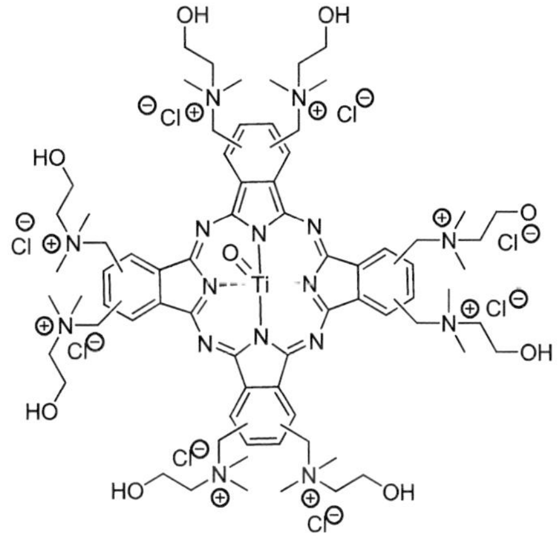

33

A series of cationic pyridyloxy Pcs (**34**) bearing either Zn(II) or Si(IV) coordinated metal ions and methyl or short PEG groups on the pyridyl substituents was reported.[93] The Si(IV)Pcs bearing two bulky axial ligands exist mainly as monomers in both organic and aqueous solutions and show higher fluorescence quantum yields as compared to the ZnPcs. The most hydrophobic Pc **34a** showed the highest uptake by HEp2 cells and best phototoxic activity. Among the SiPcs, the amphiphilic **34b** and **34c** were also shown to be highly phototoxic with preferential lysosomal uptake.

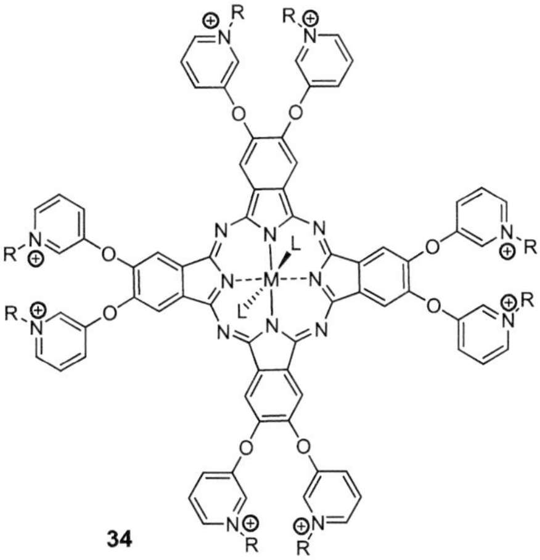

34

a: M = Zn; R = CH_3 or $(CH_2CH_2O)_3CH_3$
b: M = Si; R = CH_3; L = $OSi(CH_2CH_2CH_3)_3$
c: M = Si; R = CH3; L = $OSi[CH_2(CH_3)_2]_3$
d: M = Si; R = CH_3; L = $OSi(C_6H_5)_2C(CH_3)_3$
e: M = Si; R = $(CH_2CH_2O)_3CH_3$; L = $Si(C_6H_5)_2C(CH_3)_3$

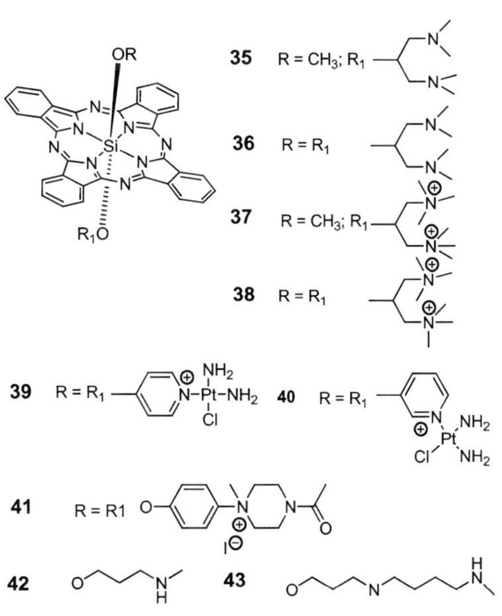

Cationic moieties were attached as axial ligands to the central metal ion of Si(IV)Pc. Among the one and two 1,3-bis(dimethylamino)-2-propoxy substituted **35** and **36**, and their two dicationic and tetracationic analogs **37** and **38**, the unsymmetrical and amphiphilic analog **35** was shown to be highly phototoxic against several cell lines.[94] The interactions of BSA and maleylated BSA with unsymmetrically substituted **35** and **37** and symmetrically substituted **36** and **38** by fluorescence spectroscopy revealed that the unsymmetrically analogs bind stronger to BSA. Non-covalent binding to BSA or mBSA enhanced the photocytotoxicity of **36** and **37** up to eight fold.[95]

The combination of PDT and chemotherapy can lead to synergistic and enhanced anticancer effects. In particular Pt(II)-based chemotherapeutic drugs that target DNA are interesting candidates for conjugation with PS. Two complexes with a SiPc core, bearing two *cis*-$[Pt(NH_3)_2]^{2+}$ units at axial positions on *meta*- (**40**) or *para*- (**39**) substituted pyridine rings, have been reported.[96] The combination of PDT and Pt(II)-based chemotherapy exhibited remarkable photocytoxicity suggesting the potential for DNA-targeting PDT as well as photo-chemotherapy. Bis[4-(4-acetylpiperazine)phenoxy]) and their *N*-methylated cationic SiPc (**39**) were non-covalently bound to BSA (**41**-BSA) and shown to exhibited high PD activities toward B16 melanoma cancer cells.[97] A series of amphiphilic, axially substituted SiPc bearing one or two 1,3-bis(dimethylamino)-2-propoxyl groups were prepared by ligand substitution and alkoxyl exchange reactions.[98] The dicationic and tetracationic analogs were obtained by methylation. In particular the unsymmetrical and amphiphilic SiPc$[C_3H_5(NMe_2)_2O]$(OMe) showed good cell uptake, selective affinity for mitochondria and strong *in vitro* PD activities. Increasing the positive charge of the Pc appears to hinder cell uptake, as evidenced by lower intracellular fluorescence. Other dicationic Pc that have been reported include bis(4-*N*-methylpyridiniumoxy)SiPc.[99] The *in vitro* PD activity of this compound was greatly enhanced by conjugation with BSA. A series of SiPc substituted axially with different polyamine moieties showed low fluorescence quantum yields in DMF due to reductive quenching by the amino moieties.[100] The values are significantly increased in aqueous media as a result of protonation of the amino substituents. All compounds are highly photocytotoxic against human colon adenocarcinoma HT29 cells and Chinese hamster ovary cells. Compounds **42** and **43** showed high affinity towards the lysosomes, but not mitochondria. The *in vivo* PD activity was also investigated using HT29 tumor bearing nude mice. Both compounds effectively inhibited the growth of the tumor without causing apparent injury to the liver of the animals.

3.3. *Non-ionic Ps*

Non-substituted Pc, particularly ZnPc, have been studied extensively as PS for PDT. The symmetrical ZnPc is highly insoluble in most solvents and various formulation strategies have been developed in order to conduct *in vivo* and *in vitro* essays.[101] Administration of ZnPc as unilamellar liposomes of l-α-Dipalmitoyl phosphatidylcholine (DPPC) to tumor bearing mice revealed that the dye is transported by lipoproteins and eliminated rapidly from most tissues, with the tumor reaching maximum tumor/muscle ratios of about seven within 1 and 2 days post-injection.[102] The same authors showed that ZnPc complexation with low-density lipoprotein (LDL) and the addition of 15% mole cholesterol to the DPPC liposomes resulted in improved tumor/muscle ratios.[103,104] A liposomal ZnPc formulation in a 9:1 mixture of 1-palmitoyl-2-oleoylphosphatidylchlorine (POPC) and 1,2-dioleoylphosphatidylserine (DOPS) showed favorable biodistribution pattern in animal models suggesting its potential for clinical use.[105,106] Administration (*i.v.*) to tumor bearing animals of BSA loaded non-covalently with ZnPc showed that the ZnPc readily redistributed over the high-density lipoprotein (HDL) fraction rendering the tumor sensitive to PDT at 24 h post-injection.[107]

A variety of gold nanoparticles including nanoshells, nanocages and nanorods have been extensively investigated for photothermal therapy (PTT). The nanoparticles are biocompatible and exert high photothermal conversion efficiencies at wavelengths where tissues show optimal transparency. Combining PDT and PTT by incorporating PS into single nanoparticle platforms appears to offer optimal tumor control. Plasmonic liposomes (PLs) loaded with ZnPc into the liposomal membrane and coated with a gold nano film resulted in efficient light absorption at wavelengths were ZnPc can be activated for PDT.[108] Combining PD and PT using ZnPc–PLs as a single agent exhibited a remarkably enhanced *in vitro* therapeutic efficacy as compared to PDT or PTT alone.

The hexadecafluorinated $ZnPcF_{16}$ (**44**) was prepared to improve solubility while retaining the symmetry and single-isomeric nature of ZnPc.[109] The fluorine atoms facilitate the formation of stable aqueous emulsions using an emulsifier[110] or polyethyleneglycol (PEG) coated poly(lactic acid) nanoparticles.[111] The added fluor atoms resulted in higher and prolonged tumor retention providing good tumor response to PDT as late as 72 h post-injection. The synthesis of a series of asymmetrically substituted dodecafluorinated Pc derivatives (**45**) has been achieved[112] via the Kobayashi ring expansion[113] of dodecafluorinated sub-PcB(III)Br. The latter reacts with numerous

1,3-diiminoisoindolines in DMSO leading to the selective formation of a series of 3:1 asymmetrically substituted dodecafluorinated ZnPc (**45**) that showed improved *in vitro* PD activity as compared to that of the symmetrically substituted $ZnPcF_{16}$. The synthesis of a non-planar, biconcave, $ZnPcF_{64}$ (**46**) has also been reported.[114] The electronic spectrum showed a bathochromic shift of the Q-band, while increased solubility, lack of aggregation and good quantum yield for singlet oxygen render this Pc interesting for PDT. Preliminary *in vivo* evaluations suggested that the introduction of bulky perfluoroalkyl groups on ZnPc results in increased efficacy for tumor PDT. The perfluorinated derivative hexa-(2,2,2-trifluoroethoxy)Pc has been synthesized,[115] and spectroscopic and PS properties have been reported.[116] The twelve periphereal trifluoroethoxy substituents in the Pc side chain improve PDT efficacy by preventing aggregation, increasing lipophilicity and inducing a red-shift of the *Q*-band absorption.

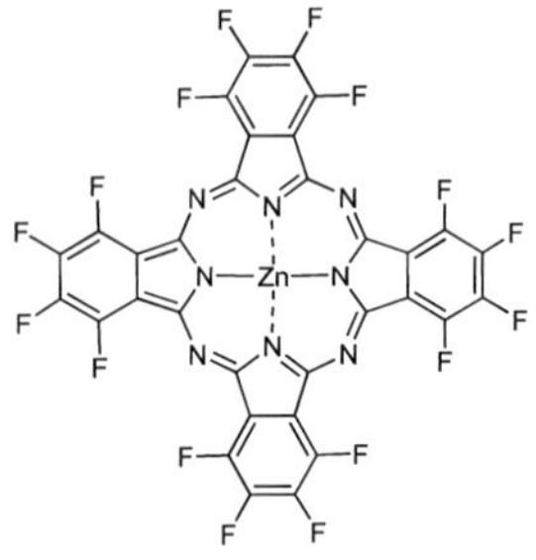

44

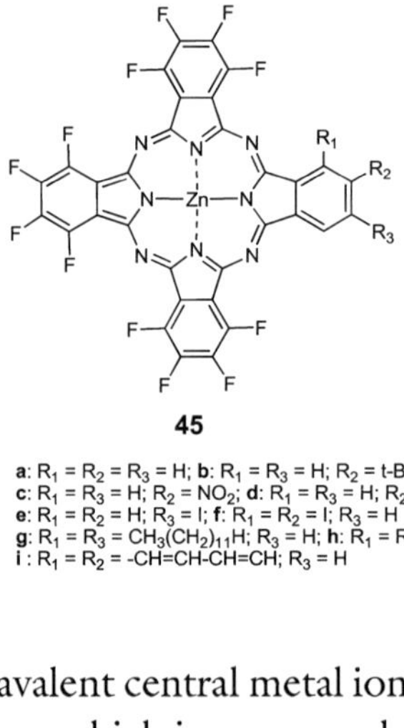

45

a: $R_1 = R_2 = R_3 = H$; **b**: $R_1 = R_3 = H$; R_2 = t-Bu
c: $R_1 = R_3 = H$; $R_2 = NO_2$; **d**: $R_1 = R_3 = H$; $R_2 = I$
e: $R_1 = R_2 = H$; $R_3 = I$; **f**: $R_1 = R_2 = I$; $R_3 = H$
g: $R_1 = R_3 = CH_3(CH_2)_{11}H$; $R_3 = H$; **h**: $R_1 = R_2 = CN$; $R_3 = H$
i : $R_1 = R_2$ = -CH=CH-CH=CH; $R_3 = H$

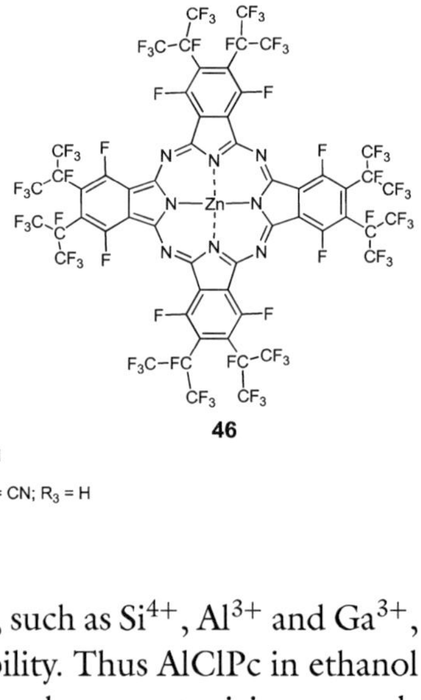

46

MPc with tri- or tetravalent central metal ions, such as Si^{4+}, Al^{3+} and Ga^{3+}, allow for axial substitution, which improves solubility. Thus AlClPc in ethanol can be dispersed in cell culture media showing photocytotoxicity towards tumor cells.[117] *In vivo* studies using an AlClPc Cremophor emulsion was also effective for *in vivo* PDT of murine tumors.[118] A single application of an AlClPc liposome preparation of Ehrlich tumors followed 24 h later by PDT induced necrosis suggesting its potential for the treatment of oral cancers.[119] Axial conjugation of AlPc with PEG or polyvinylalcohol (PVA) gave water-soluble preparations that induced complete tumor regression in an animal tumor model at similar drug and light doses as that observed with AlClPc-Cremophor emulsions.[120]

Substitution of the benzene rings of MPc, particularly at the 3 and 6-positions, induces a red-shift of the *Q* band. Octadecyl ZnPc (**47**) showed a red-shift to 715 nm and was investigated for PDT.[121–123] Topical application to an experimental mouse tumor of octakis-dodecyl and octakis-pentyl analogs of **47** in a tetrahydrofuran–azone formulation revealed good tumor uptake. The octakis-dodecyl compound showed significantly better skin-photosensitising properties than the octakis-pentyl analog and was proposed as a potential candidate drug for the treatment of psoriasis. A Cremophor emulsion of 1,4,8,11,15,18,22,25-OctadecylZnPc showed phototherapeutic activity against an intramuscularly transplanted MS-2 fibrosarcoma in mice.[124] Octa-alkoxy-Pc substituted at 3- and 6-positions showed further red-shifted *Q* bands in the λ_{max} 710–770 nm range with markedly improved properties to function as a triplet PS.[125]

ZnPc substituted with one to four hydroxyl groups have been synthesized[21] and their structure activity relationship in relation to PDT has been studied. $ZnPc(OH)_4$ (**48**) lacking spacer chains was inactive in both *in vivo* and *in vitro* assays. *In vitro* ZnPc(2-OH) was the most active PS followed by 2,3- and 2,9-$ZnPc(OH)_2$, while 2,9,16-ZnPc-$(OH)_3$ exhibiting the least activity. *In vivo*, both the monohydroxy and the 2,3-dihydroxy derivatives induced a poor PDT response, while the 2,9-dihydroxy isomer induced 75% tumor regression under the same experimental conditions. ZnPc substituted with OH groups via aliphatic spacer chains showed increased solubility and *in vitro* PD activity when compared to the ZnPc analogs lacking spacer chains.[126] Compounds with four OH-groups attached via three and six carbon chains (**49** and **50**) were tested to establish the effect of the length of the spacer chain.[127,128] In a mouse tumor model the tetra(hydroxypropyl) compound **49** showed the highest PDT efficacy while *in vitro* the differences in PD efficacy were further accentuated by an order of magnitude.

Water soluble, non-ionic ZnPcs substituted with glycerol were synthesized and their spectroscopic and PD investigated *in vitro*. Compound **51** which differs in the position by which the glycerol moiety is attached to the *iso*indoline subunits of the Pc, was shown to be significantly less aggregated, which correlated to cellular uptake and PD efficacy.[129] Three glycerol substituted ZnPc, (tetra peripherally, tetra non-peripherally and mono iodinated tri non-peripherally) were evaluated for their spectroscopic properties, capacity for singlet oxygen generation, partition coefficients, intracellular uptake, photo-induced cytotoxicity and vascular occlusion efficiency. The peripherally substituted compounds exhibited 10–100 nm red-shifted absorption peaks with higher molar absorptivity, and at least two-fold greater singlet oxygen generation rates compared to the tri non-peripherally substituted analog. Compound **51** and its peripherally substituted analog are phototoxic against various cancer cell-lines. These findings provide evidence that glycerol substitution at peripheral positions improves the PD properties of ZnPc.[130]

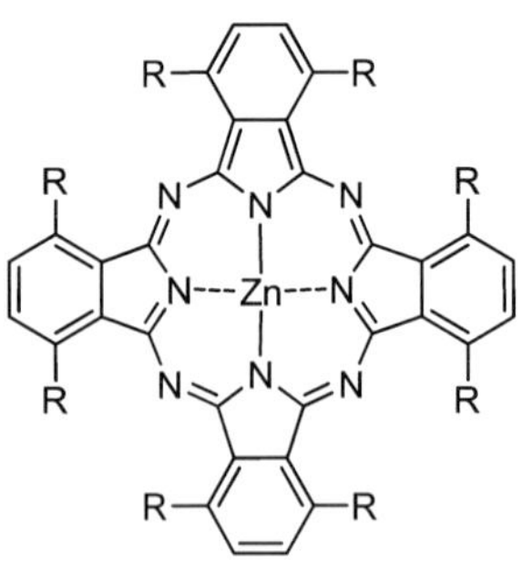

47

R = $(CH_2)_nH$; n = 1-10

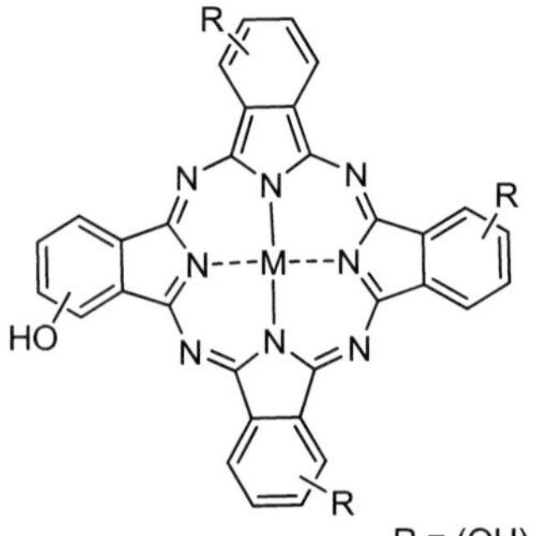

R = $(OH)_n$

n = 0-3

48a: 2H
48b: Zn

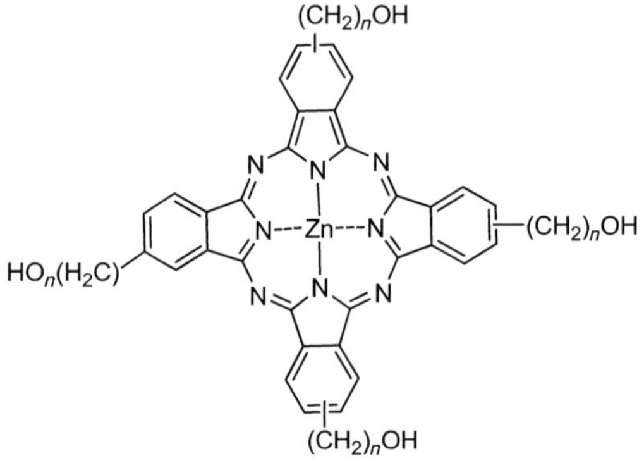

49: n = 4
50: n = 6

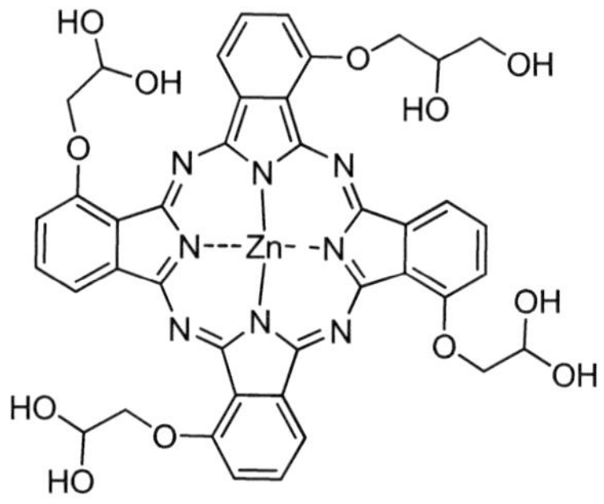

51

Three water-soluble ZnPc substituted with hydroxyl-terminated tetraethylene glycol chains (**52-54**) were evaluated for *in vitro* phototoxicity and several photophysical parameters. The less aggregated tetra non-peripherally substituted derivative **52** exhibited the best capacity to generate singlet oxygen and proved the most efficient to induce an *in vitro* tumor cell response.[131] The effect of varying chain lengths of di-substituted 1,4-dipegylated ZnPc, e.g., $ZnPc[O(CH_2CH_2O)_nMe]_2$ (n = 2, 4, *ca.* 12) and $ZnPc[O(CH_2CH_2O)_nMe]_2$ (n = 6, 8), showed that PD activities varied with chain length, and in particular, that derivatives with medium-length substituents showed the strongest *in vitro* photocytotoxicity.[132] All derivatives showed red-shifted *Q* bands (689 nm versus 670 nm), weaker fluorescence emission, higher singlet oxygen quantum yields and good cell uptake. The synthesis and photophysical properties of a series of ZnPcs bearing multiple hydroxy and tert-butyl groups has been reported.[133] All hydroxy-substituted ZnPcs show low fluorescence quantum yields in DMSO and complete fluorescence quenching in aqueous solutions. However relative high singlet oxygen quantum yields in DMSO suggests their potential for PDT. A series of Pt(II) solketal substituted Pc was synthesized and shown to be good singlet oxygen generators, however biological activities were not reported.[134]

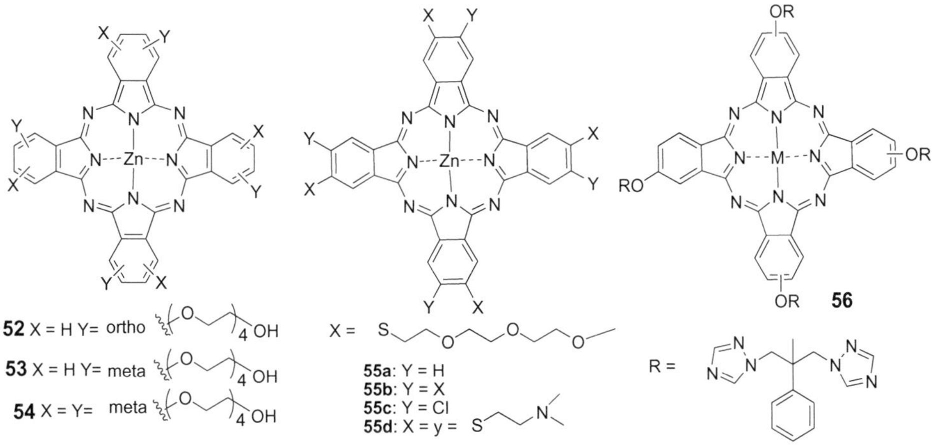

Tetra- and octa-triethyleneoxysulfonyl substituted ZnPc **55a** and **55b** are soluble in polar organic solvents, with **55a** showing better *in vitro* PD properties as compared to the octa-substituted **55b**.[135,136] Compound 2,3,9,10,16,17,23,24-octakis[(*N*,*N*-dimethylamino) ethylsulfanyl]ZnPc (**55d**) formulated as S1-l-α-dimyristoyl-phosphatidyl choline liposomes induced early, direct damage of malignant cells in an animal model leading to random necrosis.[137] ZnPc substituted with four fluconazole (**56**) antifungal groups showed antimycotic activity against *Candida albicans* in the dark, while subsequent light exposure resulted in increased cell inactivation.[138]

Group IV metallo-Pc are readily modified via the attachment of axially ligands to yield single isomeric products of varying solubility, with reduced tendencies to aggregate. The key intermediate $PcSiCl_2$ was first reported in

1965[139] and subsequently a more convenient synthesis to prepare a variety of SiPc containing two axial groups was proposed.[140] Kenney *et al.* reported the synthesis of SiPc (**57**) bearing one or two amino-substituted ligands.[141] The $HOSiPcOSiR_3$ containing a single siloxy ligand and $SiPc(OSiR_3)_2$ containing two axial ligands showed good PD activities both in cell culture and in a mouse tumor model.[142] The mono-substituted $HOSiPcOSi(CH_3)_2(CH_2)_3N(CH_3)_2$ (**57**, Pc-4) showed particularly promising activities and was submitted for clinical evaluation.[143] The presence of two aminosiloxy ligands provided a PS with similar PD activities as compared to the analog featuring one such ligand. *In vivo* studies on tumor bearing mice indicated that both derivatives were at least as effective as Photofrin in PDT protocols.[144] Compound **57b** was also shown to be effective for inactivating HIV and other enveloped viruses as well as blood borne parasites, rendering this a candidate drug for the photosterilization of red blood cells and platelet concentrates.[145] La Jolla Blue™ is a commercially available SiPc dye that contains free carboxyl groups which allow linkage via a NHS ester to biomolecules.[146,147] The dye has absorption and fluorescence emission peaks at 680 and 705 nm, respectively, and does not aggregate in solution.

A novel series of non-ionic SiPc substituted axially with one or two 1,3-bis(dimethylamino)-2-propoxy group(s) (**58**) are essentially non aggregated in common organic solvents and show a weak fluorescence emission. The methylated derivatives are also non-aggregated, even in aqueous media, and exhibit a strong fluorescence emission. These new Pc, in particular the unsymmetrical and amphiphilic analogues, are highly potent against HepG2 human hepatocarcinoma cells and J774 mouse macrophage cells. The PD efficacy is related to their cellular uptake and efficiency to generate singlet oxygen. A higher positive charge at the Pc hinders the uptake, resulting in diminished intracellular fluorescence intensity. Fluorescence microscopic studies revealed that the unsymmetrical **57b** has high and selective affinity for the mitochondria of HepG2 cells.[148] The two axial PEG chains of SiPc **59** and halogenated SiPc **60** showed *in vitro* photocytotoxicity correlating to their tendencies to aggregate rather than the presence of a halogen atom. Non-covalent conjugation of **60** to BSA augmented the *in vitro* photocytotoxicity.[149] The synthesis of a series of both axial and peripheral octa-substituted GePc has also been reported.[150] Further substitution of the hydroxyl groups with various substituted alcohols gave a series of octa-alkoxy GePc as possible PS for PDT.[151]

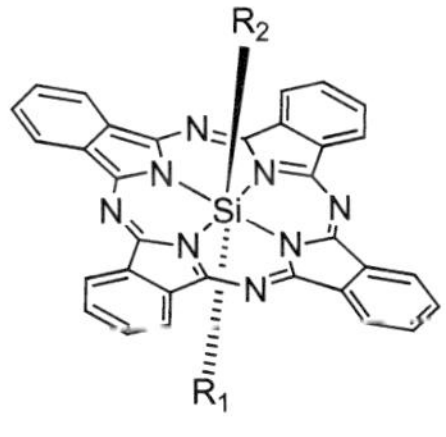

57

57a: R_1 = $OSi(CH_3)_2(CH_2)_3N(CH_3)_2$; R_2 = CH_3
57b: R_1 = $OSi(CH_3)_2(CH_2)_3N(CH_3)_2$; R_2 = OCH_3
57c: R_1 = $OSi(CH_3)_2(CH_2)_3N(CH_3)_2^{+}I^{-}$; R_2 = OH
57d: R_1 = R_2 = $OSi(CH_3)_2(CH_2)_3N(CH_3)_2^{+}I^{-}$
57e: R_1 = $OSi(CH_3)_2(CH_2)_3N(C_2H_5)(CH_2)_2N(CH_3)_2$; R_2 = OH
57f: R_1 – R_2 = $OSi(CH_3)_2(CH_2)_3N(CH_3)_2$
57g: R_1 = R_2 = $OSi(CH_3)_2(CH_2)_3N(C_2H_5)(CH_2)_2N(CH_3)_2$; R_2 = OH
57h: R_1 = R_2= $OSi(CH_3)_2C(CH_2)_3CH(CH_3)_2$
59 :R_1 = R_2= $O(CH_2CH_2O)_nCH_3$

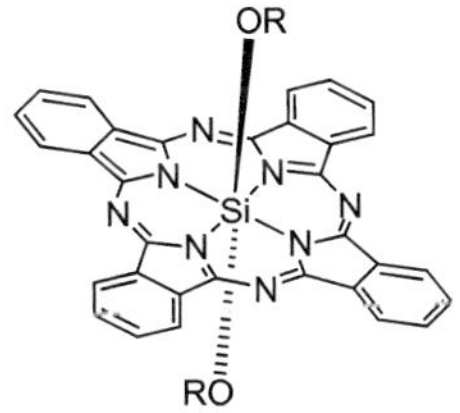

58a: R = CH_2CH_3
58b: R = $(CH_2)_3CH_3$
58c: R = $(CH_2)_{11}CH_3$
58d 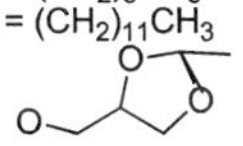
58e 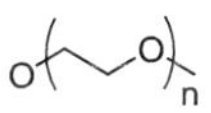

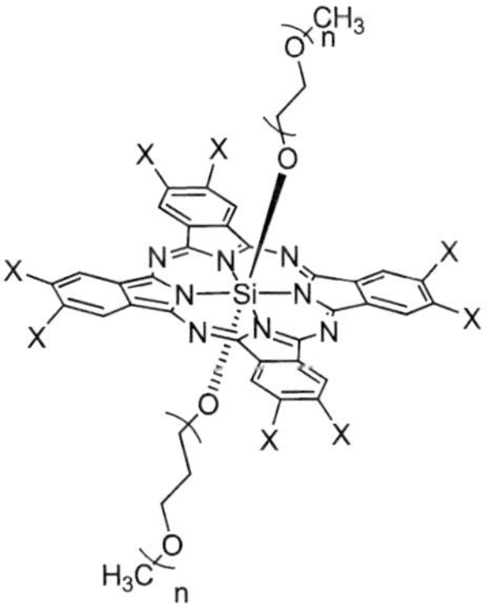

60a: X = H(with PEG_{550})
60b: X = H(withPEG_{750})
60c: X = Cl (with PEG_{750})
60d: X = Br(with PEG_{750}

SiPc substituted with a solketal group, $Si(sol)_2Pc$ (**58d**), e.g., a hydrophobic unit that can be converted *in vivo* by acid-catalyzed hydrolysis to hydrophilic glyceryl, was synthesized and shown to possess high *in vitro* PD activity.[152] Tetra-*t*-Bu-SiPc covalently linked to one or two 2,2,6,6-tetramethyl-1-piperidinyloxyl radicals showed significant *in vitro* phototoxicity when compared to $SiPc(OH)_2$.[153] Amphiphilic Si(IV)Pc axially ligated with rhodamine-B (Rh-SiPc-Rh) localizes specifically in mitochondria and showed strong *in vitro* cytotoxicity upon exposure to ambient light. Limited toxicity in nude mice and good tumor xenograft growth inhibition involving necrosis suggests its potential for PDT.[154]

Gold nanoparticles with an average size of 2–4 nm have been advanced as carriers of hydrophobic PS to improve cell delivery. To permit attachment to the gold surface, Pc featuring thiol substituents were used, providing a direct linkage *via* self-assembly.[155] The stabilized Pc gold nanoparticles are formulated using a gold salt, a phase transfer reagent (TOAB), the thiolated Pc and sodium borohydride as a reductant. The resulting preparation provides a 3-component system containing the PS and phase transfer reagent bound to the surface of the gold nanoparticles. While the free Pc derivative is only soluble in organic solvents, the phase transfer reagent facilitates solubilisation of the PS-nanoparticle conjugates in polar solvents. The *in vitro* PDT efficiency of the internalised Pc-nanoparticle conjugates were twice that observed with the free Pc derivative.[156] A thiol functionalized ZnPcSH conjugated to gold nanoparticles showed a blue shift together with an increase in the triplet quantum yield and triplet lifetime.[157]

Chalcones are vascular disrupting agents (VDA) that rapidly and specifically destroy tumour neovasculature. In order to obtain a dual antiangiogenic/PD agent, a Pc-chalcone conjugate substituted with four chalcone units (**61**) via tetraethylene glycol spacers was prepared.[158] However the attached moieties were insufficiently hydrophilic to balance the strong hydrophobicity of the Pc ring. To overcome this drawback a Pc substituted with one chalcone unit was prepared (**62**), with the potential to release the chalcone and Pc selectively at the tumor site.[159] The PS showed a highly enhanced phototoxicity, presumably due to the increased amphiphilic character of the PS favoring cellular uptake. Alternatively, the chalcone may be released upon irradiation, contributing to the cytotoxicity. These results suggest that combining a PS and VDA molecule is a highly promising strategy to develop PS for the PDT of solid tumors.

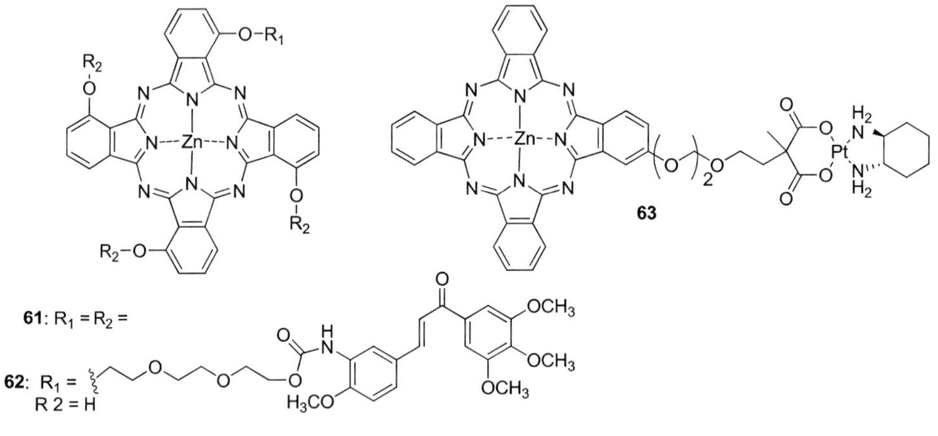

There has been considerable interest in combining PDT with other anticancer therapies. In some cases, additive or synergistic effects have been observed, which could permit reduction of the effective doses of the anti-cancer drug. Thus a combination of oxaliplatin, a platinum-based anticancer chemotherapeutic, and ZnPc substituted with an oxaliplatin derivative via a triethylene glycol linker, has been synthesized.[160] This PS linked through a triethylene glycol spacer group (**63**), permitted a five-fold reduction in drug dose based on *in vitro* essays. The cytotoxic effects of conjugates, formulated as a Cremophor emulsion, revealed that *in vitro* PD activities are enhanced due to increased cellular uptake and intracellular ROS generation.

ZnPc substituted with a 2,4-dinitrobenzenesulfonate group fluoresce and produce reactive oxygen in cultured cells, which can be enhanced by the addition of glutathione in phosphate-buffered saline.[161] Mg 2,9(10),16(17),23(24)-PcI_4 is a strong generator of $^1O_2$ although reported quantum yields vary substantially.[162] This discrepancy led to the suggestion that the photooxidation process proceeds via a mixed Type I/II and/or Type III mechanism. The dye binds to liposomes, exhibits a low fluorescence quantum yield and readily penetrates cell or lipid membranes.

4. Pz

Pz (or tetraazaporphyrins) are tetrapyrrole macrocycles, similar to porphyrins and Pc, pioneered by Sir R. Patrick Linstead in 1952.[163] Pz differ from porphyrins in that they contain *meso* nitrogen atoms, rather than carbon atoms, and differ from Pc in that their β-pyrrole positions are open for substitution. These differences confer physical properties that are distinct from both porphyrins and Pc.[164] The free dye has absorption properties similar to those of porphyrin (λ_{max} 620 nm). Attachment of electron-donating substituents results in a red-shift, whereas metallation results in a blue-shift (λ_{max} 591 nm). The red-shifted analogues are particularly attractive PS for PDT.

A series of alkylated ZnPz (**64**) were synthesized and evaluated for their PD efficacy.[165,166] Visible absorption spectra of Pz **64e–g** show an absorption maximum around 705 nm while substitution with electron-withdrawing cyano substituents (**64b**) induces a red shift (λ_{max} 750 nm), which is even more pronounced in the case of amino substituents.[167] Evaluation of **64 a–h** for their ability to photoinactivate V-79 Chinese hamster fibroblasts *in vitro* showed that the *N*,*N*-diethylamino **64f** has the highest PDT potential. Coordination of Zn in the center cavity results in little change in the biological activity, but long aliphatic chains or cyclic amines attached to the periphery of the cyano-Pz causes loss in activity while the removal of the cyano moiety also has a detrimental effect on PD efficacy. Pz functionalized with carbohydrate-, galactopyranose (**65a**) and methyl-ribose (**65b**), featuring either H_2, Zn or Ni cores, were solubilized for biological studies via encapsulation with DCM-based PEG poly(ethylene glycol)-DSPE 5000-PBS.[168] The phototoxicity of the ZnPz derivatives against MCF-7 breast cancer cells was far superior to that of the H_2Pz and NiPz analogs, although the phototoxicity of the latter could be improved using the encapsulation procedure. The condensation of 2,3-dicyano-5,6-diethoxycarbonylpyrazine with $Zn(OAc)_2$ followed by mild hydrolysis yields the water-soluble octacarboxy Zn complex that was advanced as a candidate for PDT of cancer.[169]

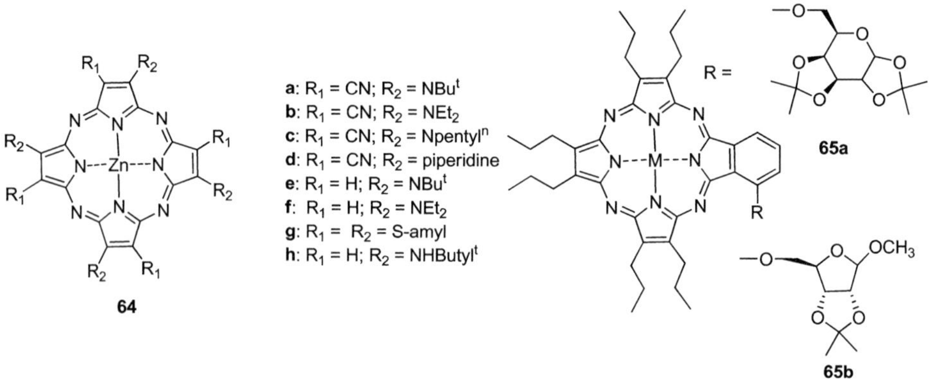

A series of Pz bearing an overall charge from neutral (**66a**), negative (**66b**) to positive (**66c**) were synthesized and evaluated for toxicity, cellular uptake and intracellular distribution pattern.[170] The anionic **66b** showed antitumor activity and to establish structure-function relationships between various anionic compounds, two additional analogs (**67b** and **68b**) featuring a fused $\beta_1\beta_1$-diisopropyloxy benzo and carboxylate group, were prepared.[171] The progressive variation in overall charge affects the balance between hydrophobic/hydrophilic properties and partitioning over cell membranes. Thus **67b** and **68b** exhibited significant phototoxicity in both tumor and normal cells, while **66b** showed selective activity against tumor cells. Anionic Pz with fewer carboxylates and larger hydrophobic core, particularly those with the largest overall negative charge, target more specifically tumor cells.[172] The addition of hydrophilic thiotriethoxy(ethoxy)methyl groups to the Pz periphery, balances the hydrophobicity of the Pz macrocycle. With the exception of **68d** ($M = Zn$), **67d** and **66d** were found to be toxic to normal and tumor cells at higher concentrations, particularly the free-base compounds. Selective tumor cell uptake and lack of toxicity suggest that **68d** is particularly promising as a dual emission probe for biomedical imaging.[173,174]

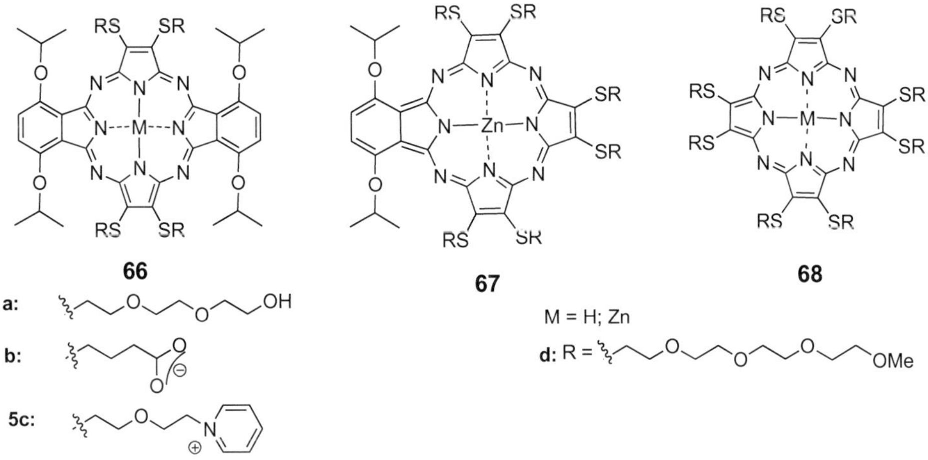

Syntheses and general physicochemical properties of free-base Pz tetrakis-2,3-[5,6-di(2-pyridyl)pyrazino]Pz, [Py_8TPyzPzH_2],[175] their metal derivatives[176] [Py_8TPyzPzM] (M) MgII(H_2O), MnII, CoII, CuII, ZnII) and the related octacations [(2-Mepy)$_8$-TPyzPzM]$^{8+}$,[177] pentanuclear species [($PdCl_2$)$_4$-Py_8TPyzPzPd], the monopalladated complex [Py_8TPyzPzPd], and its related octacation [(2-Mepy)$_8$TPyzPzPd]$^{8+}$ have been reported in detail.[178] Tetrakis-2,3-[5,6-di-(2-pyridyl)pyrazino]porphyrazinatoPdII) [Py_8 TPyzPzPd] and the corresponding pentapalladated species [($PdCl_2$)$_4$ Py_8TPyzPzPd] are potent PS for the production of singlet oxygen.[179] The water-soluble ZnPz [($PtCl_2$)(CH_3)$_6$LZn]$^{6+}$, bearing an exocyclic N_2(pyr)$PtCl_2$ coordination site, was shown capable of generating $^1O_2$ in DMF.[180] Hexacation was also shown to result in interaction with single-stranded telomeric DNA, containing the guanine-rich sequence 5′-d[AGGG (TTAGGG)3]-3, present as a G-quadruplex-type structure.[181] Binding of the hexacated Pz to the G4 structure influences the conformational equilibrium suggesting that the heterobimetallic cation can function as a potential multimodal agent for cancer therapy.

Nanostructured materials are gaining increasing interest as carriers for bioactive agents. Multicompartment micelles derived from polybutadiene-block-poly(1-methyl-2-vinyl pyridinium methyl sulfate)-block-poly(metha crylic acid) (BVqMAA) triblock terpolymers have been studied for delivering hydrophobic Pz to improve PDT efficacy.[182] The Pz derivative was incorporated into the hydrophobic core during self-assembly. The initial micelle corona is formed by negatively charged PMAA and gradually changed to

PEG through interpolyelectrolyte complex formation with positively charged PEG-block-poly(L-lysine) (PLL-b-PEG) diblock copolymers. By changing the corona composition from a highly negatively charged PMAA to neutral PEG in a controlled manner, the strong influence of the micellar corona on the interaction with biological systems was demonstrated. Fully PEGylated micelles efficiently suppressed tumor growth for up to 20 days post-PDT. The noncovalent interaction of a bimetallic water-soluble (Zn(II)/Pt(II)) Pz hexacation product, $[(PtCl_2)(CH_3)(6)LZn]^{6+}$(**69**) and its octacationic analogue $[(CH_3(8)LZn]^{8+}$(**70**) lacking the cis-platin-like functionality, with a 21-mer double strand DNA showed the presence of complexes of various number of macrocycles per DNA molecule. Comparing hexacation to octacation revealed that the former induced a higher degree of monomerization.[183] The binding of an octacationic $ZnPz^{8+}$ ligand to a human telomeric sequence showed good affinity for the G-quadruplex warranting further investigations on its photochemical behaviour.[184] A perfluorous polyphenyl dendritic shell was prepared via Pz periphery substitution with thirty-two pentafluorophenyl groups providing a compound with resistance towards photo bleaching.[185]

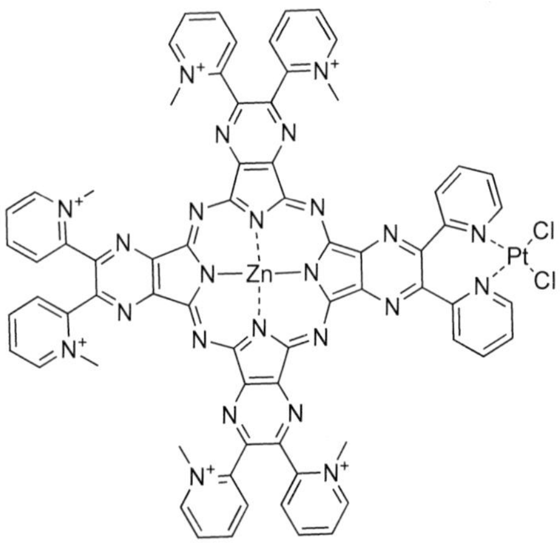

69

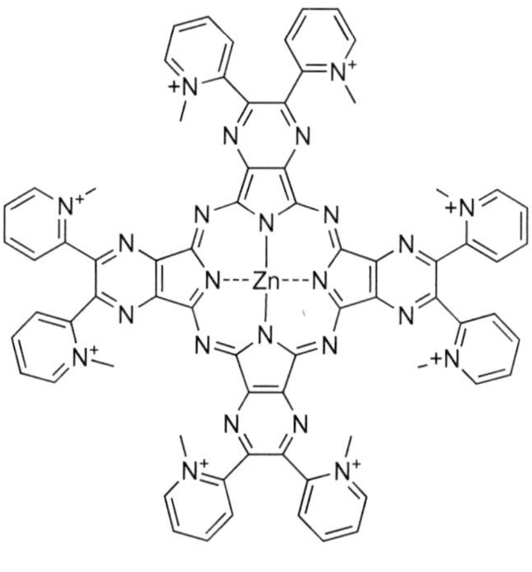

70

A near-IR (NIR) fluorescing Pz featuring pendant hydroxyl groups has been advanced as a carrier to deliver conjugated chemotherapeutic agents to tumor cells. Conjugation to Doxorubicin *via* an acid labile linker gave a complex that undergoes active endocytosis by tumor cells, while retaining strong intracellular NIR fluorescence.[186] Lipo- and hydrophilic derivatives of octaalkynyl-tetra-[6,7]-quinoxalino Pz and their Zn- and MgNPc analogs **71** exhibit intense NIR absorptions at 770 nm and show high quantum yields for singlet oxygen generation. The tri-*iso*propylsilyl-3,5-di-tert-butylphenyl- and 4-tri-*iso*propylsilyloxyphenyl-acetylenic derivatives dissolve in organic solvents only, however the polyethylene glycol derivatives can be formulated in ethanol/water mixtures.[187] Quaternized pyridine Pz **72** carry their positive charges on the macrocycle.[188] The interaction of the AlCl complex **72** with DNA suggested that the molecule binds to the outside of the DNA helix, involving interactions that are not limited to electrostatic forces only.[189] Interactions of cationic tetrakis N,N′,N″,N‴-tetramethyltetra-3,4-pyridino ZnPz (Zn tmtppa) with synthetic polynucleotides and calf thymus DNA suggest binding to external regions in the G-C step of DNA. The complex shows low tendency to aggregate, which would be advantageous for PDT applications.[190] Tetra(3-pyridyl)Pz has been modified with four peripheral $[Ru(bipy)_2Cl]^+$ groups, to yield a complex efficient in generating singlet oxygen.[191] Amphiphilic ZnPc derivatives containing non-peripheral alkyl-substituted benzenoid rings and pyridine rings have also been synthesized.[192] The triplet lifetime increased with an increasing number of pyridines attached to the molecule. The lowest symmetric isomer (**73**) with two pyridine rings showed the longest triplet state lifetime.[193]

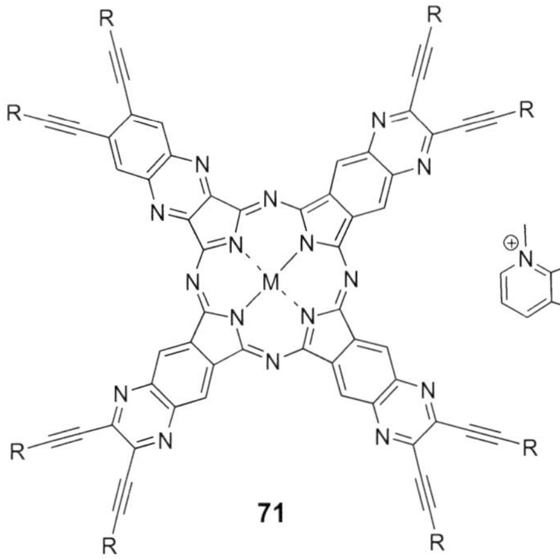
71

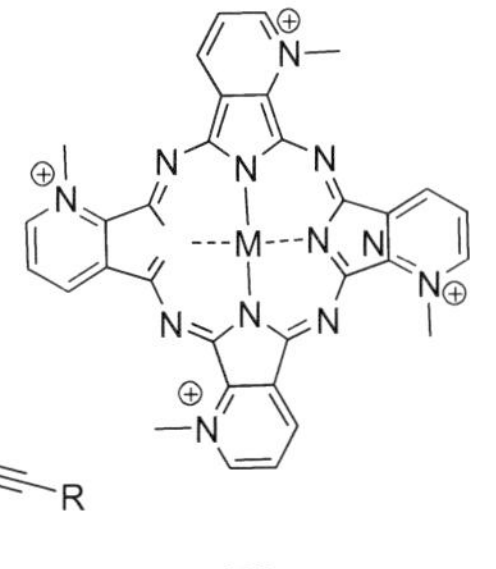
72

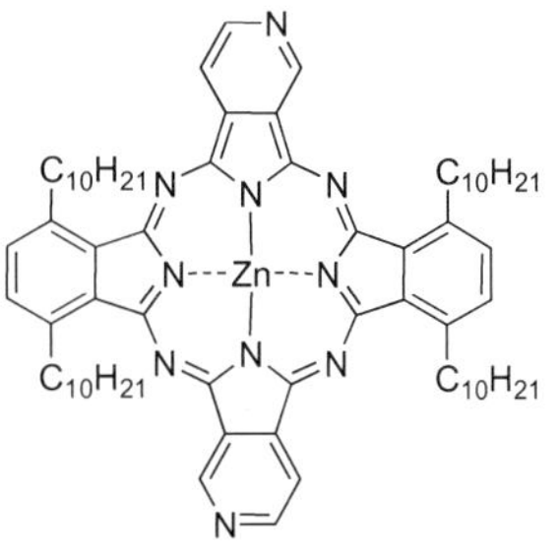
73

R = Si(*i*-pr)$_3$; 3,5-di(t-Bu)phenyl); 4-(*i*-pr)$_3$siO)Phenyl

The MYC gene plays an important role in the regulation of cell proliferation and growth and is overexpressed in a wide variety of human cancers. The interaction of water-soluble tetrapyridino ZnPc with the c-MYC gene shows a large red shift and hyperchromicity in the *Q* band providing a potential for targeted PDT applications.[194] A series of positively charged Zn(II) naphthobenzoPz (Zn-NBPz), bearing between one and eight positively charged methylpyridinium substituents, produce singlet oxygen similar to that of ZnPc, however their photostability decreases with the number of annelated naphthalene rings.[195]

Pz possessing fluoroalkylthio and dietherthio peripheral substituents can produce singlet oxygen and are moderately photoactive against human squamous carcinoma cells. Particularly Mg(II)Pz substituted with peripheral 4-fluorobutylthio groups revealed tumor cell selectivity and moderate activity.[196]

5. NPs

The first preparation of metallo NPc dates back to 1936.[197] However, compared to their Pc analogs, only a limited number of studies concerning their potential use in PDT are available. NPc are dark green crystalline compounds that resemble Pc complexes with extended aromatic naphthalene rings. NPc are of interest because of the extension of the π-electron system resulting in a red shift and intense absorption around 750–780 nm where light transmission in tissues are optimal. The common method for their preparation involves the cyclotetramerization of 1,2- and 2,3-naphthalene dicarboxylic acid and their derivatives in the presence of metal salt. Singlet oxygen yields obtained with NPc are comparable to those reported with Pc,[198] however NPc are less stable than Pc and more easily photobleach in the presence of light and oxygen.[199] To render absorbency and solubility compatible with PDT applications, NPc have been prepared featuring different central metal ions and peripheral substituents. MNPc which lack axial ligands tend to aggregate in solution resulting in loss of PD properties.[200] Due to their NIR absorption, NPc have been advanced as potential PS for highly pigmented melanomas.[201] MNPc can be rendered water soluble by introducing sulfonate groups. Similar to MPc, MNPc can be treated with oleum to furnish $MNPcS_n$ ($n = 1-4$; M = Al, Zn) (**74**), however lower temperatures are required. Products were fractionated by HPLC and studied for their PD activity.[202,203] The capacity to generate $^1O_2$ and cell uptake were similar for both sulfonated AlPc and AlNPc, however the latter were found to be photo-inactive *in vitro*. In the case of the sulfonated ZnNPc only the least sulfonated derivatives produced some phototoxicity *in vitro*, whereas the higher sulfonated derivatives were inactive, which was attributed to intracellular aggregation.

Unsubstituted ZnNPc **75a** and substituted ZnNPc, tetraamino-ZnNPc **75b**, and amido-ZnPc **75c–h** showed absorbance maxima around

760–770 nm, with quantum yield for $^1O_2$ from 0.3 to 0.36, which is about 1.4 times lower than that observed for the corresponding ZnPc. Various ZnNPc derivatives were formulated in DPPC liposomes showing different *in vivo* PDT activities depending on the nature of the substituents.[204,205] Among a series of tetraamido-substituted 2,3-ZnNPc, the tetrabenzamido-substituted derivative **75e** appeared to be the most promising PS for PDT in the series.[206] Two octa-(4-*tert*-butylphenoxy) substituted Ga- and In-NPc (**76**) were synthesized and shown to possess higher triplet state quantum yields and triplet-minus-ground state extinction coefficients than comparable Pc. Their lipo-solubility was improved by the addition of bulky substituents.[207]

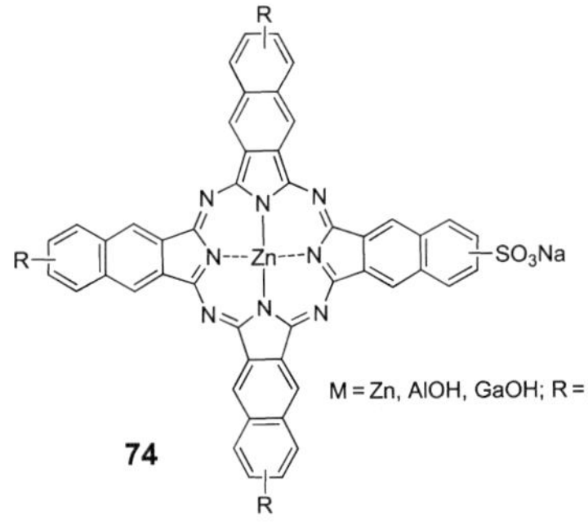

M = Zn, AlOH, GaOH; R = $(SO_3Na)_n$

74

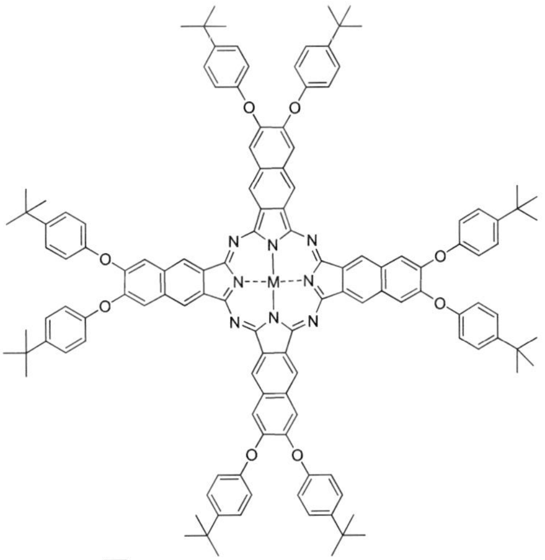

M = Ga In

76

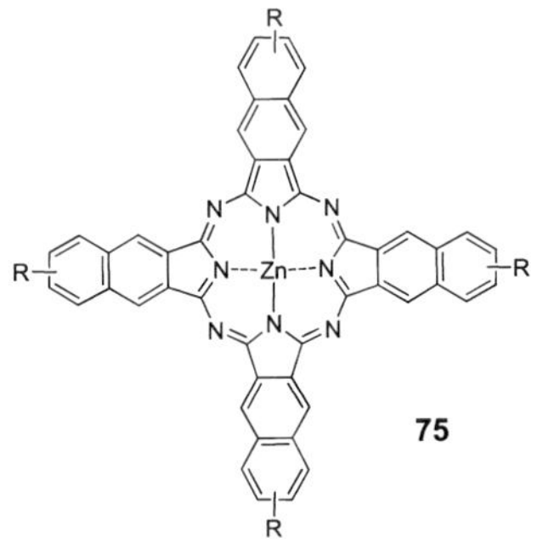

75

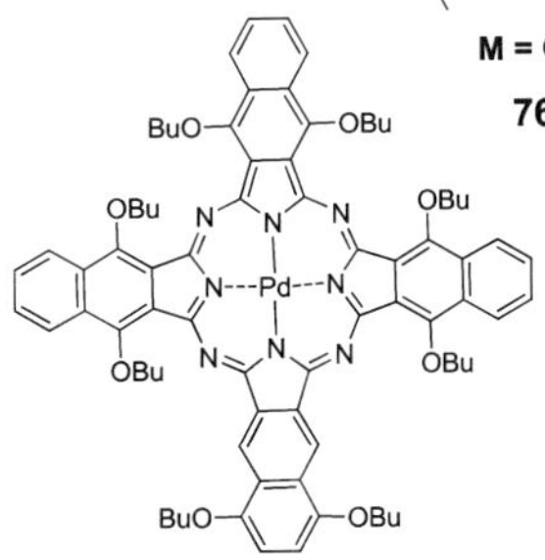

77

a: R = H; **b** R = NH_2; **c**: OCH_3, **d**: $NHCOCH_3$, **e**: $NHCOC_6H_5$, **f**: $NHCOC_6H_4OCH_3$, **g**: $NHCOC_5H_{11}$, **h**: $NHCOC_{11}H_{23}$.

Attachment of alkoxyl and alkyl groups to the macrocycle of symmetrical NPc induces a bathochromic shift of the Q-band. Substitutions closest to the point of fusion of the rings cause the largest shift, to 810–870 nm. Furthermore, alkoxyl groups induce a larger spectral shift than alkyl groups.[208,209] A

PdNPc(OBu)$_8$ (**77**) that absorbs strongly in the NIR region, without potential PD activity, was proposed as a potential agent for photothermal therapy (PTT).[210] Ge(IV)- and Si(IV)-NPc show strong, monomeric absorption peaks in the NIR region (779 nm),[211] and in particular SiNPc have been studied in detail as PS for PDT.[212,213] The tetra-valence of silicon allows for the substitution with two axial ligands, rendering the complex more stable towards photooxidation than the analogous Al- and ZnNPc, enhancing its suitability for PDT applications. Furthermore, addition of bulky substituents reduces their tendency to aggregate, favoring the monomeric and photoactive form of the dye.[214] A series of SiNPc substituted with different axial ligands, *bis*(t-butyldimethylsiloxy) (**78a**), *bis*(dimethylthexylsiloxy) (**78b**), *bis*(tri-n-hexylsiloxy) (**78c**) and *bis*(dimethyloctadecylsiloxy) (**78d**), were prepared *via* substitution of the SiNPc(OH)$_2$ precursor with the corresponding chlorosilane ligands.[215] They show strong absorption around 780 nm where tissues exhibit optimal transparency and are capable of producing singlet oxygen. The use of such far-red-absorbing PS is particularly promising for PDT of highly pigmented tumors since absorption of excitation light is not limited by melanin absorption. All four dyes showed limited phototoxicity *in vitro*, combined with substantial dark toxicity. However, in spite of the this, they exhibited excellent *in vivo* PD activity, particular the *bis*(dimethylthexylsiloxy) derivative **78b**.[216] Ligand exchange of the axial-Cl of SiNPcCl$_2$ with NaO-*n*-C_8H_{17} yields the symmetrical, disubstituted SiNPc[OSi(*i*-Bu)$_2$-*n*-$C_{18}H_{37}$]$_2$ or *iso*BOSiNPc (**78e**), which has been extensively studied as a PS for PDT.[217–219] Comparison of **78e** to an unsymmetric analog SiNPc[OSi(*n*-C_{10}H21)$_3$][OSi-$(CH_3)_2N(CH_3)_2$] or DAPSiNPc (**78f**) in melanoma cells showed that **78f** is a more efficient PS, likely reflecting higher cell uptake and less intracellular aggregation.[220] A SiNPc (**78g**) bearing two methoxyethylenglycol axial ligands formulated in Cremophor induced only a minor PDT response in the melanoma tumor model in mice, but was active in other, non-melanoma tumor models.[221]

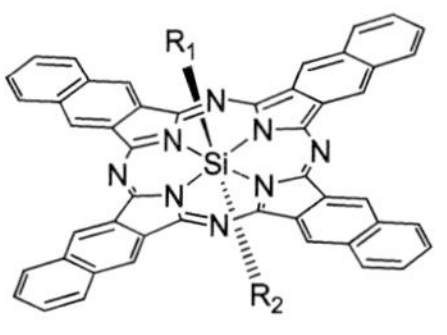

a: $R_1 = R_2 = OSi(CH_3)_2C(CH_3)_3$
b: $R_1 = R_2 = OSi(CH_3)_2C(CH_3)_2CH(CH_3)_2$
c: $R_1 = R_2 = OSi(C_6H_{13})_3$
d: $R_1 = R_2 = OSi(CH_3)_2C_{18}H_{37}$
4: $R_1 = R_2 = OSi(i\text{-}C_4H_9)_2C_{18}H_{37}$
f: $R_1 = OSi(CH_2)_2(CH_2)_3NCH_3)_2$; $R_2 = OSi(n\text{-}C_{10}H_{22})_3$
g: $R_1 = R_2 = OCH_2CH_2OCH_3$

78

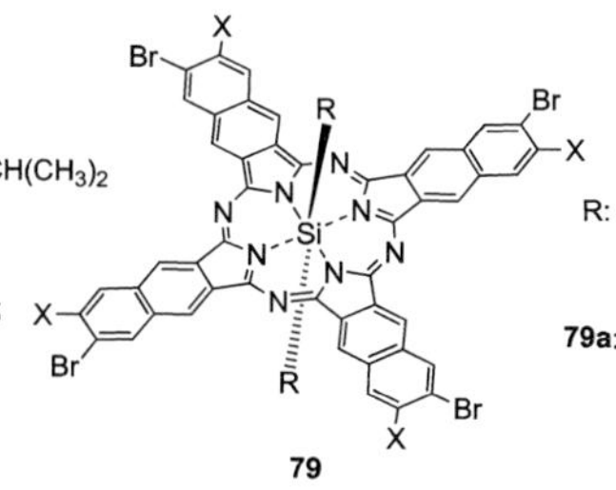

79

R: = Si–O–Si(n-C_6H_{13})$_3$

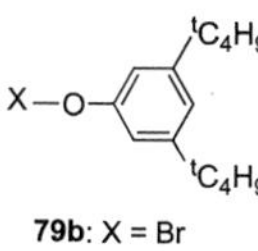

79a

79b: X = Br

The effect of the number of bromide substituents on the photophysical and nonlinear optical properties of the tetrabrominated $Br_4(t\text{-}Bu_2PhO)_4$ $NPcSi[OSi(Hex)_3]_2$ (**79a**) and the octabrominated $Br_8NcSi[OSi(Hex)_3]_2$ (**79b**) have been investigated.[222] There is a moderate increase in the quantum yield of the triplet excited-state formation and a decrease of the triplet excited-state lifetime due to the heavy atom effect of the large number of peripheral bromide substituents, particularly in **79b**. The latter was investigated for PDT of the ciliary body in pigmented rabbits. The authors confirm that tissue penetration is good at 770 nm and that thermal effects from the exciting laser alone were minimal.[223]

6. Bioconjugated Ps

Bioconjugation involves coupling of two molecules whereby one of the moieties is of biological origin, resulting in a complex with improved therapeutic or diagnostic properties. Targeted properties include: (i) enhanced cellular uptake, (ii) advantageous sub-cellular localization, (iii) targeting receptors or enzymes over-expressed in target tissues, (iv) increasing oxidative stress in the target tissue, (v) enhancing therapeutic efficacy, (vi) reducing side effects, and (vii) diminishing drug resistance. Several strategies have been explored to functionalize Pc with the aim to achieve such targeted action and to exert control over the regioselectivity of the conjugate.

5-Aminolevulinic acid (ALA) is a naturally occurring amino acid and key intermediate in the biosynthesis of protoporphyrin IX, which functions as a PS.[224] In an attempt to improve Pc bioavailability, ALA has been linked to Pc periphery to yield ALA conjugates (**80**).[225] Other amino acid derivatives that have been prepared include 2,9,16,23-tetra(2-amino-2-carboxyethyl) ZnPc, which exhibited only poor phototoxicity whereas the serine derivative 2,9,16,23-tetra(1-carboxy-2-hydroxyethyl of aminocarbonyl) ZnPc (**6**) showed unwanted dark toxicity.[226]

Synthetic approaches for the conjugation of porphyrins and Pc to peptides and proteins to target receptors in cancer cells have recently been reviewed.[227] For the attachment to peptides and proteins, carboxylic acid derivatives of Pc are of particular interest since they permit facile coupling *via* bio-compatible peptide bonds. A water-soluble cationic penta-lysine β-carbonyl complex, $ZnPc\text{-}(Lys)_5$, retained a high quantum yield for singlet oxygen, good cellular uptake and phototoxicity toward tumor cells.[228] Pharmacokinetics in tumor-bearing mice showed high tumor-to-skin ratios and PDT induced tumor growth inhibition. Mono-substituted ZnPc-COOH has been linked to lysine

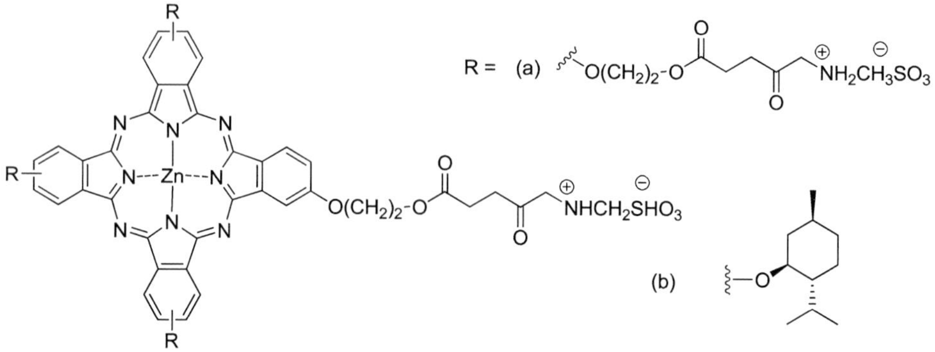

80

chains of different lengths to yield a series of water soluble ZnPc-$(Lys)_n$ ($n = 1 - 9$) conjugates. Screening for PDT efficacy showed that the ZnPc-$(Lys)_7$ possessed the highest activity.[229] A series of ZnPc-oligolysine conjugates ($n = 2, 4$, and 8) was evaluated for their PD antimicrobial activities and shown to be particularly active against Gram-positive bacteria. The photocytotoxicity increased with the length of the oligolysine chain and particularly the octa-lysine derivative showed high PD potency towards enveloped viruses.[230] ZnPc has been coupled to gonadotropin-releasing hormone (GnRH), which is over-expressed in breast tumors. The cellular uptake of the ZnPc-GnRH conjugate was specific for cells expressing GnRH surface receptors, which coincided with selective phototoxicity.[231] A Pc-bombesin conjugate targeting a subtype of gastrin releasing peptide receptor (GRPR), over-expressed in early stage prostate cancer, was prepared from $AlPcS_4$.[232] A mono-carbohexyl derivative of $AlPcS_4$ was attached to 8-Aoc-bombesin(7-14)NH_2 *via* an amide bridge to yield a bombesin-$AlPcS_4$ conjugate linked by a C-14 spacer chain. The conjugate provides a modest gain in binding affinity to the GRPR, which resulted in improved PD potency against GRPR-rich prostate cancer cells. Pz dyes containing an *iso*thiocyanate function for covalent linking have been prepared.[233] The dyes contain one carboxylate to facilitate solubilisation while another dye features a nitro group for functionalization and labeling. Both dyes were successfully conjugated to oligonucleotide primers, and showed potential for use as NIR-based DNA diagnostic agents.

ZnPc conjugated with a short, nuclear localizing peptide, Gly-Gly-Pro-Lys-Lys-Lys-Arg-Lys-Val, was synthesized by click chemistry.[234] Conjugation enhanced the cellular uptake and photocytotoxicity against HT29 human colorectal carcinoma cells. The synthesis of a series of ZnPc-peptide conjugates (**81**) targeting the gastrin-releasing peptide (GRP) and integrin receptors has also been reported.[235] Two alternative synthetic methods based on Sonogashira cross-coupling of ZnPcI with acetylenic bombesin or arginine–glycine–aspartic acid (RGD) derivatives, either in solution or on solid phase, were used. The water-soluble conjugates were screened for their PD efficacy against several cancer cell lines expressing different levels of GRP and integrin receptors. Variations in photocytotoxicity between the conjugates correlate with differences in hydrophobicity as well as receptor-mediated cell uptake. In the case of the ZnPc-bombesin conjugate, competition experiments confirmed the involvement of the GRP receptor in both the phototherapeutic activity as well as intracellular localization. A ZnPc cyclic-RGD conjugate was recently prepared and shown to target mitochondria of carcinoma cells over-expressing $\alpha v\beta 3$ integrin receptors.[236]

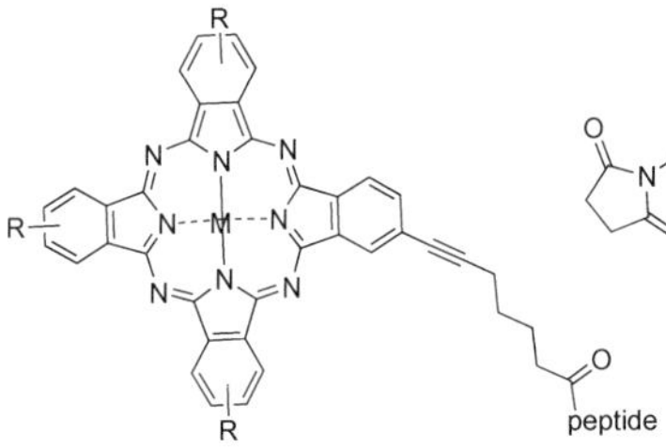

81a: M = Zn, R = t-Bu
81b: M = H_2, R = t-Bu
81c:, M = Zn, R = SO_3Na
81d: peptide = -Tyr(t-Bu)-Val-Gly-NH- **(YVG)**
81e: peptide = -Arg(Pbf)-Gly-Asp(Ot-Bu)-NH- **(RGD)**
81f:, peptide = -NH($CH_2CH_2O)_2CH_2C$(O)-D-Tyr(t-Bu)-Gln(Trt)-Trp(Boc)-Ala-Val-b-Ala-His(Trt)-3-Ala(2-thienyl)-Nle-NH- **(BBN)**

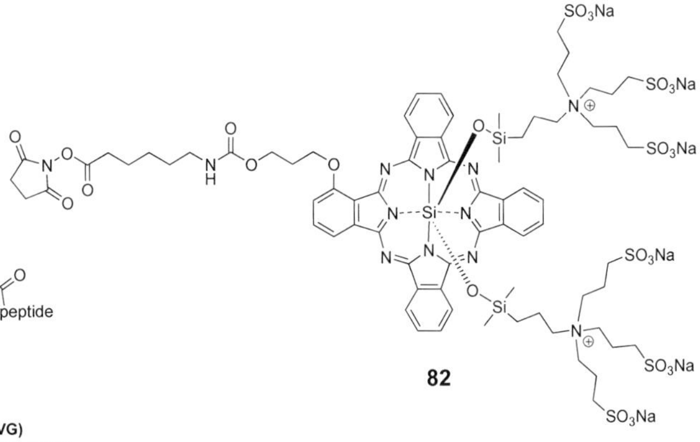

82

Two ZnPc-peptide conjugates bearing either a short linker or a long PEG-linker between the macrocycle and the fusogenic peptide sequence corresponding to the human immunodeficiency virus I transcriptional activator (HIV-1 Tat 48–60) were synthesized to increase cell-targeting and to evaluate the effect of the linker. The presence of the peptide chain increased the water solubility of the Pc and, consequently, its fluorescence in aqueous solutions. Both conjugates were found to have low dark cytotoxicity but were highly phototoxic toward human HEp2 cells. The conjugate bearing the long PEG-linker accumulated preferentially within the cells.[237] $AlPcS_4$ was covalently coupled to maleylated BSA *via* one or two sulfonamide-hexanoic-amide spacer chains.[238] Competition studies of the conjugates with ^{125}I-mal-BSA showed that coupling of $AlPcS_4$ to BSA resulted in recognition of the conjugate by the scavenger receptor, whereas coupling to mal-BSA further enhanced its binding affinity. Phototoxicity on macrophage-like J774 cells was found to be highest for the conjugate with two carboxyhexyl chains, while poor activity was shown on the non-phagocytic EMT-6 cell line. All conjugates were, however, less phototoxic than free disulfonated $AlPcS_2$.

LDLs are biological particles containing an apolipoprotein which recognizes, with high affinity, specific receptors on many cell types, including cancer cells. Their potential as vehicle to deliver $AlPcS_4$ to tumor cells for PDT was investigated.[239] $AlPcS_4$ was loaded with LDL in two different manners. In the first procedure, $AlPcS_4$ was covalently bound to the protein part of the LDL via amide bonds to 6-carboxypentylaminosulfonyl spacer chains attached to two of the four sulfonate groups, i.e. $AlPcS_4A_2$-LDL. In the second procedure, $AlPcS_4$ was substituted with a linear dodecylaminosulfonyl chain, i.e. $AlPcS_4C_{12}$ and non-covalently inserted into the phospholipid monolayer of the LDL. Association of the $AlPcS_4C_{12}$ with LDL increased *in vitro* phototoxicity of the dye substantially while the covalently loaded $AlPcS_4A_2$-LDL preparation showed little cytotoxicity. It was postulated that the covalent labeling of the protein moiety with the Pc greatly reduced LDL receptor recognition. PDT of EMT-6 tumor-bearing mice showed that the free and LDL-associated $AlPcS_4C_{12}$ exhibited similar activities. The authors conclude that the attached aliphatic chain of the $AlPcS_4C_{12}$ greatly enhances the phototoxicity of the parent $AlPcS_4$, while pre-association of the $AlPcS_4C_{12}$ with LDL does not further augment its *in vivo* PD efficacy.

Phototherapy using a cancer-targeting antibody loaded with PS has been referred to as photoimmunotherapy (PIT). Various combinations of conventional PS and MAbs have been tested showing limited success to improve *in vivo* therapeutic effects. However, PIT using the commercially available SiPc IR700 (**82**) and an epidermal growth factor receptor (EGFR) targeting MAb (mAb–IR700 conjugate) affected EGFR-expressing cells only, producing significant *in vivo* tumor shrinkage.[240] An optimized regimen of effective PIT involved fractionated light treatment guided by IR700 fluorescence.[241] A cocktail injection of two different antibody-IR700 conjugates created a more homogeneous micro distribution of the conjugates within the tumor, resulting in enhanced therapeutic effect.[242] ZnPc was also conjugated with a MAb directed against carcinoembryonic antigen (CEA), using in situ activation of the carboxylic acid on the ZnPc. The ZnPc-CEA bioconjugate with the highest ZnPc/MAb ratio of 3 was about 37 times more efficient in targeting HT-29 cells, as compared to unconjugated ZnPc.[243]

To target estrogen receptors overexpressed in breast cancer cells ZnPc derivatives were coupled via the 17α-ethynyl group of an estradiol derivative

(**83**) and evaluated for their estrogen receptor binding affinity (RBA) and *in vitro* photocytotoxicity.[244] The highest RBA values were observed with lipophilic conjugates coupled via a relative long spacer group, while the sulfonated analogues showed little binding affinities (RBA <2). The highest photocytotoxicity was observed with the sulfonated conjugates, while the nature of the spacer group appeared not have a pronounced effect.

To promote interaction with membrane structures several axially substituted GePc, including cholesterol derivatives, were prepared as possible PS for PDT.[148] Spacer groups were introduced to increase overall amphiphilicity of the complex. Although good quantum yields for singlet oxygen were observed the complex required liposomal formulation for *in vivo* PDT. Octa-substituted GePc featuring both axial and peripheral substituents including cholesterol derivatives have also been reported[152] as possible PS for PDT.[151] Two types of SiPc, featuring axial cholesterol moieties either directly linked to the central silicon ion, or via SiO-*o*-bridges, have been reported.[246]

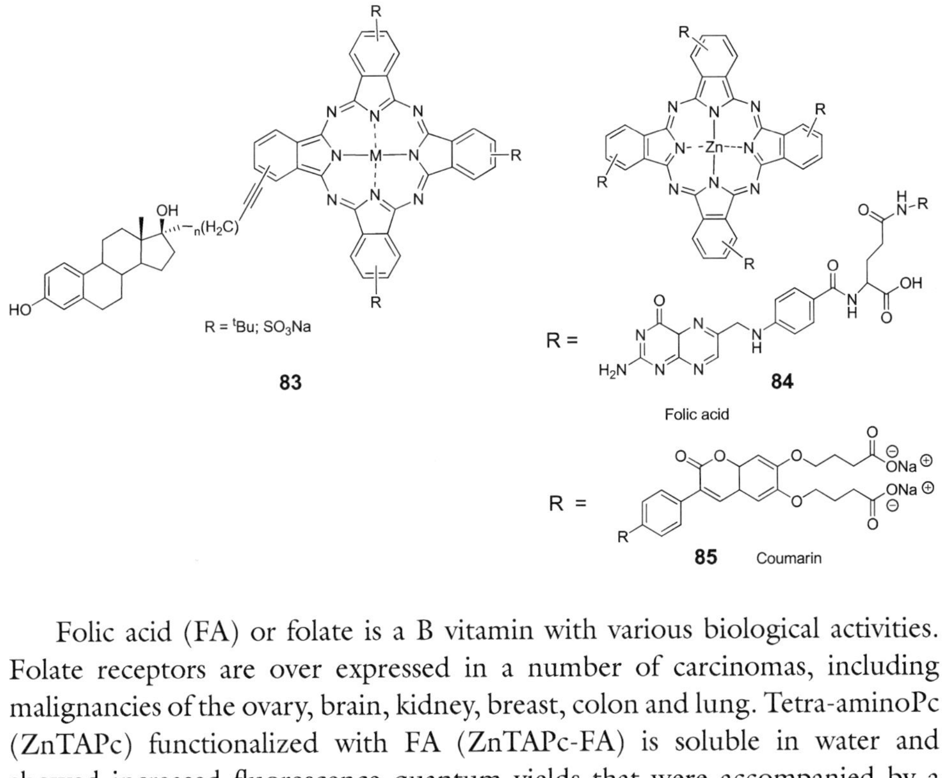

83 **84** **85**

Folic acid (FA) or folate is a B vitamin with various biological activities. Folate receptors are over expressed in a number of carcinomas, including malignancies of the ovary, brain, kidney, breast, colon and lung. Tetra-aminoPc (ZnTAPc) functionalized with FA (ZnTAPc-FA) is soluble in water and showed increased fluorescence quantum yields that were accompanied by a decrease in singlet oxygen quantum yields.[247] Covalent and non-covalent bonding between monoamino ZnPc (ZnMAPc) and either single walled carbon nanotubes (SWCNTs) or folic acid (FA) were prepared and showed a slight increase in the triplet and singlet oxygen quantum yields. Adsorption of the ZnMAPc–FA (**84**) conjugate to SWCNTs (ZnMAPc-FA-SWCNT) decreased the photophysical and photochemical parameters, although the complex was capable of generating singlet oxygen. The authors suggest that the complex may be employed in combination therapy, where ZnMAPc acts as the PDT agent, while the SWCNTs act as the photothermal (PT) agent as well as drug carrier, with the folic acid providing increased biocompatibility.[248]

Coumarin and its derivatives occur widely in nature. The diverse biological activities of natural and synthetic coumarin derivatives as anticoagulants and antithrombotics are well known. ZnPcs substituted with the coumarin framework showed excellent fluorescence properties with high triplet quantum yields.[249–251] Subsequently tetra-substitution with 7,8-(12-crown-4)-3-(4-oxyphenyl)coumarin at peripheral and non-peripheral positions was reported,[252] giving complexes with good singlet oxygen quantum yields. The presence of tetra-(3-ferrocenyl-7-oxycoumarin)-substituents strongly affected the fluorescence yield.[253] Water soluble coumarin ZnPc (**85**) complexes were obtained using sodium salts of tetra butanoic acid, to yield complexes that produce singlet oxygen and that strongly bind to BSA.[254]

The perfluorinated derivative hexa-(2,2,2-trifluoroethoxy)Pc has been synthesized,[255] and spectroscopic and PD properties have been reported.[117] The twelve periphereal trifluoroethoxy substituents improve PDT efficacy by preventing aggregation, increasing lipophilicity and inducing a red-shifted *Q* band. Dodecakis(trifluoroethoxy)ZnPc conjugated with uridine (**86**) and adenosine (**87**) via an ethynyl group was prepared as a site-selective agent for the photocleavage of DNA.[256]

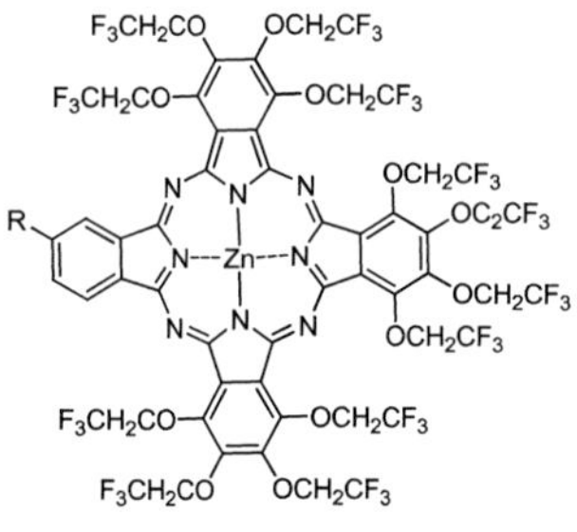

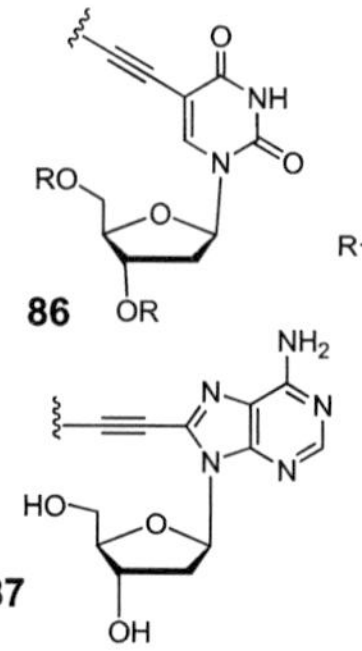

86

87

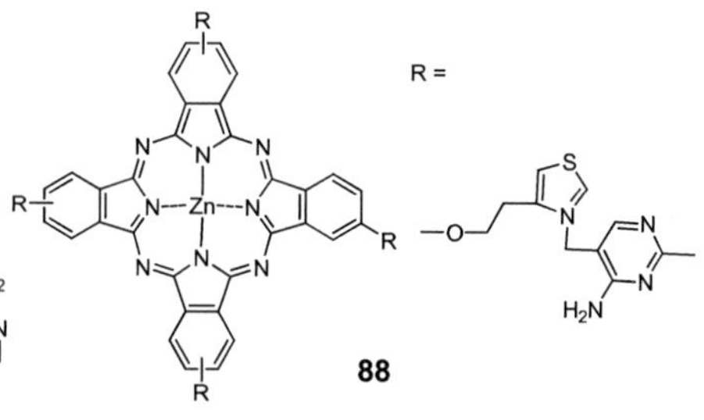

88

Gold nanoparticles (AuNPs) have been studied extensively for biological applications. They are non-toxic and were shown to be effective for photothermal therapy (PTT).[257] In the presence of gold nanoparticles, the fluorescence lifetimes and quantum yields of ZnPc tetrakis(7-coumarinthio-4-methyl) decreased, whereas the singlet oxygen quantum yield increased.[258] Conjugation of low symmetry carboxy Pc (Ge-, Ti- and Sn-MCPc) with glutathione capped gold nanoparticles (GSH-AuNPs) has been reported to induce a blue-shift of the *Q* band and increase in triplet life times.[259] The highest triplet quantum yield of 0.75 and the longest triplet lifetimes were observed for the GeMCPc–GSH-AuNPs conjugate. Water soluble tetra thiamine substituted ZnPc (**88**) was conjugated to gold nanospheres and nanorods resulting in increased fluorescence quantum yields.[260]

Asymmetric amino-functionalized Pc (AB_3, A_2B_2, A_3B) differ in molecular symmetry and water solubility, they have high singlet oxygen quantum yields and bind strongly to BSA facilitating their transport in blood.[261] Increased glucose uptake via the over-expressed glucose transporter protein in cancer cells[262] suggested that the addition of carbohydrates to Pc could give water soluble complexes with potential affinity for cancer cells.[263] The first carbohydrate substituted ZnPc was reported by Maillard *et al.*[264] Subsequently several groups developed different methods for the synthesis of glycol-conjugated ZnPc and reported on their effectiveness for PDT.[265] Covalently linked β-cyclodextrin-Pc dyads form stable, aqueous-compatible, neutral Pc.[266] Hanack and co-workers[267] described the preparation of water-soluble, symmetrically substituted Pc with four D-glucose units linked via hydroxyl groups at C-3 or C-1, respectively. The synthesis of a single isomeric, asymmetrical glycol-Pc with four D-galactose units linked via the C-6 hydroxyl group, has also been reported.[268] Mono and tetra-glycosylated ZnPc were evaluated for their *in vitro* photocytotoxicity revealing that the tetra-glycosylated Pc exhibited significantly lower activity as compared to the mono-glycosylated analogs. The lower photocytotoxicity was attributed to lower cellular uptake and higher aggregation as evidenced by weaker intracellular fluorescence.[269]

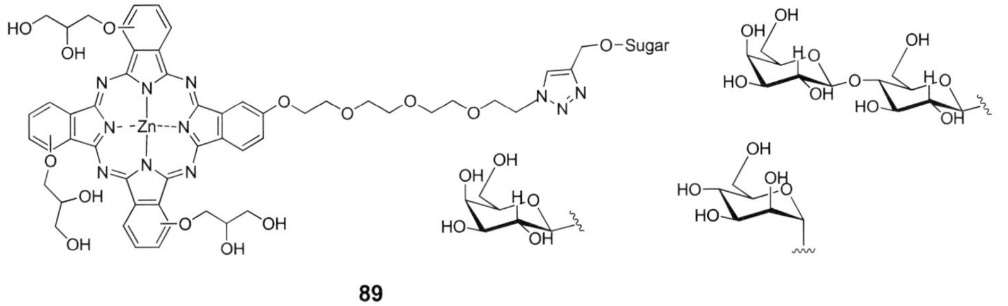

89

Click chemistry can be advantageously to graft carbohydrates on Pc (**89**). Two sets of water-soluble mono-glyco-conjugated (galactose, mannose or lactose) Pc were prepared using both conventional and click chemistry, and their cellular uptake and PDT activity was investigated using HT-29 human colon adenocarcinoma cells. The triazole linkage formed by click conjugation lowered the biological efficiency for mannose and galactose, as compared to the analogous products obtained via classical glycosylation.[270] Protected tetra-β-glycosylated ZnPc showed 10-fold higher uptake by MCF-7 cells and 2-fold greater phototoxicity than equimolar concentrations of $AlPcS_2$.[271] GlycoPc with four or eight unprotected D-galactose units were synthesised.[272] The asymmetric glycoPc with four D-galactose on the same ring induces high photocytotoxicity against cancer cells.[273] A series of asymmetrically substituted Pc conjugated to four different carbohydrate units,[274] bearing four or eight D-galactose units, were prepared as possible PS for recognition by selected cancer cells. Pc bearing four or eight D-galactose units have been reported,[275] as well as water soluble, non-ionic complexes substituted with unprotected carbohydrates.[276] Substitution of deoxyribose at *para* positions of Pc via C-C linkages improves water solubility and the lipophilicity/hydrophobicity balance, which combined with strong absorption bands in the red part of the spectrum suggested their suitability for PDT.[277] Among a series of tetra-ethylene-glycol-linked glucosylated ZnPc the tetra-glucosylated Pc was inactive whereas the mono- and di-glucosylated analogs exhibited substantial photocytotoxicity.[278] The different activities were attributed to differences in cellular uptake, aggregation, and singlet oxygen generating efficiency.

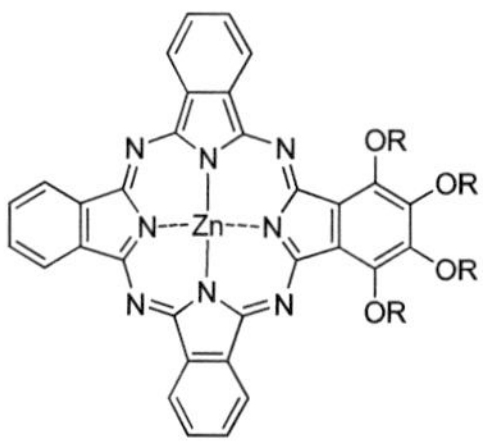

90 R =

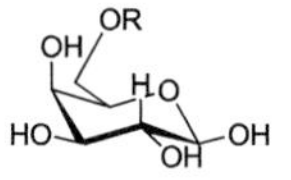

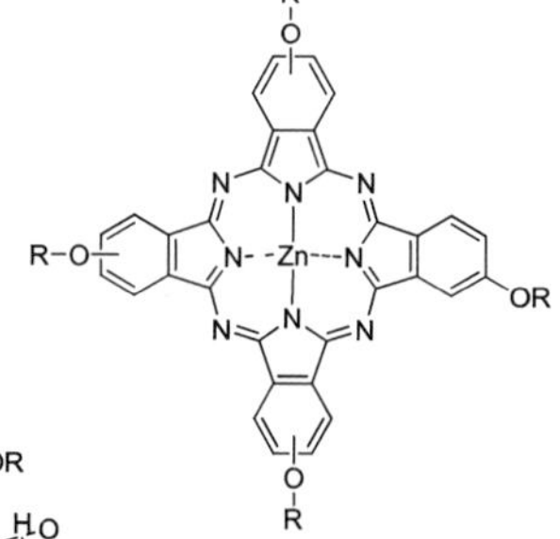

91

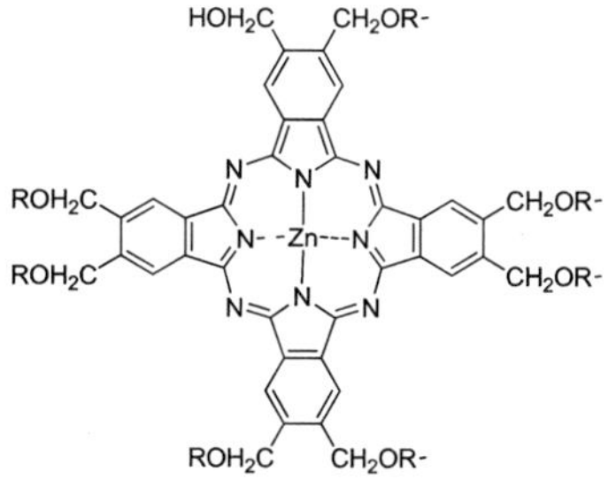

92

Tetraethyleneglycol-substituted ZnPc derivative (TEG-ZnPc) induced PDT cytotoxicity in a rat tumor model.[279] To target specific DNA sequences, MPc were conjugated with complementary deoxyribooligonucleotides and shown capable of recognizing the selected DNA sequence and to photo-induce DNA modifications. A possible role of Pc-oligonucleotide conjugates as novel artificial regulators of gene expression and as agents for cancer treatment has been advanced.[280]

A water soluble, symmetrically appended ZnPc featuring eight thioglucose units ($ZnPcGlc_8$), was synthesized from hexadecafluoroPc by controlled nucleophilic substitution of peripheral fluoro groups. Fluorescence microscopy confirmed *in vitro* cell uptake by breast cancer cells.[281] Fluorinated Pc conjugated with α-, β- and γ-cyclodextrins (CDs) were also prepared and *in vitro* photocytotoxicity against human bladder cancer cells suggested that Pc-α-CD and Pc-γ-CD have potential as PDT agents.[282] Pc–β-cyclodextrin dyads are water-soluble and possess photophysical properties suitable for applications in PDT.[266] Pc surrounded by a carbohydrate shell of sixteen galactose units distributed in a dendritic manner ($PcGal_{16}$) were prepared as potential PS for PDT of bladder cancer.[283] A series of carbohydrate-functionalized Pz (Pz-galactopyranose/methyl-ribose) derivatives bearing either H_2, Zn or Ni cores were prepared and solubilized using DCM-based PEG-$DSPE_{5000}$-PBS encapsulation. The highest degree of encapsulation was obtained with the ZnPz derivatives, which coincided with their superior phototoxicity against MCF-7 breast cancer cells.[169] Si(IV)Pc (**93**) with two axial *iso*propylidene-protected galactose moieties and one, two or eight chloro group(s) on the periphery of the macrocycle were synthesized to evaluate the heavy-atom effect on PD efficacy.[284] Although the added chloro substituents increased the quantum yield for singlet oxygen production, the non-chlorinated analog was more active due to more selective targeting of photosensitive lysosomes. Glucoconjugated Si(IV)Pc complexes featuring two axial 1,2:5,6-di-*O*-isopropylidene-α-d-glucofuranose substituents linked through a tetraethylene glycol chain were prepared with and without peripheral halogens, likewise to evaluate the heavy atom effect.[285] The non-chlorinated derivative exhibited higher *in vitro* PD activity due to lower tendency to aggregation, resulting in higher efficiency for singlet oxygen generation. Galactose-containing Si(IV)Pc derivatives with one or two axial acetal-protected galactose substituents earlier reported, also showed *in vitro* PD activity.[286]

Monoamine functionalized Pcs, featuring two different types of peripheral substituents, an amino group for covalent conjugation to biomolecules and short PEG chains to improve solubility, were prepared.[287] A novel synthetic route was developed utilizing a hydrophilic, PEG-based support with an acid labile Rink Amide linker. A microwave-assisted solid-phase synthetic method was successfully applied to obtain pure asymmetrically substituted

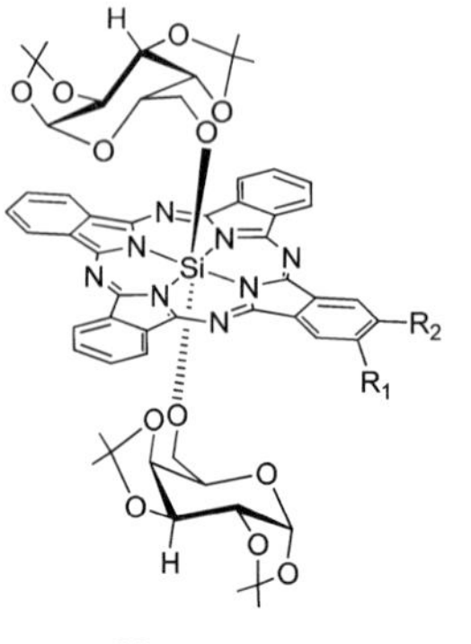

93a: $R_1 = R_2 = H$
93b: $R_1 = H$; $R_2 = Cl$
93c: $R_1 = R_2 = Cl$

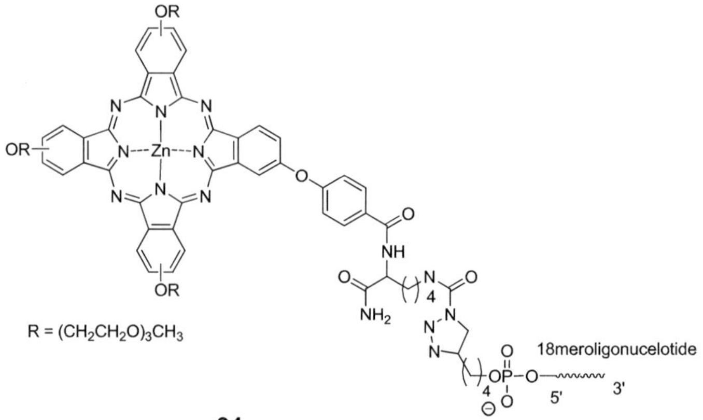

94

monoamine functionalized Pcs. Three different bioconjugation techniques, reductive amination, amidation, and Huisgen cycloaddition, were employed for covalent conjugation of Pcs to oligonucleotides (**94**).

7. Carboranyl Ps and Pz

Boron neutron capture therapy (BNCT) is based on the nuclear capture and fission reactions that occur when non-radioactive boron-10 is irradiated with neutrons to yield high energy alpha particles and high energy ^{7}Li nuclei. The nuclear reaction is:

$$^{10}\mathrm{B} + n\mathrm{th} \rightarrow [^{11}\mathrm{B}] \rightarrow \alpha + {}^{7}\mathrm{Li} + 2.31\ \mathrm{MeV}.$$

Both alpha particles and lithium ions produce closely spaced ionizations in the immediate vicinity of the boron atoms, with a range of approximately 5–9 μm (approximately the diameter of a single cell). Thus lethality is limited to boron containing cells with the success of BNCT depending on the selective delivery of sufficient amounts of ^{10}B to the tumor such that the surrounding normal tissues can be spared from the fission reactions. A wide variety of boron delivery agents have been synthesized with the aim to develop new agents that preferentially target tumor cells. Since porphyrins and related Pcs and Pzs accumulate to some extend in tumor tissues,[288] their potential use as boron-carriers for BNCT has been evaluated.[289] The synthesis of highly boronated Pc and Pz, together with an analysis of their relevant physical and chemical properties has recently been reviewed.[290] Liposomes of variable formulation and polarity can be efficiently loaded with carboranyl-Pz, opening a perspective to use supramolecular adducts as boron carriers to enter the membrane of cancerous cells for selective BNCT. *In vitro* assays have shown that particularly cationic liposomal formulations are captured by tumor cells. This behavior, together with their low toxicity, renders polycarboranyl-Pz suitable for application in BNCT.

Two strategies are used for the preparation of a boronated Pc: (1) attachment of the boron at an early stage in the synthesis by placing the boron moiety on the phthalonitrile, and (2) adding the boron group onto a reactive functional group of a preformed Pc derivative. The first water-soluble boronated Pc, bearing only one *closo*-carborane cage and three $SO_3^-Na^+$ substituents was synthesized by Soloway *et al.*[291] although little details on purification and characterization was provided. The synthesis of the first fully characterized water soluble covalently linked boronated Pc **95a,b** advanced as a BCNT agent, features four closo carborane cages containing 40 boron atoms and was reported in 1996 by Kahl and Li.[292] A similar procedure was used by Jori and co-workers[293] to synthesize the neutral ZnPc derivative **96a** and they reported photophysical properties, cell accumulation as well as *in vivo* PDT efficacy and potential for BNCT.

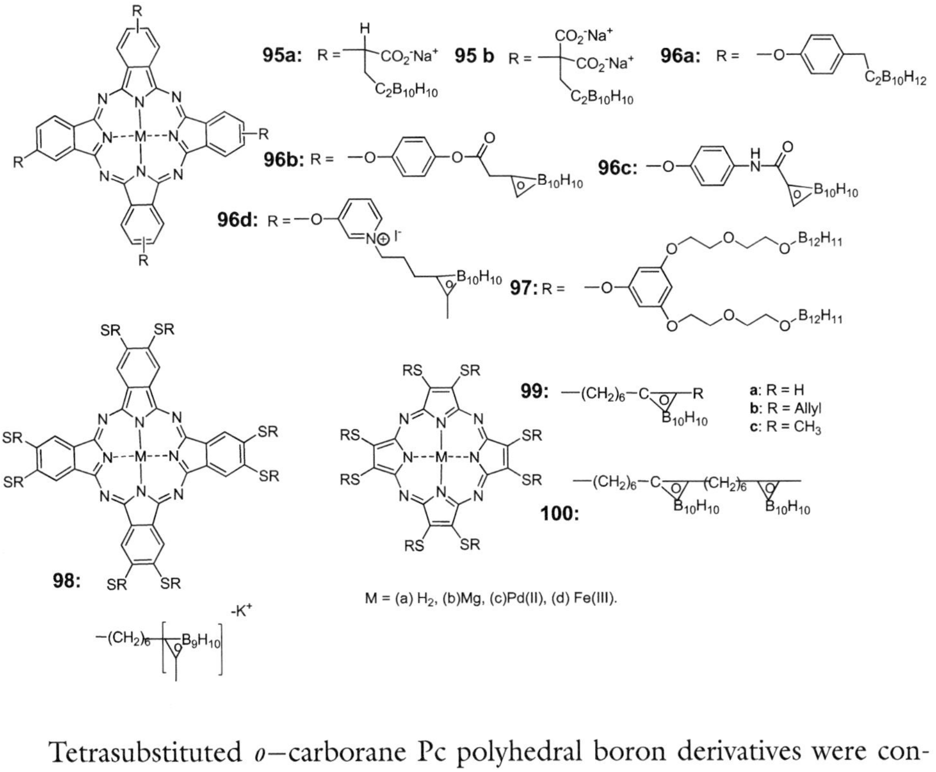

Tetrasubstituted *o*–carborane Pc polyhedral boron derivatives were connected *via* an oxyphenylene spacer to a Pc macrocycle to give complexes **96b–d**

featuring comparable photoactivity to that of $ZnPcS_4$.[294] The Pc derivative **96d** with a propylene unit between the positively charged Pc and the *o*-carboranyl groups was obtained in a yield of 34% through N-alkylation of 2,9,16,23-tetra-(3-pyridyloxy) ZnPc with 1-iodopropyl-2-methylcarborane.[245,295] ZnPc with eight closo-dodecarborane units (96 boron atoms) (**97**) was prepared via a multistep reaction sequence and shown to accumulate in human lung adenocarcinoma cells.[296] Fabris *et al.*[297] prepared a peripherally substituted ZnPc bearing dodecarborane groups that appeared to be efficient sensitizers for PDT. Adding o-carborane units to ZnPc improved singlet oxygen quantum yields, photo-oxidative stability and photocatalytic activities rendering them of interest for cancer therapy.[298] Anionic and cationic cobaltacarborane Pc conjugates have blue-shifted absorption and fluorescence bands in polar organic solvents.[299] The anionic conjugate is mainly monomeric in polar organic solvents and has fluorescence quantum yields 0.2–0.3. The synthesis of two A_3B-type ZnPc containing either one or two cobaltacarborane residues is described as potential bimodal sensitizers for PDT and BNCTPcs.[300] The direct synthesis and relevant physicochemical properties of a Pc decorated with eight thioalkyl-*nido*-carborane functions (**98**), where the anionic polyhedral are in the form of K^+ salt, are reported by Pietrangeli *et al.* as potential PD agents.[301]

A multistep synthesis of Pzs symmetrically functionalized with different carboranyl (alkylthio)-pendants, including the 1-methyl-o-closocarboranyl-2-hexylthio ZnPz, were prepared and incorporated into liposomes bearing different surface charges.[302] *In vitro* evaluation of **99c** loaded into liposomes using rat colon carcinoma and murine melanoma models showed that the use of cationic liposomes increased the cellular concentration of ^{10}B by at least 30 times more than that achieved by borophenylalanine. A highly boronated PzMgII (**100b**) and its free base **100a**, were synthesized and characterized.[303] These "spider-like" molecules bear eight dicarboranyl-substituted carborarod-like thioalkyl pendants and represent a rare example of a tetrapyrrolic systems bearing 160 boron atoms per molecule, corresponding to ca 40% boron content by weight. Transition-metal complexes PdMCHESPz (**100c**) and ClFeMCH-ESPz (**100d**), derived from lipophilic carboranyl(alkylthio)Pz H_2MCHESPz (**100a**), were also synthesized.[304] Neutral **100a** is insoluble both in water and biological fluids, which makes the use of liposomal carriers necessary for biomedical applications. To enhance solubility the carboranyl-Pz were converted in their water-soluble analogs through mild deboronation of the closo-polyedral.

8. Radiolabelled Ps for Nuclear Imaging

Nuclear medicine involves administrating a radioactive substances to a patient for diagnose and/or therapeutic use. Radionuclides are attached to carrier compounds, including known pharmaceuticals, permitting the resulting radiopharmaceuticals to visualize behavior of the parent molecule in biological systems. Depending on the mode of decomposition of the radionuclide, the radioactive decay can be used for *in vitro* and *in vivo* imaging or local radiotherapy. Compounds labelled with β-emitting radionuclides are appropriate for *in vitro* studies while γ- or positron-emitting radionuclides are particularly interesting for *in vivo* imaging, using either single photon emission tomography (SPECT) or positron emission tomography (PET).

The potential use of porphyrins for tumor delineation was suggested by Pollicard in 1924.[305] Subsequently, several porphyrin derivatives were shown to accumulate preferentially in tumor tissues. To capitalize on this phenomenon different radionuclides were used to label porphyrins and to assess their potential as tumor imaging agents.[306–308] Similarities with Pc lead to the preparation of radiolabeled Pc as potential radiopharmaceuticals for medical applications. A first account of the synthesis of Pc labeled with ^{64}Cu ($t_{1/2}$ = 12.7 h) was reported by Wrenn *et al.* in 1951.[309] The radioactive complex showed good selectivity for experimental brain tumors in an animal model. The sulfonated Al-, Ga- and Zn-PcS all are effective PDT agents. To correlate their tumor retention and PDT response, radiolabeled analogs were prepared using ^{65}Zn ($T_{1/2}$ = 244 d). Mixtures of sulfonated $ZnPcS_{1-4}$ were radiolabeled via direct metallation of the corresponding metal-free H_2PcS_n with [^{65}Zn]$ZnCl_2$.[310] Although useful for *in vitro* cell uptake studies, the long half-life of ^{65}Zn renders the tracer unsuitable as radiopharmaceutical for nuclear imaging. Attempts to insert ^{67}Ga ($T_{1/2}$ = 78 h) directly into the cavity of the macrocycle of a water-soluble sulfonated Pc resulted in poor labeling yields. Differently sulfonated radiolabeled [^{67}Ga]$GaPcS_n$ were obtained by mixed condensation of phthalic and 4-sulfophthalic acid using [^{67}Ga]$GaCl_3$ as metal template.[311] Radioactive ^{67}Ga-labeled [^{67}Ga]chlorogallium-tetraoctadecyloxy Pc and its sulfonation derivatives were prepared and evaluated for tumor uptake in a mouse tumor model.[312] In an alternative approach, DTPA was attached to the periphery of ZnPc to give ZnPc(DTPA), which was readily labelled with ^{67}Ga and ^{111}In ($T_{1/2}$ = 2.8 d). Biodistribution in tumor bearing mice showed that ^{67}Ga-DTPA-ZnPc complexes gave higher tumor uptake values than either ^{67}Ga-DTPA alone or the clinically used tumor scanning agent ^{67}Ga-citrate. Metal-free sulfonated H_2PcS_{mix} are readily labeled with $^{64}Cu^{++}$ to yield a mixture of [^{64}Cu]$CuPcS_{mix}$ by direct insertion of the central metal radionuclide.[313] ^{64}Cu-labeled Pc were tested as a mixture, since neither the metal-free H_2PcS_n nor the corresponding ^{64}Cu-labeled PcS could be obtained in high purities. PET imaging in tumor-bearing rats showed low tumor uptake. Using a new demetallation procedure,[314] a series of pure $ZnPcS_n$ was de-metallated and subsequently converted to the corresponding pure [^{64}Cu]$CuPcS_n$ in 40–50% isolated yields and >98% radiochemical purities.[315] Biodistribution patterns in tumor bearing mice showed significant differences between the various derivatives. Tumor uptake was detected only for the amphiphilic [^{64}Cu]$CuPcS_2$ and [^{64}Cu]$CuPcS_3C_6$ (1–1.5 %ID/g) suggesting their potential to establish structure–PDT efficacy relationships using micro-PET imaging.

The synthesis of a series of bimodal probes combining water-soluble ZnPcS as fluorescence imaging moiety, and either Ga-1,4,7, 10-tetraazocyclododecane-N,N′N″,N‴-tetraacetic acid (DOTA) or Cu-1,4,7-triazacyclononane-1,4,7-triacetic acid (NOTA) labeled with ^{68}Ga ($T_{1/2}$ = 68 min) or ^{64}Cu, respectively for PET-imaging, have been reported.[316] DOTA- or NOTA-ZnPcS conjugates, linked through aliphatic chains of different lengths to modulate amphiphilicity, were labeled with ^{68}Ga (**101**) and ^{64}Cu (**102**) in high efficiency >90 and high *in vitro* plasma stability suggested their potential for both fluorescence and PET imaging.

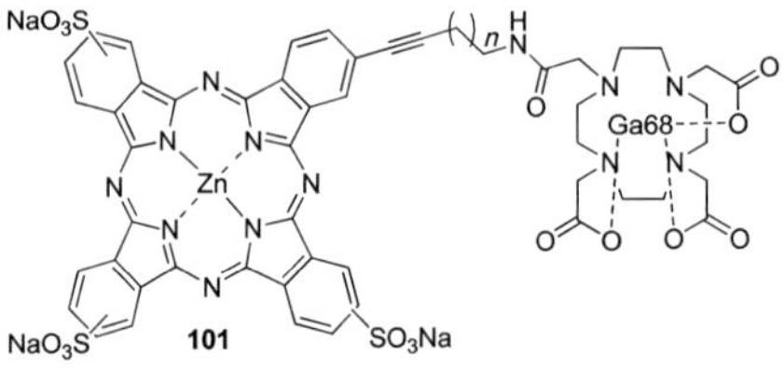

101

n = 1 68Ga/DOTA-C4-ZnPcS$_3$, 14
n = 3 68Ga/DOTA-C6-ZnPcS3, 15
n = 5 68Ga/DOTA-C8-ZnPcS$_3$,16

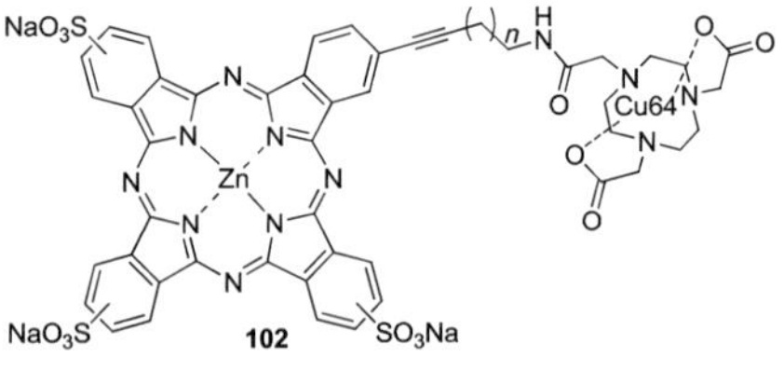

102

n = 1 64Cu/NOTA-C4-ZnPcS$_3$,17
n = 2 64Cu/NOTA-C6-ZnPcS$_3$,18
n = 3 64Cu/NOTA-C8-ZnPcS$_3$, 19

A 1962 U.S. patent[317] claims that water-soluble sulfonated $^{235}UO_2PcS$, prepared from the inorganic uranyl salt and a sulfonated Pc, is strongly retained in mouse brain tumors at 50:1 ratios compared to surrounding healthy brain tissue. The structure of $^{235}UO_2PcS$ was subsequently elucidated by Day *et al.*[318] and defined as a super-Pc consisting of a cyclic five-subunit pentakis(2-iminoisoindoline) complex. Fluoro-18 ($t_{1/2}$ = 118 min) is the most widely used PET isotope. A series of fluorinated ZnPc was prepared via attachment of aliphatic chains of different lengths bearing a terminal tosylate.[319] Nucleophilic substitution permitted introduction of cold fluorine or ^{18}F. The resulting ^{18}F-tracer (**105**) rendered the complex suitable for both fluorescence and PET imaging.

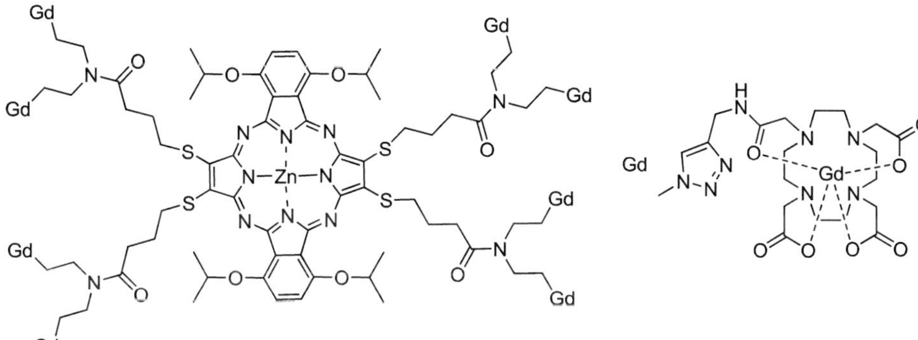

103

Labeling of Pc with ^{3}H or ^{14}C for biodistribution studies to optimize PDT protocols has the advantage that pharmacological properties of the drug are not altered.[320] Tritium labelling of Pc provides material suitable for autoradiography,[321] however *in vivo* hydrogen-exchange renders these compounds insufficiently stable.[322] A series of $GaPcS_{1-4}$ labeled with ^{14}C were prepared for *in vivo* biodistribution studies in tumor bearing mice using conventional dissection and counting revealing that the highest tumor uptake was obtained with the trisulfonated derivative at 48 h post injection.[323] Radioiodine was also used to label Pc via cyclotetramerization of [^{125}I]iodophthalonitrile and $Zn(OAc)_2$ to yield [^{125}I]$_4$ZnPc, which was formulated in different vehicles (DMSO, serum, nanoparticles) for pharmacokinetic studies.[323]

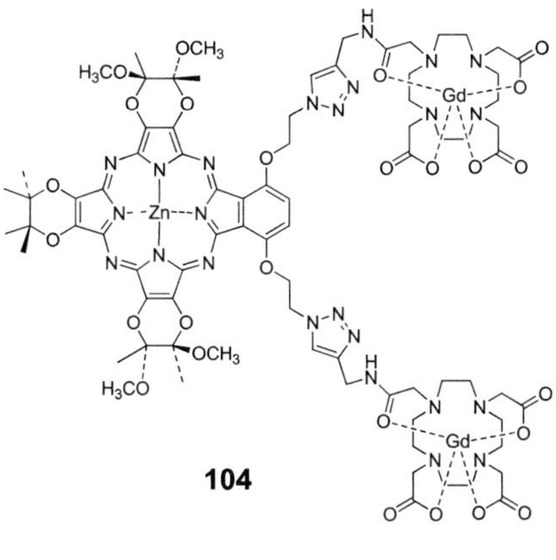

104

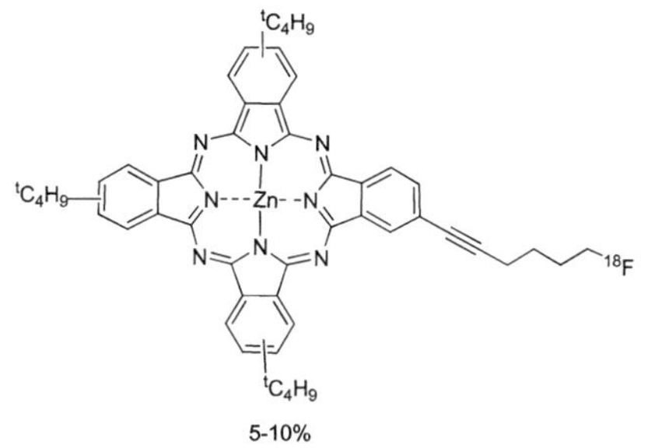

105

Technetium-99*m* ($t_{1/2}$ = 6 h; 140 keV), available from a $^{99}Mo/^{99m}Tc$ generator, is the most widely used radioisotope in conventional nuclear medicine and SPECT. Attempts to label PcS_4 with the long-lived ^{99}Tc ($t_{1/2}$ = 2.1×10^5 y) isotope gave a major product corresponding to an out-of-plane complex with the Tc-atom attached to three imidazole N-atoms and a minor complex to which a sandwich-type $^{99}Tc(PcS_4)_2$ complex was assigned.[324] Subsequently PcS_4 was also labeled with the short-lived ^{99m}Tc and scintigraphic studies in rabbits confirmed the *in vivo* stability of the ^{99m}Tc-complex.[325]

Magnetic resonance imaging (MRI) is a routine diagnostic clinical tool for three-dimensional, high-resolution imaging of tissue anatomy. The images can be enhanced by the presence of contrast agents such as paramagnetic Gd(III) complexes. The preparation of a series of agents in which 1-8 paramagnetic Gd(III) complexes are conjugated to a single fluorescent Pz has been reported.[326] Among these conjugates, the ZnPz-$[Gd(III)]_8$ (**103**) exhibits an ionic relaxivity four times that of the monomeric Gd(III) agent, while retaining the intense electronic absorption and fluorescence properties of the unmodified Pz. Combined with good cell uptake these results suggest their potential as multimodel cancer imaging agents, using both MRI and NIR fluorescence, while simultaneously acting as a PDT agent. Conjugate **103** was taken up by tumor cells *in vitro*, but poor synthetic yields precluded further development for *in vivo* testing. An amphiphilic Zn–Pz–Gd(III) conjugate (**104**) has been proposed as an MRI contrast agent.[327] The complex was prepared via azide–alkyne Huisgen cycloaddition, or 'click' chemistry, between an azide functionalized Pz and alkyne functionalized DOTA–Gd(III) analog. The complex exhibits solution-phase ionic relaxivity approximately four times higher than that of the clinically used contrast agent, DOTA–Gd(III). Breast tumor cells incubated with Cu–Pz–Gd(III) *in vitro*, showed significant contrast enhancement (9 ± 0.3 contrast-to-noise ratio) as observed in phantom cell pellet MR images. An *ex vivo* biodistribution study confirmed that Cu–Pz–Gd(III) accumulates in tumors and is rapidly cleared from the animal, primarily through the kidney pathway. Combined, differential accumulation and T1 enhancement evidence, suggests that this agent could find application in MRI imaging of necrotic tissues.

A method to prepare stable Protein Cage Nanoparticles (PCN), containing ZnPc-loaded paramagnetic micelles in their interior, has recently been reported.[328] The synthesis of these PCNs proceeds through a hierarchical self-assembly process in which an amphiphilic ZnPc dye is encapsulated in micelles formed by Gd-DOTAC10 (1,4,7,10-tetraaza-1-(1-carboxymethylundecane)-4,7,10-triacetic acid cyclododecane) and the resulting micelles, in turn, trigger the self-assembly of a virus (CCMV) coat protein (CP) around them. The authors suggest that these optically active, paramagnetic protein cages have a potential use as self-assembled protein cages for multimodal imaging and therapy.

Acknowledgments

Supported by the Jeanne and J.-Louis Lévesque Chair in radiobiology of the Université de Sherbrooke. JEvL is a member of the Research Center of the CHUS (CR-CHUS), Sherbrooke, QC, Canada, supported by the Fonds de la Recherche du Québec – Santé.

References

1. Cancer Fact sheet N°297. *World Health Organization*. Retrieved (10 June 2014).
2. Defining Cancer. *National Cancer Institute*. Retrieved (10 June 2014).
3. World Cancer Report. *World Health Organization*. Chapter 1.1. (2014).
4. A. Jemal, F. Bray, M. M. Center, J. Ferlay, E. Ward and D. Forman, *CA Cancer J. Clin.* **61**, 69 (2011).
5. B. R. Cassileth and G. Deng, *Oncologist* **9**, 80 (2004).
6. S. H. Sleigh and C. L. Barton, *Pharm. Med.* **24**, 151 (2010).
7. H. Winther and J. T. Jorgensen, *Pharm. Med.* **24**, 363 (2010).
8. *Cholesterol Autoxidation*, Ed. L. L. Smith (Plenum Press, New York, 1981).
9. P. G. Geiger, W. Korytowski, F. Lin and A. Girotti, *Free Rad. Biol.* **23**, 57 (1997).
10. R. Langlois, H. Ali, N. Brasseur, R. Wagner and J. J. van Lier, *Photochem. Photobiol.* **44**, 117 (1986).
11. J. L. Ravanat, M. Berger, G. W. Buchko, J. F. Bénard, J. E. van Lier and J. Cadet, *J. Chim. Phys.* **88**, 1069 (1991).
12. *The Porphyrins*, Vol. III. *Physical Chemistry*, Part A., Ed. D. Dolphin (Academic Press, 1978).
13. *The Porphyrin Hand Book*. Eds. K. M. Kadish, K. M. Smith and R. Guilard, Vol. 16, *Phthalocyanines: Spectroscopic and Electrochemical Characterization*. (Elsevier Science, San Diego, CA, USA, 2003).
14. A. Braun and J. Tcherniac, *Berichte der Deutschen Chemischen Gesellschaft*, **40**, 2709 (1907).
15. H. de Diesbach and E. von der Weid, *Helevtica Chimica Acta* **10**, 886 (1927).
16. (a) R. P. Linstead, *Br. Assoc. Adv. Sci. Rep.*, 465 (1933). (b) G. T. Byrne, R. P. Linstead and A. R. Lowe, *J. Chem. Soc.*, 1017 (1934). (c) C. E. Dent and R. P. Linstead, *J. Chem. Soc.*, 1027 (1934). (d) C. E. Dent, R. P. Linstead and A. R. Lowe, *J. Chem. Soc.*, 1033 (1934).
17. W. M. Sharman and J. E. van Lier, *Phthalocyanines: Synthesis*, Vol. 15, in *The Porphyrin Handbook*. Eds. K. M. Kadish, K. M. Smith and R. Guilard (Elsevier Science, San Diego, CA, USA, 2003), chapter 97, pp. 1–60.
18. A. Meller and A. Ossko, *Monatsh. Chem.* **103**, 150 (1972).

19. N. Kobayashi, R. Kondo, S.-C. Nakajima and T. Osa, *J. Am. Chem. Soc.* **112**, 9640 (1990).
20. M. Geyer, F. Plenzig, J. Rauschnabel, M. Hanack, B. del Rey, A. Sastre and T. Torres, *Synthesis*, 1139 (1996).
21. R. Minnes, H. Weitman, H.-J. Lee, S. M. Gorun and B. Ehrenberg, *Photochem. Photobiol.* **82**, 593 (2006).
22. M. P. Cormick, M. Rovera and E. N. Durantini, *J. Photochem. Photobiol. A: Chem.* **194**, 220 (2008).
23. H. Ali, R. Langlois, J. R. Wagner, N. Brasseur, B. Paquette and J. E. van Lier, *Photochem. Photobiol.* **47**, 713 (1988).
24. *Microwave in Organic Synthesis*, Eds. J. Tierney and P. Lidstrom (Wiley-VCH Verlag GmbH & Co. KGaA, Weinheim, 2002).
25. K. S. Jung, J. H. Kwon, S. M. Son, J. S. Shin, G. D. Lee and S. S. Park, *Synth. Met.* **141**, 259 (2004).
26. Z. Li, J. E. van Lier and C. C. Leznoff, *Can. J. Chem.* **77**, 138 (1999).
27. N. Brasseur, H. Ali, R. Langlois and J. E. van Lier, *Photochem. Photobiol.* **47**, 705 (1988).
28. A. Feofanov, A. Grichine, T. Karmakova, N. Kazachkina, E. Pecherskih, R. Yakubovskaya, E. Lukyanets, V. Derkacheva, M. Egret-Charlier and P. Vigny, *Photochem. Photobiol.* **75**, 527 (2002).
29. C. M. Allen, R. Langlois, W. M. Sharman, C. La Madeleine and J. E. van Lier, *Photochem. Photobiol.* **76**, 208 (2002).
30. R. de Paula Rodrigues, I. R. P. Tini, C. P. Soares and N. S. da Silva, *Cell Biol. Int.* **38**, 716 (2014).
31. M. Goeppert-Mayer, *Ann. Phys.* **401**, 273 (1931).
32. M. Pawlicki, H. A. Collins, R. G. Denning and H. L. Anderson, *Angew. Chem. Int. Ed.* **48**, 3244 (2009).
33. Y. Mir, D. Houde and J. E. van Lier, *Photochem. Photobiol. Sci.* **5**, 1024 (2006).
34. Y. Mir, D. Houde and J. E. van Lier, *Lasers Med. Sci.* **23**, 19 (2008).
35. Y. Mir, J. E. van Lier, J.-F. Allard, D. Morris and D. Houde, *Photochem. Photobiol. Sci.* **8**, 391 (2009).
36. H. Ali, N. Cauchon and J. E. van Lier, *Photochem. Photobiol. Sci.* **8**, 868 (2009).
37. (a) K. Sakamoto and F. Shibamiya, *J. Jpn. Soc. Colour Mater.* **59**, 517 (1986). (b) K. Sakamoto, E. Ohno-Okumura, T. Kato, M. Watanabe and M. J. Cook, *Metal-Based Drugs*, **39**, 2090 (2008). (c) K. Sakamoto, E. Ohno-Okumura, T. Kato, M. Watanabe and M. J. Cook, *Dyes Pigments* **78**, 213 (2008).
38. (a) B. Paquette, R. W. Boyle, H. Ali, A. H. MacLennan, T. G. Truscott and J. E. van Lier, *Photochem. Photobiol.* **53**, 323 (1991). (b) R. W. Boyle, B. Paquette and J. E. van Lier, *Br. J. Cancer* **65**, 813 (1992).
39. H.-F. Huang, Y.-Z. Chen, Y. Wu, N.-N. Li and P. Chen, *Photodiagnosis Photodynamic Therapy* **4**, 100 (2007).
40. P. R. Sekhejane, N. N. Houreld and H. Abrahamse, *Int. J. Photoenergy*, 383027 (2014).
41. S. V. Kudrevich, S. Gilbert and J. E. van Lier, *J. Org. Chem.* **61**, 5706 (1996).
42. S. V. Kudrevich, N. Brasseur, C. La Madeleine, S. Gilbert and J. E. van Lier, *J. Med. Chem.* **40**, 3897 (1997).
43. M. Idowu, J.-Y. Chenab and T. Nyokong, *New J. Chem.* **32**, 290 (2008).
44. M. Brasch, A. de la Escosura, Y. Ma, C. Uetrecht, A. J. R. Heck, T. Torres and J. J. L. M. Cornelissen, *J. Am. Chem. Soc.* **133**, 6878 (2011).

45. H. Tian, H. Ali and J. E. van Lier, *Tetrahedron Lett.* **41**, 8435 (2000).
46. N. Cauchon, H. Tian, R. Langlois, C. La Madeleine, S. Martin, H. Ali, D. Hunting and J. E. van Lier, *Bioconjugate Chem.* **16**, 80 (2005).
47. H. Ali, O. St-Jean, J. P. Tremblay-Morin and J. E. van Lier, *Tetrahedron Lett.* **47**, 8275 (2006).
48. C. M. Allen, J. M. Weber and J. E. van Lier, *Photochem. Photobiol.* **62**, 184 (1995).
49. J. E. van Lier, H. Tian, H. Ali, N. Cauchon, H. M. Hasséssian, *J. Med. Chem.* **52**, 4107 (2009).
50. S. V. Kudrevich, H. Ali, and J. E. van Lier, *J. Chem. Soc. Perkin Trans I* **19**, 2767 (1994).
51. T. Goslinski and J. Piskorz, *J. Photochem. Photobiol. C* **12**, 304 (2011).
52. E. Alléman, N. Brasseur, S. V. Kudrevich, C. La Madeleine and J. E. van Lier, *Int. J. Cancer*, 72, 289 (1997).
53. K. Fukushima, K. Tabata and I. Okura, *J. Porphyrins Phthalocyanines* **2**, 219 (1998).
54. E. Güzel, A. Atsay, S. Nalbantoglu, N. Saki, A. L. Dogan, A. Gül and M. B. Koçak, *Dyes Pigm.* **97**, 238 (2013).
55. (a) E. I. Yslas, V. Rivarola and E. N. Durantini, *Bioorg. Med. Chem.* **13**, 39 (2005). (b) L. N. Milla, E. I. Yslas, A. V. Cabral, E. N. Durantini, S. Romanini, V. Rivarola and M. Bertuzzi, *Bio. Med. Pharmacotherapy* **63**, 209 (2009).
56. S. Gaspard, P. Margaron, C. Tempete and T.-H Tran-Thi, *Photochem. Photobiol.* **4**, 419 (1990).
57. P. Margaron, S. Gaspard, and J. E. van Lier, *J. Chromatography* **634**, 57 (1993).
58. P. Margaron, R. Langlois, J. E. van Lier and S. Gaspard, *Photochem. Photobiol.* **14**, 187 (1992).
59. J. B. Liu, Y. Zhao, F. Q. Zhao, F. S. Zhang, X. Q. Song and F. T. Chao, *J. Photochem. Photobiol. A* **99**, 115 (1996).
60. L. Huang, C. Zhong, F. Zhang and C.-H. Tung, *Bioorg. Med. Chem. Lett.* **18**, 2152 (2008).
61. W. M. Sharman, S. V. Kudrevich and J. E. van Lier, *Tetrahedron Lett.* **37**, 5831 (1996).
62. R. W. Boyle and J. E. van Lier, *Synthesis* **9**, 1079 (1995).
63. R. W. Boyle and J. E. van Lier, *Synlett.*, 351 (1993).
64. M.-R. Ke, J.-D. Huang and S.-M. Weng, *J. Photochem. Photobiol. A: Chem.* **201**, 23 (2009).
65. C.-F. Choi, P.-T. Tsang, J.-D. Huang, E. Y. M. Chan, W.-H. Ko, W.-H. Fong and D. K. P. Ng, *Chem. Commun.*, 2236 (2004).
66. K. Oda, S.-I. Ogura and I. Okura, *J. Photochem. Photobiol. B: Biol.* **59**, 20 (2000).
67. K. E. Sekhosana, E. Antunes and T. Nyokong, *Polyhedron* **54**, 294 (2013).
68. Y. Peng, F. Huang, J. Wen, B. Huang, X. Ma and Q. Wang, *J. Coord. Chem.* **61**, 1503 (2008).
69. T. Yang, Chen, C. Dajun, *J. Mater. Sc.* **49**, 8116 (2014).
70. N. Malinga, E. Antunes and T. Nyokong, *Polyhedron* **55**, 121 (2013).
71. R. A. Bulgakov, N. A. Kuznetsova, O. V. Dolotova, E. N. Shevchenko, A. D. Plyutinskayab, O. L. Kaliya and T. Nyokong, *Macroheterocycles* **5**, 350 (2012).
72. H. Ihmels and D. Otto, *Top. Curr. Chem.* **258**, 161 (2005).
73. C. C. Leznoff, S. Vigh, P. I. Svirskaya, S. Greenberg, D. M. Drew, E. Ben-Hur and I. Rosenthal, *Photochem. Photobiol.* **49**, 279 (1989).

74. D. Wöhrle, N. Iskander, G. Graschew, H. Sinn, E. A. Friedrich, W. Maier-Borst, J. Stern, P. Schlag, *Photochem. Photobiol.* **51**, 351 (1990).
75. Z. Bykloglua, M. Durmus and H. Kantekina, *J. Photochem. Photobiol. A: Chem.* **222**, 87 (2011).
76. M. Idowu and T. Nyokong, *Polyhedron* **28**, 416 (2009).
77. J. Marino, M. C. Garcia Vior, L. E. Dicelio, L. P. Roguin and J. Awruch, *Eur. J. Med. Chem.* **45**, 4129 (2010).
78. G. A. Gauna, J. Marino, M. C. Garcia Vior, L. P. Roguin and J. Awruch, *Eur. J. Med. Chem.* **46**, 5532 (2011).
79. N. C. L. Zeballos, G. A. Gauna, M. C. Garcia Vior, J. Awruch and L. E. Dicelio, *J. Photochem. Photobiol. B: Biol.* **136**, 29 (2014).
80. V. Kussovski, V. Mantareva, I. Angelov, P. Orozova, D. Woehrle, G. Schnurpfeil, E. Borisova and L. Avramov, *FEMS Microbiol. Lett.* **294**, 133 (2009).
81. P.-C. Lo, B. Zhao, W. Duan, W.-P. Fong, W.-H. Ko and D. K. P. Ng, *Bioorg. Med. Chem. Lett.* **17**, 1073 (2007).
82. M.-R. Ke, J. M. Eastel, K. L. N. Ngai, Y.-Y. Cheung, P. K. S. Chan, M. Hui, D. K. P. Ng and P.-C. Lo, *Eur. J. Med. Chem.* **84**, 278 (2014).
83. I. Scalise, E. N. Durantini, *Bioorg. Med. Chem.* **13**, 3037 (2005).
84. M. Durmus¸ H. Yaman, C. Gol, V. Ahsen and T. Nyokong, *Dyes Pigm.* **91**, 153 (2011).
85. S. Moeno and T. Nyokong, *J. Photochem. Photobiol. A: Chem.* **203**, 204 (2009).
86. S. Moeno, R. W. M. Krause, E. A. Ermilov, W. Kuzyniak and M. Höpfner, *Photochem. Photobiol. Sci.* **13**, 963 (2014).
87. W. Duan, Z. Wang and M. J. Cook, *J. Porphyrins Phthalocyanines*, **13**, 1255 (2009).
88. C. Uslan and B. S. Sesalan, *Dyes Pigm.*, *94*, 27 (2012).
89. B. Turanlı-Yıldız, T. Sezgin, Z. P. Cakar, C. Uslan, B. S. Sesalan and A. Gul, *Synth. Met.* **161**, 1720 (2011).
90. A. M. Zhang, J. Huang, X. C. Weng, J. X. Li, L. G. Ren, Z. B. Song, X. Q. Xiong, X. Zhou, X. P. Cao and Y. Zhou, *Chem. Biodivers.* **4**, 215 (2007).
91. M. Ozçeşmeci, Ö. B. Ecevit, S. Sürgün and E. Hamuryudan, *Dyes Pigm.* **96**, 52 (2013).
92. J. B. Pereira, E. F. A. Carvalho, M. A. F. Faustino, R. Fernandes, M. G. P. M. S. Neves, J. A. S. Cavaleiro, N. C. M. Gomes, A. Cunha, A. Almeida and J. P. C. Tome, *Photochem. Photobiol.* **88**, 537 (2012).
93. N. Kuznetsova, D. Makarov, O. Yuzhakova, A. Strizhakov, Y. Roumbal, L. Ulanova, A. Krasnovsky and O. Kaliya, *Photochem. Photobiol. Sci.* **8**, 1724 (2009).
94. H. Li, T. J. Jensen, F. R. Fronczek and M. G. H. Vicente, *J. Med. Chem.* **51**, 502 (2008).
95. P.-C. Lo, J.-D. Huang, D. Y. Y. Cheng, E. Y. M. Chan, W.-P. Fong, W.-H. Ko and D. K. P. Ng, *Chem. Eur. J.* **10**, 4831 (2004).
96. J.-D. Huang, P.-C. Lo, Y.-M. Chen, J. C. Lai, W.-P. Fong and D. K. P. Ng, *J. Inorg. Biochem.* **100**, 946 (2006).
97. J. Mao, Y. Zhang, J. Zhu, C. Zhang and Z. Guo, *Chem. Commun.*, 908 (2009).
98. X.-J. Jiang, J.-D. Huang, Y.-J. Zhu, F.-X. Tang, D. K. P. Ng and J.-C. Sun, *Bioorg. Med. Chem. Lett.* **16**, 2450 (2006).
99. P.-C. Lo, S. C. H. Leung, E. Y. M. Chan, W.-P. Fong, W.-P. Ko and D. K. P. Ng, *Photodiagnostic Phtodyanamic Ther.* **4**, 117 (2007).
100. J.-D. Huang, W.-P. Fong, E. Y. M. Chan, M. T. M. Choi, W.-K. Chan, M.-C. Chan and D. K. P. Ng, *Tetrahedron Lett.* **44**, 8029 (2003).

101. X.-J. Jiang, S.-L. Yeung, P.-C. Lo, W.-P. Fong and D. K. P. Ng, *J. Med. Chem.* **54**, 320, (2011).
102. E. Reddi, G. LoCastro, R. Biolo and G. Jori, *Br. J. Cancer* **56**, 597 (1987).
103. E. Reddi, C. Zhou, R. Biolo, E. Menegaldo and G. Jori, *Br. J. Cancer* **61**, 407 (1990).
104. E. Reddi, S. Cernuschi, R. Biolo and G. Jori, *Lasers Med. Sci.* **5**, 339 (1990).
105. M. Shopova, V. Mantareva, K. Krastev, H. Hadjiolov, A. Milev, K. Spirov, G. Jori and F. Ricchelli, *J. Photochem, Photobiol. B Biol.* **16**, 83 (1992).
106. U. Isele, K. Schieweck, R. Kessler, P. van Hoogevest and H. G. Capraro, *J. Pharm. Sci.* **84**, 166 (1995).
107. K. Schieweck, H. G. Capraro, U. Isele, P. van Hoogevest, M. Ochsner, T. Maurer and E. Batt, *Proc. Int. Soc. Opt. Eng.* **2078**, 107 (1994).
108. C. Larroque, A. Pelegrin and J. E. van Lier, *Br. J. Cancer* **74**, 1886 (1996).
109. J. Oh, H.-J. Yoon and J.-H. Park, *J. Mater. Chem. B* **2**, 2592 (2014).
110. J. M. Birchall, R. N. Haszeldine and J. O. Morley, *J. Chem. Soc. (C)* **9**, 2667 (1970).
111. R. W. Boyle, J. Rousseau, S. V. Kudrevich, M. O. K. Obochi and J. E. van Lier, *Br. J. Cancer* **73**, 49 (1996).
112. E. N. Allemann, N. Brasseur, O. Benrezzak, J. Rousseau, S. V. Kudrevich, R. W. Boyle, J.-C. Leroux, R. Gurny and J. E. van Lier, *J. Pharm. Pharmacol.* **47**, 382 (1995).
113. W. M. Sharman and J. E. van Lier, *Bioconjugate Chem.* **16**, 1166 (2005).
114. (a) N. Kobayashi, *Coord. Chem. Rev.* **99**, 219 (2001). (b) M. Kimura, T. Muto, H. Takimoto, K. Wada, K. Ohta, K. Hanabusa, H. Shirai and N. Kobayashi, *Langmuir* **16**, 2078 (2000).
115. B. A. Bench, A. Beveridge, W. M. Sharman, G. J. Diebold, J. E. van Lier and S. M. Gorun, *Angew. Chem. Int. Ed.* **41**, 8163 (2002).
116. M. Tian, T. Wada and J. Sasabe, *Heterocyclic Chem.* **37**, 41193 (2000).
117. H. Weitman, S. Schatz, H. E. Gottlieh, N. Kobayashi and B. Ehrenberg, *Photochem. Photobiol.* **72**, 472 (2001).
118. (a) E. Ben-Hur and I. Rosenthal, *Int. J. Radiat. Biol.* **47**, 145 (1985). (b) W.-S. Chan, J. F. Marshal, R. Svensen, D. Phillips and I. R. Hart, *Photochem. Photobiol.* **45**, 757 (1987). (c) E. Ben-Hur and I. Rosenthal, *Photochem. Photobiol.* **43**, 615 (1986).
119. W.-S. Chan, N. Brasseur, C. La Madeleine and J. E. van Lier, *Eur. J. Cancer* **33**, 1855 (1997).
120. J. P. F. Longo, S. P. Lozzi, A. R. Simioni, P. C. Morais, A. C. Tedesco and R. B. J. Azevedo, *Photochem. Photobiol. B: Biology* **94**, 143 (2009).
121. N. Brasseur, R. Ouellet, C. La Madeleine and J. E. van Lier, *Br. J. Cancer* **80**, 1533 (1999).
122. M. J. Cook, A. J. Dunn, S. D. Howe, A. J. Thomson and K. J. Harrison, *J. Chem. Soc. Perkin Trans. I*, 2453 (1988).
123. M. J. Cook, I. Chambrier, S. J. Cracknel, D. A. Mayes and D. A. Russel, *Photochem. Photobiol.* **62**, 543 (1995).
124. L. Kaestner, M. Cesson, K. Kassab, T. Christensen, P. D. Edminson, M. J. Cook, I. Chambrier and G. Jori, *Photochem. Photobiol. Sci.* **2**, 660 (2003).
125. C. Ometto, C. Fabris, C. Milanesi, G. Jori, M. J. Cook and D. A. Russell, *Br. J. Cancer*, **74**, 1891 (1996).
126. W. E. Ford, B. D. Rihter, M. E. Kenney and M. A. Rodgers, *J. Photochem. Photobiol.* **50**, 277 (1989).

127. R. W. Boyle, C. C. Leznoff and J. E. van Lier, *Br. J. Cancer* **67**, 1177 (1993).
128. C. C. Leznoff, M. Hu, C. R. McArthur, Y. Qin van J. E. van Lier, *Cancer J. Chem.* **72**, 1990 (1994).
129. M. Hu, N. Brasseur, S. Z. Yildiz, J. E. van Lier and C. C. Leznoff, *J. Med. Chem.* **41**, 1789 (1998).
130. Y. Zorlu, M. A Ermeydan, F. Dumoulin, V. Ahsen, H. Savoie and R. W. Boyle, *Photochem. Photobiol. Sci.* **8**, 312 (2009).
131. Y. Chin, S. H. Lim, Y. Zorlu, V. Ahsen, L. V. Kiew, L. Y. Chung, F. Dumoulin and H. B. Lee, *PloS ONE* **9**, e97894 (2014).
132. S. Tuncel, F. Dumoulin, J. Gailer, M. Sooriyaarachchi, D. Atilla, M. Durmus¸ D. Bouchu, H. Savoie, R. W. Boyle and V. Ahsen, *Dalton Trans.* **40**, 4067 (2011).
133. J.-Y.-J. Liu, X. Jiang, W.-P. Fong and D. K. P. Ng, *Org. Biomol. Chem.* **6**, 4560 (2008).
134. H. Li, N. Nguyen, F. R. Fronczek and M. G. H. Vicente, *Tetrahedron* **65**, 3357 (2009).
135. Y. Zorlu, F. Dumoulin, M. Durmus and V. Ahsen, *Tetrahedron* **66**, 3248 (2010).
136. D. Atilla, N. Saydan, M. Durmuş, A. G. Gürek, T. Khan, A. Rück, H. Walt, T. Nyokong and V. Ahsen, *J. Photochem. Photobiol. A: Chem.* **186**, 298 (2007).
137. A. Ogunsipe, M. Durmus, D. Atilla, A. G. Gurek, V. Ahsen and T. Nyokong, *Synthetic Met.* **158**, 839 (2008).
138. N. B. R. Vittar, C. G. Prucca, C. Strassert, J. Awruch and V. A. Rivarola, *Int. J. Biochem. Cell Biol.* **40**, 2192 (2008).
139. P. M. Cormick, M. Rovera and E. N. Durantini, *J. Photochem Photobiol. A: Chem.* **194**, 220 (2008).
140. M. K. Lowery, A. J. Starshak, J. N. Esposito, P. C. Krueger and M. E. Kenney, *Inorg. Chem.* **4**, 128 (1965).
141. B. L. Wheeler, G. Nagasubramanian, A. J. Bard, L. A. Schechtman, D. R. Dininny and M. E. Kenney, *J. Am. Chem. Soc.* **106**, 7407 (1984).
142. N. L. Oleinick, A. R. Antunez, M. E. Clay, B. D. Rihter and M. E. Kenney, *Photochem. Photobiol.* **57**, 242 (1993).
143. S. I. A. Zaidi, R. Agarwal, G. Eichler, B. D. Rihter, M. E. Kenney and H. Mukhtar, *Photochem. Photobiol.* **58**, 204 (1993).
144. J. He, H. E. Larkin, Y.-S. Li, B. D. Rihter, S. I. A. Zaidi, M. A. J. Rodgers, H. Mukhtar, M. E. Kenney and N. L. Oleinick, *Photochem. Photobiol.* **65**, 581 (1997).
145. C. Y. Anderson, K. Freye, K. A. Tubesing; Y.-S. Li, M. E. Kenney, H. Mukhtar and C. A. Elmets, *Photochem. Photobiol.* **67**, 332 (1998).
146. S. Rywkin, E. Ben-Hur, Z. Malik, A. M. Prince, Y.-S. Li, M. E. Kenney, N. L. Oleinick and B. Horowitz, *Photochem. Photobiol.* **60**, 165 (1994).
147. Signal™ Technology, Diatron Diagnostic, Inc.: Miami, FL, 33186, USA.
148. H. G. Capraro, K. Schieweck, R. Hilfiker, M. Ochsner, U. Isele, P. van Hoogevest, R. Naef and M. Baumann, *Proc. Int. Symp. SPIE* (1993).
149. P.-C. Lo, J.-D. Huang, D. Y. Y. Cheng, E, Y. M. Chan, W.-P. Fong, W.-H. Ko and D. K. P. Ng, *Chem. Eur. J.* **10**, 4831 (2004).
150. J.-D. Huang, P.-C. Lo, W.-P. Fong, W.-H. Ko and D. K. P. Ng, *New J. Chem.* **28**, 348 (2004).
151. K. Schieweck, H. G. Caparo, U. Isele, E. Batt, M. Ochsner, P. van Hoogevest and W. G. Love, *Proc. Int. Symp. SPIE* (1993).

152. P. Alani, A. H. Soloway, B. V. Bapat, R. F. Barth and D. M. Adams, *Basic Life Sci.* **50**, 107 (1989).
153. J.-W. Hoffman, F. van Zeeland, S. Turker, H. Talsma, S. A. G. Lambrechts, D. V. Sakharov, W. E. Hennink and F. van Nostrum, *J. Med. Chem.* **50**, 1485 (2007).
154. K. Ishii, A. Takayanagi, S. Shimizu, H. Abe, K. Sogawa and N. Kobayashi, *Free Radical Biol. Med.* **38**, 920 (2005).
155. Z. Zhao, R. Gambari, K. K.-H. Lee, S. H.-L. Kok, R. S.-M. Wong, F.-Y. Lau, J. C.-O. Tang, K.-H. Lam, C.-H. Cheng, D. K. P. Hau, C.-H. Chui, W.-Y. Wong and W.-K. Wong, *Bioorg. Med. Chem. Lett.* **23**, 2373 (2013).
156. D. C. Hone, P. I. Walker, R. Evans-Gowing, S. FitzGerald, A. Beeby, I. Chambrier, M. J. Cook and D. A. Russell, *Langmuir* **18**, 2985 (2002).
157. M. E. Wieder, D. C. Hone, M. J. Cook, M. M. Handsley, J. Gavrilovic and D. A. Russell, *Photochem. Photobiol. Sci.* **5**, 727 (2006).
158. T. P. Mthethwa, S. Tuncel, M. Durmuş and T. Nyokong, *Dalton Trans.* **42**, 4922 (2013).
159. S. Tuncel, J. Fournier-dit-Chabert, F. Albrieux, V. Ahsen, D. S. Sylvie and F. Dumoulin, *Org. Biomol. Chem.* **10**, 1154 (2012).
160. S. Tuncel, A. Trivella, D. Atilla, K. Bennis, H. Savoie, F. Albrieux, L. Delort, H. Billard, V. Dubois, V. Ahsen, F. Caldefie-Cheìzet, C. Richard, R. W. Boyle, S. Ducki and F. Dumoulin, *Mol. Pharmaceutics* **10**, 3706 (2013).
161. J. T. F. Lau, P.-C. Lo, W.-P. Fong and D. K. P. Ng, *J. Med. Chem.* **55**, 5446 (2012).
162. H. He, P.-C. Lo and D. K. P. Ng, *Chem. Eur. J.* **20**, 6241 (2014).
163. L. Lapok, M. Cyza, A. Gut, M. Kepcznski, G. Szewczyk, T. Sarna and M. J. Nowakowska, *Photochem. Photobiol. A: Chem.* **286**, 55 (2014).
164. R. P. Linstead and M. Whalley, *J. Chem. Soc.*, 4839 (1952).
165. A. Ghosh, J. Fitzgerald, P. G. Gassman and J. Almof, *Inorg. Chem.* **33**, 6057 (1994).
166. A. R. Morgan, N. H. Petousis and J. E. van Lier, *Eur. J. Med. Chem.* **32**, 21 (1997).
167. V. N. Kopranenkov, L. S. Goncharova and E. A. Lukyanets, *Zhuranl Obshechei Khimii* **479**, 2143 (1977).
168. A. R. Morgan, G. M. Garbo, T. Krivak, M. Mastroianni, N. Petousis, T. St. Clair, M. Wiesenberger and J. E. van Lier, *SPIE* **1426**, 350 (1991).
169. T. K. Horne and M. J. Cronje, *Photochem. Photobiol.* **90**, 648 (2014).
170. S. V. Kudrevich, M. G. Galpern and J. E. van Lier, *Synthesis* **8**, 779 (1994).
171. N. D. Hammer, S. Lee, B. J. Vesper, K. M. Elseth, B. M. Hoffman, A. G. M. Barret and J. A. Radosevich, *J. Med. Chem.* **48**, 8125 (2005).
172. A. Scholto, S. Lee, B. M. Hoffman, A. G. M. Barrett and B. Ehrenberg, *Photochem. Photobiol.* **84**, 764 (2008).
173. B. J. Vesper, S. Lee, N. D. Hammer, K. M. Elseth, A. G. M. Barrett, B. M. Hoffman and J. A. Radosevich, *J. Photochem. Photobiol. B* **82**, 180 (2006).
174. S. Lee, B. J. Vesper, H. Zong, N. D. Hammer, K. M. Elseth, A. G. M. Barrett, B. M. Hoffman and J. A. Radosevich, *Metal-Based Drugs*, 391418 (2008).
175. M. Motyka, R. P. Steer, C. C. Williams, S. Lee and K. P. Ghiggino, *Photochem. Photobiol. Sci.* **12**, 1086 (2013).
176. M. P. Donzello, Z. Ou, F. Monacelli, G. Ricciardi, C. Rizzoli, C. Ercolani and K. M. Kadish, *Inorg. Chem.* **43**, 8626 (2004).

177. (a) M. P. Donzello, Z. Ou, D. Dini, M. Meneghetti, C. Ercolani and K. M. Kadish, *Inorg. Chem.* **43**, 8637 (2004). (b) M. Villano, V. Amendola, G. Sandona, M. P. Donzello, C. Ercolani and M. Meneghetti, *J. Phys. Chem. B* **110**, 24534 (2006).
178. (a) C. Bergami, M. P. Donzello, C. Ercolani, F. Monacelli, K. M. Kadish and C. Rizzoli, *Inorg. Chem.* **44**, 9852 (2005). (b) C. Bergami, M. P. Donzello, F. Monacelli, C. Ercolani and K. M. Kadish, *Inorg.Chem.* **44**, 9862 (2005).
179. M. P. Donzello, E. Viola, X. Cai, L. Mannina, C. Rizzoli, G. Ricciardi, C. Ercolani, K. M. Kadish and A. Rosa, *Inorg. Chem.* **47**, 3903 (2008).
180. M. P. Donzello, E. Viola, L. Mannina, M. Barteri, Z. Fu and C. Ercolani, *J. Porphyrins Phthalocyanines*, **15**, 984 (2011).
181. M. P. Donzello, D. Vittori, E. Viola, I. Manet, L. Mannina, L. Cellai, S. Monti and C. Ercolani, *Inorg. Chem.* **50**, 7391 (2011).
182. F. Manoli, M. P. Donzello, C. Ercolani, D. Vittori, L. Cellai, A. Masi and S. Monti, *Inorg. Chem.* **50**, 7403 (2011).
183. C. V. Synatschke, T. Nomoto, H. Cabral, M. Fortsch, K. Toh, Y. Matsumoto, K. Miyazaki, A. Hanisch, F. H. Schacher, A. Kishimura, N. Nishiyama, A. H. E. Muller and K. Kataoka, *acsnano. org.* **8**, 1161 (2014).
184. I. Manet, F. Manoli, M. P. Donzello, E. Viola, A. Masi, G. Andreano, G. Ricciardi, A. Rosa, L. Cellai, C. Ercolani and S. Monti, *Inorg. Chem.* **52**, 321 (2013).
185. I. Manet, F. Manoli, M. P. Donzello, E. Viola, G. Andreano, A. Masi, L. Cellai and S. Monti, *Org. Biomol. Chem.* **9**, 684 (2011).
186. P. Loser, A. Winzenburg and R. Faust, *Chem. Commun.* **49**, 9413 (2013).
187. E. R. Trivedi, C. M. Blumenfeld, T. Wielgos, S. Pokropinski, P. Dande, T. T. Hai, A. G. M. Barrett and B. M. Hoffman, *Tetrahedron Lett.* **53**, 5475 (2012).
188. F. Mitzel, S. FitzGerald, A. Beeby and R. Faust, *Chem. Eur. J.* **9**, 1233 (2003).
189. D. Woehrle, J. Gitzel, I. Okura and S. Aona, *J. Chem. Soc. Perkin Trans.* II, 1171 (1985).
190. T. G. Gantchev, H. Ali and J. E. van Lier, *Eur. J. Biochem.* **217**, 371 (1993).
191. H. Dezhampanah and S. Fyzolahjani, *Analytical Cellular Pathology* **36**, 125 (2013).
192. H. E. Tom, *An. Acad. Bras. Ciênc.* Rio de Janeiro 72, 1 (2000).
193. K. Sakamoto, T. Kato, E. Ohno-Okumura, M. Watanabe and M. J. Cook, *Dyes Pigm.* **64**, 63 (2005).
194. K. Sakamoto, T. Kato, T. Kawaguchi, E. Ohno-Okumura, T. Urano, T. Yamaoka, S. Suzuki and M. J. Cook, *J. Photochem. Photobiol. A: Chem.* **153**, 245 (2002).
195. L. Hassani, Z. Fazeli, E. Safaei, H. Rastega and M. Akbari, *J. Biol. Phys.* **40**, 275 (2014).
196. U. Michelsen, H. Kliesch, G. Schnurpfeil, A. K. Sobbi and D. Wohrle, *Photochem. Photobiol.* **64**, 694 (1996).
197. J. Piskorz, P. Skupin, S. Lijewski, M. Korpusinski, M. Sciepura, K. Konopka, S. Sobiak, T. Goslinski and J. Mielcarek, *J. Fluorine Chem.* **135**, 265 (2012).
198. E. F. Bradbrook and R. P. Linstead, *J. Chem. Soc.*, 1744 (1936).
199. P. A. Firey and M. A. J. Rodgers, *Photochem. Photobiol.* **45**, 535 (1987).
200. I. McCubbin and D. Philips, *J. Photochem.* **34**, 187 (1986).
201. M. L. Kaplan, A. J. Lovinger, W. D. Reents Jr. and P. H. Schmidt, *Mol. Cryst. Liq. Cryst.* **112**, 345 (1984).
202. M. Shopova, M. Peevai, N. Stoichkova, G. Jori, D. Wohrle and G. Petrov, *J. Porphyrins Phthalocyanines*, **5**, 798 (2001).
203. B. Paquette, H. Ali, R. Langlois and J. E. van Lier, *Photochem. Photobiol.* **51**, 313 (1990).

204. N. C. Yates, J. Moan and A. Western, *J. Photochem. Photobiol. B: Biol.* **4**, 379 (1990).
205. D. Woehrle, M. Shopova, S. Muller, A. D. Milev, V. N. Mantareva, K. K. Krastev, *J. Photochem. Photobiol. B: Biol.* **21**, 155 (1993).
206. M. Shopova, D. Woehrle, N. Stoichkova, A. Milev, V. Mantareva, S. Muller, K. Kassabov and K. Georgiev, *J. Photochem. Photobiol. B: Biol.* **23**, 35 (1994).
207. S. Muller, V. Mantareva, N. Stoichkova, H. Kliesch, A. Sobbi, D. Woehrle and M. Shopova, *J. Photochem. Photobiol. B: Biol.* **35**, 167 (1996).
208. J. Xu , J. Chen, L. Chen, R. Hu, S. Wang, S. Li, J. S. Ma and G. Yan, *Dyes Pigm.* **109**, 144 (2014).
209. R. Polley and M. Hannack, *J. Org. Chem.* **60**, 8278 (1995).
210. A. N. Cammidge, I. Chambrier, M. J. Cook, A. D. Garland, M. J. Heeney and K. Welford, *J. Porphyrins Phthalocyanines* **1**, 77 (1997).
211. M. Bucking, E. F. G. Dickson, M. Farahani, F. Fischer, D. Holmes, G. Jori, J. C. Kennedy, M. E. Kenney, X. Peng, R. H. Pottier and G. Weagle, *J. Photochem. Photobiol. B: Biol.* **58**, 87 (2000).
212. Y. Iawakabe, S. Numata, N. Kinjo and A. Kakuta, *Bull. Chem. Soc. Jpn.* **63**, 2734 (1990).
213. W. E. Ford, M. A. J. Rodgers, L. A. Schechtman, J. R. Sounik, B. D. Rihter and M. E. Kenney, *Inorg. Chem.* **31**, 3371 (1992).
214. S. Tai, S. Hayashida and N. Hayashi, *J. Chem. Soc. Perkin Trans.* **2**, 1637 (1991).
215. B. L. Wheeler, G. Nagasubramanian, A. J. Bard, L. A. Schechtman, D. R. Dininny and M. E. Kenney, *J. Am. Chem. Soc.* **106**, 7404 (1984).
216. N. Brasseur, T.-L. Nguyen, R. Langlois, R. Ouellet, S. Marengo, D. Houde and J. E. van Lier, *J. Med. Chem.* **37**, 415 (1994).
217. N. Brasseur, R. Ouellet, K. Lewis, W. R. Potter and J. E. van Lier, *Photochem. Photobiol.* **62**, 1058 (1995).
218. V. Cuomo, G. Jori, B. Rihter, M. E. Kenney and M. A. J. Rodgers, *Br. J. Cancer* **62**, 966 (1990).
219. R. Biolo, G. Jori, M. Soncin, R. Pratesi, U. Vanni, B. Rihter, M. Kenney and M. A. J. Rodgers, *Photochem. Photobiol.* **59**, 362 (1994).
220. (a) M. M. Zuk, B. D. Rihter, M. E. Kenney, M. A. J. Rodgers and M. Kreimer-Birnbaum, *Photochem. Photobiol.* **59**, 66 (1994). (b) M. M. Zuk, B. D. Rihter, M. E. Kenney, M. A. J. Rodgers and M. Kreimer-Birnbaum, *Photochem. Photobiol.* **63**, 132 (1996).
221. M. Soncin, A. Busetti, R. Biolo, G. Jori, G. Kwag, Y.-S. Li, M. E. Kenney and M. A. J. Rodgers, *J. Photochem. Photobiol. B: Biol.* **42**, 202 (1998).
222. V. Mantareva, M. Shopova, G. Spassova, D. Wöhrle, S. Muller, G. Jori and F. Ricchelli, *J. Photochem. Photobiol.* **40**, 258 (1997).
223. Y. Li, D. Dini, M. J. F. Calvete, M. Hanack and W. Sun, *J. Phys. Chem. A* **112**, 472 (2008).
224. R. A. Hill, J. Garrett, S. Reddi, T. Esterowitz, L. H. Liaw, J. Ryan, J. Shirk, M. Kenney, S. Shimuzu and M. W. Berns, *Lasers Surg. Med.* **18**, 86 (1996).
225. H. Ali and J. E. van Lier, in *Handbook of Porphyrin Science: Phototherapy, Radioimmunotherapy and Imaging*, Vol. 4, Eds. K. M. Kadish, K. M. Smith and R. Guilard, (World Scientific Publishing Co. Pte. Ltd., Hackensack, NJ, USA, 2010), p. 1.
226. K. T. de Oliveira, F. F. de Assis, A. O. Ribeiro, C. R. Neri, A. U. Fernades, M. S. Baptista, N. P. Lopes, O. A. Serra and Y. Iamamoto, *J. Org. Chem.* **74**, 7962 (2009).
227. U. Drechsler, M. Pfaff and M. Hanack, *Eur. J. Org. Chem.*, 3441 (1999).
228. F. Giuntini, C. M. A. Alonso and R. W. Boyle, *Photochem. Photobiol. Sci.* **10**, 759 (2011).

229. Z. Chen, S. Zhou, J. Chen, Y. Deng, Z. Luo, H. Chen, M. R. Hamblin and M. Huang, *Chem. Med. Chem.* **5**, 890 (2010).
230. L. Li, Z. Luo, Z. Chen, J. Chen, S. Zhou, P. Xu, P. Hu, J. Wang, N. Chen, J. Huang and M. Huang, *Bioconjugate Chem.* **23**, 2168 (2012).
231. M.-R. Ke, J. M. Eastel, L. K. Karry, N. Y.-Y Cheung, P. K. S. Chan, M. Hui, D. K. P. Ng and P.-C. Lo, *Chem. An. Asian J.* **9**, 1868 (2014).
232. P. Xu, J. Chen, Z. Chen, S. Zhou, P. Hu, X. Chen and M. Huang, *PloS ONE.* 7, e37051 (2012).
233. C. Dubuc, R. Langlois, F. Benard, N. Cauchon, K. Klarskov, P. Tone and J. E. van Lier, *Bioorg. Med. Chem. Lett.* **18**, 2424 (2008).
234. R. P. Hammer, C. V. Owens, S. H. Hwang, C. M. Sayes and S. A. Soper, *Bioconjugate Chem.* **13**, 1244 (2002).
235. M.-R. Ke, S.-L. Yeung, W.-P. Fong, D. K. P. Ng and P.-C. Lo, *Chem. Eur. J.* **18**, 4225 (2012).
236. E. Ranyuk, N. Cauchon, K. Klarskov, B. Guérin and J. E. van Lier, *J. Med.Chem.* **56**, 1520 (2013).
237. L. Luan, L. Ding, J. Shi and W. Liu, Abstract, ICCP 8 (2014).
238. M. Sibrian-Vazquez, J. Ortiz, I. V. Nesterova, F. Fernandez-Lazaro, A. Sastre-Santos, S. A. Soper and M. G. H. Vicente, *Bioconjugate Chem.* **18**, 410 (2007).
239. N. Brasseur, R. Langlois, C. La Madeleine, R. Ouellet and J. E. van Lier, *Photochem. Photobiol.* **69**, 345 (1999).
240. P. Urizzi, C. M. Allen, R. Langlois, R. Ouellet, C. La Madeleine and J. E. van Lier, *J. Porphyrins Phthalocyanines* **5**, 154 (2001).
241. M. Mitsunaga, M. Ogawa, N. Kosaka, L. T. Rosenblum, P. L. Choyke and H. Kobayashi, *Nat. Med.* **17**, 1685 (2011).
242. M. Mitsunaga, T. Nakajima, K. Sano, P. L. Choyke and H. Kobayashi, *Bioconjugate Chem.* **23**, 604 (2012).
243. T. Nakajima, K. Sano, P. L. Choyke and H. Kobayashi, *Theranostics* **3**, 357 (2013).
244. I. Sehgal, H. Li, B. Ongarora, D. Devillier and M. G. H. Vicente, *J. Porphyrins Phthalocyanines* **17**, 150 (2013).
245. E. H. Khan, H. Ali, H. Tian, J. Rousseau, G. Tessier, Shafiullah and J. E. van Lier, *Bioorg. Med. Chem. Letters.* **13**, 1287 (2003).
246. R. Decreau, M. Chanon and M. Julliard, *Synlett.* 375 (1998).
247. P. Khoza, E. Antunes, J.-Y. Chen and T. Nyokong, *J. Luminescence* **134**, 784 (2013).
248. R. O. Ogbodu, E. Antunes and T. Nyokon, *Polyhedron* **60**, 59 (2013).
249. M. Camur and M. Bulut, *Dyes Pigm.* 77, 165 (2008).
250. M. Camur, M. Bulut, M. Kandaz and O. Guney, *Supramol. Chem.* **21**, 624 (2009).
251. M. Camur, M. Bulut, M. Kandaz and O. Guney, *Polyhedron*, **28**, 233 (2009).
252. M. Camur, M. Durmus and M. Bulut, *J. Photochem. Photobiol. A* **222**, 266 (2011).
253. M. Çamur, A. A. Esenpınar, A. R. Özkaya and M. J. Bulut, *Organometallic Chem.* **696**, 1868 (2011).
254. M. Camur, M. Durmus and M. Bulut, *Polyhedron* **41**, 92 (2012).
255. M. Tian, T. Wada and H. Sasabe, *J. Heterocyclic Chem.* **37**, 1193 (2000).
256. M. R. Reddy, N. Shibata, Y. Kondo, S. Nakamura and T. Toru, *Angew. Chem. Int. Ed.* **45**, 8163 (2006).

257. E. B. Dickerson, E. C. Dreaden, X. Huang, I. H El-Sayed, H. Chu, S. Pushpanketh, J. F. McDonald and M. A. El-Sayed, *Cancer Lett.* **269**, 57 (2008).
258. S. Tombe, E. Antunes and T. Nyokong, *New J. Chem.* **37**, 679 (2013).
259. N. Masilela and T. Nyokong, *J. Photochem. Photobiol. A*, **223**, 124 (2011).
260. T. Mthethwa, E. Antunes and T. Nyokong, *Dalton Trans* **43**, 8230 (2014).
261. M. Göksel, M. Durmus and D. Atilla, *J. Photochem.Photobiol. A: Chem.* **266**, 37 (2013).
262. O. Warburg, *Science* **123**, 309 (1956).
263. R. E. Airley and A. Mobasheri, *Chemotherapy* (*Basel, Switz.*) **53**, 309 (2007).
264. P. Maillard, J.-L. Guerquin-Kern, M. S. Momenteau and S. Gaspard, *J. Am. Chem. Soc.* **111**, 9125 (1989).
265. A. Lyubimtsev, Z. Iqbal, G. Cruciusa, S. Syrbua, T. Zieglera and M. Hanack, *J. Porphyrins Phthalocyanines* **16**, 434 (2012).
266. A. O. Ribeiro, J. P. C. Tome, M. G. P. M. S. Neves, A. C. Tome, J. A. S. Cavaleiro, O. A. Serra and T. Torres, *Tetrahedron Lett.* **47**, 6129 (2006).
267. X. Alvarez-Mico, M. J. F. Calvete, M. Hanack and T. Ziegler, *Tetrahedron Lett.* **47**, 3283 (2006).
268. (a) A. O. Ribeiro, J. P. C. Tome, M. G. P. M. S. Neves, A. C. Tome, J. A. S. Cavaleiro, Y. Iamamotob and T. Torres, *Tetrahedron Lett.* **47**, 9177 (2006).
269. C.-F. Choi, J.-D. Huang, P.-C. Lo, W.-P. Fong and D. K. P. Ng,*Org. Biomol. Chem.* **6**, 2173 (2008).
270. D. Lafonta, Y. Zorlub, H. Savoie, F. Albrieux, V. Ahsen, R. W. Boyle and F. Dumoulin, *Photodiagnosis Photodyn. Ther.* **10**, 252 (2013).
271. S. G. Kimani, T. A. Shmigol, S. Hammond, J. B. Phillips, J. I. Bruce, A. J. MacRobert, M. V. Malakhov and J. P. Golding, *Photochem. Photobiol.* **89**, 139 (2013).
272. A. R. M. Soares, P. C. Tomé, M. G. P. M. S. Neves, A. C. Tomé, J. A. S. Cavaleiro and T. Torres, *Carbohydr. Res.* **344**, 507 (2009).
273. A. R. M. Soares, M. G. P. M. S. Neves, A. C. Tome, M. C. Iglesias-de la Cruz, A. Zamarron, E. Carrasco, S. Gonzalez, J. A. S. Cavaleiro, T. Torres, D. M. Guldi and A. Juarranz, *Chem. Res. Toxicol.* **25**, 940 (2012).
274. Y. Zorlu, F. Dumoulin, D. Bouchu, V. Ahsen and D. Lafont, *Tetrahedron Lett.* **51**, 6615 (2010).
275. (a) Z. Iqbal, M. Hanack and T. Ziegler, *Tetrahedron Lett.* **50**, 873 (2009). (b). A. R. M. Soares, J. P. C. Tome, M. G. P. M. S. Neves, A. C. Tome, J. A. S. Cavaleiro and T. Torres, *Carbohydr. Res.* **344** , 507 (2006).
276. Y. Zorlu, M. A. Ermeydan, F. Dumoulin, V. Ahsen, H. Savoie and R. W. Boyle, *Photochem. Photobiol. Scs.* **8**, 312 (2009).
277. R. M. Reddy, N. Shibata, H. Yoshiyama, S. Nakamura and T. Toru, *Synlett.*, 2186 (2007).
278. J. Y. Liu, J. Yong, P.-C. Lo, W.-P. Fong and D. K. P. Ng, *Org. Biomol. Chem.* 7, 1583 (2009).
279. M. R. Aykota, S. Tuncel, N. Kabay, C. Yenisey, V. Ahsen, F. Dumoulin and B. Kabay, Abstract, ICPP-8 (2014).
280. A. A. Chernonosov, V. V. Koval, D. G. Knorre, A. A. Chernenko, V. M. Derkacheva, E. A. Lukyanets and O. S. Fedoroval, *Bioinorg. Chem. Applications*, 1 (2006).
281. A. Aggarwal, S. Singh, Y. Zhang, M. Anthes, D. Samaroo, R. Gao and C. M. Drain, *Tetrahedron Lett.* **52**, 5456 (2011).

282. L. M. O. Lourenço, P. M. R. Pereira, E. Maciel, M. Valega, F. M. J. Domingues, M. R. M Domingues, M. G. P. M. S. Neves, J. A. S. Cavaleiro, R. Fernandes and J. P. C. Tome, *Chem. Commun.*, 8363 (2014).
283. P. M. R. Pereira, S. Silva, J. A. S. Cavaleiro, C. A. F. Ribeiro, J. P. C. T. Mail and R. F. Mail, *PloS One.*, e95529 (2014).
284. P.-C. Lo, W. Fong and D. K. P. Ng, *Chem. Med. Chem.* **3**, 1110 (2008).
285. P.-C. Lo, C. M. H. Chan, J.-Y. Liu, W.-P. Fong and D. K. P. Ng, *J. Med. Chem.* **50**, 2100 (2007).
286. P. P. S. Lee, P.-C. Lo, E. Y. M. Chan, W.-P. Fong, W.-H. Ko and D. K.P. Ng, *Tetrahedron Lett.* **46**, 1551 (2005).
287. S. S. Erdem, I. V. Nesterova, S. A. Soper and R. P. Hammer, *J. Org. Chem.* **74**, 9280 (2009).
288. N. D. Hammer, S. Lee, B. G. Vesper, K. M. Elseth, B. M. Hoffman, A. G. M. Barrett and J. A. Radosevich, *J. Med. Chem.* **48**, 8125 (2005).
289. M. G. H. Vicente, *Curr. Med. Chem. Anti-Cancer Agents* **1**, 175 (2001).
290. (a) S. Ozcelik and A. Gul, *Turk J. Chem.* **1** (2014). (b). D. Pietrangeli, A. Rosa, S. Ristori, A. Salvati, S. Altieri and G. Ricciardi, *Coord. Chem. Rev.* **257**, 2213 (2013).
291. P. Alam, A. H. Soloway, B. V. Bapat, R. F. Barth and D. M. Adams, *Basic Life Sci.* **50**, 107 (1989).
292. S. B. Kahl and J. Li, *Inorg. Chem.* **35**, 3878 (1996).
293. F. Giuntini, Y. Raoul, D. Dei, M. Municchi, G. Chiti, C. Fabris, P. Colautti, G. Jori and G. Roncucci, *Tetrahedron Lett.* **46**, 2979 (2005).
294. O. Tsaryovaa, A. Semioshkinb, D. Wöhrle and V. I. Bregadzeb, *J. Porphyrins Phthalocyanines* **9**, 268 (2005).
295. D. Woehrle, N. Iskandar, G. Graschew, H. Sinn, E. A. Friedrich, W. Maier-Borst, J. Stern and P. Schlag, *Photochem. Photobiol.* **51**, 351 (1990).
296. B. Birsoz, A. V. Efremenko, A. A. Ignatova, A. Gul, A. V. Feofanov, I. B. Sivaev and V. I. Bregadze, *Biochem. Biophys. J. Neutron Ther. Cancer Treat.* **1**, 8 (2013).
297. C. Fabris, G. Jori, F. Giuntini and G. Roncucci, *J. Photochem. Photobiol. B: Biol.* **64**, 1 (2001).
298. D. Woehrle, O. Tsaryova, A. Semioshkin, D. Gabel and A. Suvorova, *J. Organome. Chem.* **747**, 98 (2013).
299. H. Li, F. R. Fronczeck and M. G. H. Vicente, *J. Organomet. Chem.* **694**, 1607 (2009).
300. H. Li, F. R. Fronczeck and M. G. H. Vicente, *Tetrahedron Lett.* **49**, 4828 (2009).
301. D. Pietrangeli, A. Rosa, A. Pepe and G. Ricciardi, *Inorg. Chem.* **50**, 4680 (2011).
302. (a) S. Altieri, S. Bortolussi, P. Bruschi, L. Ciani, A. M. Clerici, P. Faraoni, C. Ferrari, M. A. Gadan, L. Panza, D. Pietrangeli, G. Ricciardi and S. Ristori, *J. Med. Chem.* **52**, 7829 (2009). (b) S. Ristori, A. Salvati, G. Martini, O. Spalla, D. Pietrangeli, A. Rosa and G. Ricciardi, *J. Am. Chem. Soc.* **129**, 2728 (2007). (c) A. Salvati, S. Ristori, J. Oberdisse, O. Spalla, G. Ricciardi, D. Pietrangeli, M. Giustini and G. Martini, *J. Phys. Chem. B* **111**, 10357 (2007). (d) A. Salvati, S. Ristori, D. Pietrangeli, J. Oberdisse, L. Calamai, G. Martini and G. Ricciardi, *Biophys. Chem.* **131**, 43 (2007). (e) D. Pietrangeli and G. Ricciardi, *Appl. Radiat. Isot.* **67**, 97 (2009).
303. D. Pietrangeli, S. Ristori, A. Rosa and G. Ricciardi, *J. Porphyrins Phthalocyanines* **14**, 678 (2010).
304. D. Pietrangeli, A. Rosa and G. Ricciardi, *J. Porphyrins Phthalocyanines* **15**, 1024 (2011).

305. D. A. Policard, *C.R. Soc. Biol.* **91**, 1123 (1924).
306. Y. K. Ho, R. K. Pandey, J. R. Missert, D. A. Bellnier and T. J. Dougherty, *Photochem. Photobiol.* **48**, 445 (1988).
307. J. Winkelman, G. Slater and J. Grossman, *Cancer Res.* **27**, 2060 (1967).
308. D. S. Lawrence, S. L. Gibson, M. L. Nguyen, K. R. Whittmore, D. G. Whitten and R. Hilf, *Photochem. Photobiol.* **90**, 61 (1995).
309. F. R. Wrenn Jr., M. L Good and P. Handler, *Science* **18**, 687 (1951).
310. V. Scasnár, R. W. Boyle, J. Rousseau and J. E. van Lier (Unpublished).
311. V. Scasnár and J. E. van Lier, *Nucl. Med. Biol.* **20**, 257 (1993).
312. J. Rousseau, R. W. Boyle, A. H. Maclennan, T. G. Truscott and J. E. van Lier, *Nucl. Med. Biol.* **18**, 777 (1991).
313. A. Soucy-Faulkner, J. A. Rousseau, R. Langlois, V. Berard, R. Lecomte, F. Benard and J. E. van Lier, *J. Porphyrins Phthalocyanines* **12**, 49 (2008).
314. J. Alzeer, P. J. C. Roth and N. W. Luedtke, *Chem. Commun.*, 1970 (2009).
315. E. R. Ranyuk, N. Cauchon, H. Ali, R. Lecomte, B. Guérin and J. E. van Lier, *Bioorg. Med. Chem. Lett.* **21**, 7470 (2011).
316. E. Ranyuk, R. Lebel, Y. Beìrubeì-Lauzieìre, K. Klarskov, R. Lecomte, J. E. van Lier and B. Gueìrin, *Bioconjugate Chem.* **24**, 1624 (2013).
317. N. A. Frigerio, *U.S. Patent* 3,027, 391 (1962).
318. V. W. Day, T. J. Marks and W. A. Wachter, *J. Am. Chem. Soc.* **97**, 4519 (1975).
319. E. R. Ranyuk, H. Ali, B. Guérin and J. E. van Lier, *J. Porphyrins Phthalocyanines* **17**, 850 (2013).
320. H. Ali and J. E. van Lier, in *Phthalocyanines: Properties and Applications*, Eds. C. C. Leznoff and A. B. P. Lever (VCH Publishers, New York, 1996) Chapter 11, p. 427.
321. M. Shimoni, I. Rosenthal and E. J. Ben-Hur, *J. Labelled Compds. Radiopharm.* **22**, 863 (1985).
322. C. J. Gomer and T. J. Dougherty, *Cancer Res.* **39**, 146 (1979).
323. V. Lenaerts, A. Labib, F. Chouinard, J. Rousseau, H. Ali and J. E. van Lier, *Eur. J. Pharm. Biopharm.* **41**, 38 (1995).
324. J. Rousseau, D. Autenrieth and J. E. van Lier, *Int. J. Appl. Radiat. Isot.* **34**, 571 (1983).
325. J. Rousseau, H. Ali, G. Lamoureux, E. Lebel and J. E. van Lier, *Int. J. Appl. Radiat. Isot.* **36**, 709 (1985).
326. Y. Song, H. Zong, E. R. Trivedi, B. J. Vesper, E. A. Waters, A. G. M. Barrett, J. A. Radosevich, B. M. Hoffman and T. J. Meade, *Bioconjugate Chem.* **21**, 2267 (2010).
327. E. R. Trivedia, Z. Maa, E. A. Waterse, K. W. Macrenarise, R. Subramaniana, A. G. M Barrettf, T. J. Meadea and B. M. Hoffman, *Contrast Media Mol. Imaging* **9**, 313 (2014).
328. J. G. Millán, M. Brasch, E. Anaya-Plaza, A. de la Escosura, A. H. Velders, D. N. Reinhoudt, T. Torres, M. S. T. Koay and J. J. L. M. Cornelissen, *J. Inorg. Biochem.* **136**, 140 (2014).

Chapter 6

Molecular Phthalocyanine-Based Photosensitizers for Photodynamic Therapy

Pui-Chi Lo,[†,‡] *Sun Y. S. Chow*[†] *and Dennis K. P. Ng*[†,*]

[†]*Department of Chemistry, The Chinese University of Hong Kong, Shatin, N.T., Hong Kong*
dkpn@cuhk.edu.hk
[‡]*Department of Biomedical Sciences, City University of Hong Kong, Tat Chee Avenue, Kowloon Tong, Hong Kong*

Abstract

Phthalocyanines represent a versatile class of functional dyes. Owing to their desirable spectroscopic and photophysical properties, high photostability, and ease of chemical and physical modification, these compounds have emerged as promising photosensitizers for photodynamic therapy. This chapter aims to summarize the recent development of molecular phthalocyanine-based photosensitizers being reported over the last five years, including hydrophilic analogues with non-ionic, cationic, or anionic moieties, and bioconjugated derivatives substituted with carbohydrates, peptides, nucleic acids, polyamines, or folic acid. Novel phthalocyanines that exhibit dual functions and can be activated by tumor-associated stimuli are also described. The review will focus on their molecular design, synthesis, photophysical characteristics, and *in vitro* and *in vivo* photodynamic activities.

1. Introduction

Photodynamic therapy (PDT) is a clinically approved modality for treatment of a range of cancers and non-cancerous conditions, such as psoriasis and age-related macular degeneration.[1,2] It involves the combined action of a photosensitizer, light of appropriate wavelength, and molecular oxygen to produce cytotoxic reactive oxygen species (ROS), particularly singlet oxygen,

*Corresponding author.

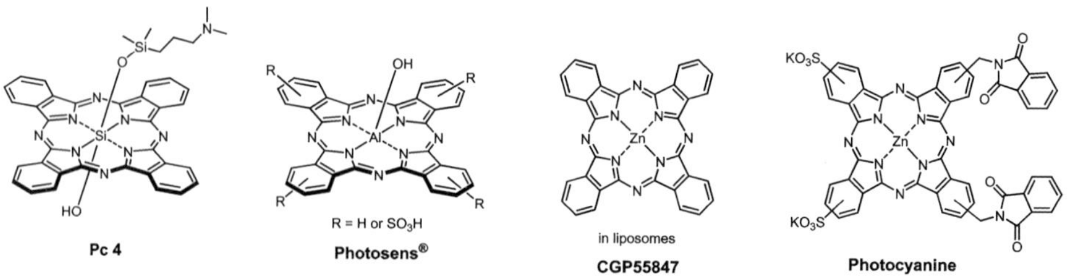

Fig. 1. Structures of the four phthalocyanine-based photosensitizers which are in clinical use or trial.

through a chain of photochemical reactions. Owing to the high reactivity and short lifetime of these species, the localization of photosensitizers usually controls the site of initial action that may affect the biochemical cell death pathways and ultimately the therapeutic efficacy.[3] Hence, rational design and modification of photosensitizers have received considerable attention.[4–7] The goals are not only to optimize their photophysical properties, but also to enhance their selectivity and phototoxicity toward malignant cells and tissues.

Among the various classes of photosensitizers, phthalocyanines are of particular interest and potent owing to their strong absorption in the tissue-penetrating red visible region (>670 nm), balance of fluorescence emission and singlet oxygen generation that can facilitate fluorescence-guided PDT, and ease of chemical and physical modification.[8] In fact, their many intriguing and superior optical and electronic properties have enabled them to extend their applications from traditional industrial dyes to many technological and practical applications. For their application in PDT, the studies were mainly stimulated by the promising therapeutic efficacy of Pc 4 developed by Kenney *et al.*[9] Over the last two decades, a vast number of phthalocyanine-based photosensitizing systems have been reported, and to our knowledge four of these systems, namely Pc 4, Photosens®, CGP55847, and Photocyanine (Fig. 1) are currently in clinical use or trial for various indications in different countries.[10]

Two intrinsic problems that hinder the use of undecorated phthalocyanines as photosensitizers for PDT stem from their poor solubility and high aggregation tendency in aqueous media, which prevent the delivery and generation of fluorescence and ROS. To circumvent these problems, a number of formulation or fabrication approaches have been utilized. For example, encapulation of these phthalocyanines in emulsions,[11] liposomes,[12] polymeric micelles,[13,14] mesoporous silica nanoparticles,[15] or layered double hydroxides[16] can enhance their water solubility, reduce their molecular stacking, and retain their photosensitizing properties. In addition, these nanoplatforms can impart additional features, such as preferential localization in tumor due to the enhanced permeability and retention (EPR) effect, active cellular uptake and subcellular targeting via conjugation to a targeting group, and activation control of their photosensitizing properties. Interestingly, changing the form of the dyes can also remedy these deficiencies. Zinc(II) phthalocyanine (ZnPc) nanowires growth from ZnPc powder by vaporization-condensation-recrystallization process, for example, exhibit good water dispersibility and highly efficient dual photodynamic and photothermal effects, both *in vitro* and *in vivo*, upon irradiation with a near-infrared laser (808 nm).[17] These multifunctional nanophotosensitizers have recently received tremendous interest.[18–20]

As an alternative strategy, phthalocyanines can be chemically modified to introduce various functionalities. Having a well-defined structure, these molecular photosensitizing systems can be studied more readily to elucidate their structure-property-activity relationships. A substantial progress has also been made in the development of molecular phthalocyanine-based photosensitizers as described in several review articles.[21–26] This chapter aims to summarize the most recent development in this field, focusing on the results published over the last five years and those photosensitizers of which their photobiological activities have been studied.

2. Hydrophilic Phthalocyanines

2.1. *With Non-ionic Moieties*

Oligo- and polyethylene glycols (OEG and PEG, respectively) are one of the well-known examples of this class. These hydrophilic and biocompatible chains have been widely used as pharmaceutical carriers that can prolong the drug circulating half-life. Conjugation of these moieties to phthalocyanines has also been documented. This can generally enhance the water solubility of the hydrophobic core, reduce the stacking tendency, and promote the cellular uptake of the resulting conjugates. Recently, Ahsen *et al.* prepared three zinc(II) phthalocyanines substituted with four or eight tetraethylene glycol chains with a terminal hydroxyl group (compounds **1–3**, Fig. 2) by typical cyclotetramerization of the corresponding phthalonitriles.[27] These compounds exhibited face-to-face aggregation in water as shown by the broad and blue-shifted *Q* band. Addition of Triton X-100 promoted disaggregation of **1** and **2**, and could restore the sharp monomeric *Q* band for the latter. The UV-Vis spectrum of the octa-β-substituted counterpart **3**, however, showed negligible

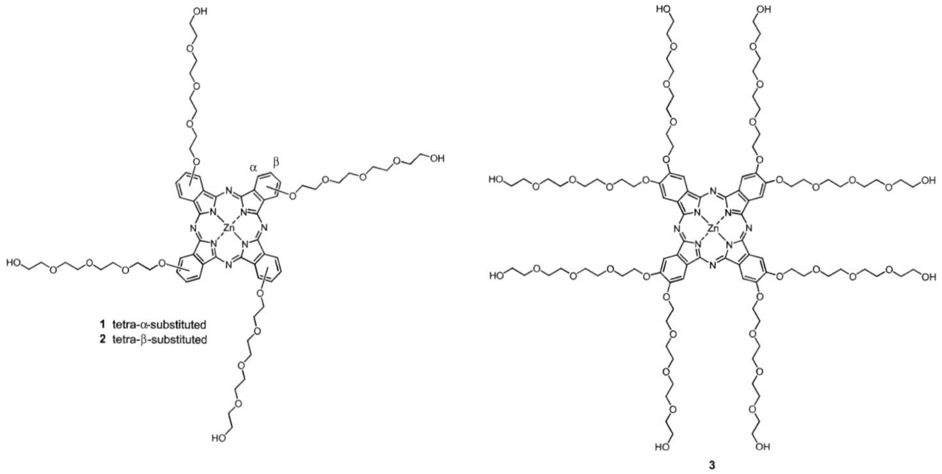

Fig. 2. Structures of tetraethylene-glycol-subtituted phthalocyanines **1**-**3**.

effect toward Triton X-100, showing that it has a higher aggregation tendency. The surfactant could also enhance the fluorescence emission and singlet oxygen generation of these compounds, especially for the least aggregated tetra-α-substituted analogue **2**. Among the three compounds, it showed the highest photocytotoxicity with an IC_{50} value of 85 μM against HT29 human adenocarcinoma cells.

By using the methoxycarbonyl-containing phthalonitrile **4** or **5** as a precursor, mixed cyclization with an excess of 1,3-diiminoisoindoline (**6**) in the presence of $Zn(OAc)_2 \cdot 2H_2O$ and 1,8-diazabicyclo[5.4.0]undec-7-ene (DBU) in tri- or polyethylene glycol monomethyl ether resulted in the formation of the corresponding "1+3" products, in which the tri- or polyethylene glycol chains had been introduced via base-promoted transesterification (Scheme 1).[28] This one-pot procedure provided a facile approach to prepare PEGylated phthalocyanines. These compounds in Cremophor EL emulsions were photocytotoxic toward HT29 cells and HepG2 human hepatocarcinoma cells with IC_{50} values in the range of 0.25–3.72 μM. Compound **8** ($n = 3$) was the most potent photosensitizer and localized preferentially in the mitochondria of the cells. Compound **7** ($n =$ ca. 12) could form surfactant-free nanoparticles in aqueous media, of which the photocytotoxicity was comparable with that of the system formulated with Cremophor EL.

The PEG chains can also be introduced after the cyclization process. For example, the mono-PEGylated ZnPc **9** were prepared by treating the hydroxyphthalocyanine **10** with the corresponding PEG tosylates (with an average molecular weight of ca. 550, 2000, or 5000 g mol^{-1}) in the

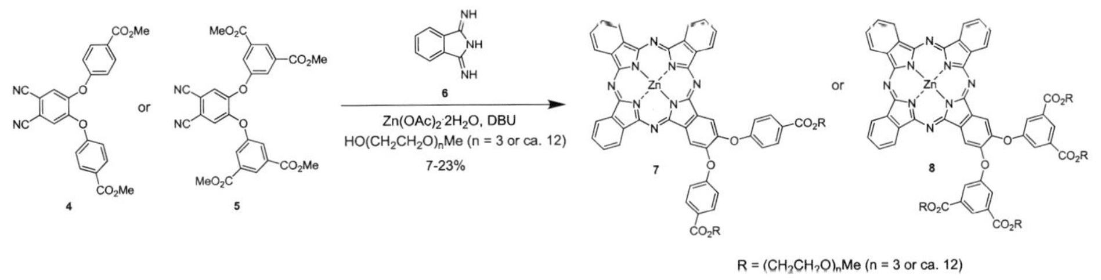

Scheme 1. Preparation of tri- and polyethylene glycol substituted phthalocyanines 7 and **8**.

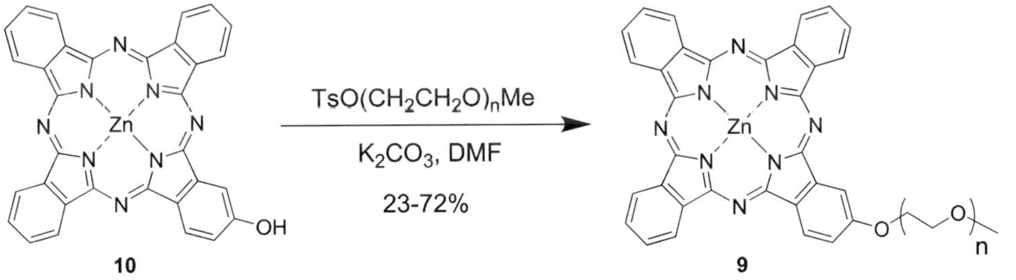

Scheme 2. Preparation of mono-PEGylated phthalocyanines **9**.

presence of K_2CO_3 in N,N-dimethylformamide (DMF) (Scheme 2).[29] All these PEGylated phthalocyanines could form self-assembled micelles in water, both in the absence and presence of Cremophor EL. Compound **9** (with PEG_{550}) was also encapsulated in organically modified silica-based nanoparticles which were formed by controlled hydrolysis of triethoxyvinylsilane in micelles. The hydrodynamic radii of all these nanoparticles were measured by dynamic laser light scattering. It was found that Cremophor EL could stabilize the nanoparticles and significantly reduce their size, and the silica-based nanoparticles showed a much narrower distribution with a hydrodynamic radius of 25 nm. Although the silica environment was not as effective as Cremophor EL in reducing the aggregation of the phthalocyanine, increasing the fluorescence intensity, and promoting the singlet oxygen generation, the *in vitro* photodynamic activities of the two systems were comparable ($IC_{50} = 0.2\,\mu M$ against HepG2 cells). Generally, the photocytotoxicity decreased as the length of the PEG chain increased as a result of the lower cellular uptake.

Huang *et al.* prepared two crown-ether-fused phthalocyanines (**11** and **12**, Fig. 3) and studied their self-assemble and metal-binding properties, and *in vitro* photodynamic activities.[30] Compound **11** was much more potent than

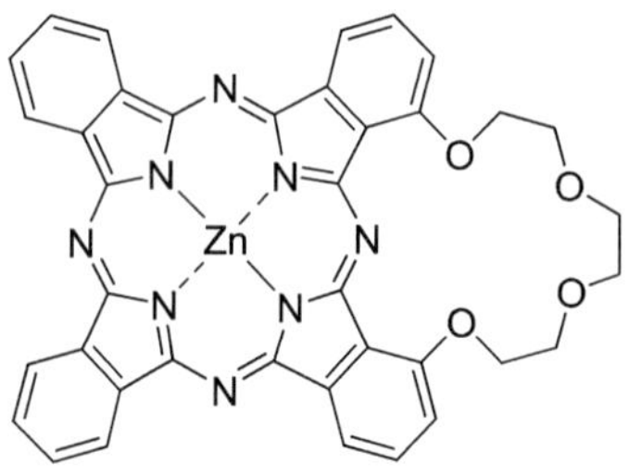

11

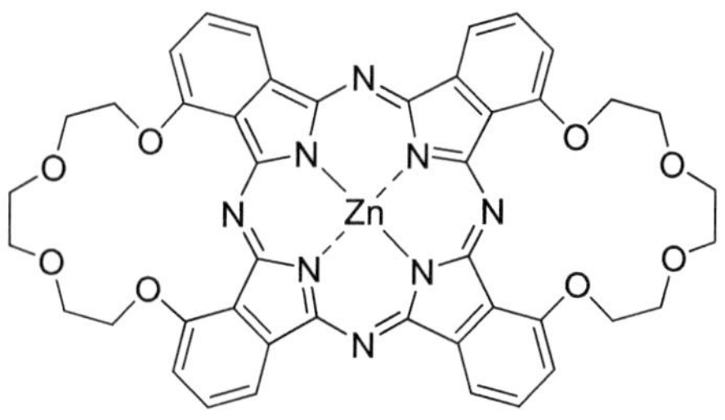

12

Fig. 3. Structures of crown-ether-fused phthalocyanines **11** and **12**.

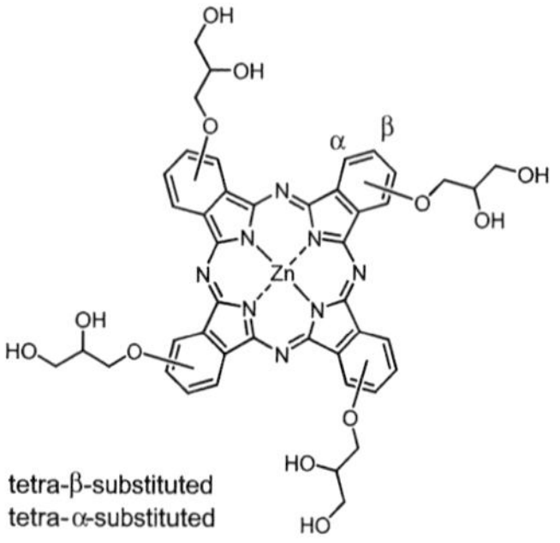

13 tetra-β-substituted
14 tetra-α-substituted

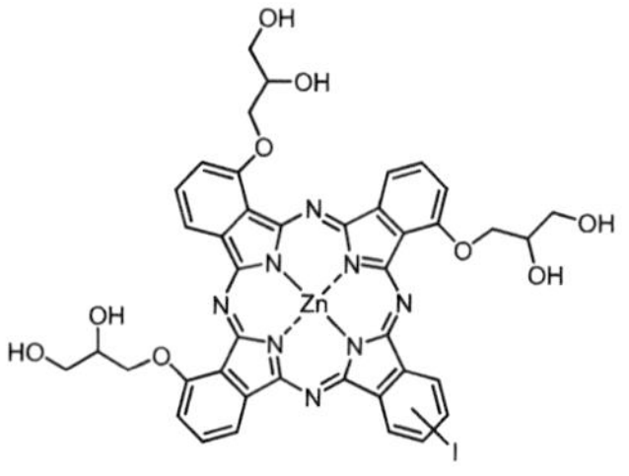

15

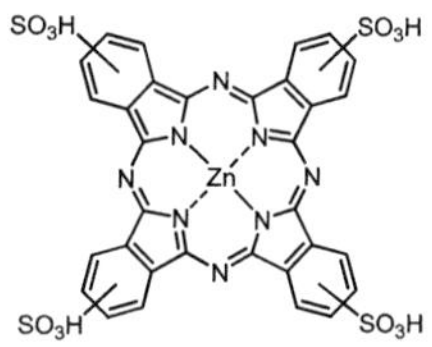

16

Fig. 4. Structures of Phthalocyanines **13-16**.

12 (IC_{50} = 0.08 μM versus 3.31 μM against HepG2 cells) as a result of its higher cellular uptake and singlet oxygen generation efficiency.

Glycerol is another oxygen-rich moiety that can impart similar properties to phthalocyanines. Compounds **13–15** were prepared by Lim *et al.* as mixtures of structural isomers,[31] which showed a red-shifted *Q* band (by 10–100 nm) and enhanced singlet oxygen generation efficiency (>two-fold) compared to the tetrasulfonated analogue **16** (Fig. 4). As expected, the tetraglycerol-substituted **13** and **14** gave a lower log P (partition coefficient) value than the iodo derivative **15** (log P = 0.9 versus 1.3) due to their higher hydrophilic character, but the latter could be taken up by cells more readily. Both **14** and **15** were phototoxic toward MCF-7 human breast carcinoma, HCT-116 human colon carcinoma, and HSC-2 human oral squamous carcinoma cells with IC_{50} values in the range of 2.8–3.2 μM and 0.04–0.06 μM, respectively, whilst **13** and **16** were essentially non-cytotoxic. The iodo analogue **15** was also more efficient in inducing vasculature damage in the chick embryo chorioallantoic membrane model, which was attributed to its lower aggregation tendency and enhanced cellular uptake. A related series of compounds with one, three, or four glycerol unit(s) were also reported by Ribeiro *et al.*, who reaffirmed the importance of the hydrophobic/hydrophilic balance in the cellular uptake.[32]

Amine is another functionality which is commonly used to increase the water solubility of photosensitizers. Xue *et al.* prepared a series of zinc(II) phthalocyanines substituted with one or four OEG appended quinoline unit(s) at the α or β position, and examined their phototoxicity against HepG2 cells.[33] The monosubstituted analogues (e.g., **17**) were much more photocytotoxic than the tetrasubstituted counterparts with IC_{50} values of ca. 0.02 μM, which again was attributed to their lower aggregation tendency. Similarly, the mono-piperazinyl phthalocyanine **18** was also found to be more potent in photoinactivation of *Candida albicans* compared with the tetra-α-substituted analogues, having an IC_{90} value of 9 μM.[34] Rivarola *et al.* studied the PDT mechanism induced by the octasubstituted phthalocyanine **19** on the transfected MCF-7c3 cells.[35] It was found that the compound caused caspase-independent apoptosis and lysosomes were the primary sites of action (Fig. 5).

Pc 4 is another example that contains an amino group (Fig. 1). Stimulated by its promising photodynamic efficacy, there has been considerable interest in studies of its various properties. Recently, there have been studies to compare the intratumor and intravenous administration of this compound to BALB/c mice bearing intradermal EMT6 tumors,[36] to assess the safety and tolerability of topically applied Pc 4 in patients with cutaneous neoplasms,[37] to evaluate the *in vitro* efficacy of Pc 4 in PEG-poly-ε-caprolactone micelles,[38] to determine its cutaneous penetration in live animals and to assess the effective absorption through the skin barrier using intravital two-photon laser scanning microscopy,[39] to examine the combined cytotoxic effect of Pc 4 and the multikinase inhibitor dasatinib on SCCVII mouse squamous cell carcinoma cells,[40]

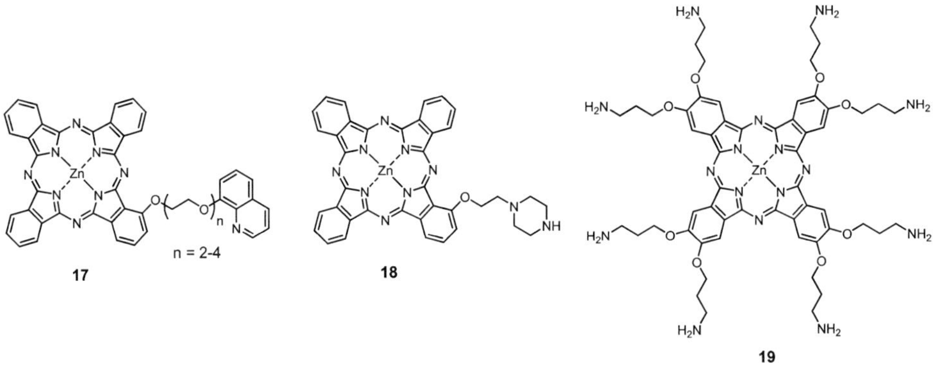

Fig. 5. Structures of amino phthalocyanines **17**-**19**.

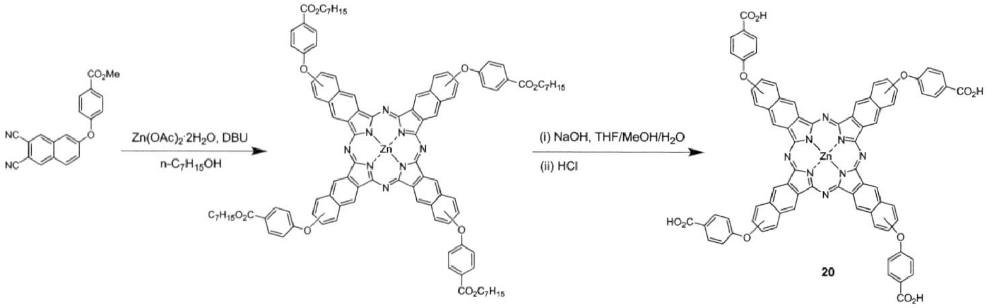

Scheme 3. Preparation of tetracarboxy naphthalocyanine **20**.

and to investigate the photocytotoxic mechanism of Pc 4 in the pathogen *Trichophyton rubrum*.[41]

The tetracarboxy naphthalocyanine **20** was another water-soluble derivative prepared according to Scheme 3.[42] Compared with the phthalocyanine analogue, this compound showed a stronger and red-shifted *Q* band (by ca. 90 nm to 763 nm) in DMF, but higher aggregation tendency in the culture medium and lower cellular uptake. However, its phototoxicity against HeLa human cervical carcinoma cells was almost the same as that of the phthalocyanine analogue ($IC_{50} = 3.7\ \mu M$ and $3.5\ \mu M$, respectively).

2.2. *With Cationic Moieties*

Awruch *et al.* prepared a series of tetracationic zinc(II) phthalocyanines (compounds **21** and **22**, Fig. 6) and evaluated their phototoxicity against KB human nasopharynx carcinoma cells.[43,44] It was found that the sulfur-linked tetra-β-substituted derivatives had higher potency and preferential localization in lysosomes. The antitumor mechanism of **21** (X = S) was subsequently examined in detail. It was found that the ROS produced disrupted the lysosomes, which was followed by mitochondrial depolarization, cytosolic release of cytochrome C, and caspases activation. The photosensitizer induced a caspase-dependent apoptotic response through the release of lysosomal proteases.[45]

Another related series with two or eight 2-dimethylaminoethylthio substituents at the β-positions, as well as their N-alkylated derivatives were also prepared.[46] The disubstituted non-cationic analogue **23** (Fig. 6) showed the highest phototoxicity against HepG2, HT29, and T84 human colon adenocarcinoma cells with IC_{50} values as low as $0.08\ \mu M$. It exhibited selective affinity to the mitochondria of HT29 cells, while its di-N-methylated counterpart showed preferential localization in the cell membrane. Wei *et al.* prepared the tetra-β-substituted phthalocyanine **24** (Fig. 6), which has eight amino groups, and

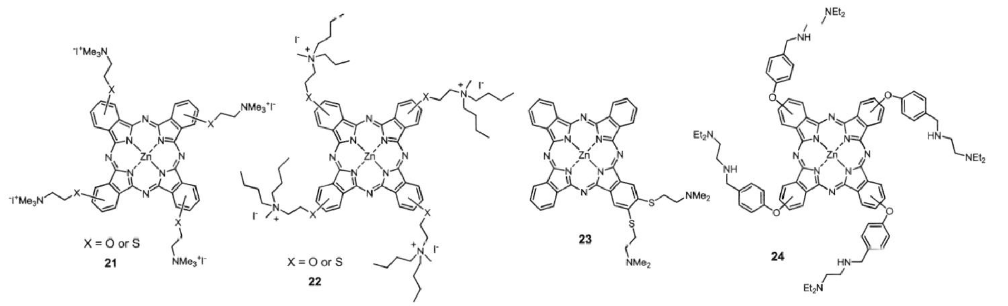

Fig. 6. Structures of phthalocyanines **21-24**.

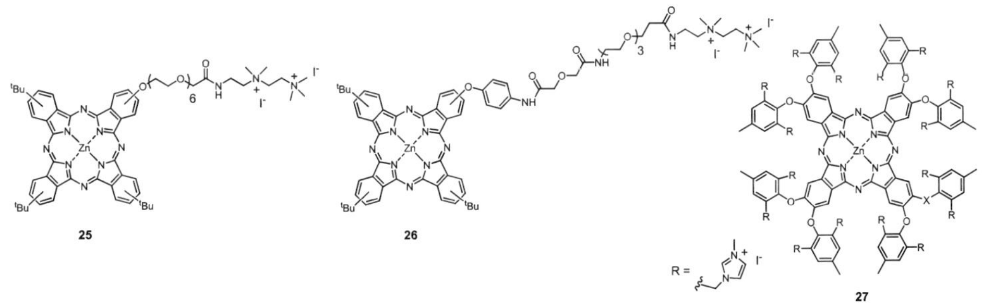

Fig. 7. Structures of phthalocyanines **25-27**.

performed *N*-methylation or protonation to increase its water solubility.[47,48] These octacationic derivatives, particularly the protonated species, were found to exhibit higher phototoxicity against HeLa cells. As protonation also proceeds more readily than *N*-methylation, this strategy was found to be of particular interest.

Vicente *et al.* synthesized a series of A_3B-type PEGylated dicationic phthalocyanines and investigated their cytotoxicity, cellular uptake, and subcellular distribution in HEp2 human carcinoma cells.[49] All the α-substituted derivatives were found to be significantly more photocytotoxic (>five-fold) than the β-substituted counterparts. Compounds **25** and **26** (Fig. 7) exhibited the highest potency with IC_{50} values of ca. 10 μM and preferred localization in the endoplasmic reticulum of the cells. Both the PEG and cationic units were found to be crucial in their photodynamic activities.

As an alternative strategy to prevent aggregation, Zimcik *et al.* prepared a novel zinc(II) phthalocyanine with 16 quaternized imidazolyl moieties (compound **27**, Fig. 7).[50] This compound was free from aggregation in water even up to 20 mM and could generate singlet oxygen effectively with a quantum yield Φ_Δ of >0.33. It also exhibited high phototoxicity against HeLa cells

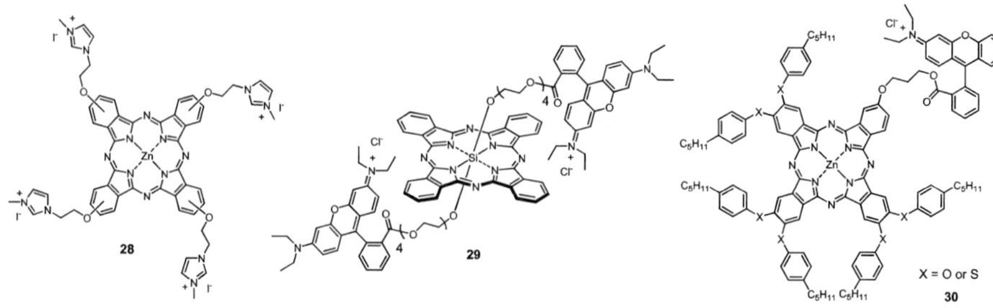

Fig. 8. Structures of mitochondrial-targeting phthalocyanines **28-30**.

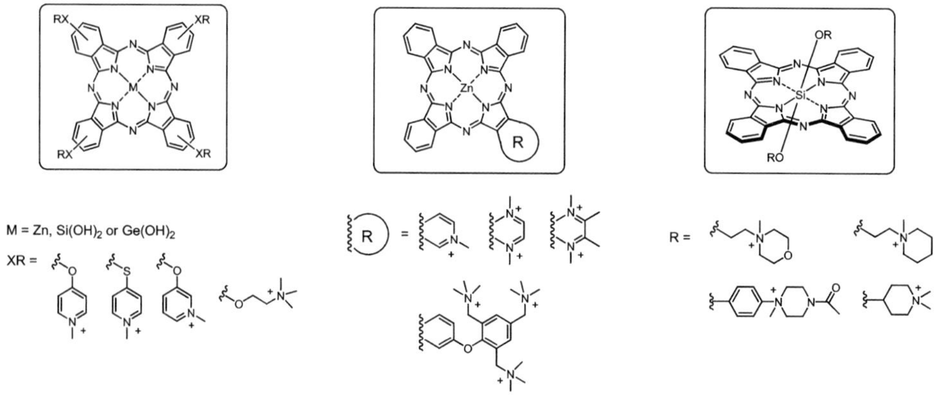

Fig. 9. Recent examples of cationic phthalocyanines with antimicrobial photodynamic activities.

(EC_{50} = 36.7 nM), but low dark toxicity to HeLa and 3T3 non-malignant mouse fibroblast cells. The respective TC_{50} values, which define the inherent toxicity in the dark, were found to be 628 μM and 395 μM.

A number of cationic phthalocyanines were found to have a mitochondrial localizing property. Some recent examples include the tetraimidazolium zinc(II) phthalocyanine **28**[51] and the rhodamine B conjugated silicon(IV) and zinc(II) phthalocyanines **29**[52] and **30** (Fig. 8).[53] Compounds **28** and **29** were also found to induce cell death mainly through apoptosis.

Cationic photosensitizers are also well-documented for their antimicrobial photodynamic activities partly due to their strong binding with the negatively charged outer membrane of most microbial cells through electrostatic interactions. A large number of such photosensitizers based on phthalocyanines have been prepared and their efficacy in photoinactivation of a range of bacteria, fungi, and viruses have also been studied. Fig. 9 shows some typical examples which have been reported recently, including the tetracationic derivatives of various metallophthalocyanines,[51–58] A_3B-type zinc(II) analogues,[59,60] and silicon(IV) phthalocyanines with axial cationic moieties.[61,62] Their photodynamic activities are generally higher than those of the neutral and anionic analogues as a result of their different binding affinity to the microbial cells.[57,60–63] Dutta *et al.* explored the potential of a series of 17 phthalocyanines for *in vitro* PDT and vaccination using a host-parasite model.[64] It was found that the cationic analogues which were localized in the mitochondria were more photolytic against cultured *Leishmania* than those in the endocytic compartments. They were useful not only for PDT against this trypanosomatid protozoa, but also for delivery of antigens for photodynamic vaccination.

2.3. *With Anionic Moieties*

Sulfonated phthalocyanines are the most common anionic phthalocyanines. The photodynamic activities of a series of these compounds with different number of sulfonate group and different metal center have been examined using a range of cancer cell lines and compared with those of some common photosensitizers.[65–69] Chen *et al.* conjugated sulfonated aluminum phthalocyanine to doxorubicin through electrostatic interaction.[70] The complexation led to a three-fold increase in uptake of the phthalocyanine by both the human hepatocellular carcinoma cell line 7701 and rat basophilic leukemia cell line. The toxicity of the conjugate in the presence of light was also found to be higher than that induced by the individual components, showing that they functioned as a synergistic manner. In addition, they also found that sulfonated aluminum phthalocyanine is a potential two-photon-absorbing photosensitizer.[71] By using femtosecond laser-induced fluorescence, its two-photon absorption cross section was measured to be 19 GM when excited at 800 nm, but increased to 855 GM when excited at 750 nm. Under the latter condition, the fluorescence images of the phthalocyanine in KB cells, its singlet oxygen generation efficiency (by 13-fold), and photocytotoxicity (by 9–10 times) were significantly higher. By using Raman microspectroscopy and confocal Raman imaging combined with confocal fluorescence microscopy, Abramczyk *et al.* could distinguish the distribution and aggregation of this phthalocyanine in cancerous and non-cancerous human breast cancer tissue, suggesting that Raman imaging could be a powerful tool for PDT studies.[72]

Photocyanine (Fig. 1) is a dianionic phthalocyanine which is under clinical trial (Phase II) in China. The drug dose and light dose were optimized for PDT in mice with murine sarcoma S180 tumors with a view to determining the effective dose range of this drug for clinical application.[73] The photocytotoxic mechanism of this photosensitizer on HepG2 cells was also investigated.[74] Being localized predominantly in the mitochondria and lysosomes, Photocyanine reduced the mitochondrial membrane potential and increased the caspase-3 activity in a concentration-dependent manner after irradiation, leading to early apoptosis followed by necrosis.

Carboxylate is another anionic moiety that can be introduced to phthalocyanines to enhance their water solubility. Torres *et al.* prepared a series of dendritic zinc(II) phthalocyanines with up to 32 carboxylate groups.[75] As expected, the degree of aggregation and singlet oxygen generation efficiency depended on the size of the dendritic wedges. A related series was actually reported by us earlier,[76] in which the second-generation analogue (compound **31**, Fig. 10) has been employed for photochemical internalization for delivery of genes and anticancer drugs.[77–79]

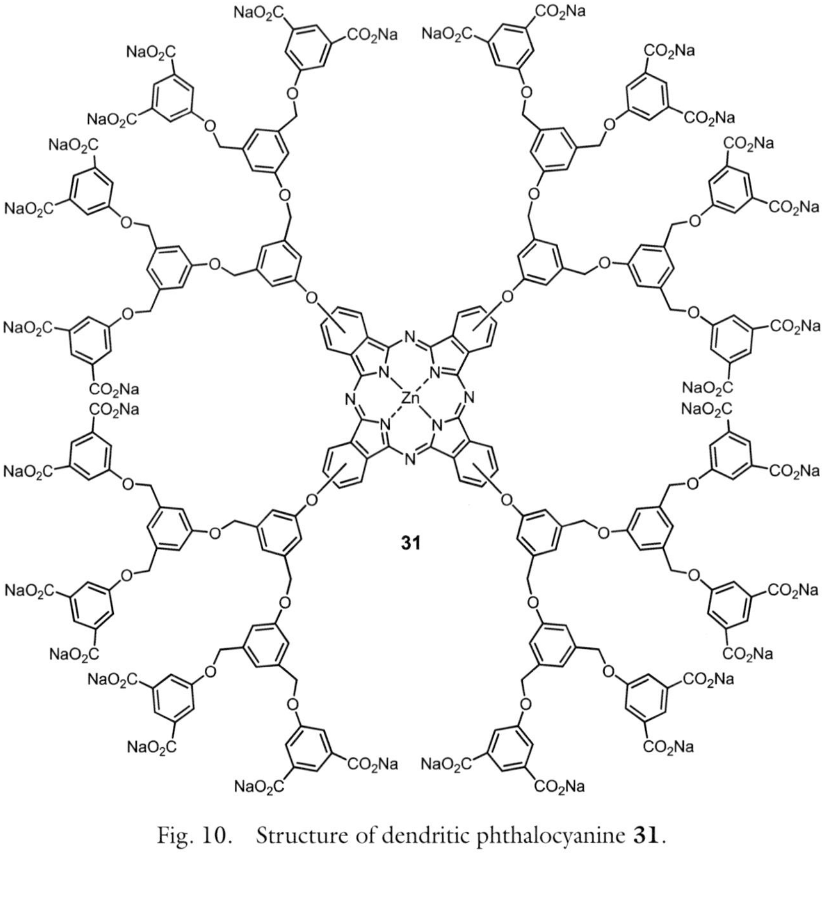

Fig. 10. Structure of dendritic phthalocyanine **31**.

3. Bioconjugated Phthalocyanines

Conjugation of phthalocyanine derivatives with various biologically relevant moieties has recently been reviewed.[80] We describe in this section the most recent examples, focusing on their photobiological properties.

3.1. *With Carbohydrates*

Glycoconjugated phthalocyanines have received considerable interest for application in PDT. In addition to their hydrophilic property that can enhance the water solubility of phthalocyanines, the sugar moieties may bind to the glucose transporters overexpressed in cancer cells leading to selective and active uptake. A large number of these derivatives have been reported of which the synthetic methods have recently been summarized.[81] Juarranz *et al.* examined the photodynamic activities and mechanism of a series of zinc(II) phthalocyanines substituted with four or eight galactopyranose moieties (compounds **32–34**, Fig. 11) against HeLa cells.[82] Having four of this hydrophilic moiety on one of the isoindole units, **34** has an amphiphilic character. It showed a lower aggregation tendency, enhanced cellular uptake, and higher photocytotoxicity compared with the other two analogues. Depending on the treatment conditions, it could induce cell death by either a necrotic or an apoptotic mechanism. At lethal dose 100 (LD100) where the cell viability dropped to almost zero (1 μM, 11.16 J cm^{-2}), the cells died by necrosis immediately after treatment. In contrast, at LD80 where there was an almost 80% decrease of the cell viability (0.1 μM, 11.16 J cm^{-2} or 0.5 μM, 5.58 J cm^{-2}), the treatment was able to induce metaphase arrest of the cells, leading to multiple spindle poles, mitotic catastrophe, and eventually apoptosis.

The tumor localization property of compound **32** and its mono-, di-, and trisubstituted analogues, in which the galactopyranose moiety is attached to a β-position of different isoindole units, in HepG2 tumor-bearing nude mice was examined.[83] By monitoring the fluorescence intensity of the tumor and

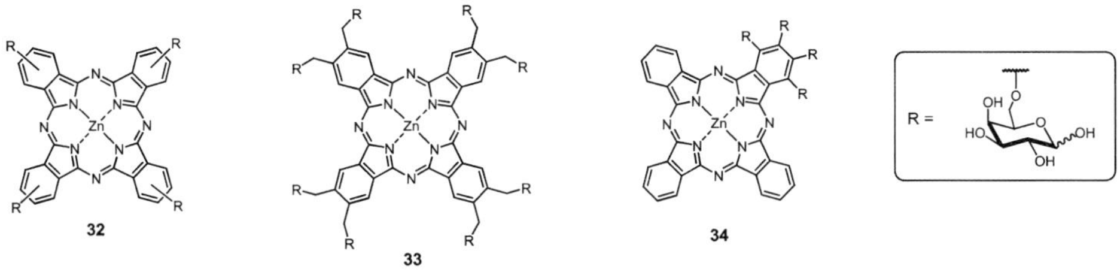

Fig. 11. Structures of galactosylated phthalocyanines **32-34**.

comparing with that of other organs, both*in vivo* and *ex vivo*, it was found that the tri- and tetrasubstituted analogues showed a certain degree of tumor-targeting ability. In addition, the antimicrobial photodynamic activities of this series of compounds were also studied.[84] They were highly potent against methicillin-resistant *Staphylococcus aureus* (MRSA), but exerted little effect on *Escherichia coli* (*E. coli*) and *Pseudomonas aeruginosa* (*P. aeruginosa*). The photoinactivation efficacy on MRSA decreased as the number of galactopyranose moiety increased. The monosubstituted analogue killed >99% of MRSA in a concentration range of 25–50 μM at a light dose of 6 J cm^{-2} as a result of its higher bacterial uptake (4–30-fold) compared with the other three analogues. The isopropylidene-protected glucose analogue of **32** was also reported, which exhibited enhanced cellular uptake (10-fold), higher PDT-induced intracellular ROS concentration (five-fold), and greater photocytotoxicity (two-fold) than disulfonated aluminum phthalocyanine.[85]

Two other series of glycerol-substituted monoglycoconjugated zinc(II) phthalocyanines (compounds **35** and **36**, Fig. 12) were also prepared to study the effects of the sugar moiety and the linker on the photodynamic activity against HT29 cells.[86] The mannose-substituted analogues exhibited the highest photocytotoxicity. Compared with the derivatives linked by direct glycosylation (compound **35**), the triazole-linked counterparts (compound **36**) showed a

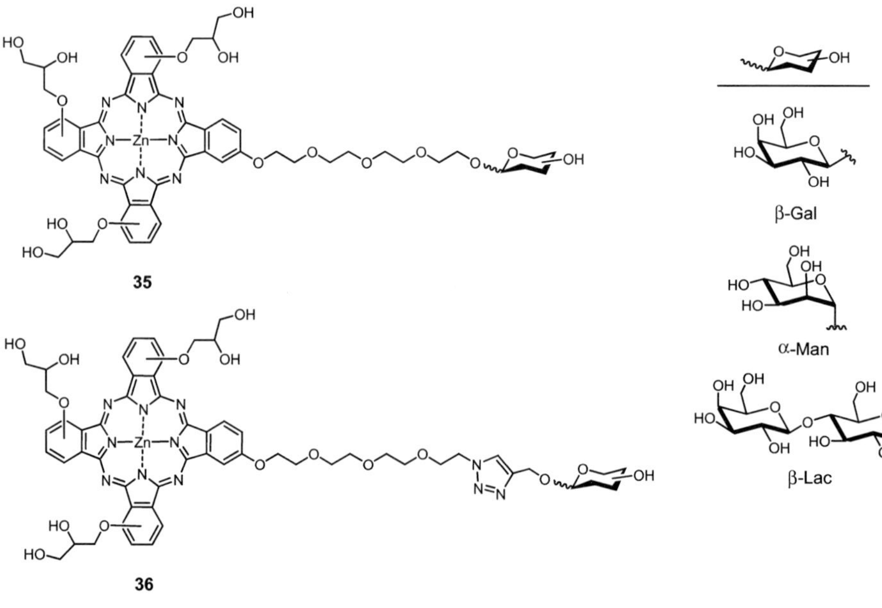

Fig. 12. Structures of glycoconjugated phthalocyanines **35** and **36**.

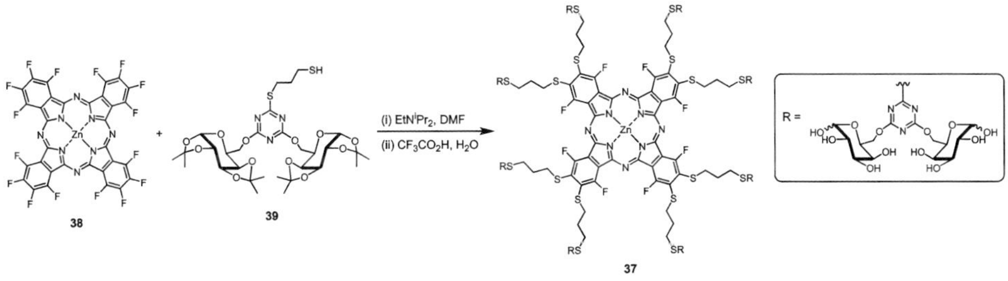

Scheme 4. Preparation of galactosylated phthalocyanine **37**.

lower photocytotoxicity for the mannose and galactose conjugates, while the effect on the lactose counterparts was negligible.

Compound **37**, which contains 16 galactopyranose moieties on the periphery of the phthalocyanine ring, was prepared by substitution reaction of the commercially available perfluorinated phthalocyanine **38** with the thiolated bis(galactose) **39** (Scheme 4)[87] and its PDT effects on two bladder cancer cell lines, namely HT-1376 and UM-UC-3, were investigated.[88] It was found that knockdown of galectin-1 (a carbohydrate-binding lectin) and GLUT1 (a well-known glucose transporter) by small interfering RNA reduced the uptake and phototoxicity of this photosensitizer, suggesting that the galactose moieties play an important role in promoting the cellular uptake through interaction with these receptors.

Apart from these sugar moieties, the oligosaccharides cyclodextrins have also been conjugated to phthalocyanines to enhance their photodynamic activities.[80] Recently, Tomé *et al.* performed substitution reaction of **38** with α-, β-, or γ-cyclodextrin in the presence of K_2CO_3, and examined the effects of this substituent on the various properties of the resulting conjugates (compounds **40**, Fig. 13).[89] Based on the mass spectrometric data, it was found that only two fluorine atoms were replaced, leading to the proposed structures as shown below. The β-cyclodextrin-substituted analogue showed a lower solubility in phosphate buffered saline (PBS), which might be due to its higher aggregation tendency. This deficiency might also account for its lower singlet oxygen generation efficiency in DMF/H_2O (9:1), reduced intracellular ROS production, lower cellular uptake, and negligible phototoxicity against UM-UC-3 cells. In contrast, the α- and γ-substituted analogues showed improved properties and were significantly more potent with IC_{50} values in the range of 26–87 μM.

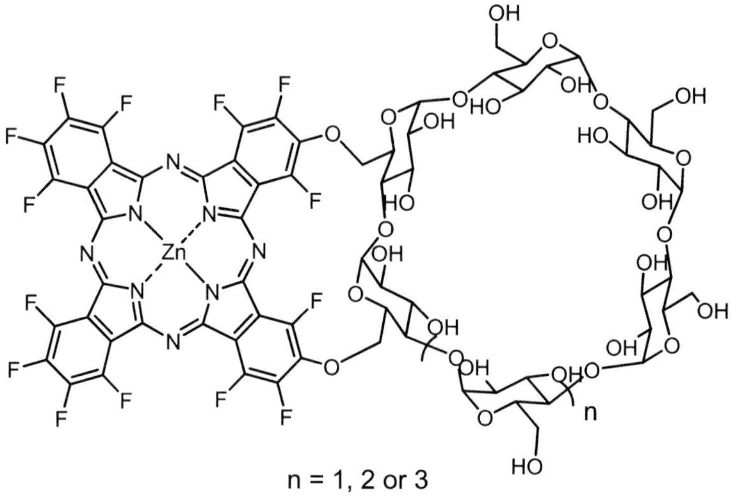

Fig. 13. Structures of cyclodextrin-conjugated phthalocyanines **40**.

3.2. *With Peptides*

Cell-penetrating peptides are effective drug carriers that can enhance the biocompatibility and cellular uptake of the drugs. A number of their conjugates with phthalocyanines have been reported for application in PDT. Chen *et al.* prepared a non-covalent conjugate of the Tat peptide, an arginine cationic peptide derived from the HIV-1 Tat protein, with sulfonated aluminum phthalocyanine by simply stirring a mixture of these compounds.[90] Its fluorescence intensity inside KB and HeLa cells as monitored by confocal laser scanning microscopy was about two-fold higher compared with the non-peptide conjugate.

Oligolysines are also cationic peptides at the physiological conditions that have been used to conjugate to phthalocyanines. Huang *et al.* prepared a series of these conjugates **41** (Fig. 14), in which an oligolysine chain ($n = 1, 3, 5, 7$ or 9) is attached to one of the β-positions of zinc(II) phthalocyanine.[91] Their photodynamic activities were studied and compared using BGC-823 human stomach adenocarcinoma cells and human embryo lung fibroblasts (HELF). These photosensitizers exhibited phototoxicity against the carcinoma cells, but minimal toxicity to the fibroblast cells. By using a sensitive in situ assay based on an electric cell-substrate impedance sensing technique, the heptalysine conjugate was found to have the highest photocytotoxicity. In another study, the pentalysine analogue was also found to exhibit antimicrobial effects against *Porphyromonas gingivalis* and *E. coli*, and in an animal infection model in which

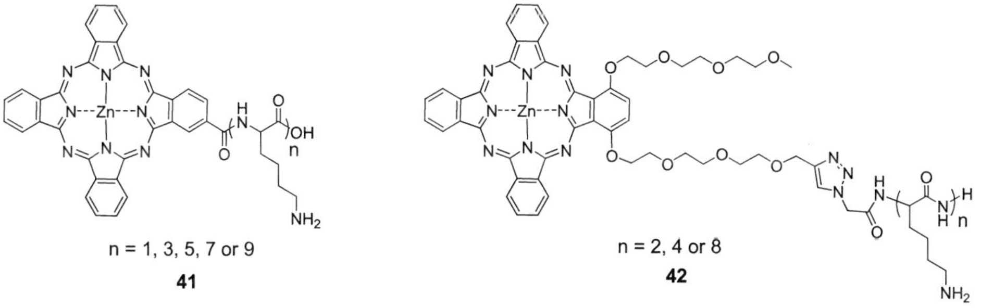

n = 1, 3, 5, 7 or 9
41

n = 2, 4 or 8
42

Fig. 14. Structures of oligolysine-conjugated phthalocyanines **41** and **42**.

a genetically engineered *E. coli* was inoculated into an excisional wound on a rabbit.[92]

Our group also reported a related series of 1,4-disubstituted zinc(II) phthalocyanines with an oligolysine chain ($n = 2$, 4 or 8) (compounds **42**, Fig. 14).[93] The compounds were slightly aggregated in water with 0.05% Cremophor EL, but they could still generate singlet oxygen effectively. Their photoinactivation efficacy was examined using a range of bacterial strains, including the Gram-positive methicillin-sensitive *Staphylococcus aureus* (*S. aureus*) and MRSA, and the Gram-negative *E. coli* and *P. aeruginosa*. The compounds were much more potent toward the Gram-positive bacteria and only 15–90 nM was required to induce a 4log reduction in the cell viability of these strains (i.e. 99.99% cell death). The photocytotoxicity toward *E. coli* increased with the length of the oligolysine chain. The most potent octalysine analogue had a 4log reduction concentration of 0.8 μM, while *P. aeruginosa* was found to be most resistant to the photo-treatment. The potency of the tetralysine analogue against a series of clinical strains of *S. aureus* and *E. coli* was found to be comparable with that toward the non-clinical counterparts. In addition, these compounds also exhibited a very high photoinactivation efficacy toward the enveloped viruses influenza A virus (H1N1) and herpes simplex virus type 1 (HSV1) with IC_{50} values as low as 0.05 nM, but their phototoxicity against the non-enveloped viruses adenovirus type 3 (Ad3) and coxsakievirus (Cox B1) was negligible.

Another important function of peptides is their ability to recognize various tumor-specific receptors, which can promote selective uptake via receptor-mediated endocytosis. A number of peptide-phthalocyanine conjugates with this property have been reported. Huang *et al.* employed 2-carboxy zinc(II) phthalocyanine as a starting material and conjugated it with a gonadotropin-releasing hormone (GnRH), which is a hypothalamic decapeptide with the sequence EHWSYGLRPG.[94] The cellular uptake and photocytotoxicity of this

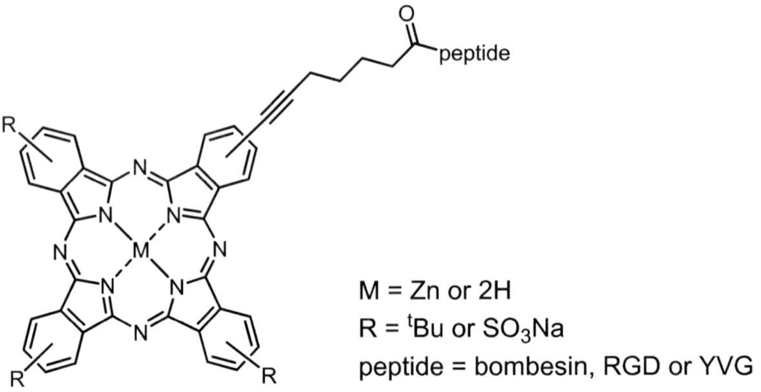

43

Fig. 15. Structures of peptide-conjugated phthalocyanines **43**.

conjugate was found to be significantly enhanced on human breast cancer cell lines MDA-MB-231 and MCF-7, which overexpress GnRH receptors, compared with HELF and HepG2, which have low levels of GnRH receptors. The uptake by MCF-7 was also found to be reduced in the presence of the GnRH receptor blocker Cetrorelix. The results suggested that the selective uptake of this conjugate was mediated by the GnRH receptors on the cell surface.

By using the Sonogashira cross-coupling reactions either in solution or on solid phase, van Lier *et al.* synthesized a series of phthalocyanines conjugated with a bombesin, which is a 14-amino acid peptide with the sequence YQRLGNQWAVGHLM, or an RGD moiety (compounds **43**, Fig. 15).[95] These two peptides are known for their high affinity to the gastrin-releasing peptide (GRP) and integrin receptors, respectively. For comparison, a non-specific peptide (YVG) that does not target any cancer-associated receptors was also conjugated to the phthalocyanines. The sulfonated conjugates were then examined for their photodynamic activities against several cell lines which have different expression levels of GRP and integrin receptors. As revealed by competition experiments, the uptake of the bombesin conjugate was found to involve the GRP receptors.

Another RGD-conjugated phthalocyanine was prepared by our group using a different strategy.[96] As shown in Scheme 5, click reaction of the alkynyl phthalocyanine **44** with the resin **45** gave the cycloadduct **46**. With a solid support, this product could be purified readily by simply washing with organic solvents. The Fmoc protecting group and the resin of **46** were then removed followed by intramolecular condensation to afford **47**. Finally, all the remaining

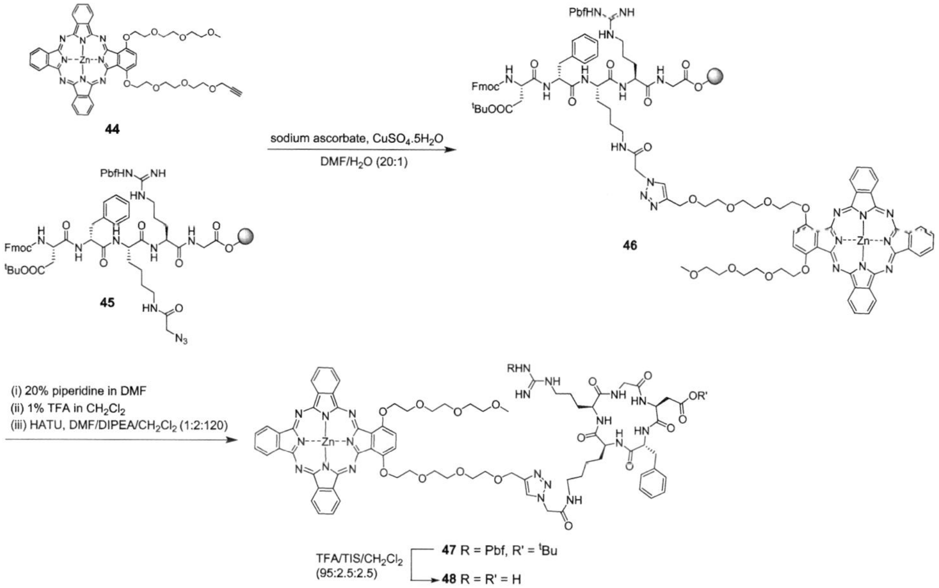

Scheme 5. Preparation of cyclic RGD-phthalocyanine conjugate **48**. DIPEA = *N*,*N*-diisopropylethylamine, Fmoc = fluorenylmethyloxycarbonyl, HATU = *O*-(7-azabenzotriazol-1-yl)-*N*,*N*,*N*′,*N*′-tetramethyluronium hexafluorophosphate, Pbf = 2,2,4,6,7-pentamethyldihydro benzofurane-5-sulfonyl, TFA = trifluoroacetic acid, TIS = triisopropylsilane.

protecting groups in **47** were removed to give the cyclic RGD conjugated phthalocyanine **48**. This short cyclic sequence is a well-known antagonist for the integrin $\alpha_v\beta_3$. In fact, it was found that **48** exhibited significantly higher cellular uptake toward the $\alpha_v\beta_3$-positive U87-MG human glioblastoma cells compared with the $\alpha_v\beta_3$-negative MCF-7 cells. However, its phototoxicity against the two cell lines was comparable owing to the similar efficiency in generating ROS inside the cells. Compound **48** was also found to localize preferentially in the lysosomes of the U87-MG cells.

In attempt to target the nucleus of cancer cells, a zinc(II) phthalocyanine conjugated with a short peptide with a nuclear localization sequence GGPKKKRKV was also prepared (compound **49**, Fig. 16).[97] The synthesis also involved **44** as a precursor and click reaction, but the peptide chain was introduced through amide coupling. Conjugation of the peptide could enhance the uptake by HT29 cells, efficiency in generating intracellular ROS, and photocytotoxicity with an IC_{50} value of 0.21 μM, despite the compound was found to localize in the cell membrane instead of the nucleus. Interestingly, the conjugate also exhibited an enhanced tumor-retention property in tumor-bearing nude mice. After 72 h post-injection, the dye concentration in the tumor was significantly higher than that in other organs.

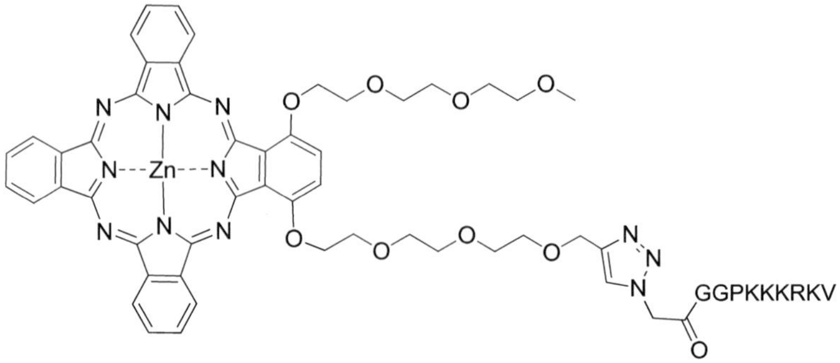

49

Fig. 16. Structure of peptide-conjugated phthalocyanine **49**.

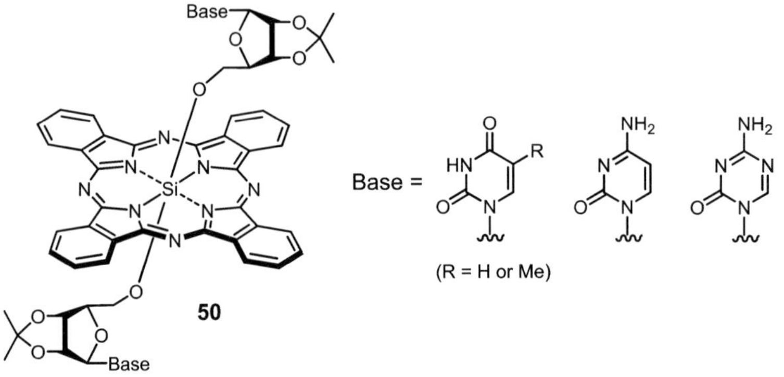

50

Fig. 17. Structures of nucleoside-conjugated phthalocyanines **50**.

3.3. *With Nucleic Acids*

Nucleic acids are also very important biomolecules, but conjugates of these compounds with phthalocyanines remain rare.[80] By simple substitution reactions, Huang *et al.* prepared a series of silicon(IV) phthalocyanines with two axial nucleoside moieties (compounds **50**, Fig. 17).[98] The uridine analogue showed the highest phototoxicity against HepG2 cells with an IC_{50} value of 6 nM as a result of its high cellular uptake and non-aggregated nature in the culture media. The compound also showed high affinity toward the mitochondria of the cells causing cell death mainly through apoptosis upon irradiation.

Gorun *et al.* employed a microwave-promoted method to prepare the perfluorinated carboxy phthalocyanine **51**, which was further conjugated with an antisense oligonucleotide targeting the glucose regulating protein 78 (GRP78), which represents a universal therapeutic target for treating

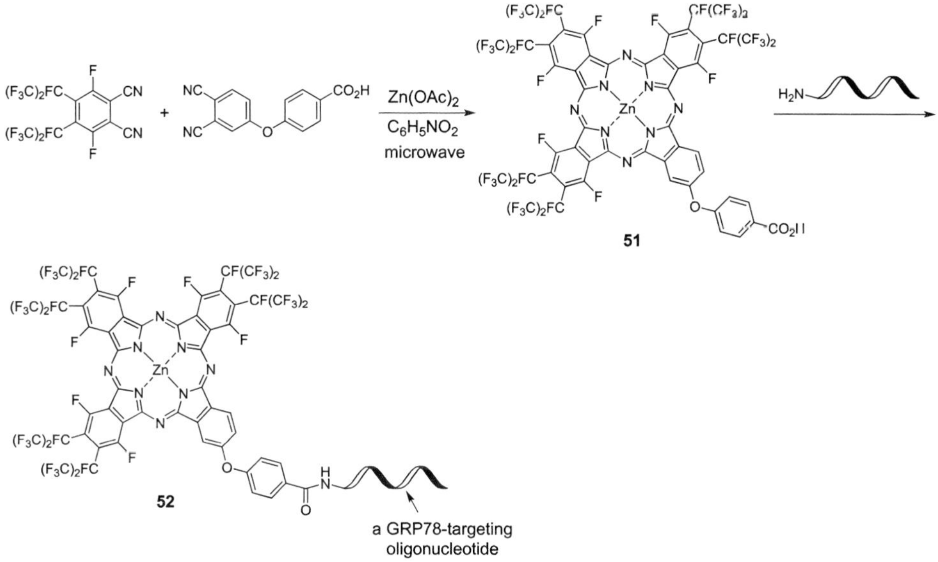

Scheme 6. Preparation of oligonucleotide-conjugated phthalocyanine **52**.

cancer (Scheme 6).[99] The fluorinated alkyl substituents could enhance the stability of the compound toward nucleophilic attack and oxidation by the ROS produced during photosensitization. The resulting conjugate **52** could hybridize with complementary GRP78 DNA and mRNA sequences giving the corresponding thermally stable duplexes. Upon irradiation followed by hot piperidine treatment, the duplexes underwent site-specific oxidative cleavage along the oligonucleotide backbone. The strategy reported therein might be potentially useful for oncogene-targeting PDT.

3.4. *With Polyamines*

Polyamines are naturally occurring compounds that are essential for cancer cells to sustain their rapid cell division. Most of these materials are imported from exogenous sources through active and specific polyamine transporters.[100] This feature has led to the development of various polyamine-based therapeutic agents.[101] Owing to this characteristic and the hydrophilic character of polyamines, our group synthesized a series of silicon(IV) phthalocyanines substituted with a range of polyamines with different lengths and spacers between the nitrogen centers (compounds **53**, Fig. 18), and examined their *in vitro* photodynamic activities.[102] All the compounds were highly photocytotoxic against HT29 and Chinese hamster ovary cells. Their IC_{50} values were as low as 1.1 nM. In a competitive study with spermidine and by adding the

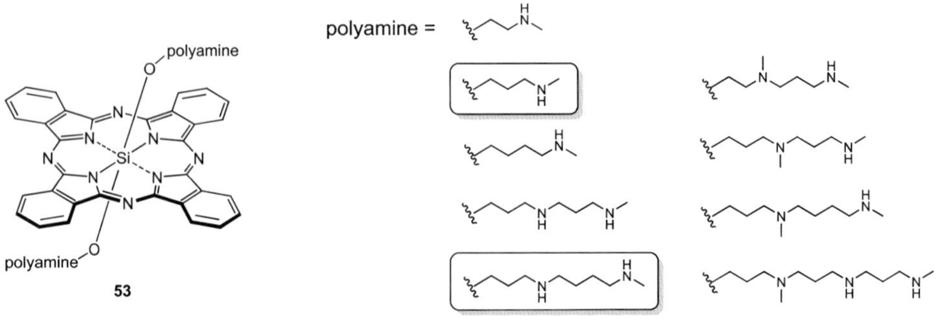

Fig. 18. Structures of polyamine-conjugated phthalocyanines **53**.

inhibitor α-difluoromethylornithine during drug incubation, it was found that the photocytotoxicity of selected conjugates was not significantly affected. The results suggested that the cellular uptake of these compounds might not involve the polyamine transporters. Further investigation of two of these compounds (highlighted below) found that they showed high affinity toward the lysosomes of HT29 cells, induced apoptosis extensively, and could inhibit the growth of tumor in HT29 tumor-bearing nude mice without causing apparent injury to the liver of the mice.

3.5. *With Folic Acid*

Folic acid is also a common tumor-targeting ligand which has strong affinity to the folate receptors overexpressed in cancer cells. While a large number of folate-conjugated drugs have been reported,[103] to our knowledge, only two phthalocyanine conjugates have been reported so far for application in PDT. The first covalent folate-phthalocyanine conjugate was prepared by treating tetraamino zinc(II) phthalocyanine with folic acid in the presence of dicyclohexylcarbodiimide and *N*-hydroxysuccinimide.[104,105] The conjugate was proposed to have only one folate group linked via the γ-carbonyl group to the phthalocyanine core (compound **54**, Fig. 19), and found to be soluble in water. The uptake of this conjugate by the folate-receptor-positive KB cells was much higher than that of the tetraamino analogue, while its uptake by the folate-receptor-negative A549 human lung epithelial carcinoma cells remained very low. These observations together with the result of a competition experiment suggested that the folate receptors played an important role in promoting the cellular uptake of this conjugate. Interestingly, the conjugate also exhibited two-photon absorption. Upon excitation by a 780 nm femtosecond laser, the conjugate caused fluorescence inside the KB cells and the singlet oxygen production was found to be proportional to the light dose. Its

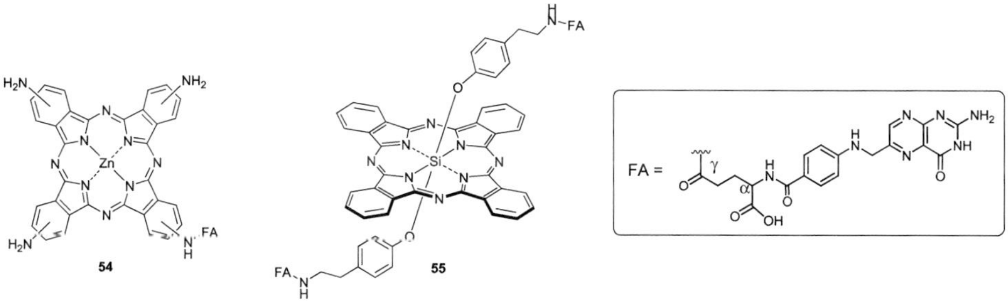

Fig. 19. Structures of folate-conjugated phthalocyanines **54** and **55**.

phototoxicity toward KB cells was ten-fold higher than that of the sulfonated aluminum phthalocyanine under two-photon excitation.

Recently, another folate-phthalocyanine conjugate (compound **55**, Fig. 19) has also been reported,[106] in which two folate moieties are axially linked to a silicon(IV) phthalocyanine core. The conjugate also showed high photocytotoxicity against the folate-receptor-positive HeLa cells ($IC_{50} = 0.07\,\mu M$), but it also had substantial potency toward the folate-receptor-negative HepG2 cells ($IC_{50} = 0.12\,\mu M$).

4. Dual Functional Phthalocyanines

By introducing another theranostic unit, the scope of biomedical applications of phthalocyanines can be extended. A number of such dual functional phthalocyanines have been reported over the last few years. van Lier *et al.* developed a series of bimodal imaging probes for fluorescence and positron emission tomography (PET) imaging.[107] These probes contain a sulfonated zinc(II) phthalocyanine and either a ^{68}Ga complex of 1,4,7,10-tetraazacyclododecane-1,4,7,10-tetraacetate or a ^{64}Cu complex of 1,4,7-triazacyclononane-1,4,7-triacetate linked via an aliphatic chain and amide linkage (compounds **56** and **57**, Fig. 20). By using *in vitro* plasma stability assays, it was found that the ^{64}Cu-labeled conjugates were highly stable, which remained intact over 24 h. The stability of the ^{68}Ga-labeled conjugates was relatively lower, which released up to 7% of free ^{68}Ga over a period of 3 h. The biodistribution of two of the ^{64}Cu-labeled conjugates in nude mice being implanted with MC7-L1 and EMT-6 tumors was also investigated, which confirmed that the probes could exhibit dual imaging functions. In a related study, the same research group replaced the metal center of several sulfonated zinc(II) phthalocyanines with ^{64}Cu via demetallation and remetallation.[108] The resulting phthalocyaninato ^{64}Cu complexes **58** (Fig. 20), which had >98% radiochemical purities, were injected

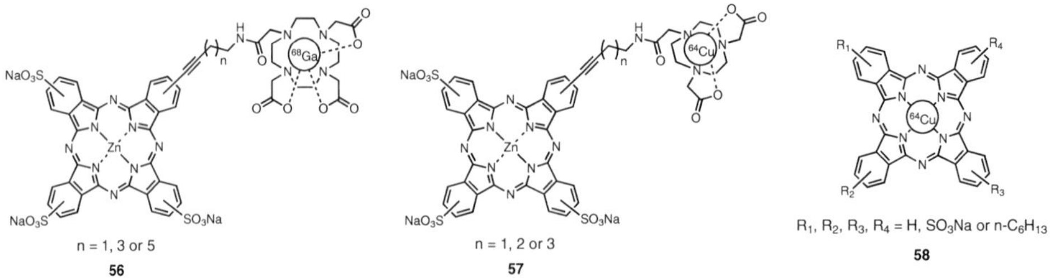

Fig. 20. Structures of phthalocyanines **56-58**.

into tumor-bearing nude mice, which were then subjected to 3 h dynamic PET imaging studies. Only the amphiphilic disulfonated and trisulfonated (with an *n*-hexyl group at the remaining isoindole unit) analogues were taken up by tumor and gave the highest tumor-to-background ratios.

With a view to achieving dual chemotherapy and PDT, a zinc(II) phthalocyanine conjugated with an oxaliplatin derivative was prepared according to Scheme 7.[109] The synthesis involved mixed cyclization of the diethyl methylmalonate-substituted phthalonitrile **59** and the unsubstituted phthalonitrile to give phthalocyanine **60**, which was then hydrolyzed followed by complexation to afford the conjugate **61**. It was cytotoxic in the dark due to the cytostatic oxaliplatin moiety and the cytotoxicity was greatly enhanced upon irradiation due to the photodynamic effect of the phthalocyanine unit. The IC_{50} value of the conjugate against HT29 cells (0.11 μM), which was five-fold lower than that of the reference compound without the platinum complex. The higher photocytotoxicity of the conjugate was attributed to the higher cellular uptake and efficiency in generating intracellular ROS. Hence, the two components in the conjugate work in a cooperative manner in the antitumor action. The conjugate was also found to localize in the lysosomes of the cells and induce cell death mainly through apoptosis.

Erlotinib is another anticancer drug that functions as a reversible tyrosine kinase inhibitor acting on the epidermal growth factor receptor (EGFR). Two conjugates of this drug and phthalocyanine (compounds **62**, Fig. 21) were also reported.[110] Their photocytotoxicity against the EGRF-positive HepG2 cells ($IC_{50} = 0.01$ and 0.04 μM) was comparable with that of the reference compound without the erlotinib moiety. However, the fluorescence intensities of the conjugates in HepG2 cells were about three-fold higher than those in EGRF-negative HELF cells, while there was no observable difference for the reference compound. In an *in vivo* fluorescence imaging study using A431 tumor-bearing nude mice, it was also found that the conjugate accumulated quickly in the tumor and the tumor/skin ratio was about five-fold higher than

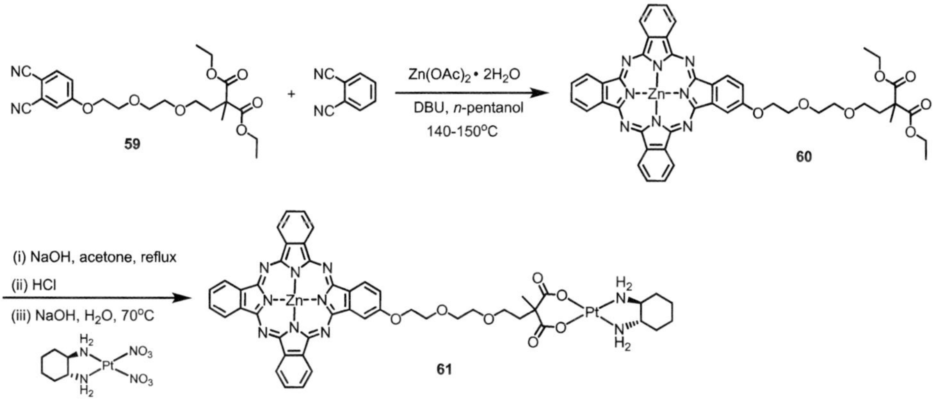

Scheme 7. Preparation of oxaliplatin-phthalocyanine conjugate **61**.

that for the reference at 2.5 h after administration. The results showed that introduction of the erlotinib moiety could impart tumor-targeting property to phthalocyanines.

You *et al.* extended the study of this class of conjugates by preparing a multifunctional prodrug (compound **63**, Fig. 21), in which the cytotoxic drug combretastatin A-4 (CA4) could be released upon cleavage of the electron-rich aminoacrylate linker by the singlet oxygen generated upon irradiation of the phthalocyanine core.[111] The conjugate showed a 19-fold lower dark toxicity against MCF-7 cells compared with CA4 ($IC_{50} = 173$ nM and 9 nM, respectively), but its cytotoxicity was greatly enhanced upon irradiation ($IC_{50} = 6$ nM). Injection of this conjugate, formulated as PEG-poly(D,L-lactide) polymeric micelles, to BALB/c mice with SC tumors (colon-26 cells) showed fluorescence spots in the tumors with a peak at around 24 h post-injection, which persisted longer than those observed by using Cremophor EL as the formulating agent, showing that the use of polymeric micelles as carriers could delay the clearance of the prodrug from the body. The *in vivo* antitumor effect of this conjugate was also found to be significantly higher than that of CA4 and the analogue with a non-cleavable linker. The results suggested that the two cytotoxic components in **63** function in a synergistic fashion.

On the basis that chalcones are able to destroy tumor vasculature, a chalcone-phthalocyanine conjugate (compound **64**, Fig. 22) was also prepared as a potential dual photodynamic and antiangiogenic agent.[112] This monochalcone conjugate showed improved hydrophilicity compared with the tetrachalcone analogue.[113] The conjugation of chalcone did not affect the singlet oxygen generation efficiency, but could enhance the photocytotoxicity (by 6–8 times) of the phthalocyanine probably due to the higher amphiphilic

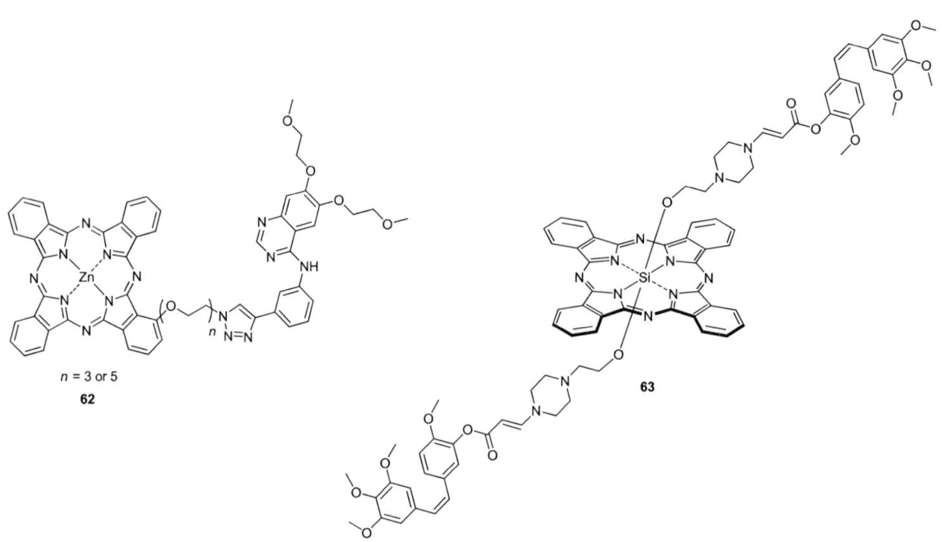

Fig. 21. Structures of phthalocyanines **62** and **63**.

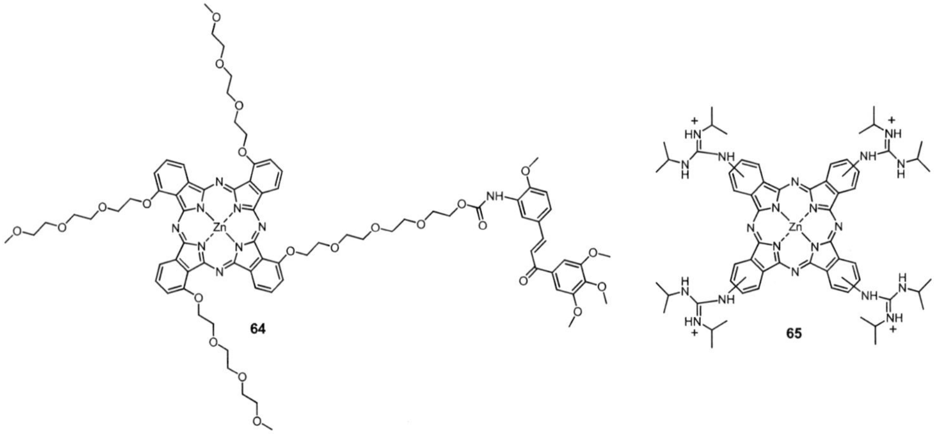

Fig. 22. Structures of phthalocyanines **64** and **65**.

character of the conjugate that could promote cellular uptake. However, as shown by migration assay using human umbilical vein endothelial cells, the conjugate **64** showed only a weak to moderate antivascular effect.

The tetrakis(diisopropylguanidino) zinc(II) phthalocyanine **65** (Fig. 22) is another promising dual functional theranostic agent.[114] Apart from being served as a photosensitizer for PDT, it also exhibited *in vivo* antimetastatic activities. Whilst the compound showed low dark toxicity against the murine metastatic B16F10 melanoma cells ($EC_{50} > 80\,\mu M$) and was well tolerated upon its intravenous injection into mice (at 8 mg/kg), its photocytotoxicity was greatly enhanced by more than 400-fold upon irradiation with laser at 660 nm ($EC_{50} \approx 0.16\,\mu M$). Compared with the well-studied tetrasulfonated zinc(II) phthalocyanine, it was more potent by about 20-fold. In addition, the compound could also inhibit the formation of metastatic tumor colonies in the lung tissues of C57BL/6 mice injected with B16F10 cells. It was suggested that the antimetastatic activities of **65** might be due to its ability to disrupt the signaling between chemokine CXCL10 and the G protein-coupled receptor CXCR3. Therefore, compound **65** could destroy the primary tumor by PDT while simultaneously inhibiting the formation of metastases by a light-independent mode of action.

5. Activatable Phthalocyanine-Based Photosensitizers

As a promising approach to address the tumor specificity, activatable photosensitizers are of much current interest.[115] In these systems, the photosensitizers are either self-quenched due to aggregation or deactivated by the neighboring quenchers. Upon interactions with tumor-associated stimuli, the photosensitizers are disaggregated or detached from the quenchers, resulting in restoration of their fluorescence and photosensitizing properties. These "smart" theranostic agents can therefore overcome the drawback of non-specific activation of phototoxicity in PDT.

Recently, our group has developed a number of activatable phthalocyanine-based photosensitizers. The tetraamino silicon(IV) phthalocyanine **66** (Fig. 23) is one of the early examples that is pH-responsive.[116] Its fluorescence emission and ROS generation in aqueous media were greatly enhanced by lowering the pH from 7.4 to 5.0. The intracellular fluorescence intensity was also much stronger at pH 5.0 and 6.0 than that at 7.4 (by about 16-fold). These results could be attributed to protonation of the amino groups that inhibited the quenching by photoinduced electron transfer. The responsive region ($pH \approx 5$–7) could roughly differentiate the extracellular region of tumors and normal tissues. Upon irradiation, compound **66** was also highly toxic toward HepG2 and HT29 cells with IC_{50} values of 0.01 μM. It was significantly more potent (>500-fold) than the classical photosensitizer Photofrin.

A related series of silicon(IV) phthalocyanines with axial aryl amino moieties (compounds **67**, Fig. 23) were also prepared.[117] Changing the pH environment altered the extent of protonation and the aggregation tendency of the compounds, which in turn affected the fluorescence emission and

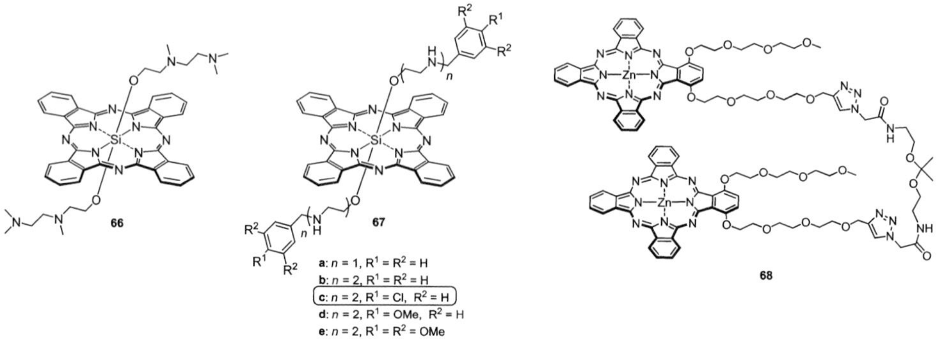

Fig. 23. Structures of phthalocyanines **66-68**.

ROS production. The compound with a terminal 4-chlorophenyl group (highlighted below) exhibited the most desirable pH-responsive properties giving the greatest changes in the pH range of 6.5–7.4. As revealed by confocal microscopy, it also showed high affinity toward the lysosomes of HT29 cells. These excellent properties together with its high photocytotoxicity (IC_{50} = $0.05-0.06$ μM) suggested that it is a promising pH-controlled photosensitizer for PDT.

Phthalocyanine dimer **68** (Fig. 23) is another pH-responsive photosensitizer.[118] Having an acid-sensitive ketal linker, it underwent cleavage at this site in an acidic environment. The separation of the phthalocyanine units prevented their self-quenching thereby restoring their fluorescence emission and singlet oxygen production. The dimer could also be activated inside the cells causing strong fluorescence throughout the cytoplasm. The intracellular fluorescence intensity was about five-fold stronger than that induced by the non-cleavable analogue, and four-fold stronger at pH 5.0 than at pH 7.4 using the ionophore nigericin to equilibrate the intra- and extracellular pH. The IC_{50} value of **68** against HT29 cells (0.35 μM) was slightly lower than that of the non-cleavable analogue (0.59 μM). The substantial photocytotoxicity of the latter suggested that the quenching of singlet oxygen production is less effective than the quenching of fluorescence inside the cells.

By using ferrocene as a quencher, which was linked to the silicon(IV) phthalocyanine core via a hydrazone linker, compound **69** (Fig. 24) could also be activated in an acidic environment.[119] The fluorescence intensity of this compound in PBS increased by about six-fold and the singlet oxygen generation efficiency was also enhanced when the pH decreased from 7.4 to 4.5. Its fluorescence intensity inside MCF-7 cells also increased by about ten-fold when the pH of the culture medium decreased from 7.4 to 5.0. Upon

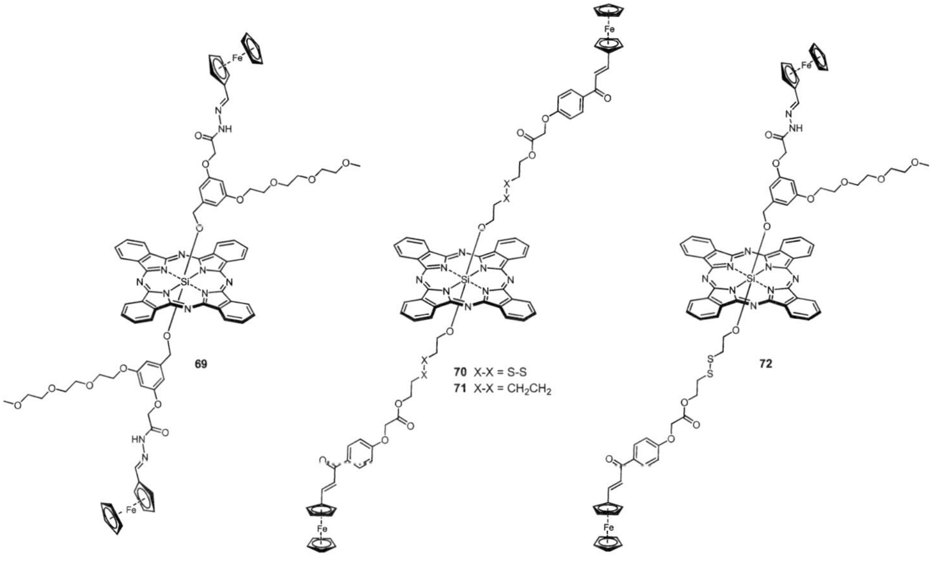

Fig. 24. Structures of phthalocyanines **69**-**72**.

irradiation, the compound was also highly potent toward these cells with IC_{50} values in the range of 73–76 nM.

Apart from these pH-responsive photosensitizers, we also prepared several thiol-sensitive counterparts. Compound **70** (Fig. 24) is an analogue of **69**, in which the acid-sensitive hydrazone linker was replaced with a thiol-cleavable disulfide bond.[120] The compound could be activated by dithiothreitol (DTT) in mM range, but not in μM range, which mimic the intracellular and extracellular reducing environment, respectively, leading to enhanced fluorescence emission and singlet oxygen production in PBS, and stronger intracellular fluorescence intensity. The photocytotoxicity of this compound against MCF-7 cells was also enhanced by preincubation with 4 mM of DTT. The IC_{50} value decreased from 124 nM to 50 nM. A similar enhancement was not observed for the non-cleavable analogue **71** (Fig. 24), in which the disulfide bond was replaced with an ethylene linker.

As an extension of these works, we also prepared a novel dual activatable analogue (compound **72**, Fig. 24).[119] Having both an acid-cleavable hydrazone linker and a DTT-cleavable disulfide linker, this compound functioned as a dual pH- and redox-responsive photosensitizer. The fluorescence intensity, singlet oxygen generation efficiency, intracellular fluorescence, and *in vitro* photodynamic activity of this compound were enhanced in a slightly acidic environment (pH = 4.5–6.0) or in the presence of DTT (in mM range). The greatest enhancement could be attained when both of these two conditions

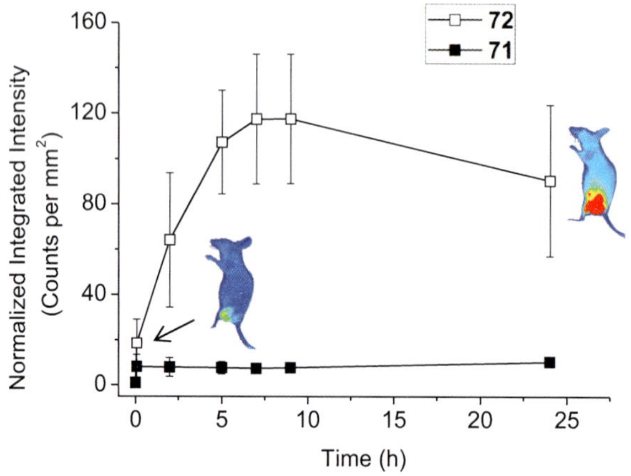

Fig. 25. Changes in fluorescence intensity of **72** and the non-cleavable analogue **71** (injected intratumorally with a dose of 1 μmol per kg body weight) per unit area of the tumor over 24 h. Five mice were used for each compound.

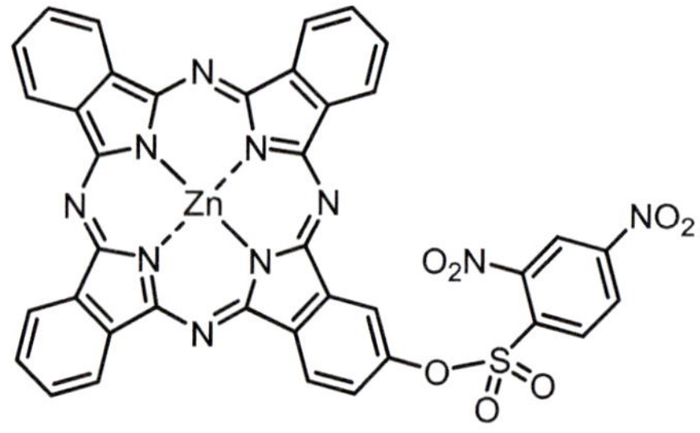

73

Fig. 26. Structure of thiol-activatable phthalocyanine **73**.

occurred, which were used to roughly mimic the acidic and reducing environment of tumor tissues. The activation of this compound was also demonstrated in tumor-bearing nude mice. As shown in Fig. 25, the intratumoral fluorescence of **72** was greatly increased within 10 h after injection, while that of the non-cleavable control **71** was negligible, showing that the fluorescence of **72** could be restored inside the tumor. In addition, this compound could also effectively inhibit the growth of tumor upon irradiation.

Compound **73** (Fig. 26) is another simple yet highly efficient thiol-activatable photosensitizer.[121] Upon interaction with glutathione, the 2,4-dinitrobenzenesulfonyl quencher was detached from the phthalocyanine. Its fluoresence emission and ROS generation were greatly enhanced in PBS and inside MCF-7 cells. The fluorescence activation was also demonstrated *in vivo*. On the basis that glutathione is the most abundant cellular thiol and its levels are often elevated in tumor tissues as compared with normal tissues, this compound can serve as a promising tumor-selective fluorescent probe and photosensitizer for targeted PDT.

6. Outlook

As an intriguing class of functional dyes, phthalocyanines have demonstrated their high versatility as advanced photosensitizers for PDT. Apart from the molecular systems described above, various nanophotosensitizing systems based on phthalocyanines have also been extensively studied,[11–20] which also show very promising results. The simultaneous administration of two phthalocyanine-based photosensitizers has been shown to induce synergistic phototoxicity, both at the cellular level and in tumor-bearing mice.[122] This approach could improve the curative rates compared with the conventional PDT with one photosensitizer alone. Recent studies have also found that phthalocyanines can also exhibit photothermal[123] and sonodynamic[124] properties, and act as contrast agents for photoacoustic imaging,[125,126] allowing them to be used for dual therapy and/or imaging of cancer. All these promising findings have opened up many possibilities for further development of multifunctional phthalocyanine-based theranostic agents. It is envisaged that this field of studies will remain as a research focus in the year ahead.

References

1. D. E. J. G. J. Dolmans, D. Fukumura and R. K. Jain, *Nat. Rev. Cancer* **3**, 380 (2003).
2. S. B. Brown, E. A. Brown and I. Walker, *Lancet Oncol.* **5**, 497 (2004).
3. L. M. Sanabria, M. E. Rodríguez, I. S. Cogno, N. B. R. Vittar, M. F. Pansa, M. J. Lamberti and V. A. Rivarola, *Biochim. Biophys. Acta* **1835**, 36 (2013).
4. A. E. O'Connor, W. M. Gallagher and A. T. Byrne, *Photochem. Photobiol.* **85**, 1053 (2009).
5. M. J. Garland, C. M. Cassidy, D. Woolfson and R. F. Donnelly, *Future Med. Chem.* **1**, 667 (2009).
6. P. Mroz, S. K. Sharma, T. Zhiyentayev, Y.-Y. Huang and M. R. Hamblin, in *Drug Delivery in Oncology: From Basic Research to Cancer Therapy*, Eds. F. Kratz, P. Senter and H. Steinhagen (Wiley-VCH, Weinheim, 2012), p. 1569.
7. A. B. Ormond and H. S. Freeman, *Materials* **6**, 817 (2013).
8. D. K. P. Ng, *Future Med. Chem.* **6**, 1991 (2014).
9. S. I. A. Zaida, R. Agarwal, G. Eichler, B. D. Rihter, M. E. Kenney and H. Mukhtar, *Photochem. Photobiol.* **58**, 204 (1993).
10. Z. Jiang, J. Shao, T. Yang, J. Wang and L. Jia, *J. Pharm. Biomed. Anal.* **87**, 98 (2014).
11. J. Shao, Y. Dai, W. Zhao, J. Xie, J. Xue, J. Ye and L. Jia, *Cancer Lett.* **330**, 49 (2013).

12. P. Skupin-Mrugalska, J. Piskorz, T. Goslinski, J. Mielcarek, K. Konopka and N. Düzgüneş, *Drug Discov. Today* **18**, 776 (2013).
13. S. Marrache, S. Tundup, D. A. Harn and S. Dhar, *ACS Nano* **7**, 7392 (2013).
14. R. Liang, L. Ma, L. Zhang, C. Li, W. Liu, M. Wei, D. Yan, D. G. Evans and X. Duan, *Chem. Commun.* **50**, 14983 (2014).
15. X. Ma, S. Sreejith and Y. Zhao, *ACS Appl. Mater. Interfaces* **5**, 12860 (2013).
16. R. Liang, R. Tian, L. Ma, L. Zhang, Y. Hu, J. Wang, M. Wei, D. Yan, D. G. Evans and X. Duan, *Adv. Funct. Mater.* **24**, 3144 (2014).
17. H. K. Moon, M. Son, J. E. Park, S. M. Yoon, S. H. Lee and H. C. Choi, *NPG Asia Mater.* **4**, e12 (2012).
18. J. U. Menon, P. Jadeja, P. Tambe, K. Vu, B. Yuan and K. T. Nguyen, *Theranostics* **3**, 152 (2013).
19. A. Master, M. Livingston and A. Sen Gupta, *J. Control. Release* **168**, 88 (2013).
20. C.-K. Lim, J. Heo, S. Shin, K. Jeong, Y. H. Seo, W.-D. Jang, C. R. Park, S. Y. Park, S. Kim and I. C. Kwon, *Cancer Lett.* **334**, 176 (2013).
21. H. Ali and J. E. van Lier, *Chem. Rev.* **99**, 2379 (1999).
22. C. M. Allen, W. M. Sharman and J. E. van Lier, *J. Porphyrins Phthalocyanines* **5**, 161 (2001).
23. S. Ogura, K. Tabata, K. Fukushima, T. Kamachi and I. Okura, *J. Porphyrins Phthalocyanines* **10**, 1116 (2006).
24. J.-p. Taquet, C. Frochot, V. Manneville and M. Barberi-Heyob, *Curr. Med. Chem.* **14**, 1673 (2007).
25. A. Wang, L. Long and C. Zhang, *J. Incl. Phenom. Macrocycl. Chem.* **71**, 1 (2011).
26. N. Sekkat, H. van den Bergh, T. Nyokong and N. Lange, *Molecules* **17**, 98 (2012).
27. S. Tuncel, F. Dumoulin, J. Gailer, M. Sooriyaarachchi, D. Atilla, M. Durmuş, D. Bouchu, H. Savoie, R. W. Boyle and V. Ahsen, *Dalton Trans.* **40**, 4067 (2011).
28. M. Bai, P.-C. Lo, J. Ye, C. Wu, W.-P. Fong and D. K. P. Ng, *Org. Biomol. Chem.* **9**, 7028 (2011).
29. B. Zhao, W. Duan, P.-C. Lo, L. Duan, C. Wu and D. K. P. Ng, *Chem. Asian J.* **8**, 55 (2013).
30. X.-W. Chen, M.-R. Ke, X.-S. Li, W.-L. Lan, M.-F. Zhang and J.-D. Huang, *Chem. Asian J.* **8**, 3063 (2013).
31. Y. Chin, S. H. Lim, Y. Zorlu, V. Ahsen, L. V. Kiew, L. Y. Chung, F. Dumoulin and H. B. Lee, *PLoS ONE* **9**, e97894 (2014).
32. F. B. do Nascimento, T. M. Manieri, G. Cerchiaro, A. O. Ribeiro, *Dyes Pigm.* **99**, 316 (2013).
33. X. Jia, F.-F. Yang, J. Li, J.-Y. Liu and J.-P. Xue, *J. Med. Chem.* **56**, 5797 (2013).
34. B.-Y. Zheng, H.-P. Zhang, M.-R. Ke and J.-D. Huang, *Dyes Pigm.* **99**, 185 (2013).
35. N. B. R. Vittar, J. Awruch, K. Azizuddin and V. Rivarola, *Int. J. Biochem. Cell Biol.* **42**, 1123 (2010).
36. T. H. Foster, B. R. Giesselman, R. Hu, M. E. Kenney and S. Mitra, *Translat. Oncol.* **3**, 135 (2010).
37. E. D. Baron, C. L. Malbasa, D. Santo-Domingo, P. Fu, J. D. Miller, K. K. Hanneman, A. H. Hsia, N. L. Oleinick, V. C. Colussi and K. D. Cooper, *Lasers Surg. Med.* **42**, 888 (2010).

38. A. M. Master, M. E. Rodriguez, M. E. Kenney and N. L. Oleinick and A. Sen Gupta, *J. Pharm. Sci.* **99**, 2386 (2010).
39. A. Y. Huang, J. T. Myers, D. Barkauskas, S. J. Howell, N. L. Oleinick, T. S. McCormick, K. D. Cooper, E. D. Baron and M. Lam, *Photodiag. Photodyn. Ther.* **9**, 225 (2012).
40. D. Separovic, P. Breen, N. B. Boppana, E. van Buren, N. Joseph, J. M. Kraveka, M. Rahmaniyan, L. Li, T. I. Gudz, A. Bielawska, A. Bai, J. Bielawski, J. S. Pierce and M. Korbelik, *Int. J. Oncol.* **43**, 2064 (2013).
41. M. Lam, M. L. Dimaano, P. Oyetakin-White, M. A. Retuerto, J. Chandra, P. K. Mukherjee, M. A. Ghannoum, K. D. Cooper and E. D. Baron, *Antimicrob. Agents Chemother.* **58**, 3029 (2014).
42. L. Luan, L. Ding, W. Zhang, J. Shi, X. Yu and W. Liu, *Bioorg. Med. Chem. Lett.* **23**, 3775 (2013).
43. J. Marino, M. C. García Vior, L. E. Dicelio, L. P. Roguin and J. Awruch, *Eur. J. Med. Chem.* **45**, 4129 (2010).
44. G. A. Gauna, J. Marino, M. C. García Vior, L. P. Roguin and J. Awruch, *Eur. J. Med. Chem.* **46**, 5532 (2011).
45. J. Marino, M. C. García Vior, V. A. Furmento, V. C. Blank, J. Awruch and L. P. Roguin, *Int. J. Biochem. Cell Biol.* **45**, 2553 (2013).
46. W. Duan, P.-C. Lo, L. Duan, W.-P. Fong and D. K. P. Ng, *Bioorg. Med. Chem.* **18**, 2672 (2010).
47. A. Wang, L. Zhou, K. Fang, L. Zhou, Y. Lin, J. Zhou and S. Wei, *Eur. J. Med. Chem.* **58**, 12 (2012).
48. A. Wang, Y. Ma, L. Zhou, S. Lu, Y. Lin, J. Zhou and S. Wei, *Dyes Pigm.* **99**, 348 (2013).
49. B. G. Ongarora, X. Hu, S. D. Verberne-Sutton, J. C. Garno and M. G. H. Vicente, *Theranostics* **2**, 850 (2012).
50. S. Makhseed, M. Machacek, W. Alfadly, A. Tuhl, M. Vinodh, T. Simunek, V. Novakova, P. Kubat, E. Rudolf and P. Zimcik, *Chem. Commun.* **49**, 11149 (2013).
51. Y. Ge, X. Weng, T. Tian, F. Ding, R. Huang, L. Yuan, J. Wu, T. Wang, P. Guo and X. Zhou, *RSC Adv.* **3**, 12839 (2013).
52. Z. Zhao, P.-S. Chan, H. Li, K.-L. Wong, R. N. S. Wong, N.-K. Mak, J. Zhang, H.-L. Tam, W.-Y. Wong, D. W. J. Kwong and W.-K. Wong, *Inorg. Chem.* **51**, 812 (2012).
53. D. K. Muli, P. Rajaputra, Y. You and D. V. McGrath, *Bioorg. Med. Chem. Lett.* **24**, 4496 (2014).
54. M. B. Spesia, M. Rovera, E. N. Durantini, *Eur. J. Med. Chem.* **45**, 2198 (2010).
55. V. Mantareva, I. Angelov, V. Kussovski, R. Dimitrov, L. Lapok and D. Wöhrle, *Eur. J. Med. Chem.* **46**, 4430 (2011).
56. J. B. Pereira, E. F. A. Carvalho, M. A. F. Faustino, R. Fernandes, M. G. P. M. S. Neves, J. A. S. Cavaleiro, N. C. M. Gomes, Â. Cunha, A. Almeida and J. P. C. Tomé, *Photochem. Photobiol.* **88**, 537 (2012).
57. V. N. Mantareva, I. Angelov, D. Wöhrle, E. Borisova and V. Kussovski, *J. Porphyrins Phthalocyanines* **17**, 399 (2013).
58. M. A. Di Palma, M. G. Alvarez, A. L. Ochoa, M. E. Milanesio and E. N. Durantini, *Fungal Biol.* **117**, 744 (2013).
59. T. C. Tempesti, M. G. Alvarez and E. N. Durantini, *Dyes Pigm.* **91**, 6 (2011).
60. M.-R. Ke, J. M. Eastel, K. L. K. Ngai, Y.-Y. Cheung, P. K. S. Chan, M. Hui, D. K. P. Ng and P.-C. Lo, *Eur. J. Med. Chem.* **84**, 278 (2014).

61. B.-Y. Zheng, T. Lin, H.-H. Yang and J.-D. Huang, *Dyes Pigm.* **96**, 547 (2013).
62. B.-Y. Zheng, X.-J. Jiang, T. Lin, M.-R. Ke and J.-D. Huang, *Dyes Pigm.* **112**, 311 (2015).
63. P. Mikula, L. Kalhotka, D. Jancula, S. Zezulka, R. Korinkova, J. Cerny, B. Marsalek and P. Toman, *J. Photochem. Photobiol. B: Biol.* **138**, 230 (2014).
64. S. Dutta, B. G. Ongarora, H. Li, M. G. H. Vicente, B. K. Kolli and K. P. Chang, *PLoS ONE* **6**, e20786 (2011).
65. E. Krestyn, H. Kolarova, R. Bajgar and K. Tomankova, *Toxicol. in Vitro* **24**, 286 (2010).
66. J. Berlanda, T. Kiesslich, V. Engelhardt, B. Krammer and K. Plaetzer, *J. Photochem. Photobiol. B: Biol.* **100**, 173 (2010).
67. K. Maduray, A. Karsten, B. Odhav and T. Nyokong, *J. Photochem. Photobiol. B: Biol.* **103**, 98 (2011).
68. S. L. Manoto, P. R. Sekhejane, N. N. Houreld and H. Abrahamse, *Photodiag. Photodyn. Ther.* **9**, 52 (2012).
69. E. Güzel, A. Atsay, S. Nalbantoglu, N. Şaki, A. L. Dogan, A. Gül and M. B. Koçak, *Dyes Pigm.* **97**, 238 (2013).
70. Y.-L. Qin, X. Huang and J.-Y. Chen, *Anti-Cancer Drugs* **23**, 1047 (2012).
71. J. Wang, W. Li, H. B. Yu, N. H. Cheung and J. Y. Chen, *Laser Phys.* **24**, 035602 (2014).
72. H. Abramczyk, B. Brozek-Pluska, J. Surmacki, J. Musial and R. Kordek, *Analyst* **139**, 5547 (2014).
73. J. Xue, C. Li, H. Liu, J. Wei, N. Chen and J. Huang, *Photodiag. Photodyn. Ther.* **8**, 267 (2011).
74. J. Shao, J. Xue, Y. Dai, H. Liu, N. Chen and J. Huang, *Eur. J. Cancer* **48**, 2086 (2012).
75. F. Setaro, R. Ruiz-González, S. Nonell, U. Hahn and T. Torres, *J. Inorg. Biochem.* **136**, 170 (2014).
76. A. C. H. Ng, X.-Y. Li and D. K. P. Ng, *Macromolecules* **32**, 5292 (1999).
77. N. Nishiyama, A. Iriyama, W.-D. Jang, K. Miyata, K. Itaka, Y. Inoue, H. Takahashi, Y. Yanagi, Y. Tamaki, H. Koyama and K. Kataoka, *Nat. Mater.* **4**, 934 (2005).
78. H.-L. Lu, W.-J. Syu, N. Nishiyama, K. Kataoka and P.-S. Lai, *J. Control. Release* **155**, 458 (2011).
79. T. Nomoto, S. Fukushima, M. Kumagai, K. Machitani, Arnida, Y. Matsumoto, M. Oba, K. Miyata, K. Osada, N. Nishiyama and K. Kataoka, *Nat. Commun.* **5**, 3545 (2014).
80. P.-C. Lo, J. T. F. Lau and D. K. P. Ng, in *Handbook of Porphyrin Science*, Vol. 18, Eds. K. M. Kadish, K. M. Smith and R. Guilard (World Scientific, New Jersey, 2012) Chapter 84.
81. L. M. O. Lourenço, M. G. P. M. S. Neves, J. A. S. Cavaleiro and J. P. C. Tomé, *Tetrahedron* **70**, 2681 (2014).
82. A. R. M. Soares, M. G. P. M. S. Neves, A. C. Tomé, M. C. Iglesias-de la Cruz, A. Zamarrón, E. Carrasco, S. González, J. A. S. Cavaleiro, T. Torres, D. M. Guldi and A. Juarranz, *Chem. Res. Toxico.* **25**, 940 (2012).
83. F. Lv, Y. Li, B. Cao and T. Liu, *J. Mater. Sci. Mater. Med.* **24**, 811 (2013).
84. Z. Zhao, Y. Li, S. Meng, S. Li, Q. Wang and T. Liu, *Lasers Med. Sci.* **29**, 1131 (2014).
85. S. G. Kimani, T. A. Shmigol, S. Hammond, J. B. Phillips, J. I. Bruce, A. J. MacRobert, M. V. Malakhov and J. P. Golding, *Photochem. Photobiol.* **89**, 139 (2013).
86. D. Lafont, Y. Zorlu, H. Savoie, F.Albrieux, V. Ahsen, R. W. Boyle and F. Dumoulin, *Photodiag. Photodyn. Ther.* **10**, 252 (2013).

87. S. Silva, P. M. R. Pereira, P. Silva, F. A. A. Paz, M. A. F. Faustino, J. A. S. Cavaleiro and J. P. C. Tomé, *Chem. Commun.* **48**, 3608 (2012).
88. P. M. R. Pereira, S. Silva, J. A. S. Cavaleiro, C. A. F. Ribeiro, J. P. C. Tomé and R. Fernandes, *PLoS ONE* **9**, e95529 (2014).
89. L. M. O. Lourenço, P. M. R. Pereira, E. Maciel, M. Válega, F. M. J. Domingues, M. R. M. Domingues, M. G. P. M. S. Neves, J. A. S. Cavaleiro, R. Fernandes and J. P. C. Tomé, *Chem. Commun.* **50**, 8363 (2014).
90. J.-F. Zhao, J.-Y. Chen, L. Mi, P.-N. Wang and Q. Peng, *Ultrastruct. Pathol.* **35**, 119 (2011).
91. L. Li, Z. Luo, Z. Chen, J. Chen, S. Zhou, P. Xu, P. Hu, J. Wang, N. Chen, J. Huang and M. Huang, *Bioconjugate Chem.* **23**, 2168 (2012).
92. Z. Chen, S. Zhou, J. Chen, L. Li, P. Hu, S. Chen and M. Huang, *J. Lumin.* **152**, 103 (2014).
93. M.-R. Ke, J. M. Eastel, K. L. K. Ngai, Y.-Y. Cheung, P. K. S. Chan, M. Hui, D. K. P. Ng and P.-C. Lo, *Chem. Asian J.* **9**, 1868 (2014).
94. P. Xu, J. Chen, Z. Chen, S. Zhou, P. Hu, X. Chen and M. Huang, *PLoS ONE* **7**, e37051 (2012).
95. E. Ranyuk, N. Cauchon, K. Klarskov, B. Guérin and J. E. van Lier, *J. Med. Chem.* **56**, 1520 (2013).
96. M.-R. Ke, D. K. P. Ng and P.-C. Lo, *Chem. Asian J.* **9**, 554 (2014).
97. M.-R. Ke, S.-L. Yeung, W.-P. Fong, D. K. P. Ng and P.-C. Lo, *Chem. Eur. J.* **18**, 4225 (2012).
98. X.-M. Shen, B.-Y. Zheng, X.-R. Huang, L. Wang and J.-D. Huang, *Dalton Trans.* **42**, 10398 (2013).
99. P. Patel, H. H. Patel, E. Borland, S. M. Gorun and D. Sabatino, *Chem. Commun.* **50**, 6309 (2014).
100. A. A. Abdulhussein and H. M. Wallace, *Amino Acids* **46**, 655 (2014).
101. T. Senanayake, H. Amunugama, T. D. Boncher, R. A. Casero Jr. and P. M. Woster, *Essays Biochem.* **46**, 77 (2009).
102. X.-J. Jiang, S.-L. Yeung, P.-C. Lo, W.-P. Fong and D. K. P. Ng, *J. Med. Chem.* **54**, 320 (2011).
103. M.-R. Ke, S.-L. Yeung, D. K. P. Ng, W.-P. Fong and P.-C. Lo, *J. Med. Chem.* **56**, 8475 (2013) and the references cited therein.
104. P. Khoza, E. Antunes, J.-Y. Chen and T. Nyokong, *J. Lumin.* **134**, 784 (2013).
105. S. Wang, J. Wang and J.-Y. Chen, *J. Mater. Chem. B* **2**, 1594 (2014).
106. Y.-W. Zheng, S.-F. Chen, B.-Y. Zheng, M.-R. Ke and J.-D. Huang, *Chem. Lett.* **43**, 1701 (2014).
107. E. Ranyuk, R. Lebel, Y. Bérubé-Lauzière, K. Klarskov, R. Lecomte, J. E. van Lier and B. Guérin, *Bioconjugate Chem.* **24**, 1624 (2013).
108. E. R. Ranyuk, N. Cauchon, H. Ali, R. Lecomte, B Guérin and J. E. van Lier, *Bioorg. Med. Chem. Lett.* **21**, 7470 (2011).
109. J. T. F. Lau, P.-C. Lo, W.-P. Fong and D. K. P. Ng, *J. Med. Chem.* **55**, 5446 (2012).
110. F.-L. Zhang, Q. Huang, K. Zheng, J. Li, J.-Y. Liu and J.-P. Xue, *Chem. Commun.* **49**, 9570 (2013).
111. M. Bio, P. Rajaputra, G. Nkepang and Y. You, *J. Med. Chem.* **57**, 3401 (2014).

112. S. Tuncel, A. Trivella, D. Atilla, K. Bennis, H. Savoie, F. Albrieux, L. Delort, H. Billard, V. Dubois, V. Ahsen, F. Caldefie-Chézet, C. Richard, R. W. Boyle, S. Ducki and F. Dumoulin, *Mol. Pharmaceutics* **10**, 3706 (2013).
113. S. Tuncel, J. Fournier-dit-Chabert, F. Albrieux, V. Ahsen, S. Ducki and F. Dumoulin, *Org. Biomol. Chem.* **10**, 1154 (2012).
114. B. R. Vummidi, F. Noreen, J. Alzeer, K. Moelling and N. W. Luedtke, *ACS Chem. Biol.* **8**, 1737 (2013).
115. J. F. Lovell, T. W. B. Liu, J. Chen and G. Zheng, *Chem. Rev.* **110**, 2839 (2010).
116. X.-J. Jiang, P.-C. Lo, S.-L. Yeung, W.-P. Fong and D. K. P. Ng, *Chem. Commun.* **46**, 3188 (2010).
117. X.-J. Jiang, P.-C. Lo, Y.-M. Tsang, S.-L. Yeung, W.-P. Fong and D. K. P. Ng, *Chem. Eur. J.* **16**, 4777 (2010).
118. M.-R. Ke, D. K. P. Ng and P.-C. Lo, *Chem. Commun.* **48**, 9065 (2012).
119. J. T. F. Lau, P.-C. Lo, X.-J. Jiang, Q. Wang and D. K. P. Ng, *J. Med. Chem.* **57**, 4088 (2014).
120. J. T. F. Lau, X.-J. Jiang, D. K. P. Ng and P.-C. Lo, *Chem. Commun.* **49**, 4274 (2013).
121. H. He, P.-C. Lo and D. K. P. Ng, *Chem. Eur. J.* **20**, 6241 (2014).
122. P. Acedo, J. C. Stockert, M. Cañete and A. Villanueva, *Cell Death Disease* **5**, e1122 (2014).
123. J. Peng, L. Zhao, X. Zhu, Y. Sun, W. Feng, Y. Gao, L. Wang and F. Li, *Biomaterials* **34**, 7905 (2013).
124. H.-N. Xu, H.-J. Chen, B.-Y. Zheng, Y.-Q. Zheng, M.-R. Ke and J.-D. Huang, *Ultrasonics Sonochem.* **22**, 125 (2015).
125. C. J. H. Ho, G. Balasundaram, W. Driessen, R. McLaren, C. L. Wong, U. S. Dinish, A. B. E. Attia, V. Ntziachristos and M. Olivo, *Sci. Reports* **4**, 5342 (2014).
126. Y. Zhang, M. Jeon, L. J. Rich, H. Hong, J. Geng, Y. Zhang, S. Shi, T. E. Barnhart, P. Alexandridis, J. D. Huizinga, M. Seshadri, W. Cai, C. Kim and J. F. Lovell, *Nat. Nanotech.* **9**, 631 (2014).

Chapter 7

Porphyrin Nanoparticles for Cancer Imaging and Phototherapy

Liyang Cui,[†,‡,§] *Juan Chen*[†] *and Gang Zheng*[†,‡,*]

[†]*Princess Margaret Cancer Center, University Health Network*
[‡]*Department of Medical Biophysics, University of Toronto*
[§]*Medical Isotope Research Center, Peking University*
gang.zheng@uhnres.utoronto.ca

Abstract

Phototherapies including photodynamic therapy (PDT) and photothermal therapy (PTT) are extensively investigated as potent alternative therapeutic modalities for cancer treatment. Porphyrins, the nature existed chromophores, have a rich history of phototherapy, especially for PDT. Paramount efforts have been involved in the development of porphyrin-based photosensitizers to enhance therapeutic efficacy and decrease toxicity. The union of porphyrin and nanotechnology offers new perspective to improve the safety and efficacy of porphyrin-based PDT. Various nanovectors, both organic and inorganic, have been investigated on porphyrin delivery for high payload photosensitizers, protection from pre-mature release and tumor-selective targeting. Most recently, the development of porphyrin-based "one-for-all" nanoparticles has opened up a brand new avenue for rational design of multifunctional nanoconstruct with multimodal imaging and therapeutic capabilities. When porphyrin is utilized as built-in elements to form the nanostructure, the high densely packed or orderedly aggregated porphyrin produce new photonic properties beyond porphyrin monomer, such as photoacoustic and photothermal properties. Marching on the frontline of cancer theranostics, porphyrin nanotechnology is turning one of nature's most fascinating molecules into powerful imaging and therapeutic platforms.

*Correponding author.

1. Porphyrin in Phototherapy

Phototherapies including photodynamic therapy (PDT) and photothermal therapy (PTT) are involved in cancer therapy as alternatives to conventional cancer treatment modalities. Both therapies are preformed non-invasively and selectively by local laser irradiation, leading to relative short recovery time, low risk of complications, and precise treatment under microsurgery range.[1,2] Although, PDT and PTT both require the administration of light-absorbing molecules, together with laser irradiation at specific wavelength in visible or near-infrared (NIR) range, they differ in the laser intensity requirement for performing different energy-transducing mechanisms.[1,3] PDT is highly dependent on molecular oxygen for sufficient singlet oxygen production to damage the region restricted to photosensitizer accumulation, while PTT, relying on the physical process of heating, is an alternative phototherapy that remains effective even under hypoxia condition. Porphyrins, the nature existed chromophores, have a rich history on phototherapy, especially on PDT. The first generation of photosensitizers is porphyrin-based, including hematoporphyrin (Hp) and its derivatives (HpDs).[4,5] Photofrin®is a commercialized HpD-based photosensitizer composed of porphyrin monomers, dimers and oligomers, which is the gold standard in clinical cancer treatment with PDT. The second generation of photosensitizers refers to porphyrin derivatives, chlorophyll derivatives and dyes that achieve higher chemical purity, absorb light at longer wavelength, and cause significantly reduced side effects such as skin photosensitization.[6,7] New photosensitizers are continuously being developed to enhance their safety, selectivity and therapeutic efficiency.[8,9] Conjugation porphyrins with targeting moieties, such as antibody,[10–12] peptide,[13,14] aptamer,[15,16] and small targeting ligands (e.g., folate,[17–19] 2-deoxyglucose[20]) etc., could improve their delivery to cancer cells, vasculatures or interstitial milieu. Activatable photosensitizers have also been introduced to provide an unprecedented level of control of PDT selectivity, by switching on/off the photosensitizer's photoreactivity in response to a variety of disease-associated activation mechanisms, such as the enzymetic-activation,[21–29] nucleic acid-activation,[30–34] and environmental activation.[35,36]

Porphyrins are well known for their medical applications in PDT and fluorescence imaging. They are also excellent metal chelators, forming highly stable metallo-complexes with many radionuclides (e.g., ^{64}Cu, $^{68/67}Ga$, ^{166}Ho) and paramagnetic metals (e.g., Mn^{3+}), thus unlocking their potential application on positron emission tomography (PET), single-photon emission computed tomography (SPECT),[37–41] and magnetic resonance imaging (MRI).[42–44] However, porphyrins are rarely explored for PTT application due to major drawbacks commonly associated with organic dyes:[45,46] (1) super high powerful lasers may be required to produce thermal ablation temperature due to their relatively low absorption (the extinction coefficient of porphyrins is in the range of 10^4–$10^5 M^{-1}cm^{-1}$) and (2) repeated exposure to the laser could cause organic molecules to breakdown due to photobleaching.

2. Porphyrin Nanotechnology

The convergence of porphyrin and nanotechnology offers new opportunities to improve the safety and efficacy of porphyrin-based PDT. Nanocarriers could provide high payload photosensitizer delivery and their enriched tumor accumulation through passive targeting (via enhanced permeability and retention) or active targeting (via cell surface receptor). Porphyrins have been incorporated into both inorganic and organic nanoparticles for various applications. Good examples of porphyrin-inorganic nanoparticle systems include porphyrin-fullerene complex,[47,48] porphyrin coated on gold,[49,50] iron oxide,[51] or carbon nanotube,[52–55] porphyrin conjugated on quantum dots,[56,57] and porphyrin-doped mesoporous silica nanoparticle,[58,59] in which many special photonic, photovoltaic and phototherapeutic properties are produced. The characteristics and applications of these systems have been extensively reviewed[60] and will not be included in this chapter. Many organic nanoparticles including micelles, polymers, lipoproteins, and liposomes,[61–63] are extensively investigated for porphyrin delivery. This book chapter focuses on the review of these porphyrin-organic nanoparticle systems for cancer imaging and phototherapy.

2.1. *Micelle and Polymer for Porphyrin Delivery*

Cremophor-EL was early used to disperse porphyrin to enhance their PDT effect *in vivo*.[64,65] However, this formulation could cause acute anaphylactic reactions.[66,67] To minimize the toxicity caused by the delivery system, many biocompatible polymeric micelles have been developed. Shieh *et al.* reported an PEOz-*b*-PLA diblock copolymeric micelle for delivery 5,10,15,20-tetrakis(meso-hydroxyphenyl)porphyrin (m-THPC), exhibiting similar antitumor effect to free m-THPC, but significantly reduced skin toxicity.[68] PEGylation was used to improve delivery pharmacokinetics. For example, poly(ethylene glycol)-*b*-poly(d,llactide) (PEG-PLA), was used for delivery of different molecular porphyrins including protoporphyrin IX[69] and 5,10,15,20-tetrakis(meso-hydroxyphenyl)porphyrin (m-THPP),[70,71] resulting in enhanced PDT efficacy. To improve the drug release specificity, pH-sensitive micelles were also developed for porphyrin delivery. Tsai *et al.* used two pH-sensitive micelles poly(N-vinylycaprolactam)-g-poly(D,L-lactide) and poly(N-vinyly caprolactam-*co*-N-vinyl imidazole)-*g*-poly($_{\mathrm{d,l}}$-lactide) for protoporphyrin IX delivery and both showed selective release of protoporphyrin IX in acidic organelles such as lysosomes, causing drastically enhanced PDT efficacy in comparison to free protoporphyrin IX.[72]

Polymers have also been widely used for photosensitizer delivery due to their high drug loading capacity.[73–75] For example, Photolon® is a molecular complex of chlorin e_6 with low molecular polyvinylpirrolidone, which has been officially approved for clinical PDT application in Republic of Belarus and Russian Federation.[76] PEGylated polymers prolong blood circulation of the system to further improve tumor accumulation and therapeutic effect. Alle'mann *et al.* developed a PEG-coated PLA polymer loaded with $ZnPcF_{16}$, which showed a continuous tumor accumulation up to 48 h post administration[77] and resulted in complete tumor ablation after treatment. An m-THPC loaded poly(ethylene glycol)-*b*-poly(lactic-*co*-glycolic acid) (PEG-PLGA) polymeric system also showed favorable blood clearance, excellent biodistribution profile, and significant decrease of toxicity compared to free m-THPC.[78,79] In some cases, targeting ligands are incorporated into the polymer delivery system for improving tumor specific delivery of porphyrin. For example, folate-conjugated polymeric nanoparticles are proved to specifically deliver 5-aminolevulinic acid (5-ALA) into folate receptor-positive cancer cells.[80,81] Activatable moieties such as reducible disulfide bonds switch are also employed[82] to trigger the photoreactivity of porphyrin molecules in response to cancer cell-specific reductive cleavage, showing significantly enhanced phototoxicity in both *in vitro* and *in vivo*.

Apart from conventional polymer delivery systems, such as PLA and PLGA, some novel polymeric constructs are also developed for porphyrin delivery. Lim *et al.* reported a ultra-small (~10 nm) heavy-atom-affected Pluronic (F-127) based nanoparticle for PDT.[83] These nanoparticles contain densely integrated iodinated aromatic molecules that form a structurally rigid core matrix, which not only encapsulate stably photosensitizers in a monomeric form, but also enhance the singlet oxygen generation by the heavy-atom effect. Another example is a supramolecular nanoconstruct built by the noncovalent integration of porphyrins and cucurbit[n]uril (CB[n]) that is a family of barrel-shaped macrocyclic host.[84] This porphyrin complex is easy to produce, rich in host-guest chemistry and potent in photodynamic activity.

2.2. Lipoprotein for Porphyrin Delivery

Porphyrins have a tendency to associate with serum proteins (e.g., albumin, hemopexin, globulins and some cytosolic proteins)[85–88] and lipoproteins.[89–91] Lipoprotein, especially low-density lipoprotein (LDL), is considered the major carrier of porphyrin in blood stream and subsequently delivers these photosensitizers to tissues through their receptors mediated pathway or passive diffusion.[89,92] In addition to this natural affinity between lipoproteins and porphyrins, lipoproteins have many other advantages favorable for porphyrin delivery: (1) as endogenous carriers, lipoproteins could escape recognition as foreign entities by the human immune system and escape absorption by the reticuloendothelial system (RES);[93] (2) lipoproteins have excellent blood circulation characteristics allowing for favorable systemic payload delivery without the need of PEGylation; (3) the desirable size of lipoprotein (7–30 nm) not only avoids rapid elimination by kidneys,[94] but also allows for superior penetration through interfibrillar openings in tumors (>40 nm);[95–99] (4) their intracellular uptake usually governed by receptor-mediated interactions, which allows for targeted delivery of payloads through lipoprotein receptors (LDL receptor for LDL delivery and scavenger receptor class B member 1 (SR-BI) for high-density lipoprotein (HDL) delivery), which are highly expressed in a number of cancer types.[100–103] Taken together, these characteristics make lipoproteins desire delivery strategies of porphyrins.

Initial approach to load porphyrins into lipoprotein was the simple mixing of the two components. LDL and HDL doped with Photofrin II and benzoporphyrin (BPD) were formulated and found to be PDT-active both *in vitro*[104] and *in vivo*.[105,106] Nevertheless, porphyrins transfer onto non-tumor-homing plasma proteins may occur due to unstable interaction between lipoprotein and porphyrin by the simple mixing method. To improve the stable delivery, Hamblin *et al.* reported a method to conjugate Hp to HDL or LDL,[107] enabling targeting delivery of porphyrin through lipoprotein-associated pathways for the selective PDT. Covalently attaching chlorin e_6 to LDL also demonstrated a specific photocytotoxicity to LDL receptor-positive retinoblastoma cells.[108] The methods for conjugation of porphyrins to the peptides or proteins of lipoproteins have been reviewed by Giuntini *et al.*[109] Since Krieger *et al.* developed a robust LDL reconstitution method to replace the endogenous core-lipids with exogenous hydrophobic cholesteryl linoleate, a number of porphyrin and porphyrin analogs have been loaded into the core of lipoproteins for stable targeting delivery including pyropheophorbide,[110] phthalocyanines[111] and naphthalocyanine[112] and bacteriochlorin *e6*.[113,114] In addition, Zheng *et al.* developed a versatile lipoprotein rerouting method, further broadening the purview of using lipoprotein nanoparticles for cancer targeting.[115,116]

Other serum-associated proteins are also employed for porphyrin delivery. Human serum albumin (HSA)-based nanoparticles, are extensively studied on delivery of pheophorbide,[117] 5,10,15-triphenyl-20-(3-N-methylpyridinium-yl)porphyrin,[118] m-THPC and m-THPP,[119,120] for enhanced PDT. Zhen *et al.* reported using RGD targeting ferritin protein to deliver zinc hexadecafluorophthalocyanine (ZnF16Pc).[121] The resulted ferritin nanocages with high drug loading rate (~60 wt.%) and super small size (~18 nm) demonstrated favorable tumor accumulation, potent PDT efficacy and minimal skin toxicity.

2.3. *Liposome for Porphyrin-Delivery*

Another attractive nanoparticle delivery system of porphyrin is liposome, which has biocompatible composition, excellent drug loading ability and amicable nature for easy structural modification. Liposome formulation for delivery of porphyrin photosensitizers including Photofrin, 5-ALA, pheophorbide, chlorin e_6 etc. have been extensively investigated.[7] Several liposomal photosensitizers have been clinically approved for the treatment of cancer and other diseases. For example, Visudyne®, a liposome structure loaded with benzoporphyrin derivative monoacid (BPD-MA), is the first liposomal photosensitizer introduced into clinical practice. PDT treatment with Visudyne® was found to effectively suppress tumor growth[122] via damaging endothelial cells of the tumor neovasculature.[123] Another example is liposomal Temoporfin (m-THPC). Temoporfin was approved in 2001 as Foscan® for the palliative treatment of advanced head and neck squamous cell carcinoma. The liposomal forms of Foscan®, named Foslip® and Fospeg®(PEGylated liposome), were found to reduce the tumor burden effectively in various preclinical cancer models including lung,[124] prostate,[125] squamous cell carcinoma[126] and oral carcinoma.[127] There are many comprehensive reviews on the liposomes for PDT.[128–131] Herein, we will summarize the recent improvement on liposomal porphyrin-mediated PDT.

2.3.1. *Increasing porphyrin payload*

A well-known loading method is to embed porphyrin in the lipid bilayer of liposome[132] to solubilize lipophilic porphyrin for intravenous administration. However, the loading capability in liposome needs to be kept low so that its structure and functions won't be compromised. To improve the porphyrin loading efficiency, cationic liposome formulations have been developed for effective delivery of negatively charged porphyrins. Shim *et al.* developed an 1,2-dioleoyl-sn-glycero-3-ethylphosphocholine-based cationic liposome that can deliver chlorin e_6 with loading efficiency up to 92% (comparing to that of 12% in neutral formulations) attributed to the electrostatic interaction between lipid and porphyrin molecules.[133] Suitable molecular modification of porphyrin could also enhance porphyrin loading in liposome. For example, conjugating protoporphyrin IX with hydrophobic oleylamine arms could facilitate the stable embedment of lipophilic porphyrins in liposome.[134] Alternative to loading lipophilic porphyrin into lipid compartment, pre-modifying porphyrin to be water soluble and then encapsulating it into the aqueous core of liposomes could also increase the porphyrin loading payload.[135]

2.3.2. *Improving delivery pharmacokinetics*

Delivery of porphyrins using liposomes was found to effectively improve their half-life in blood circulation. Such improved pharmacokinetics can be achieved by incorporating PEG or glycolipid into the formulation to preserve the liposomes from instant recognition and rapid clearance by RES. Buchholz *et al.* validated that PEGylated liposomal m-THPC (Fospeg®) had much higher median maximal plasma concentration (1523.30 ng/ml) compared to the free m-THPC (458.81 ng/ml).[136] The porphyrin intrinsic fluorescence imaging confirmed that the tumor uptake of liposomal m-THPC were nearly twice as that of the free m-THPC. Bovis *et al.* showed similar favorable pharmacokinetic properties of Fospeg® formulations over the standard m-THPC formulation (Foscan®), where longer biological half-life (99 h for 2% PEGylated formulation, Fospeg 2%) and higher tumor accumulation was observed.[124] The enhanced tumor uptake and blood plasma concentration of Fospeg 2% caused more efficient PDT treatment-induced tumor necrosis while less skin phototoxicity in comparison to free Foscan®.

2.3.3. *Enhancing treatment selectivity*

Enhanced selective porphyrin delivery in tumor can be achieved by incorporating targeting ligands, such as folate, antibody, peptide, and aptamer *etc.* to liposomal structure through surface modification.[129,137,138] Targeting tumor vasculature is widely investigated for anti-angiogenesis treatment through PDT. For example, a pentapeptide, APRPG, specific targeted to angiogenic endothelial cells, was introduced into a PEGylated liposome-BPD for neovessel-targeted tumor therapy.[139] A significantly enhanced tumor inhibition effect and vascular damage were observed upon PDT treatment with APRPG-liposome-BPD when compared to non-targeted liposome-BPD or free BPD.

Many monoclonal antibodies have been reported for the targeted delivery of liposomal PDT agents for improved treatment outcome. With the additional immuno-therapeutic function of antibody, Mir *et al.* demonstrated a synergistic effect of PDT and immune-treatment with cetuximab conjugated liposome-BPD. The resulted liposome not only specifically bound to EGFR-positive ovarian cancer cells for selective PDT, but also triggered the inhibition of EGFR signaling that further enhanced PDT-mediated ovarian cancer cell death.[140] Besides liposome surface modification, targeting can also be achieved by loading pre-conjugated targeted-photosensitizer. Rahmanzadeh *et al.* designed a photoimmunoconjugate-encapsulating liposome (PICEL) where anti-Ki-67 antibody is conjugated to photosensitizers first and loaded into the aqueous core of the liposome.[135] Once uptake by cells, PICEL can effectively release the conjugates, which will subsequently relocalize and direct the photosensitizer into the nucleus of cancer cells where the Ki-67 is highly expressed.

Controlled porphyrin release is another strategy to enhance the treatment specificity of PDT. Liposomes could be designed to specifically release porphyrins in response to tumor-associated factors, such as enzyme, pH, temperature and light explosion. This part has been reviewed elsewhere.[7,129]

2.4. *Porphysome: A Porphyrin-Assembled Nanostructure*

In 2011, porphysome, an all-organic nanoparticle self-assembled from porphyrin-phospholipid conjugates with the liposome-like structure but with intrinsic multimodal photonic properties, was discovered.[141] In this particular nanostructure, porphyrin-lipids are introduced as building blocks rather than payloads to form the nanostructure with the porphyrin-lipid content up to 100 mol% of the total lipids. Other types of lipids, such as PEG-lipid, for improving *in vivo* pharmacokinetics, and cholesterol for enhancing loading and circulation half-life, are also incorporated in the nanostructure to improve its *in vivo* performance. Transmission electronic microscopy revealed porphysome as a spherical vesicle with diameters of 100–150 nm, with a bilayer of porphyrin-lipid separated by a two nm gap (Fig. 1). The extremely high density of porphyrin-phospholipid (>80,000/particle) packed in porphysome bilayer results in the 'super'-absorption and 'super'-quenching of photoactivity (fluorescence and singlet oxygen production), which allows the conversion from light energy to thermal effect with extremely high efficiency (extinction coefficient $\sim 2.9 \times 10^9 M^{-1} cm^{-1}$, comparable to gold nanorods), giving ideal photothermal and photoacoustic activity that are unprecedented for organic nanoparticles. Upon porphysome nanostructure dissociation, fluorescence and

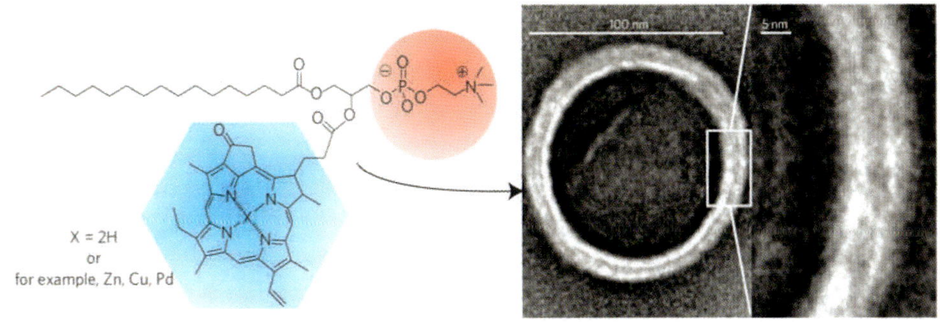

Fig. 1. Porphysome shows a liposome-like structure assembled by porphyrin-phospholipid conjugates (Readapted figure from Ref. 141. Copyright Nature Publishing Group).

singlet oxygen generation capacity of free porphyrin molecules are restored, which can be used for low background fluorescence imaging and activatable PDT.[142] In addition, porphyrin-lipid retains the natural ability of porphyrins to chelate metals (e.g., radioisotope ^{64}Cu or paramagnetic Mn), offering their potential for PET and MRI.[143–145] In a similar manner to liposomes, the large aqueous core of porphysomes could be passively or actively loaded with drugs.[141] Therefore, the porphysome nanostructure built from the intrinsically multifunctional porphyrin building block provides a "one particle to rule them all" ("one-for-all") approach for phototherapy, multi-modality imaging, and drug transport.[141]

2.4.1. *Porphysome for phototherapies (nanostructure-driven PTT and PDT)*

Porphysome nanostructure drives the complete conversion of the porphyrins' photoreactivity from a singlet oxygen generation mechanism, useful for PDT, to thermal mechanism, ideal for PTT enhancement (Fig. 2). Jin *et al.* reported the first demonstration of porphysome-enabled PTT for hypoxic tumor ablation.[146] Porphysomes were able to produce sufficient photothermal effect under PTT laser irradiation (671 nm, 750 mW, 63.75 J), allowing complete tumor ablation in both hyperoxic and hypoxic condition, while their PDT activity was eliminated under both conditions. Porphysome-enabled PTT was proved to be advantageous over Photofrin-PDT on the treatment of hypoxic tumor. Recently, an activatable strategy was introduced for switch-back-on of the PDT activity of porphysome by introducing targeting ligands, such as folate-lipid, into the porphysome formulation.[142] Folate-porphysomes exhibited targeting-enhanced intracellular uptake and efficient nanostructure disruption. This receptor-mediated porphysome dissociation allows for the release the

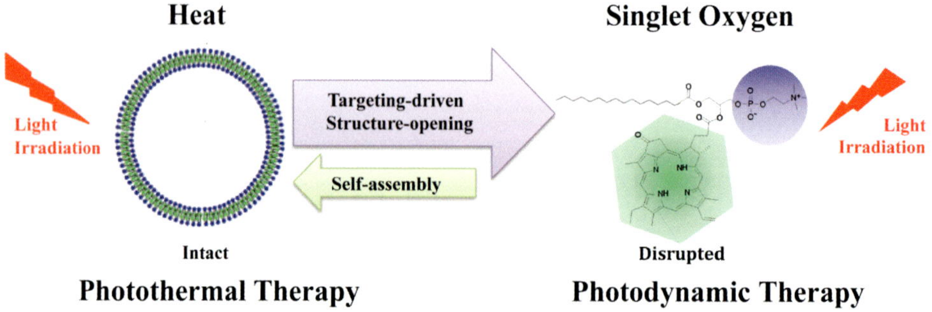

Fig. 2. Schematics of nanostructure-driven conversion between PTT and PDT. (Reprinted with permission from Ref 142. Copyright John Wiley and Sons).

suppressed PDT activity of porphyrin for highly selective and effective PDT. Therefore, the nanostructure-dependent phototherapy conversion between PTT and PDT could be manipulated selectively towards personalized cancer treatment. Due to its natural chlorophyll origin, porphysome itself is biodegradable and has very low toxicity *in vivo* (1000 mg/kg *i.v.* in mice caused no detectable functional, hematological or histological effects),[141] indicating that the porphysome-enabled phototherapies (PDT & PTT) are safe treatment approaches.

2.4.2. *Porphysome for multimodal imaging*

Porphysome converts the light absorption to heat due to efficient fluorescence quenching, allowing for photoacoustic imaging (PAI) of lymphatic network with high sensitivity.[141] The lymphatic network was clearly illustrated by PAI, showing the first draining lymph node, the inflowing lymph vessels, and the surrounding lymph vessels of rats, at 15 min post intradermal injection of porphysomes at a dose as low as 2.3 pmol (Fig. 3A). Liu *et al.*[144] also demonstrated that porphysomes could clearly delineate orthotopic prostate tumor by fluorescence imaging *in situ* at 24 h post-injection. Quantification of the fluorescence signals showed that tumor had significantly high fluorescence intensity over healthy reproductive organs, enabling low background fluorescence imaging of the deep-seated tumor during open peritoneal cavity surgery (Fig. 3B).

Due to the intrinsic metal chelation nature of porphyrin, porphysomes can be directly labeled with ^{64}Cu for PET imaging and *in vivo* pharmacokinetics evaluation.[143] The labeling procedure is fast, efficient and robust, with high reaction yield (>98%) and the highest radioactivity of nanoparticle (2800 mCi/μmol nanoparticle) ever reported for a ^{64}Cu-labeled nanoparticle.[147] The resulted ^{64}Cu-porphysomes revealed selective accumulation in orthotopic

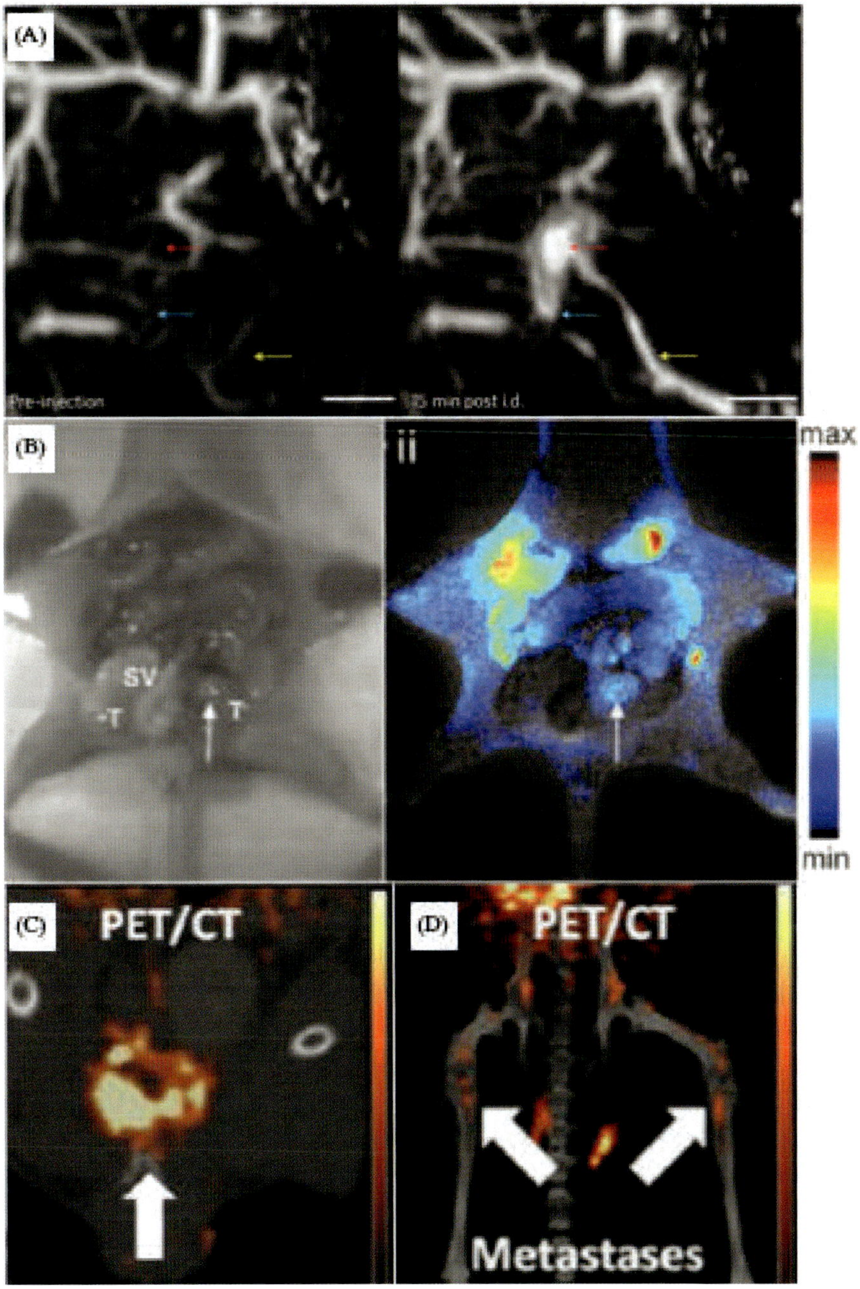

Fig. 3. (A) PAI of lymphatic network before (left) and 15 min (right) after porphysome injection. (B) *In situ* monochrome (left) and fluorescence (right) imaging of PC3 orthotopic prostate tumor-bearing mice at 24 h post-injection with porphysome. (C) PET/CT image of ^{64}Cu-porphysome in orthotopic prostate cancer model. (D) Bone metastasis detected by ^{64}Cu-porphysome using PET/CT imaging. (Reprinted with permission from Ref. 141, Copyright Nature Publishing Group and Ref. 144, copyright ACS Publications).

prostate tumors compared to the low accumulation in surrounding healthy prostate (6.83 ± 1.08 %ID/g in tumor and 1.23 ± 0.202 %ID/g in normal prostate at 24 h post-administration), allowing for clear tumor delineation (Fig. 3C). ^{64}Cu-porphysomes were also able to sensitively detect small (<2 mm) prostate cancer-derived bone metastases (Fig. 3D). Apart from PET and optical imaging, MRI contrast can be promptly added by forming the structure with Mn-chelated porphyrin-lipid.[145] Mn-porphysomes present a MRI contrast at a level comparable to the clinically used Gd- diethylenetriamine pentacetate (Gd-DTPA) while maintaining the photoacoustic property of porphysome. Recently, Lee *et al.* successfully labeled porphysome with ^{99m}Tc, showing the potential application for SPECT imaging.[148]

2.4.3. *Stimuli-responsive porphysomes*

More recently, a phase change porphyrin-based nanostructures through ordered dye aggregation has been developed, which is capable of detecting temperature change through alterations in photoacoustic signal.[149] This thermoresponsive photoacoustic nanoswitch contains 15 mol% of bacterichorophyll-lipid (Bchl-lipid) in a temperature sensitive liposome formulation. The ordered porphyrin aggregation showed a red-shifted Qy-band absorption by 74 nm with 2.7-fold increase of intensity in comparison to that of monomer and the corresponding change in photoacoustic spectrum (Fig. 4). The spectral feature is responsive to a host lipid-dependent temperature threshold. Above the melting temperature of the host lipid, the nanoswitches lose the aggregated

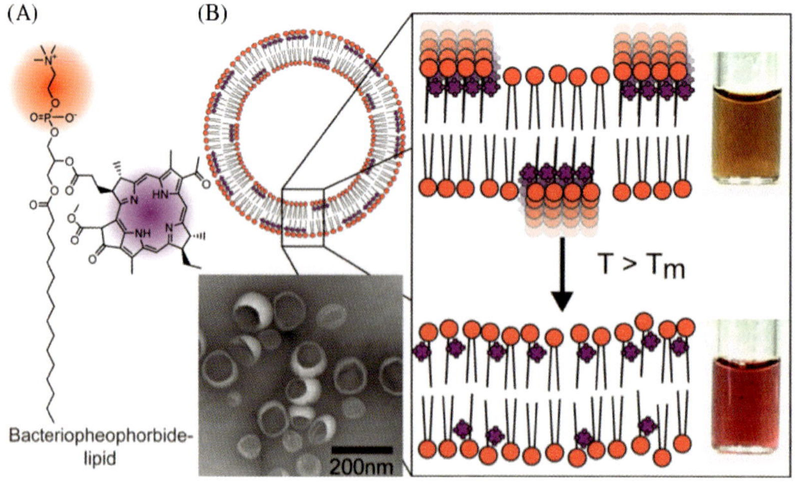

Fig. 4. Schematic of thermo-responsive nanoswitch (Reprinted with permission from Ref. 149. Copyright ACS Publications).

absorption peak and the corresponding photoacoustic signal, and restore the monomer signals, thus useful for temperature monitoring in thermal therapies such as PTT. This smart photoacoustic sensor demonstrates a great potential to make photoacoustic as versatile as fluorescence to report environment stimuli while overcoming the optical diffusion limit at depth.

Lovell *et al.* reported another stimuli-responsive porphysome for the control of drug release by light irradiation.[150] This so-called PoP-liposome was formed by doping 10% of devinyl hexyloxyethy-pyropheophorbide-lipid in liposome formulation, offering a superior porphyrin bilayer structure for thermostable cargo retention. Upon exposure to the beams of laser, PoP-liposomes released the cargo, such as doxorubicin, with robust spatial and temporal control, permitting a light-triggered selective antitumour chemotherapy.

As many environmental factors can be used to differentiate tumor cells from normal cells, stimuli-responsive porphysome could be easily designed in response to tumor microenvironment, such as tumor acidity or enzyme activation, for selective diagnosis, treatment, and payload delivery.

2.4.4. *Porphyrin-based nanoconstruct beyond porphysome*

Porphysome nanoparticle with the concept of "one-for-all" design shines the light on developing new generation of multifunctional nanoparticles. By engaging porphyrin-lipid technology into lipoprotein structure, Ng *et al.* reported a porphyrin nanodisc with ultra-small size (<40 nm) for activatable theranostics. The nanodisc is assembled by 100 mol% of porphyrin-lipid and constructed with apolipoproteins to stabilize the particle.[97] Similar to porphysome, the intact nanodisc was fluorescence and singlet oxygen generation-quenched, but became fully photodynamically active upon SR-BI receptor mediated internalization. In addition, sub-40nm porphyrin nanodisc demonstrated a greater capability on diffusing (five-fold increase in diffusion coefficient) through a tumor's collagen rich matrix compared to large-sized particles. With the potential of direct incorporation of ^{64}Cu or Mn ions, porphyrin nanodisc is very promising for phototherapy and multimodal imaging.

A naphthalocyanine-based micelle, termed nanonap, was developed by Zhang *et al.* for the non-invasive imaging of the gastrointestinal tract.[151] Due to the high hydrophobicity, the nanonaps can withstand the harsh conditions of the stomach and intestine, minimize systemic absorption, and provide good contrast for PAI following oral administration in mice. Co-registration with ultrasound, PAI imaging of nanonaps provided real-time, low-background, high-resolution visualization of the intestinal tract.

3. Conclusion and Future Perspective

Moving from small molecule-based agent to nano-sized architecture, porphyrin nanomedicine shows better performance on prolonging systemic circulation, improving tumor specificity, and reducing side effect. Porphyrin nanotechnology also allows for integrating multifunctional properties in a single platform to achieve targeted delivery, multimodality (diagnosis plus therapies), and controlled/activated release to pursue maximum therapeutic effects and minimal side effects. The inherent multifunctionality of porphyrin molecules, such as fluorescence generation, photodynamic activity and stable chelation of metal ions makes porphyrin nanoparticle ideal for theranostic application. The intrinsic imaging modalities also provide an accurate assessment on drug distribution, intratumoral dose, and even intracellular uptake pathway to guide treatment planning.

When porphyrin as building units, such as porphyrin-lipid, involved in nanoparticle construction, the high densely packed or ordered aggregated porphyrin produce new photonic properties of the nanostructure beyond porphyrin monomer. The high densely packed porphyrin in porphysome converts the fluorescence and singlet oxygen generation mechanism of porphyrin to completely thermal mechanism, enabling PTT and PAI. The ordered aggregation of bacteriochlorophyll in nanoparticles produced a special red-shifted photoacoustic spectrum, allowing for the design of a new class of smart photoacoustic sensors to report environment stimuli (temperature, pH, enzyme *etc.*). These nanostructure-dependent photonic properties are useful for monitoring nanoparticles stability and drug release. MRI and PET modalities can be intrinsically added after nanoparticle synthesis[143,145] is also excellent for clinical translation as nanoparticles can be shipped pre-formed and then radiolabelled on-site.

Porphysome nanotechnology enables the modulation of the phototherapies between PDT and PTT. The nanostructure drives conversion of PDT to PTT, but could be easily turned back by targeting-triggered activation where nanoparticles internalize cell via enhanced targeting delivery and rapidly disruption in tumor. Replacing a fraction of the porphyrin-lipid with non-fluorescent lipids to reduce the quenching efficiency could also permit fluorescence imaging and PDT in the intact nanostucture. Direct incorporation of paramagnetic Mn into porphysome not only provides an additional MRI modality, but also enhances porphysome's photostability for high efficient PTT. Therefore, the porphyrin nanoparticles can be designed by taking advantage of PTT or PDT modalities towards personalized cancer treatment.

The porphysome discovery has turned one of nature's most fascinating molecules, porphyrins, into all-organic multimodal photonic nanomaterials, subsequently opened the door for exploration of different porphyrin nanostructures based on the intrinsic multimodal porphyrin building blocks. Recently, Li *et al.* reported "nanoporpyrin", a self-assembled nanostructure from a single porphyrin polymer that integrates a variety of imaging modalities, such as NIR-fluorescence imaging and dual modal PET-MRI imaging, and therapeutic functions associated with the nanostructure such as chemotherapeutics delivery, photothermal and PDT.

Nano-scaled porphyrin aggregates are widely existed in nature. For example, chlorosomes are the primary light-harvesting devices in photosynthetic organism.[152,153] They contain nano-sized bacteriochlorophyll *c*, *d* or *e* molecules' aggregates[153] which give many advanced photonic properties, such as high light sensitivity, nonlinear optics, extraordinary energy transfer ability and capability to adapt to extreme conditions. Artificial biomimetic chlorosomal devices have been profoundly investigated and applied in solar energy materials.[154] However, most porphyrin assemblies are unstable in plasma, causing the loss of their structure-dependent photonic properties. A DNA assembled network has been introduced to stabilize hydrophobic porphyrins, where DNA network assembled on the hydrophobic porphyrin anchors via a complementary hydrogen bonding interaction.[155] Recently, Ghosh *et al.* demonstrated that a DNA assembled porphyrin nanocomplex induced both light-mediated cellular toxicity and nucleic acid-driven therapeutic efficacy,[156] thus bringing another way in design of porphyrin-nanoplatform.

Acknowledgement

The authors acknowledge funding support from the Natural Sciences and Engineering Research Council of Canada, Canadian Institutes of Health Research, Terry Fox Research Institute, Major International (Regional) Joint Research Project from National Science foundation of China, Canada Foundation for Innovation, the Prostrate Cancer Canada, and the Princess Margaret Cancer Foundation.

References

1. B. C. Wilson and M. S. Patterson, *Phys. Med. Biol.* **53**, R61 (2008).
2. D. P. O'Neal, L. R. Hirsch, N. J. Halas, J. D. Payne and J. L. West, *Cancer Lett.* **209**, 171 (2004).
3. X. Huang, P. K. Jain, I. H. El-Sayed and M. A. El-Sayed, *Lasers Med. Sci.* **23**, 217 (2008).

4. A. E. O'Connor, W. M. Gallagher and A. T. Byrne, *Photochem. Photobiol.* **85**, 1053 (2009).
5. R. R. Allison and C. H. Sibata, *Photodiagnosis Photodyn. Ther.* 7, 61 (2010).
6. I. J. Macdonald and T. J. Dougherty, *J. Porphyrins Phthalocyanines* **5**, 105 (2001).
7. B. Chen, B. W. Pogue and T. Hasan, *Expert Opin. Drug. Deliv.* **2**, 477 (2005).
8. T. W. Liu, J. Chen and G. Zheng, *Amino Acids* **41**, 1123 (2011).
9. J. F. Lovell, T. W. Liu, J. Chen and G. Zheng, *Chem. Rev.* **110**, 2839 (2010).
10. N. Kameyama, S. Matsuda, O. Itano, A. Ito, T. Konno, T. Arai, K. Ishihara, M. Ueda and Y. Kitagawa, *Cancer Biother. Radiopharm.* **26**, 697 (2011).
11. A. O. Abu-Yousif, A. C. Moor, X. Zheng, M. D. Savellano, W. Yu, P. K. Selbo and T. Hasan, *Cancer Lett.* **321**, 120 (2012).
12. A. R. Oseroff, D. Ohuoha, T. Hasan, J. C. Bommer and M. L.Yarmush, *Proc. Natl. Acad. Sci. USA* **83**, 8744 (1986).
13. C. L. Conway, I. Walker, A. Bell, D. J. Roberts, S. B. Brown and D. I. Vernon, *Photochem. Photobiol. Sci.* 7, 290 (2008).
14. L. Tirand, C. Frochot, R. Vanderesse, N. Thomas, E. Trinquet, S. Pinel, M. L. Viriot, F. Guillemin and M. J. Barberi-Heyob, *J. Control. Release* **111**, 153 (2006).
15. P. Mallikaratchy, Z. Tang and W. Tan, *Chem. Med. Chem.* **3**, 425 (2008).
16. C. Pestourie, B. Tavitian and F. Duconge, *Biochimie* **87**, 921 (2005).
17. J. Gravier, R. Schneider, C. Frochot, T. Bastogne, F. Schmitt, J. Didelon, F. Guillemin and M. Barberi-Heyob, *J. Med. Chem.* **51**, 3867 (2008).
18. D. Li, P. Li, H. Lin, Z. Jiang, L. Guo and B. Li, *J. Photochem. Photobiol. B* **127**, 28 (2013).
19. R. Schneider, F. Schmitt, C. Frochot, Y. Fort, N. Lourette, F. Guillemin, J. F. Muller and M. Barberi-Heyob, *Bioorg. Med. Chem.* **13**, 2799 (2005).
20. M. Zhang, Z. Zhang, D. Blessington, H. Li, T. M. Busch, V. Madrak, J. Miles, B. Chance, J. D. Glickson and G. Zheng, *Bioconjug. Chem.* **14**, 709 (2003).
21. J. Chen, K. Stefflova, M. J. Niedre, B. C. Wilson, B. Chance, J. D. Glickson and G. Zheng, *J. Am. Chem. Soc.* **126**, 11450 (2004).
22. M. Dong, W. H. Tepp, E. A. Johnson and E. R. Chapman, *Proc. Natl. Acad. Sci. USA* **101**, 14701 (2004).
23. J. Lin, Z. Zhang, J. Yang, S. Zeng, B. F. Liu and Q. Luo, *J. Biomed. Opt.* **11**, 024011 (2006).
24. J. Lu, Z. Zhang, J. Yang, J. Chu, P. Li, S. Zeng and Q. Luo, *Biochem. Biophys. Res. Commun.* **362**, 25 (2007).
25. G. Zheng, J. Chen, K. Stefflova, M. Jarvi, H. Li and B. C. Wilson, *Proc. Natl. Acad. Sci. USA* **104**, 8989 (2007).
26. Z. Zhang, J. Yang, J. Lu, J. Lin, S. Zeng and Q. Luo, *J. Biomed. Opt.* **13**, 011006 (2008).
27. J. Chen, T. W. Liu, P. C. Lo, B. C. Wilson and G. Zheng, *Bioconjug. Chem.* **20**, 1836 (2009).
28. P. C. Lo, J. Chen, K. Stefflova, M. S. Warren, R. Navab, B. Bandarchi, S. Mullins, M. Tsao, J. D. Cheng and G. Zheng, *J. Med. Chem.* **52**, 358 (2009).
29. T. M. Mawn, A. V. Popov, N. J. Beardsley, K. Stefflova, M. Milkevitch, G. Zheng, E. J. Delikatny, *Bioconjug. Chem.* **22**, 2434 (2011).
30. E. Clo, J. W. Snyder, N. V. Voigt, P. R. Ogilby and K. V. Gothelf, *J. Am. Chem. Soc.* **128**, 4200 (2006).
31. J. Chen, J. F. Lovell, P. C. Lo, K. Stefflova, M. Niedre, B. C. Wilson and G. Zheng, *Photochem. Photobiol. Sci.* 7, 775 (2008).

32. J. F. Lovell, J. Chen, M. T. Jarvi, W. G. Cao, A. D. Allen, Y. Liu, T. T. Tidwell, B. C. Wilson and G. Zheng, *J. Phys. Chem. B* **113**, 3203 (2009).
33. I. V. Nesterova, S. S. Erdem, S. Pakhomov, R. P. Hammer and S. A. Soper, *J. Am. Chem. Soc.* **131**, 2432 (2009).
34. J. F. Lovell, J. Chen, E. Huynh, M. T. Jarvi, B. C. Wilson and G. Zheng, *Bioconjug. Chem.* **21**, 1023 (2010).
35. J. Arnbjerg, M. Johnsen, C. B. Nielsen, M. Jørgensen and P. R. Ogilby, *J. Phys. Chem. A* **111**, 4573 (2007).
36. S. O. McDonnell, M. J. Hall, L. T. Allen, A. T. Byrne, W. M. Gallagher and D. F. O'Shea, *J. Am. Chem. Soc.* **127**, 16360 (2005).
37. R. Bases, S. S. Brodie and S. Rubenfeld, *Cancer* **11**, 259 (1958).
38. Y. Fazaeli, A. R. Jalilian, M. M. Amini, K. Ardaneh, A. Rahiminejad, F. Bolourinovin, S. Moradkhani and A. Majdabadi, *Nucl. Med. Mol. Imaging* **46**, 20 (2012).
39. F. Zoller, P. J. Riss, F. P. Montforts, D. K. Kelleher, E. Eppard and F. Rosch, *Nucl. Med. Biol.* **40**, 280 (2013).
40. N. Vahidfar, A. R. Jalılıan, Y. Fazaeli, A. Bahrami-Samani, D. Beiki and A. Khalaj, *J. Radioanal. Nucl. Chem.* **301**, 269 (2014).
41. J. Shi, T. W. Liu, J. Chen, D. Green, D. Jaffray, B. C. Wilson, F. Wang and G. Zheng, *Theranostics* **1**, 363 (2011).
42. T. Lee, X. A. Zhang, S. Dhar, H. Faas, S. J. Lippard and A. Jasanoff, *Chem. Biol.* **17**, 665 (2010).
43. D. Pan, S. D. Caruthers, A. Senpan, A. H. Schmieder, S. A. Wickline and G. M. Lanza, *Wiley Interdiscip. Rev. Nanomed. Nanobiotechnol.* (2010).
44. X. A. Zhang, K. S. Lovejoy, A. Jasanoff and S. J. Lippard, *Proc. Natl. Acad. Sci. USA* **104**, 10780 (2007).
45. G. Jori and J. D. Spikes, *J. Photochem. Photobiol. B: Biol.* **6**, 93 (1990).
46. D. Kessel and Morgan, *Photochem. Photobiol.* **59**, 547 (1994).
47. N. Armaroli, G. Marconi, G., Echegoyen, L., Bourgeois, J. P. and Diederich, F. *Chemistry* **6**, 1629,(2000).
48. M. B. Ballatore, M. B. Spesia, M. E. Milanesio and E. N. Durantini, *Eur. J. Med. Chem.* **83**, 685 (2014).
49. D. Conklin, S. Nanayakkara, T. H. Park, M. F. Lagadec, J. T. Stecher, M. J. Therien and D. A. Bonnell, *Nano Lett.* **12**, 2414 (2012).
50. H. Yamada, H. Imahori, Y. Nishimura, I. Yamazaki, T. K. Ahn, S. K. Kim, D. Kim and S. Fukuzumi, *J. Am. Chem. Soc.* **125**, 9129 (2003).
51. G. R. Reddy, M. S. Bhojani, P. McConville, J. Moody, B. A. Moffat, D. E. Hall, G. Kim, Y. E. Koo, M. J. Woolliscroft, J. V. Sugai, T. D. Johnson, M. A. Philbert, R. Kopelman, A. Rehemtulla and B. D. Ross, *Clin. Cancer Res.* **12**, 6677 (2006).
52. F. D'Souza, R. Chitta, A. S. Sandanayaka, N. K. Subbaiyan, L. D'Souza, Y. Araki and O. Ito, *Chemistry* **13**, 8277 (2007).
53. I. Hijazi, T. Bourgeteau, R. Cornut, A. Morozan, A. Filoramo, J. Leroy, V. Derycke, B. Jousselme and S. Campidelli, *J. Am. Chem. Soc.* **136**, 6348 (2014).
54. H. Ozawa, X. Yi, T. Fujigaya, Y. Niidome, T. Asano and N. Nakashima, *J. Am. Chem. Soc.* **133**, 14771 (2011).
55. J. Yu, S. Mathew, B. S. Flavel, M. R. Johnston and J. G. Shapter, *J. Am. Chem. Soc.* **130**, 8788 (2008).

56. E. Vaishnavi and R. Renganathan, *Analyst* **139**, 225 (2014).
57. C. M. Lemon, E. Karnas, M. G. Bawendi and D. G. Nocera, *Inorg. Chem.* **52**, 10394 (2013).
58. T. Yui, T. Tsuchino, T. Itoh, M. Ogawa, Y. Fukushima and K. Takagi, *Langmuir* **21**, 2644 (2005).
59. T. Zhao, H. Wu, S. Q. Yao, Q. H. Xu and G. Q. Xu, *Langmuir* **26**, 14937 (2010).
60. B. Sara, G. Damiano, J. Riccardo, M. Fabrizio, M. Marco, P. Luca, R. Enrico and Z. Nelsi in *Handbook of Porphyrin Science*, Vol. 12 (World Scientific Publishing Company, 2011) p. 349.
61. M. Bryszewska and B. Klajnert, *Curr. Med. Chem.* **19**, 4895 (2012).
62. K. K. Ng, J. F. Lovell and G. Zheng, *Acc. Chem. Res.* **44**, 1105 (2011).
63. A. Maham, Z. Tang, H. Wu, J. Wang and Y. Lin, *Small* **5**, 1706 (2009).
64. S.H. Selman. G. M. Garbo, R. W. Keck, M. Kreimer-Birnbaum and A. R. Morgan, *J. Urol.* **137**, 1255 (1987).
65. A. R. Morgan, G. M. Garbo, R. W. Keck and S. H. Selman, *Cancer Res.* **48**, 194 (1988).
66. D. L. Howrie, R. J. Ptachcinski, B. P. Griffith, R. J. Hardesty, J. T. Rosenthal, G. J. Burckart and R. Venkataramanan, *Drug Intell. Clin. Pharm.* **19**, 425 (1985).
67. R. T. Dorr, *Ann. Pharmacother.* **28**, S11 (1994).
68. M. J. Shieh, C. I. Peng, W. I Chiang, C. Wang, C. Hsu, S. Wang and P. S. Lai, *Mol. Pham.* 7, 1244 (2010).
69. H. Ding, B. D. Sumer, C. W. Kessinger, Y. Dong, G. Huang, D. A. Boothman and J. Gao, *J. Control. Release* **151**, 271 (2011).
70. E. M. Cohen, H. Ding, C. W. Kessinger, C. Khemtong, J. Gao and B. D. Sumer, *Otolaryngol. Head Neck Surg.* **143**, 109 (2010).
71. H. Ding, R. Mora, J. Gao and B. D. Sumer, *Otolaryngol. Head Neck Surg.* **145**, 612 (2011).
72. H. C. Tsai, C. H. Tsai, S. Y. Lin, C. R. Jhang, Y. S. Chiang and G. H. Hsiue, *Biomaterials* **33**, 1827 (2012).
73. N. Brasseur, D. Brault and P. Couvreur, *Int. J. Pharm.* **70**, 129 (1991).
74. A. Labib, V. Lenaerts, F. Chouinard, J. C. Leroux, R. Ouellet and J. E. van Lier, *Pharm. Res.* **8**, 1027 (1991).
75. O. Bourdon, V. Mosqueira, P. Legrand and J. Blais, *J. Photochem. Photobiol. B* **55**, 164 (2000).
76. S. V. Shliakhtsin, T. V. Trukhachova, H. A. Isakau and Y. P. Istomin, *Photodiagnosis Photodyn. Ther.* **6**, 97 (2009).
77. E. Allemann, J. Rousseau, N. Brasseur, S. V. Kudrevich, K. Lewis and J. E. van Lier, *Int. J. Cancer* **66**, 821 (1996).
78. K. Low, T. Knobloch, S. Wagner, A. Wiehe, A. Engel, K. Langer and H. von Briesen, *Nanotechnology* **22**, 245102 (2011).
79. M. Rojnik, P. Kocbek, F. Moret, C. Compagnin, L. Celotti, M. J. Bovis, J. H. Woodhams, A. J. Macrobert, D. Scheglmann, W. Helfrich, M. J. Verkaik, E. Papini, E. Reddi and J. Kos, *Nanomedicine (Lond)* 7, 663 (2012).
80. S. J. Yang, F. H. Lin, K. C. Tsai, M. F. Wei, H. M. Tsai, J. M. Wong and M. J. Shieh, *Bioconjug. Chem.* **21**, 679 (2010).
81. S. J. Yang, C. F. Lin, M. L. Kuo and C. T. Tan, *Biomacromolecules* **14**, 3183 (2013).

82. I. H. Oh, H. S. Min, L. Li, T. H. Tran, Y. K. Lee, I. C. Kwon, K. Choi, K. Kim and K. M. Huh, *Biomaterials* **34**, 6454 (2013).
83. C. K. Lim, J. Shin, Y. D. Lee, J. Kim, H. Park, I. C. Kwon and S. Kim, *Small* 7, 112 (2011).
84. K. Liu, Y. Liu, Y. Yao, H. Yuan, S. Wang, Z. Wang and X. Zhang, *Angew. Chem. Int. Ed. Engl.* **52**, 8285 (2013).
85. W. T. Morgan, A. Smith and P. Koskelo, *Biochem. Biophys. Acta* **624**, 271 (1980).
86. M. Rotenberg, S. Cohen and R. Margalit, *Photochem. Photobiol.* **46**, 689 (1987).
87. V. Rosenberger and R. Margalit, *Photochem. Photobiol.* **58**, 627 (1993).
88. G. S. Timmins and M. J. Davies, *J. Photochem. Photobiol. B* **24**, 117 (1994).
89. A. Barel, G. Jori, A. Perin, P. Romandini, A. Pagnan and S. Biffanti, *Cancer Lett.* **32**, 145 (1986).
90. D. Kessel, *Cancer Lett.* **33**, 183 (1986).
91. G. Jori and E. Reddi, *Int. J. Biochem.* **25**, 1369 (1993).
92. J. C. Maziere, R. Santus, P. Morliere, J. P. Reyftmann, C. Candide, L. Mora, S. Salmon, C. Maziere, S. Gatt and L. Dubertret, *J. Photochem. Photobiol. B* **6**, 61 (1990).
93. P. C. Rensen, R. L. de Vrueh, J. Kuiper, M. K. Bijsterbosch, E. A. Biessen and T. J. van Berkel, *Adv. Drug Deliv. Rev.* **47**, 251 (2001).
94. M. E. Davis, Z. Chen, D. M. Shin, *Nat. Rev. Drug Discov.* 7, 771 (2008).
95. A. Pluen, Y. Boucher, S. Ramanujan, T. D. McKee, T. Gohongi, E. di Tomaso, E. B. Brown, Y. Izumi, R. B. Campbell, D. A. Berk and R. K. Jain, *Proc. Natl. Acad. Sci. USA* **98**, 4628 (2001).
96. H. Cabral, Y. Matsumoto, K. Mizuno, Q. Chen, M. Murakamil, M. Kimura, Y. Terada, M. R. Kano, K. Miyazono, M. Uesaka, N. Nishiyama and K. Kataoka, *Nat. Nano* **6**, 815 (2011).
97. K. K. Ng, J. Lovell, A.Vedadi, T. Hajian and G. Zheng, *ACS Nano* 7, 3484 (2013).
98. Z. Zhang, J. Chen, L. Ding, H. Jin, J. F. Lovell, I. R. Corbin, W. Cao, P. C. Lo, M. Yang, M. S. Tsao, Q. Luo, G. Zheng, *Small* **6**, 430 (2010).
99. N. Sabnis and A. G. Lacko, *Ther. Deliv.* **3**, 599 (2012).
100. M. S. Brown and J. L. Goldstein, *Science* **191**, 150 (1976).
101. M. S. Brown and J. L. Goldstein, *Proc. Natl. Acad. Sci. USA* 76, 3330 (1979).
102. D. L. Williams, M. A. Connelly, R. E. Temel, S. Swarnakar, M. C. Phillips, M. de la Llera-Moya and G. H. Rothblat, *Curr. Opin. Lipidol.* **10**, 329 (1999).
103. D. Steinberg, *Science* **271**, 460 (1996).
104. C. Candide, P. Morliere, J. C. Maziere, S. Goldstein, R. Santus, L. Dubertret, J. P. Reyftmann and J. Polonovski, *FEBS Lett.* **207**, 133 (1986).
105. B. A. Allison, E. Waterfield, A. M. Richter and J. G. Levy, *Photochem. Photobiol.* **54**, 709 (1991).
106. U. Schmidt-Erfurth, F. Tj, E. S. Gragoudas, K. Schomacker, R. Birngruber and T. Hasan, *Exp. Eye Res.* **62**, 1 (1996).
107. M. R. Hamblin and E. L. Newman, *J. Photochem. Photobiol. B* **26**, 147 (1994).
108. U. Schmidt-Erfurth, H. Diddens, R. Birngruber and T. Hasan, *Br. J. Cancer* **75**, 54 (1997).
109. F. Giuntini, C. M. Alonso and R. W. Boyle, *Photochem. Photobiol. Sci.* **10**, 759 (2011).
110. G. Zheng, H. Li, M. Zhang, S. Lund-Katz, B. Chance and J. D. Glickson, *Bioconjug. Chem.* **13**, 392 (2002).

111. H. Li, D. E. Marotta, S. Kim, T. M. Busch, E. Wileyto and G. Zheng, *J. Biomed. Opt.* **10**, 41203 (2005).
112. L. Song, H. Li, U. Sunar, J. Chen, I. Corbin, A. G. Yodh and G. Zheng, *Int. J. Nanomedicine* **2**, 767 (2007).
113. W. Cao, K. K. Ng, I. Corbin, Z. Zhang, L. Ding, J. Chen and G. Zheng, G. *Bioconjug. Chem.* **20**, 2023 (2009).
114. D. E. Marotta, W. Cao, E. P. Wileyto, H. Li, I. Corbin, E. Rickter, J. D. Glickson, B. Chance, G. Zheng and T. M. Busch, *Nanomedicine (Lond)* **6**, 475 (2011).
115. G. Zheng, J. Chen, H. Li and J. D. Glickson, *Proc. Natl. Acad. Sci. USA* **102**, 17757 (2005).
116. J. Chen, I. R. Corbin, H. Li, W. Cao, J. D. Glickson and G. Zheng, *J. Am. Chem. Soc.* **129**, 5798 (2007).
117. K. Chen, A. Preuss, S. Hackbarth, M. Wacker, K. Langer and B. Roder, *J. Photochem. Photobiol. B* **96**, 66 (2009).
118. D. K. Deda, A. F. Uchoa, E. Carita, M. S. Baptista, H. E. Toma and K. Araki, *Int. J. Pharm.* **376**, 76 (2009).
119. M. Wacker, K. Chen, A. Preuss, K. Possemeyer, B. Roeder and K. Langer, *Int. J. Pharm.* **393**, 253 (2010).
120. A. Preuss, K. Chen, S. Hackbarth, M. Wacker, K. Langer and B. Roder, *Int. J. Pharm.* **404**, 308 (2011).
121. Z. Zhen, W. Tang, C. Guo, H. Chen, X. Lin, G. Liu, B. Fei, X. Chen, B. Xu and J. Xie, *ACS Nano* **7**, 6988 (2013).
122. K. Kurohane, A. Tominaga, K. Sato, J. R. North, Y. Namba and N. Oku, *Cancer Lett.* **167**, 49 (2001).
123. B. Chen, C. Crane, C. He, D. Gondek, P. Agharkar, M. D. Savellano, P. J. Hoopes and B. W. Pogue, *Int. J. Cancer* **123**, 695 (2008).
124. M. J. Bovis, J. H. Woodhams, M. Loizidou, D. Scheglmann, S. G. Bown and A. J. Macrobert, *J. Control Release* **157**, 196 (2012).
125. E. B. Gyenge, S. Hiestand, S. Graefe, H. Walt and C. Maake, *Photodiagnosis Photodyn. Ther.* **8**, 86 (2011).
126. E. Besic Gyenge, P. Forny, D. Luscher, A. Laass, H. Walt and C. Maake, *Photodiagnosis Photodyn. Ther.* **9**, 321 (2012).
127. S. A. de Visscher, M. J. Witjes, B. van der Vegt, H. S. de Bruijn, A. van der Ploeg-van den Heuvel, A. Amelink, H. J. Sterenborg, J. L. Roodenburg and D. J. Robinson, *Lasers Surg. Med.* **45**, 668 (2013).
128. A. S. Derycke and P. A. de Witte, *Adv. Drug. Deliv. Rev.* **56**, 17 (2004).
129. C. S. Jin and G. Zheng, *Lasers Surg. Med.* **43**, 734 (2011).
130. L. A. Muehlmann, G. A. Joanitti, J. R. Silva, J. P. Longo and R. B. Azevedo, *Braz J. Med. Biol. Res.* **44**, 729 (2011).
131. N. Dragicevic-Curic and A. Fahr, *Expert Opin. Drug Deliv.* **9**, 1015 (2012).
132. Y. N. Konan, R. Gurny and E. Allemann, *J. Photochem. Photobiol. B* **66**, 89 (2002).
133. G. Shim, S. Lee, Y. B. Kim, C. W. Kim and Y. K. Oh, *Nanotechnology* **22**, 365101 (2011).
134. E. Temizel T. Sagir, E. Ayan, S. Isik and R. Ozturk, *Photodiagnosis Photodyn Ther.* **11**, 537 (2014).
135. R. Rahmanzadeh, P. Rai, J. P. Celli, I. Rizvi, B. Baron-Luhr, J. Gerdes and T. Hasan, *Cancer Res.* **70**, 9234 (2010).

136. J. Buchholz, B. Kaser-Hotz, T. Khan, C. Rohrer Bley, K. Melzer, R. A. Schwendener, M. Roos and H.Walt, *Clin. Cancer Res.* **11**, 7538 (2005).
137. S. A. Sibani, P. A. McCarron, A. D. Woolfson and R. F. Donnelly, *Expert Opin. Drug Deliv.* **5**, 1241 (2008).
138. E. Paszko, C. Ehrhardt, M. O. Senge, D. P. Kelleher and J. V. Reynolds, *Photodiagnosis Photodyn. Ther.* **8**, 14 (2011).
139. K. Ichikawa, T. Hikita, N. Maeda, S. Yonezawa, Y. Takeuchi, T. Asai, Y. Namba and N. Oku, *Biochimica et Biophysica Acta (BBA) - Biomembranes* **1669**, 69 (2005).
140. Y. Mir, S. A. Elrington and T. Hasan, *Nanomedicine (Lond)* **9**, 1114 (2013).
141. J. F. Lovell, C. S. Jin, E. Huynh, H. Jin, C. Kim, J. L. Rubinstein, W. C. Chan, W. Cao, L. V. Wang and G. Zheng, *Nat Mater* **10**, 324 (2011).
142. C. S. Jin, L. Cui, F. Wang, J. Chen and G. Zheng, *Adv. Healthc. Mater.* **3**, 1240 (2014).
143. T. W. Liu, T. D. MacDonald, J. Shi, B. C. Wilson and G. Zheng, *Angew. Chem. Int. Ed.* **51**, 1 (2012).
144. T. W. Liu, T. D. Macdonald, C. S. Jin, J. M. Gold, R. G. Bristow, B. C. Wilsonm and G. Zheng, *ACS Nano* 7, 4221 (2013).
145. T. D. MacDonald, T. W. Liu and G. Zheng, *Angew. Chem. Int. Ed. Engl.* **53**, 6956 (2014).
146. C. S. Jin, J. F. Lovell, J. Chen and G. Zheng, *ACS Nano* 7, 2541 (2013).
147. D. Zeng, N. S. Lee, Y. Liu, D. Zhou, C. S. Dence, K. L. Wooley, J. A. Katzenellenbogen and M. J. Welch, *ACS Nano* **6**, 5209 (2012).
148. J. H. Lee, S. Shao, K. T. Cheng, J. F. Lovell and C. H. Paik, *J. Liposome Res.*, 1 (2014).
149. K. K. Ng, M. Shakiba, E. Huynh, R. A. Weersink, A. Roxin, B. C. Wilson and G. Zheng, *ACS Nano* **8**, 8363 (2014).
150. K. A. Carter, S. Shao, M. I. Hoopes, D. Luo, B. Ahsan, V. M. Grigoryants, W. Song, H. Huang, G. Zhang, R. K. Pandey, J. Geng, B. A. Pfeifer, C. P. Scholes, J. Ortega, M. Karttunen and J. F. Lovell, *Nat. Commun.* **5**, 3546,(2014).
151. Y. Zhang, M. Jeon, L. J. Rich, H. Hong, J. Geng, Y. Zhang, S. Shi, T. E. Barnhart, P. Alexandridis, J. D. Huizinga, M. Seshadri, W. Cai, C. Kim and J. F. Lovell, *Nat. Nanotechnol* **9**, 631 (2014).
152. D. A. Bryant, A. M. Costas, J. A. Maresca, A. G. Chew, C. G. Klatt, M. M. Bateson, L. J. Tallon, J. Hostetler, W. C. Nelson, J. F. Heidelberg and D. M. Ward, *Science* **317**, 523 (2007).
153. N.-U. Frigaard and D. Bryant, in *Complex Intracellular Structures in Prokaryotes Microbiology Monographs*, Vol. 2, Ed. J. M. Shively (Springer Berlin Heidelberg, 2006), Chapter 21, p. 79.
154. C. J. Medforth and J. A. Shelnutt, in *Handbook of Porphyrin Science,* Vol. 12 (World Scientific Publishing Company, 2011) p. 181.
155. K. Börjesson, E. P. Lundberg, J. G. Woller, B. Nordén and B. Albinsson, *Angew. Chem. Int. Ed.* **50**, 8312 (2011).
156. S. Ghosh, K. B. Ucer, Jt. R. D'Agostino, K. Grant, J. Sirintrapun, M. J. Thomas, R. Hantgan, M. Bharadwaj and W. H. Gmeiner, *Nanomedicine (Lond)* **10**, 451 (2014).

Chapter 8

Photodynamic Therapy Dosimetry: A TO Z

Michael S. Patterson

Juravinski Cancer Centre and McMaster University,
699 Concession St., Hamilton, Ontario, Canada, L8V 5C2
mpatters@hhsc.ca

Abstract

This chapter uses a glossary format to introduce the reader to fundamental concepts in the dosimetry of photodynamic therapy (PDT). Each term is defined in the context of PDT dosimetry and its importance is discussed in the light of current knowledge and recent developments. The goal of the chapter is to make the literature in the field accessible to those without an extensive background in optics, photophysics, or radiation transport. The chapter concludes with a discussion of the importance of dosimetry in the future of PDT.

1. Introduction

The dosimetry of photodynamic therapy (PDT) is a complex and sometimes confusing subject. While several useful review articles and protocols have been published, they can be difficult to read without an extensive background in optics, photochemistry, and radiation transport. The purpose of this chapter is to provide a bridge to these formal treatments for those without this preparation. The chapter is written in the form of a glossary but is intended to be more than that. Each term is defined in the context of PDT dosimetry and its importance is discussed in the light of current knowledge and recent developments in the field. Rigor in the definitions has sometimes been sacrificed for the sake of simplicity and readability.

While the chapter can be read on its own, it is best used as a companion to the publications mentioned above. For a general introduction to PDT the article by Agostinis *et al.*[1] is recommended. Wilson and Patterson[2] have reviewed the dosimetry problem from a biophysics standpoint and specific PDT dosimetry protocols have been disseminated by the American Association of Physicists in Medicine.[3,4]

2. Glossary of Dosimetry Terms

2.1. *Absorption Coefficient*

If light travels a short distance dx in a medium such as tissue, a fraction of that light $\mu_a dx$ is absorbed where μ_a is the absorption coefficient. The absorption coefficient is also equal to $N\sigma_a$ where N is the concentration of a specific molecule and σ_a is its molecular absorption cross section. For a medium like tissue that contains a variety of molecular species, the total absorption coefficient is just the sum of the individual contributions. In the wavelength region of interest for PDT, important absorbers in tissue include hemoglobin, melanin, and exogenous photosensitizers. Note that the product of the fluence rate and the absorption coefficient $E_0\mu_a$ is the rate of energy absorption per unit volume per unit time. If μ_a is the absorption coefficient of the photosensitizer in the target tissue, this product drives the ensuing photochemical reactions.

2.2. *Aminolevulinic Acid (ALA)*

ALA is used to sysnthesize Protoporphyrin IX (PpIX) as a part of normal cell metabolism. PpIX is a PDT sensitizer and is produced in excess when ALA is administered systemically or topically. ALA is not a photosensitizer but can be regarded as a prodrug that is metabolized to produce an active compound. This process adds an additional complexity to dosimetry of PDT with ALA.

2.3. *Anisotropy Parameter*

The angular dependence of light scattering in tissue is fully described by the phase function. The anisotropy parameter g is the mean cosine of the scattering angle. If the phase function is isotropic so that scattering in any direction is equally probable then $g = 0$. In tissue, scattering through small angles, referred to as forward scattering, is much more probable so that g is typically greater than 0.9.

2.4. *Attenuation Coefficient*

The attenuation coefficient μ_t is the sum of the absorption and scattering coefficients.

2.5. *Bolometer*

A bolometer is a device used to measure the total power output of a laser source. Its operation is based on detecting the temperature increase in a light absorbing target. The spectral response of a bolometer is very flat compared to that of a photodiode but its sensitivity and time response are inferior.

2.6. *Continuous Wave (cw)*

This term refers to light sources that emit light continuously as opposed to pulsed operation. For PDT treatment, diode lasers, LEDs, and lamps are commonly used cw light sources. A pulsed source will produce the same biological effect as a cw source of the same average power as long as the energy per pulse is not too high. If it exceeds 1 mJ, there is the potential for depletion of the photosensitizer ground state during the pulse.[2] This reduces the absorption of light by the photosensitizer and hence the PDT efficiency.

2.7. *Daylight PDT*

Refers to PDT where the sun is used as the light source. Studies have compared daylight PDT with ALA to conventional PDT with ALA for actinic keratoses[5] and basal cell carcinoma.[6] Daylight PDT was found to be as effective and less painful. In these trials the solar exposure was measured but it is not clear how critical this was to treatment success.

2.8. *Diffuse Reflectance*

This term refers to light that is reflected from a surface over a range of directions even if the incident light is collimated in a specific direction. This is in contrast to specular (or mirror-like) reflection that obeys Snel's Law. Light can be diffusely reflected by a rough surface or it can arise from sub-surface scattering, i.e. light enters the medium, is scattered one or more times, and then escapes. Diffuse reflectance is important in superficial or intracavitary PDT because it reduces the light that is available for sensitizer activation. For example, more than half of the incident light is diffusely reflected by Caucasian skin in the optical window. The diffuse reflectance depends on the optical properties of the tissue and its noninvasive measurement provides a means for estimating these *in vivo*.[7] The total diffuse reflectance is the ratio of the power of all the diffusely reflected light to the power of the incident light.

2.9. *Diffusion Theory*

Diffusion theory provides the basis for a useful equation for the fluence rate in a turbid medium like tissue. This diffusion equation can be developed heuristically by assuming that the propagation of multiply scattered light is analogous to mass diffusion or heat conduction, i.e. the net flux of light is proportional to the gradient of the fluence rate (Fick's Law). It can also be derived directly from the radiation transport equation by assuming that the radiance is almost isotropic. In tissue this condition is met when the absorption coefficient is much less than the transport scattering coefficient ($\mu_a \ll \mu_s'$) and the point of interest is not too close to sources or boundaries. The former condition is valid in the optical window where PDT is usually performed. The diffusion equation can be solved analytically for simple cases such as point, line, or external beam sources and a uniform turbid medium.[8] When the tissue is heterogeneous, numerical methods such as finite elements can be used to solve for the fluence rate.

2.10. *Direct Dosimetry*

This term was coined by Wilson *et al.*[9] in their paper that described three possible strategies for PDT dosimetry. If singlet oxygen is the main cytotoxic agent produced by PDT, then its quantitation would be a "direct" measure of the PDT dose. Singlet oxygen does emit near infrared light but with very low probability. Nonetheless, it has been detected *in vivo* and the signal has been shown to correlate with biological effects such as erythema[10,11] and tumor response.[12,13] Images of relative singlet oxygen concentration have also been generated by scanning systems[14,15] and NIR cameras.[16] There are a number of challenges in measuring and interpreting the singlet oxygen luminescence (SOL): (a) background fluorescence and phosphorescence require spectral and (ideally) temporal discrimination to select the SOL peak at 1270 nm (b) tissue optical properties and detector instrumentation affect the collection efficiency and resulting signal amplitude (c) the SOL amplitude and kinetics depend on oxygen concentration and possibly on quencher concentration in the tissue. This complication reflects the fact that luminescence arises from singlet oxygen that does not react whereas the PDT damage is caused by singlet oxygen that does react. These problems have limited the adoption and testing of direct dosimetry but it remains an appealing concept.

2.11. *Dose*

In general, dose is the amount of a therapeutic agent administered. Typically this is corrected for body weight or surface area so systemic drug doses might be expressed in mg per kg. PDT is primarily a local therapy so a measure of the local amount of the therapeutic agent would be more useful. An analogy is radiation therapy where accurate calculations or measurements of the energy absorbed per unit mass of tissue ($J\,kg^{-1}$) at any location provide a reliable prediction of eventual biological effect. There is no consensus as to the appropriate quantity for PDT dose, but a reasonable candidate would be the total amount of singlet oxygen that reacts per unit mass or volume of tissue (units moles per liter, M). Studies with a wide range of photosensitizers and biological systems have concluded that a therapeutic dose of singlet oxygen is about 1 mM.[17–19] Three approaches to estimating the singlet oxygen dose during PDT have been proposed (see the definitions of direct, explicit, and implicit dosimetry in this chapter), but this is rarely done in the clinical setting. Instead, the systemic dose of sensitizer ($mg\,kg^{-1}$) is reported along with the radiant exposure (for superficial delivery), the delivered energy per unit length of optical fiber (for treatment with line sources), or the total delivered energy (for point sources).

These quantities are relatively straightforward to measure[3,4] but it is easy to see why they may not provide a good indication of the local PDT effect. Individual variation in pharmacokinetics can affect the concentration of photosensitizer in the target tissue at the time of irradiation. The optical properties of tissue are wavelength, site, and subject dependent. Tissues may be hypoxic at the outset or may become hypoxic due to oxygen consumption or vascular shutdown. Finally, all three components of PDT (drug, light, and oxygen) may change dynamically during treatment in an interdependent manner — for example, photobleaching of the sensitizer could reduce the rate of oxygen consumption and allow reoxygenation to occur. Clearly, achieving accurate local dosimetry equivalent to that available for radiation therapy is a difficult assignment.

The magnitude of the therapeutic singlet oxygen dose (about 1 mM) has some interesting implications. Given that the average concentration of the photosensitizer is in the μM range, for an effective treatment each molecule must participate in hundreds of excitation-ISC-energy transfer cycles. The quantum yield of photobleaching must be low enough that the photosensitizer can survive long enough to participate in these repeated cycles. In well oxygenated tissue the oxygen concentration is around 40 μM. In a successful treatment this oxygen must be replaced at least 25 times in order to provide the necessary reacting singlet oxygen. This may not be the case if the vasculature itself is the target of the PDT treatment.

2.12. *Dose Rate Effects*

The rate of photodynamic chemical reactions can be readily modified by changing the fluence rate. Treatment efficacy can be impaired if the fluence rate is high enough to induce hypoxia[20] and the trend in recent years has been to reduce the fluence rate. The lower limit is imposed by practical considerations of overall treatment time, although the use of implantable light sources[21] could facilitate the exploration of very low fluence rates. Lower irradiance has also been found to reduce pain in ALA-PDT of the skin.[22]

2.13. *Effective Attenuation Coefficient*

As distance d from a light source increases in a turbid medium like tissue, the fluence rate falls off approximately as $f(d)\ exp-(\mu_{eff}d)$ where μ_{eff} is the effective attenuation coefficient and $f(d)$ accounts for geometric factors specific to the source — for example, $f(d) = 1/r$ for a point source. In the spectral region where diffusion theory is valid in tissue $\mu_{eff} = (3\mu_a\mu_s')^{1/2}$. Note that increasing the absorption coefficient by a certain percentage will have the same effect on the effective attenuation coefficient as increasing the scattering coefficient by the same percentage.

2.14. *Explicit Dosimetry*

This term was proposed by Wilson *et al.*[9] to describe one of three strategies for PDT dosimetry. The basic idea is that the singlet oxygen dose at any point can be calculated if the history of the local fluence rate, photosensitizer concentration, and oxygen concentration are known along with the rate constants for the relevant photochemical reactions. Detailed information regarding oxygen may not be necessary as long as it is known that tissue oxygenation does not decrease enough to impair PDT efficiency. In this case, the singlet oxygen dose will be proportional to the time integral of the product of the fluence rate and the photosensitizer concentration. In the absence of significant photobleaching the integral is simply the product of photosensitizer concentration and fluence. This "drug-light" product form of explicit dosimetry has predicted biological effect in some *in vivo* studies,[23,24] but explicit dosimetry has also failed in other experiments.[25]

2.15. *Fluence Rate*

The radiance is the fundamental descriptor of the light distribution in tissue but it provides more information than needed for PDT dosimetry. Light absorption by a photosensitizer molecule does not depend on the direction of propagation of that light so it is sufficient to consider only the integral of the radiance over 4π solid angle. This definition is equivalent to considering a small spherical volume, measuring the radiant energy that passes through that volume per unit time, and dividing that power by the cross sectional area of the sphere. This quantity is called the fluence rate E_0 and the SI units are $\mathrm{Wm^{-2}}$. Calculating or measuring the fluence rate is usually much easier than the radiance. The integral of the fluence rate over time is the fluence H_0 and has SI units $\mathrm{J\,m^{-2}}$.

2.16. *Fluorescence*

Fluorescence is the emission of light with transition of a molecule from an excited electronic state to a lower state with the same overall electron spin. Fluorescence is often observed after excitation of a photosensitizer and results from transition from the first excited singlet state to the ground state. This process competes with intersystem crossing (ISC) so fluorescence does not contribute to photodynamic action. On the other hand, fluorescence can be a very useful tool to monitor the concentration of photosensitizer in tissue and many photosensitizers have a fluorescence quantum yield of a few per cent.

2.17. *Fractionation*

This term is used in ionizing radiation therapy to describe treatment with a series of small doses or fractions instead of a single large dose. Empirical formulas have been devised to adjust the dose per fraction so that the desired treatment effect is achieved. In PDT, fractionation usually refers to multiple irradiations for a single photosensitizer application. The dosimetry of fractionated PDT is relatively unexplored except for ALA mediated treatment of skin lesions where superior response has been reported for a two fraction irradiation scheme.[26]

2.18. *Hemoglobin*

In addition to its role as an oxygen transporter in the blood, hemoglobin is the main contributor to the tissue absorption coefficient in the PDT spectral range. The rapid decrease in its absorption above 600 nm is the main impetus for "second generation" photosensitizers with absorption peaks at longer wavelengths. The absorption spectra of deoxyhemoglobin and oxyhemoglobin are quite different — a fact that is exploited in routine pulse oximetry of arterial blood. It can also be used to measure the average oxygenation of hemoglobin in tissue and hence the average tissue oxygenation during PDT.[27]

2.19. *Hypoxia*

The term hypoxia refers to a condition in which the oxygen concentration in tissue is lower than normal. Tissue oxygenation is crucial to successful

PDT — without oxygen there can be no singlet oxygen! At oxygen concentration of about 10 μM, photosensitizer excited state triplets are as likely to de-excite by competing mechanisms as they are by energy transfer to ground state oxygen. This means that the singlet oxygen quantum yield will be only half of what it is under normal oxygenation. It is well known that tumors can be hypoxic because of an inadequate or abnormal vascular supply. Hypoxia can also be induced by PDT itself as a result of vasoconstriction or simply by a PDT oxygen consumption rate that exceeds the local perfusion rate. The significance of the latter mechanism was not appreciated in the early days of PDT but many studies have confirmed that the efficacy of PDT can be reduced if the excitation fluence rate is too high.[20]

2.20. *Immune Response*

PDT is known to induce a local inflammatory response as well as systemic effects that may be important in successful treatment.[1] It is an open question whether these effects can be correlated with the local PDT dose discussed in this chapter.

2.21. *Implicit Dosimetry*

This term was also coined by Wilson *et al.*[9] to refer to the idea that a surrogate for PDT effect may be measurable. If the photosensitizer is bleached by reactions with singlet oxygen, then the reduction in its fluorescence during treatment should be an indicator of singlet oxygen dose.[17] Similarly, the accumulation of the photoproduct of such reactions would also provide this information.[19,28] In principle, implicit dosimetry is simple because it requires only a measure of photosensitizer (or photoproduct) fluorescence rather than separate measurements of fluence rate, photosensitizer concentration, and oxygen. In practice, implicit dosimetry has had mixed success because photobleaching may be caused by reactions that do not involve singlet oxygen.[29,30]

2.22. *In vitro PDT Dosimetry*

Fundamental studies of PDT mechanisms and cell response depend on treatment of cell cultures. Cells may be treated as adherent monolayers, as suspensions, or as three dimensional cultures including multicellular spheroids. It is relatively easy to deliver reproducible treatments because the variables of photosensitizer concentration in the incubation medium, time of irradiation after incubation, and irradiance can be precisely measured and controlled.

Because the cell cultures do not scatter light strongly at typical cell concentrations, the fluence rate is equal to the irradiance. This is not the case in turbid media like tissue where the fluence rate can be several times greater than the irradiance because of multiple light scattering. PDT can induce hypoxia in cell monolayers and 3D cultures at surprisingly low irradiances and researchers must be aware of this when planning experiments and interpreting results.[31] On the other hand, gently stirred cell suspensions are virtually immune to hypoxia unless the irradiance is extremely high.

2.23. *Integrating Sphere*

An integrating sphere is a device that simplifies the measurement of total power output from an optical fiber intended for use as an interstitial or intracavitary source. It consists of a hollow sphere with an inner surface of high diffuse reflectance. This can be achieved by coating the surface or manufacturing the sphere from a material with low intrinsic absorption and high scattering. The sphere also has at least two ports: one allows the fiber source to be inserted and the other permits measurement of the irradiance at the sphere surface. For an ideal integrating sphere with a diffuse reflectance of unity, the irradiance will be the same everywhere on the internal surface. Hence the total power emitted by the fiber can be obtained by multiplying the measured irradiance by the internal surface area of the sphere. In practice corrections are required for the actual diffuse reflectance of the surface and loss of light at the ports. Some clinical PDT light sources have built-in integrating spheres to facilitate the measurement of optical fiber output.

2.24. *Interstitial PDT*

For an interstitial treatment, optical fibers are inserted directly into the tissue to be treated. These can be either line sources or point sources. Interstitial PDT is performed when the tumor volume is too large for superficial treatment or when the treatment site is not directly accessible, for example, the prostate gland.

2.25. *ISC*

Refers to the process in which a molecule undergoes rearrangement of electron spin and converts from a singlet state to a triplet state (or vice versa) without emission of radiation. ISC occurs with high probability (>0.5) for Type II photosensitizers and results in the production of a long-lived excited triplet state from the initial excited singlet state. This triplet then transfers energy to ground state oxygen.

2.26. *Intracavitary PDT*

For intracavitary treatments such as those for bladder cancer or Barrett's esophagus, optical fibers are placed within a natural body cavity. The term can also be used to refer to treatment of a surgical cavity, for example, adjuvant PDT of residual brain tumor following resection. The irradiation of a cavity introduces another complication to dosimetry.[32] Diffusely reflected light is not lost as it is in irradiation of the skin, but can reenter the tissue at a different point. This effect can significantly increase the fluence rate (compared to irradiation of a flat surface) and is thought to have contributed to unexpectedly high toxicity in some trials of PDT for bladder cancer.

2.27. *Irradiance*

The radiant energy incident on a surface per unit time per unit area is called the irradiance E. The SI units of irradiance (Wm^{-2}) are the same as those for the fluence rate but they are physically distinct quantities. The irradiance can be directly measured on the skin surface with a calibrated power meter of known detector area. This constitutes the simplest form of light dosimetry for PDT but it is important to realize that it does not directly yield the fluence rate inside the tissue. The integral of the irradiance over time is called the radiant exposure H (units $J\,m^{-2}$).

2.28. *Jablonski Diagram*

A pictorial representation of the electronic energy levels for a molecule and the transitions between those states.

2.29. *Laser Diode*

A semiconductor junction that emits light when the electrical current through the junction exceeds a threshold. Diode lasers are the most widely used light sources for PDT because they provide high power (several Watts) in narrow spectra that can be selected to match specific sensitizers, their output can be efficiently coupled to multimode optical fibers, and they are reliable, energy efficient, and relatively inexpensive.

2.30. *Light Emitting Diode (LED)*

A semiconductor device similar to a diode laser but it has lower output power, a broader emission spectrum, and cannot be coupled as efficiently to optical fibers. LED's are well suited to superficial PDT treatments and are cheaper than laser diodes.

2.31. *Monte Carlo Method*

The Monte Carlo method is a means to simulate a physical process like the propagation of light in a multiple scattering medium. It is particularly useful for situations with complex geometries or other difficulties that defy analytical or numerical solutions. A bundle of light is launched into the medium at a specific starting point and the history of that bundle is traced until a predetermined end point is reached. The location and type of interactions undergone by the bundle is determined by sampling the appropriate probability distributions. For example, if scattering occurs, the new direction of propagation is sampled from the phase function. Because sampling is a stochastic process, every history will be different (up to the limitations of pseudorandom number generators). During the history certain events will be scored — for PDT dosimetry that event might be the absorption of part of the bundle by the photosensitizer. After many histories, a spatial map of light absorption in the tissue will be produced. The accuracy of this map will depend on the number of histories — one of the disadvantages of Monte Carlo simulations is that a lot of computer time can be necessary to get useful estimates of relatively rare events. This can be overcome to some extent by using biased sampling techniques.

2.32. *Noise Equivalent Power (NEP)*

The NEP of a light detector is the power incident on the detector that produces a signal equal to the noise. The lower the NEP, the more sensitive the detector.

2.33. *Optical Fiber*

An optical fiber is essentially a waveguide that allows light to be transmitted with low losses over considerable distances. It consists of a glass or plastic core surrounded by a cladding of lower refractive index. The fibers used in PDT are multimode fibers that support several transmission modes within a core that is typically a few hundred microns in diameter. In PDT fibers are used to deliver light from the source to the treatment site. One end of the fiber is usually terminated with a standard fiber connector (such as an SMA connector) suitable for coupling to the light source and the other end is modified for light delivery. For superficial irradiation a microlens fixed to the fiber produces a uniform circular beam. Interstitial or intracavitary irradiation use fibers that act as line or point sources. The former can be fabricated by stripping the cladding from the fiber and coating with a suitable scattering material. Point sources consist of a small scattering sphere fixed to the end of an optical fiber. These devices can also serve as nearly isotropic detectors for the *in situ* measurement of fluence rate.

2.34. *Optical Window*

The optical window is the spectral range for which the penetration depth in tissue is maximized. The range is somewhat arbitrary (typically quoted as 650–950 nm), but the wavelength at which maximum penetration occurs is around 700 nm. Although the scattering coefficient of tissue continues to decrease above this wavelength, this is offset by an increase in absorption by tissue water and lipids.

2.35. *Penetration Depth*

The reciprocal of the effective attenuation coefficient μ_{eff} is called the penetration depth δ. It is roughly the distance in tissue over which the fluence rate decreases by a factor of e. In the optical window where PDT is usually performed the penetration depth is 1–5 mm. The penetration depth is sometimes incorrectly thought of as the maximum depth of light penetration. In fact, there is no such thing — there is a finite probability that light will penetrate to any depth.

2.36. *Phantom*

In dosimetry, a phantom is an object that has the same properties as tissue. For PDT these properties are the absorption coefficient, the scattering coefficient, and the refractive index. Many materials have been proposed as tissue substitutes[33] but three are commonly used in PDT. A liquid phantom composed of a scattering emulsion (typically an intravenous feeding liquid) and an absorber such as ink or a dye is inexpensive and easy to fabricate. Optical fibers can easily be placed within these phantoms and moved from location to location. Solid phantoms can be molded into complex shapes, can comprise more than one tissue “type” and can be stable over time. Typically a gel or epoxy is combined with a powder such as titanium dioxide to provide the scattering and ink or dyes are added to adjust the absorption.

Phantoms serve two main purposes in PDT dosimetry. The first is to provide a realistic mockup of a proposed treatment plan. Optical fiber detectors can then be inserted into the phantom to map out the resulting fluence rate distribution and confirm the treatment plan. The second role is to facilitate the calibration of fluence rate detectors before they are used for *in vivo* measurements. These typically consist of a scattering or fluorescent bulb affixed to an optical fiber so that a fraction of light striking the bulb from any direction (except in the shadow of the fiber itself) will be coupled into the fiber and transmitted to a photodiode. The resulting near-isotropic response means that the photodiode signal will be proportional to fluence rate but absolute calibration requires that the fiber tip be placed at a location with known fluence rate. One way to do this is to fabricate a phantom with well characterized optical properties, irradiate the surface of the phantom with a measured irradiance, and calculate the fluence rate within the phantom using the diffusion equation or some other solution of the radiation transport equation.

2.37. *Phase Function*

The phase function for a medium describes the probability that light travelling in initial direction Ω will be scattered into direction Ω'. In PDT dosimetry it is usually assumed that this function does not depend on the initial direction and is a function only of the angle between Ω and Ω'. This may not be the case for tissue with directional structure such as collagen fibers. The phase function can be directly measured for tissue samples thin enough that multiple scattering can be ignored (10 microns or less). For calculation of the fluence rate, measured phase functions are usually replaced by simpler analytic expressions such as the Henyey–Greenstein function.[34] As long as the anisotropy parameter is the same, the resulting fluence rate distribution will not be significantly different.[35]

2.38. *Phosphorescence*

Phosphorescence is the emission of light with transition of a molecule from an excited triplet state to a lower singlet state or (as in the case of oxygen) from an excited singlet state to a lower triplet state. Because these transitions are spin forbidden they occur with much lower probability than the spin allowed transitions associated with fluorescence. The lifetimes of the excited states are also correspondingly longer — typically microseconds as opposed to nanoseconds. Phosphorescence of singlet oxygen is the basis for direct dosimetry or singlet oxygen luminescence dosimetry (SOLD).

2.39. *Photobleaching*

In the context of PDT dosimetry, photobleaching refers to the irreversible loss of photosensitizer molecules due to photochemical reactions. This could result from reactions of the sensitizer ground state with singlet oxygen or by

Type I reactions involving excited states of the sensitizer. Irreversible loss of the sensitizer has important implications for dosimetry. First, photobleaching imposes an upper limit on the total amount of singlet oxygen that can be generated for a given initial concentration of sensitizer regardless of the fluence. If this amount is less than the threshold dose of singlet oxygen required for a given effect, say necrosis, then that endpoint will not be achieved. It has been suggested that this provides a means to enhance the difference in response between normal tissues and tumors.[36] If the initial concentration of sensitizer in the normal tissue is low enough to ensure it is protected (regardless of the fluence) and the initial concentration in the tumor is high enough to achieve necrosis then dosimetry is almost superfluous. Second, photobleaching reduces the rate of local PDT oxygen consumption. Regions of the tumor that became hypoxic due to PDT might be reoxygenated later in the treatment. This complicates the dosimetry problem and could result in surprisingly heterogeneous doses of singlet oxygen as shown in modeling studies by Liu *et al.*[37]

2.40. *Photodiode*

A semiconductor junction that produces an electrical current when irradiated with light. It can be used to measure power or, if the detector area is known, irradiance. Silicon photodiodes are useful for PDT measurements because they have good sensitivity in the visible and near infrared. A commercial power meter incorporating a silicon photodiode is the most useful instrument for a PDT facility. The meter must have a calibration that is traceable to the relevant national standards laboratory (e.g., National Institute of Standards and Technology, NIST, in the U.S.). The response of the diode, and therefore the calibration factor, is wavelength dependent. This correction is often programmed into commercial instruments.

2.41. *Photoproduct*

Refers to the product of a photochemical reaction. In PDT, the photoproducts of photosensitizer bleaching can themselves be effective photosensitizers.[38] Quantitation of these photoproducts may also be the basis for implicit dosimetry.[19,28]

2.42. *Photosensitizer*

In general, a photosensitizer is a molecule that initiates photochemical reactions by the absorption of light. Photodynamic sensitizers transfer energy to oxygen resulting in cytotoxic singlet oxygen. Many compounds can produce singlet oxygen with high quantum yield in solution but the ideal clinical PDT sensitizer would have several other attributes. It should be non-toxic without light, should preferentially accumulate in the target tissue, should be soluble or readily administered in a delivery vehicle such as liposomes, should be cleared rapidly to prevent longterm photosensitization, should be easily synthesized and purified, should absorb light strongly in the optical window to maximize penetration of the excitation light, should be fluorescent (but not too fluorescent!) to aid dosimetry and tumor demarcation, and should not photobleach too rapidly. With all of these constraints it is not surprising that only a handful of drugs has been approved for routine clinical use[1] and that none of them is ideal. Another consideration that may not be immediately obvious is that the ideal drug should probably be protected as intellectual property so that the costs of clinical trials to gain approval can eventually be recouped.

2.43. *Power*

The amount of radiant energy per unit time (SI unit W). The measurement of power is an essential part of PDT dosimetry. This usually consists of a measurement of the total power emitted from a treatment fiber or the power per unit area (i.e., the irradiance). A power meter with calibration at the treatment wavelength traceable to a national standards laboratory is required. Several types of meters are available but the most useful is probably a silicon photodiode.

2.44. *Quantum Yield*

Refers to the probability of a specific event per absorbed quantum. For example, if one molecule of singlet oxygen is produced for every two quanta absorbed by the photosensitizer, the singlet oxygen quantum yield is 0.5.

2.45. *Quencher*

A quencher is a molecule that interacts with the excited state of the molecule of interest and reduces its concentration. This reduction can be due to chemical reactions, energy transfer, or collisional de-excitation.

2.46. *Radiance*

The radiance $L\ (r, \Omega, t)$ is the fundamental descriptor of the radiation distribution in tissue. It is the radiant power propagating in direction Ω per unit solid angle per unit area at location r at time t. The SI unit is $W\ sr^{-1}\ m^{-2}$.

2.47. *Radiation Transport Equation*

The radiation transport equation is the integrodifferential equation that describes the radiance as a function of space, direction, and time. It is essentially a statement of conservation of energy. It can rarely be solved analytically but is amenable to numerical methods such as discrete ordinates or expansion of the radiance as a series of basis functions. Monte Carlo methods are also effective tools to provide approximate solutions.

2.48. *Scattering Coefficient*

If light travels a short distance dx in tissue a fraction of that light $\mu_s dx$ will be scattered or redirected where μ_s is the scattering coefficient. In tissue, scattering results from changes in the refractive index associated with its complex structure. These inhomogeneities include those smaller than visible wavelengths (small organelles) and those several wavelengths in size (cells). A recent review by Jacques[39] explains how the former structures contribute to Rayleigh scattering while the latter cause the forward scattering component. The scattering coefficient of soft tissue in the PDT spectral range is 5–50 mm^{-1}. Because the scattering is forward peaked, several events are required to randomize the direction of light propagation. A useful concept is the effective isotropic scattering coefficient, also referred to as the reduced or transport scattering coefficient, defined by $\mu_s' = (1 - g)\mu_s$ where g is the anisotropy parameter. Typical values of μ_s' are 0.5–5 mm^{-1}. The reciprocal of the reduced scattering coefficient is the distance over which the direction of light propagation is "randomized".

2.49. *Second Generation Photosensitizer*

The first clinically approved photosensitizer was hematoporphyrin derivative (HPD), a mixture of porphyrins derived from blood. HPD (and a purified version with the trade name Photofrin) has a relatively weak absorption peak at 630 nm. Newer, second generation drugs, are synthetic (sometimes pure) compounds with strong absorption peaks at longer wavelengths up to 800 nm.

2.50. *Singlet Oxygen*

Oxygen is a very unusual molecule in that its ground state is a triplet. There is more than one excited singlet state but the term "singlet oxygen" usually refers to the $^1\Delta_g$ state. In PDT it is produced by energy transfer from the excited triplet state of the photosensitizer to ground state oxygen. The forbidden transition from this state to the ground state results in the emission of near infrared light of wavelength 1270 nm. Detection of this weak emission is the basis of "direct" PDT dosimetry. Singlet oxygen is highly reactive and its lifetime in tissue is estimated to be 0.05–0.20 μs.[40] For comparison, the lifetime in water is about 3.7 μs.

2.51. *Singlet State*

In a singlet state all of the electron spins are paired. This is typically the case for the ground state of photosensitizers and for the excited state that results from light absorption. ISC then produces an excited triplet state that efficiently transfers energy to ground state oxygen.

2.52. *Specular Reflectance*

Refers to the mirror-like reflection of light from smooth surfaces. In contrast to diffuse reflectance, specular reflectance is directional with the angle of reflection equal to the angle of incidence. The fraction of light that is reflected at the surface depends on the index of refraction of the material and the angle of incidence as expressed in Snel's Law. Specular reflection can pose an eye hazard during laser procedures such as PDT and should be considered in establishing laser safety protocols.

2.53. *Superficial PDT*

This term refers to PDT of an accessible surface such as the skin. Typically this is performed with an optical fiber and micolens combination that produces a uniform circular beam. If the area is very large it may be more efficient to use a lamp or LED source. Irregular fields can be produced by "cut-outs" of aluminum foil or similar materials.

2.54. *Threshold Dose*

The observation of a sharp boundary between necrotic and normal tissue on histology following PDT led to the concept of a minimum threshold dose required for a therapeutic effect.[23] Studies with a variety of sensitizers and tissues have shown that a cumulative dose of about 1 mM of singlet oxygen will cause necrosis. *In vitro* studies with cell cultures show that this dose will also reduce clonogenic cell survival by about a factor of *e*.

2.55. *Treatment Planning*

Refers to actions taken prior to or during a PDT treatment to achieve a specific outcome. At its simplest, treatment planning would be the design of an irradiation scheme (e.g., length of interstitial optical fibers, or diameter of superficial beam) to ensure adequate coverage of a lesion. At its most complex, treatment planning incorporates measurement of photosensitizer concentration, fluence rate, and oxygenation to construct a three dimensional map of reacting singlet oxygen and to plan a treatment that achieves a threshold dose everywhere in the target volume.[41,42]

2.56. *Triplet State*

A triplet state has two unpaired electrons. Historically the term triplet arose from the observation that a magnetic field allowed the three previously degenerate states (both spins parallel to the field, both spins antiparallel to the field, spins opposite) to be differentiated in the spectroscopic signature. In PDT the first excited triplet state of the photosensitzer is efficiently produced by ISC from the first excited singlet state.

2.57. *Two Photon PDT*

If a photosensitizer can be excited by a photon with energy $h\nu$ it can also be excited by the simultaneous absorption of two photons each with energy $h\nu/2$. This two photon absorption leads to exactly the same photochemistry as standard one photon absorption, however the rate of the initial excitation will depend on the square of the fluence rate. This could result in better spatial confinement of the PDT effect — in fact, two photon PDT can target subcellular structures.[43]

2.58. *Type I Reaction*

In a Type I reaction the excited photosensitizer molecule participates directly in an electron transfer reaction with a substrate. Such reactions can produce reactive oxygen species (e.g., superoxide radical anion) that cause biological damage or the photosensitizer itself can react with a biological target. A potential advantage of Type I sensitizers is that they can be effective even in hypoxic environments. It is generally agreed that PDT sensitizers currently approved for clinical use cause damage mainly by Type II reactions.

2.59. *Type II Reaction*

In a Type II reaction the excited photosensitizer molecule transfers energy to molecular oxygen and returns to its ground state. Reactive singlet oxygen then causes the biological effects. In the absence of oxygen Type II reactions are not possible.

2.60. *Vascular Effects*

It has long been known that PDT can cause vasoconstriction and temporary or permanent reduction in blood flow.[20] This could be a desirable outcome if the goal is to destroy the vasculature as in PDT for macular degeneration. Temporary blood flow reduction could be undesirable if it results in hypoxia and protects the target tissue from PDT damage.

2.61. *Zinc Phthalocyanine*

A second generation photosensitizer with strong absorption at about 675 nm and first letter z.

3. Conclusions

The title of this chapter might imply that everything is known about PDT dosimetry but this is not the case. Since the advent of the modern age of PDT in the mid 1970's, great progress has been made in developing the tools and concepts of dosimetry. Among those accomplishments are the identification of singlet oxygen as the initial cytotoxic product of PDT, the concept of a threshold dose, the recognition of PDT-induced hypoxia as a critical limitation, the development of noninvasive techniques for the measurement of tissue optical properties, the testing and implementation of methods for measuring and calculating fluence rate in tissue, and the introduction of absorption and fluorescence spectroscopy for the quantitation of photosensitizers *in vivo*. Researchers are now in a position where the paradigms of explicit, implicit, and direct dosimetry can be tested in pre-clinical animal models and in human trials. The goal of this work in the near future will be to establish the dose-response relationships of malignant and normal tissue and to determine how important accurate dosimetry is for successful PDT.

An analogy could be drawn between the current state of PDT dosimetry and dosimetry for ionizing radiation therapy about forty years ago. In that field, the concept of absorbed dose in tissue had been firmly established and methods for its measurement and rapid calculation had been widely implemented in clinical practice. This led to the understanding that a change in absorbed dose of about 5% could lead to clinically significant changes in toxicity or tumor control. This finding provides a rational basis for the ongoing development of new treatment methods, image-guided radiation therapy, and faster, more accurate algorithms for dose calculation and treatment planning. It is reasonable to suggest that the future of PDT will be more secure and successful if it is built on an equally solid foundation of practical dosimetry.

References

1. P. Agostinis, K. Berg, K. A. Cengel, T. H. Foster, A. W. Girotti, S. O. Gollnick, S. M. Hahn, M. R. Hamblin, A. Juzeniene, D. Kessel, M. Korbelik, J. Moan, P. Mroz, D. Nowis, J. Piette, B. C. Wilson and J. Golab, *CA Cancer J. Clin.* **61**(4), 250–281 (2011). doi: 10.3322/caac.20114.
2. B. C. Wilson and M. S. Patterson, *Phys. Med. Biol.* **53**, R61–R109 (2008).
3. F. Hetzel, M. Patterson, L. Preuss and B. Wilson, Recommended nomenclature for physical quantities in medical applications of light. *American Association of Physicists in Medicine* (American Institute of Physics, Woodbury, N.Y, 1996), Report No. 57.
4. T. C. Zhu, C. Bonnerup, V. C. Colussi, M. L. Dowell, J. C. Finlay, L. Lilge, T. W. Slowey and C. Sibata, *Med. Phys.* **40**(8), 081501 (2013). doi: 10.1118/1.4813897.
5. D. M. Rubel, L. Spelman, D. F. Murrell, J. A. See, D. Hewitt, P. Foley, C. Bosc, D. Kerob, N. Kerrouche, H. C. Wulf and S. Shumack, *Br. J. Dermatol.* (2014). doi:10.1111/bjd.13138.
6. S. R. Wiegell, V. Skødt and H. C. Wulf, *J. Eur. Acad. Dermatol. Venereol* (2013). doi:101111/jdv.12076.
7. T. J. Farrell, M. S. Patterson and B. C. Wilson, *Med. Phys.* **19**, 874–888 (1992).
8. S. L. Jacques, *Photochem. Photobiol.* **67**(1), 23–32 (1998).
9. B. C. Wilson, M. S. Patterson and L. Lilge, *Lasers Med. Sci.* **12**, 182–199 (1997).
10. M. J. Niedre, C. S. Yu, M. S. Patterson, B. C. Wilson, *Br. J. Cancer* **92**, 298–304 (2005).
11. H. J. Laubach, S. K. Chang, S. Lee, I. Rizvi, D. Zurakowski, S. J. Davis, C. R. Taylor and T. Hasan, *J. Biomed. Opt.* **13**(5), 050504 (2008). doi:10.1117/1.2981813.
12. J. Yamamoto, S. Yamamoto, T. Hirano, S. Li, M. Koide, E. Kohno, M. Okada, C. Inenaga, T. Tokuyama, N. Yokota, S. Terakawa and H. Namba, *Clin. Cancer Res.* **12**(23), 7132–7139 (2006).
13. B. C. Wilson, M. S. Patterson, B. Li and M. T. Jarvi, *J. Innov. Opt. Health Sci.* 1540006 (2014).
14. M. J. Niedre, M. S. Patterson, A. Giles and B. C. Wilson, *Photochem. Photobiol.* **81**, 941–943 (2005).
15. J. C. Schlothauer, J. Falckenhayn, T. Perna, S. Hackbarth and B. R?der, *J. Biomed. Opt.* **18**(11), 115001 (2013). doi: 10.1117/1. JBO.18.11.115001.
16. S. Lee, K. Galbally-Kinney, M. F. Hinds, J. A. O'Hara, B. W. Pogue, A. Liang, T. Hasan and S. J. Davis, *Proc. SPIE* **7380**, 738046 (2013). doi:10.1117/12.823030.
17. J. S. Dysart, G. Singh and M. S. Patterson, *Photochem. Photobiol.* **81**, 196–205 (2005).
18. J. S. Dysart and M. S. Patterson, *Phys. Med. Biol.* **50**, 2597–2616 (2005).
19. M. A. Weston and M. S. Patterson, *Photochem. Photobiol.* **87**, 1129–1137 (2011).
20. T. M. Busch, *Lasers Surg. Med.* **38**(5), 494–499 (2006).
21. G. S. Rogers, *J. Natl. Compr. Canc. Netw.* **1**(10), Suppl 2, S14–S17 (2012) .
22. W. J. Cottrell, A. D. Paquette, K. R. Keymel, T. H. Foster and A. R. Oseroff, *Clin. Cancer Res.* **15/14**(14), 4475–4483 (2008). doi: 10.1158/1078-0432.CCR-07-5199.

23. M. S. Patterson, B. C. Wilson and R. Graff, *Photochem. Photobiol.* 51, 343–349 (1990).
24. T. J. Farrell, B. C. Wilson, M. S. Patterson, M. C. Olivo, *Photochem. Photobiol.* **68**, 394–399 (1998).
25. K. K. Wang, S. Mitra and T. H. Foster, *Med. Phys.* **35**(8), 3518–3526 (2006).
26. W. M. Star, A. J. van't Veen, D. J. Robinson, K. Munte, E. R. de Haas and H. J. Sterenborg, *Acta Derm. Venereol.* **86**(5), 412–417 (2006).
27. J. Tyrrell, C. Thorn, A. Shore, S. Campbell and A. Curnow, *Br. J. Dermatol.* **165**(6), 1323–1331 (2011). doi: 10.1111/j.1365-2133.
28. J. S. Dysart and M. S. Patterson, *Photochem. Photobiol. Sci.* **5**, 73–81 (2006).
29. I. Georgakoudi and T. H. Foster, *Photochem. Photobiol.* **67**, 612–625 (1998).
30. J. C. Finlay, S. Mitra, M. S. Patterson and T. H. Foster, *Phys. Med. Biol.* **49**, 4837–4860 (2004).
31. M. A. Weston and M. S. Patterson, *Photochem. Photobiol.* **90**, 878–888 (2014).
32. W. M. Star, H. P. Marijnissen, H. Jansen, M. Keijzer and M. J. van Gemert, *Photochem. Photobiol.* **46**(5), 619–624 (1987).
33. B. W. Pogue and M. S. Patterson, *J. Biomed. Opt.* **11**, 041102 (2006).
34. L. G. Henyey and J. L. Greenstein, *Astrophys. J.* **93**, 70–83 (1941).
35. D. R. Wyman, M. S. Patterson and B. C. Wilson, *Appl. Opt.* **28**, 5243–5249 (1989).
36. W. R. Potter, T. S. Mang and T. J. Dougherty, *Photochem. Photobiol.* **46**(1), 97–101 (1987).
37. B. Liu, T. J. Farrell and M. S. Patterson, *Phys. Med. Biol.* **55**, 5913–5932 (2010).
38. E. F. Gudgin Dickson, R. H. Pottier, *J. Photochem. Photobiol. B*, **29**(1), 91–93 (1995).
39. S. L. Jacques, *Phys. Med. Biol.* **58**(11), R37–R61 (2013). doi: 10.1088/0031-9155/58/11/R37.
40. M. T. Jarvi, M. J. Niedre, M. S. Patterson and B. C. Wilson, *Photochem. Photobiol.* **82**, 1198–1210 (2006).
41. J. Swartling, J. Axelsson, G. Ahlgren, K. M. Kälkner, S. Nilsson, S. Svanberg, K. Svanberg and S. Andersson-Engels, *J. Biomed. Opt.* **15**(5), 058003 (2010). doi:10.1117/1.3495720.
42. T. C. Zhu, J. C. Finlay and S. M. Hahn, *J. Photochem. Photobiol. B* 79, 231–241 (2005).
43. A. Gollmer, F. Besostri, T. Breitenbach and P. R. Ogilby, *Free Radic. Res.* **47**(9), 718–730 (2013). doi: 10.3109/10715762.2013.817670.

Part 2

Chapter 9

PDT: Death and Survival Pathways

David Kessel

Department of Pharmacology, Wayne State University School of Medicine, Detroit MI 48201, USA
dhkessel@med.wayne.edu

Abstract

Photodynamic therapy (PDT) elicits a variety of signals that can lead to both cell death and cytoprotection. The irreversible process of apoptosis is readily initiated by mechanisms involving mitochondrial or lysosomal photodamage. An often competing process is autophagy, a pathway that can lead to recycling of photodamaged organelles before they can initiate apoptosis. There can be crosstalk between these processes with sometimes unpredictable results. The net effect is that a relatively minor degree of photodamage can result in a significant level of cell death. More recent studies point to the possibility that combinations of PDT with chemotherapy, or even involving multiple photosensitizers can promote PDT efficacy.

1. Introduction: What is PDT?

Photodynamic therapy (PDT) is the name given to a form of cancer treatment that requires a photosensitizing (PS) agent, oxygen and light. The PS will absorb light, then convert this energy into both fluorescence and an atomic rearrangement that can pass energy along to oxygen molecules. The latter are converted to reactive oxygen species (ROS) that rapidly interact with any nearby biologic structures, where the PS has become localized. This effect can lead to the initiation of cell death pathways, depending on the target(s) of photodamage. There have been several reviews on this topic; among the more comprehensive is one from the late 1990s.[1] A recent update summarizes progress during the past 10 years.[2]

A unique feature of PDT is that nothing happens until the irradiation of photosensitized cells and tissues, assuming no dark toxicity of the PS being used. This permits a highly-selective form of therapy with the direction of light being an additional factor in selectivity. Treatment of malignant disease relies on the localization of PS in the tumor and/or tumor vasculature. Because of the requirement for light, protocols can involve waiting after drug administration for an optimal interval before irradiation. This permits optimizing the ratio of PS in neoplastic versus normal tissues. Where vascular shut-down is the primary goal, this interval may be quite short.

Determinants of tumor localization by PS molecules are not yet fully known. There is some evidence that lipoprotein binding is involved,[3] but this does not provide a comprehensive explanation of localization patterns. An important element of PDT is the concomitant shutdown of the tumor blood supply.[4] Determinants of this process have yet to be fully explained, but this effect clearly plays a role in cancer control. PS with proven efficacy tend to localize in mitochondria or lysosomes, sites especially sensitive to photodamage that has lethal consequences.[5]

2. PDT and ROS

In the context of PDT, direct photokilling arises from the production of ROS during the irradiation of photosensitized cells. The predominant species is singlet oxygen ($^1O_2$), although hydroxyl radical ($^{\bullet}OH$) can also be formed and is the predominant species generated by some photosensitizers, e.g., the palladium bacteriopheophorbide termed Tookad.[6] Superoxide radical ($^{\bullet}O_2^-$) can also be formed. An interaction between $^{\bullet}O_2^-$ and the enzyme superoxide dismutase leads to the formation of hydrogen peroxide. There is also evidence for formation of some reactive nitrogen species.[7] When assessing the different ROS formed during irradiation of photosensitized cells, it is necessary to do this during irradiation because the highly-reactive species have a very short half-life and will not be detected after irradiation has ceased. The ROS detected after irradiation are usually formed by dysfunctional mitochondria, an effect often observed in dying cells. Moreover, the responsiveness of various fluorescent probes to ROS varies substantially.[8] For example, the often-used probe DCFDA is relatively insensitive to $^1O_2$.[9]

Involvement of ROS as the proximate cause of phototoxicity has been appreciated for many years.[10] The determinants of ROS formation are complex. A distracting factor is the common use of cell-culture experimental conditions where the oxygen level is in the vicinity of 20%, i.e., the composition of the atmosphere. Malignant tissues generally show an oxygenation level of 5% or below.[11,12] Under conditions of low oxygenation, PDT efficacy will decrease, but this appears to depend on the PS chosen. For example, we have reported that, as the oxygen level was decreased to near zero, •OH formation derived from irradiation of WST11 was diminished, while $^1O_2$ formation from NPe6 remained.[13] These results suggest that the photodynamic efficacy of NPe6 will be minimally affected by hypoxic conditions.

3. Death Pathways

The major determinants of success include sites of photosensitizer localization, the sensitizer concentration, the light dose delivered and the cell phenotype. A major cellular death pathway involves apoptosis, an irreversible process. While apoptosis can be initiated by a variety of stimuli, photokilling appears to rely mainly on damage to anti-apoptotic proteins or to lysosomes. The intrinsic pathway to apoptosis can therefore be triggered by a relatively minor degree of photodamage that ultimately results in release of cytochrome c from mitochondria. Pertinent mechanisms are discussed below.

As death pathways are initiated, a variety of other signaling will occur, leading to a variety of downstream responses.[14–16] Inhibition of these responses may be feasible, but once begun, apoptosis is an irreversible pathway to death. Lockshin has commented on cells being able to choose any available death pathway once the process is initiated.[17] A possible mitigating factor is the ability of autophagy to protect cells from apoptotic death, e.g., by sequestering photodamaged mitochondria before they can trigger apoptosis via release of cytochrome c. A report on the role of autophagy in the context of PDT has been published.[18]

4. The Role of Localization Patterns

The delineation of death pathways elicited by PDT has been facilitated by the availability of PS agents with well-defined localization patterns. It is not very helpful to be examining effects of PS that target every intracellular membrane and organelle. Of the agents in current use, the more useful have included the chlorin NPe6 — targets lysosomes,[19] the benzoporphyrin derivative BPD — targets mitochondria,[20] m-tetrahydroxyphenylchlorin (mTHPC) — targets mitochondria, ER or both.[21,22] The phthalocyanine Pc 4 has a broad spectrum of affinity although mitochondria are reported to be a primary target.[23] Other PS can have multiple targets.[5,24] The protoporphyrin produced by treatment with aminolevulinic acid (ALA) is initially localized in mitochondria, but may then diffuse to other cellular regions.[25]

Affinity of photosensitizers for different sub-cellular loci appears to be a factor in PDT efficacy. Localization is a function of both the time of exposure of cells to a photosensitizer as well as hydrophobicity and charge distribution.[26–28] Since singlet oxygen does not migrate far from its site of formation,[29] the initial pattern of sensitizer localization is an important factor in the outcome of PDT. Localization patterns can be determined by comparing patterns of photosensitizer fluorescence with results obtained using fluorescent probes with well-characterized sites of localization.[30,31] Another useful procedure involves use of fluorescent probes for detecting sites of photodamage.[28,31] Alterations in cellular morphology caused by the initiation of apoptosis can be prevented by maintaining the temperature of irradiation and subsequent operations below 15°C.[32] This will prevent apoptosis so that direct consequences of photodamage can be assessed, avoiding secondary effects initiated by subsequent processes.

5. Apoptosis versus Autophagy

It was initially believed that photodamage was primarily directed against the plasma membrane and its assorted functions.[33] Irradiation of cells immersed in solutions of photosensitizers will often yield results consistent with this interpretation. Later work demonstrated that, when the plasma membrane was the primary target, membrane-bound sensitizers migrated into the cytoplasm during irradiation and catalyzed photodamage to procaspases, thereby inhibiting the apoptotic response.[34] Moreover, the high efficacy of PDT, where substantial photokilling results from a low light dose, was not consistent with the much higher light doses needed to kill cells when the plasma membrane is the primary target. Later evidence revealed that one of the critical targets for photodamage is the Bcl-2 family of anti-apoptotic proteins which is located in mitochondria and the ER.[35–37] Oleinick's group reported that while Bcl-2 photodamage was associated with photokilling,[37] Bcl-2 mutants that were not membrane bound were unaffected by PDT.[38] Since photosensitizers used for PDT are relatively hydrophobic and associate with sub-cellular membranes, the targeting of membrane-bound proteins is not surprising.

It is important to remember that apoptotic death can be triggered by a relatively low level of photodamage at pertinent sites. The substantial magnification of the photodynamic effect by the triggering of apoptosis likely accounts for the ability of PDT to bring about a degree of tumor eradication that would require a much higher energy dose where cell destruction is required, e.g., using a carbondioxide laser. All that is required to activate the 'intrinsic' pathway to apoptosis is release of the protein cytochrome c from mitochondria.[39]

Since cytochrome c translocation from mitochondria to the cytoplasm is an initiator of apoptosis, it is not unexpected that the mitochondrion would be a sensitive PDT target.[40–42] There can also be a migration of other pro-apoptotic factors from photodamaged mitochondria into the cytoplasm, e.g., Smac/Diablo and AIF, additional initiators of the apoptotic program.[37,38,43] Although the porphycene termed CPO[44] initially localizes in the ER,[30] photodamage also leads to an apoptotic response associated with Bcl-2 photodamage[36] This effect is likely derived from 'cross-talk' pathways that can evoke a pro-apoptotic response in mitochondria from signals that originate in the ER.[45,46] Photodamage to lysosomes results in a somewhat circuitous apoptotic cell death pathway beginning with release of lysosomal proteases into the cytoplasm.[47] These enzymes then cleave the inactive form of the pro-apoptotic protein Bid to a truncated form (t-Bid) that interacts with the mitochondrial membrane. This leads to the opening of pores and release of cytochrome c.[48] While this may be an indirect procedure for initiating apoptosis, a study involving two phthalocyanine photosensitizers revealed that photodamage to lysosomes was more efficacious than when mitochondria were the target.[49]

6. Other Targets

There may be additional sensitive PDT targets that have yet to be identified and characterized. One example is the reported impairment of the endocytic pathway responsible for lipoprotein uptake.[50] In this regard, it is interesting to note that one hypothesis for the preferential accumulation of photosensitizers by neoplastic tissues is based on indications that many show affinity for plasma lipoproteins.[51] Low-density lipoprotein (LDL) receptors tend to be over-expressed in neoplastic tissues.[52] This does not appear to uniformly apply since, e.g., NPe6 tends to bind to HDL.[53]

An example of the roles of autophagy and apoptosis in PDT-induced cell death is shown in Fig. 1. The murine L1210 leukemia was employed since, in this cell line, these processes occur promptly after photodamage. Panel A shows untreated cells. After an LD_{90} PDT dose, autophagic vacuoles appear within 15 min (Panel B). This likely represents a survival response, but apoptotic morphology is visible 60 min after irradiation (Panel C). This becomes more predominant 4 h later (Panel D). After 24 h, only a few cells remain, many with autophagic vacuoles (Panel E).

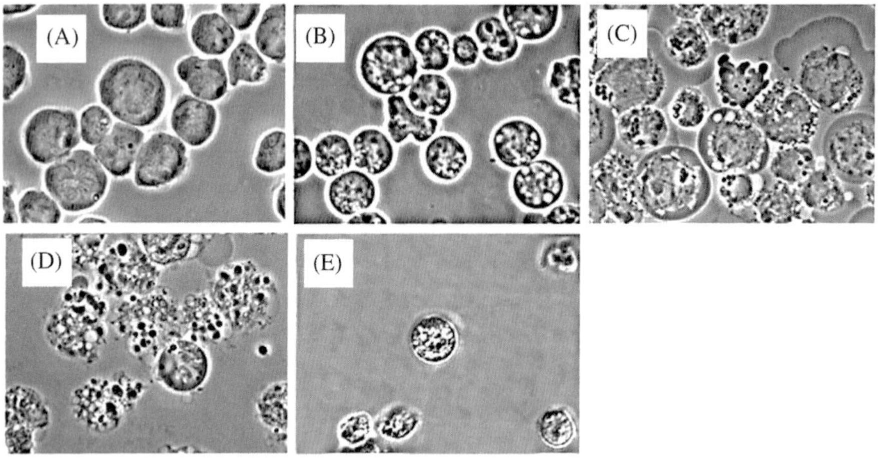

Fig. 1. Effects of an LD_{90} PDT dose using the porphycene termed CPO on murine leukemia L1210 cells. A. Controls; B, 15 min after PDT showing early autophagic vacuoles; C, 60 min later showing beginning of an apoptotic morphology; D, more significant numbers of apoptotic cells 4 h later; E, surviving cells 24 h later.

The rapidity with which pro-apoptotic signals can be generated after PDT is shown in Fig. 2, again involving L1210 cells. Fragmentation of nuclei as a result of apoptosis is visible 60 min after irradiation, as indicated by fluorescence labeling patterns with HO33342 (Panel A). Genomic DNA damage can be seen as early as 10 min after irradiation with pulse-field electrophoresis (Panel B). The DNA 'ladders' characteristic of apoptosis appear within 60 min (Panel C).

With regard to autophagy, this is usually associated with promotion of cell and tissue survival. A portion of the intracellular space containing proteins, organelles and other biological products is engulfed by a three-dimensional double membrane termed the autophagosome. This structure then fuses with lysosomes, the contents are digested by lysosomal enzymes and the resulting mixture of amino acids, lipids and other basic components is released back into the cytoplasm for re-utilization. The net effect is the recycling of biologic molecules that can alleviate starvation effects.[54] Under some conditions, autophagy has been reported to lead to cell death, but this has been disputed.[55,56] It has been proposed that a more accurate description of the process is 'death with autophagy'.[57] In the context of PDT, it is at this point safe to conclude that the outcome of PDT will depend on a collection of variables that have yet to be fully delineated. As noted above, the general response is formation of autophagic vacuoles, followed by an apoptotic morphology.[58] The role of autophagy in efficacy of cancer treatment has also received considerable

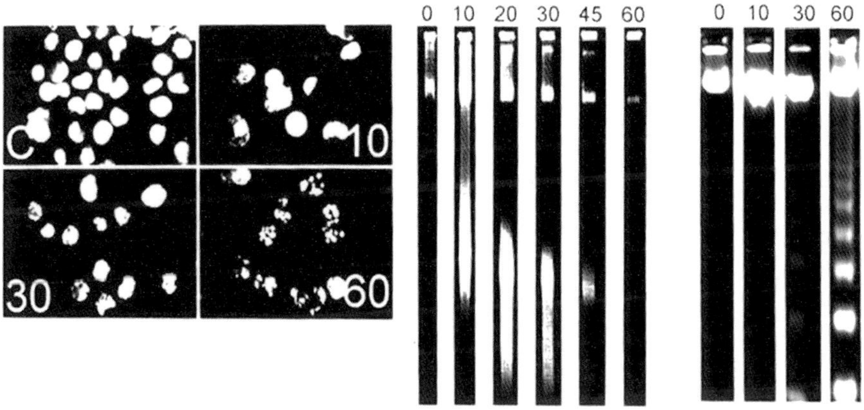

Fig. 2. Appearance of apoptosis in L1210 cells after an LD_{90} PDT dose using the tin etiopurpurin termed SnET2. A, appearance of apoptotic morphology after 60 min as visualized by the fluorescent nuclear label HO33342; B, fragmentation of genomic DNA within 10 min detected by pulse-field electrophoresis; C, appearance of DNA 'ladders' indicating endonuclear cleavage within 60 min.

attention. Survival of cancer 'stem cells' has been associated with enhanced autophagy.[59]

7. New Roles for Autophagy

Another series of studies with L1210 cells demonstrates the 'cross-talk' between apoptosis and autophagy (Fig. 3). Treatment with a Bcl-2 antagonist led to appearance of a few apoptotic nuclei 60 min after irradiation (panel B). Inhibition of autophagy results in increased apoptosis (C) while inhibition of apoptosis resulted in an enhanced number of autophagic vacuoles (D).

Cell lines lacking the capacity for autophagy can be prepared by deleting or knocking-down a critical autophagy gene. Such cells lose the 'shoulder' on the PDT dose-response curve.[60] This is consistent with a 'survival-promoting' effect of autophagy. A similar finding has been made during exposure of cells to ionizing radiation.[61] There are reports that autophagy can recycle damaged mitochondria.[62,63] In the context of PDT, this would prevent the release of cytochrome c that results in apoptosis. An example of this effect is seen in Fig. 4 panel A. Murine hematoma 1c1c7 cells were photosensitized with BPD (targeting predominantly mitochondria) and irradiated with varying light doses to produce a dose-response curve. Effects are shown with wild-type (WT) cells and a subline lacking the ATG7 or ATG5 genes, thereby preventing an autophagic response. In either case, the shoulder on the dose-response curve

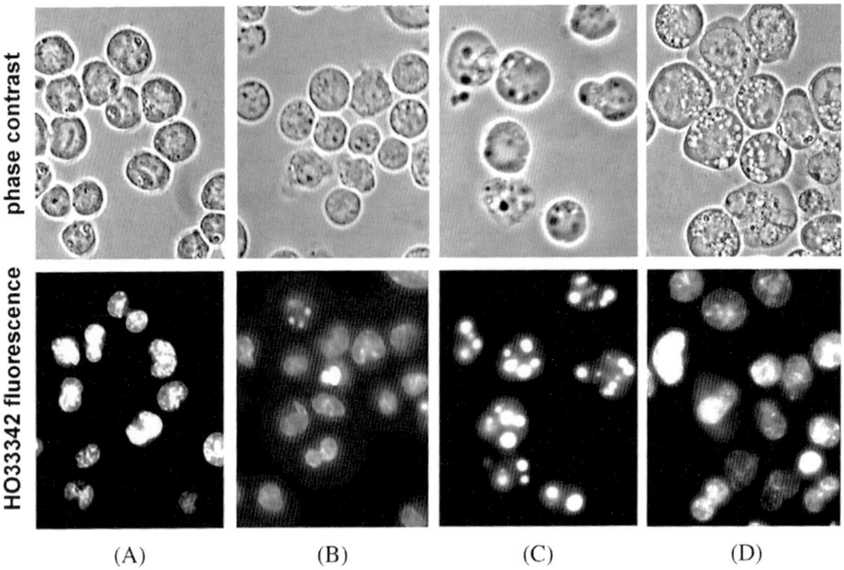

Fig. 3. Interdependence of apoptosis and autophagy. L1210 cells were exposed to a Bcl-2 antagonist and appearance of apoptotic nuclei and autophagic vacuoles followed by microscopy. A, control cells; B, 60 min after treatment with the Bcl-2 antagonist ABT-737; C, promotion of apoptosis by the autophagy inhibitor wortmannin; D, promotion of autophagy by the caspase antagonist zDEVD-fmk.

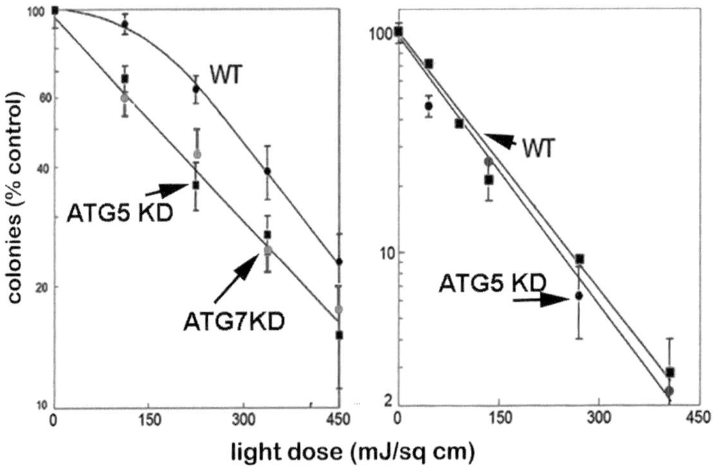

Fig. 4. Left panel: dose-response curve of WT murine hepatoma 1c1c7 cells photosensitized with BPD and treated with varying light doses. The 'shoulder' on the dose-response curve is abolished when ATG5 or ATG7 knockdowns are similarly treated. Right panel, results of a similar study using NPe6 as the PS agent. See text for further details. Right panel, effects of ATG5 KD on PDT efficacy of NPe6.

was abolished. This is consistent with the proposal that autophagy can act to suppress low-level of photodamage that would otherwise lead to cell death.

Panel B shows a similar study carried out with the photosensitizer NPe6 where the targets are the lysosomes. Since lysosomes are an important element of the autophagy program, photodamage produced by NPe6 is going to prevent an autophagic response. As a result, there was no shoulder on the dose-response curve and silencing of an autophagy gene (ATG5) had no effect (Fig. 4 panel B).

Autophagy may also represent an important part of the overall PDT response relating to immunologic effects. The recycling process can result in class II presentation of antigens derived from cytosolic proteins.[64] The autophagic response to PDT may also provide an explanation for the finding that treatment of tumor cells with low-dose PDT can yield anti-tumor vaccines. Oleinick's group recently reported that cells unable to undergo apoptosis tend to die with an elevated level of autophagy and that Bcl-2 was unable to protect cells from this autophagic response.[66]

A recent review in Cancer Research did, however, point out that inhibiting autophagy, so as to prevent the potentially cytoprotective role of the process, may not necessarily promote the efficacy of chemotherapy.[67] Four potential effects of autophagy were identified: cytoprotective, cytotoxic, cytostatic and non-protective. In the context of PDT, it appears that autophagy can be cytoprotective, but the ATG5 and ATG7 effects outlined above indicate that lack of autophagy-related genes can also impair PDT responses.

8. Tweaking the System: Combinations

Hasan's group has proposed new strategies for the promotion of PDT efficacy (94). It is often feasible to target sensitizers to specific proteins on tumor cell surfaces via antibody-antigen interactions.[68,69] The extension of this work to the field of nanotechnology has also been reported.[70] Another possibility is the use of combinations involving PDT + conventional chemotherapy. We have shown that concurrent treatment with PDT and Bcl-2 antagonists resulted in enhanced photokilling.[71] Hydrogen peroxide formed during PDT may also play a role in photokilling.[72] Hydrogen peroxide has also been implicated as a factor in the initiation of autophagy.

We observed (see Fig. 4A) that loss of the autophagy-related protein ATG7 promoted photokilling when the principal target was mitochondria.[60] A different result was obtained with lysosomal targets: PDT efficacy was significantly impaired;[74] two photosensitizers (NPe6 and the palladium derivative termed WST11) directed against lysosomes were found to be ineffective (Fig. 5). This

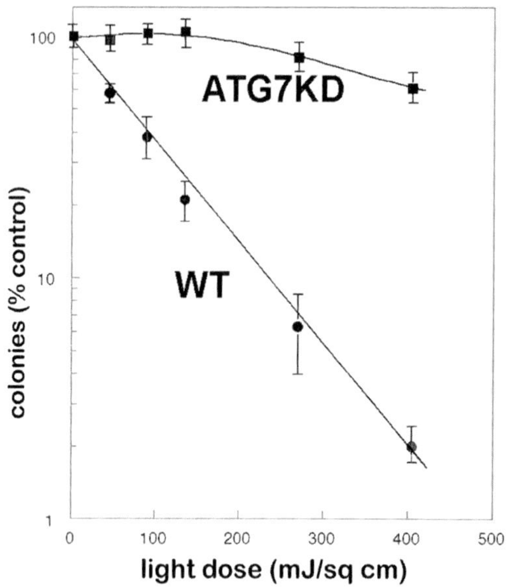

Fig. 5. PDT dose-response curve of 1c1c7 cells and an ATG7 KD to two lysosomal photosensitizers: NPe6 and WST11.

was traced to a failure of lysosomes to release the proteases needed to initiate the apoptotic process as described in Ref. 47. Initiation of apoptosis after lysosomal photodamage derives from the release of proteases that cleave cytoplasmic Bid to a pro-apoptotic molecule termed t-Bid. After ATG7 depletion, lysosomal photodamage resulted in a transient loss of the permeability gradient responsible for retention of fluorescent probes, e.g., the LysoTrackers. But loss of larger molecules, e.g., fluorescent dextran particles, was prevented. It was concluded that the presence of ATG7 was necessary for lysosomal photodamage to lead to opening of large pores necessary for protease release.

An additional indication of the role of ATGs in PDT death pathways was indicated by studies when low-dose lysosomal photodamage preceded mitochondrial photodamage.[75] This was feasible since photosensitizers that target lysosomes (NPe6) and mitochondria (BPD) have sufficiently different excitation spectra so that they can be independently activated. Using the murine hepatoma 1c1c7 cells, it was observed that under experimental conditions employed in this work, a minor degree of lysosomal photokilling (~5%) could significantly promote the efficacy of a subsequent PDT dose directed at mitochondria. A possible explanation for this result was obtained when an ATG5 knockdown was tested and no such potentiation was observed (Fig. 6). The increased efficacy of this PDT combination may be explained by a study showing that ATG5 can be cleaved by the protease calpain. This yields an

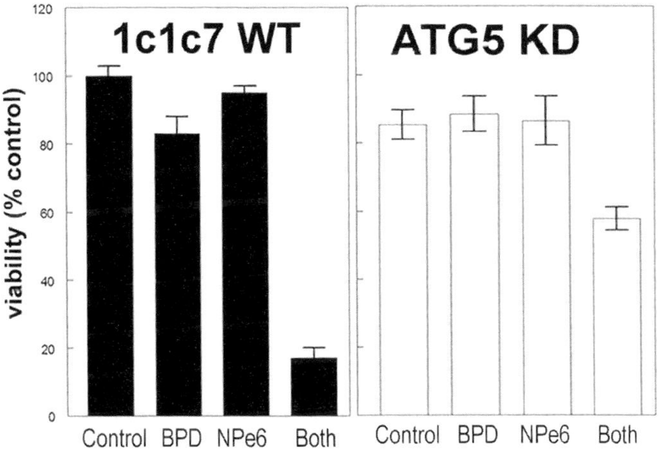

Fig. 6. Clonogenic assays of 1c1c7 murine hepatoma cells and an ATG5 knockdown. From left: controls, low-dose BPD, low-dose NPe6, the combination (NPe6 activated first).

ATG5 fragment that interacts with mitochondria to promote pro-apoptotic signals.[76] Consistent with this mechanism, we found that treatment with the calpain antagonist E64d, along with the calcium chelation agent BAPTA-AM prevented the ability of the PDT combination to synergistically elicit an apoptotic response (not shown).

The discovery that low-dose lysosomal photodamage can promote any pro-apoptotic signal that activates the intrinsic pathway, i.e., involving activation of caspase 3/7, indicates potential usefulness with any PS that target mitochondria, e.g., Photofrin, protoporphyin induced by ALA administration, and assorted other Photofrin analogs. Moreover, the procedure might be used to promote response to conventional anti-tumor agents that initiate apoptosis, e.g., doxorubicin. It is unknown whether other pro-apoptotic pathways can be amplified. Loss of lysosomal permeability barriers has also been associated with amplification of cell death signals.[77] Other avenues of crosstalk between lysosomes and mitochondria have been proposed.[78]

It has been noted that cancer 'stem cells' tend to be less responsive to PDT than other malignant cell types; this has been attributed to an overexpression of genes involved in autophagy.[59] Proof of this observation was, in part, obtained by promoting lysosomal permeability and thereby decreasing the ability of these organelles to digest cell components during the autophagy cycle. It is not clear whether this result was derived from inhibition of autophagy or promotion of ferrous iron uptake by mitochondria, thereby promoting ROS formation.[79]

9. Modeling: 2D versus 3D and Beyond

While many pertinent death mechanisms have been identified using 2D cell culture systems (cells growing in monolayers on glass slides), investigators are now turning to 3D culture systems that more closely approximate the tumor microenvironment.[80] It is feasible to incorporate endothelial cells, cancer-associated fibroblasts and other components of more complex biological systems. Many studies have also been carried out in animal models. Orthotopic tumor implants are often used to better approximate the clinical situation.

Results from experiments carried out with 2D culture conditions are not necessarily replicated in 3D cultures. The latter are believed to more closely resemble the situation that prevails in vivo where there is significant tissue thickness and presence of other cell types may offer protection from PDT photokilling.[81] When the photosensitizer BPD was shown to inhibit autophagy in the dark, we carried out a study in 2D culture and demonstrated that photokilling was enhanced when the BPD level was elevated and the light-dose correspondingly reduced. The dose-response curves were consistent with loss of the cytoprotective effects of autophagy, i.e., the shoulder on the curve was lost.[60] When high-dose BPD was tested in 3D culture, PDT efficacy was greatly reduced.[81] It appears likely that the high extinction coefficient of BPD prevented light from penetrating deep into the 3D structure. The 3D culture systems are also proving useful in the delineation of increased efficacy when PDT + conventional chemotherapy are employed.[82]

The question of which cell lines are appropriate for PDT studies has prompted some attention. Many of the early studies that explored PDT mechanisms used murine leukemias, a cell type that is not an obvious PDT target. But the essentials of apoptosis, autophagy, signaling pathways, effects of ROS formation and other details were discovered. More complex models have been developed with a view toward better assessing selectivity of PDT and potential pitfalls. These include orthotopic animal implants, use of tumors of human origin in nude mice and similar attempts to more closely approximate the clinical situation. These will necessarily be much more expensive models than simple cell-culture techniques, but can provide information not accessible in simpler models.

Even the more realistic 3D systems lack important elements of clinical PDT: a vascular system, immune responses and lymphatics. Determinants of selective tumor localization remain difficult to identify. Many of the pertinent elements of PDT have, however, been successfully deduced from studies in cell culture. Moreover, studies involving small animals, e.g., mice, do not always predict for successful clinical outcomes. It appears that continued progress in PDT will involve a broad approach: synthesis of new and more selective agents, perhaps taking advantage of nanotechnology to promote targeting, identification of cell death mechanisms and utilization of combinations that may significantly promote PDT efficacy.

References

1. T. Dougherty, C. Gomer, B. Henderson, G. Jori, D. Kessel, M. Korbelik, J. Moan and Q. Peng, *J. Natl. Cancer Inst.* **90**, 889 (1998).
2. P. Agostinis, K. Berg, K. Cengel, Y. Foster, A. Girotti, S. Gollnick, S. Hahn, M. Hamblin, A. Juzeniene, D. Kessel, M. Korbelik, J. Moan P. Mroz, D. Nowis, J. Piette, B. Wilson and J. Golab, *CA Cancer J. Clin.* **61**, 250 (2011).
3. G. Jori and E. Reddi, *Int. J. Biochem.* **25**, 1369 (1983).
4. V. Fingar, *J. Clin. Laser. Med. Surg.* **14**, 323 (1996).
5. D. Kessel, Y. Luo, Y. Deng and C. Chang. *Photochem. Photobiol.* **65**, 422 (1996).
6. I. Ashur, R. Goldschmidt, I. Pinkas, Y. Salomon, G. Szewczyk, T. Sarna and A. Scherz, *J. Phys. Chem. A* **113**, 8027 (2009).
7. D. Kessel and J. Reiners, *Isr. J. Chem.* **52**, 674 (2012).
8. M. Price and D. Kessel, *J. Biomed. Opt.* **15**, 051605 (2010).
9. K. Setsukinai, Y. Urano, K. Kakinuma, H. Majima and T. Nagano, *J. Biol. Chem.* **278**, 3170 (2003).
10. K. Weishaupt, C. Gomer and T. Dougherty. *Cancer Res.* **36**, 2326 (1976).
11. Z. Yun and Q. Lin. *Adv. Exp. Med. Biol.* **772**, 41 (2014).
12. G. Semenza, *Ann. Rev. Pathol.* **9**, 47 (2014).
13. M. Price, L. Heilbrun and D. Kessel, *Photochem. Photobiol.* **89**, 683 (2013).
14. L. Xue, J. He and N. Oleinick, *Cell Death Differ.* **6**, 855 (1999).
15. E. Buytaert, J. Matroule, S. Durinck, P. Close, S. Kocanova, J. Vandenheede, P. de Witte, J. Piette and P. Agostinis, *Oncogene.* **27**, 1916 (2008).
16. M. Dewaele, T. Verfaillie, W. Martinet and P. Agostinis, *Meth. Mol. Biol.* **635**, 7 (2010).
17. Z. Zakeri and R. Lockshin, *Adv. Exp. Med. Biol.* **615**, 1 (2008).
18. J. Reiners, P. Agostinis, K. Berg, N. Oleinick and D. Kessel, *Autophagy* **6**, 7 (2010).
19. W. Roberts, L. Liaw and M. Berns, *Lasers Surg. Med.* **9**, 102 (1989).
20. T. Peng, C. Chang, M. Guo, Y. Wang, J. Yu, H. Wu and M. Jou, *Ann. NY Acad. Sci.* **042**, 419 (2005).
21. M. Teiten, L. Bezdetnaya, J. Merlin, C. Bour-Dill, M. Pauly, M. Dicato and F. Guillemin, *Photochem. Photobiol. B* **62**, 146 (2001).
22. J. Chen, N. Mak, C. Yow, M. Fung, L. Chiu, W. Leung and N. Cheung, *Photochem. Photobiol.* **72**, 541 (2000).
23. N. Trivedi, H. Wang, A. Nieminen, N. Oleinick and J. Izatt, *Photochem. Photobiol.* **71**, 634 (2000).
24. D. Kessel, *J. Porph. Phthalo.* **8**, 1009 (2004).
25. R. Chen, Z. Huang, G. Chen, Y. Li, X. Chen, J. Chen and H. Zeng, *Int. J. Oncol.* **32**, 861 (2008).
26. T. Jensen, M. Vicente, R. Luguya, J. Norton, F. Fronczek and K. M. Smith, *J. Photochem. Photobiol. B* **100**, 100 (2010).

27. M. Dellinger, *Photochem. Photobiol.* **64**, 82 (1996).
28. D. Kessel and R. Poretz, *Photochem. Photobiol.* **71**, 94 (2000).
29. J. Moan and K. Berg, *Photochem. Photobiol.* **53**, 549 (1991).
30. D. Kessel, M. Conley, M. Vicente and J. Reiners, *Photochem. Photobiol.* **81**, 569 (2005).
31. D. Kessel and Y. Luo, *J. Porphy. Phthalo.* **5**, 181 (2001).
32. J. Pryde, A. Walker, A. Rossi, S. Hannah and C. Haslett, *J. Biol. Chem.* **275**, 33574 (2000).
33. D. Kessel, *Biochemistry* **16**, 3443 (1977).
34. D. Kessel, *Photochem. Photobiol. Sci.* **1**, 837 (2002).
35. H. Kim, Y. Luo, G. Li and D. Kessel, *Cancer Res.* **59**, 3429 (1999).
36. D. Kessel and M. Castelli, *Photochem. Photobiol.* **74**, 318 (2001).
37. L.-Y. Xue, S.-M. Chiu and N. Oleinick, *Oncogene* **20**, 3420 (2001).
38. J. Usuda, S. Chiu, E. Murphy, M. Lam, A. Nieminen and N. Oleinick, *J. Biol. Chem.* **278**, 2021 (2003).
39. P. Li, D. Nijhawan, I. Budihardjo, S. Srinivasula, M. Ahmad, E. Alnemri and X. Wang, *Cell* **91**, 479 (1997).
40. N. Oleinick, R. Morris and I. Belichenko, *Photochem. Photobiol. Sci.* **1**, 1 (2002).
41. J. Morgan and A. Oseroff, *Adv. Drug Deliv. Rev.* **49**, 71 (2001).
42. D. Kessel and Y. Luo, *Cell Death Differ.* **6**, 28 (1999).
43. M. Lam, N. Oleinick and A. Nieminen, *J. Biol. Chem.* **27**, 47379 (2001).
44. H. Toledano, R. Edrei and S. Kimel, *J. Photochem. Photobiol. B* **42**, 20 (1998).
45. V. Lamarca and L. Scorrano, *EMBO J.* **28**, 1681 (2009).
46. J. Su, L. Zhou, M. Xia, Y. Xu, X. Xiang and L. Sun, *Biomed. Res. Int.* **2014**, 234370 (2014).
47. J. Reiners, J. Caruso, P. Mathieu, B. Chelladurai, X. Yin and D. Kessel, *Cell Death Differ.* **9**, 934 (2002).
48. V. Stoka, B. Turk, S. L. Schendel, T. H. Kim, T. Cirman, S. J. Snipas, L. M. Ellerby, D. Bredesen, H. Freeze, M. Abrahamson, D. Bromme, S. Krajewski, J. C. Reed, X. M. Yin, V. Turk and G. Salvesen, *J. Biol. Chem.* **276**, 3149 (2001).
49. M. Rodriguez, B. Turk, S. L. Schendel, T. H. Kim, T. Cirman, S. J. Snipas, L. M. Ellerby, D. Bredesen, H. Freeze, M. Abrahamson, D. Bromme, S. Krajewski, J. Reed, X. Yin, V. Turk and G. Salvesen, *Photochem. Photobiol.* **85**, 1189 (2009).
50. G. Ouédraogo, P. Morlière, C. Mazière, J. Mazière and R. Santus, *Photochem. Photobiol.* **72**, 458 (2000).
51. G. Jori and E. Reddi, *Int. J. Biochem.* **25**, 1369 (1993).
52. C. Peterson, S. Vitols, M. Rudling, H. Blomgren, F. Edsmyr and L. Skoog, *Med. Oncol. Tumor Pharmacother.* **2**, 143 (1985).
53. D. Kessel, *Photochem. Photobiol.* **61**, 646 (1995).
54. Z. Yang and D. Klionsky, *Curr. Opin. Cell Biol.* **22**, 124 (2010).
55. S. Shen, O. Kepp and G. Kroemer, *Autophagy* **8**, 1 (2012).
56. B. Levine and G. Kroemer, *Cell. Death Differ.* **16**, 1 (2009).
57. G. Kroemer and B. Levine, *Nat. Rev. Mol. Cell. Biol.* **9**, 1004 (2008).
58. D. Kessel and N. Oleinick, *Meth. Mol. Biol.* **635**, 35 (2010).
59. M. Wei, M. Chen, K. Chen, P. Lou, S. Lin, S. Hung, M. Hsiao, C. Yao and M. Shieh, *Autophagy* **10**, 1179 (2014).
60. M. Andrzejak, M. Price and D. H. Kessel, *Autophagy* **7**, 979–984 (2001).

61. B. Liang, D. Kong, Y. Liu, N. Liang, M. He, S. Ma and X. Liu, *Radiat Oncol.* 7, 213 (2012).
62. A. Terman, H. Dalen, J. Eaton, J. Neuzil and U. Brunk, *Exp. Gerontol.* **38**, 863 (2003).
63. H. Lu, G. Li, L. Liu, L. Feng, X. Wang and H. Jin, *Autophagy* **9**, 1720 (2013).
64. V. Crotzer and J. Blum, *Immunology* **131**, 9 (2010).
65. M. Korbelik, *Photochem. Photobiol. Sci.* **10**, 664 (2011).
66. L. Xue, S. Chiu, K. Azizuddin, S. Joseph and N. Oleinick, *Autophagy* **4**, 125 (2008).
67. D. Gewirtz, *Cancer Res.* **74**, 647 (2014).
68. B. Goff, J. Blake, M. Bamberg and T. Hasan, *Br. J. Cancer* **74**, 1194 (1996).
69. R. Rahmanzadeh, P. Rai, J. P. Celli, I. RizviI, B. Baron-Lühr, J. Gerdes and T. Hasan, *Cancer Res.* **70**, 9234 (2010).
70. Y. Mir, S. Elrington and T. Hasan, *Nanomedicine* **9**, 1114 (2013).
71. D. Kessel, M. Castelli and J. Reiners, *Photochem. Photobiol.* **76**, 314 (2002).
72. M. Price, S. Terlecky and D. Kessel, *Photochem. Photobiol.* **85**, 491 (2009).
73. R. Scherz-Shouval, E. Shvets, E. Fass, H. Shorer, L. Gil and Z. Elazar, *EMBO J.* **26**, 1749 (2007).
74. D. Kessel, M. Price and J. Reiners, *Autophagy* **8**, 1333 (2012).
75. D. Kessel and J. Reiners, *Photochem. Photobiol.* **90**, 889 (1994).
76. S. Yousefi, R. Perozzo, I. Schmid, A. Ziemiecki, T. Schaffner, L. Scapozza, T. Brunner and H. U. Simon, *Nat. Cell Biol.* **8**, 1124 (2006).
77. U. Repnik, M. Hafner Cesen and B. Turk, *Mitochondrion* **19**, 49 (2014).
78. B. Turk and U. Repnik, *Mitochondrion* **10**, 662 (2010).
79. S. Saggu, H. Hung, G. Quiogue, J. Lemasters and A. Nieminen, *Photochem. Photobiol.* **88**, 461 (2012).
80. J. Rothberg, M. Sameni, K. Moin and B. Sloane, *Biochim. Biophys. Acta.* **1824**, 123 (2012).
81. I. Rizvi, S. Anbil, N. Alagic, J. Celli, L. Zheng, A. Palanisami, M. Glidden, B. Pogue and T. Hasan, *Photochem. Photobiol.* **89**, 942 (2013).
82. I. Rizvi, J. P. Celli, C. L. Evans, A. O. Abu-Yousif, A. Muzikansky, B. Pogue, D. Finkelstein and T. Hasan, *Cancer Res.* **70**, 9319 (2010).

Chapter 10

Vascular Effects of Photodynamic Therapy for Tumors

*Keith A. Cengel, Charles B. Simone and Theresa M. Busch**

Department of Radiation Oncology, Perelman School of Medicine
University of Pennsylvania, Philadelphia, PA 19104, USA
buschtm@mail.med.upenn.edu

Abstract

The localized damage of tumor vasculature contributes to the therapeutic effect of photodynamic therapy (PDT) in many of its oncologic uses. As a function of the treatment protocol, vascular effects can manifest as vasoconstriction, vasodilation, hemostasis, and/or hyperpermeability, among other hemodynamic changes. The extent and nature of vascular damage will depend on treatment-specific factors, such as the photosensitizer used (targeted or untargeted, pharmacokinetics and dosing), drug-light interval, and illumination fluence and fluence rate. Moreover, microenvironmental factors specific to the pre-existing structure (e.g., stromal composition) and function (e.g., flow velocity) of tumor blood vessels may determine their sensitivity to PDT. PDT-created vascular damage is integral to the inflammatory response that can accompany it, and may also play a role in the development of anti-tumor immunity. Simultaneous with vascular and immune effects, survival signaling can be activated by PDT, potentially detracting from tumor response. Indeed, combination therapies that include molecular targeting of PDT-upregulated pathways, such as those involved in angiogenesis, have proven beneficial to long-term tumor response. The advantages and limitations of PDT-triggered vascular damage are also evident in its clinical application. Looking ahead, the incorporation of hemodynamic monitoring in clinical trials of PDT may provide a means for individualized optimization of treatment.

*Corresponding author.

1. Introduction

The aberrant vascularization that is characteristic of tumor-associated angiogenesis is one of the major hallmarks of cancer.[1] Many approaches have been developed and continue to be sought for the eradication of these blood vessels toward the treatment of solid malignancies. Photodynamic therapy (PDT) can produce localized damage of tumor blood vessels as one component of its multiple mechanisms of cytotoxicity.[2] Many factors contribute to the vascular reactions of PDT, and the response can be manifested in many ways. Choice of treatment condition, such as photosensitizer type, drug-light interval, fluence, and fluence rate can determine the nature and timing of a vascular effect. The sensitivity of blood vessels to insult by PDT can be a function of their structural composition and is mediated by the release of vasoactive factors in the tumor microenvironment. Damage is revealed through changes in tumor blood flow due to vasoconstriction, hyperpermeability, and occlusion. These changes are accompanied by alterations in molecular signaling that can contribute to or impede a long-term therapeutic effect. In practice, PDT-created vascular damage can be both beneficial and detrimental to efficacy. Benefits are obtained when PDT leads to the progressive and irreversible shutdown of the tumor vasculature, but ischemia during illumination can be therapy-limiting when the photosensitizing agent relies on oxygen to produce cytotoxic reactive oxygen species. Vascular damage can further lead to the activation of survival signaling pathways that enable the recovery and regrowth of malignancies and their vasculature. Nevertheless, there is an important role for vascular damage in the success of many PDT treatments. Outcome can hinge on vascular damage both when the vessels are intended as the primary target, as in the case of PDT at short drug-light intervals, as well as when vascular damage is a secondary effect, for example in treatments at longer drug-light intervals. In the following pages, the causes and consequences of PDT-induced vascular damage are considered.

2. Vascular Damage as a Selective Response

2.1. *Passively Targeting Aberrant Vasculature*

Several applications of PDT are centered specifically on its ability to damage the blood vessels of solid malignancies or other pathologic conditions. Prominent examples of this "vascular-targeting" approach include the FDA-approved indication for treatment of age-related macular degeneration (AMD), as well as clinical trial investigations of PDT for treatment of prostate cancer.[3] PDT can be directed toward an anti-vascular therapy through the use of

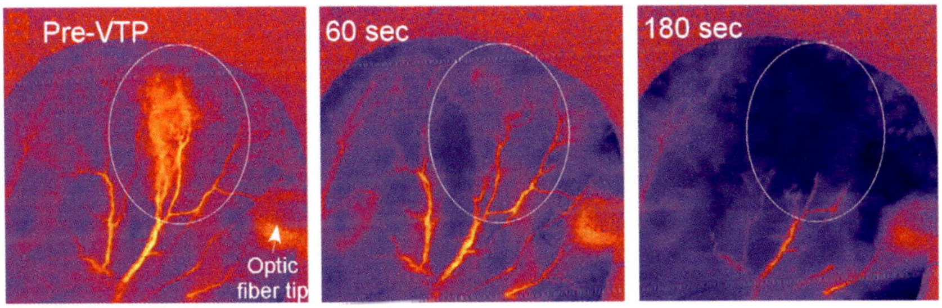

Fig. 1. Vascular-targeted PDT (VTP) rapidly shuts down tumor blood vessels. An ellipse outlines the margins of a tumor grown on the earlobe of a nude mouse. PDT consisted of the i.v. injection of Tookad (10 mg/kg) during the delivery of 763 nm light. Tumor blood flow was evaluated using dynamic light scattering imaging at the indicated times during light delivery. Selective and progressive damage of the tumor vasculature can be observed. Reprinted with permission from *PLoS One* **5**(4), e10282 (2010).

short time intervals between intravenous administration of a photosensitizing drug and the exposure of tissue to illumination at drug-exciting wavelengths (Fig. 1).[4] Essentially, this results in excitation of the photosensitizer when it is highly concentrated in the circulating blood, thereby leading to localized damage of endothelial cells, blood cells, and other components of a tissue's vascular network. The extent to which damage is dependent upon circulating levels of photosensitizer is well exemplified by studies of PDT with Tookad (WST-09), a palladium-metalated bacteriopheophorbide that rapidly clears from the blood. Using Tookad, PDT of the hamster cheek pouch is highly effective at a drug-light interval of 10 min, but this efficacy rapidly falls off as the drug-light interval is increased to 30, 120 and then 240 min; at a drug-light interval of 240 min, PDT is ineffective at all tested drug and light doses.[5] A more recently developed soluble form of the drug, Tookad-Soluble (WST-11; palladium bacteriopheophorbide monolysotaurine) is also rapidly cleared (murine $t_{1/2}$ of 1.65 min in the blood) and demonstrates great efficacy when light is delivered immediately following drug injection.[6]

The binding characteristics of photosensitizers can also determine the extent and nature of PDT-mediated vascular damage. The hydrophobicity of many photosensitizers leads them to associate with plasma proteins. In the case of Tookad-Soluble, binding of the drug to serum proteins can impede cellular uptake of the photosensitizer and subsequent PDT-mediated cytotoxicity (e.g., in endothelial cells).[6,7] This is expected to result in vascular occlusion independent of endothelial cell damage.[7] Indeed, Tookad-PDT of a rodent tumor resulted in extremely rapid decreases in blood flow and the initiation of clot formation.[4]

Another example of the importance of photosensitizer binding is evident in applications of verteporfin (Visudyne or benzoporphyrin derivative monoacid ring A; BPD) for PDT in ophthalmology. Verteporfin-mediated PDT can be used to treat the aberrant neovascularization that is secondary to AMD, a leading cause of blindness in the elderly of developed countries. Typical PDT protocols for AMD stipulate that light be delivered at ~15 min after the intravenous administration of verteporfin, while plasma levels are high.[8] However, the anti-vascular effect is additionally mediated by BPD binding to low density lipoproteins (LDL),[9] in conjunction with the presence of high numbers of LDL receptors on the proliferating cells of angiogenic blood vessels.[10] It has also been shown that LDL binding is enhanced when BPD is formulated with liposomes, as done in the production of verteporfin.[9] As a result of its LDL binding, verteporfin-mediated PDT at short drug-light intervals will damage endothelial cells, and it can lead to thrombosis in the absence of hemorrhage.[11]

Even when longer intervals between photosensitizer administration and light delivery are employed, the localization characteristics of photosensitizers may contribute to the anti-vascular effects of PDT. Some photosensitizers have an affinity for collagenous structure, which has led to observations of their association with the basement membrane of vessels.[12] Because PDT of collagen is thrombogenic,[13] it is not surprising that the co-localization of photosensitizer with vascular basement membrane is associated with PDT-induced vascular damage. Importantly, this mechanism of vascular damage is dependent upon there being sufficient time for the photosensitizing agent to localize to the vessel-lining collagenous structure. Consequently, a short drug-light interval (i.e., illumination prior to extra-vascular photosensitizer localization) can be highly effective in promoting vessel damage irrespective of collagen deposition in tumors. In contrast, when Photofrin-mediated PDT is employed at longer drug-light intervals, it is more effective in tumors with a higher stromal collagen content.[14]

2.2. *Active Targeting of Blood Vessels*

In addition to the above-described passive targeting of blood vessels, active targeting of photosensitizer to blood vessels provides another means of creating vascular damage. Numerous approaches have been tested toward active targeting of blood vessels. Hu *et al.*[15] conjugated verteporfin to factor VII as a means of directing photosensitizer delivery to tissue factor, an important protein in the coagulation pathway that is commonly expressed on angiogenic vessels and some tumor cells. Using the same ligand, chlorin e6 was also targeted as a photosensitizer and found to produce significant increases in tumor necrosis and apoptosis compared to untargeted PDT.[16] In other vascular-directed approaches, PDT was targeted to neuropilin-1, a receptor that is associated with the vascular endothelial growth factor (VEGF) receptor,[17] or the somatostatin 2 receptor, which is expressed by proliferating endothelium and many tumor cells.[18] Both of these targeting mechanisms were effective in promoting vascular damage and slowing tumor growth.[17,18]

In the development of active targeting platforms, it is important to consider how PDT response may be impacted by changes in tumor uptake and cellular association of photosensitizer due to its targeting. In work to this point, Ichikawa *et al.*[19] showed that targeting could be used to overcome limitations in therapeutic effect that stemmed from reductions in the bioavailability of BPD when it was delivered in the form of pegylated liposomes. Modification of the liposomes with a peptide that is specific to angiogenic vessels (Ala-Pro-Arg-Pro-Gly), provided for its effective delivery to the tumor, led to a strong anti-vascular effect from PDT (tested in a dorsal air sac angiogenesis model), and could prolong the survival of tumor-bearing mice treated with PDT.

2.3. *Microenvironmental Determinants of Vessel Response*

In many applications of PDT, vascular damage is thought to develop when damage of endothelial cells leads to their retraction and the thrombosis-promoting exposure of underlying basement membrane.[20] This effect may be mediated by numerous factors that include the direct cytotoxic insult of PDT on the cells themselves, rearrangement of cytoskeletal proteins, or a PDT-related decrease in the function of integrins that are necessary for maintaining cell adhesion to basement membrane and other components of extracellular matrix.[21,22] Additionally, PDT effect on the production or release of soluble factors that mediate hemodynamic and inflammatory reactions could contribute to the propensity for treatment to result in vascular damage.

Notwithstanding the above factors, it is also clear that the stromal composition of a tumor can contribute to its sensitivity to PDT. Specific to the tumor vasculature, the subendothelial basement membrane, composed mostly of collagen IV, represents an important target. We have shown that tumors with a more highly collagenated vascular structure are more susceptible to PDT-induced vascular damage, and that this increase in vascular damage translates into better long term outcome in murine studies (Fig. 2).[14] Known effects of PDT on stroma could contribute to this sensitivity. Firstly, and not

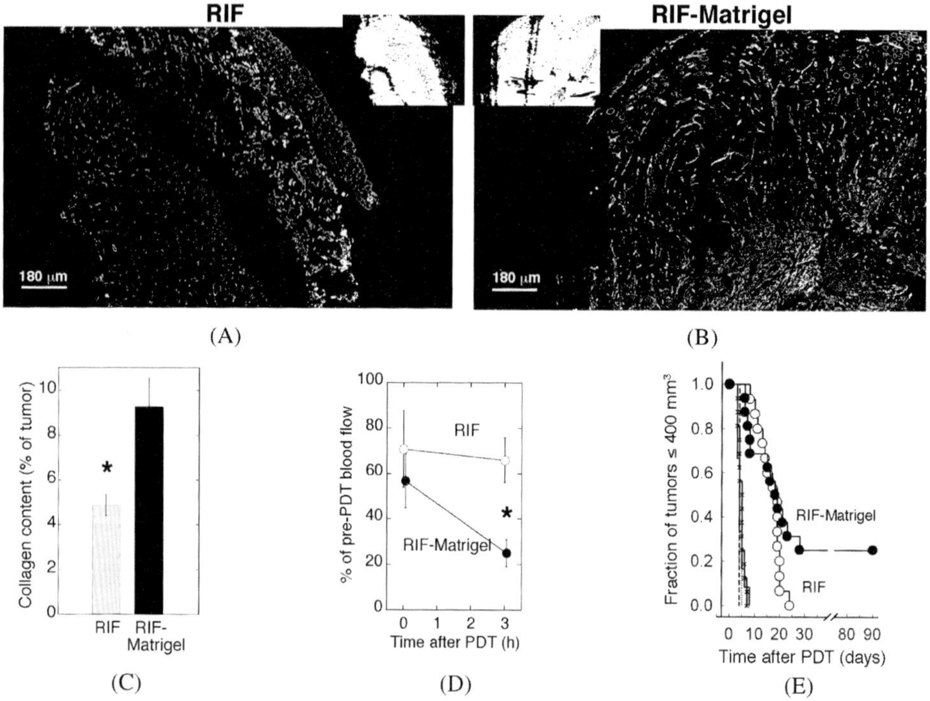

Fig. 2. Tumors that contain more collagen IV are more sensitive to PDT-mediated vascular damage and more responsive to treatment. Murine tumors supplemented with Matrigel Basement Membrane Matrix (RIF-Matrigel) at the time of tumor inoculation were characterized by a greater deposition of collagen IV in a vascular-associated pattern when compared to unsupplemented RIF tumors (compare panels A and B for collagen staining; insets show tumor area on the section). RIF-Matrigel tumors contained more collagen (C), and in response to PDT they experienced larger decreases in tumor blood flow (D) and better long-term response (E). PDT performed with Photofrin (5 mg/kg) and 632 nm illumination at 75 mW/cm^2 to a dose of 135 J/cm^2. *indicates $p < 0.05$ for differences between RIF and RIF-Matrigel tumors. Reprinted with permission from *Cancer Res.* 72(8), 2079 (2012).

surprisingly, given the tendency for photosensitizers to localize to collagens, it has been shown that PDT can directly damage the basement membrane of vessels.[12] Furthermore, it is known that PDT of a collagen matrix will increase the adhesion and aggregation of platelets to its surface. This may be a result of structural changes that occur secondary to PDT-induced crosslinking of collagen, and an effect of these changes on the direct interaction of platelets with collagen.[13] Ultimately, any increases in platelet binding consequent to PDT of collagen would be expected to contribute to more vascular congestion and deteriorating vascular function in vessels with higher collagen content.[14]

It is also important to consider how irregularities in vascular composition, which are common in tumors, come into play during PDT. Gaps or inconsistencies in the endothelial cell lining of tumor blood vessels can inherently make these vessels at greater risk for vascular events, perhaps by "lowering the bar" in terms of the extent of vascular damage needed to initiate the process of thrombosis. In this regard, it is revealing that use of a short drug-light interval as a vascular targeting mechanism lends little from a pharmacological standpoint to provide for selectivity of the PDT effect. This is because illumination takes place when circulating drug levels are high, providing little basis for differentials in PDT effect on diseased versus normal tissues in the illumination field. Nevertheless, selectivity has been observed.[9,23] In the case of PDT with verteporfin, some of this selectivity may be accounted for by the higher expression of receptors for LDLs on proliferating endothelial cells (given the affinity of BPD for LDLs). However, it is additionally posited that the heightened sensitivity of tumor blood vessels to vascular-targeting PDT stems from their poorly developed structure and a resulting greater propensity for thrombosis.[4]

Irregularities in vascular composition and other abnormalities of the tumor microenvironment lead to heterogeneities in tumor hemodynamics that can also play a role in vessel response to PDT. Mesquita *et al.*[24] showed that baseline hemodynamics differ between tumors of the same model when they are propagated in different murine strains, and these discrepancies in vasosensitivity are reflected in hemodynamic response to PDT. Overall, tumors with more regular patterning in baseline oscillatory blood flow were more sensitive to vascular stresses. This included greater reductions in blood flow over the course of illumination for PDT.[24] Other studies support these observations at the level of the individual blood vessel. Using intravital microscopy, He *et al.*[25] established a direct correlation between the initial velocity of cells in a tumor blood vessel and the time elapsed after PDT until vessel shutdown. Vessels with more sluggish flow at the initiation of PDT experienced a more rapid stoppage of blood flow in response to PDT. The implications of these heterogeneities in vascular response to the cytotoxic effects of PDT are undeniable. In a study of intratumoral heterogeneities in PDT response, regional variations in PDT-induced ischemia clearly had a negative impact on cytotoxicity. Intratumoral heterogeneities in vessel response to PDT led to greater hypoxia at the base of intradermal tumors. In turn, PDT-created hypoxia at the tumor base protected this area from PDT-mediated direct cytotoxicity.[26]

3. Functional Characteristics of Vascular Damage

3.1. *General Considerations of the Vascular Reaction*

The response of tumor blood vessels to PDT is multifaceted. Among its many characteristics, the reaction has been documented to include vasoconstriction, hyperpermeability, ischemia, and occlusion. The prevalence of one reaction compared to another may be a function of the treatment parameters, determined by conditions such as the photosensitizer and its dose, the drug-light interval, and the total fluence and fluence rate. For example, He *et al.*[25] showed that higher doses of BPD lead to vascular occlusion, while lower photosensitizer doses favor the development of a hyperpermeable response. In general, increasing the treatment fluence leads to more durable vascular response,[4,27] with the potential to result in irreversible vascular shutdown. The longer treatment times that are associated with lower fluence rates have also been show to increase the extent of PDT-created vascular damage.[28,29]

Investigations by Robinson and colleagues support a role for direct cytotoxicity in promoting vascular damage.[30] They documented that the extent of PDT-created vascular damage in the murine skin was associated with levels of the photosensitizer protoporphyrin IX in the wall of its vessels.[30] Others have studied whether vascular damage is dependent on PDT dose, which is the product of the administered photosensitizer dose and incident light dose.[31] In this work, reciprocity could be established between drug and light dose in the creation of vascular occlusion that was associated with endothelial cell damage. Thus, the collective interpretation of these studies points toward the direct effect of PDT as a key contributor to the vascular reaction.

3.2. *PDT-induced Vasoconstriction*

A vasoconstrictive response to PDT can been seen very early during treatment (Fig. 3).[32] This response is more evident in arterioles compared to venules, which is an expected consequence of better vasomodulatory capabilities by arterial-derived vessels.[30,33] One hypothesis is that PDT damage of the endothelial cells lining these vessels leads to reductions in the local availability of nitric oxide, thereby reducing the ability of these vessels to maintain tone and manifesting in vasoconstriction. Indeed, studies have suggested that nitric oxide can protect from PDT-created vascular effects. For example, it has been shown in murine fibrosarcomas that inhibition of nitric oxide during PDT can promote vasoconstriction.[34] Also, the thrombogenic response that is triggered by PDT when damaged endothelial cells expose the underlying basement membrane can lead to the release of vasoconstrictive eicosanoids such as thromboxane.[35]

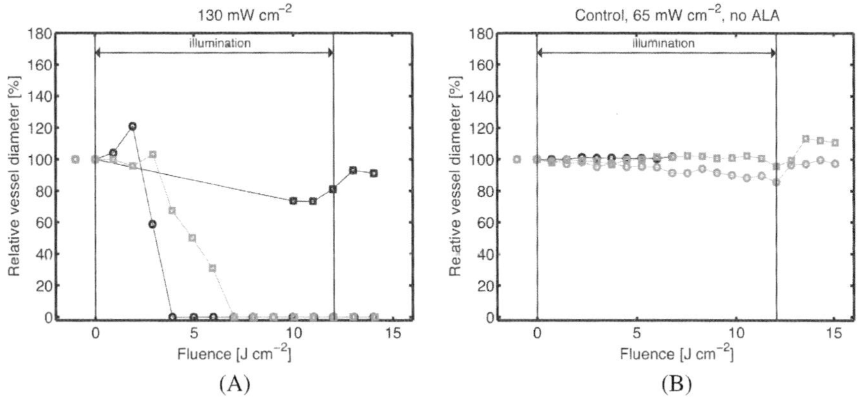

Fig. 3. PDT leads to rapid vasoconstriction during illumination. Plots depict vessel diameters in the subcutaneous vasculature of individual animals as measured by intravital confocal imaging. After topical application of aminolevulinic acid (ALA), light was delivered to a total fluence of 13 J/cm^2 at a fluence rate of 130 mW/cm^2 (A). Control animals did not receive ALA (B). Reprinted with permission from *Photodiagnosis Photodyn. Ther.* **11**(2), 71 (2014).

3.3. *Vascular Hyperpermeability as a Response to PDT*

PDT damage on the venous side of tissue vasculature can be manifested by large increases in vessel permeability and thrombus formation.[33] PDT-induced hyperpermeability is associated with the release of numerous factors that modulate vascular permeability, such as histamine and VEGF.[36,37] At a cellular level, hyperpermeability likely results from the PDT-created disruption of endothelial cell barrier function when treated endothelial cells retract from the basement membrane, leading to the formation of intracellular gap junctions.[38] Moreover, the PDT-triggered binding of leukocytes to endothelial cells is also important to treatment-induced permeability. In one study of this phenomenon, PDT of murine muscle tissue was performed in a dorsal skinfold chamber, allowing for visualization of leukocyte rolling and sticking to the surface of endothelial cells.[39] Prior to PDT, mice were inoculated with an antibody cocktail that targeted the selectin family of cell adhesion molecules for the purpose of inhibiting leukocyte interaction with the endothelium. Leukocyte rolling was significantly inhibited in these animals, and this was accompanied by decreases in permeability detected by reductions in the vascular leakage of macromolecules.

Prominent intratumoral heterogeneities exist in the microvasculature's permeability response to PDT. In a rodent model of prostate cancer, the periphery of tumors exhibited larger increases in vascular permeability and better maintenance of blood flow after PDT than did the vasculature at the center of the tumors.[40] The authors speculated that this may have been related to the presence of high interstitial fluid pressure in tumors, which could make the central vasculature more susceptible to collapse. They also suggested that the larger size of peripheral blood vessels may have made them less prone to collapse, obstruction or shutdown. Such an explanation is in line with other studies that found tumor response to vascular stress to differ between tumors grown in different murine hosts. Vessel size was a distinguishing, host-dependent characteristic of the tumors, and smaller treatment-induced changes in blood flow were detected in tumors with larger blood vessels.[24] Others have also observed that PDT-induced vessel occlusion, or even destruction, is more complete in smaller blood vessels.[41,42]

The characteristic hyperpermeability that accompanies PDT provides opportunity for the design of novel combination therapies. For example, hyperpermeability can be exploited as a means for increasing the delivery of other agents into the treatment site.[43,44] In studies along these lines, some investigators have explored the potential to combine vascular-permeabilizing PDT protocols with chemotherapy or other drugs toward the goals of increasing drug accumulation and thus, its effect on the target disease.[45]

3.4. *PDT-initiated Hemostasis*

The anti-vascular effects of PDT frequently cumulate in activation of the coagulation cascade and thrombosis of blood vessels that are contained within the illumination field. The PDT-initiated retraction of endothelial cells from their underlying matrix facilitates the binding of platelets to the exposed site. This triggers the release of coagulation-stimulating factors such thromboxane and von Willebrand factor. Thromboxane production by activated platelets promotes vasoconstriction and can potentiate platelet aggregation, while von Willebrand factor mediates platelet adhesion by providing a structural link between platelets and collagen in the basement membrane matrix.[13,22] As discussed above, binding of leukocytes to the PDT-damaged endothelium has also been observed.[33] This is associated with the production of inflammatory mediators such as leukotriene B4. Moreover, the formation of mixed platelet-leukocyte aggregates (for example, through P-selectin receptors on each cell type) provides for a point of interface between inflammatory and coagulant responses.[46] Leukocyte binding to PDT-treated endothelial cells occurs via B_2 integrins on the leukocytes,[20] and may be aided by the PDT-induced upregulation of integrin expression on these cells. For example, PDT of polymorphonuclear leukocytes leads to increases in the expression of Mac-1 integrin on these cells, as well as results in their generation of leukotriene B4.[47]

Functional evidence for a role of neutrophils in PDT-created coagulation can be gleaned from studies of thrombosis in transplanted vein grafts that were subsequently treated with PDT. This work established that the pre-PDT administration of an anti-neutrophil antibody to animals that had received a vein graft bypass could completely inhibit the PDT-induced development of thrombosis.[48]

Further contributing to the coagulant effects of PDT is the ability of treatment to directly alter the balance of pro-and anticoagulant factors through which endothelial cells promote or impede thrombosis. PDT can reduce expression of the anticoagulant, glycoprotein thrombomodulin on the surface of endothelial cells. It can simultaneously increase the expression and release of tissue factor, a procoagulant initiator of the clotting cascade.[49] The PDT-induced production of endothelial-cell derived microparticles also stands to play a role in treatment-produced thrombosis.[19,50]

3.5. *Hemodynamic Effects*

As one would expect, blood flow stasis accompanies the progressive development of thrombus formation in PDT. Using heparin to inhibit coagulation in PDT-treated mammary tumors, Dolmans *et al.*[51] showed that interfering with thrombus formation can simultaneously lead to significant prolongation in the maintenance of tumor perfusion after PDT. Furthermore, it is evident that vascular congestion can develop rapidly during PDT delivery, with some investigators observing the intravascular accumulation of platelets within seconds of initiating PDT.[42] Vasoconstriction also can occur early during light delivery for PDT.[4,52] Whether operating in concert or independently, vascular congestion and vasoconstriction can contribute to measureable alterations in blood flow during PDT.[32,53] Indeed, early studies using intravital microscopy provided evidence of changes in tumor blood flow that sometimes, but not always, were associated with decreases in vessel diameter.[33,54,55] In the absence of changes in vessel diameter, a PDT effect on lumen diameter could be found, which was thought to result from physical obstruction of the vessel lumen by aggregating cells. Notably, transient vasodilation has also been found during some PDT applications, thereby providing an additional cause for PDT-triggered alterations in blood flow.[4]

Using present day optical technologies, it is possible to noninvasively monitor blood flow during PDT in the clinical, as well as the preclinical setting. Continuous monitoring of blood flow during light delivery reveals that it is highly dynamic, characterized by both increases and decreases in flow over the course illumination. Despite this seeming randomness, specific patterns do emerge, and there is evidence of photosensitizer-specific trends in vascular dynamics. Blood flow tends to increase at the beginning of illumination for Photofrin and ALA-mediated PDT, followed by decreasing and then oscillatory flow.[56,57] PDT with motexafin lutetium (MLu) at a short drug-light interval resulted in large initial decreases in blood flow that recovered and then declined for the remainder of treatment.[28] Not unlike these results, small positive and negative fluctuations were overridden by progressive decreases in tumor blood flow during illumination for vascular-targeted, Tookad-mediated PDT.[4] Clinical studies confirm the presence of similar trends in tumor hemodynamics for PDT of human disease, providing direct evidence that PDT-created ischemia is worthy of consideration in the planning of patient treatment.[57–60]

The consequences of ischemia during PDT are undeniable. It can contribute to the development of hypoxia during illumination, the presence of which has long been known to negatively impact PDT outcomes with most (oxygen-dependent) photosensitizers.[61,62] Also of importance, hemodynamic monitoring can be used as a tool to predict therapy response. Several investigators have shown that PDT effect on tumor oxygenation and/or blood flow is highly correlated with treatment outcomes.[56,63–66] For example, treatment-induced changes in blood flow during Photofrin-PDT in rodents is predictive of tumor response, measured as either the time-to-tumor regrowth[56] or the extent of necrosis.[66] Similarly, the assessment of tumor oxygenation after PDT can provide for an early indication of therapeutic effectiveness.[64,65,67] Ultimately, the adaption of hemodynamic monitoring to the clinic could provide a means for real-time monitoring of tumor physiology during PDT, thereby facilitating individualized treatment of patients based on functional assessment of the vascular response of their tumor to PDT.

4. Vascular Damage in Inflammation and Immune Reactions

4.1. *Complement Activation and Inflammation of the Innate Immune Response*

The vascular changes initiated by PDT are intrinsic to the activation of innate and adaptive immune responses in animals undergoing treatment. Innate immunity involves the stimulation of immune defenses that lack a memory of the instigating antigen. PDT-initiated damage of tumor and stromal cells, such as endothelial cells, can result in the activation of the innate immune response.[68] This occurs when cellular damage leads to the release of molecules that are normally maintained intracellularly. These molecules then act as damaged-associated molecular patterns (DAMPS) that initiate an immune reaction. Known DAMPS with prominent roles in PDT include the heat shock proteins (HSPs), phosphatidylserine, calreticulin, and ATP/UTP. Ultimately, the type of DAMP that is detected after PDT may be a function of the extent of oxidative stress that is inflicted at the site of DAMP origin, and in this manner, it may be related to intracellular localization of photosensitizer.[69]

Upon the recognition of vascular damage, the complement cascade can be activated. The complement system is an important component of the innate immune response that, among other functions, serves to mark damaged cells for destruction. This pathway is a distinctly different system from the coagulation cascade that is activated by vascular damage. Nevertheless, there are points of interaction between the pathways that result in crosstalk between them. Observations that thrombin can proteolytically activate complement component 5 (C5) demonstrate that complement activation can be induced by factors of the coagulation cascade. Inversely, the activation products of certain complement components can induce platelet activation and aggregation, and the complement-induced expression of tissue factor on endothelial cells contributes to activation of the coagulation cascade in the presence of compromised endothelium.[70]

PDT activates multiple components of the complement cascade. Stott and Korbelik[71] reported that Photofrin PDT of Lewis lung carcinoma increased gene expression of complement proteins C3, C5, and C9. This occurred as a function of the PDT-triggered release of HSP because complement gene expression was decreased when PDT was performed in the presence of antibodies against HSP70. Upon their production, complement proteins can recognize and bind to the surfaces of PDT-damaged cells, contributing to the full activation of the cascade[72] and the potential to induce vascular effects.[73] Evidence of complement-mediated vascular effects in PDT can be gleaned from studies of the tumor microenvironment in knockout mice deficient for complement. Photofrin-PDT of these knockout mice resulted in less tumor hypoxia than in wild type animals. This was expected to be a result of less PDT-created vascular damage in the knockout animals. Since vascular damage resulted in hypoxia during the delivery of PDT, it was possible to increase cytotoxicity and improve therapeutic outcomes by mitigating complement activation. Under circumstances in which complement activation was much less pronounced (i.e., with BPD-PDT at a 3 h drug-light interval), there was no differential in PDT responses between the knockout and wildtype mice.[73]

Table 1. Murine complete blood counts (mean ± SD) after intrathoracic PDT.[a,b]

Treatment condition	WBC[c] (10^3/ml)	HgB (g/dl)	PLT (10^3/ml)	(%WBC) NEU	LYM	MON
Untreated	2.1 ± 0.8	12.1 ± 0.8	1003 ± 84	27.32 ± 4.27	65.68 ± 6.54	1.10 ± 0.44
HPPH-only	3.4 ± 1.7	10.8 ± 3.7	855 ± 156	27.83 ± 8.50	67.63 ± 8.94	0.90 ± 0.28
Light-only	4.8 ± 1.2	12.4 ± 0.6	918 ± 88	23.25 ± 5.54	70.82 ± 9.55	1.00 ± 0.43
PDT	4.7 ± 1.8	7.9 ± 3.4	345 ± 225[d]	86.50 ± 7.79[d]	7.10 ± 5.37[d]	4.03 ± 0.06[d]

[a]Adapted from *Lasers Surg. Med.* **43**-(7), 663(2011).
[b]Blood collected from tumor-free animals 24 hours after intrathoracic (2-11-Hexyloxyethyl)-2-devinyl pyropheophorbide-a) (HPPH)-PDT to 50 J/cm or the indicated controls; $n = 3$–6 animals.
[c]WBC, While blood cell count; HgB, hemoglobin; PLT, platelet count; NEU, neutrophils; LYM, lymphocytes; MON, monocytes.
[d]$p < 0.05$ for comparison to untreated samples.

Inflammation is another characteristic of innate immunity that is commonly triggered by PDT. Inflammation involves the release of cytokines, most prominently IL-6, as well as initiates the infiltration of leukocytes into the injured area. PDT has a particularly large effect on the circulation and migration of neutrophils.[74] Large increases in the number of circulating neutrophils, known as neutrophilia, can rapidly develop when PDT is delivered to either tumor or normal tissues. In studies of PDT-treated subcutaneous tumor models, neutrophilia was noted to peak at times 3 h and 8–10 h after PDT, and elevated levels of circulating neutrophils were detectable through at least 24 h after illumination.[75] Similarly, after PDT of the (normal) murine thoracic cavity, prominent increases in circulating neutrophils and monocytes were detected at 24 hours[76] (Table 1). Moreover, in these investigations, PDT also led to significant decreases in the levels of circulating platelets.[76] Such thrombocytopenia has also been observed in clinical applications of PDT, and can potentially even pose as a dose-limiting toxicity.[77,78]

PDT affect on circulating and local accumulation of immune cells is strongly integrated with its vascular effects. In studies of vascular biology, thrombocytopenia can be linked to hemorrhage at local sites of inflammation.[79] The specific consequences of PDT-triggered thrombocytopenia on the local tumor response have been little studied; nevertheless, it is noteworthy that the induction of thrombocytopenia *prior* to PDT resulted in significantly less vasoconstriction and thrombus formation during PDT, which detracted from vascular response to Photofrin-PDT.[80] More work has been performed on the effects of neutrophilia in PDT, and studies have identified neutrophils as positive effectors of tumor response.[81] Interestingly, mice depleted of neutrophils at the time of PDT exhibited slower recovery of tumor blood flow in the post-treatment period.[82] Thus, one effect of local neutrophil accumulation after PDT could be to alter the kinetics of vascular response, perhaps through release of vasodilators such as nitric oxide from activated neutrophils. It has been hypothesized that such a prolongation of ischemia could negatively impact treatment outcomes by reducing the contribution of ischemia-reperfusion injury to the PDT response.[82] Additionally, an extended hypoxic state that results from prolonged ischemia could exacerbate the production of angiogenic factors, potentially contributing to tumor recurrence and growth. Finally, it is important to mention that neutrophils can play a role in promoting anti-tumor immunity, thereby providing additional mechanisms through which the inflammation-associated influx of leukocytes into the treatment field serves to mediate PDT response.[83]

4.2. *The Adaptive Immune Response*

There is a major role for the PDT-triggered acute phase reaction, including its vascular components, in the post-treatment induction of an adaptive immune response. The adaptive immune response leads to anti-tumor immunity through the production of antibodies that are specific to tumor antigens. Although it does not contribute to the initial damage of a tumor by PDT, adaptive immunity can be important to controlling tumor regrowth or inhibiting recurrence.[84] Many aspects of the innate immune response contribute to the development of antitumor immunity. It is thought that the simultaneous presence of both complement proteins and DAMPS, especially HSPs, on the surface of PDT-damaged cells may provide for the enhanced presentation of tumor antigen, an essential step in the process of developing antibody-mediated immunity.[84] In fact, it has been shown that the level of PDT-induced expression of HSP70 is related to the maturation of dendritic cells, which in turn enables these cells to provide for antigen-presentation to T-cells.[85] The migration of neutrophils into PDT-treated tumors also has a role in instigating a T-cell response to therapy, and mice depleted of neutrophils exhibit a blunted $CD8^{+}$ T-cell response to PDT.[83] More recently, it has been shown that PDT-induced production of IL-17 is an important factor in mediating neutrophil migration to tumor-draining lymph nodes, a key step in the development of adaptive immunity.[86] Perhaps the clearest evidence that PDT-created vascular damage can lead to the development of anti-tumor immunity comes from studies of vascular-targeted PDT. An intact immune system was essential for vascular-targeted PDT with Tookad-Soluble (WST-11) to effectively control tumor regrowth.[87] Furthermore, in these studies, the immunity that was generated by PDT against one tumor cell line could serve to protect mice from challenge with a different tumor cell line. This leads to the intriguing possibility that immunity may have been produced against antigens that originated from cells of the PDT-damaged vasculature, instead of from the tumor cells themselves, which makes it an area that is worthy of further investigation.[87]

5. The Role for Stromal-associated Signaling

5.1. *Stromal-associated Signaling and Inflammation*

The molecular response of vasculature and other stromal components to PDT-induced damage involves activation of inflammatory and vasoactive cytokine/growth factor pathways. As noted above, the influx of neutrophils and the subsequent recruitment of $CD8^+$ T-cells are critical for the success of PDT, especially for its non-local effects. However, inflammation is a highly complex and diverse process and many inflammatory mediators have both "good" and "bad" effects. For example, PDT-induced stromal (in particular vascular) damage results in production of IL-6, a pleotropic inflammatory cytokine that can enhance PDT efficacy by promoting neutrophil migration and suppression of CD4+CD25+CD127FoxP3+ regulatory T-cell (Treg) function, but may also paradoxically inhibit long term anti-tumor responses. In studies of Photofrin-mediated PDT in mouse mammary tumors, antibodies targeting IL-6 completely abrogated neutrophilia. In contrast, the inhibition of neutrophilia could be achieved only at certain post-treatment timepoints using antibodies against another inflammation-associated cytokine, TNFα and the anti-inflammatory cytokine IL-10.[82] Similarly, augmentation of the PDT-induced IL-6 response by delivering a soluble IL-6R-IL-6 fusion protein (hyper-IL-6) significantly enhanced HPPH-mediated PDT efficacy in mice bearing Colo-26 xenografts.[88] In patients receiving Photofrin-mediated PDT for Barrett's esophagus, PDT-induced increases in systemic IL-6 expression was correlated with diminished Treg activity.[89] However, inhibition of PDT-induced IL-6 using anti-IL-6 antibodies did not significantly diminish the influx of neutrophils into treated tumors.[90] Moreover, PDT-mediated immune memory was enhanced by abrogation of host IL-6 production after HPPH-mediated PDT of Colo26 tumors in an IL-6 mutant mouse model.[91] Interestingly, while systemic IL-6 levels were (unsurprisingly) nearly undetectable in the mutant mice of these studies, there was still a small, statistically significant increase in tumor IL-6 levels following PDT, presumably relating to tumor cell production of this cytokine. This observation suggests the possibility that IL-6 and other cytokines produced by PDT-mediated stromal damage may have qualitatively different (positive or negative) effects on the overall anti-tumor efficacy of PDT, depending on either the level or origin of cytokine production.

5.2. *PDT-induced Activation of Stromal versus Tumor Cell Growth Factor Pathways*

As with cytokine production, damage to both tumor cells and tumor stroma can induce growth factor pathway activation, but the effects of tumor cell versus stromal growth factor production can be challenging to separate. This is exemplified by the multiple, intertwined mechanisms by which PDT can stimulate production of VEGF. One mechanism for stromal growth factor production is through the creation of hypoxia following vascular damage/shutdown. Under hypoxic conditions, the oxygen-dependent tagging of hypoxia inducible factor 1α (HIF-1α) for proteosomal degradation does not occur, leading to formation of the heterodimeric HIF1α/β transcription factor and expression of a wide array of hypoxia inducible genes, including VEGF.[92] Indeed, a number of investigators have correlated PDT-mediated HIF-1α stabilization and VEGF expression in a variety of cancer models, including mammary, bladder, head and neck, prostate and glial carcinomas.[89,92–98] In addition, stromal damage-mediated production of VEGF and other growth factor pathways can produce cytokine-dependent activation of the STAT-3 (signal transducer and activator of transcription-3) transcription factor. STAT-3 activation by cytokine receptors such as IL-6R and growth factor receptors such as epidermal growth factor receptor (EGFR) can promote VEGF production either directly by activating its transcription, or indirectly by increasing HIF-1α expression.[99–102] Another mechanism by which PDT-mediated stromal damage can produce VEGF is through activation of cyclooxygenase-2 (Cox-2). PDT-induced expression of Cox-2 has been demonstrated in a wide array of tumor types, including sarcomas and cervical, skin, head and neck and prostate carcinomas.[103–106]

The EGFR pathway is also activated by PDT in both tumor and stromal cells. For example, EGFR is activated in tumor cells in response to BPD- or Photofrin-mediated PDT in both non-small cell lung cancer (NSCLC) and ovarian cancer cells *in vitro*, and promotes cancer cell survival by inhibiting apoptotic death.[107] Using immunofluorescence, we have also found BPD-mediated EGFR activation (phosphorylation) in H460 human NSCLC

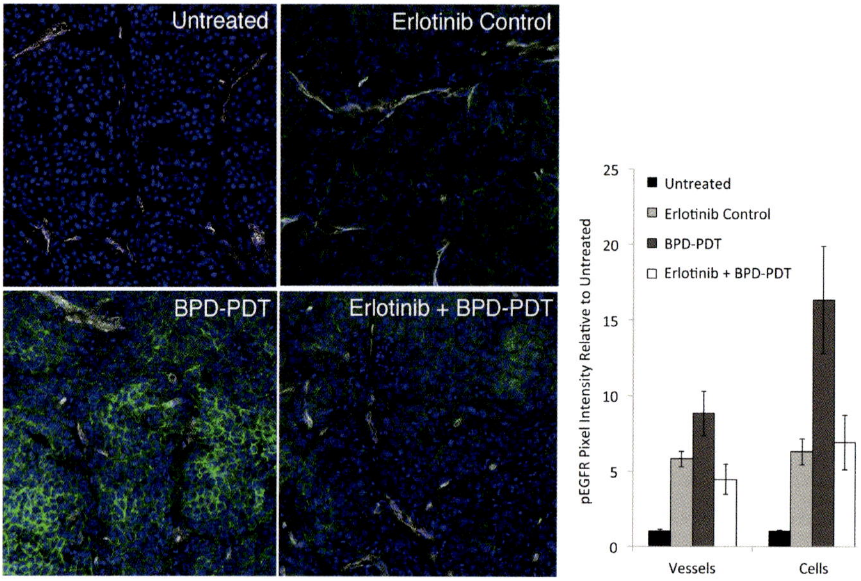

Fig. 4. BPD-PDT leads to increases in EGFR activation that can be attenuated by pre-treatment with Erlotinib. Confocal images (425 μm × 425 μm) of phosphorylated EGFR (green), nuclei (blue), and blood vessels (white) in H460 tumors. PDT with BPD (1 mg/kg, 3 h) and 135 J/cm^2 (75 mW/cm^2 at 690 nm). When indicated, Erlotinib (50 mg/kg) was administered by oral gavage at 48 h, 24 h, and 3 h before light delivery.

xenografts and their vasculature (Fig. 4). Interestingly, Cox-2/prostanoids can be produced in response to damage of both tumor cells and stroma (e.g., lymphatic/vascular endothelium), and this, in turn, can lead to activation of VEGF, EGFR, and STAT3 signaling pathways.[108–110] Conversely, PDT can induce nuclear translocation of EGFR and STAT3,[107] and EGFR/STAT3 nuclear translocation can increase expression of Cox-2.[111] Thus, while PDT-induced signaling pathways are complex and diverse, STAT-3 activation seems to be a common theme in PDT-mediated signaling by growth/survival, cytokine and angiogenesis/hypoxia response pathways.

With the complexity of these mechanisms, it can be very difficult to understand the relative roles of PDT tumor cell versus stromal damage in the induction of growth factors such as VEGF or EGFR. However, models with human tumor xenografts in rodents provide one opportunity for such study since Bevacizumab binds selectively to human (but not murine) VEGF. In experiments using H460 human NSCLC tumors in athymic mice, BPD-mediated PDT produced strong inductions of both human and murine VEGF (hVEGFR/mVEGF) that could be inhibited by either anti-hVEGF (bevacizumab) or anti-mVEGF antibodies, respectively.[112] In these studies, anti-hVEGF therapy did not improve PDT efficacy, but the combination of anti-h and anti-m VEGF therapies increased the PDT cure rate from 8.3 to 33.3%. Thus, it seems likely that the PDT-damage response of both tumor cells and stroma are important contributors to the molecular responses of tumors to PDT.

5.3. *Molecular Targets that Augment Anti-vascular Effects of PDT*

The molecular and physiologic responses to PDT-induced vascular damage include both positive and negative elements, the balance of which will help to decide patient outcome. On the positive side, vascular-targeted PDT can combine the complete destruction of both tumor cells and the cannibalized host components that make up the tumor stroma with a robust host anti-tumor immune reaction; this serves to simultaneously prevent local recurrence and treat any distant micro (or macro) metastases. In this case, vascular-targeted PDT represents a potentially ideal cancer treatment modality to be used in the primary therapy (single agent or multimodality) or adjuvant/second line setting. However, the anti-vascular and pro-inflammatory features that can make vascular-targeted PDT successful bring with it potential negative consequences that result in renewal of the pro-angiogenic, chronic inflammatory state of the original tumor. Currently, vigorous efforts are underway to meet this challenge by developing new targeted PDT agents or improved light delivery/dosimetry systems to fine tune the PDT damage so that only the positive aspects remain. However, this challenge may also be addressed by using currently available molecularly-targeted therapies to augment the positive impact while minimizing the negative consequences of PDT.

Preclinical studies have demonstrated that increased PDT efficacy can be achieved in multiple cancer types by targeting PDT-induced VEGF release from tumor and stroma using bevacizumab and other anti-angiogenic therapies. Initial studies used either Endothelial-Monocyte Activating Polypeptide-II (EMAP-II) or a l-glutamyl-l-tryptophan dipeptide to inhibit angiogenesis following Photofrin-mediated PDT of murine BA mammary carcinoma cells and tumors.[113] These studies demonstrated improvements of PDT efficacy both *in vitro* and *in vivo* that correlated with decreases in tumor HIF-1α and VEGF levels. Further studies by multiple investigators have demonstrated the reproducibility and robustness of this combination using a variety of photosensitizers, tumor models and anti-angiogenic agents.[93–97,112,114] Given the role of the proteasome in regulating HIF-1α-mediated VEGF transcription, another approach to limit the pro-angiogenic effects of vascular-targeted PDT is to use the proteasome inhibitor bortezomib, which was demonstrated to improve PDT efficacy using a murine prostate cancer model.[115]

Other strategies to limit the pro-angiogenic effects of vascular-targeted PDT have also met with significant preclinical success. For example, targeting EGFR activation, either alone or in combination with VEGF inhibition, has shown promise. In a study of BPD-mediated PDT in a murine ovarian carcinomatosis model, combining the anti-EGFR antibody cetuximab with PDT resulted in improvements in both objective tumor response (tumor burden) and overall survival, with no increase in treatment toxicity.[116] Further studies inhibiting EGFR pathway signaling using either cetuximab or the PI 3-Kinase inhibitor BEZ235 confirmed these results, and demonstrated that superior enhancement of PDT could be achieved using the combination of cetuximab and bevacizumab as compared to PDT with either single agent.[96,117] Similarly, targeting Cox-2 using selective NSAIDs such as NS-398, celecoxib, or rofecoxib can decrease PDT-induced VEGF secretion and increase PDT efficacy, although the mechanism may involve a combination of anti-vascular and immunemodulatory effects.[103–105]

While Cox-2 inhibition may target both pro-angiogenic and pro-inflammatory PDT effects, directly targeting pro-inflammatory cytokines using molecularly targeted therapies is significantly less well explored. As noted above, blocking IL-6 activation using a soluble decoy receptor decreased PDT efficacy, while performing PDT in an IL-6 mutant mouse improved PDT efficacy. Nevertheless, there are now a large variety of cytokine-targeted therapies that could be explored as potential agents to "fine tune" the immune responses from vascular-targeted PDT. This is exemplified by the use of tocilzumab in patients receiving CD19 directed Chimeric Antigen Receptor T-cell (CART-19) therapies. Despite highly promising clinical results, these patients can experience severe, life threatening toxicity from cytokine release syndrome (CRS).[118] In patients experiencing severe CRS, where non-directed anti-inflammatory agents such as corticosteroids were not effective, short duration delivery of anti-IL-6 therapy (tocilzumab) produced dramatic reductions in the acute toxicity. Clinical trials are ongoing and additional study is needed, but patients treated with tocilzumab + CART-19 therapy appear to have intact immunologic treatment responses. Further work is also necessary to see if such agents can augment the "good" and inhibit the "bad" pro-inflammatory effects of PDT, but this would appear to be a highly promising new therapeutic approach.

6. Vascular Effects in Clinical Oncology

6.1. *Advantages and Limitations of Vascular Damage in Clinical PDT*

Just as found in preclinical studies, PDT exerts its anti-cancer effect in clinical application through several different mechanisms. In many treatments, vascular damage from PDT is prevalent. This provides both advantages and disadvantages in the clinical setting. Among the advantages, PDT effect on the vasculature is one factor that facilitates its repeatability. Namely, it allows for PDT to be administered multiple times to the same site without compromising its efficacy. The ability of PDT to damage the tumor-supporting vasculature also enables it to avoid some of the effects that are imposed by cancer chemoresistance and radioresistance.[119]

Among its limitations, vascular PDT is not easily exportable to all treatment sites. Care must especially be taken when vascular-damaging PDT is applied to large surface areas. There is a technical difficulty to reproducibly and homogeneously deliver PDT to large internal surfaces, like what is required for treatment to the entire intrapleural or intraperitoneal spaces, and there is a lack of light-directed targeting for PDT administration to these surfaces. As discussed in the preceding sections, hyperpermeability is common with PDT, and the treatment-induced venule leakage of macromolecules can begin shortly after the start of light administration and continue for an extended period of observation.[80] While this can be clinically asymptomatic, a more systemic capillary leakage can be induced by PDT delivery to large surfaces like that which is required for intraperitoneal or intrapleural PDT. In such cases, symptoms can be profound and life threatening.

To this point, evidence of vascular effects was found during a phase II clinical trial of intraoperative PDT for disseminated intraperitoneal cancer. Among 65 patients treated with surgical debulking and PDT, the mean operative time was 9.8 h and the mean blood loss was nearly 1.5 L. Patients required crystalloid and blood product replacement in the intraoperative and postoperative period. The median ICU stay postoperatively was four days, and 69% of patients developed at least one clinically significant complication, including 9% with acute respiratory distress syndrome (ARDS).[120] This clinical picture is in keeping with the known mechanisms of systemic inflammatory response syndrome. Systemic inflammatory response syndrome is characterized by increased vascular permeability, which results in intravascular volume depletion and tissue edema. It can lead to ARDS and death.[121] Of note, the constellation of massive fluid requirements, increased need for prolonged mechanical ventilation, and high incidence of pulmonary complications seen with PDT of intraperitoneal cancer is rarely observed after similar major abdominal surgeries when not delivering PDT.[120] Given these potential vascular effects, patients receiving PDT to large internal surface areas should undergo careful ICU monitoring and aggressive volume resuscitation.

6.2. *Clinical Monitoring of Vascular Effects*

Due to the known clinical occurrence of vascular damage and its documented relevance to preclinical outcomes, there is great interest in measuring tissue physiologic properties in patients who are undergoing PDT. Overall, the ability to obtain quantitative measures of optical parameters, photosensitizer concentrations, blood flow, and tissue oxygenation allows for improved and more accurate PDT dosimetry. New methods to obtain optical measurements are minimally invasive to patients and can allow for assessment during light delivery. Thus, these technologies can facilitate real-time adjustments to be made during PDT, which could ultimately lead to improved clinical outcomes.[60]

To test the feasibility of incorporating tissue physiologic measurements during the process of light delivery for PDT, Yu *et al.*[58] employed a dual-purpose probe that employed diffuse reflectance spectroscopy for measuring tissue hemoglobin oxygen saturation and diffuse correlation spectroscopy for measuring blood flow. This probe was applied clinically to measure the physiologic properties of three patients with locally recurrent prostate cancer being treated with MLu-mediated interstitial prostate PDT. MLu was given at 2 mg/kg with a 3-h drug-light interval, and treatment was administered to a total light dose of $100\,J/cm^2$. Results showed that prostrate hemoglobin oxygen saturation decreased slightly (approximately 3%) after treatment, whereas prostate blood flow and total hemoglobin concentration decreased more appreciably over the course of treatment, by 50% and 15%, respectively. Therefore, this study demonstrated that PDT created a vascular effect that could be measured over the course of light delivery.

Similar clinical studies assessing diffuse optical measurements of blood flow and tissue hemoglobin oxygen saturation in conjunction with PDT have been performed with other photosensitizers. In a phase I trial of HPPH-mediated PDT for head and neck cancer patients, blood flow, oxygenation, and HPPH drug photobleaching were analyzed before and after PDT delivery using fast, non-invasive diffuse optical methods.[122,123] HPPH-PDT was found to induce significant drug photobleaching and reduce blood flow and oxygenation, thus suggesting significant vascular and cellular reactions. Of note, these changes were accompanied by cross-linking of STAT3, a molecular measure for the oxidative photoreaction.[124]

Physiologic measurements could also prove useful during ALA-PDT. Given its short half-life and depth of penetration, ALA has emerged as a leading choice of photosensitizer for early-stage invasive and in situ non-melanoma skin cancers. However, difficulties with topical and intra-lesional applications have been described,[125] which has led to the suggestion that vascular disruption early in treatment should be identified and prevented. In patients with basal cell carcinoma, topical ALA-mediated PDT induced early blood flow changes assessed by diffuse correlation spectroscopy that were irradiance dependent. Since there can be significant variations in the blood flow responses in patients from lesion to lesion, the ability to monitor blood flow in real time can aid clinicians in evaluating ALA-PDT response and treatment delivery.[57]

In conjunction with Photofrin-mediated PDT of large surface areas, optical measurements of tissue properties are also being pursued. Such measurements were previously performed as part of the above-mentioned intraperitoneal trials,[126] and are presently being conducted in the now accruing multi-center phase II randomized trial, which compares survival for a lung-sparing surgery with or without intraoperative Photofrin-mediated PDT for patients with malignant pleural mesothelioma. In the prospective trial, probes are placed intraoperatively on or immediately above the tissue surface in order to provide optical-based monitoring of the PDT field throughout light delivery. One set of probes uses frequency-modulated red and near-IR light to evaluate relative changes in hemoglobin volume and saturation, and concomitantly, to quantify changes in blood flow. Another set of probes uses the PDT treatment light or frequency-modulated light sources, and either both optical and electronic filters or optical filters alone to assess fluorescence emission during light delivery, thereby reporting on the extent and rate of photobleaching of the sensitizer.[127,128] It is a long term goal of this work to compare the acquired optical data to patient outcome for the purpose of testing the clinical predictive value of physiologic and photosensitizer (Photofrin) measurements.

6.3. *Clinical Examples of Vascular-targeted PDT for Tumors*

The phenomenon of vascular-targeted PDT in oncology has perhaps been best demonstrated, and is most easily conceptualized, through the examination of prostate PDT as a case in point. Vascular-targeted PDT of prostate cancer has been performed in conjunction with the previously mentioned vascular-occluding photosensitizers: Tookad (palladium bacteriopheophorbide; WST-09), and its water-soluble derivative Tookad-Soluble (palladium bacteriopheophorbide monolysotaurine; WST-11). The properties of these photosensitizers allow them to remain ideally confined to the vascular bed after administration. In the first report of Tookad in humans, 24 patients with histologically-proven recurrent prostate cancer following definitive radiotherapy were treated in cohorts receiving drug dose escalation (0.1 mg/kg to 2.0 mg/kg) at a fixed 100 J/cm per interstitial source fiber, or at a fixed drug dose of 2 mg/kg with escalating light doses (100 J/cm to 360 J/cm). This study demonstrated that vascular-targeted PDT was technically feasible, was not associated with serious adverse events, and appeared effective, with all six patients at the highest light dose responding to treatment.[129] In a phase II study of escalating light doses, 28 patients with recurrent localized prostate cancer were given whole-prostate ablation treatment after the failure of external beam radiotherapy. Trachtenberg *et al.*[130] found that treatment of the whole prostate was possible with limited damage to surrounding organs. Furthermore, tumor response, as assessed on MRI, improved with increasing light dose. Of the 13 patients treated at the highest light dose levels, 8 had a complete and modestly durable response to therapy, as assessed by a negative repeat biopsy six months after treatment, and a decreased level or inability to detect PSA in peripheral blood after treatment.

In an attempt to avoid the intraoperative hypotension with cardiovascular effects that were seen in these early studies,[129,130] Tookad-Soluble has more recently been employed for clinical use. In a recent phase II prospective, multicenter, open-label, multiple-arm trial, 85 men with localized low-risk prostate cancer, not previously receiving therapy, were treated with palladium Tookad-Soluble-mediated vascular-targeted PDT. Patients received 4 mg/kg or 6 mg/kg of photosensitizer administered intravenously, and activated with 200 J/cm or 300 J/cm of light, delivered by light-diffusing fibers within the prostate through the perineum. At six months following treatment, 74% of patients who underwent repeat prostate biopsy had histopathology that was negative for malignancy. While 87% of patients reported at least one treatment-related adverse event, most events were mild or moderate in intensity and thought to be most related to the technical procedures of the study, and no clinical hypotension was observed.[131] Based on these encouraging results, a randomized phase III trial using Tookad-Soluble in prostate cancer patients was designed and is currently accruing.

7. Conclusions

More often than not, vascular response is central to the clinical and preclinical therapeutic actions of PDT. Vascular damage can result from protocols that use short drug-light intervals or targeted antibodies to direct PDT effect against the blood vessels. Yet, it also develops as a result of damage to the endothelium when PDT conditions are designed to promote cellular damage. Under both sets of circumstances, the vascular reaction consists of a complex interplay of hemodynamic, inflammatory and signaling responses that interact through multiple pathways and mediators to create what collectively amounts to vascular damage. Our knowledge of how its action on blood vessels leads to other aspects of PDT response serves to facilitate new methods toward optimizing treatment delivery. For example, it can support the rationale for design of combination therapies that pair PDT with molecular targeting drugs so as to augment PDT effect through vascular/immune pathways. It provides impetus to monitor vascular changes during the clinical delivery of PDT toward the development of real-time hemodynamic-based dosimetry for light delivery. In short, the centralized role of blood vessels in many mechanisms of tissue response to PDT provides an abundance of opportunities ready for exploration that could serve to develop a more individualized approach to PDT.

Acknowledgments

Support during the preparation of this text was provided by NIH Grants CA-085831, CA-129554, and CA-087971. We thank Ms. Chiharu Sako for administrative assistance, and Ms. Shirron Carter and Ms. Joann Miller for their critical reading of the document.

References

1. D. Hanahan and R. A. Weinberg, *Cell* **144**(5), 646 (2011).
2. R. R. Allison and K. Moghissi, *Photodiagnosis Photodyn. Ther.* **10**(4), 331 (2013).
3. C. M. Moore, A. R. Azzouzi, E. Barret, A. Villers, G. Muir, N. Barber, J. Trachtenberg, N. Arumainayagam, B. Gaillac, C. Allen, A. Schertz and M. Emberton, *BJU Int.* **116**(6), 888 (2015).
4. N. Madar-Balakirski, C. Tempel-Brami, V. Kalchenko, O. Brenner, D. Varon, A. Scherz and Y. Salomon, *PLoS One* **5**(4), e10282 (2010).
5. F. Borle, A. Radu, P. Monnier, H. van den Bergh and G. Wagnieres, *Photochem. Photobiol.* **78**(4), 377 (2003).
6. O. Mazor, A. Brandis, V. Plaks, E. Neumark, V. Rosenbach-Belkin, Y. Salomon and A. Scherz, *Photochem. Photobiol.* **81**(2), 342 (2005).

7. A. Brandis, O. Mazor, E. Neumark, V. Rosenbach-Belkin, Y. Salomon and A. Scherz, *Photochem. Photobiol.* **81**(4), 983 (2005).
8. B. Chen, B. W. Pogue, P. J. Hoopes and T. Hasan, *Crit. Rev. Eukaryot. Gene Expr.* **16**(4), 279 (2006).
9. U. Schmidt-Erfurth and T. Hasan, *Surv. Ophthalmol.* **45**(3), 195 (2000).
10. U. Schmidt-Erfurth, T. Hasan, K. Schomacker, T. Flotte and R. Birngruber, *Lasers Surg. Med.* **17**(2), 178 (1995).
11. U. Schmidt-Erfurth, T. J. Flotte, E. S. Gragoudas, K. Schomacker, R. Birngruber and T. Hasan, *Exp. Eye Res.* **62**(1), 1 (1996).
12. Q. Peng and J. M. Nesland, *Ultrastruct. Pathol.* **28**(5–6), 333 (2004).
13. P. Fungaloi, R. Statius van Eps, Y. P. Wu, J. Blankensteijn, P. de Groot, H. van Urk, R. van Hillegersberg and G. LaMuraglia, *Photochem. Photobiol.* **75**(4), 412 (2002).
14. A. L. Maas, S. L. Carter, E. P. Wileyto, J. Miller, M. Yuan, G. Yu, A. C. Durham and T. M. Busch, *Cancer Res.* **72**(8), 2079 (2012).
15. Z. Hu, B. Rao, S. Chen and J. Duanmu, *BMC Cancer* **10**, 235 (2010).
16. J. Duanmu, J. Cheng, J. Xu, C. J. Booth and Z. Hu, *Br. J. Cancer* **104**(9), 1401 (2011).
17. D. Bechet, L. Tirand, B. Faivre, F. Plenat, C. Bonnet, T. Bastogne, C. Frochot, F. Guillemin and M. Barberi-Heyob, *Pharm. Res.* **27**(3), 468 (2010).
18. J. R. Starkey, E. M. Pascucci, M. A. Drobizhev, A. Elliott and A. K. Rebane, *Biochim. Biophys. Acta.* **1830**(10), 4594 (2013).
19. K. Ichikawa, T. Hikita, N. Maeda, S. Yonezawa, Y. Takeuchi, T. Asai, Y. Namba and N. Oku, *Biochim. Biophys. Acta.* **1669**(1), 69 (2005).
20. W. J. de Vree, A. N. Fontijne-Dorsman, J. F. Koster and W. Sluiter, *Br. J. Cancer* **73**(11), 1335 (1996).
21. M. Pazos and H. B. Nader, *Braz. J. Med. Biol. Res.* **40**(8), 1025 (2007).
22. V. H. Fingar, *J. Clin. Laser Med. Surg.* **14**(5), 323 (1996).
23. Z. Huang, Q. Chen, K. C. Dole, A. B. Barqawi, Y. K. Chen, D. Blanc, B. C. Wilson and F. W. Hetzel, *Photochem. Photobiol. Sci.* **6**(12), 1318 (2007).
24. R. C. Mesquita, S. W. Han, J. Miller, S. S. Schenkel, A. Pole, T. V. Esipova, S. A. Vinogradov, M. E. Putt, A. G. Yodh and T. M. Busch, *PloS One* 7(5), e37322 (2012).
25. C. He, P. Agharkar and B. Chen, *Pharm. Res.* **25**(8), 1873 (2008).
26. T. M. Busch, X. Xing, G. Yu, A. Yodh, E. P. Wileyto, H. W. Wang, T. Durduran, T. C. Zhu and K. K. Wang, *Photochem. Photobiol. Sci.* **8**(12), 1683 (2009).
27. S. Dick, I. Khludeyev, L. Vasilevskaya and V. Zorin, *Z Med. Phys.* **19**(2), 90 (2009).
28. T. M. Busch, H. W. Wang, E. P. Wileyto, G. Yu and R. M. Bunte, *Radiat. Res.* **174**(3), 331 (2010).
29. M. Seshadri, D. A. Bellnier, L. A. Vaughan, J. A. Spernyak, R. Mazurchuk, T. H. Foster and B. W. Henderson, *Clin. Cancer Res.* **14**(9), 2796 (2008).
30. T. A. Middelburg, H. S. de Bruijn, L. Tettero, A. van der Ploeg van den Heuvel, H. A. Neumann, E. R. de Haas and D. J. Robinson, *J. Photochem. Photobiol. B* **126**, 26 (2013).
31. M. Khurana, E. H. Moriyama, A. Mariampillai, K. Samkoe, D. Cramb and B. C. Wilson, *J. Biomed. Opt.* **14**(6), 064006 (2009).
32. F. van Leeuwen-van Zaane, H. S. de Bruijn, A. van der Ploeg-van den Heuvel, H. J. Sterenborg and D. J. Robinson, *Photodiagnosis Photodyn. Ther.* **11**(2), 71 (2014).
33. V. H. Fingar, T. J. Wieman, S. A. Wiehle and P. B. Cerrito, *Cancer Res.* **52**(18), 4914 (1992).

34. K. J. Reeves, M. W. Reed and N. J. Brown, *J. Photochem. Photobiol. B* **101**(3), 224 (2010).
35. V. H. Fingar, K. A. Siegel, T. J. Wieman and K. W. Doak, *Photochem. Photobiol.* **58**(3), 393 (1993).
36. R. C. Brooke, A. Sinha, M. K. Sidhu, R. E. Watson, M. K. Church, P. S. Friedmann, G. F. Clough and L. E. Rhodes, *J. Invest. Dermatol.* **126**(10), 2296 (2006).
37. A. Ferrario, A. M. Fisher, N. Rucker and C. J. Gomer, *Cancer Res.* **65**(20), 9473 (2005).
38. B. Chen, B. W. Pogue, J. M. Luna, R. L. Hardman, P. J. Hoopes and T. Hasan, *Clin. Cancer Res.* **12**(3), 917 (2006).
39. E. Debefve, F. Mithieux, J. Y. Perentes, Y. Wang, C. Cheng, S. C. Schaefer, C. Ruffieux, J. P. Ballini, M. Gonzalez, H. van den Bergh, H. B. Ris, H. A. Lehr and T. Krueger, *Lasers Surg. Med.* **43**(7), 696 (2011).
40. B. Chen, C. Crane, C. He, D. Gondek, P. Agharkar, M. D. Savellano, P. J. Hoopes and B. W. Pogue, *Int. J. Cancer* **123**(3), 695 (2008).
41. A. C. Kubler, W. Stenzel, M. Ruhling, B. Meul and J. H. Fischer, *Lasers Surg. Med.* **33**(4), 247 (2003).
42. E. Debefve, B. Pegaz, H. van den Bergh, G. Wagnieres, N. Lange and J. P. Ballini, *Angiogenesis* **11**(3), 235 (2008).
43. E. Debefve, C. Cheng, S. C. Schaefer, H. Yan, J. P. Ballini, H. van den Bergh, H. A. Lehr, C. Ruffieux, H. B. Ris and T. Krueger, *J. Photochem. Photobiol. B* **98**(1), 69 (2010).
44. Z. Zhen, W. Tang, Y. J. Chuang, T. Todd, W. Zhang, X. Lin, G. Niu, G. Liu, L. Wang, Z. Pan, X. Chen and J. Xie, *ACS Nano* **8**(6), 6004 (2014).
45. J. W. Snyder, W. R. Greco, D. A. Bellnier, L. Vaughan and B. W. Henderson, *Cancer Res.* **63**(23), 8126 (2003).
46. M. O. Senge and M. W. Radomski, *Photodiagnosis Photodyn. Ther.* **10**(1), 1 (2013).
47. W. Kobayashi, Q. Liu, T. Matsumiya, H. Nakagawa, H. Yoshida, T. Imaizumi, K. Satoh and H. Kimura, *Oral Oncol.* **40**(5), 506 (2004).
48. G. R. Nigri, S. Kossodo, P. Waterman, P. Fungaloi and G. M. LaMuraglia, *J. Vasc. Surg.* **39**(4), 843 (2004).
49. P. Fungaloi, P. Waterman, G. Nigri, R. Statius-van Eps, W. Sluiter, H. van Urk and G. LaMuraglia, *Photochem. Photobiol.* **78**(5), 475 (2003).
50. L. Zhou, X. L. Qi, M. X. Xu, Y. Mao, M. L. Liu and H. M. Song, *Acta Pharmacol. Sin.* **35**(9), 1103 (2014).
51. D. E. Dolmans, A. Kadambi, J. S. Hill, C. A. Waters, B. C. Robinson, J. P. Walker, D. Fukumura and R. K. Jain, *Cancer Res.* **62**(7), 2151 (2002).
52. M. Khurana, E. H. Moriyama, A. Mariampillai and B. C. Wilson, *J. Biomed. Opt.* **13**(4), 040502 (2008).
53. M. W. Reed, F. N. Miller, T. J. Wieman, M. T. Tseng and C. G. Pietsch, *J. Surg. Res.* **45**(5), 452 (1988).
54. K. S. McMahon, T. J. Wieman, P. H. Moore and V. H. Fingar, *Cancer Res.* **54**(20), 5374 (1994).
55. V. H. Fingar, P. K. Kik, P. S. Haydon, P. B. Cerrito, M. Tseng, E. Abang and T. J. Wieman, *Br J. Cancer* **79**(11–12), 1702 (1999).
56. G. Yu, T. Durduran, C. Zhou, H. W. Wang, M. E. Putt, H. M. Saunders, C. M. Sehgal, E. Glatstein, A. G. Yodh and T. M. Busch, *Clin. Cancer Res.* **11**(9), 3543 (2005).
57. T. L. Becker, A. D. Paquette, K. R. Keymel, B. W. Henderson and U. Sunar, *Biomed. Opt. Express* **2**(1), 123 (2010).

58. G. Yu, T. Durduran, C. Zhou, T. C. Zhu, J. C. Finlay, T. M. Busch, S. B. Malkowicz, S. M. Hahn and A. G. Yodh, *Photochem. Photobiol.* **82**(5), 1279 (2006).
59. D. M. Milstein, A. M. van Kuijen, M. P. Copper, B. Karakullukcu, I. B. Tan, J. A. Lindeboom, W. J. Fokkens and C. Ince, *Photodiagnosis Photodyn. Ther.* **9**(1), 69 (2012).
60. U. Sunar, *World J. Clin. Cases* **1**(3), 96 (2013).
61. T. M. Busch, *Lasers Surg. Med.* **38**(5), 494 (2006).
62. T. J. Dougherty, C. J. Gomer, B. W. Henderson, G. Jori, D. Kessel, M. Korbelik, J. Moan and Q. Peng, *J. Natl. Cancer Inst.* **90**(12), 889 (1998).
63. J. Dong, H. J. Toh, P. S. Thong, C. S. Tee, R. Bi, K. C. Soo and K. Lee, *J. Photochem. Photobiol. B* **140**, 163 (2014).
64. H. W. Wang, M. E. Putt, M. J. Emanuele, D. B. Shin, E. Glatstein, A. G. Yodh and T. M. Busch, *Cancer Res.* **64**(20), 7553 (2004).
65. T. H. Pham, R. Hornung, M. W. Berns, Y. Tadir and B. J. Tromberg, *Photochem. Photobiol.* **73**(6), 669 (2001).
66. B. A. Standish, K. K. Lee, X. Jin, A. Mariampillai, N. R. Munce, M. F. Wood, B. C. Wilson, I. A. Vitkin and V. X. Yang, *Cancer Res.* **68**(23), 9987 (2008).
67. H. W. Wang, E. Rickter, M. Yuan, E. P. Wileyto, E. Glatstein, A. Yodh and T. M. Busch, *Photochem. Photobiol.* **83**(5), 1040 (2007).
68. M. Firczuk, D. Nowis and J. Golab, *Photochem. Photobiol. Sci.* **10**(5), 653 (2011).
69. A. D. Garg, D. V. Krysko, P. Vandenabeele and P. Agostinis, *Photochem. Photobiol. Sci.* **10**(5), 670 (2011).
70. K. Oikonomopoulou, D. Ricklin, P. A. Ward and J. D. Lambris, *Semin. Immunopathol.* **34**(1), 151 (2012).
71. B. Stott and M. Korbelik, *Cancer Immunol. Immunother.* **56**(5), 649 (2007).
72. I. Cecic and M. Korbelik, *J. Environ. Pathol. Toxicol. Oncol.* **25**(1–2), 189 (2006).
73. I. Cecic, A. I. Minchinton and M. Korbelik, *Photochem. Photobiol.* **83**(5), 1049 (2007).
74. S. O. Gollnick, *J. Natl. Compr. Canc. Netw.* **10** Suppl 2, S40 (2012).
75. I. Cecic, B. Stott and M. Korbelik, *Int. Immunopharmacol.* **6**(8), 1259 (2006).
76. C. E. Grossman, S. Pickup, A. Durham, E. P. Wileyto, M. E. Putt and T. M. Busch, *Lasers Surg. Med.* **43**(7), 663 (2011).
77. T. W. Bauer, S. M. Hahn, F. R. Spitz, A. Kachur, E. Glatstein and D. L. Fraker, *Ann. Surg. Oncol.* **8**(3), 254 (2001).
78. M. A. Rosenthal, B. Kavar, J. S. Hill, D. J. Morgan, R. L. Nation, S. S. Stylli, R. L. Basser, S. Uren, H. Geldard, M. D. Green, S. B. Kahl and A. H. Kaye, *J. Clin. Oncol.* **19**(2), 519 (2001).
79. T. Goerge, B. Ho-Tin-Noe, C. Carbo, C. Benarafa, E. Remold-O'Donnell, B. Q. Zhao, S. M. Cifuni and D. D. Wagner, *Blood* **111**(10), 4958 (2008).
80. V. H. Fingar, T. J. Wieman and P. S. Haydon, *Photochem. Photobiol.* **66**(4), 513 (1997).
81. W. J. de Vree, M. C. Essers, H. S. de Bruijn, W. M. Star, J. F. Koster and W. Sluiter, *Cancer Res.* **56**(13), 2908 (1996).
82. I. Cecic and M. Korbelik, *Cancer Lett.* **183**(1), 43 (2002).
83. P. C. Kousis, B. W. Henderson, P. G. Maier and S. O. Gollnick, *Cancer Res.* **67**(21), 10501 (2007).
84. M. Korbelik, *Lasers Surg. Med.* **38**(5), 500 (2006).
85. C. M. Brackett and S. O. Gollnick, *Photochem. Photobiol. Sci.* **10**(5), 649 (2011).

86. C. M. Brackett, J. B. Muhitch, S. S. Evans and S. O. Gollnick, *J. Immunol.* **191**(8), 4348 (2013).
87. D. Preise, R. Oren, I. Glinert, V. Kalchenko, S. Jung, A. Scherz and Y. Salomon, *Cancer Immunol. Immunother.* **58**(1), 71 (2009).
88. L. H. Wei, H. Baumann, E. Tracy, Y. Wang, A. Hutson, S. Rose-John and B. W. Henderson, *Br J. Cancer* **97**(11), 1513 (2007).
89. E. Reginato, J. Lindenmann, C. Langner, N. Schweintzger, I. Bambach, F. Smolle-Juttner and P. Wolf, *Photochem. Photobiol. Sci.* **13**(9), 1281 (2014).
90. S. O. Gollnick, S. S. Evans, H. Baumann, B. Owczarczak, P. Maier, L. Vaughan, W. C. Wang, E. Unger and B. W. Henderson, *Br J. Cancer.* **88**(11), 1772 (2003).
91. C. M. Brackett, B. Owczarczak, K. Ramsey, P. G. Maier and S. O. Gollnick, *Lasers Surg. Med.* **43**(7), 676 (2011).
92. G. L. Semenza, *Trends Pharmacol. Sci.* **33**(4), 207 (2012).
93. A. Ferrario, K. F. von Tiehl, N. Rucker, M. A. Schwarz, P. S. Gill and C. J. Gomer, *Cancer Res.* **60**(15), 4066 (2000).
94. B. Kosharskyy, N. Solban, S. K. Chang, I. Rizvi, Y. Chang and T. Hasan, *Cancer Res.* **66**(22), 10953 (2006).
95. R. Bhuvaneswari, G. Y. Yuen, S. K. Chee and M. Olivo, *Photochem. Photobiol. Sci.* **6**(12), 1275 (2007).
96. R. Bhuvaneswari, G. Y. Yuen, S. K. Chee and M. Olivo, *Lasers Surg. Med.* **43**(7), 651 (2011).
97. Q. Zhou, M. Olivo, K. Y. Lye, S. Moore, A. Sharma and B. Chowbay, *Cancer Chemother. Pharmacol.* **56**(6), 569 (2005).
98. Q. Zhan, W. Yue and S. Hu, *Photodiagnosis Photodyn. Ther.* **8**(4), 314 (2011).
99. S. H. Jee, S. C. Shen, H. C. Chiu, W. L. Tsai and M. L. Kuo, *Oncogene* **20**(2), 198 (2001).
100. T. Cohen, D. Nahari, L. W. Cerem, G. Neufeld and B. Z. Levi, *J. Biol. Chem.* **271**(2), 736 (1996).
101. Q. Xu, J. Briggs, S. Park, G. Niu, M. Kortylewski, S. Zhang, T. Gritsko, J. Turkson, H. Kay, G. L. Semenza, J. Q. Cheng, R. Jove and H. Yu, *Oncogene* **24**(36), 5552 (2005).
102. G. Niu, K. L. Wright, M. Huang, L. Song, E. Haura, J. Turkson, S. Zhang, T. Wang, D. Sinibaldi, D. Coppola, R. Heller, L. M. Ellis, J. Karras, J. Bromberg, D. Pardoll, R. Jove and H. Yu, *Oncogene* **21**(13), 2000 (2002).
103. A. Ferrario, K. Von Tiehl, S. Wong, M. Luna and C. J. Gomer, *Cancer Res.* **62**(14), 3956 (2002).
104. N. Hendrickx, C. Volanti, U. Moens, O. M. Seternes, P. de Witte, J. R. Vandenheede, J. Piette and P. Agostinis, *J. Biol. Chem.* **278**(52), 52231 (2003).
105. Y. Akita, K. Kozaki, A. Nakagawa, T. Saito, S. Ito, Y. Tamada, S. Fujiwara, N. Nishikawa, K. Uchida, K. Yoshikawa, T. Noguchi, O. Miyaishi, K. Shimozato, S. Saga and Y. Matsumoto, *Br J. Dermatol.* **151**(2), 472 (2004).
106. N. Solban, P. K. Selbo, A. K. Sinha, S. K. Chang and T. Hasan, *Cancer Res.* **66**(11), 5633 (2006).
107. C. Edmonds, S. Hagan, S. M. Gallagher-Colombo, T. M. Busch and K. A. Cengel, *Cancer Biol. Ther.* **13**(14), 1463 (2012).
108. T. Karnezis, R. Shayan, S. Fox, M. G. Achen and S. A. Stacker, *Oncotarget* **3**(8), 893 (2012).

109. K. S. Chun and R. Langenbach, *Mol. Carcinog.* **50**(6), 439 (2011).
110. C. Han, A. J. Demetris, D. B. Stolz, L. Xu, K. Lim and T. Wu, *J. Biol. Chem.* **281**(34), 24831 (2006).
111. H. W. Lo, X. Cao, H. Zhu and F. Ali-Osman, *Mol. Cancer Res.* **8**(2), 232 (2010).
112. S. M. Gallagher-Colombo, A. L. Maas, M. Yuan and T. M. Busch, *Isr. J. Chem.* **52**, 681 (2012).
113. A. Ferrario, K. F. von Tiehl, N. Rucker, M. A. Schwarz, P. S. Gill and C. J. Gomer, *Cancer Res.* **60**(15), 4066 (2000).
114. Q. Zhan, W. Yue and S. Hu, *Photodiagnosis Photodyn. Ther.* **8**(4), 314 (2011).
115. Z. Li, P. Agharkar and B. Chen, *Cancer Lett.* **339**(1), 128 (2013).
116. M. G. del Carmen, I. Rizvi, Y. Chang, A. C. Moor, E. Oliva, M. Sherwood, B. Pogue and T. Hasan, *J. Natl. Cancer Inst.* **97**(20), 1516 (2005).
117. B. Fateye, W. Li, C. Wang and B. Chen, *Photochem. Photobiol.* **88**(5), 1265 (2012).
118. S. A. Grupp, *Best Pract. Res. Clin. Haematol.* **27**(3–4), 222 (2014).
119. P. Agostinis, K. Berg, K. A. Cengel, T. H. Foster, A. W. Girotti, S. O. Gollnick, S. M. Hahn, M. R. Hamblin, A. Juzeniene, D. Kessel, M. Korbelik, J. Moan, P. Mroz, D. Nowis, J. Piette, B. C. Wilson and J. Golab, *CA Cancer J. Clin.* **61**(4), 250 (2011).
120. R. J. Canter, R. Mick, S. B. Kesmodel, D. J. Raz, F. R. Spitz, J. M. Metz, E. J. Glatstein, S. M. Hahn and D. L. Fraker, *Ann. Surg. Oncol.* **10**(5), 514 (2003).
121. R. C. Bone, *Crit Care Med.* **24**(1), 163 (1996).
122. D. J. Rohrbach, N. Rigual, E. Tracy, A. Kowalczewski, K. L. Keymel, M. T. Cooper, W. Mo, H. Baumann, B. W. Henderson and U. Sunar, *Biomed. Opt. Express* **3**(9), 2142 (2012).
123. U. Sunar, D. Rohrbach, N. Rigual, E. Tracy, K. Keymel, M. T. Cooper, H. Baumann and B. H. Henderson, *Opt. Express* **18**(14), 14969 (2010).
124. N. Rigual, G. Shafirstein, M. T. Cooper, H. Baumann, D. A. Bellnier, U. Sunar, E. C. Tracy, D. J. Rohrbach, G. Wilding, W. Tan, M. Sullivan, M. Merzianu and B. W. Henderson, *Clin. Cancer Res.* **19**(23), 6605 (2013).
125. R. F. Donnelly, P. A. McCarron and A. D. Woolfson, *Perspect. Medicin. Chem.* **1**, 49 (2008).
126. H. W. Wang, T. C. Zhu, M. E. Putt, M. Solonenko, J. Metz, A. Dimofte, J. Miles, D. L. Fraker, E. Glatstein, S. M. Hahn and A. G. Yodh, *J. Biomed. Opt.* **10**(1), 14004 (2005).
127. J. C. Finlay and T. H. Foster, *Appl. Opt.* **44**(10), 1917 (2005).
128. J. C. Finlay, T. C. Zhu, A. Dimofte, D. Stripp, S. B. Malkowicz, T. M. Busch and S. M. Hahn, *Photochem. Photobiol.* **82**(5), 1270 (2006).
129. J. Trachtenberg, A. Bogaards, R. A. Weersink, M. A. Haider, A. Evans, S. A. McCluskey, A. Scherz, M. R. Gertner, C. Yue, S. Appu, A. Aprikian, J. Savard, B. C. Wilson and M. Elhilali, *J. Urol.* **178**(5), 1974 (2007).
130. J. Trachtenberg, R. A. Weersink, S. R. Davidson, M. A. Haider, A. Bogaards, M. R. Gertner, A. Evans, A. Scherz, J. Savard, J. L. Chin, B. C. Wilson and M. Elhilali, *BJU Int.* **102**(5), 556 (2008).
131. A. R. Azzouzi, E. Barret, C. M. Moore, A. Villers, C. Allen, A. Scherz, G. Muir, M. de Wildt, N. J. Barber, S. Lebdai and M. Emberton, *BJU Int.* **112**(6), 766 (2013).

Chapter 11

EtNBS in Photodynamic Therapy

*Nicholas H. Nowell**, *Hsin-I Hung*†, *Lauren A. Austin*†
and Conor L. Evans†

**Harvard University, Cambridge, MA 02138*
†*Wellman Center for Photomedicine, Harvard Medical School, Massachusetts General Hospital, Charlestown, MA 02129*

Abstract

A derivative of Nile Blue, a small molecule dye which selectively stains tumors, EtNBS has in recent decades become a popular photosensitizer in photodynamic therapy (PDT). Its structure has been designed to maximize its photosensitization capability and to improve upon the shortcomings of earlier porphyrin-based photosensitizers. EtNBS rapidly localizes to the lysosome and efficiently absorbs light at a wavelength of 652 nm. Unlike many photosensitizers, it can operate via both oxygen-dependent and oxygen-independent pathways with remarkable efficiency. As a result, PDT with EtNBS has shown particular promise in the eradication of entire tumors, including the hypoxic, acidic cores of nodules that have traditionally been unresponsive to most therapeutic regimens. EtNBS has also been investigated, with success, as a possible treatment for Gram-negative and Gram-positive bacterial infections, leishmaniasis, tuberculosis, and fungal infections, among many other diseases.

1. Introduction

Photodynamic therapy (PDT) has traditionally relied on three indispensable components — a light-activatable molecule known as a photosensitizer, tissue oxygenation to enable formation of destructive reactive oxygen species (ROS), and excitation with a specific wavelength of light to set the photosensitizing process in motion.[1] The uptake and subcellular localization profiles of a given photosensitizer significantly affect the mechanism by which it ultimately kills cells and govern its overall PDT efficacy.[2] Over the past few decades, much work has gone toward elucidating these structure–function relationships across and within diverse classes of photosensitizers.

Photosensitizers based on porphyrins have played a large role in the development of PDT. The first photosensitizer studied clinically was hematoporphyrin derivative (HPD), which emerged in the 1970s and saw use in breast, skin, and lung cancer patients.[3–6] However, both HPD, which exists as a mixture of oligomers, and the more purified Photofrin (PII) suffered from several shortcomings.[7–11] Among these were:

- Low absorption in the so-called "therapeutic window" or "optical window" from approximately 600–1000 nm in which red or near-infrared light penetrates deep into tissue.
- Complicated uptake profiles and photoactivation conditions, as both HPD and PII are mixtures of several porphyrin ethers and ester oligomers.
- Prolonged retention in the skin, causing undesirable, delayed, long-term photosensitization.

The desire to overcome these limitations drove a number of investigations to design next generation photosensitizers based on porphyrins, chlorins, pupurins, phthalocyanines, and other photoactive structures.[12–14]

One promising line of research began with Lewis *et al.*,[15–17] who found that Nile Blue dyes, a family of cationic benzo[*a*]phenoxazines, stained tumors selectively and retarded their growth in mice with xenografted sarcomas while inducing only minimal systemic toxicity. Subsequent experiments reproduced this nontoxic staining on several occasions in the original model and in spontaneous adenocarcinomas.[16] Unfortunately, when tested as part of a PDT regimen, Nile Blue dyes offered no photoactivity. Nevertheless, in pursuit of a novel, non-porphyrin photosensitizer that could overcome the aforementioned obstacles, several groups continued to investigate the Nile Blue family and its derivatives.[18–21] Through a few modifications, including substitution of the central oxygen atom of benzophenoxazines with a sulfur atom, a promising new derivative, 5-ethylamino-9-diethylaminobenzo[*a*]phenothiazinium (EtNBS), was created.[9] The structure and delocalized positive charge of EtNBS made it both lipophilic and water-soluble. Both of these traits, which seem to be shared by most Nile Blue derivatives, were found essential to its potency as a photosensitizer.

Ethylation of the 5-amino group of Nile Blue leads to the creation of ethylamino-9-diethylaminobenzo[*a*]phenoxazinium (EtNBA). Replacement of the central oxygen atom of EtNBA with sulfur or selenium yields EtNBS or EtNBSe, respectively. The introduction of atoms heavier than oxygen has a dramatic effect on singlet oxygen production and on overall phototoxicity. Under standard cell culture conditions, the order of efficacy is EtNBA < EtNBS ≪ EtNBSe.[9,11] All three compounds are rapidly taken up by cells and have similar uptake and efflux profiles. In contrast to HPD and PII, which bind to the plasma membrane and vascular stroma, these three Nile Blue derivatives localize intracellularly, in the lysosome.[9,21,22] From here, they cause direct cellular damage through either radical- or singlet oxygen-mediated mechanisms. Much attention has been given to this series of compounds, particularly EtNBS, which has functioned well in various PDT regimens across a breadth of organisms and in both normoxic and hypoxic conditions, unlike the majority of clinically available PDT regimens.[23–26]

Here, we present what we feel is a comprehensive review of the photosensitizer EtNBS, including its synthesis, photochemical properties, derivatives, antimicrobial and antineoplastic activity, and in particular its versatility with respect to environmental oxygenation. The chapter will conclude with some thoughts on the future of EtNBS in PDT. Although we strive to provide the best possible summary and analysis, the present understanding of EtNBS and the relevant chemistry and biology is undoubtedly imperfect and we fully expect that ongoing research will come to revise existing knowledge and provide new insight into this potent photosensitizer.

2. Synthesis and Chemical Modifications

Since EtNBS is derived from Nile Blue A (NBA), any discussion of the synthesis of EtNBS should begin with that of its precursor. NBA (**3**) was first created in the late 1800s as part of larger investigations into novel industrial dyes made from naphthalenes.[27] The original scheme involved the condensation of 1-naphthylamine (**2**) with 5-diethylamino-2-nitrosophenol (**1**) in acetic acid (Scheme 1A). Although the positive charge of NBA is delocalized throughout the molecule, the imino tautomers of the amino groups evidently retain enough electrophilicity that they experience a finite hydrolytic lifetime. In fact, hydrolysis of NBA with a hot, aqueous solution of dilute sulfuric acid was an early method for obtaining the phenoxazone dye Nile Red.[28] Kanitz and Hartmann[29] demonstrated that hydrolysis of the imino tautomers of NBA could be minimized by converting the amino substituents to fused bridges. While they did not work with EtNBS itself, its overwhelming similarity to NBA suggests that it too could be hydrolyzed.

They also expanded upon the original synthesis of NBA by using arylazo compounds (**5**) with any of a variety of bridged amine precursors (**4**) to generate

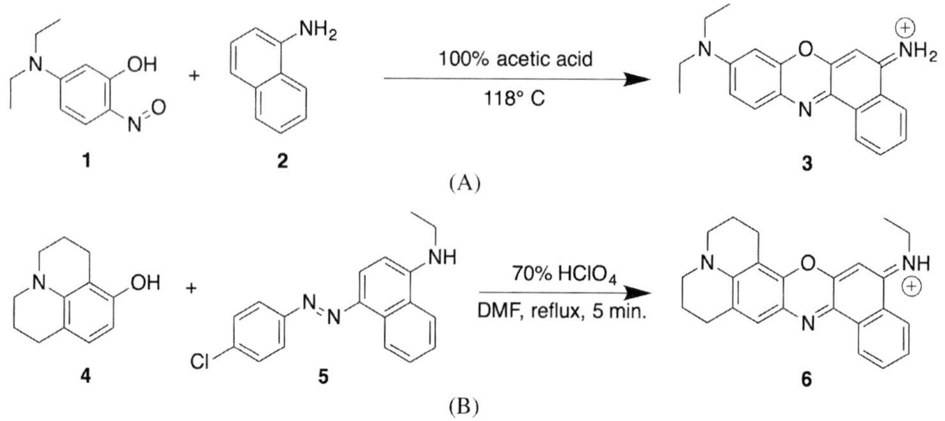

Scheme 1. Nile Blue synthetic routes. (A) Original synthesis of Nile Blue A. (B) Synthesis of bridged amine derivative of Nile Blue A.

an array of benzophenoxazinium salts with bridged amines (**6**) (Scheme 1B). Of note is their use of perchloric acid, in lieu of the weaker acetic acid, to generate the final benzophenoxazinium salts with respectable yields. Especially since Cincotta *et al.*[19] observed that simple tricyclic phenoxazines required some sort of exocyclic "D ring" to be practical photosensitizers, future work with bridged amine EtNBS derivatives offers an intriguing and possibly more efficient avenue of study.

Bridged amines represent just one category of potential derivatives of NBA and, by extension, EtNBS. Other possibilities, many of which are reviewed by Jose and Burgess,[30] include halogenation, conjugation to biomolecules, and modification of the amine substituents. Recent work in the realm of pure synthesis includes new routes to benzophenoxazines through sulfur insertion into *N*-phenyl-1-naphthylamine with elemental sulfur and iodine, as well as through benzenediamines with novel bismuth catalysts.[31,32] To say the field of benzophenoxazine synthesis is ripe for innovation would be a gross understatement.

As mentioned in the introduction, Nile Blue itself offers no photosensitizing activity. The tumor specificity of Nile Blue has nonetheless made benzophenoxazines an attractive starting point for a rationally designed photosensitizer. For clarity, the numbering scheme of benzo[*a*]phenoxazines is shown in Fig. 1. A retrospective analysis of the work of Lewis *et al.*[15–17] showed that a 9-dialkylamino substituent and at least one proton attached to the 5-amino nitrogen atom markedly improved the tumor-localizing ability of benzophenoxazines dyes.[19] Additionally, incorporation of heavier atoms was predicted to increase spin-orbit coupling, in turn generating more

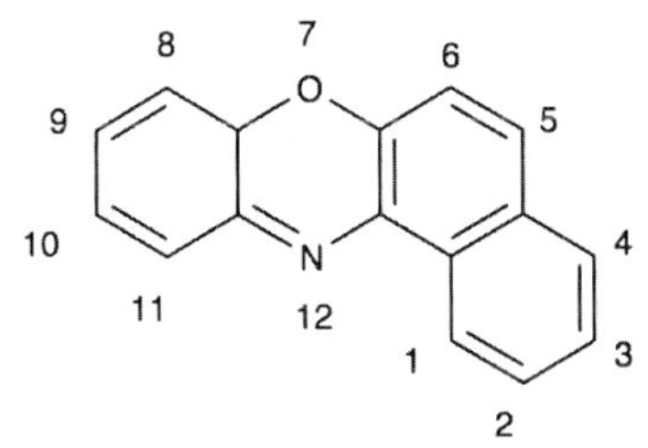

Fig. 1. Numbering scheme of benzo[*a*]phenoxazines.

singlet oxygen and thereby yielding a photosensitizer much more potent than NBA.

To this end, a sulfur-substituted version of NBA, known as NBS, was synthesized around 1989 by Cincotta *et al.*,[33] and was followed shortly thereafter by EtNBS (**9**) itself.[34,35] Their protocol greatly resembled the original synthesis of NBA by Mohlau and Uhlamnn. It has been optimized slightly over the past few decades, but the most recent synthesis entails mixing Bunte salt **7** with *N*-ethyl-1-aminonaphthalene (**6**) in the presence of a mild oxidizing agent (Scheme 2A).[35,36] Ion exchange is often done to ensure chloride is the counterion.[11] To prepare the Bunte salt, aluminum sulfate, sodium thiosulfate, and zinc chloride are added to *N*, *N*-diethyl-*p*-phenylenediamine (**10**) (Scheme 2B).[37–39] While this straightforward procedure has gone essentially unchanged for over a century, the synthesis of EtNBS, particularly with its poorly understood final condensation, remains much more open to refinement.

EtNBSe, the next product in the series of chalcogen analogs of EtNBA, follows a more complicated synthetic protocol (Scheme 2C).[11,35] The paucity of organoselenium knowledge and the so-called "double bond rule", which suggests that selenium generally does not form pi bonds, likely contribute to the comparative difficulty of this synthesis. The multistep pathway to EtNBSe consists of a Grignard reaction between selenium powder and *N*, *N*-diethyl-3-iodoaniline (**11**) to afford a diselenide complex (**12**), which is then nitrosated to obtain **13**. A final condensation with *N*-ethyl-1-napthylamine (**8**) yields EtNBSe (**14**). Although it was first created at more or less the same time as EtNBS, EtNBSe has not enjoyed the enduring popularity in the PDT community that EtNBS has experienced. For example, unlike EtNBS, EtNBSe has not yet been tested in naturally occurring tumors in large mammals or in modern 3D cell cultures.

In light of the promise EtNBS has shown in recent years as a photosensitizer in PDT, some groups have investigated derivatives of EtNBS, adding

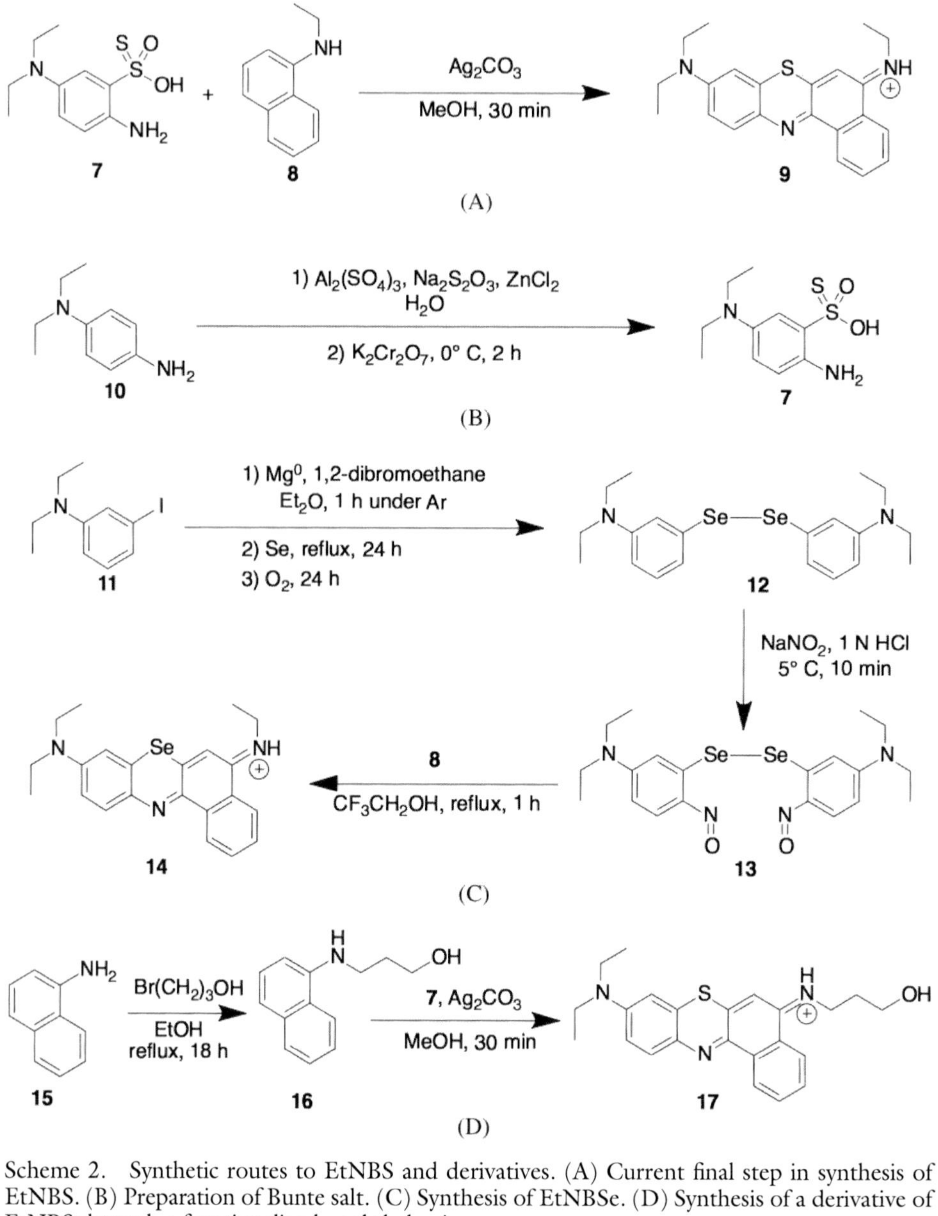

Scheme 2. Synthetic routes to EtNBS and derivatives. (A) Current final step in synthesis of EtNBS. (B) Preparation of Bunte salt. (C) Synthesis of EtNBSe. (D) Synthesis of a derivative of EtNBS through a functionalized naphthylamine.

substituents or changing sidechains in hopes of optimizing its photosensitizing ability or altering its uptake and localization patterns.[26,36,38] Most of these derivatives arise either from modification of the original naphthylamine or by later addition to one of the benzene rings. A common motif is an S_N2-like *N*-alkylation with a functionalized alkyl bromide to yield a derivatized napthylamine, which then condenses with the Bunte salt as usual to yield a benzophenothiazinium, now with an altered 5-amino moiety. An example from Verma *et al.*[38] is provided in Scheme 2D, in which 1-naphthylamine (**15**) reacts with a primary bromide to afford **16**, which then condenses with the ordinary Bunte salt to yield **17**, a 5-amino sidechain-functionalized derivative of EtNBS. Derivatives of EtNBS are far too numerous to explore each in extensive detail, but several merit mentions in later sections.

3. Photochemistry

While Nile Blue first garnered attention for its unique tumor-staining properties, EtNBS is now more known for its role in PDT as a photosensitizer. With this in mind, a brief review of the photochemistry of such photosensitizers is warranted. The ground state of EtNBS is a singlet state, designated S_0. Upon absorption of a photon, it can be promoted to an excited state, S_1. Multiphoton excitation is possible, but higher excited states tend to decay to the lowest energy excited state, following Kasha's rule.[40] From S_1, a number of processes can occur.[12] Among other things, S_1 can:

- Relax to S_0 radiatively, emitting a photon through fluorescence,
- Relax to S_0 non-radiatively, generating local heating,
- React with surrounding moieties, or
- Undergo intersystem crossing, whereby the spin of one electron is flipped, leading to a triplet state T_1.

Like S_1, T_1 can also decay radiatively — via phosphorescence — or non-radiatively. More interesting, though, are situations in which a photosensitizer transfers its energy to another molecule. In the context of PDT, these reactions are classified into two categories: type I and type II. In type I mechanisms, the excited state initiates a cascade of radical reactions, such as homolytic C-H abstraction, with surrounding molecules. In type II mechanisms, T_1 relaxes to T_0 by transferring its energy to nearby ground state (triplet) molecular oxygen ($^3O_2$), which proceeds to become singlet oxygen ($^1O_2$). All of these processes are shown in a Jablonski diagram (Fig. 2).

A recent review by O'Connor *et al.*[12] of photosensitizers in PDT contends that the ideal photosensitizer should be a single, stable compound, with minimal dark toxicity, that absorbs in the red or infrared "therapeutic window" of roughly 600–1000 nm that offers maximum light penetration, has a high

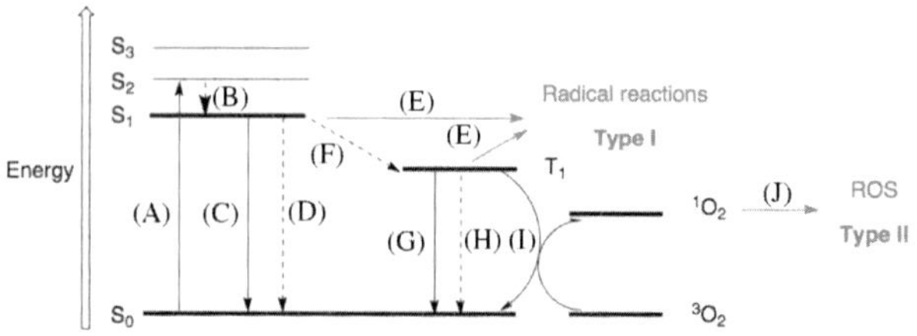

Fig. 2. Jablonski diagram for photosensitizers. (A) Promotion of the photosensitizer to an excited state by absorption of a photon. (B) Relaxation to the lowest energy excited state. (C) Fluorescence photon emission. (D) Non-radiative decay. (E) Initiation of a cascade of electron transfer reactions. (F) Intersystem crossing. (G) Phosphorescence photon emission. (H) Non-radiative decay. (I) Energy transfer to triplet oxygen and its promotion to singlet oxygen. (J) ROS formation and oxidation of biomolecules.

singlet oxygen quantum yield, and is easy to synthesize. Additionally, it has been noted that stable, planar systems around which pi electrons can travel freely, and where any heteroatoms can comfortably bear a positive charge, maximize intersystem crossing.[40] Furthermore, cationic dyes in particular, as opposed to neutral or anionic molecules, have been shown to localize to mitochondria and lysosomes within cancer cells and offer better cellular penetration in some types of bacteria.[41–43] To retain this positive charge at physiological pH, a pKa of at least 9 is required. These constraints greatly thin the crowd of possible photosensitizers.

In addition to meeting the criteria previously laid out by Foley *et al.*[19] EtNBS satisfies nearly all of these broader criteria. At room temperature, it is a stable blue powder.[35] Its absorption maximum rests comfortably in the red at 652 nm.[11] Unlike in NBA, with its unfavorable oxonium resonance structure, the sulfur atom of EtNBS can bear a partial positive charge with relative ease, allowing electrons to freely circulate about all four rings. Its singlet oxygen quantum yield of 0.03, while small, is still effective and several times greater than that of NBA or EtNBA.[11,44] That said, as will be discussed later, having a high singlet oxygen quantum yield may not always be ideal, as a balance between type I and type II mechanisms can lead to more versatile photosensitization overall. A study that varied the chain length of the phenothiazine Methylene Blue showed that dyes with ethyl substituents by far had the best phototoxicity to dark toxicity ratios; it is likely that these findings can be generalized to benzophenothiazines like EtNBS.[45] Lastly, as discussed above, the synthesis of EtNBS is fairly facile, especially compared to that of porphyrins like PII or benzoporphyrin derivative (BPD). For reference, some photophysical values of EtNBS and related molecules are included in Table 1.

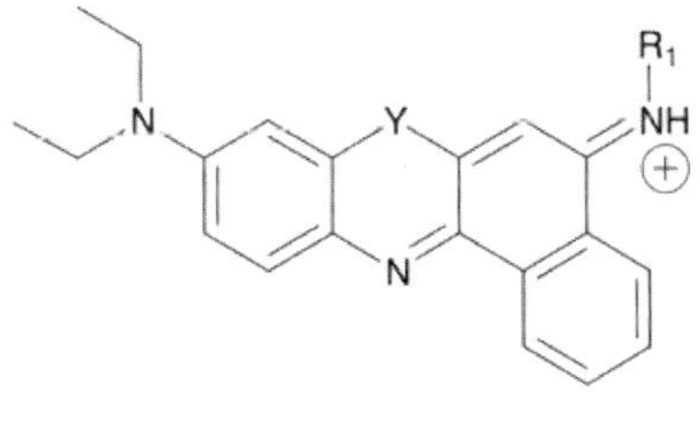

Table 1. Photophysical data for EtNBS and related photosensitizers. (a) Absorption maximum, in methanol or ethanol with 0.1–1.0% hydrochloric acid or acetic acid. (b) Fluorescence emission maximum in ethanol or methanol. (c) Absolute quantum yield of fluorescence. (d) Absolute quantum yield of singlet oxygen formation. (e) Log of partition coefficient between octanol and phosphate-buffered saline at pH 7.4. (f) NBS-6I contains an additional iodine substituent at the 6 position of the benzophenothiazine ring. Data adapted with permission from multiple sources.[9,11,18,33,38,44]

Name	Y	R_1	$\lambda_{abs}(nm)^a$	$\lambda_{fl}(nm)^b$	ϕ_{fl}^c	$\varphi^1O_2^d$	pKa	log P^e
NBA	O	H	623	660	0.27	0.005	9.9	2.23
NBS	S	H	645	—	—	0.025	10.0	2.48
NBS-6I[f]	S	H	660	—	—	0.170	6.5	3.80
EtNBA	O	C_2H_5	632	—	—	0.005	—	2.69
EtNBS	S	C_2H_5	652	693	0.21	0.025	—	2.76
EtNBSe	Se	C_2H_5	661	703	0.03	0.65	—	2.08
EtNBS-OH	S	C_3H_6OH	655	685	0.12	0.026	—	2.03
EtNBS-COOH	S	C_4H_8COOH	655	685	0.12	0.032	—	1.51

The "heavy atom effect" dictates that larger atoms, including sulfur, selenium, and halogens, provide much greater spin-orbit coupling and enable otherwise forbidden quantum mechanical transitions, such as the one from a singlet to a triplet state, thereby promoting intersystem crossing.[40] As a consequence, exchange of oxygen for heavier atoms, such as sulfur and selenium, increases the rate of intersystem crossing and overall singlet oxygen formation. While EtNBSe is predictably a much more potent type II photosensitizer than EtNBS, it also suffers from nontrivial dark toxicity, displays just a low fluorescence quantum yield that makes it difficult to detect, requires a more laborious synthesis, is beleaguered by a low partition coefficient, likely has weaker type I photosensitizing ability, and is easily reduced to its harmless leuko form *in vitro* by NAD(P)H.[9,11] Halogenated derivatives of EtNBS have been synthesized, but they are plagued by their increased acidity and extremely high hydrophobicity, both of which are problematic for ensuring uptake and photosensitivity.[21] In short, whereas EtNBSe is unbalanced and halogenated products also have significant flaws, EtNBS presents a better balance of the properties that make for a good photosensitizer.

Although EtNBS may on paper appear to be close to an "ideal" photosensitizer, its finicky behavior in solution can complicate experimental PDT studies. Despite its usage *in vitro* and even *in vivo*, EtNBS can aggregate, dimerize, and self-quench in water.[11] To counter this tendency, most spectroscopic measurements must be carried out exclusively in acidified methanol or ethanol. Further complicating matters is the fact that excited benzophenothiazine dyes have a markedly increased basicity, often on the order of 10,000 times greater than that of the ground state, which leads to transient changes in charge and absorption.[46] Moreover, EtNBS paradoxically is both more lipophilic and more water-soluble than its primary amine cousin NBS.[9] The introduction of the hydrophobic ethyl group increases overall lipophilicity, yet it also changes the surface area and polarity of the molecule, minimizing its aggregation in water. Such odd behavior has come to be expected from EtNBS; for example, its parent, NBA, is noted for its wild solvatochromic shifts in absorption and even for changes in absorption in response solely to changes in the counterion.[30]

All cationic benzophenothiazines with a pKa near physiological pH exist *in vitro* in an equilibrium of three states: the cationic form around which discussion thus far has centered, a neutral pink deprotonated form, and a neutral, colorless, reduced "leuko" form (Fig. 3). The weak absorption of the deprotonated form, whose absorption maximum lies outside the therapeutic window, renders it a poor photosensitizer.[9] The leuko form similarly lacks photosensitizing capability. Reductases and cofactors such as NADH abound *in cells*, however, and have been shown to reduce EtNBS and its analogs to their leuko forms in the absence of O_2.[9] While re-oxygenation does return the leuko form to its familiar and more potent protonated state, reduction of EtNBS could certainly pose difficulties for the treatment of hypoxic regions of tumors.

Although it can sometimes be frustrating to study, the indisputable utility of EtNBS as a photosensitizer has made it a popular platform for further exploration of benzophenothiazines in PDT. Substituents at the 1 position (and, to a lesser extent, the 4 position) interfere with the nitrogen atom in the 12 position, distorting the planar extended pi system and impairing photoactivity.[36] Broadly speaking, sidechain derivatives of EtNBS have photophysical properties in line with those of their parent molecule; for example, their dark toxicities are similar. Neutral or cationic sidechains likewise have a negligible effect on singlet oxygen quantum yield, as they do not alter the delocalized ring system. Absorption maxima rarely shift by more than a few nanometers.[26,36]

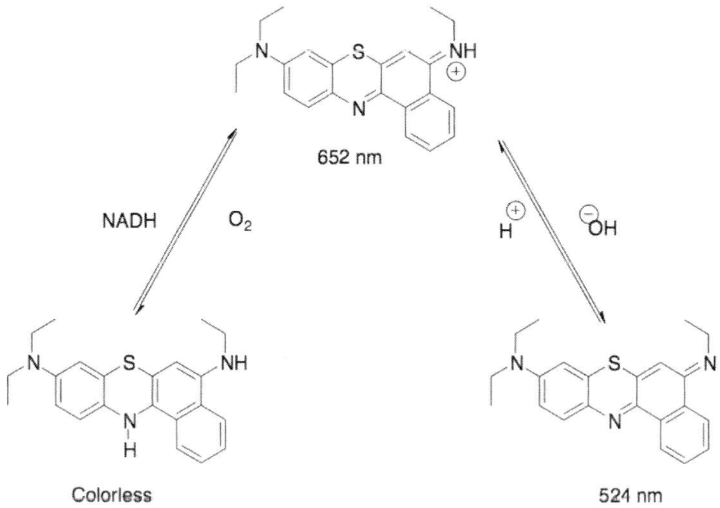

Fig. 3. Protonation and redox states of EtNBS. EtNBS is typically protonated (652 nm absorption), but can be deprotonated (524 nm absorption) in mildly basic conditions (approx. pH 10). It can also be reduced, e.g., by NADH, to the colorless leuko form.

Carboxyl-terminated sidechains, while useful for further functionalization, yield zwitterionic species that see reduced retention and, consequently, heavily reduced phototoxicity in certain applications.[26] However, these and other 5-amino derivatives, e.g., EtNBS-OH and EtNBS-COOH, retain singlet oxygen quantum yields on par with that of EtNBS itself.[38] Therefore, the decrease in PDT efficacy of the carboxyl-terminated side chain derivatives appears to stem not from any inherent reduction in photoactivity, but rather from unfavorable changes in uptake and retention. Provided a new polar group elsewhere on the photosensitizer counterbalanced this increase in lipophilicity, an ester or amide linkage to another molecule to yield a photosensitizer conjugate could perhaps restore PDT efficacy. Of note is such a conjugation, of EtNBS to a cephalosporin antibiotic which was recently reported, with exciting implications.[47]

That sidechain effects are largely divorced from the photochemistry of the parent benzophenothiazine is quite promising, since this orthogonality leaves the door open for further derivatization of EtNBS without any loss of photosensitizing potency. Thus, attachment to bioactive molecules such as antibodies, drugs, or targeting ligands; tagging with fluorophores for ratiometric imaging; or any of the countless other modifications in the toolbox of modern organic synthesis all remain available as part of the continued exploration of EtNBS as a treatment for a variety of cancers and microbes.

4. Uptake and Localization of EtNBS and Related Nile Blue Derivatives

4.1. *Eukaryotes, 3D Cultures, and Animal Models*

As early as 1946, Nile Blue (NBA) was shown to preferentially stain malignant sarcoma cells over surrounding healthy tissue when administered to mice via feed.[15,16] This foundational study reported a diffuse staining pattern within the stained cells, along with the apparent retardation of malignant cell growth. Since these initial findings, the uptake mechanism and the intracellular localization pattern of Nile Blue analogs, including EtNBS and its derivatives, have been investigated both *in vitro* and *in vivo*.

Unlike many exogenous agents that enter a cell through endocytosis or a carrier-mediated mechanism, NBA accumulates intracellularly through an energy-dependent ion trapping mechanism.[20,21] NBA first diffuses across the cell membrane in its neutral, deprotonated form. On the way, some molecules, due to their lipophilicity, become lodged in the membrane. Those that manage to cross the membrane accumulate in the lysosome. Here, the low pH — lysosomal pH is around 5, compared to the cytosolic pH of 7.4 — ensures that virtually all molecules exist in their protonated, cationic form. The positive charge prevents NBA from crossing the membrane again and returning to the cytosol; thus, it remains trapped in the lysosome.

Since the pKa of NBA, EtNBS, and related photosensitizers is close to 10, only a small fraction will be neutral and able to enter the cell across the cell membrane at a given time. However, the subcellular trapping of molecules drives the equilibrium forward, toward accumulation first in the cell and ultimately within the lysosome. In the lysosome, only a few molecules in a million will exist in their deprotonated form. As a result, very few ever diffuse out of the lysosome. By altering these pH gradients, and by disrupting membrane potentials and oxidative phosphorylation (i.e. disrupting energy), Lin *et al.*[20,21] provided conclusive evidence for this energy-dependent ion trapping mechanism and refuted competing hypotheses based on photosensitizer self-aggregation or accumulation in lipids.

As the charge of a molecule becomes more delocalized, the molecule itself more closely resembles a neutral species, and the ease with which it could cross the nonpolar portion of the cell membrane increases. Several lipophilic ions have demonstrated successful internalization due to their delocalized charge,[48,49] and that possibility should not be ruled out here, with EtNBS and related photosensitizers. Regardless of their protonation state, these analogs enter the cell by passive diffusion across the cell membrane, aided by their lipophilicity, followed by their irreversible sequestration in the low pH compartments of the lysosome, just as with NBA.

While all Nile blue derivatives preferentially localize to the lysosome, localization rates and residual staining vary.[21] For NBA and iodinated derivatives, for example, lysosomal localization occurred within five minutes of incubation, with no residual staining of the cytosol or mitochondria. NBS and D ring-saturated derivatives, however, showed some diffuse cytosolic and mitochondrial staining on top of the normal perinuclear lysosomal staining. These compounds also required 15–40 min to show particulate localization. Broader cellular distribution (i.e. mitochondrial and cytomembrane staining) was achieved as incubation doses rose from 10^{-8} M to 10^{-3} M. In these scenarios, localization to lipophilic regions such as the endoplasmic reticulum and plasma membrane may be explained by stabilizing hydrophobic interactions or by membrane trafficking within cells.

The rapid localization seen in iodinated derivatives of NBA, compared to that of sulfur and D-ring saturated derivatives, was attributed to their increased acidity and hydrophobicity, which likely accelerated permeation of the cell and lysosomal membranes and resulted in the inability of the iodinated derivatives to cross back over the lysosomal membrane after protonation.

Since more photoactive analogs of NBA — like EtNBA, EtNBS, and EtNBSe — still retain its cationic, lipophilic, aromatic skeleton, their uptake resembles that of NBA.[9] Immediately after administration, EtNBA, EtNBS, and EtNBSe all show rapid intracellular uptake that levels out only once the extracellular media is depleted. In light of concentration-dependent studies, this plateau suggests that uptake of EtNBA, EtNBS, and EtNBSe is linearly proportional to the treatment concentration. The incubation dose also has a strong influence on the intracellular localization of these ethyl-substituted analogs of NBA. When dye concentrations ranged from 10–100 μM, punctate lysosomal sequestration dominated (Fig. 4), while concentrations above 100 μM induced cytomembrane and mitochondrial staining. Because EtNBS is primarily used now as a photosensitizer, its localization behavior during PDT must also be discussed. The subcellular distribution of EtNBS has indeed been shown to change after irradiation. Fluorescence images revealed that, after 200 s of 50 mW/cm^2 irradiation by light at 662 nm, EtNBS was redistributed to the cytosol.[50] Redistribution was also reported in 3D spheroids as a second wave of triplet oxygen consumption was observed after 10 J/cm^2 fluence was reached. As expected from its substantially greater singlet oxygen quantum yield, EtNBSe required approximately 10% of the fluence that EtNBS did to trigger redistribution in spheroid models.

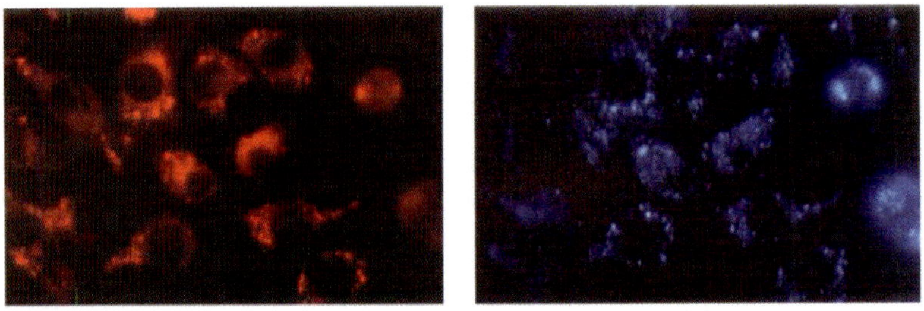

Fig. 4. Colocalization studies of EtNBS in breast cancer (EMT6) cells. EtNBS showed perinuclear staining (A), resembled the localization pattern of carboxylate-modified blue fluorescent spheres (B) that are known to accumulate in the lysosome. Adapted with permission from Georgakoudi *et al.*[50] Copyright 1998 John Wiley and Sons.

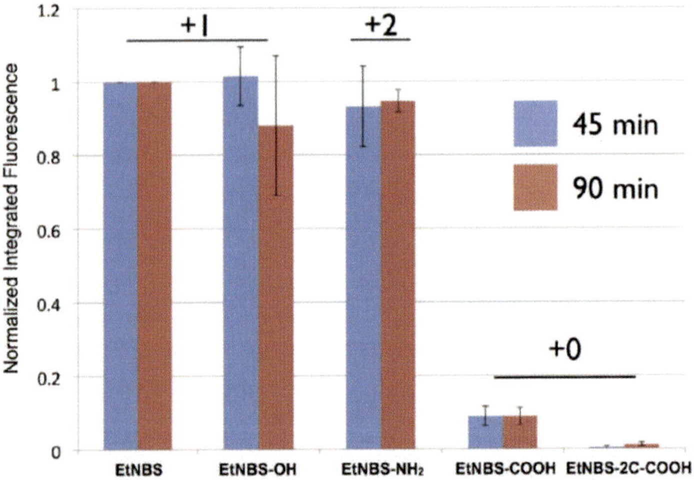

Fig. 5. Relationship between uptake and overall net charge of EtNBS and its derivatives in ovarian cancer (OVCAR5) monolayer cells. Cationic compounds (EtNBS, EtNBS-OH and EtNBS-NH_2) exhibited increased cellular uptake when compared to zwitterionic compounds (EtNBS-COOH and EtNBS-2C-COOH). Adapted with permission from Klein *et al.*[26] Copyright 2012 American Chemical Society.

Altering the net charge of EtNBS in physiological conditions helped elucidate the relationship between charge and uptake (Fig. 5). In more recent work, Klein *et al.*[26] specifically investigated the uptake and localization behaviors of EtNBS derivatives, including those with hydroxyl (EtNBS-OH), "long" carboxylic acid (EtNBS-COOH), "short" carboxylic acid (EtNBS-2C-COOH), and primary amine (EtNBS-NH_2) sidechains attached to the 5-amino substituent of EtNBS. When the derivative was neutral, as with the zwitterionic EtNBS-COOH and EtNBS-2C-COOH, it saw significantly reduced uptake.

Since the pKa of carboxylic acids is slightly below the lysosomal pH of 5, a large proportion of these carboxylic acid EtNBS derivatives are overall neutral in the lysosome. With no net charge and a generally lipophilic structure, these molecules easily leave the lysosome and exit the cell. As a result, they demonstrate only transient uptake and minimal retention. EtNBS-2C-COOH especially suffered, since its point charges are so close that they even more closely approximate a neutral species. EtNBS, EtNBS-OH, and EtNBS-NH_2 — all cationic molecules below pH 10 — show essentially identical rates of uptake.

While understanding uptake and localization profiles in monolayer cell culture provides clues into the nature of EtNBS, visualizing the distribution of the photosensitizer in biologically relevant 3D cultures is more helpful in better understanding its behavior *in vivo*. Therefore, several works have investigated its behavior in spheroid models.[25,26,50] As previously mentioned, redistribution of EtNBS was seen in spheroid models; however, this work indirectly determined localization through triplet oxygen consumption measurements, not by more direct microscopic approaches. In 2011, Evans *et al.*[25] showed that EtNBS was able to penetrate ovarian cancer spheroids *in vitro* and concentrate in the hypoxic cores of these nodules. As hypoxic regions are also more acidic, its localization within these regions agrees well with the previously proposed ion trapping mechanism from monolayer studies. Section 6.1 covers these interrelated phenomena in more detail.

The derivatives by Klein *et al.*[26] introduced earlier were also screened in 3D spheroid cultures (Fig. 6). Using time-lapse confocal imaging, the uptake of EtNBS-OH was shown to occur over 4.5 h and resulted in its accumulation at the core of the 3D spheroids. This uptake pattern is similar to that which was observed with the parent EtNBS compound, suggesting hydroxylation does not alter its tumor penetrating ability. EtNBS-NH_2 displayed initial uptake kinetics similar to those of EtNBS and EtNBS-OH, but its uptake leveled off after just 2 h and it showed a peripheral localization pattern. The shortened time required to reach the observed uptake plateau was attributed to its poor nodule penetration, while its suboptimal localization is explained by the protonation at physiological pH of EtNBS-NH_2 to afford EtNBS-NH_3^+. The point charge experiences significant attraction to anionic components of the cell membrane and extracellular matrix that hinder its penetration. In agreement with monolayer studies, EtNBS-COOH exhibited significantly reduced spheroidal uptake. The natural, delocalized +1 charge of EtNBS therefore seems ideal for most scenarios.

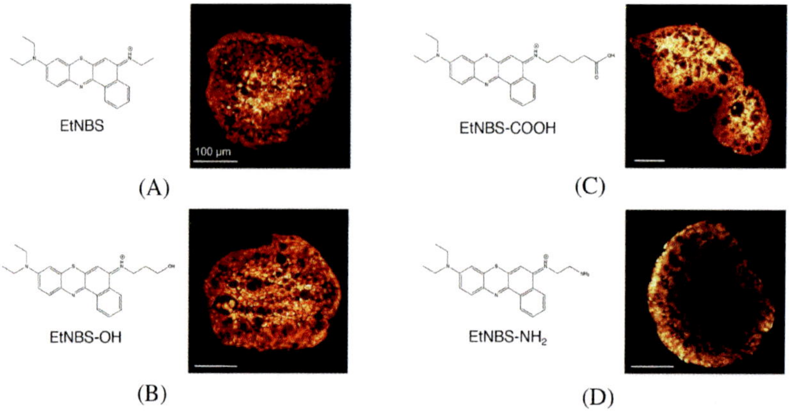

Fig. 6. Uptake and localization patterns of EtNBS and its derivatives in 3D ovarian cancer tumor nodules. EtNBS (A) and EtNBS-OH (B) displayed similar localization behavior with accumulation within the central core of the nodule, however, EtNBS-COOH (C) showed diffused uptake throughout the nodule and EtNBS-NH2 (D) accumulated only on the outer periphery. EtNBS-2C-COOH was omitted from 3D studies due to its lack of phototoxicity in monolayer experiments. Adapted with permission from Klein *et al.*[26] Copyright 2012 American Chemical Society.

To confirm the results observed via confocal imaging, the localization patterns of EtNBS derivatives were further evaluated by Klein *et al.* using cryosectioning and fluorescence microscopy. While EtNBS-OH, like EtNBS, showed preferential localization to the hypoxic core, its distribution here was more diffuse. This slightly different accumulation pattern can likely be credited to the increased hydrophilicity offered by the hydroxyl group. Cryosectioning also verified the localization patterns discussed earlier from time-lapse microscopy of EtNBS-NH_2 and EtNBS-COOH.

In order for EtNBS or any of its counterparts to be approved for human use, its pharmacokinetic and biodistribution profiles must be carefully explored using *in vivo* animal models. To that end, Cincotta and coworkers[10] screened EtNBS against EMT6 and RIF, two murine mammary sarcoma models. In EMT6, rapid uptake in the tumor, normal skin, and the gastrocnemius muscle was observed after subcutaneous administration. However, preferential accumulation was seen at the tumor site and reached an uptake plateau at 3 h post-administration. Unlike what was seen during *in vitro* studies, the uptake of EtNBS was not linear with respect to treatment concentration, but sigmoidal. This behavior was thought to be due to the initial saturation of surrounding tissue with EtNBS at the injection site before entering circulation. Interestingly, the uptake of EtNBS in a RIF model *in vivo* was just one third of its *in vitro* uptake, despite pharmacokinetics similar to and uptake greater than in EMT6.

This discrepancy between *in vitro* and *in vivo* studies was ascribed to the poor vasculature of RIF tumors as compared to EMT6.[51]

The circulating blood plasma concentration of EtNBS was also explored. The authors first observed minimal EtNBS in plasma when it was injected subcutaneously at concentrations needed to induce tumor reduction. However, when EtNBS concentrations were increased in an attempt to promote photosensitizer accumulation in the tumor, it was found that this extra EtNBS stayed in the plasma and did not accumulate in the tumor. Additionally, tumoral accumulation of EtNBS was shown to be heavily influenced by the method of administration. Likely due to electrostatic interactions in the peritoneal cavity, intraperitoneal injection resulted in 50% less tumor accumulation as compared to subcutaneous and intravenous injections. This result highlights the pressing need to understand the myriad possible interactions that could arise upon introduction of an exogenous agent, like EtNBS, to more complex biological systems.

4.2. *Prokaryotes*

As the promise of EtNBS in the PDT of mammalian malignancies became apparent, it could only be expected that EtNBS, its analogs, and its derivatives would be probed for antimicrobial activity as well. The success of many similarly charged bacterial staining agents suggests that the delocalized charge of EtNBS could prove highly advantageous in many antimicrobial applications.[52] One of the earliest investigations into the antimicrobial activity of such analogs showed that EtNBS, EtNBSe and EtNBA displayed significantly greater bacterial cell binding than traditional phenothiazinium dyes, e.g., Methylene Blue, in both Gram-positive and Gram-negative bacteria.[11] While all three analogs demonstrated superior binding compared to earlier photosensitizers, there were substantial differences among the photosensitizers and across types of bacteria with respect to bacterial cell binding. In both Gram-negative *Escherichia coli* and Gram-positive *Enterococcus faecalis*, EtNBSe displayed the greatest bacterial cell interaction, with EtNBS a close second and EtNBA lagging significantly behind. All analogs showed greater binding and uptake in *E. coli* than in *E. faecalis*. Gram-negative bacteria have a generally impermeable cell wall composed of complex, anionic lipopolysaccharides, whereas Gram-positive bacteria cell walls are much more porous. While many antimicrobial agents are hydrophilic, EtNBS and its analogs are known to be extremely lipophilic and therefore would favor penetration into and through the negatively charged, lipid rich membrane found in Gram-negative bacteria.

Just as they were studied in eukaryotic models, derivatives of EtNBS were also explored in representative Gram-positive and Gram-negative prokaryotic bacterial models to determine if structural alterations could tune their antimicrobial properties. In 2009, two EtNBS derivatives, EtNBS-OH and EtNBS-COOH, were assessed as photosensitizers in PDT of *E. coli* and *Staphylococcus aureus* 29213, a Gram-positive bacterial strain.[38] Consistent with the findings listed above, EtNBS-OH and EtNBS-COOH showed greater bacterial binding to Gram-negative bacteria than to Gram-positive bacteria (Fig. 7A). It should be noted that these sidechain modifications of EtNBS did not produce significant differences in uptake. Hamblin and colleagues[36] further explored the effect of derivatizing the parent EtNBS compound through substitutions at the 1-carbon position that introduced electron withdrawing or donating groups in hopes of altering the its photochemical properties. In the process of examining the phototoxicity of EtNBS derivatives, they concluded that additional positive charges improved association with and uptake in Gram-negative bacteria.

As witnessed in murine models with malignant tumors, EtNBS displayed rapid initial uptake into the subcutaneous site of *M. bovis* BCG-induced granuloma, a model for mycobacterial infections (Fig. 7B).[53] The total intravenous delivery scaled linearly with time, with maximum accumulation seen 60 min post-injection. Surprisingly, intravenous injection of EtNBSe resulted in rapid clearance from the blood stream with negligible accumulation seen at the infection site, requiring local site injections for all PDT studies.

While many studies have focused on bacterial destruction, Hasan and colleagues investigated the PDT efficacy of EtNBS and EtNBSe against *Leishmania* parasites, specifically *Leishmania major*.[44] EtNBSe was shown to have markedly higher accumulation inside parasites 1 h post-administration when compared to the commonly used photosensitizers BPD and protoporphyrin IX (PpIX). EtNBS and EtNBSe exhibited similar uptake dynamics, with maximal internalization occurring one hour post injection, though EtNBSe showed greater accumulation overall. After 2.5 h, both compounds showed drastic decreases — 40% for EtNBSe, 46% for EtNBS — in their respective intracellular concentrations. Since transmission electron microscopy further revealed a perinuclear localization pattern of EtNBS and EtNBSe within *Leishmania* parasites, intracellular localization has been deemed a critical factor in the destruction of these parasites.

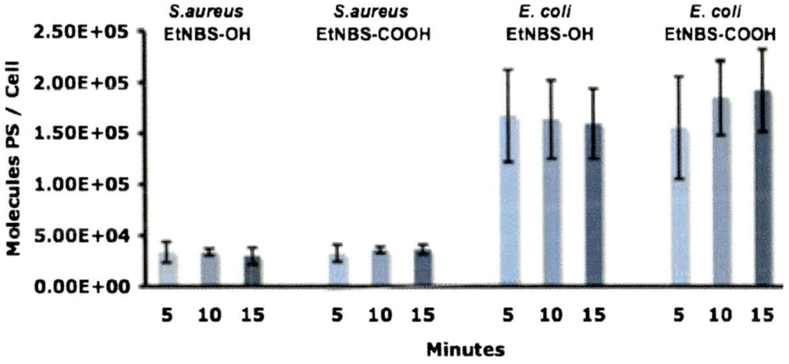

(A)

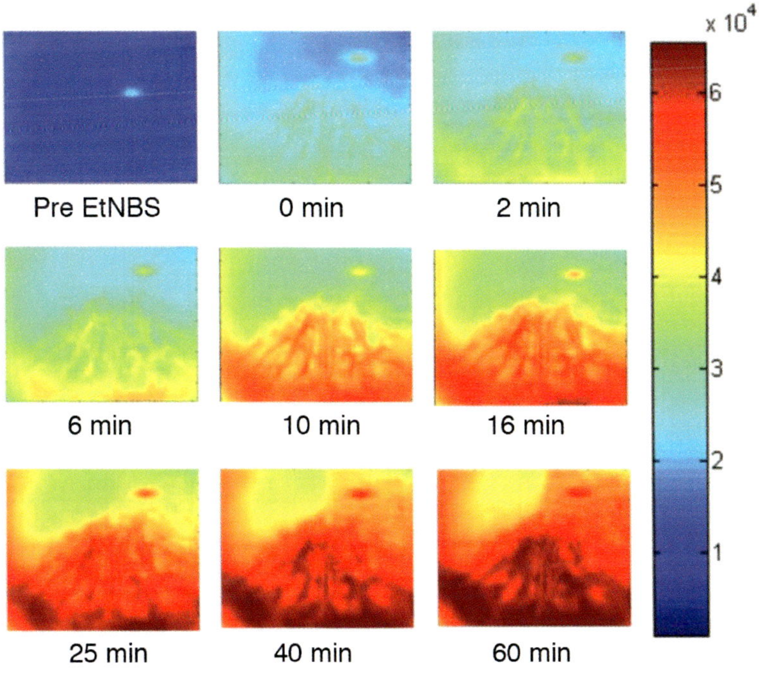

(B)

Fig. 7. Binding and accumulation dynamics of EtNBS and its derivatives in both *in vitro* and *in vivo* models. (A) EtNBS derivatives showed significantly higher binding affinities to Gram-negative bacteria, such as *E. coli*, than to Gram-positive bacteria, such as *S. aureus*, due to the lipophilic structures. (B) EtNBS showed maximum accumulation in a *M. bovis* BCG-induced granuloma model after 60 min when administered intravenously. (A) Adapted with permission from Verma *et al.*[38] Copyright 2009 John Wiley and Sons. (B) Reproduced from O'Riordan *et al.*[53] with permission from the European Society for Photobiology, the European Photochemistry Association, and The Royal Society of Chemistry.

5. Antimicrobial Applications

While most of the attention paid to EtNBS in recent years has been a consequence of its anticancer properties, it has also demonstrated great efficacy in photodynamic inactivation (PDI) — the application of PDT to the treatment of pathogens or infectious diseases. In PDI, a nontoxic photosensitizer is first applied to an affected region, either topically or by systemic administration. After a sufficient incubation period during which it accumulates in the affected region, the photosensitizer is activated by a specific wavelength of light, permitting it to generate radicals and ROS that inactivate pathogens and kill microbes. Since PDI is considered safe and is thought not to induce therapeutic resistance — it targets indirectly the entire cell, not a specific enzyme or structure, as normal antibiotics often do — it is a very attractive solution to the growing problem of multi-drug resistance, particularly in the treatment of localized infections.[11,43,47,54] To date, numerous successful instances of PDI have been reported, with several different photosensitizers treating a variety of bacterial infections, fungal infections, and parasites.[43,55,56]

Although many commonly used porphyrin-based photosensitizers have demonstrated appreciable killing of Gram-positive bacteria *in vitro*, their antimicrobial activity declines in fungal models like *Candida* and drops even further in Gram-negative bacteria.[43] As mentioned above, and in contrast to anionic or neutral molecules, cationic photosensitizers offer much more efficient PDI of both Gram-positive and Gram-negative organisms. Several groups have used heavily polycationic polypeptides to permeabilize Gram-negative membranes, with some success.[57] Others have discovered that the membrane need not be penetrated at all; provided the photosensitizer has attached to the membrane, it can generate singlet oxygen that diffuses through to the interior of the cell.[58,59] A slew of related studies suggest that cationic photosensitizers are strongly attracted to the outer membrane and damage cells through some combination of type I and type II mechanisms. The exact mechanisms of uptake and phototoxicity are not well understood, but these findings bode well for the more general application of EtNBS in antimicrobial therapies. As introduced above, Foley *et al.*[11] evaluated PDI of Gram-negative *Escherichia coli*, Gram-positive *Enterococcus faecalis*, and the fungus *Candida albicans* with EtNBA, EtNBS, and EtNBSe. They found that all derivatives were effective against the three organisms, with efficacy descending from EtNBSe to EtNBS to EtNBA (Fig. 8). These results mesh very well with the present body of knowledge in PDI — that Gram-positive bacteria and fungi are inactivated much more easily than are Gram-negative species, but that

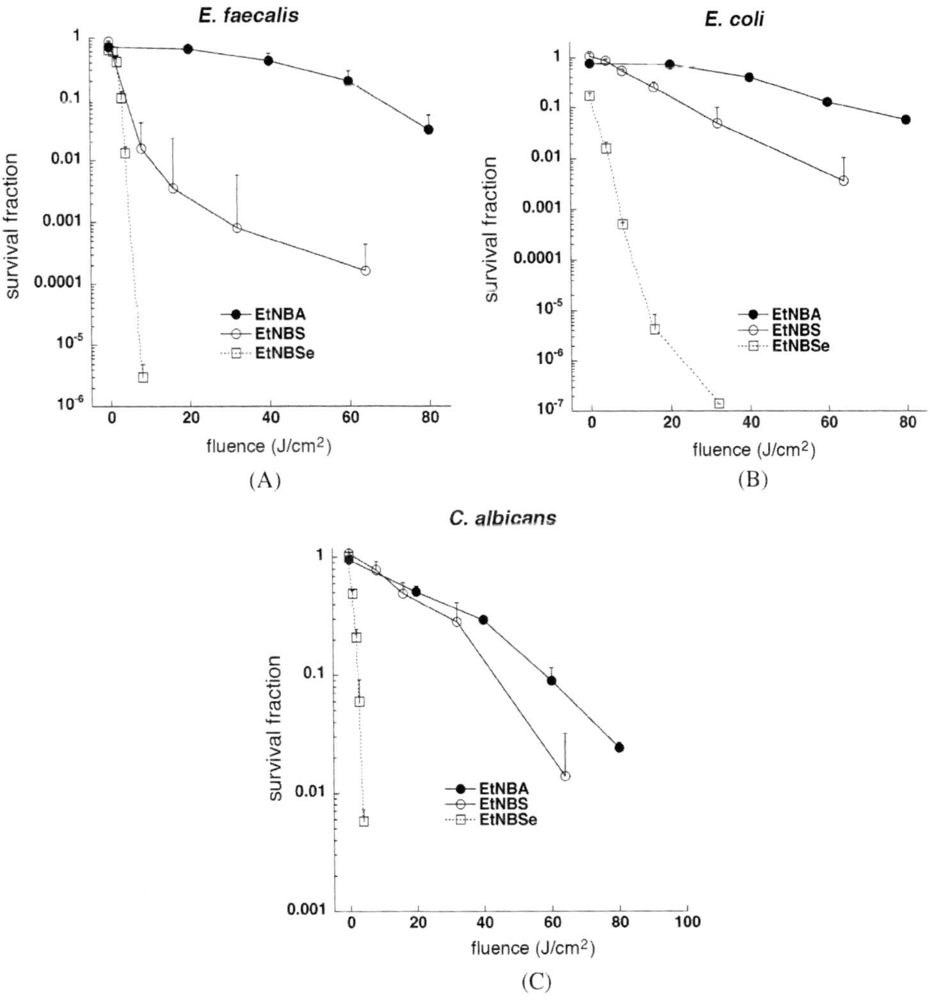

Fig. 8. Antimicrobial activity of EtNBA, EtNBS, and EtNBSe in (A) *E. faecalis*, (B) *E. coli*, and (C) *C. albicans*. Bacteria were incubated for 30 min with 2 μM, 5 μM, or 20 μM photosensitizers, respectively. The cells were then washed and irradiated with red light. Reused with permission from Foley *et al.*[11] Copyright 2006 American Chemical Society.

cationic photosensitizers significantly improve upon that discrepancy by more readily killing Gram-negative bacteria. However, EtNBSe suffered significant dark toxicity, indicating that EtNBS may be the best overall candidate for future studies.

Later work by Verma *et al.*[38] lent support to the hypothesis that the positive charge of EtNBS has a significant influence on its bacterial binding and uptake, especially in Gram-negative species. There, they saw greatly reduced phototoxicity when *E. coli* was treated with a neutral zwitterionic carboxylate derivative of EtNBS, EtNBS-COOH, but only minimal change in the Gram-positive model *Staphylococcus aureus*. A non-ionizable sidechain derivative, EtNBS-OH, produced negligible changes in photochemistry or phototoxicity.

As introduced above, PDI with EtNBS has also been evaluated as one potential treatment for cutaneous leishmaniasis.[60] As *Leishmania* possess a negative surface charge, much like Gram-negative bacteria, it was speculated that PDI with a cationic photosensitizer would be remarkably effective. Akilov *et al.*[44] reported that EtNBS and EtNBSe offered a significant improvement over the porphyrin-based photosensitizers PpIX and BPD that are currently in clinical use for the treatment of leishamniasis. This finding is rather interesting, as EtNBSe and BPD have comparable singlet oxygen quantum yields. The authors suggest that the cationic benzophenoxazine analogs see greater overall uptake and localize to the lysosome, whereas the anionic BPD localizes to the mitochondria; as a result, EtNBS and EtNBSe mediate their cytotoxicity of *Leishmania* in an inherently more effective manner. With by far the highest cellular photosensitizer concentration one hour after incubation, EtNBSe in particular exhibited excellent cell killing — the overall order of PDI efficacy was EtNBSe > EtNBS > BPD > PpIX. The morphological changes seen after PDI indicate that EtNBS and EtNBSe destroyed the nuclear membrane and scattered the nucleus, instead of disorganizing the cytoplasm, as the porphyrins did.

While the authors focused on EtNBSe, it was used only in low nanomolar concentrations well below the threshold for dark toxicity. By switching to EtNBS and optimizing the dose, fluence, and irradiance, even more impressive antiparasitic results could likely be achieved. Since *Leishmania* are quickly phagocytized by macrophages and reside there intracellularly,[61] some modifications must be made to ensure more selective parasite killing while sparing the host cell. Nevertheless, EtNBS and EtNBSe display efficient killing of *Leishmania* without stimulating resistance to treatment, making PDI with them one of the most promising antiparasitic therapies presently being studied.

Tuberculosis is another infection notorious for its long, grueling treatment regimen, typically spanning several months, and for its lethal drug-resistant forms, which demand second-line antibiotics or even invasive surgeries.[62,63] Negatively charged, long-chained fatty acids in the cell wall are a hallmark of the genus *Mycobacterium*, which is responsible for the infections of the skin and lungs that cause tuberculosis.[53] For this reason, photosensitizers that are both cationic and lipophilic appear to be well suited for antimicrobial PDI. Indeed,

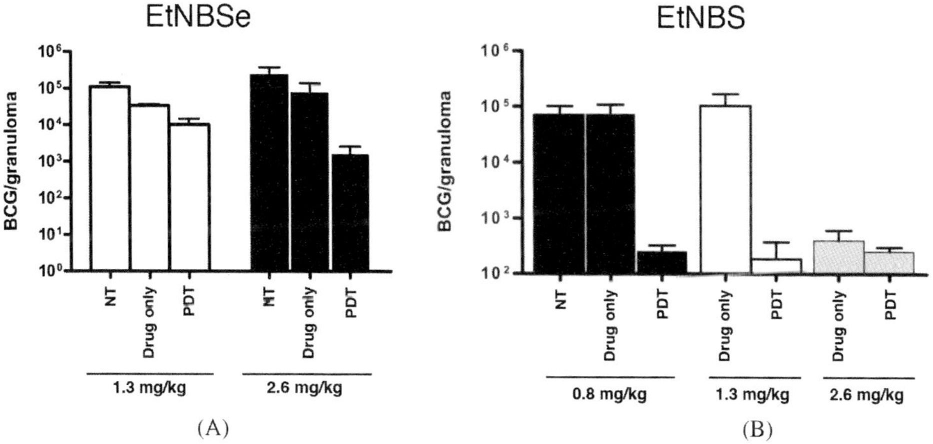

Fig. 9. PDI of *M. bovis* BCG in a murine subcutaneous granulomata model. Mice were treated with 0.8 mg/kg, 1.3 mg/kg or 2.6 mg/kg of (A) EtNBSe or (B) EtNBS, then irradiated with 60 J/cm^2 light at 635 nm. Drug only groups received no light. No treatment groups received neither light nor drug. 72 h post-treatment, the mice were sacrificed and the number of viable bacteria remaining was quantified. Reproduced from O'Riordan *et al.*[53] with permission from the European Society for Photobiology, the European Photochemistry Association, and The Royal Society of Chemistry.

EtNBS has been proven effective both *in vitro* and *in vivo* for PDI of the *M. bovis* strain BCG.[53] While the phototoxicities of EtNBS and EtNBSe were shown to be similar and dose-dependent *in vitro*, EtNBS surprisingly yielded greater efficacy in a murine subcutaneous granulomata model despite its lower singlet oxygen quantum yield as compared to EtNBSe. Within just a few hours, virtually all EtNBSe was eliminated from the mouse. EtNBSe was unfortunately plagued by dark toxicity at even the lowest doses used as well (Fig. 9A). Even with EtNBS, higher drug doses caused unwanted dark toxicity, which could hinder further applications (Fig. 9B). Nevertheless, a range of active doses of EtNBS did exist in which there was substantial cell killing without any associated dark toxicity.

With the rise of potentially lethal nosocomial infections such as methicillin-resistant *Staphylococcus aureus* (MRSA), much effort has been devoted to the search for novel treatment modalities that can overcome their resistance to ordinary antibiotics. One avenue of exploration has been PDI, which has great potential in the treatment of MRSA and other β-lactamase producing bacteria.[64] Zheng *et al.*[47] recently devised an intriguing construct, dubbed β-LEAP, in which a derivative of the β-lactam antibiotic cephalosporin is conjugated to two EtNBS-COOH molecules. Because the two EtNBS molecules are in close proximity — only 2.4 nm apart — they undergo ground state quenching. Cleavage of the β-lactam ring of the cephalosporin by β-lactamase-producing bacteria like MRSA, however, releases the EtNBS molecules from quenching, separating the two and allowing photosensitization to occur upon excitation. Thus, it was predicted that β-LEAP would exclusively target treatment-resistant bacteria.

In vitro studies of this construct with MRSA strains bore out this hypothesis beautifully. All strains were inactivated at moderate doses. As expected, the team saw an inverse relationship between susceptibility to penicillin, another β-lactam antibiotic, and susceptibility to β-LEAP. Even more encouragingly, experiments with uninfected mammalian cell lines showed reduced phototoxicity compared to EtNBS-COOH alone. This selectivity for bacteria and the orthogonality of penicillin and β-LEAP strongly suggests that combination therapy with a β-lactam antibiotic and β-LEAP-PDI could kill virtually all Gram-positive bacteria, even those that resist ordinary treatments.

While benzophenoxazines and their analogs have been tested in only some of these applications, the large body of work on PDI suggests that the premise is fundamentally sound. Admittedly, more investigations into PDI with EtNBS must first be done *in vivo*, but the promise shown by this approach hints that it may have broad applicability in treating a wide range of pathogens and infections.

6. Antineoplastic Applications

Since Lewis *et al.* discovered the preferential tumor staining of Nile Blue, many studies have investigated the anti-proliferative and cytotoxic qualities of Nile Blue and its derivatives. Of the myriad benzophenoxazine derivatives, EtNBS in particular has earned attention for its antineoplastic potency across an array of cancer cell lines. The first investigations of EtNBS *in vitro* and *in vivo* involved PDT of human squamous carcinoma (FaDu), murine mammary sarcoma (EMT6), and rat glioma (9L).[65] In all of these models, EtNBS demonstrated remarkable and, to the authors at the time, quite surprising, phototoxicity. Even more impressively, it was found that adjacent non-cancerous tissue survived treatment undisrupted and without collateral damage. Interestingly, these early studies also recognized a certain pattern of cell killing in which the cores of tumors experienced significant cell death post-PDT, while the periphery emerged unscathed. Recent work, elaborated later in this section, has potentially elucidated the mechanism behind these early results.[25,26,50]

The surprising success of EtNBS *in vivo* led Foley *et al.*[34] to investigate EtNBA and EtNBSe as well. Examination of the *in vivo* efficacy of these

Table 2. (a) Uptake of photosensitizer by EMT6 cells *in vitro*, measured in μg photosensitizer/mg total cell protein. (b) Percentage of EMT6 cells surviving after treatment with 0.15 mM photosensitizer and irradiation with 610 nm light for 10 min at 5.4 mW/cm^2. Data adapted from Foley *et al.*[34] with permission.

Photosensitizer	Uptake by EMT6[a]	% Viable cells[b]
EtNBA	3.60 ± 0.06	114.1 ± 2.0
EtNBS	4.67 ± 0.11	94.5 ± 2.0
EtNBSe	4.97 ± 0.11	2.8 ± 1.3

three analogs demonstrated EtNBSe to be a potent photosensitizer, with cytotoxicity greater than that of EtNBS and much greater than that of EtNBA (Table 2). In EMT6, Cincotta *et al.*[9] found that EtNBS and EtNBSe elicited post-irradiation cell killing significantly greater than that of PII, whose cell killing effects were undetectable in that study. As expected, EtNBA offered only minimal, but nevertheless nonzero, photodynamic efficacy. These results built on similar findings by Lin *et al.*[66] who found that, in a human bladder carcinoma cell line, some benzophenoxazines analogs were 1000 times more potent than HPD.

Further investigation of EtNBS probed its uptake and elimination patterns *in vivo*.[10] Maximum uptake in EMT6-expressing mice occurred between 3 and 8 h after subcutaneous injection. In these mice, EtNBS additionally demonstrated strong tumor specificity, with a tumor-to-tissue ratio of 4:1 in skin and 8:1 in muscle. Furthermore, EtNBS was rapidly cleared from the study mice. Plasma concentrations 24 h post-injection were a negligible 54 ng/mL, orders of magnitude lower than the 7.5 mg/kg required to provoke cytotoxicity. Absorption spectroscopy was unable to detect any residual EtNBS in the tumors or in nearby healthy tissue. All these findings pointed to a lack of undesirable long-term photosensitization in tissue.

This same study also began to tune and optimize treatment conditions in both RIF and EMT6 *in vivo* murine models. They found that decreasing the irradiance from 200 to 50 mW/cm^2 greatly improved treatment response. Additionally, increasing the total light dose from 50 to 100 or 200 J/cm^2 elicited a similar improvement, eradicating entire tumors, with no regrowth observed within 90 days in most instances. The authors noted that tumor regrowth seemed to begin from the periphery of former tumors that had been mostly destroyed. These light doses induced minor burns and edema that were noted to resolve within a few weeks of treatment. As observed previously, nearby and overlying tissue was not damaged by the photosensitizer itself. Using EtNBA as a toxicity control, it was verified that PDT-triggered cytotoxicity stemmed from type I or type II mechanisms and not from the thermal effects of non-radiative relaxation. Even at the highest doses, EtNBS was found to avoid systemic toxicity. Combination PDT of EtNBS with BPD-MA, a benzoporphyrin photosensitizer, demonstrated impressive synergistic effects in the same *in vivo* mouse model, leading to the destruction of more cells than with either photosensitizer alone.[67]

The flurry of *in vitro* studies of EtNBS in the 1990s culminated, in a sense, in the study of EtNBS-PDT of naturally occurring tumors, including an assortment of squamous cell carcinomas and mast cell tumors in dogs and cats.[68] Here, nearly all animals studied saw at least a short-term response to PDT, and several saw a complete, long-term response. None of the animals experienced any lingering photosensitization within five days after treatment. While no severe systemic toxicity or pain was observed, some minor nausea and vomiting, which resolved quickly, was observed in a few cats. The efficacy of EtNBS-PDT in these animals was exciting, and showed promise for future antineoplastic applications of the photosensitizer.

6.1. *Hypoxia and Oxygen Independence in Antioneoplastic Applications*

Building on the pioneering work by Foley and the Cincottas, recent anti-neoplastic studies utilizing EtNBS have specifically focused on the problem of hypoxia in the treatment of cancer. The rapid and haphazard growth of tumors leads to the formation of leaky and tortuous vasculature that does not adequately deliver oxygen throughout cancerous tissue. Oxygen diffusing from blood vessels is typically consumed within 70 and 150 μm; after this point, more distant cells progressively become starved of oxygen.[69] Hypoxic cancer cells undergo numerous adaptations, including the upregulation of anti-apoptotic machinery, DNA repair enzymes, and anaerobic glycolysis, the acidic byproducts of which reduce the local extracellular pH.[25,70]

It has been a challenge to fully treat these heterogeneous tumor regions. As much as 60% of tumors demonstrate severe hypoxia (pO_2 under 10 mmHg), which renders chemotherapy drugs such as carboplatin and paclitaxel less effective, often reduced to just 1% of their observed monolayer efficacy.[24,71] It can be difficult to deliver drugs into hypoxic regions due to many factors including their distance from blood vessels, the acidity of the microenvironment, and the extracellular matrix that is often present throughout tumorous tissue. Since radiation therapy and type II-based PDT regimens require O_2 to mediate their cytotoxicity, it is not surprising that these methods inadequately treat hypoxic regions, which instead can display treatment resistance, recovery, and eventual metastasis.[23,70–72] Oxygen-dependent therapies have the added downside of consuming oxygen, thereby depleting oxygen during treatment and compounding the challenges posed by hypoxia. In order to overcome these challenges, an approach is needed that can function in acidic conditions, regardless of oxygenation, and without being hampered by the extracellular matrix (Fig. 10).

It was hypothesized that EtNBS, as a small cationic molecule capable of both type I and type II photochemistry, might show antineoplastic activity in hypoxic regions. To explore this possibility, three-dimensional *in vitro* tumor models, which replicate the hypoxic and acidic cores found *in vivo*, were incubated with EtNBS.[25] EtNBS diffused down the acidic pH gradient of

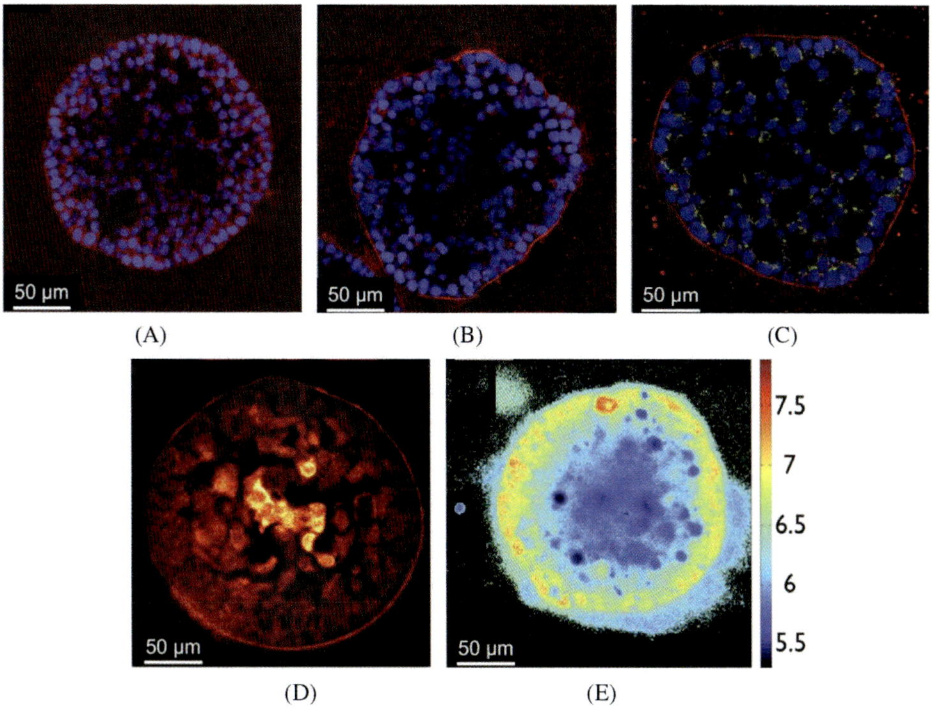

Fig. 10. Physiological properties of *in vitro* tumor nodules. (A) Stain for fibronectin. (B) Stain for collagen IV. (C) Stain for laminin V. (D) Pimonidazole stain for hypoxia. (E) pH imaging with SNARF-4F. In a generic *in vitro* tumor nodule, a scattered meshwork of extracellular matrix proteins exists, along with distinct, decreasing pO_2 and pH gradients from the periphery to the core. Adapted with permission from Evans *et al.*[25]

nodules and selectively localized to the hypoxic cores of 3D models. At low light doses of 5 mW/cm^2, EtNBS-PDT triggered cytotoxicity only in these hypoxic cells, and at higher fluences could completely kill all cells in 500 μm spheroids (Fig. 11). It is worth noting that the pattern of cytotoxicity — where the core of the tumor nodules experiences greater cell death than the periphery — matches the pattern of treatment response observed in the first antineoplastic studies with EtNBS.

Both EtNBS and the sidechain-hydroxylated EtNBS-OH were highly effective in hypoxic nodules, suggesting that tandem therapy with more traditional chemotherapy drugs already in clinical use may offer toxicity towards both hypoxic and normoxic cells alike.[26] In contrast, the porphyrin BPD was found to localize just to the periphery of large nodules (Fig. 12). Indeed, a separate imaging study found the limited localization of BPD significantly impedes its photodynamic therapy efficacy in large *in vitro* tumor spheroids.[25] The success of EtNBS against hypoxic cellular subpopulations is thought to arise from its propensity to cause type I phototoxicity, which has been demonstrated to cause cell death even in the absence of molecular oxygen. The fact that EtNBS has a singlet oxygen quantum yield of only 3% suggests that the majority of its phototoxic interactions are mediated through the direct generation of reactive radical species. The ability of EtNBS to react via two separate photochemical channels seems to make it a highly potent molecule for operation across a range of oxygen tensions. EtNBS-PDT in 3D nodule cultures was found to be irradiance-dependent; at a constant total light dose of 15 J/cm^2, the greatest level of cell killing occurred at the lowest level of irradiance, while cell viability reached an asymptote at an irradiance of 100 mW/cm^2. This result can be explained by the dual-channel phototoxicity of EtNBS — in low-irradiance scenarios, the consumption of oxygen is less than the oxygen diffusion rate, such that enough oxygen reaches EtNBS to maintain Type I and II activity; at higher irradiances, the more rapid oxygen consumption, which outpaces oxygen diffusion, starves the type II mechanism, forcing EtNBS to react via the still effective type I pathway.

Interestingly, EtNBS-PDT studies in 3D nodules carried out in both 100% nitrogen and carbon dioxide atmospheres were found to be equally effective, indicating that radical species other than those derived from molcular oxygen and nitrogen must mediate cell killing effects. These results shed some light on observations in early EtNBS studies, including the lack of strong correlation found between singlet oxygen quantum yields and overall cell killing, especially *in vivo*.[18,33,66]

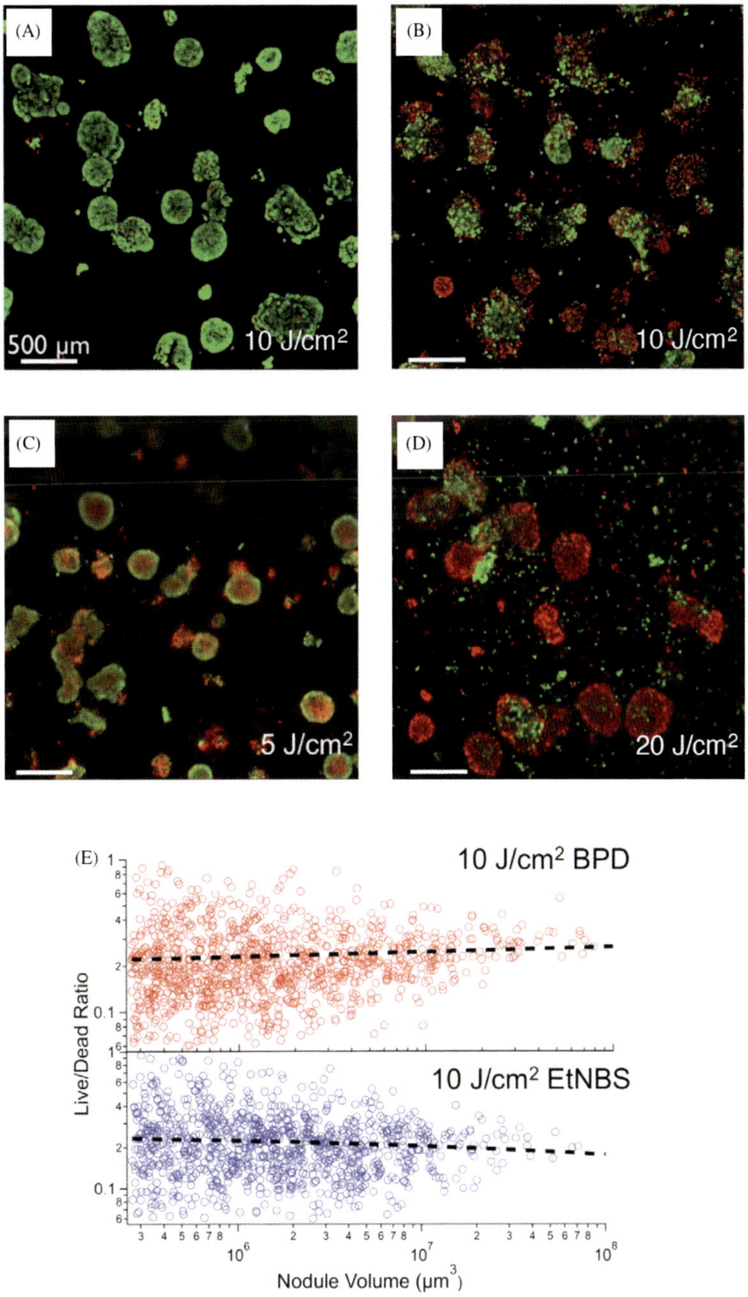

Fig. 11. Cytotoxicity of EtNBS and BPD in 3D ovarian cancer (OVCAR5) nodules using live/dead ratios. OVCAR5 cells were exposed to (A) no treatment (B) 240 nM BPD-PDT at 10 J/cm^2, (C) 500 nM EtNBS-PDT at 5 J/cm^2, or (D) 500 nM EtNBS-PDT at 20 J/cm^2. (E) Live/dead ratios of nodules of various sizes treated with EtNBS-PDT or BPD-PDT at 10 J/cm^2. While most cells survived BPD, most were killed by EtNBS. Adapted with permission from Evans *et al.*[25]

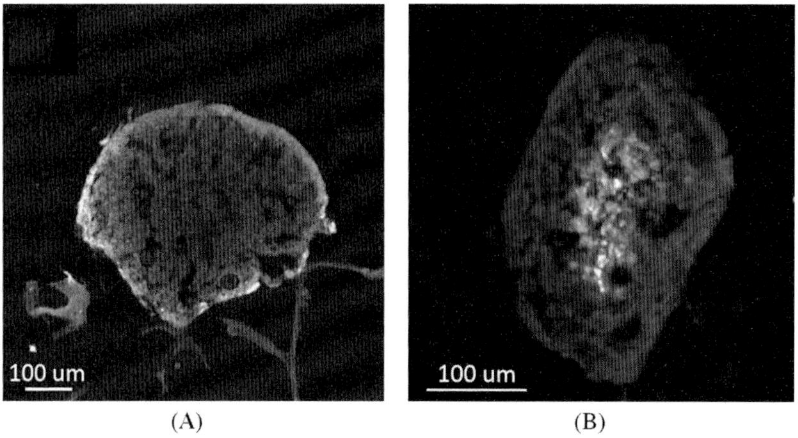

Fig. 12. Frozen section images of the uptake of BPD and EtNBS in 3D ovarian cancer nodules. (A) BPD fluorescence (gray-white) after 4.5 h incubation. (B) EtNBS fluorescence after 4.5 h incubation. Adapted with permission from Evans *et al.*[25]

7. Conclusions and Future Directions

EtNBS and its related heavy atom, ring, and sidechain derivatives form a family of photosensitizers with properties that make them promising for either stand-alone application or combination with other treatment approaches. Its ease of synthesis and the ability to readily tune its uptake, localization, and photochemical properties through straightforward chemical reactions additionally make EtNBS ideal for research and investigative studies. In many ways, EtNBS and its family of compounds can be considered a model photosensitizer system for exploring a wide range of photodynamic therapy conditions and outcomes.

As a photosensitizer, EtNBS shows promise as a first line agent in the treatment of difficult-to-treat cancers. The results of recent hypoxia studies, discussed above, suggest that reliance exclusively on the Type II mechanism of cytotoxicity in PDT may be suboptimal for achieving a complete cytotoxic response in many tumors. Although Type II photosensitizers have been proven to be highly potent in clinical studies, evidence is building that PDT with these agents, like many chemotherapeutics, leaves key cell subtypes untouched that proceed to repopulate tumors post-treatment. It is also acknowledged that type I mechanisms may not be ideal for cell killing *en masse*, as radical reactions often alter the photosensitizer irreparably. Nevertheless, photosensitizers like EtNBS, which can make use of oxygen and still operate in its absence, seem to solve a key limitation inherent in many PDT approaches. This dual efficacy is critical for future PDT approaches, as hypoxia in cancer is not a niche concern; rather, it is arguably one of the largest current barriers to effective cancer treatment.

In conjunction with other photosensitizers, EtNBS may show additional promise. Studies by Cincotta *et al.* found that EtNBS, in combination with BPD, was highly synergistic in animal models. The mechanism underlying this enhancement has not yet been identified in literature and is ripe for exploration. If EtNBS-PDT enhances BPD-PDT, it is likely synergistic with other photosensitizers as well. In this regard, EtNBS may serve as a complementary tool in PDT to kill hard-to reach hypoxic cell populations. In particular, the ability of EtNBS to penetrate into tumor regions normally untouched by porphyrins may further enhance its synergy with first generation photosensitizers such as PII. Even if synergy in cell killing is not observed for EtNBS combination therapies, it is hypothesized that *in vivo* studies utilizing EtNBS as part of a PDT combination strategy will still see greater long-term tumor shrinkage or growth retardation due to the elimination of subpopulations normally unresponsive to therapy.

EtNBS is likely to be synergistic with traditional chemotherapy approaches as well. Though no studies involving EtNBS itself have been published to date, two recent studies with other photosensitizers suggest that EtNBS could potentiate chemotherapy regimens. In a study by Rizvi *et al.*[73] BPD-PDT displayed synergy with carboplatin in a 3D spheroid model of ovarian cancer. The study revealed that BPD-PDT induced cytotoxicity that caused an "unpacking" of the tumor nodules, which may have allowed for the re-oxygenation of the tumor core and improved delivery of the platinum drug deep within spheroids.[25] The hydroxyl side-chain derivative EtNBS-OH was found to trigger similar unpacking of *in vitro* spheroids, suggesting that similar synergy may be observed with this photosensitizer.[26] Interestingly, this unpacking mechanism was not caused by other derivatives of EtNBS or by EtNBS itself.

Outside of cancer applications, the ability to work in hypoxic environments may enable EtNBS to have efficacy against infections and bacterial biofilms. Bacteria and fungi overwhelmingly tend to cluster into biofilms in wounds and on medical devices such as catheters.[74] These biofilms hinder penetration by most antimicrobial drugs, often become mildly hypoxic and acidic, and secrete dense, anionic extracellular polymers that act to increase resistance to treatment.[75] Thus, biofilms are generally far more resistant to treatment than free-floating bacteria. Because of the success EtNBS has seen in treating a variety of microbes, and because it is cationic and diffuses preferentially towards acidic, hypoxic environments, we speculate that EtNBS may show fungicidal and bactericidal properties in the treatment of difficult wound infections.

One challenge facing the application of EtNBS is its inherent dark toxicity when taken up by cells. While not as toxic as EtNBSe, the dark toxicity of EtNBS has the potential to limit its applicability and may pose regulatory barriers. There are several approaches that may reduce this potentially problematic issue. For example, EtNBS can be linked to a targeting agent, either through photoimmunoconjugates, like those developed by Hasan and colleagues, or through the use of aptamers.[76,77] Encapsulation and delivery via liposomes, which has been found helpful for molecules such as BPD, may serve to reduce the toxicity profile.[78] Furthermore, the use of nanoparticle approaches that can shield cells from interactions with EtNBS prior to phototherapy likely will also find applicability.

The future remains bright for EtNBS and its family of photosensitizers. Even though it has a rich history of studies spanning several decades, EtNBS still has a long road ahead to reach clinical use. Future research focusing on combination therapies, the utility of EtNBS against hypoxic cell populations, and its application in the treatment of microbial infections has the potential to drive the translation of EtNBS forward.

References

1. T. J. Dougherty, C. J. Gomer, B. W. Henderson, G. Jori, D. Kessel, M. Korbelik, J. Moan and Q. Peng, *J. Natl. Cancer Inst.* 90, 889–895 (1998).
2. Q. Peng, J. Moan and J. M. Nesland, *Ultrastruct. Pathol.* 20, 109–129 (1996).
3. T. J. Dougherty, G. B. Grindey, R. Fiel, K. R. Weishaupt and D. G. Boyle, *J. Natl. Cancer Inst.* 55, 115–121 (1975).
4. T. J. Dougherty, J. E. Kaufman, A. Goldfarb, K. R. Weishaupt, D. Boyle and A. Mittleman, *Cancer Res.* 38, 2628–2635 (1978).
5. T. Christensen, J. Moan, E. Wibe and R. Oftebro, *Br. J. Cancer*, 39, 64 (1979).
6. T. J. Dougherty, G. Lawrence, J. H. Kaufman, D. Boyle, K. R. Weishaupt and A. Goldfarb, *J. Natl. Cancer Inst.* 62, 231–237 (1979).
7. T. J. Kinsella, V. C. Colussi, N. L. Oleinick and C. H. Sibata, *Expert Opin. Pharmacother.* 2, 917–927 (2001).
8. D. E. Dolmans, D. Fukumura and R. K. Jain, *Nat. Rev. Cancer* 3, 380–387 (2003).
9. L. Cincotta, J. W. Foley and A. H. Cincotta, *Cancer Res.* 53, 2571–2580 (1993).
10. L. Cincotta, J. W. Foley, T. MacEachern, E. Lampros and A. H. Cincotta, *Cancer Res.* 54, 1249–1258 (1994).
11. J. W. Foley, X. Song, T. N. Demidova, F. Jilal and M. R. Hamblin, *J. Med. Chem.* 49, 5291–5299 (2006).
12. A. E. O'Connor, W. M. Gallagher and A. T. Byrne, *Photochem. Photobiol.* 85, 1053–1074 (2009).

13. C. J. Gomer, *Photochem. Photobiol.* 54, 1093–1107 (1991).
14. M. Kreimer-Birnbaum, *Semin. Hematol.* 26, 157–173 (1989).
15. M. R. Lewis, H. A. Sloviter and P. P. Goland, *Anat. Rec.* 95, 89–96 (1946).
16. M. R. Lewis, P. P. Goland and H. A. Sloviter, *Anat. Rec.* 96, 201–220 (1946).
17. M. R. Lewis, P. P. Goland and H. A. Sloviter, *Cancer Res.* 9, 736–740 (1949).
18. L. Cincotta, J. W. Foley and A. H. Cincotta, *Photochem. Photobiol.* 49, 751–758 (1987).
19. J. W. Foley, L. Cincotta and A. H. Cincotta, *SPIE Proc.* 90–95 (1988).
20. C.-W. Lin, J. R. Shulok, Y. K. Wong, L. Cincotta and J. W. Foley, *SPIE Proc.* 287–292 (1990).
21. C.-W. Lin, J. R. Shulok, S. D. Kirley, L. Cincotta and J. W. Foley, *Cancer Res.* 51, 2710–2719 (1991).
22. D. Kessel, Ed., *Photodynamic therapy of neoplastic diseases.* Vol. 2 (CRC, New York, 1990), p. 1–15.
23. M. Höckel and P. Vaupel, *J. Natl. Cancer Inst.* 93, 266–276 (2001).
24. X. Song, X. Liu, W. Chi, Y. Liu, L. Wei, X. Wang and J. Yu, *Cancer Chemother. Pharmacol.* 58, 776–784 (2006).
25. C. L. Evans, A. O. Abu-Yousif, Y. J. Park, O. J. Klein, J. P. Celli, I. Rizvi, X. Zheng and T. Hasan, *PLoS ONE*, 6, e23434 (2011).
26. O. J. Klein, B. Bhayana, Y. J. Park and C. L. Evans, *Mol. Pharmaceutics*, 9, 3171–3182 (2012).
27. R. Möhlau and K. Uhlmann, *Eur. J. Org. Chem.* 289, 90–130 (1896).
28. J. F. Thorpe, *J. Chem. Soc., Trans.* 91, 324–336 (1907).
29. A. Kanitz and H. Hartmann, *Eur. J. Org. Chem.* 1999, 923–930 (1999).
30. J. Jose and K. Burgess, *Tetrahedron*, 62, 11021–11037 (2006).
31. J. V. Crivello, *J. Polym. Sci., Part A: Polym. Chem.* 46, 3820–3829 (2008).
32. B. C. Raju, K. V. Prasad, G. Saidachary and B. Sridhar, *Org. Lett.* 16, 420–423 (2013).
33. L. Cincotta, J. W. Foley and A. H. Cincotta, *SPIE Proc.* 145–153 (1989).
34. J. W. Foley, L. Cincotta and A. H. Cincotta, *SPIE Proc.* 208–215 (1991).
35. J. W. Foley and L. Cincotta, *U.S. Patent No. 4,962,197.* (U.S. Patent and Trademark Office, Washington, D.C., 1990).
36. D. Vecchio, B. Bhayana, L. Huang, E. Carrasco, C. L. Evans and M. R. Hamblin, *Eur. J. Med. Chem.* 75, 479–491 (2014).
37. R. C. Clapp, J. P. English, C. E. Fellows, J. Forsythe, R. E. Grotz and R. G. Shepherd, *J. Am. Chem. Soc.* 74, 1994–1996 (1952).
38. S. Verma, U. W. Sallum, H. Athar, L. Rosenblum, J. W. Foley and T. Hasan, *Photochem. Photobiol.* 85, 111–118 (2009).
39. A. Bernthsen, *Justus Liebigs Ann. Chem.* 251, 1–97 (1889).
40. K. H. Drexhage, Ed., *Dye Lasers* (Springer, Berlin Heidelberg, 1973) pp. 144–193.
41. I. C. Summerhayes, T. J. Lampidis, S. D. Bernal, J. J. Nadakavukaren, K. K. Nadakavukaren, E. L. Shepherd and L. B. Chen, *Proc. Natl. Acad. Sci. U.S.A.* 79, 5292–5296 (1982).
42. A. R. Oseroff, D. Ohuoha, G. Ara, D. McAuliffe, J. Foley and L. Cincotta, *Proc. Natl. Acad. Sci. U.S.A.* 83, 9279–9733 (1986).
43. M. R. Hamblin and T. Hasan, *Photochem. Photobiol. Sci.* 3, 436–450 (2004).
44. O. E. Akilov, S. Kosaka, K. O'Riordan, X. Song, M. Sherwood, T. J. Flotte, J. W. Foley and T. Hasan, *Chem. Biol.* 13, 839–847 (2006).

45. K. J. Mellish, R. D. Cox, D. I. Vernon, J. Griffiths and S. B. Brown, *Photochem. Photobiol.* 75, 392–397 (2002).
46. R. S. Becker, S. Chakravorti and S. Das, *Photochem. Photobiol.* 51, 533–538 (1990).
47. X. Zheng, U. W. Sallum, S. Verma, H. Athar, C. L. Evans and T. Hasan, *Angew. Chem. Int. Ed.* 41, 2148–2151 (2009).
48. R. Benz, *Biophys. J.* 54, 25 (1988).
49. P. Gros, F. Talbot, D. Tang-Wai, E. Bibi and H. R. Kaback, *Biochem.* 31, 1992–1998 (1992).
50. I. Georgakoudi and T. H. Foster, *Photochem. Photobiol.* 68, 115–122 (1998).
51. V. H. Fingar and B. W. Henderson, *Photochem. Photobiol.* 46, 837–847 (1987).
52. T. Hamouda, A. Y. Shih and J. R. Baker, *Lett. Appl. Microbiol.* 34, 86–90 (2002).
53. K. O'Riordan, O. E. Akilov, S. K. Chang, J. W. Foley and T. Hasan, *Photochem. Photobiol. Sci.* 6, 1117–1123 (2007).
54. T. Maisch, *Lasers Med. Sci.* 22, 83–91 (2007).
55. C. E. Millson, M. Wilson, A. J. MacRobert and S. G. Bown, *Photochem. Photobiol. B.* 32, 59–65 (1996).
56. T. Dai, Y.-Y. Huang and M. R. Hamblin, *Photodiagn. Photodyn. Ther.* 6, 170–188 (2009).
57. Y. Nitzan, M. Gutterman, Z. Malik and B. Ehrenberg, *Photochem. Photobiol.* 55, 89–96 (1992).
58. A. Minnock, D. I. Vernon, J. Schofield, J. Griffiths, J. H. Parish and S. B. Brown, *Antimicrob. Agents Chemother.* 44, 522–527 (2000).
59. T. A. Dahl, W. R. Midden, P. E. Hartman, *Photochem. Photobiol.* 46, 345–352 (1987).
60. S. L. Croft, S. Sundar and A. H. Fairlamb, *Clin. Microbiol. Rev.* 19, 111–126 (2006).
61. E. Handman and D. V. Bullen, *Trends Parasitol.* 18, 332–334 (2002).
62. R. L. Treasure and B. J. Seaworth, *Ann. Thorac. Surg.* 59, 1405–1409 (1995).
63. C. Lienhardt, M. Raviglione, M. Spigelman, R. Hafner, E. Jaramillo, M. Hoelscher, A. Zumla and J. Gheuens, *J. Infect. Dis.* 205, S241–S249 (2012).
64. X.-J. Fu, Y. Fang and M. Yao, *BioMed. Res. Int.* 2013, 1–9 (2013).
65. A. H. Cincotta, L. Cincotta and J. W. Foley, *SPIE Proc.* 202–210 (1990).
66. C.-W. Lin, J. R. Shulok, Y.-K. Wong, C. F. Schanbacher, L. Cincotta and J. W. Foley, *Cancer Res.* 51, 1109–1116 (1991).
67. L. Cincotta, D. Szeto, E. Lampros, T. Hasan and A. H. Cincotta, *Photochem. Photobiol.* 63, 229–237 (1996).
68. A. E. Frimberger, A. S. Moore, L. Cincotta, S. M. Cotter and J. W. Foley, *Clin. Cancer Res.* 4, 2207–2218 (1998).
69. J. G. Rajendran and K. A. Krohn, *Radiol. Clin. N. Am.* 43, 169–187 (2005).
70. M. Höckel and P. Vaupel, *Semin. Oncol.* 28, 36–41 (2001).
71. S. Koch, F. Mayer, F. Honecker, M. Schittenhelm and C. Bokemeyer, *Br. J. Cancer.* 89, 2133–2139 (2003).
72. J. S. Rasey, W.-j. Koh, M. L. Evans, L. M. Peterson, T. K. Lewellen, M. M. Graham and K. A. Krohn, *Int. J. Radiat. Oncol. Biol. Phys.* 36, 417–428 (1996).
73. I. Rizvi, J. P. Celli, F. Xu, C. L. Evans, A. O. Abu-Yousif, A. Muzikansky, S. A. Elrington, B. W. Pogue, D. M. Finkelstein and U. Demirci, *SPIE BiOS*, 788609 (2011).
74. R. M. Donlan and J. W. Costerton, *Clin. Microbiol. Rev.* 15, 167–193 (2002).

75. J. Bonhomme, M. Chauvel, S. Goyard, P. Roux, T. Rossignol and C. d'Enfert, *Mol. Microbiol.* 80, 995–1013 (2011).
76. B. Q. Spring, A. O. Abu-Yousif, A. Palanisami, I. Rizvi, X. Zheng, Z. Mai, S. Anbil, R. B. Sears, L. B. Mensah and R. Goldschmidt, *Proc. Natl. Acad. Sci.* 111, E933–E942 (2014).
77. O. C. Farokhzad, J. Cheng, B. A. Teply, I. Sherifi, S. Jon, P. W. Kantoff, J. P. Richie and R. Langer, *Proc. Natl. Acad. Sci.* 103, 6315–6320 (2006).
78. M. Kramer, J. W. Miller, N. Michaud, R. S. Moulton, T. Hasan, T. J. Flotte and E. S. Gragoudas, *Ophthalmology*, 103, 427–438 (1996).

Chapter 12

Negative Impact of Tumor-Generated Nitric Oxide on Photodynamic Therapy

Albert W. Girotti

Department of Biochemistry, Medical College of Wisconsin, Milwaukee, Wisconsin, USA, 53226
agirotti@mcw.edu

Abstract

Many malignant tumors exploit endogenously generated nitric oxide (NO) as a pro-survival molecule, a role often linked to constitutive expression of inducible nitric oxide synthase (iNOS). NO is known to antagonize the antitumor effects of ionizing radiation, chemotherapeutic agents, and also photodynamic therapy (PDT), the latter effect recognized relatively recently. This chapter will focus on low level NO as a tumor-promoting signaling molecule and how it can compromise PDT efficacy by stimulating tumor microvasculature relaxation or more directly by enhancing tumor cell resistance to photokilling and also proliferation, migration, and invasiveness of surviving cells. To be discussed (along with its troubling implications) is the striking observation that for certain cancer cell types, these negative effects are attributable more to PDT stress-induced iNOS/NO than to preexisting iNOS/NO. Approaches for overcoming NO's still poorly recognized anti-PDT activity through pharmacologic use of appropriate iNOS inhibitors will also be discussed.

1. Introduction

The purpose of this chapter is to discuss the relatively recent discovery that nitric oxide (NO) generated in tumor vascular cells and/or in tumor cells themselves can reduce the anti-tumor efficacy of photodynamic therapy (PDT) or even stimulate tumor progression in cells that survive PDT stress. To the extent that they are known, some of signaling mechanisms that underlie these pro-survival effects of NO will also be discussed based on *in vitro* studies carried out on various cancer cell lines. Finally, some ideas about counteracting these negative effects of NO via pharmacologic intervention, e.g., with inhibitors of NO generation, will be presented, along with prospects for introducing such agents into the clinical PDT setting.

2. Brief Overview of Nitric Oxide's Physiological and Pathophysiological Properties

NO is a relatively short-lived free radical molecule produced naturally by enzymes of the nitric oxide synthase (NOS) family, including neuronal NOS (nNOS or NOS1), inducible NOS (iNOS or NOS2), and endothelial NOS (eNOS or NOS3).[1–4] These enzymes catalyze a 5-electron oxidation of L-arginine at the expense of NADPH and O_2 to give L-citrulline and NO as products (Fig. 1). NOS-generated NO has numerous biological functions, including involvement in blood vessel relaxation (eNOS), nerve cell signal transmission (nNOS), and eradication of pathogenic microorganisms (iNOS).[1–3] When generated at high rates, e.g., by activated macrophages or neutrophils at pro-inflammatory sites, NO can act as a cytotoxic prooxidant.[4] However, at low generation rates, NO, in addition to mediating vasorelaxation, can act as a cytoprotective antioxidant, signaling for cell survival and even accelerated growth after an oxidative challenge.[5] Accordingly, at steady state levels in the micromolar range, NO tends to be cytotoxic, whereas at levels in the nanomolar range, it can signal for cell survival and proliferation.[4–6] Low level

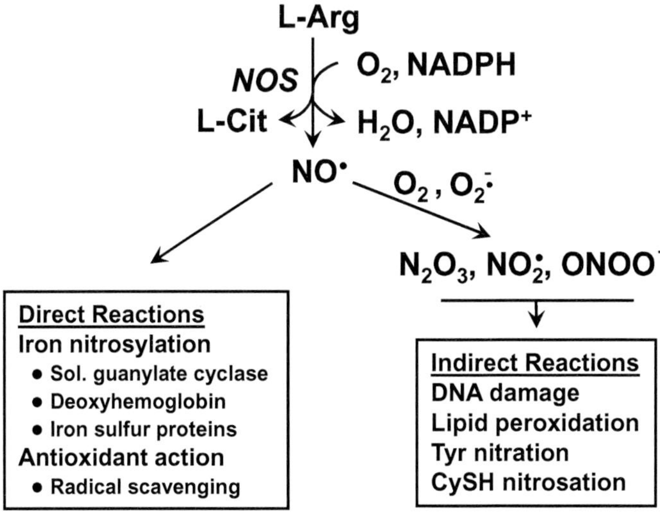

Fig. 1. Generation of nitric oxide (NO) by nitric oxide synthase (NOS) enzymes: direct *vs.* indirect reactions of NO with target molecules.

NO can react directly with heme or other ligated iron to give nitrosylated derivatives (Fig. 1), the classic example being eNOS NO stimulation of soluble guanylate cyclase (sGC) in vascular smooth muscles, resulting in cyclic GMP (cGMP) formation, the latter causing vasorelaxation and lowering of blood pressure.[3] NO itself can also act as an antioxidant by intercepting cytotoxic species such as lipid oxyl and peroxyl radicals.[7] High level NO produced by vascular macrophages, for example, usually reacts indirectly with target molecules, i.e. after conversion to some oxidized intermediate such as dinitrogen trioxide (N_2O_3), nitrogen dioxide (NO_2), or peroxynitrite ($ONOO^-$). The latter two species are strong oxidants which can damage DNA, induce lipid peroxidation, or modify tyrosine residues (Fig. 1). $ONOO^-$ arises from rapid reaction of NO with superoxide, a reactive oxygen species (ROS) also generated by activated macrophages.[8] N_2O_3, an autoxidation product of NO, is of special interest because it can reversibly modify (S-nitrosate) specialized cysteine residues in effector proteins, thereby initiating pro-survival or pro-death signaling responses.[9–11] Intracellular transnitrosation from S-nitrosoglutathione can also give rise to S-nitroso (SNO) cysteine. Although NO does not react directly with cysteine thiol groups, it can react rapidly with cysteine thiyl radicals, giving SNO derivatives.[10]

There are numerous examples of endogenous (NOS-derived) and exogenous (chemical donor-derived) NO acting in a salutary fashion on normal (non-transformed) mammalian cells. For example, studies with primary hepatocytes[12] and differentiated (neuron-like) PC12 cells[13] showed that apoptosis induced by growth factor or serum deprivation could be inhibited by exogenous NO through activation of sGC and thence cGMP-responsive protein kinase G. A more recent study demonstrated that human keratinocyte apoptosis induced by UVB radiation could be strongly enhanced by NOS and sGC inhibitors, but suppressed by NO from a chemical donor or by iNOS overexpression.[14] These findings demonstrated an anti-apoptotic role of NO that was mediated at least in part by cGMP. Similar cytoprotective effects of NO have been reported for UVA-challenged keratinocytes and endothelial cells.[15] However, whether this NO derived solely from constitutively expressed NOS or whether any photostress-induced NOS might be involved was not determined.

The many studies that have been carried out concerning NO's role in cancer have revealed a fascinating dichotomy of effects that it can have on cell proliferation, migration, invasion, metastasis, and apoptosis.[4–6,16–19] In general NO has been shown to be either pro- or anti-tumorigenic, depending on factors such as where it arises in the tumor setting, rate of generation, steady state concentration attained, and properties of the tumor microenvironment.[4,5,18]

As already suggested, NO-derived oxidants ($ONOO^-$, NO_2) produced by high NO-generating inflammatory cells can be toxic to existing tumors, which tend to be more sensitive to oxidative pressure than normal tissues.[5,18] On the other hand, NO at low levels tends to do the opposite, inhibiting apoptosis, enhancing resistance to therapeutic agents, and promoting tumor expansion via angiogenesis and metastasis.[5,19] Low level NO can also stimulate aerobic glycolysis (Warburg effect) and inhibit mitochondrial ROS-generating oxidative phosphorylation, thereby benefiting tumor survival/proliferation.[20,21] This NO may derive not only from macrophages and endothelial cells in the surrounding tumor vasculature, but also from tumor cells themselves. Diverse experimental approaches have led to a general awareness of low level NO's tumor-promoting ability, including genetic ablation or knockdown of NOS enzymes, use of competitive inhibitors of NOS activity, and use of NO scavengers.[22,23] As one notable example, Sikora *et al.*[24] found that expansion of human malignant melanoma cells *in vitro* and in a mouse model could be suppressed by the iNOS inhibitor L-NIL, which also acted synergistically with cisplatin in blocking tumor growth. These investigators subsequently reported that iNOS/NO also plays a key role in tumor-promoting immunosuppression in melanoma-bearing mice.[25] Other investigators showed that ROS-mediated apoptosis of human non-small lung carcinoma cells after exposure to ionizing radiation was strongly enhanced by the general NOS inhibitor L-NMMA and that this correlated with down-regulation of hypoxia-inducible factor-1α (HIF-1α) and pyruvate dehydrogenase kinase-1 (PDK-1).[26] This suggested a link between NO and HIF-1α and PDK-1 expression, both proteins known to play key roles in survival/progression of various tumors.[20] Of added interest is an recent study by Eyler *et al.*[27] showing that glioblastoma stem cells isolated from human brain specimens expressed much higher levels of iNOS than non-stem cell counterparts. The stem cells were found to depend on iNOS/NO for relatively rapid basal proliferation *in vitro* and in mouse xenografts because administration of a blood–brain barrier-penetrating iNOS inhibitor slowed tumor progression and extended animal survival.[27] A subsequent study by another group revealed that iNOS/NO also plays a key role in the aggressive migration and invasiveness of glioma stem cells and that this is linked to expression of proteins involved in extracellular matrix degradation, e.g., matrix metalloproteinase-9 (MMP-9), and $\alpha 9\beta 1$ integrin.[28] Of added interest vis-à-vis brain malignancies is recent work showing that fractionated ionizing radiation dramatically stimulates iNOS/NO expression and stem cell expansion in several glioma cell lines, one of which was from a patient with highly malignant GBM.[29] Importantly, iNOS knockdown greatly reduced radiation-stimulated stem cell expansion and abolished cell resistance to cisplatin- or taxol-induced apoptosis.[29] These studies,[24–29] along with many relevant others not cited here, clearly demonstrate that many malignant cells can exploit low level iNOS-derived NO not only to evade apoptosis, but also to stimulate migration and invasion as early steps in metastasis.

3. Influence of Tumor-generated NO on PDT Outcome

The question of how NO produced by tumor cells per se and/or tumor-supporting vascular cells might affect tumor responsiveness to PDT has been addressed in surprisingly few studies to date. In a report on an early study carried out with isolated rat aortas, it was suggested that PDT impairs NO production by vascular endothelial cells and that this plays a crucial role in the vasoconstrictive effects often associated with tumor ablation by PDT.[30] However, no mechanistic evidence was provided to support such a mechanism, e.g., endothelial NOS activity before and after PDT was not assessed.[30] A later *in vitro* study showed that epidermoid cancer cells rapidly induced nNOS and NO after being photodynamically stressed and subsequently underwent caspase activation and apoptosis; cell death was linked to formation of cytotoxic NO derivatives such as $ONOO^-$.[31] This was the first reported evidence for NOS induction in a cancer cell line subjected to PDT-like stress. However, it appears unlikely that PDT-induced NOS/NO could have been the cause of cell death as a result of photodynamic stress. The more likely possibility, not considered by the investigators,[31] is that the observed NOS upregulation after photostress with accompanying low flux NO was a cytoprotective or pro-survival response.

How NO might affect PDT efficacy in an *in vivo* setting was first addressed about 15 years ago in studies involving various mouse tumor models subjected to Photofrin®-based PDT. Using Photofrin®-PDT on RIF tumor-bearing mice, Henderson *et al.*[32] showed that the general NOS inhibitor L-NNA significantly improved tumor responses over those seen in non-L-NNA controls, particularly when relatively high light fluence rates were used. Slightly before and after this work, Korbelik *et al.*[33,34] carried out comparable groundbreaking studies, which demonstrated that the Photofrin®-PDT cure rate for various murine subcutaneous tumors could be significantly improved when a NOS inhibitor (L-NAME or L-NNA) was administered immediately after irradiation. Importantly, it was found that the extent of improvement correlated directly with constitutive tumor capacity to generate NO, tumors with relatively high constitutive NO output responding best.[34] Moreover, the "high" NO tumors exhibited somewhat greater resistance to PDT than the "low" NO counterparts, which suggested that NO output might be exploited as a prognostic indicator of PDT effectiveness. Reeves *et al.*[35] using murine tumors sensitized with 5-aminolevulinic acid (ALA)-induced protoporphyrin IX (PpIX) have recently confirmed that NO yield can predict extent to which PDT can be improved by NOS inhibitors. The observations described[32–35] were interpreted in terms of competing effects of PDT and NO on the tumor microvasculature, i.e. the vasodilatory effects of endothelium-derived NO acting in opposition to the known anti-tumor vasoconstrictive effects of PDT. Surprisingly little additional work has been carried out at the *in vivo* level to further characterize how endogenous NO may reduce anti-tumor PDT efficacy. Thus, important questions such as the following remain open at the present time, at least for *in vivo* PDT: (i) whether NO derives from tumor cells *per se*, tumor vasculature cells, or both; (ii) which of the three NOS isoforms is most important in supplying NO; (iii) whether the NOS/NO involved is constitutive to the tumor and/or upregulated by PDT stress; (iv) the signaling mechanisms that may underlie NO's negative effects on PDT effectiveness. In the case of *in vitro* PDT models, however, such questions have begun to be addressed over the past 5–6 years and some of the key findings will be discussed in the next section.

4. Exogenous NO as a Cytoprotective Chain-breaking Antioxidant in Tumor Cells

Early studies in the author's laboratory involved COH-BR1 cells, a human breast tumor subline that is sensitive to photodynamic killing because it lacks GPx4, a selenoperoxidase that catalyzes detoxification of membrane lipid hydroperoxides (LOOHs), including photodynamically-generated LOOHs.[36] Cells were sensitized with ALA-generated PpIX using two different protocols — one with PpIX confined to mitochondria, where it originates, and the other with most of the PpIX disseminated to other intracellular sites, including plasma membrane.[37] We found that cell death after irradiation was mainly apoptotic when PpIX was in mitochondria, but mainly necrotic after it was allowed to diffuse to the cell periphery.[37] Using cells sensitized with disseminated PpIX, we asked how NO released from a chemical donor, spermine NONOate (SPNO), would affect cell sensitivity to photokilling. When photosensitized cells were treated with non-toxic doses of SPNO (0.2–0.4 mM) immediately before irradiation (designated "NO-now"), a striking resistance to necrosis was observed compared with controls lacking SPNO or treated with decomposed SPNO.[38] Photokilling was enhanced when a lipophilic iron

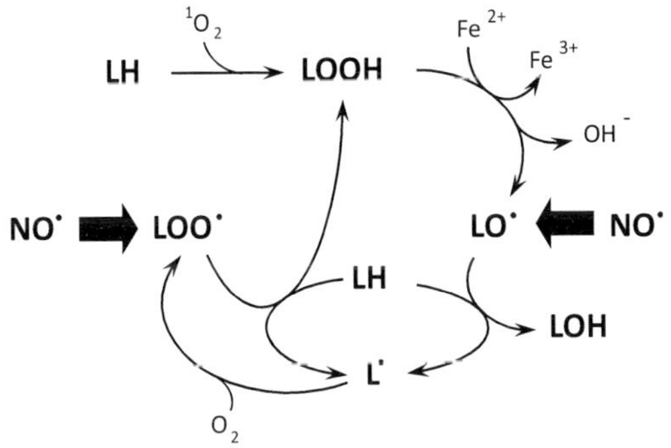

Fig. 2. Inhibition of free radical-mediated (chain) lipid peroxidation by NO. $^1O_2$ addition to an unsaturated lipid (LH) gives the lipid hydroperoxide (LOOH), which undergoes iron-catalyzed reduction to a lipid oxyl radical (LO·). Chain peroxidation induced by LO· gives lipid peroxyl radical (LOO·) intermediates. Interception of LO· and LOO· by NO suppresses chain peroxidation. (Reproduced from Ref. 39 with permission).

chelate was introduced at the outset and again SPNO was strongly inhibitory. Based on earlier findings,[39] we attributed these results to NO inhibition of plasma membrane-damaging lipid peroxidation, i.e. scavenging of chain-propagating lipid oxyl (LO·) and peroxyl (LOO·) radicals arising from one-electron reduction of photogenerated LOOHs (Fig. 2). Previous work had ruled out the possibility that NO might scavenge PpIX-generated singlet oxygen ($^1O_2$), which adds to unsaturated lipids to give LOOHs.[40] Of added interest regarding inhibition of peroxidative necrosis by exogenous NO was the subsequent discovery that NO has the remarkable ability to support nascent apoptosis, particularly at low levels of photooxidative stress.[41] Thus, in the presence of SPNO, residual photokilling of disseminated PpIX-sensitized COH-BR1 cells was found to be predominantly apoptotic rather than necrotic, and glucose supplementation enhanced this, at least in part by supporting ATP synthesis. The proffered explanation was that since NO suppressed both membrane disruption and loss of cellular ATP under photostress, this allowed apoptosis to be sustained and detected just as necrosis was strongly repressed.[41]

5. Endogenous NO as a Cytoprotective Agent in Photostressed Tumor Cells

Another early study in the author's laboratory is noteworthy because it revealed novel cytoprotective signaling effects of NO that were distinct from its ability to directly intercept lipid derived free radicals (Sect. 4). COH-BR1 cells were exposed to a non-lethal dose of SPNO (e.g., 0.2 mM) and 20 h later were sensitized with ALA-induced PpIX, which was allowed to disseminate from mitochondria. At this point, NO was no longer being liberated by SPNO because it's decomposition half-life is approximately 39 min at 37 °C.[42] The SPNO-treated cells exhibited a striking resistance to MTT- or LDH release-assessed photokilling, whereas spent SPNO-treated controls behaved like procedural controls. This resistance (also seen when cells were challenged with *tert*-butyl hydroperoxide) was attributed to "NO-then", i.e. NO's delayed or long-term signaling effects.[43,44] Substantial photoresistance was also observed when tumor cells were exposed overnight to a natural source of NO — activated RAW-264.7 macrophages on microporous inserts — before being photochallenged. Observed at two different light fluences, this resistance was abolished when cell pre-incubation with macrophages was carried out in the presence of L-NAME, thus confirming that macrophage-derived NO was acting cytoprotectively.[43,44] Immunoblot analyses on SPNO- and macrophage-preconditioned cells revealed an early upregulation of heme oxygenase-1 (HO-1) at ~2 h, followed by ferritin from 4–20 h.[43] This suggested that NO was signaling for ultimate ferritin sequestration of redox iron in free and/or heme form, which would otherwise catalyze lipid peroxidation-mediated photokilling. Supporting evidence was obtained by showing that the level of intracellular free iron in SPNO-treated cells was significantly lower than that in decomposed SPNO controls after 24 h. This study showed for the first time that NO emerging from the tumor microvasculature (endothelial cells, macrophages) could signal for PDT resistance via mobilization and then containment of redox iron.[43]

Subsequent studies focused on cell photokilling sensitized exclusively by mitochondria-localized PpIX, using an ALA treatment protocol that limited the porphyrin to this compartment.[45] As demonstrated earlier for COH-BR1 cells sensitized in this manner, photokilling occurred primarily via mitochondria-initiated (intrinsic) apoptosis, as assessed by Hoechst or Annexin V-FITC staining.[37] Similar to its effect on disseminated PpIX-sensitized necrosis, SPNO-derived NO strongly inhibited mitochondrial PpIX-sensitized apoptosis. Comparable SPNO anti-apoptotic effects were observed on two other ALA-treated cell lines, breast cancer MDA-MB-231 and cervical cancer HeLa.[45] We predicted that LOO·/LO· scavenging by NO could not be the major reason for cytoprotection in these cases because little of the incoming NO would be expected to reach mitochondrial membranes. It appeared instead that NO was acting by inducing anti-apoptotic signaling pathways, possibly including those leading to HO-1 and ferritin upregulation.[43] Initial attention was directed to mitogen-activated protein kinase (MAPK) pathways, i.e. those involving JNK and p38 MAPKs (typically associated with apoptosis) and ERK1/2 MAPKs (typically associated with cell survival and proliferation).[46] Although the role of MAPK activation in PDT had received considerable attention,[47–49] none of this focused on how NO might possibly affect this activation. As shown by Bhowmick and Girotti,[45] broad-band visible irradiation (4 J/cm^2) of ALA-treated COH-BR1 cells induced a transient phosphorylation-activation of total p38 and JNK, observed by Western analysis 15–30 min after the light dose. No signal for p-JNK or p-p38 was observed at 4 h post-irradiation. Unlike p38 and JNK, ERK1/2 was already activated prior to irradiation and underwent gradual deactivation during post-irradiation incubation. These after-light deactivations are attributed to MAPK phosphatases such as MPK-1/2.[50] Of the several p38 isoforms known to exist, p38α typically exhibits pro-apoptotic properties and p38β anti-apoptotic properties.[46,51] Greater scrutiny of p38 responses revealed that p38α was activated by photostress whereas p38β was deactivated,[45] consistent with their established characteristics. Introduced immediately before irradiation, SPNO-derived NO not only induced HO-1 and suppressed apoptosis, but strongly inhibited the observed post-irradiation MAPK responses, i.e. activation of JNK and p38α, and deactivation of p38β and ERK1/2. Downstream of photostress activation of JNK and p38α in COH-BR1 cells, there was a marked upregulation/activation of pro-apoptotic Bax and Bid, along with a down-regulation of anti-apoptotic Bcl-xL, each effect being reversed by SPNO.[45] Thus, the responses of these various pro-apoptotic/anti-apoptotic markers to exogenous NO were all consistent with the latter's demonstrated ability to enhance COH-BR1 cell resistance to photodynamic killing. This was the first in-depth investigation of how NO can modulate the stress signaling that underlies PDT.[45]

Given that exogenous NO (chemical donor- or macrophage-derived) could directly/indirectly protect against photokilling,[38,43,44] and that many cancer cells produce low constitutive levels of NO,[19,24–28] the obvious question is whether the latter might also act cytoprotectively. This question was first addressed in the author's laboratory by subjecting COH-BR1 cells containing mitochondrial ALA-induced PpIX to a photooxidative stress in the presence of a NOS activity inhibitor to limit NO generation. Both non-specific L-NAME and iNOS-specific 1400W, when present in the medium before and after photostress, caused a striking concentration-dependent increase in caspase-3/7 activation and Hoechst-assessed apoptosis, which was observed for two different light fluences. At 1 J/cm^2, for example, apoptotic count after 20 h of dark incubation went from ~22% to 50% on increasing L-NAME from 0 to 4 mM, whereas it went from ~22% to 70% on increasing 1400 W from 0 to

10 μM.[52] Neither inhibitor had any effect on its own in the dark or light. An equally striking large increase in apoptosis was observed when another human breast tumor line, MDA-MB-231, was PpIX-sensitized and irradiated in the presence of 4 mM L-NAME.[52] In subsequent work,[53] the role of iNOS is photostress resistance was examined more directly by assessing the effects of shRNA-based iNOS knockdown (kd), which was shown by Western blotting to reduce constitutively expressed COH-BR1 iNOS level by at least 80%. As shown in Fig. 3A, iNOS knockdown, like inhibition by 1400W, caused a large increase in photostress-induced apoptosis, suggesting that iNOS was playing a crucial pro-survival role in these cells, presumably because of the NO it was generating. Furthermore, photostressed iNOS-kd cells exhibited much higher caspase-9 activation than controls, and this was reversed by active SPNO, further supporting the argument for NO-elicited resistance (Fig. 3B).[53] Western analysis indicated that iNOS was, in fact, constitutively expressed by COH-BR1 cells (and also by MDA-MB-231 cells). However, its level increased dramatically after ALA/light stress, leveling off at ~2-fold above background 4 h after a 1 J/cm^2 light fluence and ~3-fold above background 4 h after a 2 J/cm^2 fluence (Fig. 4A).[52,53] These post-irradiation elevations persisted for at least another 16 h, suggesting that long-term iNOS upregulation had occurred in response to photooxidative stress. No upregulation was observed in light-only or ALA-only controls. Moreover, 1400W did not enhance apoptosis after a H_2O_2 challenge, suggesting that iNOS-dependent hyperresistance was specific to photostress.[52] ALA/light-induced iNOS upregulation has also been observed for other human tumor types, including prostate PC-3 cells[54] and glioblastoma U87 cells[55] (Fig. 4B,C), suggesting that this may be a general response to PDT stress. That constitutive NO and/or stress-induced iNOS NO was the causative agent in photokilling resistance was first established by using a Sievers analyzer to measure NO produced by ALA/light-stressed COH-BR1 cells.[52] In this highly sensitive/specific approach, oxidized NO as nitrite is reduced back to NO, which is then detected via ozone-based chemiluminescence. For cells exposed to a 4 J/cm^2 fluence, post-irradiation nitrite increased progressively in parallel with iNOS induction, the intracellular level reaching ~5-fold that of a dark control at 20 h and the extracellular level even greater. Importantly, 1400W reduced these levels nearly to background. These data[52] confirmed that NO as well as iNOS was induced by photodynamic stress. One can monitor NO more conveniently by using the probe DAF-2DA, which is nitrosated by NO-derived N_2O_3 to give DAF-2 triazole, the highly fluorescent product that is detected. This approach revealed a dramatic increase in NO-producing cells after ALA/light stress, which was almost

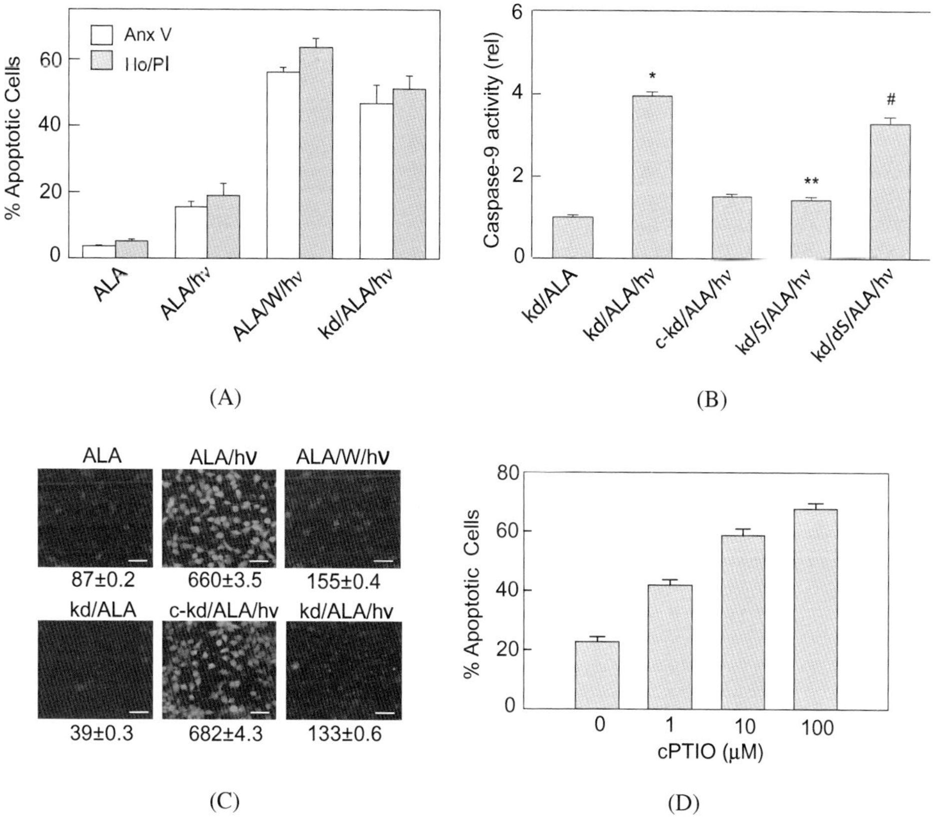

Fig. 3. Effects of iNOS inhibition, iNOS-knockdown, and a NO scavenger on manifestations of ALA/light-induced stress in COH-BR1 cells. (A) Apoptosis assessed by Annexin V-FITC or Hoechst staining (W, 1400W; kd, iNOS knockdown); (B) Caspase-9 activation (kd, iNOS knockdown; c-kd, knockdown control; S, SPNO; dS, spent SPNO); (C) DAF-2DA-assessed NO production (fluorescence yield shown below images); (D) Effect of NO scavenger cPTIO on apoptosis. (Reproduced from Ref. 53 with permission).

nullified by 1400W treatment or iNOS knockdown, control knockdown cells showing no significant effect (Fig. 3C).[53] That NO was a key causative agent in photostress-induced cytoprotection was confirmed by showing that cPTIO, a nitronyl nitroxide known to scavenge NO with a high degree of specificity,[56] increased both caspase-3/7 activation and apoptosis in a dose-dependent manner (Fig. 3D).[52]

Prostate carcinoma PC-3 cells have been found to exploit iNOS/NO similarly to breast COH-BR1 cells in response to photostress sensitized by ALA-generated mitochondrial PpIX.[54] Thus, PC-3 sensitivity to caspase-3/7 activation and apoptotic cell death was strongly enhanced by various NOS inhibitors (L-NAME, 1400W, GW274150) and also by the NO trap

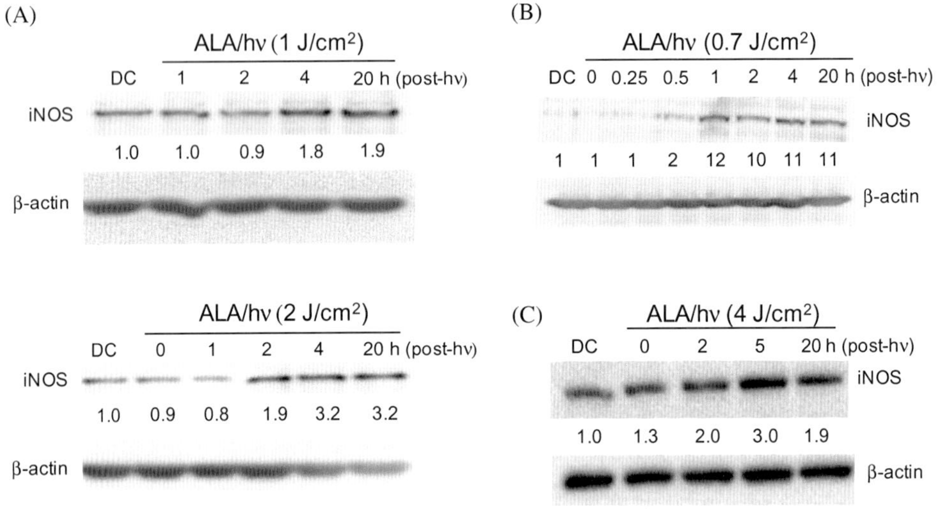

Fig. 4. Levels of iNOS protein at various post-irradiation times in ALA/light-stressed tumor cells. (A) Breast cancer COH-BR1 cells; (B) Prostate cancer PC-3 cells; (C) Glioblastoma U-87 cells. Numbers below bands indicate iNOS level relative to β-actin and normalized to an ALA-only dark control (DC). (Reproduced from Refs. 52 and 53 (A) Ref. 54 (B) with permission).

cPTIO, consistent with iNOS/NO playing a major cytoprotective role. The iNOS inhibitor GW274150, though less potent than 1400W, deserves special attention because it has already been tested in a clinical trial — not in conjunction with PDT, but as an asthma ameliorator.[57] Like breast cancer counterparts, PC-3 cells overexpressed iNOS after being photostressed, but were more robust in this response, reproducibly expressing 10-12-times more enzyme than dark controls 1–2 h after irradiation (Fig. 4B). It follows that most of the apoptotic resistance of PC-3 cells must have been due to stress-induced iNOS/NO, whereas for COH-BR1 cells, pre-existing iNOS (Fig. 4A) probably made a significant contribution. It was of interest to learn whether in addition to apoptosis resistance, iNOS/NO-mediated adaptation to photostress might involve growth stimulation, as observed previously for PDT models, but not assessed for a possible NO role.[58,59] Using two different approaches to examine live cell expansion, Bhowmick and Girotti[54] consistently found that PC-3 cells surviving an ALA/light challenge exhibited at least a 3-day growth spurt relative to dark- or light-only controls, and this was nullified by 1400W. Flow cytometric cell cycle analysis, using propidium iodide-labeled cells, provided strong confirmatory evidence for NO-dependent growth acceleration. The most compelling observation was increasingly greater S-phase (DNA doubling) occupancy over a 36 h post-irradiation period and near elimination of this by iNOS inhibition.[54] An examination of cell cycle regulators revealed, for example, that p27, a cyclin-dependent kinase inhibitor, was strongly down-regulated by ALA/light (1 J/cm^2) in 1400W-inhibitable fashion.[55] Recent experiments have shown that photostressed PC-3 cells can also exploit iNOS/NO to migrate more rapidly, as demonstrated by gap closure or "wound healing" assay.[55] For example, percent migration was 2.4-times greater at 24 h post-irradiation and 1.6-times greater at 48 h relative to a dark control, 1400W negating this accelerated migration. The question of how ALA/light stress might affect the known invasive and metastatic properties of PC-3 cells[60] was also addressed recently. A Boyden chamber device was used for testing cell ability to degrade and traverse a basement membrane layer that simulates the extracellular matrix.[55] It was found that after exposure to ALA and light (1 J/cm^2), PC-3 cells invaded ~40% faster than non-irradiated controls over 24 h. The accelerated invasion, along with an observed activation of MMP-9, which plays a key role in invasion,[61] were both strongly suppressed by 1400W.[55] That endogenous NO can enhance migration and invasion is entirely novel in the context of PDT. It has troubling implications for clinical PDT because cells escaping photodynamic eradication could potentially become more metastatic and tumorigenic.

6. Pro-survival versus Pro-death Signaling Events in Photostressed Tumor Cells

Considerable attention has been directed toward elucidation of the signaling events that underlie not only iNOS/NO induction, but also NO-dependent pro-survival versus pro-death responses in ALA/light-stressed tumor cells. Regarding iNOS induction by photostress, an immunocytological imaging approach revealed that transcription factor NF-κB in COH-BR1 cells translocated from cytosol to nucleus after an ALA/light challenge.[62] This was confirmed by Western blot analysis of the p65 subunit of NF-κB p65 in cytosolic and nuclear fractions. Importantly, Bay11-7082, an inhibitor of IKK (the kinase that activates NF-κB by phosphorylating and dissociating inhibitory subunit IκBα)[63] suppressed not only NF-κB translocation after a photochallenge, but also iNOS upregulation. Concomitantly, Bay11-7082 acted similarly to iNOS inhibition or knockdown in dramatically increasing the extent of stress-induced apoptosis.[62] These and related findings clearly indicated that NF-κB was responsible for the transcriptional upregulation of iNOS. Considering possible upstream signaling events associated with NF-κB activation, Bhowmick and Girotti found that Akt/PKB, an oncogenic (pro-survival/growth) kinase, underwent robust phosphorylation-activation

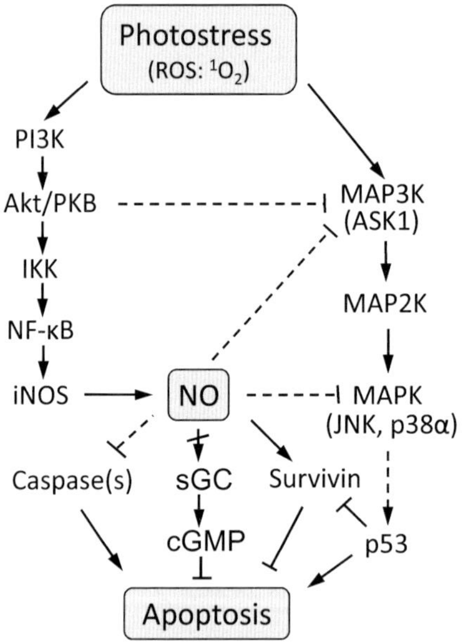

Fig. 5. Deduced pro-apoptotic *vs.* anti-apoptotic signaling pathways in ALA/light-stressed tumor cells. (Reproduced from Ref. 62 with permission).

in ALA/light-stressed COH-BR1 cells.[62] Wortmannin-induced inhibition of PI3K, the kinase involved in Akt activation through formation of plasma membrane phosphatidylinositol triphosphate,[64] was found to suppress not only Akt activation, but also iNOS upregulation while stimulating apoptosis.[62] The implication of these and related findings is that photostress activation of Akt and thence NF-*κ*B (possibly via Akt-mediated IKK phosphorylation[64]) played a key role in survival signaling by triggering iNOS/NO upregulation. A scheme for iNOS upregulation based on this evidence is shown in Fig. 5.

One of the first questions asked in considering how iNOS-derived NO signaled for stress resistance was whether cyclic-GMP (cGMP) played a role in this. NO is required for activation of soluble guanylate cyclase, resulting in synthesis of cGMP, a known activator of pro-survival/anti-apoptotic protein kinase G (PKG).[65] An early PDT-related study showed that exogenous NO from chemical donors protected lymphoblastoid cells from a phthalocyanine-sensitized photokilling and that this could be suppressed by a sGC inhibitor (ODQ) and also by a PKG inhibitor, thus implicating cGMP in the cytoprotection.[66] In contrast to this, ODQ in concentrations that antagonized NO cytoprotection in the study described,[66] had no such effect on ALA/light-challenged COH-BR1 or PC-3 cells, i.e. apoptosis was not enhanced by ODQ.[54,62] On the basis of these and consistent other findings, PKG activation by any cGMP arising from NO stimulation of sGC was ruled out as a possible cytoprotective signaling mechanism in these cells. Investigation of other possible signaling activity revealed that survivin, an inhibitor of apoptosis protein (IAP), was upregulated and phosphorylated in photostressed COH-BR1 cells, whereas p53, the pro-apoptotic tumor suppressor, appeared unaffected according to Western analysis.[62] However, when iNOS-kd cells were photostressed, total p53 increased significantly whereas survivin became barely detectable, consistent with the greater degree of apoptosis observed in these cells.[62] Based on these findings and supported by other evidence in the literature,[67] it seems likely that stress-induced NO signaled for survivin upregulation and p53 for survivin suppression, both effects being overturned by iNOS inhibition; see Fig. 5. Of added importance with regard to death versus survival signaling is the observation that pro-apoptotic MAPKs such as overall p38 and JNK in COH-BR1 cells underwent a very robust, yet transient phosphorylation-activation after ALA/light (1 J/cm^2) treatment, as evidenced by Western blot analysis, the signals appearing and then disapppearing over $\sim$1 h of post-irradiation incubation.[62] Both the extent and duration of post-irradiation JNK and p38 activation were dramatically elevated by iNOS inhibition (1400W) or shRNA-based knockdown, implying that stress-induced NO was antagonizing these MAPK activations. These results were consistent with those obtained using an exogenous NO donor (SPNO), which effectively prevented the transient photostress activation of JNK and p38.[45] Transient activation of JNK and p38 after ALA/light stress was also observed with PC-3 cells, and 1400W also caused these activations to be intensified and prolonged in this case.[54] However, the phosphorylation pattern of ERK1/2 was strikingly different. Although both ERK isoforms were strongly expressed in PC-3 cells, they exhibited little or no phosphorylation-activation until at least 20 h after irradiation, and this was abolished by 1400W.[54] This slow ERK1/2 activation may explain the post-irradiation growth spurt observed for surviving PC-3 cells.[54]

A major question at this point is how in chemical terms iNOS-derived NO can act as a pro-survival or anti-apoptotic signaling molecule in the in the PDT setting PDT setting. As already mentioned, NO does not quench $^1O_2$, but can intercept and deactivate lipid-derived free radicals.[38] However, the latter process is irreversible, putting it outside the realm of signal transduction. As also mentioned, NO signaling for cGMP formation and activation of pro-survival PKG has been ruled out for at least two types of cancer cell exposed to ALA/light stress.[54,62] Another distinct possibility is S-nitrosation of specialized (low pK_a) cysteine residues on key target proteins.[9–11] As indicated, NO cannot add directly to such residues to produce SNO groups, but may do so after conversion to N_2O_3.[10] S-nitrosated (SNO-modified) proteins can initiate signaling cascades which subside when the SNO groups are removed, e.g., by glutathione or thioredoxin.[9–11,68] Based on published data for other systems, possible molecular targets of SNO modification in the PDT models described include the following: (i) pro-apoptotic MAPKs such as JNK and ASK1, which become inhibited;[69,70] (ii) procaspase-9, whose cleavage to apoptosis-initiating caspase-9 is inhibited;[71] (iii) anti-apoptotic Bcl-2, whose proteosomal degradation is inhibited;[72] (iv) tumor suppressor PTEN, which is inactivated as a negative regulator of Akt;[73] (v) phospho-JNK/p38 phosphatase MKP-1, whose proteasomal degradation is inhibited.[74] These possibilities are being explored in ongoing studies in an effort to better understand the signaling mechanisms that underlie much of NO's anti-PDT activity. It is important to note that Rapozzi *et al.*[75,76] have recently shown that endogenous iNOS/NO also modulates pheophorbide a-sensitized photokilling of murine melanoma cells; low level stress-induced NO protected cells (attributed to upregulation of antiapoptotic NF-κB and Snail), whereas high level stress NO enhanced photokilling (attributed to Snail down-regulation and RKIP upregulation). Thus, NO signaling responses may vary with photostress intensity.

7. NOS Inhibitors as PDT Adjuvants

The ability of many malignant tumors to exploit iNOS/NO for a survival, proliferative, and metastatic advantage suggests that pharmacological use of iNOS inhibitors might attenuate these effects and model studies with tumor-bearing mice have supported this idea.[24–27] If the described negative effects of constitutive and/or stress-upregulated iNOS/NO occur in conjunction with clinical PDT, they could seriously undermine treatment efficacy. The iNOS inhibitor GW274150 was shown to stimulate caspase activation and apoptosis in ALA/light-challenged breast and prostate cancer cells,[54,62] consistent with NO-dependent resistance. GW274150 would be an excellent candidate as a PDT adjuvant for iNOS/NO suppression, since it has already been tested on humans for controlling pulmonary inflammation associated with asthma.[57] Another iNOS inhibitor, L-Nil, has been tested in similar fashion.[77] These agents are currently being evaluated by the author's group as part of an *in vivo* PDT study using a mouse breast tumor model.

8. Summary and Perspectives

This chapter has focused on the relatively recent awareness that many malignant tumors utilize NO to survive, expand, and resist various therapeutic interventions, including PDT. NO induction can now be included among the various strategies by which tumors mount a resistance to PDT, including COX-2, and survivin upregulation.[78,79] Studies in the author's laboratory have shown that several different cancer cell types rapidly overexpress iNOS and NO after being exposed to a PDT-like oxidative stress initiated by PpIX photoexcitation in mitochondria.[52,53,62] Induced, along with any background iNOS/NO, made cells more resistant to apoptotic photokilling, but when iNOS was inhibited, knocked down, or NO scavenged, more extensive apoptosis occurred. This was the first known example of cancer cells mounting a NO-dependent defense against PDT, or any therapeutic oxidative stress for that matter. Cell studies have demonstrated that NO can cause PDT resistance in several different ways, ranging from trapping of lipid-derived radicals to signaling for up-regulation of anti-apoptotic proteins (e.g., HO-1, survivin, p38β) or down-regulation of pro-apoptotic proteins (e.g., JNK, p38α, Bax).[45,62] The relative importance of these mechanisms for a given cancer cell type and photostress conditions is still not clear. Also unclear is how and at which sequence position(s) effector proteins are modified by NO, although activation or deactivation via S-nitrosation is a likely possibility being pursued in the author's laboratory. More mechanistic information along these lines is needed to better understand and control NO's negative actions in the PDT setting.

Acknowledgments

Studies by the author's group were supported by Grant CA72630 from the National Cancer Institute and by a grant from the MCW Cancer Center and Wisconsin Breast Cancer Showhouse for a Cure. Reshma Bhowmick, Jonathan Fahey, and Tyler Lee are thanked for their valuable contributions to this research. The helpful advice and suggestions of Neil Hogg and Witold Korytowski are also appreciated.

References

1. S. R. Moncada, R. M. J. Palmer and E. A. Higgs, *Pharmacol. Rev.* **43**, 109–143 (1991).
2. D. A. Wink and J. B. Mitchell, *Free Radic. Biol. Med.* **25**, 434–456 (1988).
3. L. J. Ignarro, *J. Physiol. Pharmacol.* **53**, 503–514 (2002).

4. D. D. Thomas, L. A. Ridnour, J. S. Isenberg, W. Flores-Santana, C. H. Switzer, S. Donzelli, P. Hussain, C. Vecoli, N. Paolocci, S. Ambs, C. A. Colton, C. C. Harris, D. D. Roberts and D. A. Wink, *Free Radic. Biol. Med.* **45**, 18–31 (2008).
5. L. A. Ridnour, D. D. Thomas, S. Donzelli, M. G. Espey, D. D. Roberts, D. A. Wink and J. S. Isenberg, *Antioxid. Redox Signal* **8**, 1329–1337 (2006).
6. A. Villalobo, *Letter* **273**, 2329–2344 (2006).
7. H. Rubbo, R. Radi, M. Trujillo, B. Kalyanaraman, S. Barnes and B. A. Freeman, *J. Biol. Chem.* **269**, 26066–26075 (1994).
8. J. S. Beckman, T. W. Beckman, J. Chen, P. A. Marshall and B. A. Freeman, *Proc. Natl. Acad. Sci. USA* **87**, 1620–1624 (1990).
9. H. W. Foster, D. T. Hess and J. S. Stamler, *Trends Mol. Med.* **15**, 391–404 (2009).
10. N. Hogg and K. A. Broniowska, *Antiox. Redox Signal* **17**, 969–980 (2012).
11. D. D. Thomas and D. Jord'heuil, *Antiox. Redox Signal* **17**, 924–936 (2015).
12. Y. M. Kim, M. E. de Vera, S. C. Watkins and T. R. Billiar, *J. Biol. Chem.* **272**, 1402–1411 (1997).
13. Y. M. Kim, H. T. Chung, S. S. Kim, J. A. Han, Y. M. Yoo, K. M. Kim, G. H. Lee, H. Y. Yun, A. Green, J. Li, R. L. Simmons and T. R. Billiar, *J. Neurosci.* **19**, 6740–6747 (1999).
14. R. Weller, A. Schwentker, T. R. Billiar and Y. Vodovotz, *Am. J. Physiol. Cell Physiol.* **284**, C1140–C1148 (2003).
15. C. V. Suschek, V. Krischel, D. Bruch-Gerharz, D. Berendji, J. Krutmann, K. D. Kröncke and V. Kolb-Bachofen, *J. Biol. Chem.* **274**, 6130–6137 (1999).
16. C.-Q. L. Li and G. N. Wogan, *Cancer Lett.* **226**, 1–15 (2005).
17. A. Villalobo, *FEBS J.* **273**, 2329–2344 (2006).
18. A. J. Burke, F. J. Sullivan, F. J. Giles and S. A. Glynn, *Carcinogenesis* **34**, 503–512 (2013).
19. H. Cheng, L. Wang, M. Mollica, A. T. Re, S. Wu and L. Zuo, *Cancer Lett.* 353, 1–7 (2014).
20. J. Lu, M. Tan and Q. Cai, *Cancer Lett.* (in press) (2014).
21. C. A. Caneba, L. Yang, J. Baddour, R. Curtis, J. Win, S. Hartig, J. Marini and D. Nagrath, *Cell Death Dis.* (in press) (2014).
22. D. A. Wink, Y. Vodovotz, J. Laval, G. Laval, M. W. Dewhirst and J. B. Mitchell, *Carcinogenesis* **19**, 711–721 (1998).
23. J. A. Crowell, V. E. Steele, C. C. Sigman and J. R. Fay, *Mol. Cancer Ther.* **1**, 815–823 (2003).
24. A. G. Sikora, A. Gelbard, M. A. Davies, D. Sano, S. Ekmekcioglu, J. Kwon, Y. Hailemichael, P. Jayaraman, J. N. Myers, E. A. Grimm and W. W. Overwijk, *Clin. Cancer Res.* **16**, 1834–1844 (2010).
25. P. Jayaraman, F. Parikh, E. Lopez-Rivera, Y. Hailemichael, A. Clark, G. Ma, D. Cannan, M. Ramacher, M. Kato, W. W. Overwijk, S. H. Chen, V. Y. Umansky and A. G. Sikora *J. Immunol.* **188**, 5365–5376 (2012).
26. W. Saleem, Y. Suzuki, A. Mobaraki, Y. Yoshida, S. Noda, J. I. Saitoh and T. Nakano, *Cancer Sci.* **102**, 2150–2156 (2011).
27. C. E. Eyler, Q. Wu, K. Yan, J. M. MacSwords, D. Chandler-Militello, K. L. Misuraca, J. D. Lathia, M. T. Forrester, J. Lee, J. S. Stamler, S. A. Goldman, M. Bredel, R. E. McLendon, A. E. Sloan, A. B. Hjelmeland and J. N. Rich, *Cell* **146**, 53–66 (2011).
28. T. Zhuang, B. Chelluboina, S. Ponnala, K. K. Velpula, A. A. Rehman, C. Chetty, E. Zakharian, J. S. Rao and K. K. Veeravalli, *BMC Cancer* **13**, 590–600 (2013).

29. R. K. Kim, Y. Suh, Y. H. Cui, E. Hwang, E. J. Lim, K. C. Yoo, G. H. Lee, J. M. Yi, S. G. Kang and S. J. Lee, *Cancer Sci.* **104**, 1172–1177 (2014).
30. M. J. Gilissen, L. E. A. van de Merbel-de Wit, W. M. Star, J. F. Koster and W. Sluiter, *Cancer Res.* **53**, 2548–2552 (1993).
31. S. Gupta, N. Ahmad and H. Mukhtar, *Cancer Res.* **58**, 1785–1788 (1998).
32. B. W. Henderson, T. M. Sitnik-Busch and L. A. Vaughan, *Photochem. Photobiol.* **70**, 64–71 (1999).
33. M. Korbelik, H. Shibuya and I. Cecic, *Proc. SPIE — Int. Soc. Opt. Eng.* **3247**, 98–105 (1998).
34. M. Korbelik, C. S. Parkins, H. Shibuya and I. Cecic, *Br. J. Cancer* **82**, 1835–1843 (2000).
35. K. J. Reeves, M. W. R. Reed and N. J. Brown, *J. Photochem. Photobiol. B: Biol.* **101**, 224–232 (2010).
36. R. Hurst, W. Korytowski, T. Kriska, R. S. Esworthy, F. F. Chu and A. W. Girotti, *Free Radic. Biol. Med.* **31**, 1051–1065 (2001).
37. T. Kriska, W. Korytowski and A. W. Girotti, *Free Radic. Biol. Med.* **33**, 1389–1402 (2002).
38. M. Niziolek, W. Korytowski and A. W. Girotti, *Photochem. Photobiol.* **78**, 262–270 (2003).
39. W. Korytowski, M. Zareba and A. W. Girotti, *Biochemistry* **39**, 6918–6928 (2000).
40. W. Korytowski, M. Wrona and A. W. Girotti, *Anal. Biochem.* **270**, 123–132 (1999).
41. M. Niziolek-Kierecka, A. Pilat, W. Korytowski and A. W. Girotti, *Photochem. Photobiol.* **86**, 681–686 (2010).
42. L. K. Keefer, R. W. Nims, K. M. Davies and D. A. Wink, *Methods Enzymol.* **268**, 281–293 (1996).
43. M. Niziolek, W. Korytowski and A. W. Girotti, *Free Radic. Biol. Med.* **40**, 1323–1331 (2006).
44. M. Niziolek-Kierecka, W. Korytowski and A. W. Girotti, *SPIE Proc.* **6427** (2007).
45. R. Bhowmick and A. W. Girotti, *Free Radic. Biol. Med.* **47**, 731–740 (2009).
46. T. Wada and J. M. Penninger, *Oncogene* **23**, 2838–2849 (2004).
47. L. O. Klotz, C. Fritsch, K. Briviba, N. Tsacmacidis, F. Schliess and H. Sies. *Cancer Res.* **58**, 4297–300 (1998).
48. A. Weyergang, O. Kaalhus and K. Berg, *Mol. Cancer Ther.* 7, 1740–1750 (2008).
49. P. S. Chan, H. K. Koon, Z. G. Wu, R. N. Wong, M. L. Lung, C. K. Chang and N. K. Mak, *Photochem. Photobiol.* **85**, 1207–1217 (2009).
50. W. Guan, J. Sha, X. Chen, Y. Xing, J. Yan and Z. Wang, (2012) S-Nitrosylation of mitogen activated protein kinase phosphatase-1 suppresses radiation-induced apoptosis. *Cancer Lett.* **314**, 137–146 (2012).
51. A. Porras, S. Zuluaga, E. Black, A. Valladares, A. M. Alvarez, C. Ambrosino, M. Benito and A. R. Nebreda, *Mol. Biol. Cell* **15**, 922–933 (2014).
52. R. Bhowmick, A. W. Girotti, *Free Radic. Biol. Med.* **48**, 1296–1301 (2010).
53. R. Bhowmick and A. W. Girotti, *Photochem. Photobiol.* **87**, 378–386 (2011).
54. R. Bhowmick and A. W. Girotti, *Cancer Lett.* **343**, 115–122 (2014).
55. J. Fahey and A. W. Girotti, *Nitric Oxide* **49**, 47–55 (2015).
56. T. Akaike and H. Maeda, *Methods Enzymol.* **268**, 211–222 (1996).
57. D. Singh, D. Richards, R. G. Knowles, S. Schwartz, A. Woodcock, S. Langley and B. J. O'Connor, *Am. J. Respir. Crit. Care Med.* **176**, 988–993 (2007).
58. R. D. Almeida, B. J. Manadas, A. P. Carvalho and C. B. Duarte, *Biochim. Biophys. Acta* **1704**, 59–86 (2004).

59. C. Edmonds, S. Hagan, S. M. Gallagher-Colombo, T. M. Busch and K. A. Cengel, *Cancer Biol. Ther.* **13**, 1463–1470 (2012).
60. Y. C. Lee, J. K. Jin, C. J. Cheng, C. F. Huang, J. H. Song, M. Huang, W. S. Brown, S. Zhang, L. Y. Yu-Lee, E. T. Yeh, B. W. McIntyre, C. J. Logothetis, G. E. Gallick and S. H. Lin, *Mol. Cancer Res.* **11**, 405–417 (2013).
61. S. O'Sullivan, C. Medina, M. Ledwidge, M. W. Radomski and J. F. Gilmer, *Biochim. Biophys. Acta* **1843**, 603–617 (2014).
62. R. Bhowmick and A. W. Girotti, *Free Radic. Biol Med.* **57**, 39–48 (2013).
63. H.-M. Shen and V. Tergaonkar, *Apoptosis* **14**, 348–363 (2009).
64. T. F. Franke, *Oncogene* **27**, 6473–6488 (2008).
65. F. Hofmann, R. Feil, T. Kleppisch and J. Schlossmann, *Physiol. Rev.* **86**, 1–23 (2006).
66. E. R. Gomes, R. D. Almeida, A. P. Carvalho and C. B. Duarte, *Photochem. Photobiol.* **76**, 423–430 (2002).
67. V. Fetz, C. Bier, N. Habtemichael, R. Schuon, A. Schweitzer, M. Kunkel, K. Engels, A. F. Kovács, S. Schneider, W. Mann, R. H. Stauber and S. K. Knauer, *Int. J. Cancer* **124**, 2033–2041 (2009).
68. A. K. V. Iyer, Y. Rojanasakul and N. Azad, *Nitric Oxide* **42**, 9–18 (2014).
69. H. S. Park, S.-H. Huh, M.-S. Kim, S.-H. Lee and E.-J. Choi, *Proc. Natl. Acad. Sci. USA* **97**, 14382–14387.
70. H. S. Park, J.-W. Wu, J. H. Cho, M.-S. Kim, S.-H. Huh, K. Ryoo and E. J. Choi, Inhibition of apoptosis signal-regulating kinase 1 by nitric oxide through a thiol redox mechanism. *J. Biol. Chem.* **279**, 7484–7490 (2004).
71. J.-E. Kim and S. R. Tannenbaum, *J. Biol. Chem.* **279**, 9758–9764 (2004).
72. N. Azad, V. Vallyathan, L. Wang, V. Tantishaiyakul, C. Stehlik, S. S. Leonard and Y. Rojanasakul, *J. Biol. Chem.* **281**, 34124–34134 (2006).
73. Y. D. Kwak, T. Ma, S. Diao, X. Zhang, Y. Chen, J. Hsu, S. A. Lipton E. Masliah, H. Xu and F. F. Liao, *Mol. Neurodegener.* **5**, 49 (2010).
74. W. Guan, J. Sha, X. Chen, Y. Xing, J. Yan and Z. Wang, *Cancer Lett.* **314**, 137–146 (2012).
75. V. Rapozzi, K. Umezawa and L. E. Xodo, *Lasers Surg. Med.* **43**, 575–585 (2011).
76. V. Rapozzi, P. E. Della, S. Zorzet, M. Zacchigna, B. Bonavida and L. E. Xodo, *Nitric Oxide* **30**, 26–35 (2013).
77. T. T. Hansel, S. A. Kharitonov, L. E. Donnelly, E. M. Erin, M. G. Currie, W. M. Moore, P. T. Manning, D. P. Recker and P. J. Barnes, *FASEB J.* **17**, 1298–1317 (2003).
78. A. Ferrario, K. von Tiehl, S. Wong, M. Luna and C. J. Gomer, *Cancer Res.* **62**, 3956–3961 (2002).
79. A. Ferrario, N. Rucker, S. Wong, M. Luna and C. J. Gomer, *Cancer Res.* **67**, 4989–4995 (2007).

Chapter 13

On the Use of Photodynamic Therapy, as Monotherapy or in Combination, in the Treatment of Polypoidal Choroidal Vasculopathy: An Update

Hubert van den Bergh and Patrycja Nowak-Sliwinska†*

Swiss Federal Institute of Technology (EPFL), Lausanne, Switzerland, CH-1015
**Hubert.vandenBergh@epfl.ch*
†Patrycja.Nowak-Sliwinska@epfl.ch

Abstract

The introduction of Visudyne-photodynamic therapy (PDT) to the treatment of choroidal neovascular, (CNV) age-related macular degeneration (AMD) and polypoidal choroidal vasculopathy (PCV) has rescued millions of patients from vision loss and blindness. Especially in the case of PCV, the clinical results obtained with PDT, in combination with angiogenesis inhibitors, vascular disrupting agents, or/and anti-inflammatory compounds as adjuvant therapies, may well keep PDT as one the main treatment options. In the present article we review past and very recent data on PDT in PCV, including combination therapies.

1. Introduction

Polypoidal choroidal vasculopathy (PCV) is a neovascular disease located in the choroidal part of the retina. A good recent reference to the treatment of this disease.[1] PCV has mainly been treated by (i) photodynamic therapy (PDT), (ii) anti-VEGF therapy, (iii) by a combination of PDT with anti-VEGF therapy, (iv) by triple therapy, where PDT is combined with both anti-VEGF therapy and a steroid, and (v) by thermal laser photocoagulation. Here we provide an update on the treatment of PCV and a brief discussion of some of these recent results.

2. PCV

Exudative or "wet" age-related macular degeneration (AMD) is a major blinding disease of the retina in the elderly population world-wide. It was previously divided into two classifications based on leakage characteristics observed in fluorescein angiography (FA). Lesions were thus identified as "classic" or "occult", or different combinations of the two. The first effective treatment of exudative AMD was PDT with verteporfin (Visudyne® therapy), and more than 2 million eyes have now been successfully treated by this technology.[2,3] More recently exudative AMD has been quite successfully treated by molecules that interfere with the VEGF pathway (anti-VEGFs). This approach has shown better visual acuity (VA) results on exudative AMD patients than PDT, at least in the short and medium term, but it needs monthly, or rather frequent, intravitreal injections of one of the anti-VEGFs. Furthermore, the effect of this anti-VEGF therapy seems to "wear off" after several years.[4] Seven years after ranibizumab therapy in the ANCHOR and MARINA trials, one third of the patients had lost more than 15 letters (three lines).

Over the last decades both Indocyanine Green video angiography (ICGA) and optical coherence tomography (OCT) have helped to show a more detailed spectrum of diseases involving new choroidal blood vessels in the retina than was possible before, when only fluorescein angiography was used.[5–8] Thus one can now distinguish, among others, different kinds of so-called retinal angiomatous proliferations (RAP)[9] including retinal choroidal anastomoses, and PCV as separate clinical entities from wet AMD. We are thus now confronted with the fact that the possibility of this more precise clinical identification of the different sub-types of choroidal neovascularization (CNV) can lead to their more specific, and hence probably more optimal, treatment.

In the present article we are concerned with updating previous information on the treatment of PCV,[10–14] because new clinical reports on important longer-term observations of the disease have now become available.

PCV, even though it shares many characteristics with wet AMD, was first recognized and described as a separate clinical entity by Yannuzzi in 1982.[16] Some of PCV's hallmarks are hyper fluorescent nodules that show up in the first 5 min of dynamic ICGA (Fig. 1). These nodules or "polyps" represent terminal aneurisms, i.e. bulb-like vascular structures at the end of choroidal vessels, that are often attached to a so-called branching vascular network (BVN).

In normal color fundus photography the polyps show up as orange-red nodules. PCV is also frequently accompanied by repeated serosanguinous

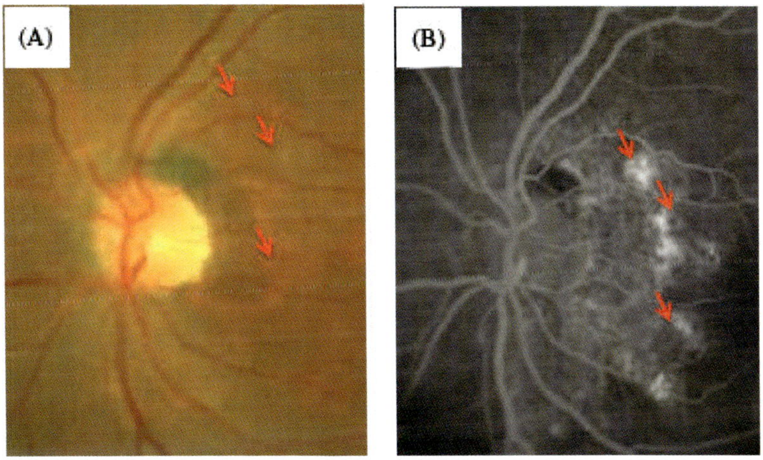

Fig. 1. Images of the first published patient, a 68-year-old African American woman, with PCV. (A) Color fundus image with a haemorrhage. (B) Late phase of a fluorescein angiogram (FA) with patches of hyperfluorescence (red arrows). Adapted from Imamura *et al.* 2010.[15]

detachment of the retinal pigmented epithelium (RPE), and even detachment of the overlying neurosensory retina, as well as sub macular haemorrhage with lipid exudates. Using only fluorescein angiography the leaky polyps were hard to distinguish from exudative AMD.

PCV is more prevalent in the pigmented races than in Caucasians. Typically, in large parts of the East Asian population 25–50% of the patients originally identified with exudative AMD by fluorescein angiography may actually have PCV. This percentage drops to about 10% in Caucasians even though that had number may increase somewhat when more ICGA will be applied in Caucasian patients in the future. The latter implies that white patients that are found to be resistant to anti-VEGF therapy, and that have large pigmented epithelium detachment (PED) and lipid exudates, should probably be considered for ICGA and more detailed diagnostics in order to check if they have PCV.[17] Furthermore, PCV as compared to wet AMD, more often can be found in younger people. Also PCV can be encountered in parts of the retina outside of the macula. We became aware of the existence of PCV following some apparent discrepancies in the PDT results obtained in the Visudyne® clinical trials. These data allowed us to realize that the significantly better VA outcome reported from Japan after PDT with Verteporfin, as compared to data from Europe and the U.S., implied that in Japan a much larger fraction of the patients whose disease had originally been identified by fluorescein angiography as exudative AMD, actually had PCV (see Fig. 2).

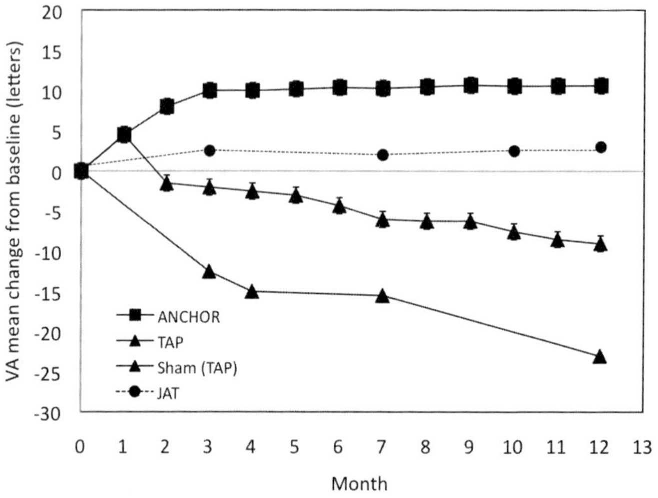

Fig. 2. Comparison of mean change of the VA with different treatments in wet AMD. ANCHOR trial (PDT + ranibizumab), JAT and TAP trials (PDT). Month zero indicates the patients visual acuity at baseline.

2.1. *Treatment Modalities for PCV*

In the naturally progressing PCV disease, the prognosis for VA seems to be better than is the case for naturally progressing exudative AMD.[18,19] In fact, about half of the PCV patients may undergo spontaneous polyp regression and will not need treatment. In the other 50%, recurrent haemorrhage and leakage may lead to permanent photoreceptor and RPE damage.

2.1.1. *PDT with verteporfin (Visudyne®)*

PDT has frequently been used to treat PCV,[20–25] mostly with Visudyne® at standard therapeutic conditions, i.e. 6 mg/m^2 of drug slowly injected over 10 min, followed at 15 min after the start of the injection by light application. 50 J/cm^2 of visible light at 689 nm is given over 83 s, i.e. with 600 mW/cm^2, focused onto the circular treatment area which includes the polyps and the BVN as visualized by ICGA, plus a margin of at least several hundred microns. This PDT will lead to local angio-occlusion which in most of the cases[10] results in very significant, or even complete, regression of the polyps, especially when considering the short-term results. PDT at standard fluence conditions in most eyes will unfortunately also lead to closure of a significant fraction of the normal choriocapillaries in the irradiated area, thus possibly inducing some collateral damage, with the possible accompanying inflammation, hypoxia and induction of VEGF. Nevertheless, most PDT of PCV leads to quite good results when one looks at the first year after treatment, with either stable or even improved VA, better than 80% reduction of the polyps, and a cessation, or at least significant reduction of leakage as assessed by FA.[20,26–28] Unfortunately, longer-term observations of the polyps, the BVN, and all-importantly, the VA after PDT on PCV, did not show as good results as what was reported over the first year.[21,24,25,29,30] In fact Miki *et al.*[25] in their 5-year study showed that whereas PCV treated by PDT alone gave improved BCVA results versus baseline after six months, after 60 months BCVA was considerably worse than what was measured at baseline. It was noted that both older age of the patients and larger lesions at baseline correlated with poorer treatment outcome. Akaza *et al.*[27] reported in their 3-year study of PDT on PCV that BCVA at 36 months dropped below baseline and that polyps recurred in 77% of the patients. Due to the small amount of collateral damage often associated with "standard" fluence Visudyne® PDT, possibly correlated with the partial closure of the normal choriocapillaries in the irradiated region, this treatment was at times accompanied by complications including sub retinal and vitreous haemorrhage. Such observations, consequently led to several investigations of PDT for PCV treatment at lower light fluencies with rather promising results (see for instance Yamashita *et al.*).[31]

2.1.2. *Anti-VEGF therapy*

Patients diagnosed with PCV were shown to have increased levels of both pro-angiogenic VEGF and anti-angiogenic pigmented epithelium derived factor (PEDF) in different parts of their eyes.[32,33] Hence anti-VEGF therapy was tested for treating PCV using either the anti-VEGF bevacizumab[23,34,35,37,38] or the anti-VEGF ranibizumab.[39–41] Whereas VA could be stabilized and even improved in most patients, as expected from the observed reduction in macular edema and sub retinal fluid, in most cases of PCV the hoped for reduction of polyps and the associated BVN was not very effective. Thus anti-VEGF monotherapy was not nearly as successful in treating PCV as it was in the case of treating exudative AMD. One may speculate that this lack of efficiency of the anti-VEGFs in treating PCV might be due to the more "mature" vasculature involved in the polyps and BVN as compared to the choroidal neovessels in wet AMD.

During the long-term anti-VEGF monotherapy of exudative AMD, a certain amount of "resistance" against this treatment modality appears, i.e. the efficiency of the anti-VEGF treatment is reduced with time.[4,37,42] In particular some evidence has been reported by Guttman Krader[43] that anti-VEGF treatment upregulated platelet derived growth factor (PDGF), as well as pericyte recruitment and neovascular membrane maturation in AMD patients.

This has led to the development of a therapy for wet AMD where an aptamer against PDGF-B (Fovista®) is combined with Lucentis® which is now being tested in a phase III clinical study. The idea behind this combination therapy is that the protection of the vasculature by pericytes, which would weaken the effect of the anti-VEGF compounds, can be reduced by the anti-PDGF molecule. We would like to mention here that this combination of anti-PDGF-B and anti-VEGF might also be tried in the treatment of PCV. An interesting study that should also be reported on when comparing anti-VEGF therapy with PDT is the so-called LAPTOP 24-month trial.[44] This trial was undertaken because the EVEREST-1 trial had mainly concentrated on polyp removal as principle outcome, and a second trial was thought to be necessary in which BCVA was at least part of the primary outcome. Another weakness of the EVEREST trial was its duration of only six months. In LAPTOP, 97 patients that had not yet been treated for PCV were randomized to either PDT with retreatment based on fluorescein leakage observed in FA, or three intravitreal ranibizumab injections of 0.5 mg at one month intervals, followed by retreatment as needed (*pro re nata*) following the defined PRN criteria. Here the results of the bevacizumab arm on VA were better than those of the PDT arm. Indeed, baseline and final logMAR visual assessment showed slightly improved vision in the case of bevacizumab treatment (from 0.48 to 0.4), whereas it showed a decrease in vision in the PDT arm (from 0.57 to 0.65). The change in logMAR observations of VA was significant between the two groups with $P = 0.004$, i.e. in the PDT arm 0.7 lines of BCVA were lost whereas in the bevacizumab 0.9 lines were gained. Unfortunately, in the LAPTOP study a third arm with the combination between anti-VEGF and PDT was not included.

2.1.3. *PDT + anti-VEGF*

Combination of the angio-occlusive effect of PDT with the anti-angiogenic and reduced leakage effects of the anti-VEGF molecules, ranibizumab (Lucentis®) or bevacizumab (Avastin®), might be expected to lead to an additive overall effect, and possibly even a synergetic treatment for PCV. A further reason behind this therapeutic combination is the observation of increased VEGF levels in eyes with PCV, as well as an increase in VEGF after photodynamic treatment.[30] Consequently a number of studies have reported on verteporfin PDT (Visudyne® therapy) combined with anti-VEGF for treating PCV, see Fig. 3. An early 12 months study by Gomi *et al.*[38] compared PDT monotherapy with the combination of PDT and anti-VEGF. The results showed that, whereas both approaches resulted in quite similar and rather good results on

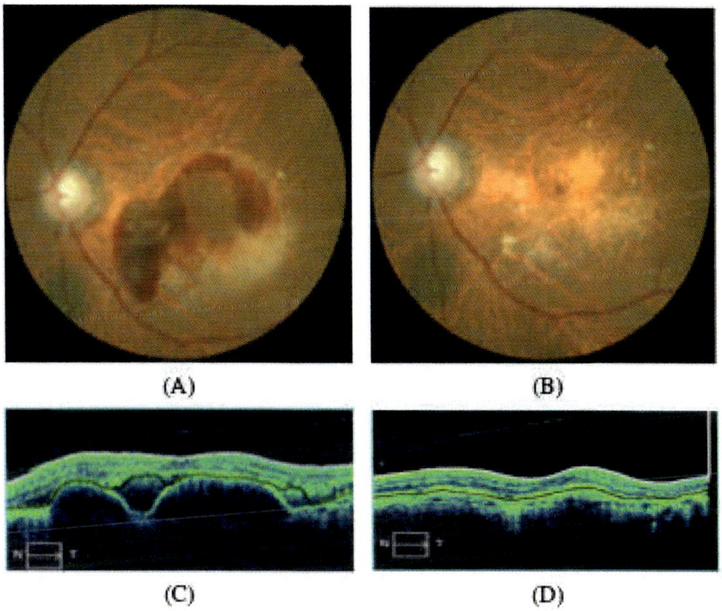

Fig. 3. Multimodal imaging of PCV, comprising color fundus photography (A,B), and OCT images (C,D). (A) Color fundus image with submacular hemorrhage and bi-lobed large hemorrhagic pigment epithelial detachments, surrounding neurosensory detachment and subretinal lipid exudation occupying almost the entire left macula. (B) Corresponding color fundus image three months after a single application of combination treatment of PDT and ranibizumab, showing complete resolution of pigment epithelial detachments, macular hemorrhage and neurosensory detachment, and restoration of the normal macular architecture with corresponding improvement in VA. Pre-treatment (C) and Post-treatment (D) OCT showing resolution of macular lesions as observed on color fundus image.

the resolution of polyps, significantly better results for the VA outcome were obtained in the combination arm of the study (PDT + anti-VEGF) as compared to PDT alone.

Other investigations along the same lines included the EVEREST-1 trial, which as mentioned above was unfortunately limited to only six months of observations.[45] EVEREST-1 was a multicenter double-masked ICGA-guided study on 61 Asian patients. This study had three arms: PDT, PDT + ranibizumab, and ranibizumab monotherapy. The primary endpoint was complete regression of polyps, secondary endpoints included mean change in best corrected VA (BCVA) and safety. Complete polyp regression, as evaluated by ICGA, was found to be 77.8% in the combination therapy, 71.4% in PDT alone, versus only 28.6% in ranibizumab monotherapy. BCVA at six months was rather good with a vision increase of +10.9 letters, +7.5 letters and +9.2 letters in the three respective study arms, as compared to baseline. Unfortunately, there was too much statistical uncertainty in these BCVA values so that these

VA outcomes could not be concluded to be significantly different. No safety problems were observed at all in the EVEREST-1 trial. One of the important conclusions was that "Longer term studies with larger patient populations are required to assess the efficacy of the treatments in achieving complete regression of polyps and its correlation with VA". A further worry was the lack of precise information on the number and kind of (anti-VEGF or PDT) retreatments needed, as the number of retreatments clearly is one criterion that one would like to minimize while maintaining optimal polyp regression and best possible VA outcome. Such shortcomings as encountered in the EVEREST trial, probably due to time and financial constraints, are unfortunately not unique in the literature. The EVEREST-2 trial will consequently improve on EVEREST-1 by employing three study arms. In this new trial, ranibizumab monotherapy will be compared with Ranibizumab in combination with Verteporfin PDT, and Verteporfin PDT alone, for the treatment of PCV. Outcomes will be evaluated at 12 and 24 months, and the primary outcome will include the change from baseline in BCVA, as well as complete regression of polyps evaluated by ICGA. The secondary outcomes will include the measurement of FA leakage, and retinal thickness, as well as the reporting of adverse events. Study completion is expected around September 2016. In the meantime reports on several longer duration studies of the combination treatments of PCV, using PDT and anti-VEGFs, have appeared in the literature. A few of these are underlined below. Saito *et al.*[46] reported on a better than 2.6 line BCVA gain after 24 months of PDT + intravitreal ranibizumab. Kim *et al.*[47] even reported on an improvement from baseline BCVA of three lines after 12 months of such a combination therapy. Unfortunately, not all "longer duration" studies of the combination of PDT with anti-VEGFs had such good results. Nemoto *et al.* in a 24-month study of the combination therapy on their PCV patients showed a nice early improvement of BCVA, that to a large extent disappeared at the end of the 2-year period.[48] Lee *et al.* also reported that early VA gains disappeared towards the end of their 24-month study of combination therapy for PCV.[49] In the review by Wong *et al.*[13] this rather poor outcome was ascribed to the fact that the anti-VEGF was given seven days before PDT, rather than after PDT, as was done in most studies. Furthermore, the intravitreal half life of bevacizumab is only about four days, so it was not sure if enough of the drug was still around at the time when PDT was applied. Kang *et al.*[50] actually did some of the longest observations in the treatment of PCV using the combination of Verteporfin therapy with intravitreal bevacizumab with a mean follow-up period of 46.8 months. They observed that three years after treatment initialization, 88.2% of the patients had avoided any VA loss, and actually showed a mean 0.2 logMAR improvement. It is interesting to note that, even though these VA results are quite satisfactory, over the nearly four years of follow up 61.8% of the patients actually showed recurrence of the treated PCV, which is not that good. During the follow-up in this study PDT was applied as needed (PRN) on the average 1.4 times, whereas a mean of 9.2 intravitreal injections of bevacizumab were found to be necessary. Observation at baseline of the largest polyp size, as well as the location of the polyps, correlated well with long term VA outcome after the combination therapy. PDT, or in some cases thermal laser coagulation, was also combined with intravitreal bevacizumab in a study by Jeon *et al.*,[51] where there was an important focus on "adjusted retreatment". The original polypoidal lesion was treated by PDT applied to the entire area including polyps and BVN in combination with anti-VEGF. The follow-up in the case of persistent or recurrent exudation was then by PDT or thermal laser, but now focused on the each of the individual polyps. Obviously the thermal laser was only used well outside of the fovea. If no polyps were found, intravitreal bevacizumab was given by itself. The reduced exposure of large parts of the eye to PDT in the follow-up treatments led to better stabilisation of VA and much reduced numbers of PDT treatments.

Another approach taken in the combination treatment with PDT and anti-VEGF of PCV was to reduce the fluence used in the PDT, most often from $50\,J/cm^2$ to $25\,J/cm^2$. This was to avoid, or at least to significantly reduce, the damage to the normal choroidal vessels, which may have contributed to the induction of sub retinal haemorrhage observed in some patients. This is done either by reducing the light intensity by a factor of two or by reducing the light exposure time by a factor of two. This reduced-fluence combination therapy between Visudyne® PDT and intravitreal anti-VEGF actually showed very nice results both for the polyp reduction as evaluated by ICGA and the VA outcome.[52–54] The factors predicting therapeutic outcome of the different therapeutic approaches mentioned in the present "update" are quite interesting and important. In a recently published one year study on the treatment of PCV by the combination of a single Verteporfin PDT with three loading doses of ranibizumab[55] the authors not only found significant VA improvement (logMAR went down from 0.828 at baseline to 0.728 at 12 months) and 81% polypoidal lesion control, but they also took great care to underline the factors at baseline that predicted the degree of success of therapeutic outcome. These factors were found to include baseline VA, absence of neovascular foveal involvement, as well as the absence of hard exudates and sub retinal fluid.

It should also be mentioned that due to the important differences between PCV patients from white and pigmented races, it is interesting to cite a report by Davis *et al.*[17] who investigated PCV in white patients presumed with exudative AMD, that did not respond well to anti-VEGF therapy. Patients with resistance to anti-VEGF, combined with large PED, lipid exudates and few drusen, are suspected of PCV and then undergo ICGA for confirmation of PCV. It is particularly noteworthy here that after PDT the patients still needed anti-VEGF but its frequency was reduced by 2/3. VA results here too were quite good as vision was maintained in 77% of the patients.

2.1.4. *PDT combined with a topical steroid*

Steroid monotherapy with 12 mg triamcinolone acetonide (TA) infused via a trans-tenon's retrobulbar injection could reduce the size and the elevation of a PCV lesion, and resolve a case of serous retinal detachment. VA improved from 20/32 to 20/20. VA remained good in the 15 months follow-up.[56] Lai *et al.*[57] reported on PDT with and without intravitreal triamcinolone acetonide (IVTA) for treating PCV. They concluded that PDT alone reduced loss of VA in the short term, but that this beneficial effect could not be observed to be sustained after one year. The adjunctive use of IVTA did not help in the visual outcome, and these patients had a tendency to develop cataract requiring surgery and ocular hypertension. Some discussion remains on the relative advantages of applying the steroid by retrobulbar infusion (subtenon) or intravitreally.

2.1.5. *Triple therapy*

Triple therapy for treating PCV using the combination of Verteporfin PDT, intravitreal anti-VEGF and intravitreal steroid, has also been tried following the success of this approach in treating exudative AMD.[58–61] Indeed, in the case of wet AMD, the triple therapy outcome published in these patients showed both very good results on the average BCVA gain, as well as excellent results on the reduced number of retreatments needed. The reasons for the triple therapy, as applied in the treatment of exudative AMD, and probably also in the treatment of PCV, are a consequence of the multifunctional pathogenesis of this disease. PDT is thus used to destroy existing choroidal neovascularization (CNV), the steroid will decrease inflammation associated with the disease, as well as inflammation induced by the PDT treatment, and the anti-VEGF will help to reduce the formation of VEGF and inhibit the formation of new CNV. For the triple therapy of PCV,[62] PDT with Verteporfin was combined with both intravitreal bevacizumab and intravitreal TA, the two of which were injected three days before PDT. In this treatment of PCV the triple therapy was found to be significantly better than PDT alone. More than 40% of the patients showed a significant improvement of VA over the PDT results, retreatment as necessary rates decreased significantly, and the triple therapy also resulted in much less haemorrhage. It should be noted here that TA has been used as monotherapy for PCV with mixed results (see above). The better results were obtained when TA was injected retrobulbarly rather than intravitreally. This should be considered when continuing the triple therapy efforts on PCV. In a second report on the outcome of triple therapy for treating PCV, again with verteporfin-PDT combined with intravitreal bevacizumab and TA, Nakata *et al.*[63] confirmed the very advantageous results obtainable with the triple therapy strategy. Relative to PDT monotherapy where improved VA was found in only one out of 40 eyes, in the triple therapy improved VA was found in 10 eyes, i.e. 41.7% of the patients. Retreatment had to be given in 75% of the PDT group and only in 37% of the triple therapy group. Post treatment haemorrhage was seen in two patients of the PDT group and none in the triple therapy group. So triple therapy seems to be a very good approach to the treatment of PCV indeed! In a recent clinical update on PCV published on the American Academy of Ophthalmology website, B. Boughton interviewed Drs. A.H.C. Koh, G.T. Kokame and R.F. Spaide as to the best therapeutic approach to PCV. Here the recommendation was that if the patient's VA is still not too bad, start with anti-VEGF. If that does not work move directly to triple therapy where PDT is combined with anti-VEGF and dexamethasone intravitreal injections. Dexamethasone was preferred to TA because it has a shorter residence time in the eye and was thought to cause fewer side effects like increased intraocular pressure and cataract. PDT in this case may be considered in particular at the lower fluences of 25 J/cm^2 or even less. It may be worth noting here that in the original clinical research on PDT of wet AMD the window of the best results included 25 J/cm^2 fluence. The higher value, 50 J/cm^2, was actually chosen because it was supposed that many of the wet AMD patients might have cataract, which would reduce the light delivered to the retina.

2.1.6. *Thermal laser coagulation*

Different types of lasers have been used for treating the extrafoveal polyps associated with PCV. Polyp regression could generally be achieved with good efficacy.[64,65] This thermal laser treatment was found to be most effective when both the polyps and the associated surrounding abnormal vasculature were coagulated together.

3. Discussion

A number of clinical studies over the past few decades have shown that what was designated previously, mainly based on fluorescein angiography, as exudative age related macular degeneration (classic and occult), really consisted of different disease subtypes including, among others, PCV. Whereas wet AMD treatment by PDT was to a large extend replaced by frequent repetitive anti-VEGF injections this treatment did not seem to work that well on PCV. Thus PDT was continued as a mainstream treatment for PCV, either as a monotherapy or in combination mainly with the anti-VEGFs. Whereas the short-term results of PDT monotherapy on PCV looked good, the longer term results were far from impressive. The combination therapy results of PDT + anti-VEGF appeared somewhat better in the short term, especially concerning VA outcome. But there too, not all of the long-term observations seem to lead to consistently good results. From what has been published up to date, triple therapy, where PDT is combined with both anti-VEGF and a steroid, seems even more promising probably using reduced fluence PDT. From the many clinical studies that have been reported on in this field (some of which have been cited above) it is as yet hard to distill a completely consistent and clear picture, mainly due to the many differences in the large number of variables in these studies.

Small numbers of patients and insufficient follow-up time contribute to the remaining uncertainties in finding optimal treatments for these patients. Furthermore, a number of new agents are arriving on the market that may be of interest in treating PCV by multi therapy approaches. One such compound is a new anti-VEGF molecule called VEGF Trap-Eye (aflibercept), which consists in part of VEGF receptor ligand fragments.

Other approaches can be used to block the VEGF pathway including the inhibition of the tyrosine kinase cascade downstream from the VEGFRs. Among these compounds are pazopanib and vatalanib. Due to the observation of both increased VEGF and PDGEF-β in the vitreous of patients with PCV, possibly using pagabtamt in combination with PDT, an anti-VEGFA and a steroid like dexamethasone, might be worth while trying.

4. Conclusion

The multifactorial character of PCV implies that combination therapies are likely to play an important role in the treatment of this disease. The present combination of reduced-fluence PDT with an anti-VEGF and a carefully dosed steroid, applied at an optimal timing sequence, looks like a very good approach for the time being. The possibility of further combination with an anti-PDGF can also be considered. It should also be realized that for these combination therapies the number of necessary retreatments can also be considerably reduced which relieves the burden in both the patients and the country's medical finances. The outcome of more (and longer duration) trials like EVEREST-2 is also awaited. Careful observation of the factors defining the patients disease and comparing this with therapeutic outcome will certainly also be helpful and finally, for the long-term future, replacing at least in part the RPE with stem-cell therapy may offer a solution in the future.

Acknowledgments

The authors acknowledge Adrian H.C Koh for PCV related images.

References

1. C. W. Wong, T. Y. Wong, C. M. G. Cheung, *J. of Clin. Med.* 4, 782–811 (2015).
2. U. Schmidt-Erfurth, J. Miller, M. Sickenberg, A. Bunse, H. Laqua, E. Gragoudas, L. Zografos, R. Birngruber, H. van den Bergh, A. Strong, U. Manjuris, M. Fsadni, A. M. Lane, B. Piguet and N. M. Bressler, *Graefe's Arch. Clin. Exper. Ophthalmol.* **236**(5), 365–374 (1998).
3. H. van den Bergh, *Semin. Ophthalmol.* **16**(4), 181–200 (2001).
4. S. Rofagha, R. B. Bhisitkul, D. S. Boyer, S. R. Sadda, K. Zhang SEVEN-up Study Group, *Ophthalmology* 120(11), 2292–2299 (2013).
5. T. H. Lim, A. Laude and C. S. Tan, *Eye* **24**(3), 483–490 (2010).
6. R. F. Spaide, J. M. Klancnik, M. J. Comey, *JAMA Ophthalmol.* 133(1), 45–50 (2015).
7. Jia Yali *et al.*, PNAS (2015), 112(18), E2395–E2402.
8. Saeed T. Alshahiani *et al.*, *Clin Ophthalmol.*, 8, 1689–1695 (2014).
9. L. A. Yannuzzi, S. Negrao, T. Iida, C. Carvalho, H. Rodriguez-Coleman, J. Slakter, K. B. Freund, J. Sorenson, D. Orlock and N. Borodoker, *Retina* **21**(5), 416–434 (2001).
10. P. Nowak-Sliwinska, M. Sickenberg, H. van den Bergh and A. H. C. Koh, *Prog. Retin. Eye Res.* **37**, 182–199 (2013).
11. S. Honda, W. Matsumiya, A. Negi, *Ophthalmolgica*, 231(2), 59–74 (2014).
12. Hae Min Kaung and Hyoung Jun Koh, *am. J. Ophthalmology* 156(4), 652–660 (2013).
13. R. L. Wong and T. Y. Lai, *J. Ophthalmic. Vis. Res.* **8**(4), 359–371 (2013).
14. B. Boughton (interviewer), H. C. Voh *et al.*, *Internet Eyenet Magasine,* AAO 2012 (Dec.) 35–37 "update on Polypoidal Choroidal Vasculopathy (PCV)".
15. Y. Imamura, M. Engelbert, T. Iida, K. B. Freund and L. A. Yannuzzi, *Surv. Ophthalmol.* **55**(6), 501–515 (2010).
16. L. A. Yannuzzi, J. Sorenson, R. F. Spaide and B. Lipson, *Retina* **10**(1), 1–8 (1990).
17. S. J. Davis, A. K. Lauer and C. J. Flaxel, *Retina.* **34**(11), 2185–2191 (2014).
18. M. Uyama, M. Wada, Y. Nagai, T. Matsubara, H. Matsunaga, I. Fukushima, K. Takahashi and M. Matsumura, *Am. J. Ophthalmol.* **133**(5), 639–648 (2002).
19. H. Bessho, S. Honda, H. Imai and A. Negi, *Retina* **31**(8), 1598–1602 (2011).

20. R. F. Spaide, M. L. Martin, J. Slakter, L. A. Yannuzzi, J. Sorenson, D. R. Guyer and K. B. Freund, *Am. J. Ophthalmol.* **134**(1), 62–68 (2002).
21. S. Leal, R. Silva, J. Figueira, M. L. Cachulo, I. Pires, J. R. de Abreu and J. G. Cunha-Vaz, *Retina* **30**(8), 1197–1205 (2010).
22. A. Yamashita, F. Shiraga, C. Shiragami, Y. Shirakata and A. Fujiwara, *Am. J. Ophthalmol.* 155(1), 96–1020 (2013).
23. J. K. Chhablani, R. Narula and R. Narayanan, *Indian J. Ophthalmol.* **61**(3), 136–138 (2013).
24. E. Akaza, M. Yuzawa and R. Mori, *Jpn. J. Ophthalmol.* **55**(1), 39–44 (2011).
25. A. Miki, S. Honda, H. Kojima, M. Nishizaki, T. Nagai, M. Fujihara, M. Uenishi, M. Kita, Y. Kurimoto, A. Negi and G. Hyogo Macular Disease Study, *Jpn. J. Ophthalmol.* **57**(3), 301–307 (2013).
26. R. M. Silva, J. Figueira, M. L. Cachulo, L. Duarte, Faria J. R. de Abreu and J. G. Cunha-Vaz, *Graefe's Arch. Clin. Exper. Ophthalmol.* **243**(10), 973–979 (2005).
27. E. Akaza, M. Yuzawa, Y. Matsumoto, S. Kashiwakura, K. Fujita and R. Mori, *Jpn. J. Ophthalmol.* **51**(4), 270–277 (2007).
28. W. M. Chan, D. S. Lam, T. Y. Lai, D. T. Liu, K. K. Li, Y. Yao and T. H. Wong, *Ophthalmology* **111**(8), 1576–1584 (2004).
29. D. Tsuchiya, T. Yamamoto, R. Kawasaki and H. Yamashita, *Retina* **29**(7), 960–965 (2009).
30. H. M. Kang, Y. M. Kim and H. J. Koh, *Am. J. Ophthalmol.* **155**(3), 438–447, e431 (2013).
31. A. Yamashita, F. Shiraga, C. Shiragami, Y. Shirakata and A. Fujiwara, *Am. J. Ophthalmol.* **155**(1), 96–102, e101 (2013).
32. M. Matsuoka, N. Ogata, T. Otsuji, T. Nishimura, K. Takahashi and M. Matsumura, *Br. J. Ophthalmol.* **88**(6), 809–815 (2004).
33. J. P. Tong, W. M. Chan, D. T. Liu, T. Y. Lai, K. W. Choy, C. P. Pang and D. S. Lam, *Am. J. Ophthalmol.* **141**(3), 456–462 (2006).
34. F. Gomi, M. Sawa, H. Sakaguchi, M. Tsujikawa, Y. Oshima, M. Kamei and Tano Y. *Br. J. Ophthalmol.* **92**(1), 70–73 (2008).
35. T. Y. Lai, W. M. Chan, D. T. Liu, F. O. Luk and D. S. Lam *Br. J. Ophthalmol.* **92**(5), 661–666 (2008).
36. Wei Wang, Mias He and Xiulan Zhang, Plos one (2014) 9(10), e110667.
37. H. J. Cho, J. W. Kim, D. W. Lee, S. W. Cho and C. G. Kim *Eye* **26**(3), 426–433 (2012).
38. F. Gomi, M. Sawa, T. Wakabayashi, Y. Sasamoto, M. Suzuki and M. Tsujikawa, *Am. J. Ophthalmol.* **150**(1), 48–54, e41 (2010).
39. G. T. Kokame, L. Yeung and J. C. Lai, *Br. J. Ophthalmol.* **94**(3), 297–301 (2010).
40. T. Hikichi, M. Higuchi, T. Matsushita, S. Kosaka, R. Matsushita, K. Takami, H. Ohtsuka, H. Kitamei and S. Shioya, *Br. J. Ophthalmol.* **97**(5), 617–621 (2013).
41. D. M. Marcus, H. Singh, M. N. Lott, J. Singh and M. D. Marcus, *Retina* **33**(1), 35–47 (2013).
42. P. Tranos, A. Vacalis, S. Asteriadis, S. Koukoula, A. Vachtsevanos, Perganta G. and I. Georgalas, *Drug Des. Devel. Ther.* 7, 485–490 (2013).
43. C. Guttman Krader, *Ophthalmol. Times* (2013).
44. A. Oishi, N. Miyamoto, M. Mandai, S. Honda, T. Matsuoka, H. Oh, M. Kita, T. Nagai, N. Bessho, M. Uenishi, Y. Kurimoto and A. Negi, *Ophthalmology* **121**(5), 1151–1152 (2014).

45. A. Koh, W. K. Lee, L. J. Chen, S. J. Chen, Y. Hashad, H. Kim, T. Y. Lai, S. Pilz, P. Ruamviboonsuk, E. Tokaji, A. Weisberger and T. H. Lim, *Retina* **32**(8), 1453–1464 (2012).
46. M. Saito, T. Iida, M. Kano and K. Itagaki, *Graefe's Arch. Clin. Exper. Ophthalmol.* **251**(9), 2099–2110 (2013).
47. M. Kim, K. Kim, G. Kim do, S. Y. Yu and H. W. Kwak, *Ophthalmologica, Journal International d'Ophtalmologie, International Journal Of Ophthalmology, Zeitschrift Fur Augenheilkunde* **226**(4), 205–213 (2011).
48. R. Nemoto, M. Miura, T. Iwasaki and H. Goto, *Clin. Ophthalmol.* **6**, 1633–1638 (2012).
49. Y. A. Lee, C. H. Yang, C. M. Yang, T. C. Ho, C. P. Lin, J. S. Huang and M. S. Chen, *Am. J. Ophthalmol.* **154**(5), 872–880, e872 (2012).
50. H. M. Kang, H. J. Koh, C. S. Lee and S. C. Lee, *Am. J. Ophthalmol.* **157**(3), 598–606, e591 (2014).
51. S. Jeon, W. K. Lee and K. S. Kim, *Retina.* **33**(6), 1193–1200 (2013).
52. F. Ricci, A. Calabrese, F. Regine, F. Missiroli and A. P. Ciardella, *Retina* **32**(7), 1280–1288 (2012).
53. M. Sagong, S. Lim and W. Chang, *Am. J. Ophthalmol.* **153**(5), 873–882, e872 (2012).
54. Y. Yoshida, T. Kohno, M. Yamamoto, T. Yoneda, H. Iwami and K. Shiraki, *Jpn. J. Ophthalmol.* **57**(3), 283–293 (2013).
55. M. Ho, E. C. Lo, A. L. Young and D. T. Liu, Eye (2014), 28(12), 1469–1476.
56. A. Okubo, M. Ito, T. Kamisasanuki and T. Sakamoto, *Graefe's Arch. Clin. Exper. Ophthalmol.* **243**(8), 837–839 (2005).
57. T. Y. Lai, C. P. Lam, F. O. Luk, R. P. Chan, W. M. Chan, D. T. Liu and D. S. Lam, *J. Ocul. Pharmacol. Ther.* **26**(1), 91–95 (2010).
58. S. J. Bakri, S. M. Couch, C. A. McCannel and A. O. Edwards, *Retina* **29**(5), 573–578 (2009).
59. W. Bethke, *Rev. Ophthalmol.* **9**(31) (2009).
60. P. Bucholtz, A. Bucholtz, J. Kirchhof and A. J. Augustin, *Retinal Physician* (3/1/2010).
61. D. Veritti, V. Sarao and P. Lanzetta, *Exp. Rev. Ophthalmol.* 5(5), 681–688 (2010).
62. T. Sakai, Y. Ohkuma, H. Kohno, T. Hayashi, A. Watanabe and H. Tsuneoka, *Br. J. Ophthalmol.* **98**(12), 1642–1648 (2014).
63. I. Nakata, A. Tsujikawa, K. Yamashiro, A. Otani, S. Ooto, Y. Akagi-Kurashige, N. Ueda-Arakawa, D. Iwama and N. Yoshimura, *Graefe's Arch. Clin. Exper. Ophthalmol.* **251**(4), 1073–1080 (2013).
64. M. W. Lee, I. Yeo, D. Wong and C. L. Ang, *Eye.* **23**(1), 145–148 (2009).
65. M. Yuzawa, R. Mori and M. A. Haruyama, *Jpn. J. Ophthalmol.* **47**(4), 379–384 (2003).

Part 3

Chapter 14

Photodynamic Therapy of Head and Neck Cancer — What's Old and What's New

Merrill A. Biel

Department of Otolaryngology — Head and Neck Surgery
University of Minnesota Medical Director, Head and Neck Oncology
Virginia Piper Cancer Institute 2211 Park Avenue South
Minneapolis, Minnesota 55404 612-871-2980
bielx001@umn.edu

Abstract

Photodynamic therapy (PDT) is a FDA approved minimally invasive medical treatment modality that utilizes light in the presence of oxygen to activate photosensitizing agents that are relatively selectively concentrated in abnormal or neoplastic cells resulting in cell death. At the present time, PDT has been approved for clinical treatment in the United States, European Union, Canada, Russia, and Japan. In the EU Foscan has been approved for the treatment of early head and neck cancers and palliative treatment of head and neck cancer. In the United States, PDT treatment of head and neck cancer is performed off label using Photofrin, as to date there are no photosensitizers approved by the U.S. FDA for the treatment of head and neck cancers.

PDT is an excellent minimally invasive modality for the treatment of carcinomas of the head and neck as the treatment preserves normal tissue. First and second generation photosensitizer PDT has been effective in the treatment of early head and neck cancers, including severe premalignant dysplasia, Tis, T1, and T2 cancers of the oral cavity, pharynx and larynx with complete response rates of 71–90%. In addition, combination therapies and targeted therapies are being investigated that may improve the therapeutic efficacy of PDT to treat later stage disease and recurrent cancers of the head and neck, including ChemoPDT and Photoimmunotherapy (PIT).

1. Introduction

Photodynamic therapy (PDT) is a FDA approved minimally invasive medical treatment modality that utilizes light in the presence of oxygen to activate photosensitizing agents that are relatively selectively concentrated in abnormal or neoplastic cells resulting in cell death. At the present time, PDT has been approved for clinical treatment in the United States, European Union, Canada, Russia, and Japan. In the United States, U.S. Food and Drug administration approval has been given for the use of PDT in the treatment of Barrett's esophagus, obstructing esophageal carcinoma and early and obstructing tracheobronchial carcinoma using the photosensitizer Photofrin; actinic keratosis using the photosensitizer Levulan (aminolevulinic acid ALA) and Metvix (methyl aminolevulinate); and macular degeneration using the photosensitizer BPD. In the EU the above noted indications have also been approved in addition to the treatment of early head and neck cancers and palliative treatment of head and neck cancer using the photosensitizer Foscan; and treatment of basal and squamous cell skin cancers using the photosensitizer Metvix. In the United States, PDT treatment of head and neck cancer is performed off label using Photofrin, as to date there are no photosensitizers approved by the U.S. FDA for the treatment of head and neck cancers.

First and second generation photosensitizer PDT has been effective in the treatment of early head and neck cancers, including severe premalignant dysplasia, Tis, T1, and T2 cancers of the oral cavity, pharynx and larynx with complete response rates of 71–90%. However, treatment of Stage III and Stage IV disease with standard PDT alone has been less promising. Therefore, combination therapies and targeted therapies are being investigated that may improve the therapeutic efficacy of PDT to treat later stage disease and recurrent cancers of the head and neck, including ChemoPDT, Photochemical Internalization (PCI), and Photoimmunotherapy (PIT). This chapter will review the past experience with PDT treatment of early cancers of the head and neck as well as discuss newer PDT strategies and therapies for treatment of head and neck cancers.

2. Mechanism of PDT

The method of PDT involves the use of a photosensitizing agent that is relatively selectively concentrated in abnormal or neoplastic cells. Depending on the type of photosensitizer, it may be injected intravenously, ingested orally, or applied topically. After application of the photosensitizer, it is relatively selectively retained by tumor cells so that after several hours to days, determined by the kinetics of the compound's distribution, there is more sensitizer in the neoplastic tissue than in the normal tissue. The photosensitizer is then activated with a specific wavelength of light matching the absorption characteristics that are unique to that specific photosensitizer, usually using a laser. This results in tumor necrosis via several mechanisms including oxygen radical production as well as vascular shutdown to the tumor.[1] Because there is less sensitizer in the adjacent normal tissue, only the neoplastic tissue necroses and the normal tissue is preserved. The advantage of PDT over the other conventional modalities of surgery, radiation and chemotherapy is that it is a minimally invasive treatment technique that lacks systemic toxicity yet results in selective tumor destruction with normal tissue preservation. This advantage is of particular importance for cancers of the head and neck, where excessive tissue loss results in significant functional debilities. In addition, since this is an entirely different process, the use of chemotherapy, ionizing radiation, or surgery does not preclude the use of PDT. Also, unlike ionizing radiation, repeated applications of the photosensitizer and activating light treatments can be performed indefinitely.

PDT requires three ingredients: A photosensitizer, a light source, and oxygen. The ideal photosensitizer is one that is highly selectively retained by tumor cells, is activated at a long light wavelength, which increases light penetration into tissue, and has no side effects.

Photofrin (dihematoporphyrin ether) is the most extensively studied and clinically used photosensitizer to date. Over 20,000 patients with various cancers have been treated with Photofrin. This photosensitizer is injected intravenously and usually 48 h after infusion the tumor tissue is activated at 630 nm nonthermal light. Photofrin is relatively selectively retained by cancer cells due to unknown mechanisms but tends to localize in mitochondria and endoplasmic reticulum. Tumor cell death is caused by oxygen radical production and tumor microvascular shutdown as well as a complex series of immunologic responses, contributed to by the intense inflammatory response induced by PDT.[1] The major side effect of Photofrin is a skin photosensitivity that lasts for an average of six weeks post-infusion. Patients must remain out of outdoor daylight for that period of time, that is, when outdoors they need to keep their skin covered with a wide brim hat, long sleeve shirt, and gloves. However, they can function normally in all indoor lighting situations without precautions. Therefore, for most people, the inconvenience of light photosensitivity on normal daily function is minimal.[1,2] Another limitation of this photosensitizer is the wavelength of light used to activate it. 630 nm light therapeutically penetrates 0.5 cm into tissue. Therefore, larger solid tumors cannot be uniformly illuminated with surface illumination only due to limited depth of penetration. Therefore, to treat larger tumors, implanted diffusers are used to distribute the light deeply into tissue.

Foscan (meta tetrahydroxyl phenyl chlorin, mTHPC) is the most potent of the second generation photosensitizers.[3] It is 100 times more active than Photofrin in animal model studies. It is activated at 664 nm wavelength light with depth of penetration of approximately one centimeter.[4] Tumor to normal tissue concentration is 20:1 in animal studies.[5] It has very intense skin photosensitivity for seven days such that one needs to be in a darkened room for four days. However, the length of photosensitivity is three to four weeks.

Vertiporfin (benzoporphyrin derivative, BPD) is an intravenously administered photosensitizer that is activated with 692 nm light 3 h after infusion. It is rapidly cleared and therefore has very limited skin photosensitivity lasting only up to one week.[6,7] It has the advantages of limited skin photosensitivity with a higher wavelength of activation resulting in deeper tissue penetration of light. It has not been evaluated for the treatment of head and neck or other cancers to date. BPD has been approved by the U.S. FDA for the treatment of macular degeneration.

ALA is an intrinsic photosensitizer. It has no photosensitizing properties but is converted *in situ* to a photosensitizer, Protoporphyrin IX. The exact rate of Protoporphyrin accumulation in any tissue is dependent upon metabolic turnover and the capacity of the cell for heme synthesis. Cancer cells and gastrointestinal mucosa accumulate Protoporphyrin IX more actively than normal cells. This differential synthesis of Protoporphyrin IX in cancer cells provides the basis for selective cancer treatment with ALA.

ALA can be administered topically,[8] orally,[9] or intravenously. Topical ALA has been used to treat basal cell and squamous cell carcinomas of the skin with cure rates up to 90%.[8,10] ALA has been approved in the United States for the treatment of actinic keratosis. Oral ALA has been used in the treatment of a small series of oral cancers in England with limited success.[9] The limitation of ALA in treating oral cancers is the limited depth of ALA accumulation, only occurring in the mucosa itself. In addition, oral ALA ingestion results in transient liver function test abnormalities that may last up to one week after ALA ingestion.[9] ALA is activated at 635 nm light with peak tumor fluorescence occurring three to five hours after administration. It is rapidly cleared from the body with skin photosensitivity lasting less than 24 h.[8] Due to the limited depth of accumulation of ALA and the limited penetration of 635 nm light, tumors of greater than 2 mm depth are not consistently cured.[8,9]

3. Light Delivery Systems

Light is delivered to the tumor tissue bed from the light source (usually a diode laser) with silicone-quartz fiberoptic cables. The distal end of the quartz fiber is modified to obtain optimal uniform geometric light delivery to the tissues. In the head and neck, the fiber tips used are a microlens and cylindrical diffuser. The microlens tip distributes the light in a uniform circle and is used to illuminate superficial tissue surfaces. Cylindrical diffuser tips are used to distribute light 360° along the axis of the fiber. This type of fiber is used to deliver light into a hollow viscus such as the tracheobronchial tree or the esophagus. As well, this type of fiber can be implanted into the tissue to distribute light more deeply into the tumor tissue bed.

4. Light Dosimetry

The calculation of the optimal amount of light delivered to the tissue to be treated is referred as dosimetry. The required exposure time for PDT tumor treatment depends on the laser power output in Watts, the fluence rate measured in Watts/cm^2 and the fluence in Joules/cm^2. These relationships are expressed in the following equations:

Light dose (Joules) = Watts × seconds
Exposure time (seconds) = Fluence (J/cm^2)
Fluence rate (mW/cm^2).

The fluence (J/cm^2) and the fluence rate (mW/cm^2) to be administered to the tumor tissue is a predetermined number based on clinical trials for each specific tissue type and tumor to be treated. The amount of light (mWatts) coming out of the end of the treatment fiber (measured by a light power/wavelength meter) needed to treat the tumor = the area to be treated in cm^2 × fluence rate (mW/cm^2).

The fluence and fluence rates will vary with each tissue treated and with each unique photosensitizer.

5. Methods

The author's clinical experience with PDT for the treatment of head and neck cancers predominantly involves the use of Photofrin. However, the general method of treatment described would be used for any photosensitizer except for parameters that are unique to the specific photosensitizer used including: drug concentration, activating wavelength and the amount of activating light to be administered. Photofrin will be used as the demonstration of the method of PDT use in the treatment of Head and Neck cancers.

Photofrin is injected intravenously at a dose of 2.0 mg/kg over a 5 min period as an outpatient procedure. 48 h after the injection, the patients undergo treatment with light from a cold light diode laser (Diomed) at 630 nm wavelength. Light is delivered to the tissue bed with a 400 μm fused silica optical fiber (Laserguide, Inc., Buelton, California). A microlens treatment is used for all tumors with a depth of less than 3 mm. These treatments are performed at a fluence of 50–75 J/cm^2 and a fluence rate of 150 mW/cm^2 in the oral cavity, nasopharynx and skin and 80 J/cm^2 and 150 mw/cm^2 in the larynx. For laryngeal tumors, the treatment is performed under standard laryngoscopy on an outpatient basis. Patients are observed in post op recovery for 2 h to ensure no airway problems and are discharged to home. There is no need for a preoperative tracheostomy unless the patient requires a stable airway due an obstructing tumor. For tumors greater than 3 mm in depth or T2/T3 laryngeal tumors with paraglottic involvement, cylindrical diffusers 0.5–2.5 cm in length are placed in the tumor bed using an 18 gauge catheter under local or general anesthesia. These treatments are performed at a fluence of 100 J/cm fiber length and a fluence rate of 400 mW/cm fiber length. For all treatments of the oral cavity, larynx and pharynx, all the normal tissues should be covered with either moist 4 $\times$ 4 sponges or in the case of laryngeal treatments, a neurosurgical cottonoid is placed to cover the normal vocal cord. Covering the normal tissues will prevent any laser light scatter from activating the photosensitizer that is present in the normal tissues. This reduces the extent of any edema of the normal tissues due to inadvertent light activation.

6. Clinical Treatment Results

Data are available for over 2500 patients treated with PDT using Photofrin, HPD, ALA, or Foscan for the treatment of head and neck cancers. These patients include a mixture of presentations including primary, recurrent, and metastatic lesions. The predominant histology is squamous cell carcinoma, but other histologies treated include mucosal melanoma, Kaposi's sarcoma, adenocarcinoma, metastatic breast carcinoma, and adenoid cystic carcinoma. Fortunately, the first multi-institutional Phase II clinical trials evaluating PDT treatment of head and neck cancers have been completed. These trials have demonstrated the efficacy of this minimally invasive therapy in the treatment of early orophayngeal primary and recurrent cancers as well as the palliative treatment of refractory head and neck cancers. Importantly, different photosensitizers have been used and continue to be investigated for the treatment of head and neck cancers with very similar clinical response and cure rates.

6.1. *Early Stage Head and Neck Cancer: Photofrin-HPD Based PDT*

Patients with early stage cancers or early recurrences in the oral cavity and larynx (Tis, Tl, T2) tend to have an excellent response to PDT. Of 749 patients treated with Tis, T1, or T2 cancers of the oral cavity, larynx, pharynx and nasopharynx, 657 (87.7%) obtained a complete clinical response after one PDT treatment.[11–33] Laryngeal cancers, comprising 235 patients in this group, obtained a durable complete response rate of 87.2% with up to a eighteen year follow-up.[12–32] The largest series of laryngeal and oral cancer PDT treatment is described by Biel. 155 patients with recurrent or primary Tis, T1N0 and T2N0 laryngeal tumors were treated from 1990 to 2013 with PDT for cure. Six patients had recurrent Tis, 132 patients had T1N0 carcinomas of the true vocal cord of which 27 were radiation failures, and 17 patients had T2N0 carcinomas of the true vocal cord of which eight were radiation failures. All patients underwent a single microlens light treatment and most T2 tumors also underwent cylindrical diffuser implants into the paraglottic space. All treatments were performed under general anesthesia with standard laryngoscopy and all patients were discharged to home the same day of PDT treatment. All patients obtained a complete histopathologic response after a single light treatment. With follow-up to 60 months (mean 58 months) there were 14 recurrences for a five-year cure rate of 90%. Importantly all the recurrences were salvaged using either PDT, surgery or radiation for a total five-year cure rate of 100%. In the entire treatment group there were no episodes of airway compromise and the degree of postoperative pain was minimal and easily controlled with oral analgesics. All patients after treatment developed an immediate breathy voice that persisted for 2–3 weeks. At 4–6 weeks after treatment, the quality of voice was universally much improved over the pretreatment state. In addition, videostroboscopy six weeks post PDT treatment demonstrated a normal vocal cord mucosal fluid wave on the treated vocal cord.

155 patients with recurrent or primary Tis and T1N0 squamous cell carcinomas of the oral cavity were treated from 1990 to 2013. Five patients had recurrent Tis and 150 patients had T1N0 lesions. All patients were treated with a microlens and if the tumors clinically were invasive greater than 3 mm, they were also treated with cylindrical diffuser implantation to distribute the light deeper into the tissues. The diffusers were spaced 1 cm apart from each other. All patients obtained a complete pathologic and clinical response after a single PDT treatment. With follow-up to 60 months (mean 57 months), there were 16 local recurrences within eight months of PDT treatment. These

were all salvaged with either repeat PDT treatment or surgical resection. Two patients with T1 tongue tumors developed regional lymph nodes within three months of PDT treatment and went on to conventional neck dissection and have remained free of disease for at least five years. Five-year cure rates for these patients therefore remains at 100%. Unfortunately, there was one peri-PDT treatment death unrelated to PDT due to a drug overdose.

83 patients with superficial T2N0 and T3N0 squamous cell carcinomas of the oral cavity were treated. All of these patients had extensive areas of tumor involvement up to 7 cm in size. The maximal depth of the tumor, however, was clinically less than 1 cm. These patients were treated with a microlens and for those areas where there was clinical invasion greater than 3 mm, cylindrical diffusers of 0.5–1.0 cm were employed to distribute the 630 nm light deeper into the tissues. All patients obtained a complete clinical and pathologic response after a single PDT treatment. With follow-up to 60 months (mean 56 months), there were seven recurrences, all at the edge of the PDT treatment field. These were all salvaged with either repeat PDT treatment or surgical resection. No patients developed regional lymph node involvement. Three-year cure rates for these patients is 100%. Importantly, the post-PDT healing of these patients demonstrated normal mobile oral mucosa without scar formation and preservation of the patency of the submandibular and parotid salivary ducts. Histologic evaluation of the post-PDT healing process demonstrated preservation of the cellular collagen matrix with repopulation of the normal mucosal cells into the preserved collagen matrix resulting in normal mucosa and submucosa.

6.2. *Early Stage Head and Neck Cancer: Foscan (mTHPC)-Mediated PDT*

Foscan (mTHPC, Biolitec, Germany) is a potent second generation photosensitizer that is activated at 652 nm light. To date, this is the only photosensitizer that has been evaluated in multi-institutional Phase II trials for the treatment of head and neck cancers. Numerous single investigator trials have also demonstrated the efficacy of Foscan PDT in treating head and neck cancers.[34–43]

The multi-institutional trial evaluating Foscan PDT for the treatment of primary oropharyngeal cancers involved 114 patients with Tis-T2 oropharyngeal cancers.[39,40] These patients received Foscan 0.15 mg/kg intravenously and underwent light activation at 652 nm light at 20 J/cm^2 at 100 mW/cm^2. Up to three light treatments were allowed under the protocol. A complete response rate of 85% (97/114) was achieved at completion of therapy. With two year follow-up there was a 77% complete response rate at two years with a disease free survival of 89% and 75% at one and two years after PDT treatment respectively. This trial demonstrated complete durable response rates that are equivalent to those obtained with conventional therapies.[39,40] Overall, of 465 patients treated with Tis, T1, or T2 cancers of the oral cavity, larynx and pharynx, 357 (74.7%) obtained a complete clinical response after one PDT treatment.[35–43] (Table 2).

Table 1. Summary of published results with Photofrin PDT of early head and neck squamous cell cancer.

Study	Patients	Lesion and site	Drug, dose, mg/kg	Response, *n*		
				Complete	Partial	None
Keller *et al.*[11]	3	T1 and T2, oral cavity	Photofrin, 1.5–2	3	0	0
Feyh *et al.*[12,13]	15	T1 and T2, oral cavity	Photosan III	13	2	0
	12	T1 and T2, larynx	Photosan III	11	1	0
Wenig *et al.*[14]	26	T1 recurrent, various sites	Photofrin, 2	20	6	0
Grossweiner *et al.*[15]	9	Early oral cavity and pharynx	Photofrin, 2	8	1	0
Freche *et al.*[16]	32	T1, larynx	HPD, 3 Photofrin, 2	25	7	0
Schweitzer[17–19]	30	T1, oral cavity	Photofrin, 2	24	6	0
	10	T1, larynx		8	2	0
Gluckman 20	13	T1, oral cavity	Photofrin, 2	11	2	0
	2	T1, larynx		2	0	2
	8	Tis, condemned mucosa		7	1	0
Grant *et al.*[21]	12	T1, oral cavity	Photofrin, 2	11	1	0
Biel[22–26]	155	Tis, T1 and T2 larynx	Photofrin, 2	141	14	0
	155	Tis, T1 and T2 oral cavity		139	16	0
	83	T2, T3 superficial		76	7	0
Zhao *et al.*[27,28]	50	Lip cancer	HPD, 3	50	0	0
Kulapeditheron[29,30]	41	Oral and nasopharynx	Photofrin, 2	30	11	0
Rigual[31]	20	Tis, T1 oral cavity	Photofrin, 2	19	1	0
	6	Tis, T1 larynx		5	1	0
Civantos[32]	24	Tis, T1–T3 oral cavity	Photofrin, 2	19	5	0
	18	Tis, T1–T2 larynx		13	5	0
Ikeda[33]	25	T1 and T2 oral cavity	Photofrin, 2	24	1	0

Table 2. Summary of published results with ALA and FOSCAN PDT of early oral and laryngeal squamous cell cancer.

Study	Patients	Lesion and site	Drug, dose, mg/kg	Response, *n*		
				Complete	Partial	None
Grant *et al.*[9]	4	Oral cavity	ALA	3	1	0
Fan *et al.*[44]	18	Dysplasia, ca oral cavity	ALA	14	4	0
Sieron *et al.*[45]	5	Larynx and hypopharynx	ALA	0	5	0
Grossman *et al.*[46]	1	Severe dysplasia larynx	ALA	1	0	0
Kubler *et al.*[47]	12	Oral leukoplakia	ALA	5	4	3
Dilkes *et al.*[35]	19	T1, T2 oral cavity and pharynx	Foscan	10	9	0
Savary *et al.*[36]	4	Tis, T1 oral cavity	Foscan	4	0	0
Fan *et al.*[37]	6	Tis, T1, T2 oral cavity	Foscan	6	0	0
Kubler *et al.*[38]	25	Tis–T2 lip cancer	Foscan	22	3	0
Hopper *et al.*[39]	114	Tis–T2, oral cavity	Foscan	97	17	0
Cooper *et al.*[40]	25	T1, T2 oral cavity and pharynx	Foscan	25	4	0
Karakullukcu *et al.*[41]	226	Tis–T2 oral cavity	Foscan	160	14	0
Jerjes *et al.*[42]	38	T1–T1 oral cavity	Foscan	26	12	0
Von Beckerath *et al.*[43]	8	T1 larynx	Foscan	7	1	0

6.3. *Early Stage Head and Neck Cancer: ALA-Mediated PDT*

Grant treated four patients with oral carcinoma following the oral administration of ALA.[9] 630 nm light of 50–100 J/cm^2 was delivered 6 h after oral ingestion of the ALA. All patients sustained tumor necrosis but normal surrounding tissue also necrosed, indicating no tumor selectivity. Fan treated 18 patients with dysplasia and malignant lesions of the oral cavity with ALA. Only two of six patients with carcinomas obtained a complete response.[44] Sieron treated five patients with larynx and hypopharynx carcinomas using ALA PDT. All five patients only achieved a partial response.[45] Transient liver function abnormalities occurred in all patients. Due to the limited depth of accumulation of ALA and the limited penetration of 635 nm light, tumors of greater than 2 mm depth are not consistently cured.[9,44,45] Grossman treated a patient with recurrent severe dysplasia of the larynx with ALA. A complete response was achieved after a single treatment at six months follow-up.[46] These limited trials have demonstrated that oral ALA PDT is not as effective as Photofrin or Foscan in the treatment of oral and laryngeal cancers.

Topical ALA has been employed for the treatment of premalignant oral leukoplakia and dysplasia. Kubler treated 12 patients with oral leukoplakia with 20% topical ALA cream and 630 nm light. Five patients demonstrated a complete response, four patients a partial response, and three patients no response. One patient with a partial response was retreated resulting in a complete response.[47] Chen treated 24 patients with oral leukoplakia using ALA topically twice a week. Eight patients had a complete response and 16 patients had a partial response. Significant pain during light treatment was noted.[48] Sieron treated 48 patients with oral leukoplakia using topical ALA PDT. Treatments were repeated 6–8 times in 2–3 week intervals. A complete response rate of 88% was achieved with the repeated therapy technique.[49] Yu treated 46 patients with oral erythroleukoplakia using topical ALA PDT. Lesions underwent repeat therapy an average of 3.7 times. A complete response rate of 91% was achieved with repeat PDT therapy.[50] Lin treated 40 patients with oral verrucous hyperplasia and 40 patients with oral erythroleukoplakia using topical ALA. A complete response was achieved in all 40 patients with oral verrucous hyperplasia after an average of 3.6 treatments. A 95% complete response rate was achieved in oral erythroleukoplakia after an average of 3.4 treatments.[51] Jerjes treated 147 patients with premalignant oral dysplasia with either topical ALA (57 patients) or IV Foscan (90 patients). An 81% complete response rate was achieved.[52] These studies demonstrate that topical ALA is effective in the treatment of oral leukoplakia and premalignant dysplasia although it requires multiple repeated treatments to achieve a complete response. In addition, pain is associated with the therapy and therefore requires local anesthesia during the light administration. The treatment however is well tolerated without systemic photosensitivity.

6.4. *Recurrent Head and Neck Cancers — Foscan PDT*

A Phase II multi-institutional clinical trial of Foscan PDT in 96 patients with recurrent or second primary carcinomas in the oral cavity was performed demonstrating the efficacy of PDT for the treatment of recurrent oral squamous cell cancers. These patients demonstrated a 50% histologically confirmed complete response rate with a 79% survival rate at one year.[53] Lorenz treated 35 patients with recurrent or second primary tumors of the oral cavity and larynx. Local control was achieved in 21 patients (60%), partial remission in 10 patients (28.5%) and four patients did not respond (11.5%).[54] Tan treated 39 patients with head and neck squamous cell carcinoma that were determined to be incurable or untreatable with surgery or radiation. 19 patients obtained a complete response, two patients a partial response, five patients stable disease and five patients had progressive disease.[55] These studies demonstrate the utility and effectiveness of Foscan PDT in the curative or palliative treatment of limited recurrent or second primary squamous cell carcinomas in patients who are untreatable with surgery or radiation therapy.

6.5. *Interstitial Image Guided PDT (iPDT) Treatment of Recurrent Head and Neck Cancer*

Lou described the method of interstitial light fiber PDT using Foscan to treat 39 patients with recurrent unresectable head and neck cancers. The overall median survival was 14 months with 72% overall palliative benefit achieved. Importantly, the local control rate at 12 months was 41%.[56] Jerjes treated 21 patients with advanced or recurrent tongue base cancers using Foscan PDT and image guided placement of the interstitial PDT light fibers. Nine of 11 patients reported an improved airway, 19/21 patients noted improved swallowing and 11/13 reported improved speech. Fifteen patients had a reduction in tumor size at six weeks post treatment. The author's noted that PDT is a successful palliative modality for the treatment of large recurrent unresectable tongue base tumors.[57] Interstitial imaged guided PDT treatment of head and neck cancers may be performed under ultrasound, CT or MRI guidance. Additionally, iPDT may be performed under endoscopic guidance (Figs. 1, 2) or Robotic guidance (Fig. 3).

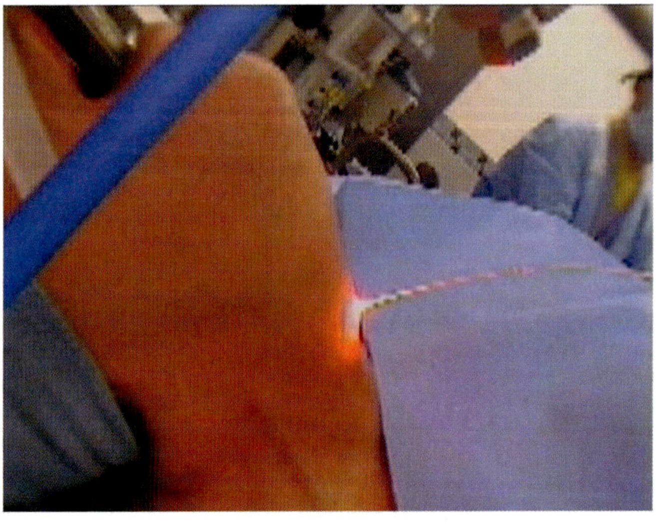

Fig. 1. Transcervical endoscopic placement of interstitial PDT fiber (iPDT) for vocal cord illumination.

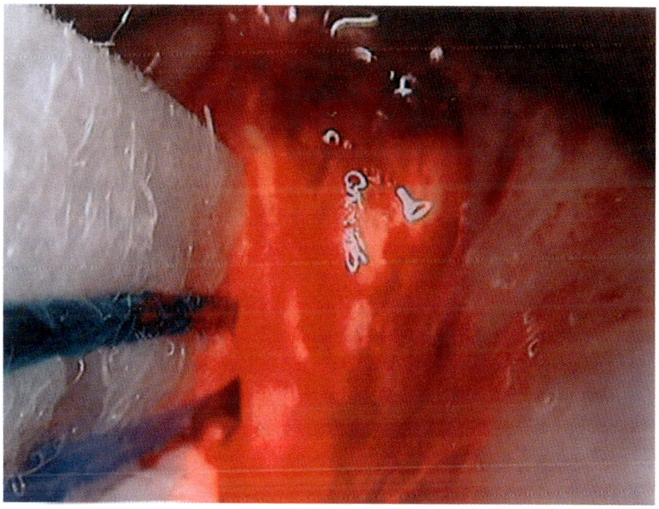

Fig. 2. Transcervical iPDT illumination of right vocal cord.

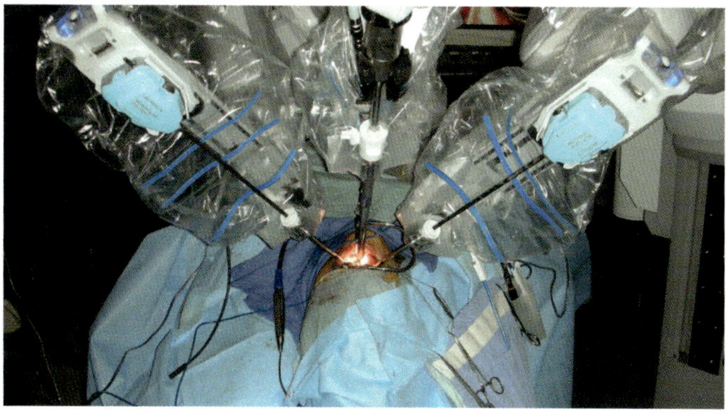

Fig. 3. Transoral robotic placement of interstitial PDT fiber (iPDT) for base of tongue PDT treatment.

6.6. *Intraoperative Adjuvant PDT for Massive Recurrent Head and Neck Cancers*

Biel reported the first human clinical trial of 35 patients with long term follow-up using PDT as an intraoperative adjuvant treatment for recurrent head and neck cancer at the time of surgical resection.[26,58] Eighteen patients with recurrent infiltrating squamous cell carcinoma of the head and neck were treated. Each patient had undergone previous treatment of the primary lesion of the head and neck with surgical resection, radiotherapy, and chemotherapy.

All patients were followed postoperatively (60 months) with only six patients developing recurrent or metastatic disease, two inside the field of PDT treatment and four outside the field of surgical and PDT therapy. In another treatment group, PDT was performed for curative intent following gross tumor debulking on 17 patients, 11 laryngeal and 6 oral cavity. Of the 11 laryngeal, eight were supraglottic and three were glottic. The oral cavity lesions were tongue and floor of mouth. For the laryngeal tumors treated, with follow-up to 60 months, there have been no recurrences. For the six oral cavity tumors, with follow-up to 60 months there was one recurrence that went on to conventional surgical resection and remains free of disease. This investigator determined that adjuvant intraoperative PDT performed after surgical resection may improve cure rates of recurrent head and neck malignancies by providing for larger tumor-free margins of resection while preserving normal structures.

Dilkes treated 14 patients with intraoperative Foscan mediated PDT. Two patients remain free of disease with up to five month follow-up. Two patients suffered carotid blow-outs.[59]

7. ChemoPDT

Patients with recurrent head and neck cancer who have failed previous surgery, chemoradiation, palliative chemotherapy, molecular targeted agents and PDT have no conventional treatment options available for further treatment. The author performed a limited clinical trial involving three patients to determine whether combining chemotherapy with PDT would enhance tumor response in this very difficult group of patients who have failed all other anticancer therapies. The treatment involved the patient receiving two courses of chemotherapy consisting of either platinum, taxol or eributux, followed within 1 week by PDT treatment and resuming the chemotherapy regimen the following week (sandwich chemo-PDT treatment). The first patient had an unresectable pharynx cancer in a previously radiated field. She received chemotherapy with no response. Was treated with PDT with a 50% reduction in tumor volume. She then underwent sandwich chemo-PDT treatment with platinum followed by Photofrin PDT followed by platinum and obtained a complete response and remained disease free for 37 months. The second patient had a recurrent oropharyngeal cancer that failed previous surgery and chemoradiation. She underwent sandwich chemo-PDT treatment with taxol followed by Photofrin PDT and obtained a complete response and has remained free of disease for 19 months. The third patient presented with a recurrent large ulcerative left lateral pharynx cancer that failed multiple previous surgical resections and chemoradiotherapy. He underwent sandwich chemo-PDT with platinum followed by Photofrin PDT with endoscopic interstitial diffuser placement. The patient obtained a complete response and at four months post chemo-PDT, the PET scan demonstrated no viable disease. The patient however developed a separate tongue base cancer at nine months post treatment and was again treated with chemo-PDT using Erbitux and Photofrin PDT with a partial response.

This limited clinical trial of chemoPDT to treat head and neck cancers that have failed all previous therapies is promising and further controlled studies are warranted.

8. PCI — Photochemical Internalization

PCI is a novel treatment modality that facilitates the delivery of macromolecules into cytoplasm. The initial mechanism and clinical application was described by Berg *et al.* in 1999.[60] In 2001, the same group performed an *in vivo* study using PCI for cancer therapy. The study demonstrated a synergistic effect of combining a photoactivated photosensitizer and gelonin to effectively destroy cancers *in vivo*.[61] In 2005, bleomycin was introduced as the chemotherapy agent to be effectively used with PCI for cancer treatment.[62] Based on these preclinical studies, a Phase I human dosing clinical trial using Amphinex (PCI Biotech, Lysaker, Norway) as the photosensitizer and bleomycin as a chemotherapeutic has been completed in Europe. Twenty two patients were treated at increasing doses of Amphinex, at a fixed bleomycin and light dose. The treatment was well tolerated with strong tumor responses noted across the higher Amphinex dose levels. A Phase II human clinical trial is presently underway in Europe to evaluate the efficacy of PCI to treat recurrent head and neck cancers at the optimal light and drug treatment dose determined in the Phase I study.

9. PIT — Photoimmunotherapy

PIT is a molecular targeted anticancer treatment that combines an antitumor receptor antibody to a photosensitizer resulting in a light activated targeted cancer specific treatment modality. This modality was first described by Mew and Levy in 1983[63] with further studies performed by others including Hasan and colleagues.[64–66] In 2011, Kobayashi and colleagues described a PIT agent consisting of a near infrared photoactive phthalocyanine dye, IR 700, conjugated to a monoclonal antibody targeting epidermal growth factor receptors. In vitro and *in vivo* studies demonstrated immediate cell death upon irradiation of the tumor with near infrared light.[67] Further studies using the PIT IR 700 conjugate drug demonstrated increased effectiveness of chemotherapeutic agents immediately after activating PIT light treatment. This was due to post PIT treatment increased vascular and cell membrane permeability allowing for the chemotherapeutic agents to accumulate in the PIT treated tumor, resulting in increased tumor necrosis compared to PIT treatment alone.[68] Based on these promising results, Aspyrian Therapeutics (San Diego, CA) has planned a Phase I human clinical trial using RM-1929, a PIT conjugate, to treat recurrent head and neck cancer that has failed prior surgical and chemotherapeutic therapies.

10. Summary

The present studies indicate the effectiveness of PDT in the treatment of specific anatomic areas in the head and neck. In particular, Tis and T1 carcinomas of the larynx appear to be particularly effectively treated with PDT. This clinical series demonstrates the efficacy of Photofrin-mediated PDT as a curative treatment for Tis, T1 (80–91%) and T2 (72%) squamous cell carcinomas of the larynx. PDT for laryngeal carcinomas results in no glottic scarring as compared to conventional laser or surgical excision or vocal cord stripping. For recurrent carcinomas of the larynx that have failed conventional radiation therapy, PDT allows excellent voice preservation and may eliminate the need for partial or total laryngectomy. Also, PDT can be repeated without additional functional laryngeal compromise that can occur from repeated conventional laser surgery or cordectomy. Importantly, PDT treatment of primary T1 and T2 laryngeal carcinomas reserves radiation therapy for treatment of recurrences or of second head and neck primaries that may occur in these high risk patients.

The side effects of PDT treatment of laryngeal carcinomas is quite minimal as compared to conventional radiotherapy or surgery. PDT treatment is performed as a single outpatient procedure as compared to 6–7 weeks of radiotherapy or the hospitalization associated with a partial or total laryngectomy. The photosensitivity of Photofrin is a temporary inconvenience not associated with systemic toxicity and is minimized by patient education and temporary changes in daily outdoor activities.

PDT for treatment of T1 and T2 laryngeal carcinomas in the present series has cure rates that are comparable to if not better than that of conventional therapies with less morbidity of treatment. PDT should be considered as a reasonable option for the treatment of primary and recurrent Tis, T1 and T2 squamous cell carcinomas of the larynx.

PDT also is effective in the treatment of Tis and T1 primary and recurrent carcinomas of the oral cavity including the palate, floor of mouth, nasopharynx and posterior pharyngeal walls. These results have been demonstrated in the single and multi-institutional Phase II Foscan PDT clinical trials[69] and in single investigator Photofrin trials in which cure rates were comparable to those of conventional therapy with less morbidity.

Photofrin and Foscan PDT have been shown to be effective in the curative and palliative treatment of recurrent head and neck squamous cell cancers. The adjuvant intraoperative use of PDT performed at the time of surgical resection to treat recurrent infiltrating carcinomas of the head and neck has been impressive to date. In this group of patients with a very high incidence of recurrence, adjuvant intraoperative PDT may improve cure rates by providing for larger tumor-free margins of resection while preserving normal structures. In addition, the use of image guided interstitial PDT for the treatment of large recurrent head and neck cancers appears to provide good local control and palliation for this difficult patient population.

Sandwiched ChemoPDT has demonstrated promise in a limited clinical study. Further formal clinical trials are needed to assess the efficacy of this combination treatment modality to treat recurrent head and neck cancers.

Newer novel therapies such as PCI and PIT are promising treatments for recurrent head and neck cancers that have failed previous treatments of surgery and radiotherapy. Human clinical trials are presently underway or planned to evaluate the efficacy of these new promising treatment modalities.

11. Conclusion

PDT is an excellent minimally invasive modality for the treatment of carcinomas of the head and neck as the treatment preserves normal tissue. Great advances have been achieved in the past few years in defining the role of PDT in the treatment of head and neck cancers. The future for the use of PDT in the treatment of head and neck cancers lies in its ability to treat early carcinomas of the head and neck with minimal morbidity, its potential use as an adjuvant therapy intraoperatively to treat surgical margins following resection for T3 and T4 head and neck and skull base tumors using superficial or interstitial light applications and as a palliative treatment for recurrent unresectable head and neck squamous cell carcinomas. PDT may also have a role in combination with radiation, chemotherapy and biotherapies. In order to further assess the effectiveness of this treatment on various areas within the head and neck, further standardized controlled studies are necessary. Human clinical trials are presently underway for the evaluation of the promising novel therapies of PCI and PIT for the treatment of recurrent head and neck cancers.

References

1. B. W. Henderson and T. J. Dougherty, *Photochem. Photobiol.* **55**, 145–157 (1992).
2. S. B. Brown, D. I. Vernon, J. A. Holroyd, S. Marcus, R. Trust, W. Hawkings, A. Shal and A. Tonelli, Pharmacokinetics of photofrin in man. Photodynamic therapy and biomedical lasers. In *Proceedings of International Conference*, Eds. Milan, P. Spinelli, F. M. Dal and R. Marchesi (Excerpta Medica, London, 1992), pp. 475–479.
3. R. Bonnet, R. D. White, U. J. Winfield and M. C. Berenbaum, *Biochem. J.* **261**(1), 277–280 (1989).
4. H.-B. Ris, J. H. Altermatt, R. Interbitzi, R. Hess, B. Nachbur, J. C. M. Stewart, Q. Wang, C. K. Lim, R. Bonnett, M. C. Berenbaum and U. Althaus, *Br. J. Cancer* **64**, 1116–1120 (1991).
5. H.-B. Ris, J. H. Altermatt, C. M. Steward, T. Schaffner, Q. Wang, C. K. Lim, R. Bonnett and U. Althaus, *Int. J. Cancer* **55**, 245–249 (1993).
6. H. Lui, N. Kollias, J. Wimberly and R. Rox Anderson, A preliminary report of cutaneous photosensitivity in patients undergoing photodynamic therapy with Benzoporphyrin derivative-monoacid ring A. In *Photodynamic Therapy and Biomedical Lasers*, Eds. P. Spinelli, P. M. Dal, R. Marchesina (Excerpta Medica, London 1992) pp. 459–460.
7. E. W. Kelly, K. K. Wang, N. N. Swanson, K. A. Hitzemann and M. R. Pittelkow, *Photodyn. Ther. Biomedi. Lasers*, 493–502 (1992).
8. K. Svanberg, T. Andersson and D. Killander, Photodynamic therapy of human skin malignancies and laser-induced fluorescence diagnostics utilizing Photofrin and 5amino levulinic acid. In *Photodynamic Therapy and Biomedical Lasers*, Eds. P. Spinelli, M. F. Dal, R. Marchesini, (Excerpta Medica, London, 1992), pp. 436–440.
9. W. E. Grant, C. Hopper, A. J. MacRobert, P. M. Speight and S. C. Bown, *Lancet*, **342**, 147–149 (1993).
10. T. Warloe, Q. Penq, J. Moan, H. L. Qvist and K. E. Giercksky, Photochemotherapy of multiple basal cell carcinoma with endogenous porphyrins induced by topical applications of 5-aminolevulinic acid. In *Photodynamic Therapy and Biomedical Lasers*, Eds. P. Spinelli, F. M. Dal and R. Marchesini (Excerpta Medica, London, 1992) pp. 436–440.
11. G. S. Keller, D. R. Doiron and C. U. Fisher *Arch. Otolaryngol.* **111**, 758–761 (1985).
12. J. Feyh, A. Goetz, W. Muller, R. Konigsberger and E. Kastenbauer, *J. Photochem. Photobiol. B:Biol.* 7, 353–358 (1990).
13. J. Feyh, A. Gutmann and A. Leunig, *Laryngo. Rhino. Otol.* **72**, 273–278 (1993).
14. B. L. Wenig, D. M. Kurtzman, L. Grossweiner, J. F. Mafee, D. M. Harris, R. V. Lobraico, R. A. Prycz and E. L. Appelbaum, *Arch. Otolaryngol. Head Neck Surg.* **116**, 1267–1270 (1990).
15. L. I. Grossweiner, J. H. Hill and R. V. Lobraico *Photochem. Photobiol.* **46**, 911–917 (1987).
16. C. Freche and S. DeCorbiere, *J. Photochem. Photobiol. B* **6**, 291–296 (1990).

17. V. G. Schweitzer, *Otolaryngol. Head Neck Surg.* **102**, 225–232 (1990).
18. V. G. Schweitzer, *Lasers Surg. Med.* **29**, 305–313 (2001).
19. V. G. Schweitzer and M. L. Somers, *Lasers Surg. Med.* **42**, 1–7 (2010).
20. J. L. Gluckman, *Laryngoscope* **101**, 36–42 (1991).
21. W. E. Grant, C. Hopper, P. M. Speight, A. D. Macrobert and S. C. Bown, *J. Laryngol. Otol.* **107**(12), 1140–1145 (1993).
22. M. A. Biel, *Laryngoscope.* **104**(4), 399–403 (1994).
23. M. A. Biel, *Semin. Surg. Oncol.* **11**, 355–359 (1995).
24. M. A. Biel, *Laryngoscope* **108**, 1259–1268 (1998).
25. M. A. Biel, Photodynamic Therapy Treatment of Early Oral and Laryngeal Cancers. *Photochem. Photobiol.* **83**, 1063–1068 (2007).
26. M. A. Biel, *Methods Mol. Biol.* **635**, 281–293 (2010).
27. F. Y. Zhao, K. H. Zhang, D. Q. Ma, Z. Q. He, Chen 5, X. M. Ni and G. Y. Yu, *Ann. Acad. Med. Singapore*, **18**, 533–536 (1989).
28. F. Y. Zhao, K. H. Zhang, H. N. Huang, K. H. Sun, Q. B. Ling and B. Xu, *Lasers Med. Sci.* **1**, 253–256 (1986).
29. B. Kulapaditharom and V. Boonkitticharoen, *J. Med. Assoc. Thai.* **83**(3), 249–258 (2000).
30. B. Kulapaditharom and V. Boonkitticharoen, *J. Med. Assoc. Thai.* **82**(11), 1111–1117 (1999).
31. M. F. Rigual, T. Krishnakumar, M. Cooper, M. A. Sullivan, T. Dougherty, S. R. Popat, T. R. Loree, M. A. Biel and B. Henderson, *Arch. Otolaryngol. Head Neck Surg.* **135**(8), 784–788 (2009).
32. F. Civantos, *J. Natl. Compr. Cancer Netw.* **10**(2), S-65–S-68 (2012).
33. H. Ikeda, T. Tobita, S. Ohba, M. Uehara and I. Asahina, *Photodiagnosis Photodyn. Ther.* **10**, 229–235 (2013).
34. M. G. Dilkes and M. L. DeJode, *Lasers Med. Sci.* **11**(1), 23–29 (1996).
35. M. G. Dilkes, E. Benjamin, S. Ovaisi and A. S. Banerjee, *J. Laryngol.* **117**(9): 713–717 (2003).
36. F. J. Savary, H. P. Monnier, C. Fontolliet *et al.*, *Arch. Otolaryngol. Head Neck Surg.* **123**(2), 162–168 (1997).
37. F. K. Fan, C. Hopper, P. M. Speight *et al.*, *Int. J. Cancer* **73**(1), 25–32 (1997).
38. A. D. Kubler, J. de Carpentier, C. Hopper, A. G. Leonard and G. Putnam, *Int. J. Oral Maxillofac. Surg.* 30, 504–509 (2001).
39. C. Hopper, A. Kubler, H. Lewis, I. B. Tan, G. Putnam, *Int. J. Cancer*, 111, 138–146 (2004).
40. M. P. Copper, B. I. Tan, H. Oppelaar, M. C. Ruevekamp and F. A. Stewart, **129**(7), 709–711 (2003).
41. B. Karakullukcu, K. van Oudenaarde, M. P. Copper, W. M. C. Klop, R. van Veen, M. Wildeman and I. B. Tan, *Eur. Arch. Otorhinolaryngol.* **268**, 281–288 (2011).
42. W. Jerjes, T. Upile, Z. Hamdoon, C. A. Mosse, M. Morcos and C. Hopper, *Lasers Surg. Med.* **43**, 468–469 (2011).
43. M. P. Von Beckerath, J. A. Reizenstein, A. L. Berner, K. W. O. Nordqvist, F. J. Landstrom, A. L. Lofgren and C. G. Moller, *Acta Oto-Laryngologica* **134**, 852–858 (2014).
44. K. F. M. Fan, C. Hopper, P. M. Speight, G. Buonaccorsi, A. J. MacRobert and S. G. Bown, *Cancer*, 78, 1374–1383 (1996).

45. A. Sieron, G. Namyslowski, M. Misiolek, M. Adamek and K. Dawczyk, *Eur. Arch. Otorhinolaryngol.* 258, 349–352 (2001).
46. C. Grossman, T. Zhu, J. Finlay, A. Dimofte, K. Malloy, B. O'Malley Jr., G. Weinstein, T. M. Busch and H. Quon, *Photodiagnosis Photodyn. Ther.* 7, 158–161 (2010).
47. A. Kubler, T. Haase, M. Rheinwald, T. Barth and J. Muhling, *Int. J. Oral Maxillofac. Surg.* **27**(6), 466–469 (1998).
48. H. Chen, C. Yu, P. Tu, C. Yeh, T. Tsai and C. Chiang, *Lasers Surg. Med.* **37**, 114–122 (2005).
49. A. Sieron, A. Kosciarz-Grzesiok, A. Kawczyk-Krupka, K. Sieron-Stoltny and A. Misiak, *Wiad. Lek.* **61**(10–12), 283–287 (2008).
50. C. H. Yu, H. P. Lin, H. M. Chen, H. Yang, Y. P. Wang and C. P. Chiang, *Lasers Surg. Med.* **41**, 628–633 (2009).
51. H. P. Lin, H. M. Chen, C. H. Yu, H. Yang, Y. P. Wang and C. P. Chiang, *J. Oral Pathol. Med.* **39**, 624–630 (2010).
52. W. Jerjes, T. Upile, Z. Hamdoon, C. A. Mosse, S. Akram and C. Hopper, *Lasers Surg. Med.* **43**,192–199 (2011).
53. A. K. D'Cruz, M. H. Robinson and M. A. Biel, *Head Neck* **26**, 232–240 (2004).
54. K. J. Lorenz and H. Maier, *Eur. Arch. Otorhinolaryngol.* **266**, 1937–1944 (2009).
55. I. B. Tan, G. Dolivet, P. Ceruse, Vander V. Poorten, G. Roest and W. Rauschning, *Head & Neck* **32**, 1597–1604 (2010).
56. P. J. Lou, H. R. Jager, L. Jones, T. Theodossy, S. G. Bown and C. Hopper, *Br. J. Cancer* **91**(3), 441–446 (2004).
57. W. Jerjes, T. Upile, Z. Hamdoon, S. Abbas, S. Akram, C. A. Mosse, S. Morley and C. Hopper, *Lasers Surg. Med.* **43**, 283–292 (2011).
58. M. A. Biel, *Arch. Otolaryngol. Head Neck Surg.* **122**, 1261–1265 (1996).
59. M. G. Dilkes, G. Alusi and B. J. Djaezeri, *Rev. Contemp. Pharmacother.* **10**, 47–57 (1999).
60. K. Berg, P. K. Selbo, L. Prasmickaite, T. E. Tjelle, Sandvig, J. Moan, G. Gaudernack, O. Fodstad, S. Kjolsrud, H. Anholt, G. H. Rodal and A. Hogset, *Cancer Res.* **59**(6), 1180–1183 (1999).
61. P. K. Selbo, G. Sivan, O. Fodstad, K. Sandvig and K. Berg, *Int. J. Cancer* **92**(5), 761–766 (2001).
62. K. Berg, A. Dietze, O. Kaalhus and A. Hogset, *Clin. Cancer Res.* **11**(23), 8476–8485 (2005).
63. D. Mew, C. Wat, G. H. N. Towers and J. Levy, *Jour. Immunol.* **130**(3), 1473–1477 (1983).
64. B. A. Goff, U. Hermanto, J. Rumbaugh, J. Blake, M. Bamberg and T. Hasan, *Br. J. Cancer* **70**, 474–480 (1994).
65. L. R. Duska, M. R. Hamblin, J. L. Miller and T. Hasan, *J. Natl. Cancer Inst.* **91**(18), 1557–1563 (1999).
66. I. Rizvi, T. A. Dinh, W. Yu, Y. Chang, M. E. Sherwood and T. Hasan, *Isr. J. Chem.* **53**(8–9), 776–787 (2012).
67. M. Mitsunaga, M. Ogawa, N. Kosaka, L. T. Rosenblum, P. L. Choyke and H. Kobayashi, *Nat. Med.* **17**(12), 1685–1691 (2011).
68. K. Sano, T. Nakajima, Choyke and H. Kobayashi, *ACS Nano* **7**(1), 717–724 (2013).
69. S. A. H. J. de Visscher, P. U. Dijkstra, I. B. Tan, J. L. N. Roodenburg and M. J. H. Witjes, *Oral Oncol.* **49**, 192–210 (2013).

Chapter 15

Endoscopic Photodynamic Therapy for Advanced Inoperable Esophageal Cancer

Joerg Lindenmann

Division of Thoracic- and Hyperbaric Surgery, Department of Surgery, Medical University of Graz, Auenbruggerplatz 29, 8036 Graz, Austria
jo.lindenmann@medunigraz.at

Abstract

Patients suffering from advanced esophageal cancer ineligible for curative surgery still have a poor prognosis. Photodynamic Therapy (PDT) represents a gentle and effective approach beneath the different palliative treatment options for this aggressive malignancy. The mechanism of function is based on the illumination of malignant esophageal tumor tissue after selective accumulation of the specific photosensitizer in these tumor cells. As a result of this photochemical reaction distinct necrosis of the endoluminal tumor load develops quickly and sustainably. In case of esophageal cancer, PDT is performed endoscopically under general anaesthesia. By means of PDT, considerable reduction of the endoluminal tumor load associated with increased quality of life can be achieved which helps conditioning the patient for further palliative treatment procedures. However, local endoluminal palliation with the aim to improve swallowing, decreament of dysphagia and therefore improvement of the patient´s quality of life are the main therapeutic goals of PDT in case of inoperable esophageal cancer. Therefore, PDT has become an accepted as well as safe, feasible and efficient endoscopic treatment option within the various multimodal treatment procedures for advanced inoperable esophageal cancer.

1. Introduction

Esophageal cancer, one of the least studied cancers, is the sixthleading cause of death from cancer worldwide.[1] There are remarkable geographical variations in the incidence of esophageal carcinoma from approximately 10 cases per 100,000 persons in Europe and the United States to 139 cases per 100,000 persons in some areas of China.[2–4] Esophageal squamous cell carcinoma represents one of the two most common tumor subtypes, which is mainly associated with tobacco smoking[5,6] and heavy alcohol use[5] but also with age >60 years, male gender, and heredity.[1–8] Esophageal adenocarcinoma represents the second histologic subtype which increasingly gains in both frequency and importance. It is mainly associated with chronic gastro-esophageal reflux disease and obesity.[9]

Up to now the conventional treatment modality for the cure of esophageal cancer is still esophageal resection either with or without radio-chemotherapy. These radical procedures are, however, associated with a high morbidity and mortality in these very patients, who are mostly in a reduced nutritional as well as performance status often because of accompanying diseases such as limited cardiac and/or pulmonary function or impaired liver function.

In general, the prognosis of patients with operable esophageal cancer depends on the depth of tumor penetration (T-stage) and the extent of lymph node metastasis (N-stage).[10] In patients with intra-epithelial cancer, the risk of lymph node metastasis is low, and resection of the carcinoma carries an excellent prognosis.[11] Nevertheless the overall 5-year survival rate of patients after curative esophagectomy for esophageal cancer is between 20% and 40% depending on the histologic subtype, respectively. However these results still represent a distressing outcome although many efforts had been undertaken during the last decades to improve this poor prognosis.[4]

In the vast majority of patients, esophagel cancer is diagnosed when these affected individuals are already symptomatic with higher levels of dysphagia. At this point in time, the disease has progressed to regional lymph node infiltration and/or distant metastases in most cases resulting in definitive ineligibility for curative resection. Considering this advanced tumor stage, these patients are provided for palliative treatment settings which aim at an improvement of swallowing, slowing of endoluminal tumor growth and treatment or prevention of tumor associated complications. The inability to swallow a regular diet is one of the most severe symptoms reducing quality of life for patients with an incurable disease and short life expectancy.[12,13] Therefore, the main goal in the palliation of patients with advanced inoperable esophageal carcinoma is a decrease in dysphagia which always results in an improvement of quality of life. Generally, the prognosis of esophageal carcinoma is poor in palliative settings by obtaining mean survival rates of only several months.[14–16]

Concerning this poor prognosis it is still required to search for further different treatment options which may help improving these disappointing results.

A variety of tumoricidal palliative measures allow at least a limited oral diet, but they require repetitive hospitalization. Many published reports discuss different combinations of tumoricidial procedures such as endoscopic brachy-radiotherapy and external beam irradiation,[17] chemotherapy,[18] endoscopic tumor desobliteration with laser technique or cautery,[19] stenting,[20] photo-dynamic therapy (PDT)[21] and combinations of these techniques. Therefore, in order to minimize the risk of procedure induced complications and increase the treatment benefit, different procedures should be combined in the treatment of these terminally ill patients. However, all physicians involved in the management of patients with advanced inoperable cancer of the esophagus know the agony and frustration these patients suffer from. For this reason PDT increasingly became the focus of attention in these patients.

In the course of time endoluminal PDT has developed to a promising non-thermal laser technique in the palliative treatment of advanced esophageal cancer. Generally the mechanism of function is based on the illumination of malignant tissue after selective accumulation of photosensitizers in tumor cells resulting in local tumor tissue necrosis.[22,23] Unlike radiotherapeutic or chemotherapeutic measures the method is not limited by cumulative doses.

Considering this mode of action, endoscopic PDT is leading to a targeted local destruction of neoplastic under preservation of normal esophageal tissue. As a result local endoluminal palliation with the aim to improve swallowing, decreament of dysphagia and therefore improvement of the patient's quality of life are the main therapeutic goals of PDT. After decades of clinical use, PDT is now considered a feasible and efficient minimally invasive endoscopic option for palliation of advanced esophageal cancer,[4,24,25] applied alone as sole treatment option or as part of a multimodal approach within the palliative treatment regimen tailored to the specific needs of the patient.[26–29]

2. Basics of PDT for Advanced Esophageal Carcinoma

2.1. *General Comments on PDT*

Basically in PDT for esophageal cancer a photosensitizer is first administered intravenously and then visible light is delivered endoscopically to the cancerous area. Light activation of the photosensitizer triggers a photochemical reaction resulting in the production of highly reactive oxygen species followed by immediate cell damage.

The detailed technique of cytotoxic action of PDT has been reviewed extensively and described elsewhere.[30–38]

The requirements for PDT are a photosensitizing drug, laser light of an adequate wavelength appropriate to the sensitizer and the presence of molecular oxygen which has shown to be crucial for the photochemical reaction.

The exact mechanism by which photosensitizers preferentially accumulate in tumor cells is unknown. Probably, a combination of multiple factors is important in causing this selective retention. Poorly developed tumor lymphatics may prevent removal of sensitizers from tumor cells. Many sensitizers are relatively hydrophobic and bind to lipoproteins within the plasma. Increased expression of lipoprotein receptors on tumor cell membranes may increase photosensitizer concentration within these cells.[39]

Other possible factors include altered pH in tumor cells, tumor neovascularization, and the increased metabolic activity found in tumors.[30,40] At this point, it is important to know that in the absence of light activation, photosensitizers have no inherent effect on tissue.

After infusion, the photosensitizer binds to the cellular plasma membrane. Over time, it is internalized and binds to organelle membranes, especially those of mitochondria. These membranes are the initial target of singlet oxygen. Damage to these membranes causes the release of cellular contents, including inflammatory mediators, which ultimately leads to cell death. The cells of the vasculature appear to be the initial target of photosensitizers: Damage to the tumor microvasculature produces localized tumor ischemia and cell death. The extent of tumor destruction that occurs from direct tumor cell killing versus ischemic death from destruction of tumor vasculature is unknown.

2.2. *Photosensitizers for PDT of Advanced Esophageal Carcinoma*

The ideal photosensitizer would have the following properties: High selectivity and concentration in tumor compared to normal tissue, selectable depth of tumor penetration, high yield production of singlet oxygen, and minimal side effects, especially a low skin photosensitivity.

This type of photosensitizer is not yet available; however active research on many different types of photosensitizer is ongoing with the goal to define a photosensitizer for early cancer and another one for advanced cancer treatment each fulfilling the criteria mentioned above.

Within the following paragraphs the three current photosensitizing drugs available for esophageal PDT are described. Considering their different properties and their use in daily routine, porfimer sodium has to be proven to serve as the most appropriate photosensitizer for advanced esophageal cancer.

2.2.1. *Porfimer sodium*

The compounds are oligomeric mixtures of dihematoporphyrin ethers and esters with very similar photochemical properties. Porfimer sodium represents a purified form of the first clinically used photosensitizer, hematoporphyrin derivative (HpD).

Peak light absorption with porfimer sodium occurs at around 400 nm; however, tissue attenuation at this wavelength is high, limiting the depth of tissue penentration of light and hence of the extent of tumour necrosis. A secondary, lower absorption peak is present around 630 nm, a longer wavelength less attenuated by tissue pigments and hemoglobin. The decreased activation at this wavelength is compensated by increasing the dose of photosensitizer or light. The depth of effective treatment with porfimer sodium at 630 nm varies between 2 and 4 mm. In case of esophageal cancer studies have demonstrated that treatment with 630 nm light can produce full-thickness inflammation and tissue injury.[41]

Although porfimer sodium is selectively retained in tumor tissue compared to normal tissue at a ratio of at least two to one, the drug does accumulate to significant levels in the skin and in the reticulo-endothelial system, especially of the liver and spleen resulting in photosensitivity lasting 4–6 weeks. The standardized dose of porfimer sodium for PDT of esophageal carcinoma is 2 mg/kg body weight by slow intravenous injection, 48 h prior to irradiation.[42]

2.2.2. *Meta-tetrahydroxyphenyl-chlorin (mTHPC)*

mTHPC, is a second generation photosensitizer with a shorter period of skin photosensitivity (2 weeks), a longer activation wavelength (652 nm), and therefore increased depth of effect, higher yields of singlet oxygen, and better tumor selectivity. This compound was typically used intravenously at a dose of 0.15–0.3 mg/kg body weight. Early results in regulatory trials, studying palliative treatment of patients with squamous cell carcinoma of the head and neck were complicated by extensive tissue necrosis, tissue breakdown and stricture formation probably related to excessive light doses. Phase IIb studies in head and neck cancer have been completed, and mTHPC has also been studied in bronchial and esophageal tumours with encouraging results.[43,44]

2.2.3. *5-Aminolaevulinic acid (5-ALA)*

5-ALA, is an early intermediary in heme biosyntheses. It is formed by the combination of glycine and succinyl CoA. The formation of 5-ALA is the rate-limiting step in the heme pathway and is catalyzed by 5-ALA synthase. 5-ALA is converted to porphobilinogen, porphyrin intermediates via porphobilinogen deaminase, then Protophorphyrin IX (PpIX). The final step in this pathway is conversion of PpIX to heme, catalyzed by ferrochelatse. Under normal conditions, 5-ALA production downregulates 5-ALA synthase, limiting heme synthesis; however, administration of large doses of 5-ALA results in intracellular accumulation of PpIX. It has been shown that PpIX accumulates in greater concentrations in adenocarcinoma of the esophagus compared to normal tissue because of a relative increase in activity of porphobilinogen deaminase compared to ferrochelatase.[45]

As a photosensitizer, 5-ALA has several advantages compared to other photosensitzers. PpIX is produced in greater levels in the mucosa compared to the submucosa or muscularis propria in the esophagus, making 5-ALA useful for treatment of superficial lesions.[40,46–48]

5-ALA may be administered orally, and the interval between drug administration and activation by laser light is only 6 h. Skin photosensitivity lasts only for 48 h. However, the oral administration of 5-ALA with necessary doses of 30–75 mg/kg body weight, may lead to nausea and vomiting. This is why the application of an antiemetic drug is recommended prior to photosensitization.[40,46]

2.3. ***Light Delivery***

Dosimetry refers to the calculation of light delivery to the tumor. Dosimetry calculations are complex and must take into account multiple factors involving the respective photosensitizer, the method of light delivery, and characteristics of the tumor itself.

One of the first steps in evaluating PDT was to analyze its light dosimetry, or the effects of various light doses on the degree of tumor necrosis. After testing a variety of light doses, a correlation between tissue dose of light (J/cm^2) and the resulting depth of tumor necrosis could be found.[49]

The tissue dose of light varies inversely with the luminal diameter of a tumor segment (when the light dose delivered from the fibre tip remains constant) and was calculated by figuring the surface area of an open cylinder. By using this method of analysis and working backward from tissue doses, the light dose of 300 J/cm at the fibre tip was determined to give a safe, yet effective, range of tumor necrosis.[27,28] This light dose was still lower than that which had been previously used. Furthermore, it was reported that the higher light doses previously used were associated with complications (e.g., pleural effusions, mediastinitis, fistula).[50] The dose of 300 J/cm at the fibre tip for palliative therapy of malignant esophageal obstruction represents a standardized light dose for PDT.

Several authors have studied the relationship between the extent of tumor necrosis and light dose: They were able to show that the results are consistent with a threshold effect.[51] Bown and co-authors revealed a logarithmic relationship between the extent of tumor necrosis and applied energy fluence rate.[52] However, in all of these studies, only experimental tumors, not human carcinomas were investigated.

In clinical studies, very high-energy fluence rates up to 4,800 J/cm^2 were used in the treatment of colorectal carcinomas.[53,54] In contrast, Gossner and co-workers showed that colorectal carcinomas react very sensitively to PDT, requiring low energy doses (25–75 J/cm^2) for effective tumour destruction. They showed that high-energy fluence rates ($>$300 J/cm^2) were associated with undesired side effects such as hyperthermia.[55] These observations were corroborated by the results of a quantitative study for colorectal cancer.[56] Similarly, high-energy fluence rates were applied to gastric (960 J/cm^2) and esophageal carcinomas (6,000 J/cm^2) in other clinical studies.[50,57] Interestingly, energy fluence higher than 100–150 J/cm^2 did not markedly increase the PDT effect. Thomas and colleagues found an increased risk for complications by using power higher than 1.5 watt but were not able to enhance the effectiveness of PDT in tumors by irradiating at the highest doses.[50]

These studies demonstrate that the prevailing clinical dosimetry for gastrointestinal carcinomas may result in inadequate treatment by under- or overdosage of the applied energy fluence rate in PDT. However, the dose dependence of light cannot be discussed on the basis of energy fluence alone. On a more fundamental scientific level, one has to keep in mind that the propagation of light in tissue depends on optical parameters: One of them is the absorption coefficient, which varies significantly in different tissues because of the presence of various light absorbing chromophores (e.g., hemoglobin and melanin).[58] Therefore tumors containing a large amount of blood and tissues with great amounts of melanine have a high absorption coefficient.

A further optical parameter is the scattering coefficient, which depends on the structural pattern of tumor tissue.[59] It is easy to understand that different organ tumors might exhibit a variety of structural matrices of differential blood flow. These factors could explain the different reactions of gastrointestinal tumors to the same light dose. The optical parameters of esophageal carcinomas vary with different tumor diameters because of changes in vascular supply and necrosis in its centre. The effects these parameters may have on the therapeutic results are expressed by different penetration depth of light at different tumor diameters.[60] Due to the changing optical properties resulting from increased tumor diameter, dosimetry needs to be adapted to achieve proper and comparable results after PDT. In other words, an increasing tumor diameter requires an increasing treatment time of a standardized light dose, depending on its correction factor.[60]

2.3.1. *Light delivery systems*

Light delivery systems are generally fibre-optic devices adapted for endoscopic use, varying from a quartz light-diffuser fibre placed through the working channel of the endoscope to specially modified diffusers that are put in place with the help of fluoroscopic guidance. For esophageal carcinoma the cylindrical diffuser fibre is commonly used. It is positioned under endoscopic or fluoroscopic guidance within the lumen of the esophagus. It is necessary to centrally align these applicator systems within the lumen of the esophagus, which in clinical practice is possible only in presence of high-grade stenosis. Furthermore or they allow light delivery only over a limited distance and thus segmental light application is needed in longer tumor segments.[61]

Additionally, this technique has the disadvantage that it is not possible to flatten the mucosal folds of the esophagus, and thus some areas may be shielded from the light as the configuration of the lumen continually changes with esophageal peristalsis and respiratory movement.[62]

This is thought to produce excessive light dosing in some areas, with the risk of stricture and/or inadequate light dosing to other areas with incomplete mucosal ablation. Modern light systems are optimized to deliver the desired light intensity distribution to the targeted region with minimal losses. In the esophagus this implies the use of a device that is adapted to the shape of the hollow organ, such as an elastic balloon catheter.[63]

Superficial light application should therefore always be done by the use of light applicator systems which guarantee a standardized distance of the irradiation source from the surface of the tumor and eliminating the shadowing phenomenon, described above, of a hill-and-valley effect caused by mucosal folds.

Considering the variables for proper calculation of light dosimetry and the irregular endoluminal surface of the tumor, it is impossible to treat the tumor site accurately only by using the light diffuser inserted through the working channel of the endoscope. Based on the findings of optical parameters and considering all the above mentioned aspects, a very simple and save dosimetry formula was developed and approved in clinical practice of more than 300 esophageal PDT's.[60] The dosimetry formula is based on the use of a standardized light applicator of different length and includes the following variables: The diameter of the tumor stenosis and distance of the irradiation source from the tumor surface, the length of the irradiation source and appropriate light applicator system, the optical parameters of esophageal carcinoma and finally, the desired light dose (J/cm^2).

For interstitial illumination the optical fibre can be introduced through the biopsy channel of the endoscope and the cylindrical diffusing end section is placed into the tumor itself under direct vision.[64] Despite sufficient tumor eradication especially in tumor masses that protrude into the lumen, the risk of perforation and/or penetration might be higher compared to superficial treatment.

2.4. *Oxygen Supply for Esophageal PDT*

Both the photochemical reaction and the efficacy of PDT mainly depend on the presence of molecular oxygen. There seems to be a decreased cell sensitivity to PDT in the presence of low oxygen, which means that hypoxic cells are resistant to PDT.

Furthermore, animal tumor models have demonstrated a decreased effect to PDT under hypoxemic conditions, because hypoxemic tumor cells seem to be less affected by porphyrins and light.[65,66]

PDT-induced cytotoxicity most likely occurs through photo-oxidative reactions, and there are two major reaction pathways.[67] Type I photo-oxidation involves a direct reaction of the excited sensitizer with a substrate by a mechanism involving hydrogen or electron transfer to yield transient radicals that further react with oxygen. In the Type II photo-oxidative reaction, energy transfer occurs from the excited triplet state of the sensitizer to molecular oxygen producing singlet oxygen, which can further react with substrates susceptible to oxidation.

Uncommonly, electron transfer from the sensitizer generates a superoxide radical through the activation of hypoxanthine/xanthine oxidase system by catabolism of high-energy phosphate damage of mitochondria and a calcium-dependent protease conversion of xanthine dehydrogenase to xanthine oxidase.[68]

Considering these effects, too high fluence rates of exposure light will lead to oxygen depletion: Using transcutaneous oxygen electrodes, Tromberg and co-workers reported that tumor oxygen tension becomes irreversibly low with large PDT fluences (fluence is a measure of time-integrated particle flux, expressed as particles per square centimeter).[69]

On the other hand, too low fluence rates need a long exposure time and lead to vascular shutdown which also cause hypoxia of the tumor tissue.[35] Therefore, the presence of molecular oxygen in tumor tissue is crucial for the effectiveness of PDT.

In clinical practice different approaches are used to enhance tissue oxygen saturation and thereby the effect of PDT. In literature oxygen administration during PDT was described as simple nasal oxygen support of only 3–4 l/min,[70] general anaesthesia with intubation and FiO_2 of 0.21–1.0[27,28,40,46,71] and additional hypberbaric oxygenation with enhanced effectivity.

At this time the general recommendation for clinical practice in PDT application is general anesthesia with oro-tracheal intubation and oxygen supply using 100% pure oxygen.[6–28,71]

3. Practical Considerations for PDT of Advanced Esophageal Carcinoma

3.1. *Indication for PDT*

PDT as an important component of the multimodality treatment in anatomically and/or oncologically non-resectable esophageal cancer with palliative intention. Therefore, PDT is used in a meaningful combination with other endoluminal treatment options, with chemo- and/or radiotherapy, respectively.

3.2. *Contra-indications for PDT*

- Porphyria,
- Renal insufficiency,
- Liver insufficiency,
- Pre-existent fistula formation,
- Allergy against photosensitizing drug,
- Tumor involvement of the trachea, the carina or the main bronchi,
- Patient's incompliance regarding required protection from light after photosensitation,
- Terminally ill patients.

3.3. *Clinical Examination and Diagnostic Work-up*

Patients history, recording of dysphagia grade and determination of the Karnofsky performance status, physical examination, routine laboratory parameters including hepatic and renal function.

3.4. *Clinical Staging*

- Esophagogram using water soluble contrast medium
- Flexible esophago-gastroscopy
- Flexible tracheo-bronchoscopy
- Computed tomographic scan of the chest and abdomen
- Abdominal ultrasonography
- Endoscopic ultrasound
- Electrocardiogram
- Cardiac ultrasonography and spiro-ergomety

Before PDT can be considered as treatment of choice, both flexible esophago-gastroscopy and tracheo-bronchoscopy have to be performed to rule out possible tumor infiltration of the adjacent trachea, the carina and the main bronchi. If tumorous involvement can be confirmed, PDT must not carried out due to the considerably increased risk of perforation resulting in a malignant esophago-respiratory fistula.

3.5. *Patient Information and Education Before PDT*

Extensive and complete patient education is of supreme importance in PDT because of the persistence of photosensitivity after drug injection. Patients are instructed to wear opaque clothing that covers most of their skin, including long-sleeve shirt and long pants, gloves, wide-brimmed hat, and sunglasses, when they report for their injection. Because of the persistence of photosensitizer in the skin for several weeks the patients must wear similar clothing when going outside during this time. The daily application of a potential sunblocker with high sun protection factor is mandatory to avoid any event of sunburn. Due to the increasing intensity of the sunlight, we are currently using a special type of sunblocker with high sun protection factor (Daylong® extreme-SPF 50+, Egerkingen, Suisse).

Inside, the patients are instructed to avoid direct sunlight and bright indoor lighting, although indirect sunlight is acceptable and probably shortens the duration of photosensitivity because of photo-bleaching (removal of the photosensitizer from the skin caused by low-intensity light exposure). Topical sunscreens are generally ineffective in preventing photo-toxicity because they block only ultraviolet light.

During endoscopic laser-light treatment the patient is covered with drapes, taking care to protect also the face. As an alternative, a total sun-blocker can be applied on the face.

If a patient requires surgery after PDT, drapes must be placed over the skin, liver, and spleen because of the risk of photo-toxicity generated by bright operating room lights.

After the average time of skin photosensitivity has elapsed, the patients are instructed to expose the back of one hand to sunlight for 5 min. If no redness or swelling occurs, they can gradually increase their exposure to sunlight over time.

3.6. *Photosensitization*

Intravenous photosensitization of porfimer sodium is recommended at a dose of 2 mg/kg body weight, 48 h prior to irradiation as we commonly do at our department. Before intravenous administration the cannula should be checked with a test of 20 ml standardized physiologic saline solution because para-venous injection may cause necrosis of the surrounding tissue accompanied by severe pain. Topical application of a sunblocker immediately after photosensitization as well as the above-mentioned general recommendations for light protection are commonly sufficient to prevent sunburn.

3.7. *Endoscopic Illumination Procedure*

48 h after uneventful photosensitization, esophageal PDT should be done under general anesthesia with endotracheal intubation and appropriate oxygen support. Routine cardio-respiratory monitoring including electrocardiogram, non-invasive continuous blood pressure control and pulsoxymetry should be provided (Fig. 1).

The endoscopic placement of the light delivery system is best done with the help of a flexible endoscope. We are using balloon catheter (PhotoDynamicTherapy®, Vienna, Austria) which are inserted through the biopsy channel of the flexible endoscope (Fig. 2). In case of very advanced esophageal cancer with subsequent extensive tumor stenosis, stepwise endoscopic bouginage using the Savary-Gillard device may become necessary before PDT can be performed.

A flexible guide wire is passed under radiological guidance through the endoscope and careful and gentle bouginage is done enabling the passage

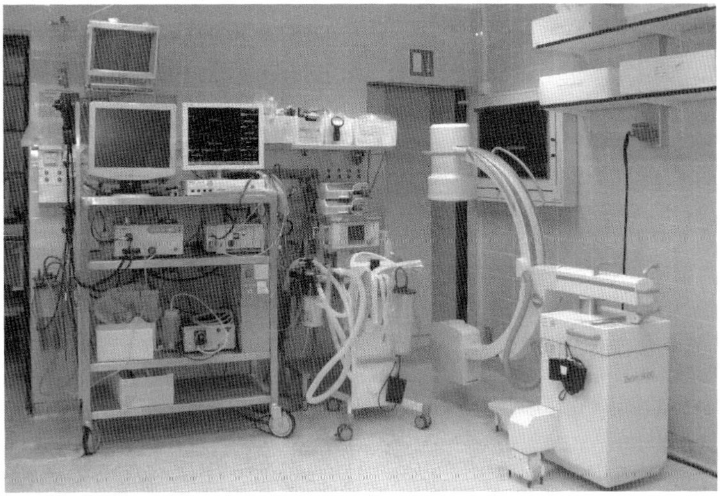

Fig. 1. Multifunctional endoscopy unit provided with anesthesiological working station and X-ray fluoroscopic equipment for endoluminal dilatation and stenting.

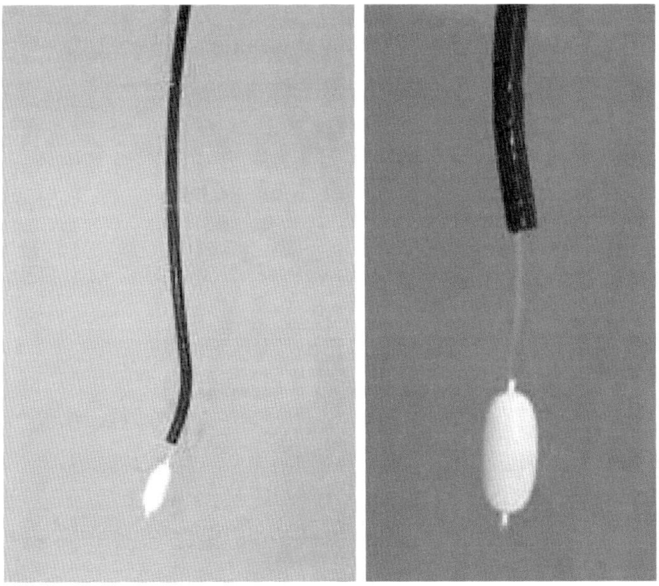

Fig. 2. Laser application system (expanded balloon catheter) inserted through the biopsy channel of the flexible endoscope.

through the tumor by the flexible gastroscope. In the same intervention suspicion of perforation can be excluded by esophagogram using a water-soluble contrast medium. Furthermore, it is necessary to pass the entire esophageal carcinoma with the endoscope in order to perform exact

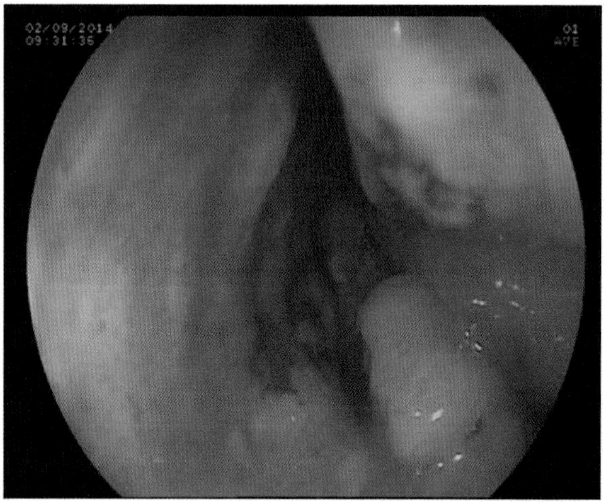

Fig. 3. Histologically proven esophageal squamous cell carcinoma causing considerable endoluminal tumor stenosis.

measurement of the entire extent of the tumor reaching from the beginning of the upper border to the end of the lower border of the tumor (Fig. 3). Meticulous measuring is essential for determination of both the type of balloon catheter and the duration of PDT, respectively.

Afterwards the flexible endoscope is placed above the carcinoma and the laser application system, using a balloon catheter is inserted in a folded condition through the biopsy channel. The folded balloon is placed at the distal aspect of the tumor. After instillation of distillated water the balloon becomes unfolded and is applied very close to the surface of the tumor allowing homogenous light distribution (Fig. 4).

Afterwards the expanded balloon is withdrawn proximally in order to treat the entire tumor length. Withdrawal of the expanded balloon occurs step by step corresponding to the length of the used balloon which amounts usually 4 cm. The required time for each placement of the balloon is approximately 20 min. which is calculated by the help of the dosimetry formula developed at our department as mentioned above.[60] The light dose is calculated as 100 J/cm. Light intensity of 630 nm is applied by a KTP-Nd:YAG laser with DYE-box (Laserscope; Surgical Systems, Gwent, UK). For safety reasons, wavelength and light dose at the tip of the light diffuser should be controlled before and after PDT. Depending on the topography and length of the tumor, single or multiple placements of the balloon may be necessary in order to illuminate the entire tumor length (Fig. 5). Therefore the required irradiation time may range from at least 20 min up to 90 min. After PDT, a conventional chest X-ray is done to rule out any pneumothorax or pleural effusion.

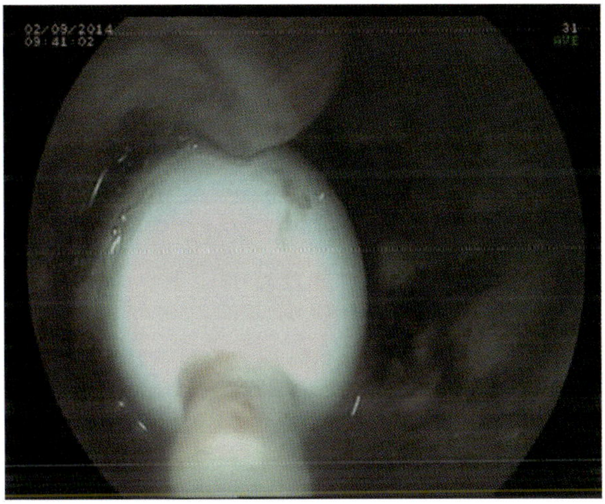

Fig. 4. Completely expanded balloon catheter with broad contact to the tumor surface immediately before illumination.

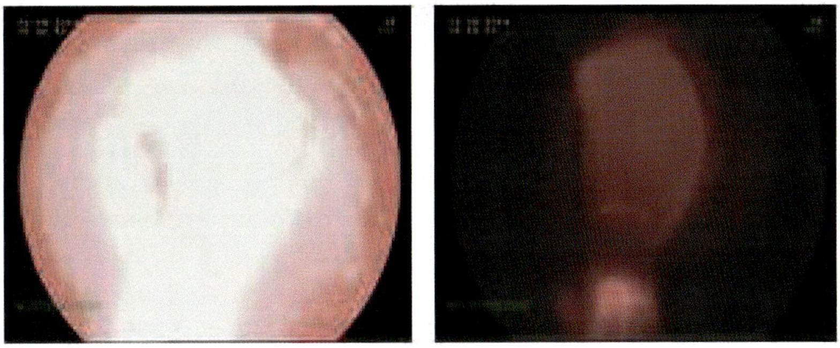

Fig. 5. Expanded balloon catheter during illumination procedure using red laser light.

5–7 days after successful PDT, flexible esophago-gastroscopy under short-term anesthesia is repeated in order to verify the therapeutic response (Fig. 6). In the course of this endoscopic intervention supernumerous necrotic tumor tissue may be removed mechanically by using the forceps or by mere suction when necessary. The depth of tumor necrosis is determined indirectly by the visible increase in luminal diameter measured at the maximal point of constriction or directly by determination using endoscopic ultrasound if available. Especially in large carcinoma with considerable endoluminal tumor stenosis a second laser-illumination after necrotic debridement may become necessary and is therefore recommended by various authors.[27,28,40,46,71] The persistent photosensitation can be used therapeutically by applying the second

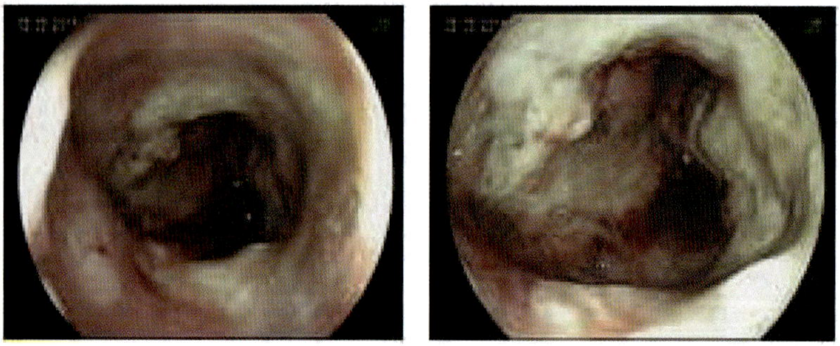

Fig. 6. Extensive endoluminal tumor necrosis one week after PDT resulting in reopening of the esophageal lumen.

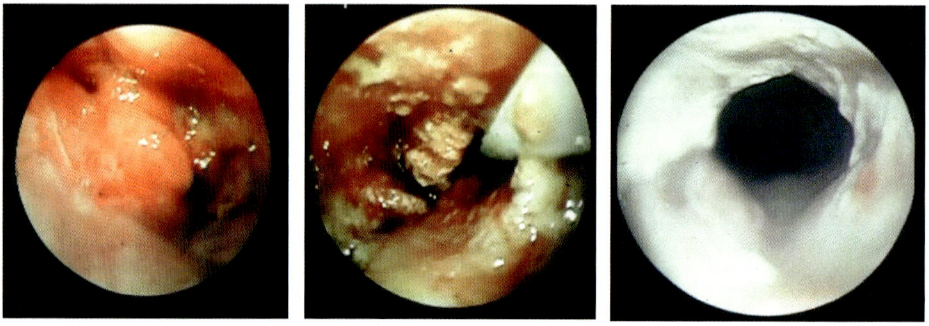

Fig. 7. Three months follow-up of advanced esophageal carcinoma before, five days and three months after PDT: Notable tumor necrosis on the 5th post-interventional day and patent esophageal lumen two months later.

illumination several days after the first without reapplication of a photosensitizing drug.

3.8. *Follow-up after Esophageal PDT*

After effective PDT the patients undergo repetitive endoscopy after one month and thereafter every three months for the first year (Fig. 7).

CT scans of the thorax and the abdomen are performed every six months.[27,28,40,46] However, increasing dysphagia at follow-up is still the indication for repetitive PDT treatment. In case of endoscopically verified local tumor progression, the photosensitizer is injected again and PDT can be repeated in the same manner mentioned above. In this context it has to be mentioned that repeated PDT should never be considered before at least one month has passed after first PDT because of the high risk of perforation due to overwhelming tumor necrosis induced by PDT.

4. PDT Related Complications and Their Management

Although PDT represents a highly safe and efficient treatment option, it carries some risks which should be always kept in mind.

Notwithstanding the used photosensitizer porfimer sodium is selectively retained in tumor tissue compared to normal tissue at a ratio of at least two to one, the drug does accumulate to significant levels in the skin and in the reticulo-endothelial system, especially of the liver and spleen. This results in photosensitivity- mainly prolonged skin photosensitivity for at least six weeks is documented. However, strict adherence to the detailed precautions mentioned above decreases the rate of sun burn to lower than 10%.[26]

Minor side effects such as nausea, vomiting, dysphagia, odynophagia, epigastric pain, chest pain, transient elevation of white blood cell count as well as C-reactive protein and temperature frequently occur in the very post-interventional period following PDT. These symptoms are reversible and can be efficiently managed by intravenous administration of analgetics, anti-inflammatory drugs, proton-pump inhibitors, and prophylactic broad spectrum antibiotics for several days. Unilateral or bilateral pleural effusions are rare and usually resolve spontaneously over several days.[27,28,71,72] In some cases insertion of a chest tube drainage followed by intravenous broad-spectrum antibiosis may become necessary. However, flexible endoscopy (esophagoscopy as well as bronchoscopy) should be added in these cases to rule out possible PDT-induced perforation.[26]

After PDT inflammation-related mucosal edema develops within the first 72 h leading to increasing mucosal swelling and therefore decrement of the esophageal lumen. This PDT-related condition may be worsened by necrotic tumor tissue secondary obstructing the esophageal lumen which may cause high grade dysphagia or complete aphagia. Besides intravenous administration of anti-inflammatory drugs, fexible esophagoscopy becomes necessary and the necrotic tissue is removed mechanically by using the forceps or by mere suction if required. Sometimes it may be helpful to push the necroses gently into the stomach resulting in quick reopening of the esophageal lumen.

However after PDT when necrotic tumor tissue finally sloughs, esophageal perforation remains a serious risk which must not be underestimated. Though the depth of tumor necrosis after porfimer sodium is limited to approximately 5 mm, the extent of tumor necrosis may develop imprecisely. Even though quick development of tumor necrosis within the first 48 h after PDT is desired and in these patients suffering from tumor-related high grade dysphagia required, there is a slightly increased risk of esophageal perforation when PDT-related tumor necrosis occurs too fast and too distinct. However, perforation rates are low at 0–8%[73,74] and.[26,75] Moreover, if PDT is conducted in centers which are experienced in this treatment, the therapy-related mortality rate is less than 3%.[26,76]

Due to their anatomical location, advanced tumors of the esophagus located at the level of the trachea and/or the bifurcation, are likely to penetrate into the left main bronchus, the bifurcation or the trachea. Therefore, the possibility of an affection of the tracheo-bronchial tree also has to be considered after PDT for esophageal cancer. In general, the risk of perforation and/or penetration increases with the T-stage of the tumor.[27,28,40,46,72] In case of persistent clinical symptoms after PDT such as fever, cough, aspiration, pneumonia, and other clinical signs of sepsis, an esophageal perforation and/or penetration into the tracheo-bronchial tree should be taken into consideration. Esophagoscopy and bronchoscopy should be performed immediately followed by computed chest tomography with oral and intravenous contrast. In presence of perforation, in patients with anatomically resectable carcinoma palliative esophagectomy should be considered provided that the patient would be fit enough for this major surgical approach.[26,77,78] In terminally ill patients and/or in case of non-resectable cancer, esophageal stenting represents an effective approach with good results enabling immediate closure of the perforation.[20,78–81] In case of distinct esophago-respiratory fistula, sandwich stenting[82] may also be a safe and feasible option to seal the target site. In this context it has to be pointed out that stenting should only be applied in well selected cases of patients with non-resectable esophageal cancer offering better chance of survival, particularly in frail patients with high comorbidity. Furthermore, supportive measures such as antibiotics, intensive care and drainage of pleural and mediastinal fluid accumulations are strictly recommended.

Esophageal stricture formation has been reported as a late complication after PDT, its incidence ranging from 4.8–53%.[27,28,71,73] The high variability depends on the tissue involved, the light intensity and the photosensitizer used. Fortunately, the majority of strictures are mild and readily respond to gentle endoscopic bouginage using the Savary–Gillard device as we commonly do at our department. Direct injection of triamcinolone into the dilated strictures may reduce stricture re-formation. Interestingly, oral prednisone did not reduce the incidence of post-PDT strictures (29%) if compared to patients who underwent PDT without oral steroid medication (16%).[83]

5. PDT in Combination with Other Treatment Modalities

Besides endoscopic PDT there are many treatment modalities, such as endoluminal procedures and external beam irradiation as well as chemotherapy which can be used in the treatment of advanced inoperable esophageal carcinoma. A combination of these different measures is frequently applied because PDT can be included into any local or systemic treatment protocol along with other modalities without interference. The guiding principles are that PDT is more selective than other methods of treatment and that it can be performed repetitively without a limitation of the cumulative dosage to the mucosa.[26,64,73,75] Statistical analysis revealed PDT as a safe and effective modality to be the most significant treatment for increasing esophageal luminal diameter and decreasing dysphagia which helps conditioning the patient for further palliative treatment procedures.[26,27,64,75,84] In contrast, PDT cannot be expected to eliminate bulky tumor outside the esophageal lumen or tumor within para-esophageal lymph nodes beyond the reach of the laser beam. Therefore additional chemotherapy and/or external beam irradiation is required according to the multimodal approach mentioned above.

5.1. *PDT Combined with Nd: YAG Laser*

Beneath the different palliative endoscopic treatment options, thermal tumor ablation using the Nd:YAG laser enables quick relief of tumor-related dysphagia but the depth of penetration is unpredictable and has therefore been associated with increased risk of esophageal perforation.[75] If compared to this thermal laser technique, which usually affects all parts of the tissue, damaging collagen fibrils and esophageal smooth muscle, PDT is more selective, sparing the collagen fibrils and enhances healing predominantly by regeneration.[85] In addition, PDT shows a significantly lower rate of perforations and a better response rate.[86] In this context, a second study confirmed that the duration of response was found to be longer with PDT, respectively.[87]

5.2. *PDT Combined with Brachy-Radiotherapy*

Moreover, PDT can also be applied in conjunction with radiotherapy to improve survival and quality of life.[27,28] In patients with advanced disease and close adherence of the tumor towards the trachea and/or the main bronchi, caution is advised.

If these tumors are treated aggressively with a combination of PDT as initial treatment followed by radiotherapy, they are at a higher risk of developing major complications such as fistula formation to the tracheo-bronchial tree. Therefore, aggressive endoluminal treatment including PDT and/or brachy-radiotherapy and external beam irradiation should not be combined as first line treatment. In such cases it is recommended to start the treatment with external beam irradiation and/or chemotherapy. Because radiotherapy induces peri-esophageal fibrosis preventing overt perforation, endoluminal PDT and brachy-radiotherapy can be done at a low risk of major complications several weeks later.

On the contrary to patients with esophageal carcinoma located distal the tracheal bifurcation, PDT as first-line treatment in combination with radiotherapy is feasible. In this case the incidence of fistula is unlikely due to fibrosis of the surrounding fat and connective tissue which form a protective layer.[4]

In a large series of 119 patients, our study group published the combination of PDT and radiotherapy.[28] 44 patients received PDT followed by brachy-radiotherapy, 75 were treated with brachy-radiotherapy alone. In both groups, some patients received external beam irradiation. At 3-months follow-up the performance status had increased in both groups. A decrease of tumor related stenosis and improvement in dysphagia was observed in all patients, with a significant improvement in those treated with additional PDT. Further it has been shown that the mean survival of patients undergoing combined treatment with PDT was significantly higher.[4]

It has to be stressed again that in patients suffering from advanced inoperable esophageal cancer an increase in quality of life always has to be the main goal of palliative therapy. Therefore sustainable improvement of swallowing and slowing of intraluminal tumor growth are the cornerstones within the treatment regimen. Regarding the considerably decreased residual life expectation in these very patients, accurate prognostication still remains a notable component for the further course of these individuals. Consecutive and appropriate decisions on interventional and non-interventional procedures during palliative care are of utmost importance for these end stage cancer patients.

Regarding these requirements mentioned above the different treatment options have to be tailored to the specific particular needs of each patient considering the loco-regional endoscopic situation, the tumor stage and the individual co-morbidity.

In this context an individualized, multimodal treatment regimen comprising a variety of interventional and non-interventional palliative procedures has been established at our department.

In a large series of 248 patients with advanced inoperable esophageal cancer, our study group wanted to evaluate the therapeutic effect of our arranged multimodal approach in due consideration of endoscopic PDT.

The therapeutic approach consisted of PDT, high-dose-rate endoluminal brachythcrapy, external beam radiotherapy, systemic chemotherapy, esophageal dilatation and stenting procedures, respectively. Feeding enterostomy and palliative esophageal resection were done in the minority of these patients. The different treatment modalities were applied in combination or subsequently in order to maintain best possible individual results for each patient with preferably sustainable tumoricidal effect. The use of PDT which causes a quick relief of dysphagia, was the central mainstay in our treatment schedule, applied in 69% of those patients. Therefore, PDT was set as first-line treatment in the majority of those patients (69%), followed by three sessions of high-dose-rate brachytherapy. External beam radiotherapy was done after brachytherapy had been completed. However the mean survival time yielded in this study collective was 34 months. Interestingly, if PDT was used as first-line treatment, median survival was significantly prolonged to 50.9 months compared to 17.3 months if other options were used as initial modality.

According to these findings, on may conclude that palliative treatment of esophageal carcinoma may result in prolonged survival time with an acceptable quality of outpatient life, if a multimodal approach is chosen starting with PDT as first-line treatment. If high-dose-rate brachytherapy is scheduled three weeks after PDT, an additive anti-tumoral effect can be achieved. Additional external beam radiotherapy will further sustain the tumoricidal effect. Chemotherapy can be beneficial in patients with distant metastasis who present with good general condition. Stenting should be reserved for complications, for patients who are unfit for other treatment modalities or for recurrent stenosis not manageable otherwise. In this study there is strong evidence that PDT, applied as initial endoluminal treatment in patients without gross tumour infiltration into the mediastinum, the great vessels or the tracheo-bronchial tree, enables a considerable beneficial effect in the palliative setting resulting in significantly improved and mainly prolonged survival.[26]

This extraordinary improvement in survival might be due to PDT's induction of immune responses which already had been demonstrated in various experimental settings and in animal-based studies.[32,88,89] However, the effects of PDT on immune function, which may depend on the type of immune response, are still controversial. Considering this presumption controlled studies on the clinical impact of PDT on the immune system of human patients suffering from (advanced) malignancies are required.

In this context we conducted a clinical pilot study where we investigated the immunological changes induced by endoscopic PDT and the effect of PDT on level and function of regulatory T cells in patients with invasive esophageal squamous cell carcinoma. Therefore we analyzed both patient's blood and tissue samples before and after PDT. We could observe that PDT downregulated the immuno-suppressive function of peripheral regulatory T cells whereas their levels did not change significantly after PDT.

While we observed no change in absolute lymphocyte numbers, PDT affected systemic leukocyte levels. We could verify statistically significant increases in neutrophil (granulocyte and monocyte) numbers, as they have been found to play a key role in PDTinduced anti-tumor immunity *in vivo* and *in vitro*.[90–93] Furthermore a statistically significant increase in serum Interleukin-6 levels could be detected. On may conclude that these results enrich the understanding of the immunological events linked to clinical PDT especially in esophageal cancer. However, further investigations are warranted to elucidate the complicated immune profile elicited by PDT and translate it into improved therapy for patients with esophageal cancer.[94]

5.3. *PDT Combined with Hyperbaric Oxygenation (HBO)*

Finally it has to be mentioned that endoscopic PDT for esophageal carcinoma can be combined with additional HBO which significantly increases the rate of tumor necrosis. Considering the interactions of photosensitizer, light and oxygen, an enhanced tumoricidal effect may be achieved by increasing the amount of oxygen available for the photochemical reaction. This concept is of crucial importance, as PDT by itself induces reduced blood flow and causes a shutdown of tumor vessels resulting in hypoxia with decreased oxygen tension leading to a decreased effect of PDT because hypoxemic tumor cells seem to be less affected by porphyrins and light.[37,38]

Therefore, the use of HBO under 2 ATA (atmospheres absolutes) in this concept increases the oxygen tension from below 75 mmHg up to 750 mmHg. Under HBO, oxygen physically dissolves in all fluid components of the body, resulting in the fact, that oxygenation is no longer dependent on the presence of red blood cells. The use of HBO in this particular field of cancer treatment could be the key to obtain higher levels of molecular oxygen in tumor tissue in order to increase cytotoxicity followed by increased development of tumor necrosis. However the results of experimental studies conducted in tumor-bearing nude mice corroborated that HBO improves the efficiency of PDT by increasing the depth of tumor cell damage.[66,95]

According to those animal-based studies, experimental clinical pilot studies in humans were conducted.[72,84,96] Those promising results suggested that supportive HBO for PDT might accelerate the photodynamic reaction by raising the transmission efficiency of light energy, increasing the quantum amount of oxygen. Furthermore, the radius of effective distance is extended resulting in increased depth of penetration, which is a very important instance to improve the clinical outcome of the patients and however, enabling prolongation of their residual survival time.

According to recent literature, several experimental studies have confirmed the positive effect of PDT combined with adjunct HBO, which may help to enhance the therapeutical effect of PDT in case of advanced inoperable esophageal cancer.[72,84,96] By the help of supportive HBO, satisfactory reduction of the endoluminal tumor load can be achieved and it helps conditioning the patient for further palliative treatment procedures.

However, despite the undisputable benefits of HBO during PDT, this technique requires additional expertise in hyperbaric medicine and is therefore unlikely to become widely adopted outside of specialized HBO centres.[39]

References

1. P. C. Enzinger and R. J. Mayer, *N. Engl. J. Med.* **349**(23), 2241–2252 (2003).
2. L. Corti, J. Skarlatos, C. Boso, F. Cardin, L. Kosma, M. I. Koukourakis, A. Giatromanolaki, L. Norberto, M. Shaffer and K. Beroukas, *Int. J. Radiat. Oncol. Biol. Phys.* **47**(2), 419–424 (2000).
3. Y. Li, H. Yang, J. Cao, *PLoS One* **6**(4), e18776 (2011).
4. J. Lindenmann, A. Maier, V. Matzi, N. Neuboeck, U. Anegg, C. Porubsky, O. Sankin, B. Fell, H. Renner, P. Swatek and F. M. Smolle-Juettner, *Eur. Surg.* **43**(6), 355–365 (2011).
5. L. M. Brown, R. Hoover, D. Silverman, D. Baris, R. Hayes, G. M. Swanson, J. Schoenberg, R. Greenberg, J. Liff, A. Schwartz, M. Dosemeci, L. Pottern and Jr. J. F. Fraumeni, *Am. J. Epidemiol.* **153**(2), 114–122 (2011).
6. E. De Stefani, E. Barrios and L. Fierro, *Eur. J. Cancer* **29A**(5), 763–766 (1993).
7. A. Ellis, J. K. Field, E. A. Field, P. S. Friedmann, A. Fryer, P. Howard, I. M. Leigh, J. Risk, J. M. Shaw and J. Whittaker, *Eur. J. Cancer B Oral Oncol.* **30B**(2), 102–112 (1994).
8. J. M. Risk, H. S. Mills, J. Garde, J. R. Dunn, K. E. Evans, M. Hollstein and J. K. Field, *Dis. Esophagus.* **12**(3), 173–176 (1999).
9. M. Pera, C. Manterola, O. Vidal and L. Grande, *J. Surg. Oncol.* **92**(3), 151–159 (2005).
10. F. H. Ellis Jr., E. Watkins Jr. and M. G. Krasna, *J. Surg. Oncol.* **52**, 231–235 (1993).
11. K. Moghissi, *Br. J. Surg.* **79**, 935–937.
12. S. E. Oliver, C. S. Robertson and R. F. A. Logan, *Br. J. Surg.* **79**, 1321–1325 (1992).
13. M. Fok, J. S. Sham, D. Chay, S. W. K. Chung and J. Wong, *Surgery* **113**, 138–147 (1993).
14. C. D. Amdal, A. B. Jacobsen, J. E. Tausjø, J. N. Wiig, T. Warloe, B. Sandstad and K. Bjordal, *Dis. Esophagus.* **24**, 502–509 (2011).
15. S. R. Lord, P. S. Hall, P. McShane, J. Brown and M. T. Seymour, *Clin. Oncol.* **22**(2), 107–113 (2010).
16. M. Javle, S. Ailawadhi, G. Y. Yang, C. E. Nwogu, M. D. Schiff and H. R. Nava, *J. Support Oncol.* **4**(8), 365–373 (2006).

17. R. K. Sur, D. P. Singh and S. C. Sharma, *Int. J. Radiat. Concol. Biol. Phys.* **22**, 1043–1046 (1992).
18. T. Sakamoto, H. Kotah and T. Shimizu, *Chest* **112**, 1487–1493 (1997).
19. P. Rutgeerts, G. Vantrappen and L. Broeckaert, *Gastrointest. Endosc.* **34**, 87–90 (1988).
20. A. Maier, H. Pinter, G. B. Friehs, H. Renner and F. M. Smolle-Juettner, *Ann. Thorac. Surg.* **67**, 781–784 (1999).
21. J. S. McCaughan Jr., Photodynamic therapy versus Nd: YAG laser treatment of endobronchial or esophageal malignancies. *Photodynamic Therapy and Biomedical Lasers*, Eds. P. Spinelli, M. Dal Fante and R. Marchesini (Elsevier, New York, 1992), pp. 23–36.
22. T. J. Dougherty, C. J. Gomer, B. W. Henderson, G. Jori, D. Kessel, M. Korbelik, J. Moan and Q. Peng, *J. Natl. Cancer Inst.* **90**(12), 889–905 (1998).
23. D. E. Dolmans, D. Fukumura and R. K. Jain, *Nat. Rev. Cancer* **3**(5), 380–387 (2003).
24. D. Wu, Z. Liu, Y. Fu, Y. Zhang, N. Tang, Q. Wang and L. Tao, *Oncol. Lett.* **6**(4), 1111-1119 (2013).
25. T. Yano, M. Muto, K. Minashi, M. Onozawa, K. Nihei, S. Ishikura, K. Kaneko and A. Ohtsu, *Endoscopy* **43**(8), 657–663 (2011).
26. J. Lindenmann, V. Matzi, N. Neuboeck, U. Anegg, E. Baumgartner, A. Maier, J. Smolle and F. M. Smolle-Juettner, *Lasers Surg. Med.* **44**(3), 189–198 (2012).
27. A. Maier, U. Anegg, B. Fell, F. Tomaselli, O. Sankin, U. Prettenhofer, H. Pinter, P. Rehak, G. B. Friehs and F. M. Smolle-Juttner, *Lasers Surg. Med.* **26**(5), 461–466 (2000).
28. A. Maier, F. Tomaselli, F. Gebhard, P. Rehak, J. Smolle and F. M. Smolle-Juttner, *Ann. Thorac. Surg.* **69**(4), 1006–1009 (2000).
29. A. Sibille, R. Lambert, J. C. Souquet, G. Sabben and F. Descos, *Gastroenterology* **108**(2), 337–344 (1995).
30. T. J. Dougherty and S. L. Marcus, *Eur. J. Cancer* **28A**, 1734–1742 (1992).
31. H. I. Pass, *J. Natl. Cancer Inst.* **85**, 443–456 (1993).
32. A. P. Castano, P. Mroz and M. R. Hamblin, *Nat. Rev. Cancer* **6**(7), 535–545 (2006).
33. A. P. Castano, T. N. Demidova and M. R. Hamblin, *Photodiagn. Photodyn. Ther.* **2**, 1–23 (2005).
34. A. P. Castano, T. N. Demidova and M. R. Hamblin, *Photodiagn. Photodyn. Ther.* **2**, 91–106 (2005).
35. J. Moan, Q. Peng, R. Sorensen, V. Iani and J. M. Nesland, *Endoscopy* **30**, 387–391.
36. N. L. Oleinick, R. L. Morris and I. Belichenko, *Photochem. Photobiol. Sci.* **1**(1), 1–21 (2002).
37. B. Krammer, *Anticancer Res.* **21**, 4271–4277 (2001).
38. D. E. Dolmans, *Cancer Res.* **62**, 2151–2156 (2002).
39. J. Lindenmann, V. Matzi, N. Neubock, A. Maier and F.M. Smolle-Juttner, *Eur. Surg.* **42**(5), 220–228 (2010).
40. A. Maier, F. Tomaselli, V. Matzi, P. Rehak, H. Pinter and F. M. Smolle-Juettner, *Lasers Surg. Med.* **29**(4), 323–327.
41. P. Grosjean, G. Wagnieres and C. Fontolliet, *Br. J. Cancer* **77**, 1989–1995 (1998).
42. C. J. Lightdale, *Gastrointest. Endosc. Clin. North America* **10**, 397–408 (2000).
43. J. F. Savary, P. Grosjean and P. Monnier, *Endoscopy* **30**, 258–265 (1998).
44. P. Grosjean, J. F. Savary and G. Wagnieres, *Lasers Med. Sci.* **11**, 227–235 (1996).
45. P. Hinnen, F. W. de Rooij and M. L. van Velthuysen, *Br. J. Cancer*, **78**, 679–682 (1998).

46. A. Maier, F. Tomaselli, V. Matzi, P. Rehak, H. Pinter and F. M. Smolle-Juettner, *Ann. Thorac. Surg.* 72, 1136–1140 (2001).
47. C. S. Loh, D. Vernon and A. J. MacRobert, *J. Photochem. Photobiol.* **20**, 47–54 (1993).
48. J. Webber, D. Kessel and D. Fromm, *J. Surg. Res.* **68**, 31–37 (1997).
49. S. K. Heier, K. Rothmann, L. M. Heier and W. S. Rosenthal, *Gastrointest. Endosc.* **38**, 279 (1992).
50. R. J. Thomas, M. Abott, P. S. Bhathal, D. J. B. St. John and G. Morstyn, *Ann. Surg.* **206**, 193–199 (1987).
51. V. H. Fingar and B. W. Henderson, *Photochem. Photobiol.* **46**, 837–841 (1987).
52. S. G. Bown, C. J. Tralu, P. D. Coleridge-Smith, D. Akdemir and T. J. Wieman, *Br. J. Cancer.* **54**, 43–52 (1986).
53. S. Karanov, M. Shopova and H. Getov, *Laser Surg. Med.* **11**, 395–398 (1991).
54. S. Z. Song, J. H. Li, J. Zou, M. Y. Shu, F. Y. Zhao, M. L. Jin and Z. H. Guo, *Laser Surg. Med.* **5**, 61–66 (1985).
55. L. Gossner, H. Wittke, A. Warzecha, R. Sroka, H. Ernst, M. Meier and C. Ell, *Eur. J. Gastroenterol. Hepatol.* **6**, 159–165 (1994).
56. H. Barr, N. Krasner, P. B. Boulus, P. Chatlani and S. G. Bown, *Br. J. Surg.* 77, 93–96 (1990).
57. H. Kato, M. Kawaguchi, C. Konaka, Y. Hayata and H. Okitsu, *Laser Med. Sci.* **1**, 67–74 (1986).
58. W. F. Cheong, S. A. Prahl and A. J. Welch, *IEEE J. Quant. Elect.* **26**, 2166–2185 (1990).
59. B. C. Wilson and M. S. Patterson, *Phys. Med. Biol.* **31**, 327–360 (1986).
60. A. Maier, D. Sullmann, U. Anegg, F. Tomaselli, P. Rehak, H. Hutten, H. Pinter and F. M. Smolle-Jüttner, *Lasers Surg. Med.* **27**, 350–357 (2000).
61. L. Gossner, A. May and R. Sroka, *Endoscopy* **31**, 370–376 (1999).
62. K. K. Wang, *Gastrointest. Endosc.* **49**, S20–S23 (1999).
63. H. van den Bergh, *Endoscopy* **30**, 392–407 (1998).
64. K. Moghissi and K. Dixon, J. A. C. Thorpe, M. Stringer and P. J. Moore, *Eur. J. Card Thorac. Surg.* **17**, 95–100 (2000).
65. C. Gomer and N. Razum, *Photochem. Photobiol.* **40**, 435–439.
66. M. Jr. Jirsa, P. Pouckiva, J. Dolezal, J. Pospísil and M. Jirsa, *Eur. J. Cancer* **27**, 109 (1991).
67. C. S. Foote, Mechanisms of photooxygenation. *Porphyrin Localization and Treatment of Tumors*, Eds. D. R. Doiron and C. J. Gomer (Alan R. Liss, New York,1984) p. 3.
68. M. Athar, C. A. Elmets and D. R. Bickers, *J. Clin. Invest.* **83**, 1137–1143 (1989).
69. B. J. Tromberg, A. Orenstein and S. Kimel, *Photochem. Photobiol.* **52**(2), 375–385 (1990).
70. J. S. McCaughan, T. E. Williams and B. H. Bethel, *Ann. Thorac. Surg.* **40**(2), 113–120 (1985).
71. K. Moghissi and K. Dixon, *Technol. Cancer Res. Treat.* **2**(4), 319–326 (2003).
72. A. Maier, F. Tomaselli, U. Anegg, P. Rehak, B. Fell, S. Luznik, H. Pinter and F. M. Smolle-Jüttner, *Eur. J. CardThorac. Surg.* **18**, 649–655 (2000).
73. J. D. Luketich, N. A. Christie, P. O. Buenaventura, T. L. Weigel, R. J. Keenan and N. T. Nguyen, *Surg. Endosc.* **14**, 653–657 (2000).
74. J. S. McCaughan Jr., E. C. Ellison, J. T. Guy, W. J. Hicks, J. J. Jones, L. R. Laufman, E. May, T. A. Nims, C. H. Spiridonidis and T. E. Williams, *Ann. Thorac. Surg.* **62**, 1005–1010 (1996).

75. V. R. Litle, J. D. Luketich, N. A. Christie, P. O. Buenaventura, M. Alvelo-Rivera, J. S. McCaughan, N. T. Nguyen and H. C. Fernando, *Ann. Thorac. Surg.* **76**(5), 1687–1692 (2003).
76. J. D. Birkmeyer, T. A. Stukel, A. E. Siewers, P. P. Goodney, D. E. Wennberg and F. L. Lucas, *N. Engl. J. Med.* **349**(22), 2117–2127 (2003).
77. J. W. Mall, H. Zuckermann-Becker, C. Pollmann, I. Opitz, P. Rogalla and M. Walter, *Thorac. Cardiovasc. Surg.* **50**, 111–113 (2002).
78. J. Lindenmann, V. Matzi, N. Neuboeck, U. Anegg, A. Maier, J. Smolle, F. M. Smolle-Juettner, *J. Gastrointest. Surg.* **17**(6), 1036–1043 (2013).
79. F. Tomaselli, A. Maier, O. Sankin, M. Woltsche, H. Pinter and F. M. Smolle-Jüttner, *Eur. J. Cardthor. Surg.* **20**, 734–738 (2001).
80. F. Di Franco, P. J. Lamb, D. Karat, N. Hayes and S. M. Griffin, *Br. J. Surg.* **95**(7), 837–839.
81. R. E. White, C. Mungatana and M. Topazian, *J. Gastrointest. Surg.* **7**, 715–719 (2003).
82. J. Lindenmann, N. Neuboeck, U. Anegg, V. Matzi, A. Maier and F. M. Smolle-Juettner, *Asian Cardiovasc. Thorac. Ann.* **17**(1), 79–81 (2009).
83. M. Panjehpour, B. F. Overhold and J. M. Haydek, *Am. J. Gastroenterol.* **95**, 2177–2184 (2000).
84. A. Maier, U. Anegg, B. Fell, P. Rehak, B. Ratzenhofer, F. Tomaselli, O. Sankin, H. Pinter, F. M. Smolle-Jüttner and G. B. Friehs, *Lasers Surg. Med.* **26**(3), 308–315 (2000).
85. H. Barr, C. J. Tralau, P. B. Boulos, A. J. MacRobert, R. Tilly and S. G. Bown, *Photochem. Photobiol.* **46**(5), 795–800 (1987).
86. C. J. Lightdale, S. K. Heier, N. E. Marcon, J. S. McCaughan Jr., H. Gerdes, B. F. Overholt, M. V. Sivak Jr., G. V. Stiegmann and H. R. Nava, *Gastrointest. Endosc.* **42**(6), 507–512 (1995).
87. S. K. Heier, K. A. Rothman, L. M. Heier and W. S. Rosenthal, *Gastroenterology* **109**, 63–72 (1995).
88. S. O. Gollnick, B. Owczarczak and P. Maier, *Lasers Surg. Med.* **38**(5), 509–515 (2006).
89. M. Korbelik, *Lasers Surg. Med.* **38**(5), 500–508 (2006).
90. P. C. Kousis, B. W. Henderson, P. G. Maier and S. O. Gollnick, *Cancer Res.* **67**(21), 10501–10510 (2007).
91. W. J. de Vree, M. C. Essers, H. S. de Bruijn, W. M. Star, J. F. Koster, W. Sluiter, *Cancer Res.* **56**(13), 2908–2911.
92. M. Korbelik, G. Krosl, J. Krosl and G. J. Dougherty, *Cancer Res.* **56**(24), 5647–5652 (1996).
93. M. Korbelik and G. Krosl, *Photochem. Photobiol.* **60**(5), 497–502 (1994).
94. E. Reginato, J. Lindenmann, C. Langner, N. Schweintzger, I. Bambach, F. Smolle-Jüttner and P. Wolf, *Photochem. Photobiol. Sci.* **13**(9), 1281–1289 (2014).
95. G. C. Dong, S. X. Hu, G. Y. Zhao, S. Z. Gao and L. R. Wu, *Chin. Med. J. Engl.* **100**(9), 697–702 (1987).
96. A. Maier, U. Anegg, F. Tomaselli, P. Rehak, O. Sankin, B. Fell, H. Renner, H. Pinter, F. M. Smolle-Jüttner and G. B. Friehs, *Endoscopy* **32**(1), 42–48 (2000).

Chapter 16

PDT for Lung Cancer

Keishi Ohtani and Norihiko Ikeda

Department of Surgery, Tokyo Medical University, Tokyo, Japan 6-7-1, Nishi-shinjuku, Shinjuku-ku, Tokyo, 160-0023, Japan
keishi_o0429@yahoo.co.jp

Abstract

Photodynamic therapy (PDT) for lung cancer is becoming increasingly common. In Japan, PDT is recommended as a treatment option for centrally located early-stage lung cancers (CLELCs). The second generation photosensitizer, Laserphyrin has strong anti-tumor effects with much less photosensitivity than Photofrin. Moreover, the laser equipment is compact and portable, and Laserphyrin is now the current mainstay of PDT for lung cancer. For successful PDT, accurate evaluation of tumor extent and bronchial invasion is crucial. Detailed examination of the tumor using autofluorescence bronchoscopy (AFB) and endobronchial ultrasonography (EBUS) or optical coherence tomography is extremely useful before PDT. At present, PDT has become the first choice of treatment for CLELC in institutions with the necessary equipment. It can also be effective for advanced lung cancer causing tracheobronchial obstruction. With such advances in PDT for CLELC, we are expanding the indications of PDT for not only CLELC, but also peripheral type lung cancer.

1. Introduction

Since the report of the first modern clinical trial of photodynamic therapy (PDT) by Dougherty *et al.*,[1] PDT using the photosensitizer, Photofrin, has been applied for the treatment of many cancers, and is now approved by the United States Food and Drug Administration (FDA) for the treatment of early stage lung cancer as well as advanced lung cancers.[2–5] At present, PDT for lung cancer is increasingly widespread. In Japan, PDT is recommended as a treatment option for centrally located early-stage lung cancers (CLELCs) in the therapeutic guidelines for lung cancer established by the Japanese Ministry of

Health, Labor and Welfare using the principles of evidence-based medicine.[6–8] The first endoscopic PDT of lung cancer in patient was performed in 1980, and since then this technique has been increasingly used to treat CLELC. Moreover, PDT allows lung function to be preserved and is recommended for CLELCs in the evidence-based clinical practice guidelines of the American College of Chest Physicians.[9]

We describe the history, current status and future prospects of PDT for lung cancer.

2. History of PDT for Lung Cancer

In our department, we started our study of PDT and photodynamic diagnosis (PDD) in 1978, using hematoporphyrin derivative (HpD) and a canine lung cancer experimental model.[10–13] In 1980, we performed the first treatment for human lung cancer using HpD and laser irradiation for tracheal and bronchial metastatic lesions from advanced primary adenocarcinoma of lung. Although sufficient local tumor responses were obtained, the patient died due to distant metastases. In March 1980, the first early stage lung cancer was treated. It was squamous cell carcinoma of the right upper lobe bronchus (B2b). This patient was a 74-year-old man who adamantly refused surgery. It was the first case in the world in which PDT was performed for a CLELC.[14,15] The tumor disappeared completely after PDT, however, he died 46 months after treatment due to COPD. In this period, the therapy was called photoradiation therapy (PRT). Complete response (CR) was also obtained in the second case of CLELC, but the patient died of cerebral infarction 31 months after PDT. A disease-free survival in excess of five years was eventually obtained in the third case. She was a 59-year-old woman. Well differentiated squamous cell carcinoma extended 2×1.5 cm from the right upper lobe bronchus to B1, B2, and B3. One week after PDT, tumor tissue had completely disappeared and tumor recurrence was not recognized.

While these first steps were being taken, PDT was gradually spread to various investigators with recognizable therapeutic effect. In Japan, phase II clinical trials using Photofrin and argon dye laser or eximer dye laser for CLELC were conducted from June 1989 to March 1992 to evaluate the activity and toxicity of PDT. In the studies, a total of 61 CLELCs (51 patients) were registered and treated using Photofrin (2.0 mg/kg) and 630 nm laser light illumination of 100 J/cm^2 using an eximer dye laser (Fig. 1) at a power of 160 mW, and 200 J/cm^2 using an argon dye laser at a power of 300 mW.[16] Of the 59 assessable CLELCs, CR was obtained in 50 (84.8%), six had partial

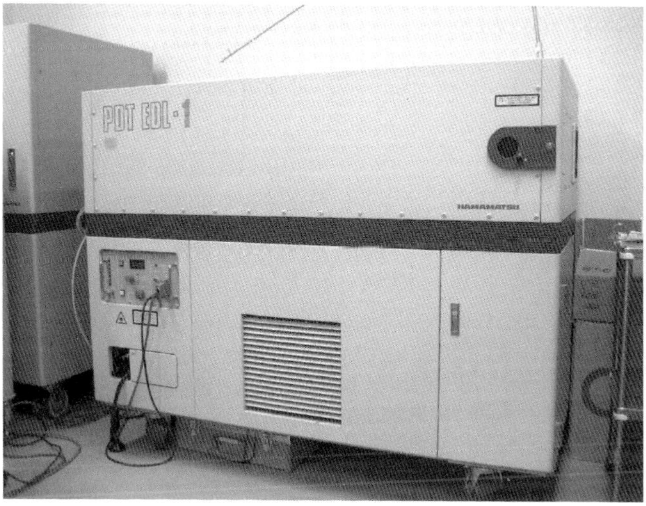

Fig. 1. Eximer dye laser for Photofrin-PDT (EDL-2: Hamamatsu Photonics, Japan).

response (PR), and three showed no response. Toxicity assessment (World Health Organization [WHO] grade 2) showed transient elevation of ALT (1.9%), pulmonary toxicity (7.7%), and allergic reaction (3.8%), as well as sunburn (1.9%). WHO grade 3 or 4 toxicities caused by PDT were not encountered. This study demonstrated that the length of longitudinal tumor extension on the bronchial surface was strongly related to the therapeutic outcome. Out of 45 CLELCs 1 cm or less, CR was obtained in 44 (97.8%), on the contrary, among 14 CLELCs that had a longitudinal extension greater than 1 cm, only six (42.9%) showed CR after PDT. In Japan, with this result, Photofrin-PDT obtained government approval in October 1994 and finally obtained National Health Insurance reimbursement status in April 1996. In 1998, the FDA approved Photofrin for the treatment of early lung cancer.[6,7] As a result, Photofrin became the most commonly used photosensitizer for PDT.

Although PDT as a treatment for lung cancer was started with great therapeutic expectations, there was still a problem: skin photosensitivity. In spite of excellent results of Photofrin-PDT, patients had to avoid sunlight for a minimum of 30 days to avoid sunburn.[17] Many investigators therefore showed interest in the synthesis of new photosensitizers, which possess lower skin photosensitivity for use in PDT. Among those photosensitizers, Laserphyrin (mono-l-aspartyl chlorin e6, NPe6) was considered as a promising photosensitizer.[18] It showed both anti-tumor efficacy in a murine tumor model and rapid clearance from skin.[19] From October 1997 to March 2000, we conducted a phase II clinical study to investigate anti-tumor effects on CLELCs as well as the safety of Laserphyrin. Laser irradiation (100 J/cm^2) using a diode laser was performed at 4 h after administration of Laserphyrin (40 mg/m^2). A total of 45 CLELCs were evaluated and CR was seen in 84.6% of lesions (33/39). Moreover, 28 of 33 patients (84.8%) in whom the first skin photosensitivity test was performed had no photosensitivity by day 14 after drug administration.[20] After this study, Laserphyrin was approved for the treatment of CLELC by the Japanese Ministry of Health, Labor, and Welfare in 2004.

At present, PDT has come to be considered as the first choice of treatment for CLELC. With recent advances in the ability to diagnose lung cancer, we continually attempt to expand the indications of PDT, not only for CLELC but also for more advanced disease or peripheral type lung cancer.

3. PDT for CLELC

In Japan, endoscopic early lung cancer is defined as 1. Normal chest X-ray and computed tomography (CT), 2. No lymph node metastasis or distant metastasis, 3. Located not beyond the level of sub-segmental bronchi, 4. Lesion peripheral margin is endoscopically visible, 5. Less than 2 cm in size, 6. Histologically squamous cell carcinoma.[21] These diagnostic criteria were established based on extensive analyses of radiologic, bronchoscopic and pathologic data by the Ikeda Study Group of the Ministry of Health and Welfare in 1975.[22] Although such tumors can be the target of PDT, the Japan Photodynamic Association and Japanese Society of Laser Surgery and Medicine established more detailed therapeutic indications for PDT in CLELC: 1. Other curative treatments such as surgical resection are impossible or the cases require preservation of lung function. 2. Overall lesions can be confirmed endoscopically. 3. Early lung cancers that can receive laser light irradiation. In particular, lesions less than 1 cm in maximal dimension and with less than bronchial cartilage layer tumor invasion on biopsy specimens are considered to be good indications for PDT.

3.1. *Photofrin-PDT*

Photofrin was the first-generation photosensitizer. It is a chemically-altered HpD porfimer sodium. Porfimer sodium is an aromatic ring complex, which achieves an excited state after exposure to 630 nm wavelength light. The excited molecule then reacts with oxygen to generate singlet oxygen and other reactive oxygen species.[23] Because of its hydrophobic nature, porfimer sodium tends to accumulate in membranes of cells and organelles, especially mitochondria, causing reactive oxygen-induced damage to the membrane as well as membrane-embedded proteins. This disruption leads to multiple intracellular changes that eventually result in cellular apoptosis, necrosis, ischemia, inflammation, and immune responses.[24] Photofrin-PDT also causes local vascular destruction, which depletes the nutrient and oxygen source for tumor cells.[25]

Clinically, Photofrin is injected intravenously at 2.0 mg/kg body weight. It accumulates in tumor tissue at a level about 4 times more than normal tissue and tumor cells need time to selectively retain more photosensitizer than normal tissue due to differences in cancerous tissue vascular supply and lymphatic drainage.[26] Therefore, bronchoscopical laser irradiation is performed 48 h after injection with topical anesthesia.[27] After Photofrin injection, the patients are instructed to avoid direct sunlight for at least four weeks because of skin photosensitivity.[28] The laser beam (630 nm wavelength) is transmitted via a quartz fiber (400 mm) inserted through the biopsy channel of a flexible bronchoscope. The total energy of the laser irradiation is 100 J/cm^2, and energy levels in this range do not cause any heat degeneration or other adverse effects. The duration of irradiation required is usually 10–20 min. Clean-up bronchoscopies to remove necrotic tissue produced by the PDT reaction are necessary at 1, 3, and 7 days after PDT.[29]

Furukawa *et al.*[29] reported their experience of PDT in 93 patients with 114 CLELC lesions. The 114 lesions were classified in two groups according to the maximum longitudinal tumor extent. The CR rates in the group of patients with lesions less than 1.0 cm in maximum dimension and 1.0 cm or more in maximum dimension were 92.8% (77 of 83 patients) and 58.1% (18 of 31 patients), respectively. Excellent efficacy was obtained in patients with lesions <1.0 cm, thus this seems a good standard in deciding on the indications of PDT. However, recurrences after CR were recognized in 9 of 77 lesions (11.7%) <1.0 cm. These local recurrences at the same site may be caused by residual tumor cells from deep layers because of inadequate laser irradiation and penetration. Therefore, to reduce the recurrence rate, it is essential to accurately grasp the tumor extent and the depth of lung cancers before performing PDT.

3.2. *Laserphyrin-PDT*

Laserphyrin is a second-generation, water-soluble photosensitizer with a molecular weight of 799.69, and it has a chlorine annulus. Its maximum absorption peak is at a wavelength of 407 nm, and there is a second peak at 664 nm.[30,31]

Because Laserphyrin has high tumor affinity and it is excited by visible red light with a longer wavelength of 664 nm, Laserphyrin-PDT is considered to have a stronger antitumor effect on deeper lesions than Photofrin-PDT.

The procedure of Laserphyrin-PDT is similar to Photofrin-PDT, but the dosage and interval time after injection to laser irradiation are different. In the initial phase I clinical study, the optimal conditions yielding high safety and efficacy were suggested to be intravenous administration of NPe6 (40 mg/m^2).[30] At 4 h after administration, a wavelength 664 nm laser beam is irradiated to the tumor site endoscopically using a quartz fiber (power density: 150 mW/cm^2, energy level: 100 J/cm^2). A small and portable diode laser system (Matsushita Electric Industrial Co.) (Fig. 2) emitting continuous-wave laser light is used as the light source.[31] The patient is prohibited from going out into direct sunlight for two weeks after administration of Laserphyrin.

Usuda *et al.*[32] showed outcome of Laserphyrin-PDT in Tokyo Medical University between June 2004 and December 2008. A total of 75 patients (91 lesions) with CLELCs were treated and CR was obtained in 85 lesions (CR rate: 93.4%). In this study, 70 lesions were ≤1.0 cm a diameter and 21 lesions were >1.0 cm in size. The CR rate of lesions ≤1.0 cm in diameter was 94% (66/70) and for those >1.0 cm in diameter, 90.4% (19/21), respectively. There was no statistically significant difference between tumor size and clinical response. This result suggests that Laserphyrin-PDT has a stronger anti-tumor effect than Photofrin-PDT against large tumors >1.0 cm in diameter. Laserphyrin-PDT has many benefits such

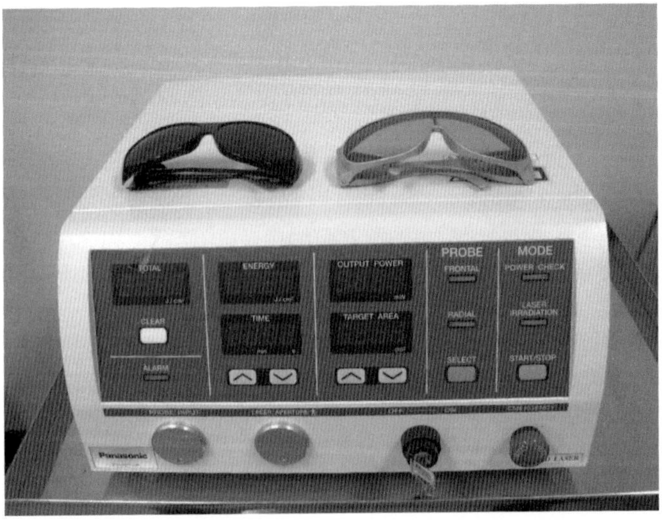

Fig. 2. Diode laser for Laserphyrin-PDT (PD Laser: Panasonic HealthCare Co., Ltd., Japan).

as strong anti-tumor effect, short photoradiation interval time and low photosensitivity.

4. Imaging

Accurate evaluation of tumor extent and tumor depth are very important to the success of PDT.[6,29,31] Therefore, we perform detail examination of tumor using various imaging equipment before PDT.[33] Our routine clinical process for PDT is shown in Fig. 3.

4.1. *Autofluorescence Bronchoscopy (AFB)*

AFB makes use of fluorescence and absorption properties to provide information about the biochemical composition and metabolic state of endogenous fluorophores in bronchial tissues.[34] Upon illumination by violet or blue light (380–460 nm), normal bronchial tissues greenly fluoresce strongly (480–520 nm). In diseased states such as lung cancer, as the bronchial epithelium changes from normal to dysplasia and then to carcinoma *in situ* (CIS) and invasive cancer, there is a progressive decrease in green autofluorescence but proportionately less decrease in the red fluorescence intensity. The excitation wavelength that produces the highest tumor to normal tissue intensity and chromatic contrast is 405 nm.[35,36] AFB allows rapid scanning of large areas of the bronchial surface for subtle abnormalities that are not visible to white-light examination. Several studies have shown that it improved the sensitivity of detection of cancerous and precancerous lesions of the airway compared to white light bronchoscopy, especially in cases in the intraepithelial stage.[37–41]

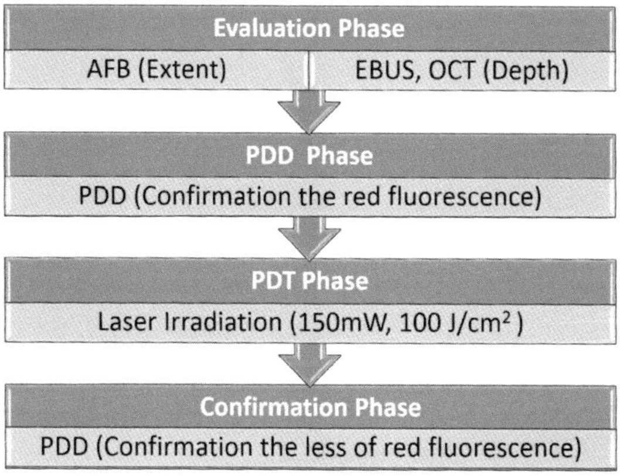

Fig. 3. The clinical process for PDT, AFB, EBUS, OCT, PDD.

Thus, AFB is essential to detect CLELC. It also makes it possible to show tumor margins clearly, therefore, we perform AFB before PDT to identify the tumor extent accurately.

Moreover, there is another important role regarding PDT: PDD. The photosensitizer accumulated in tumor tissue can be excited by a blue AFB laser and the red fluorescence can be observed. Thus PDD enables us to clearly determine the tumor margin.

SAFE-3000 (Hoya Co., Tokyo, Japan) is the videoendoscopic AFB system we use. Just before starting the PDT, we perform PDD using the SAFE-3000 with a diode laser (408 nm) to define the tumor margin based on the red fluorescence emitted by the tumor. Immediately after PDT, we again perform PDD to compare the intensity of the red fluorescence with its intensity just before PDT,[31] because the red fluorescence disappears by photobleaching after appropriate irradiation of PDT laser light. In this way, tumor extent is confirmed accurately before PDT and sufficient laser irradiation can be performed leading to the improvement of quality and efficacy of PDT (Fig. 4).

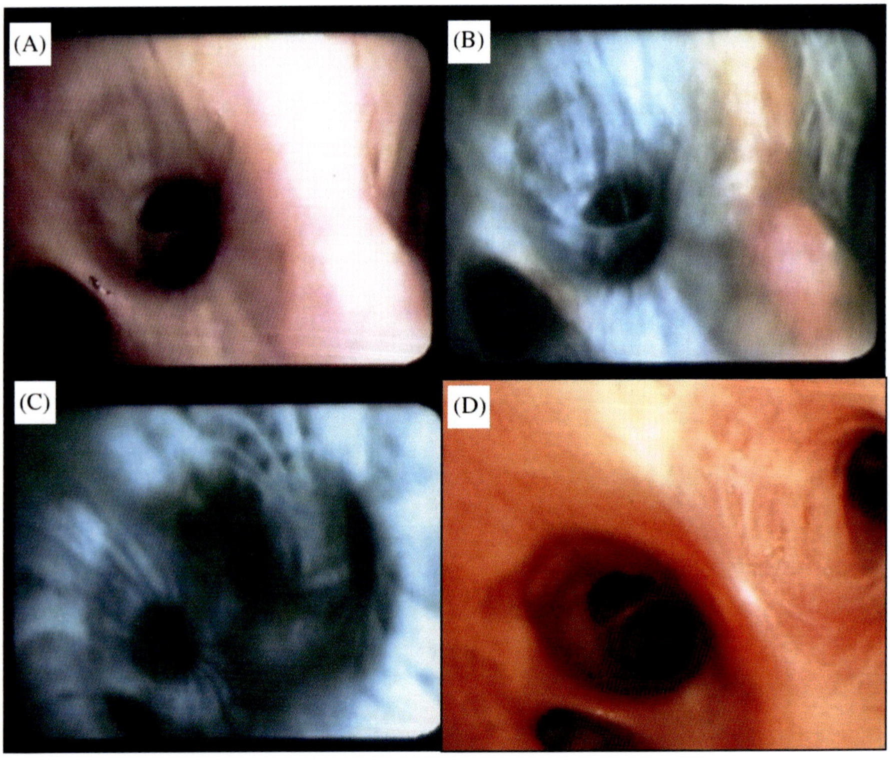

Fig. 4. Before and after PDT for CLELC (A) White light bronchoscopy image. (B) PDD by SAFE-3000 before PDT. (C) PDD image immediately after PDT. (D) Four months after PDT.

4.2. *Optical Coherence Tomography (OCT)*

Accurate evaluation of tumor depth is very important for the success of PDT. Many investigators reported the relation between tumor size and depth. Konaka *et al.*[42] showed that the size and endoscopic features correlate with its depth of penetration based on histopathological investigations of surgical specimens. Hypertrophic-type tumors only invade beyond the cartilage in 4.8% of the cases. On the other hand, nodular and polyploid lesions invade beyond the cartilage in 17.6% and 27.3% of the cases, respectively. In analysis based on the greatest dimension, 68.6% less than 10 mm in greatest dimension were either in-situ carcinomas or micro-invasive tumors within the muscle layer. This result yields useful clinical information. Kurimoto *et al.*[43] examined the usefulness of endobronchial ultrasonography (EBUS) in the determination of the depth of tumor invasion. Comparisons between the EBUS findings and the histopathologic findings in 24 lung cancer cases revealed that depth diagnosis was the same in 23 lesions (95.8%). Miyazu *et al.*[44] also showed the relationship between bronchoscopic appearance and the depth of tumor invasion estimated by EBUS and found that most lesions >1.0 cm in diameter were diagnosed as extracartilaginous lesions. They showed the usefulness of the EBUS for evaluation of tumor depth. However, the resolution of EBUS is still insufficient. Sometimes, it is difficult to perform accurate measuring because of the resolution.

OCT is an optical imaging method that can offer near histological resolution for visualizing cellular and extracellular structures at and below the tissue surface.[45–49] OCT is similar to ultrasound, but properties of light waves instead of sound waves is used for imaging. The imaging procedure is performed using fiberoptic probes that can be miniaturized to enable imaging of airways down to terminal bronchioles. OCT axial and lateral resolutions range from approximately 5–30 mm depending on imaging conditions, typically more than an order of magnitude better than ultrasound. The increase in resolution with OCT compared with ultrasound is offset by a corresponding loss in imaging depth, typically 2–3 mm in tissue for OCT. However, this combination of resolution and imaging depth is ideal for examining preneoplastic changes originating in epithelial tissues.[50] Tsuboi *et al.*[48] collected OCT images of normal bronchus and primary tumors. OCT imaging revealed a layered bronchial wall structure in normal bronchus that is lost in lung cancer. Lam *et al.*[49] performed quantitative analysis of bronchial wall using OCT images. Quantitative measurement of the epithelial thickness showed that invasive carcinoma was significantly thicker than CIS ($p = 0.004$) and dysplasia was significantly thicker than metaplasia or hyperplasia ($p = 0.002$). The nuclei became more readily visible in high-grade dysplasia or CIS. The basement membrane was still intact in CIS but became discontinuous or no longer visible with invasive cancer. OCT can show the bronchial layers clearly and the depth of tumor invasion with high resolution, so it is very useful for the evaluation of indication of PDT.

5. PDT for Advanced Lung Cancer

Palliating symptoms of tracheobronchial obstruction due to malignant tumors can be difficult, though approximately 30% of patients with lung cancer will develop central airway obstruction.[51–53] Progressive tumor growth obstructing the bronchi can lead to dyspnea, coughing, postobstructive pneumonia, and hemoptysis, all of which can be life threatening and dramatically decrease the quality of life (QOL). PDT is effective not only in early stage lung cancer, but also in such advanced lung cancer. Many investigators have shown the efficacy of PDT in palliation of symptoms in patients with advanced lung cancer.[54–57] McCaughan and Williams[55] reported their 14 years experience in 175 non-surgical lung cancer patients with lung cancers receiving bronchoscopic PDT. While the multivariate analysis of survival showed that the most significant influence on long-term survival was the stage of the tumor, their data also appear to show that, in those of their patients with advanced tumor stage (Stage IIIa–IV), the performance status became an important determinant of length of survival. They concluded that the length of palliation for advanced disease was equal or better than other reported treatment regimens. Another study of 100 patients by Moghissi and coworkers[56] reported their experience regarding advanced unresectable lung cancer in 100 cases of PDT. In this study, all patients were stage IIIa–IV and a total of 90 patients with non-small cell lung cancer and 10 patients with small cell lung cancer were treated with PDT. PDT was associated with improvement in endobronchial obstruction, ventilation, and performance status of patients 6–8 weeks after treatment. Mean endoluminal obstruction fell from 86% to 18%, mean forced vital capacity (FVC) and mean forced expiatory volume in 1 s (FEV1) increased by 430 ml and 280 ml, respectively. Moreover, a study by Diaz-Jimenez and colleagues[57] showed the results of a prospective controlled randomized trial comparing PDT with neodymium-yttrium aluminium garnet (Nd-YAG) laser resection in 31 patients with partial or complete tracheobronchial obstruction due to inoperable non-small cell lung cancer. In this trial, 14 patients were treated with PDT and 17 patients were treated with an Nd-YAG laser. Patients in both groups experienced symptomatic relief after treatment ($p = 0.03$). Patients in the PDT group had a significantly longer duration until treatment failure

($p = 0.03$), with the median time to treatment failure in the PDT group being 50 days versus 38 days in the Nd-YAG group. These studies show PDT is able to palliate the symptoms of tracheobronchial obstruction due to malignant tumors.

6. PDT for Peripheral Lung Cancer

With the extremely widespread use of CT, many more small peripheral lung cancers have been identified than before. There is a possibility that such small cancers, especially less than 2.0 cm pure ground glass nodules, can be cured by local treatment. PDT has a potential for such a treatment. Okunaka *et al.*[58] showed their experience of Photofrin-PDT for peripheral primary lung cancers and metastatic lung cancers by inserting optical fibers inside the tumor via a percutaneous needle. Nine patients were registered and the largest dimensions of the tumors were 1.2–8 cm. The rates of PR and stable disease (SD) were 77.7% (7/9 lesions) and 22.2% (2/9 lesions), respectively. Two patients had pneumothorax, one of which needed chest drainage. In this period, there was no method to insert a laser probe accurately to the target endoscopically, therefore, a percutaneous approach that has risks like pneumothorax and air embolism was chosen to perform PDT for the peripheral lung tumors.

Recently, several guided-bronchoscopy technologies have been developed to improve the yield of transbronchial biopsy for small pulmonary nodule diagnosis: electromagnetic navigation bronchoscopy (ENB), virtual bronchoscopic navigation (VBN), and radial EBUS through a guide sheath (EBUS-GS).[59] ENB and VBN are able to create a virtual bronchoscopic image and a pathway to the pulmonary nodule. ENB is an image-guided localization system, which is designed to guide bronchoscopic instruments to predetermined points within the bronchial tree. The exact position of the steerable probe that contains a position sensor when placed within the electromagnetic field is depicted on the system monitor. Thus, it has the ability to guide bronchoscopy instruments to reach lung targets.[60] VBN is a method to display three-dimensional images of the tracheal and bronchial lumens prepared from continuous volume data of helical CT depicting them as if observed on bronchoscopy.[61] A bronchoscope is guided using VBN images on the bronchial route to peripheral lung nodules.[62] On the other hand, EBUS-GS allows the practitioner to visualize the peripheral lung nodule and maintain the position in the periphery for biopsy.[63] Several investigators reported that the combination of navigation system and EBUS-GS improved the diagnostic yield of peripheral lung nodules.[64–66]

Using these techniques, we consider that it is possible to insert a laser probe for PDT to the peripheral target accurately: guiding EBUS probe with a GS to the target using a navigation system, confirming the site of the tumor by radial EBUS, and inserting a laser probe through a GS instead of EBUS probe. Now we are planning GS clinical trials of PDT for peripheral lung cancer. PDT could become one treatment option for peripheral lung cancer especially for the patients for whom neither surgery nor radiation therapy are indicated because of poor general physical condition.

7. Summary

PDT is a minimally invasive treatment with excellent anti-tumor effects. Other benefits are unlimited repeatability, limited side effects and simplicity of operation. Combination therapy with chemotherapy or surgery also holds hope. The main disadvantages are photosensitivity, the need for direct laser irradiation of the target, expensive equipment and cost of the photosensitizing agent, although it has better cost-effectiveness treatment than surgery.[67] Photosensitivity has been considerably improved by the development of the second-generation photosensitizer, Laserphyrin. Technology which can be used to guide a laser device to the target also has improved. Various types of laser devices have also been developed. The cost is still high. However, it might be possible to greatly reduce the cost by increased applications. For that purpose, close medicine and engineering collaboration is necessary. PDT has the potential to be developed more, and it could play an important role in the treatment and management of lung cancer.

Acknowledgment

The authors are indebted to Emeritus Prof. J. Patrick Barron, of Tokyo Medical University for his review of this manuscript.

References

1. T. J. Dougherty, G. Lawrence, J. H. Kaufman, D. Boyle, K. R. Weishaupt and Goldfarb A, *J. Natl. Cancer Inst.* **62**, 231–237 (1979).
2. T. J. Dougherty, C. J. Gomer, B. W. Henderson, G. Jori, D. Kessel, M. Korbelik, J. Moan and Q. Peng, *J. Natl. Cancer Inst.* **90**, 889–905 (1998).
3. M. Cheng, A. Pennathur and J. D. Luketich, *Lasers Surg. Med.* **38**, 396–402 (2006).
4. D. A. Cortese, E. S. Edell, and J. H. Kinsey, *Mayo. Clin. Proc.* **72**, 595–602 (1997).
5. A. J. M. Boxem, J. Westerga, B. J. W. Venmans, P. E. Postmus and T. Sutedja, *Lung Cancer* **31**, 31–36 (2001).

6. J. Usuda, H. Kato, T. Okunaka, K. Furukawa, H. Tsutsui, K. Yamada, Y. Suga, H. Honda, Y. Nagatsuka, T. Ohira, M. Tsuboi and T. Hirano, *J. Thorac. Oncol.* **5**, 489–493 (2006).
7. H. Kato, J. Usuda, T. Okunaka, K. Furukawa, H. Honda, N. Sakaniwa, Y. Suga, T. Hirata, K. Ohtani, T. Inoue, S. Maehara, M. Kubota, K. Yamada and H. Tsutsui, *Lasers Surg. Med.* **38**, 371–375 (2006).
8. P. N. Mathur, E. Edell, T. Sutedja and J. M. Vergnon, *Chest* **123**, 176–180 (2003).
9. J. P. Wisnivesky, R. C. Yung, P. N. Mathur and J. J. Zulueta, *Chest* **143**, e263S–77S (2013).
10. H. Kato, J. Usuda, T. Okunaka and K. Furukawa, *Photodiagnosis Photodyn. Ther.* **1**, 107–110 (2004).
11. H. Kato, C. Konaka, Y. Hayata, J. Ono, I. Iimura, Y. Matsushima, M. Tahara, J. Lei, M. Nasiell and G. Auer, *Recent Results Cancer Res.* **82**, 69–86 (1982).
12. H. Kato, C. Konaka and J. Ono, *Lung Cancer* **21**, 439–445 (1981).
13. Y. Hayata, H. Kato, C. Konaka, N. Hayashi, M. Tahara, T. Saito and J. Ono, *Cancer* **51**, 50–56 (1983).
14. Y. Hayata, H. Kato, C. Konaka, J. Ono, Y. Matsushima, K. Yoneyama and K. Nishimiya, *Chest* **81**, 269–277 (1982).
15. H. Kato, C. Konaka, N. Kawate, H. Shinohara, K. Kinoshita, M. Noguchi, S. Ootomo and Y. Hayata, *Chest* **90**, 768–770 (1986).
16. K. Furuse, M. Fukuoka, H. Kato, T. Horai, K. Kubota, N. Kodama, Y. Kusunoki, N. Takifuji, T. Okunaka and C. Konaka, *J. Clin. Oncol.* **11**, 1852–1857 (1993).
17. N. Razum, O. J. Balchum, A. E. Profio and F. Carstens, *Photochem. Photobiol.* **46**, 925–928 (1987).
18. J. D. Spikes and J. C. Bommer, *J. Photochem. Photobiol. R Biol.* **17**, 135–143 (1993).
19. T. A. Katsumi, K. Aizawa, Y. Kuroiwa, K. Saito, Y. Kurata, Y. Ii, T. Okunaka, C Konaka and H. Kato, *Photochem. Photobiol.* **64**, 671–675 (1996).
20. H. Kato, K. Furukawa, M. Sato, T. Okunaka, Y. Kusunoki, M. Kawahara, M. Fukuoka, T. Miyazawa, T. Yana, K. Matsui, T. Shiraishi and H. Horinouchi, *Lung Cancer.* **42**, 103–111 (2003).
21. The Japan Lung Cancer Society. General rule for clinical and pathological record of lung cancer. 7th edn. (Kanehara & Co. LTD., Japan, 2010), pp. 157–159.
22. S. Ikeda, Atlas of early cancer of major bronchi. (Igakushoin publisher. Japan, 1975).
23. I. MacDonald and T. Dougherty, *J. Porphyrins Phthalocyanines* **5**, 105–129 (2001).
24. G. Abbas, A. Pennathur, S. B. Keeley, R. Landreneau and J. Luketich, *Semin. Thorac. Cardiovasc. Surg.* **17**, 313–319 (2005).
25. C. J. Kelty, S. L. Marcus and R. Ackroyd, *Dis. Esophagus.* **15**, 137–144 (2002).
26. N. Christie, A. Patel and R. Landreneau, *Surg. Clin. N. Am.* **85**, 569–582 (2005).
27. D. Kessel and M. Cheng, *Photochem. Photobiol.* **41**, 277–282 (1985).
28. T. J. Dougherty, M. T. Cooper and T. S. Mang, *Lasers Surg. Med.* **10**, 485–488 (1990).
29. K. Furukawa, H. Kato, C. Konaka, T. Okunaka, J. Usuda and Y. Ebihara, *Chest* **128**, 3269–3275 (2005).
30. K. Furukawa, T. Okunaka, T. Tsuchida, C. Konaka and H. Kato, *Porphyrins* 7, 199–206 (1998).
31. J. Usuda, H. Tsutsui, H. Honda, S. Ichinose, T. Ishizumi, T. Hirata, T. Inoue, K. Ohtani, S. Maehara, K. Imai, Y. Tsunoda, M. Kubota, N. Ikeda, K. Furukawa, T. Okunaka, H. Kato, *Lung Cancer* **58**, 317–323 (2007).

32. J. Usuda, S. Ichinose, T. Ishizumi, H. Hayashi, K. Ohtani, S. Maehara, S. Ono, H. Honda, N. Kajiwara, O. Uchida, H. Tsutsui, T. Ohira, H. Kato and H. Ikeda, *Clin. Cancer Res.* **16**, 2198–2204 (2010).
33. N. Ikeda, J. Usuda, H. Kato, T. Ishizumi, S. Ichinose, K. Otani, H. Honda, K. Furukawa, T. Okunaka and H. Tsutsui, *Lasers Surg. Med.* **43**, 749–754 (2011).
34. S. Lam, The role of autofluorescence bronchoscopy in diagnosis of early lung cancer. Eds. F. R. Hirsch, P. A. Bunn, H. Kato *et al.*, In *IASLC Textbook for Prevention and Detection of Early Lung Cancer* (Taylor & Francis, London, 2006), pp. 149–158.
35. J. Hung, S. Lam, J. C. LeRiche and B. Palcic, *Lasers Surg. Med.* 11, 99–105 (1991).
36. M. Zellweger, P. Grosjean, D. Goujon, P. Monnier, H. van den Bergh and G. Wagnières, *J. Biomed. Opt.* **6**, 41–51 (2001).
37. S. Lam, T. Kennedy, M. Unger, Y. Miller, D. Gelmont, V. Rusch, B. Gipe, D. Howard, J. C. LeRiche, A. Coldman and A. F. Gazdar, *Chest* **113**, 696–702 (1998).
38. N. Ikeda, H. Honda, T. Katsumi, T. Okunaka, K. Furukawa, T. Tsuchida, K. Tanaka, T. Onoda, T. Hirano, M. Saito, N. Kawate, C. Konaka, H. Kato and Y. Ebihara, *Diagn. Ther. Endosc.* **5**, 85–90 (1999).
39. N. Ikeda, K. Kim, T. Okunaka, K. Furukawa, T. Furuya, M. Saito, C. Konaka, H. Kato and Y. Ebihara, *Diagn. Ther. Endosc.* **3**, 197–201 (1997).
40. M. Kakihana, K. Kim, T. Okunaka, K. Furukawa, T. Hirano, C. Konaka, H. Kato and Y. Ebihara, *Diagn. Ther. Endosc.* **5**, 99–104 (1999).
41. S. Lam, C. MacAulay, J. Hung, J. LeRiche, A. E. Profio and B. Palcic, *J. Thorac. Cardiovasc. Surg.* **105**, 1035–1040 (1993).
42. C. Konaka, T. Hirano, H. Kato, K. Furuse, M. Takada, Y. Saito, Y. Monden, E. Matsui and Y. Watanabe, *Br. J. Cancer* **80**, 1435–1439 (1999).
43. N. Kurimoto, M. Murayama, S. Yoshioka, T. Nishisaka, K. Inai and K. Dohi, *Chest.* **115**, 1500–1506 (1999).
44. Y. Miyazu, T. Miyazawa, N. Kurimoto, Y. Iwamoto, K. Kanoh and N. Kohno, *Am. J. Respir. Crit. Care Med.* **165**, 832–837 (2002).
45. D. Huang, E. A. Swanson, C. P. Lin, J. S. Schuman, W. G. Stinson, W. Chang, M. R. Hee, T. Flotte, K. Gregory, C. A. Puliafito, *et al.*, *Science* **254**, 1178–1181 (1991).
46. J. G. Fujimoto, M. E. Brezinski, G. J. Tearney, S. A. Boppart, B. Bouma, M. R. Hee, J. F. Southern, E. A. Swanson, *Nat. Med.* **1**, 970–972 (1995).
47. G. J. Tearney, M. E. Brezinski, B. E. Bouma, S. A. Boppart, C. Pitris, J. F. Southern, and J. G. Fujimoto, *Science* **276**, 2037–2039 (1997).
48. M. Tsuboi, A. Hayashi, N. Ikeda, H. Honda, Y. Kato, S. Ichinose and H. Kato, *Lung Cancer* **49**, 387–394 (2005).
49. S. Lam, B. Standish, C. Baldwin, A. McWilliams, J. leRiche, A. Gazdar, A. I. Vitkin, V. Yang, N. Ikeda and C. MacAulay, *Clin. Cancer Res.* **14**, 2006–2011 (2008).
50. K. Ohtani, A. M. Lee and S. Lam, *Respirology* **17**, 261–269 (2012).
51. C. S. Chin, V. Litle, J. Yun, T. Weiser and S. J. Swanson, *Ann. Thorac. Surg.* **85**, S792–S796 (2008).
52. S. Stohr and C. T. Bolliger, *Monaldi. Arch. Chest Dis.* **54**, 264–268 (1999).
53. F. Venuta, E. A. Rendina, De Giacomo T, E. Mercadante, F. Francioni, F. Pugliese, M. Moretti and G. F. Coloni, *Ann. Thorac. Surg.* **74**, 995–998 (2002).
54. J. LoCicero, M. Metzdorff and C. Almgren, *Chest* **98**, 97–100 (1990).

55. J. S. McCaughan, Jr. and T. E. Williams, *J Thorac Cardiovasc Surg.* **114**, 940–946; discussion 946–947 (1997).
56. K. Moghissi, K. Dixon, M. Stringer, T. Freeman, A. Thorpe and S. Brown, *Eur. J. Cardiothorac. Surg.* **15**, 1–6 (1999).
57. J. P. Diaz-Jiménez, J. E. Martínez-Ballarín, A. Llunell, E. Farrero, A. Rodríguez and M. J. Castro, *Eur. Respir. J.* **14**, 800–805 (1999).
58. T. Okunaka, H. Kato, H. Tsutsui, T. Ishizumi, S. Ichinose and Y. Kuroiwa, *Lung Cancer* **43**, 77–82 (2004).
59. J. S. Wang Memoli. P. J. Nietert and G. A. Silvestri, Meta-analysis of guided bronchoscopy for the evaluation of the pulmonary nodule. *Chest.* **142**, 385–393 (2012).
60. T. R. Gildea, P. J. Mazzone, D. Karnak, M. Meziane and A. C. Mehta, *Am. J. Respir. Crit. Care Med.* **174**, 982–989 (2006).
61. D. J. Vining, K. Liu, R. H. Choplin and E. F. Haponik, *Chest* **109**, 549–553 (1996).
62. F. Asano, Y. Matsuno, T. Matsushita and A. Seko, *J. Bronchol.* **9**, 108–111 (2002).
63. N. Kurimoto, T. Miyazawa, S. Okimasa, A. Maeda, H. Oiwa, Y. Miyazu and M. Murayama, *Chest* **126**, 959–965 (2004).
64. T. Ishida, F. Asano, K. Yamazaki, N. Shinagawa, S. Oizumi, H. Moriya, M. Munakata and M. Nishimura, *Thorax* **66**, 1072–1077 (2011).
65. H. Asahina, K. Yamazaki, Y. Onodera, E. Kikuchi, N. Shinagawa, F. Asano and M. Nishimura, *Chest* **128**, 1761–1765 (2005).
66. M. Tamiya, N. Okamoto, S. Sasada, T. Shiroyama, N. Morishita, H. Suzuki, E. Yoshida, T. Hirashima, K. Kawahara and I. Kawase, *Respirology* **18**, 834–839 (2013).
67. H. Kato, T. Okunaka, T. Tsuchida, H. Shibuya, S. Fujino and K. Ogawa, *Diagn. Ther. Endosc.* **6**, 9–16 (1999).

Chapter 17

New Developments in Antimicrobial PDT

Wolfgang Bäumler

Department of Dermatology, University of Regensburg,
93042 Regensburg, Germany
baeumler.wolfgang@klinik.uni-regensburg.de

Abstract

Today the increasing resistance of bacteria against antibiotics is one of the most important clinical challenges. Beside the investigation and development of new drugs, it is absolutely necessary to focus also on other antibacterial approaches like antimicrobial photodynamic therapy (PDT). In the photodynamic process, visible light is absorbed in a harmless dye molecule (photosensitizer), which transfers the absorbed light energy mainly to adjacent oxygen molecules generating highly reactive oxygen species (ROS). In case the photosensitizer is a cationic molecule, it can attach to the negatively charged surface of bacteria. Then, ROS are generated directly at the bacteria cell wall upon light irradiation. Bacteria are killed via oxidative damage and killing efficacy is independent of bacterial resistance pattern. Due to the simple oxidative mechanisms, the development of resistance against antimicrobial PDT is very unlikely. The photodynamical produced ROS can be permanently applied on surfaces to kill bacteria which settle down on such a surface.

1. Introduction

In 2009 Cesar Arias and Barbara Murray posted in the "New England Journal of Medicine" that bacteria are champions of evolution.[1] Today the increasing resistance of bacteria against antibiotics is one of the most important clinical challenges, the pace of antibiotic development to struggle against these bacteria has dramatically dropped down. Several bacterial pathogens termed ESKAPE (*Enterococcus faecium, Staphylococcus aureus, Klebsiella pneumoniae, Acinetobacter baumannii, Pseudomonas aeruginosa and Enterobacter strains*) have become worldwide resistance because they "escape" antibiotic treatment.

Especially immunosuppressed patients in hospitals are suffering from life-threatening diseases like sepsis, pneumonia and meningitis due to infection with these bacteria. On the 'Nature News Blog' was recently posted that the lack of new antibiotics and the emergence of drug-resistant bacteria pose a "catastrophic threat".[2] The author highlighted that the dry antibiotic pipeline will threat the people's health even when they have only routine infections.

In September 2014, the White House published a national strategy for combating antibiotic resistant bacteria. The U.S. government will "prevent, detect, and control illness and death related to infections caused by antibiotic resistant bacteria by implementing measures to mitigate the emergence and spread of antibiotic resistance and ensuring the continued availability of therapeutics for the treatment of bacterial infections".[3]

Methicillin resistant *S. aureus* (MRSA) is one of the most important nosocomial and community-acquired pathogens. Up to date MRSA has developed resistance not either against all β-lactam antibiotics but also against last resort antibiotics like vancomycin (VCM) or linezolid (LZD).[4] In 2011 in Germany the prevalence of MRSA runs up to 21.9 %,[5] in U.S. hospitals the proportion of *S. aureus* isolates that were methicillin resistant increased up to 64.4%.[6] The first case of VCM resistant *S. aureus* (VRSA) in Europe was reported in May 2013 in Portugal. In this case VRSA was not only resistant to VCM but also resistant against aminoglycoside, fluoroquinolone, and macrolide antibiotics.[7] Among enterococci resistance against cell enveloped drugs like VCM fluctuate from 3.9% (*Enterococcus Faecalis*) to 75% (*Enterococcus faecium*) in 2011. Recently Master and colleagues also emphasize an increasing daptomcyin resistance of *E. faecium*: from 2.5% in 2007 to 12% in 2011.[8]

The Infectious Diseases Society of America (IDSA) highlighted that, over the past several years, the number of new antibacterial drugs approved continues to decrease. Only three of the eight antibiotics confirmed within the last decade act towards a new mechanism of action, whereas all the others are only modifications of already known antibiotics.[9,10] Complete decolonisation of MRSA patients is a complicated and difficult process. In clinical applications, a successful decolonization of MRSA strongly depends on the compliance of the patients and the treatment protocol used. Such a protocol usually consists of a triple therapy for five days of twice daily nasal mupirocin ointment, chlorhexidine mouth rinse 3-times daily and daily full-body wash with chlorhexidine soap. MRSA decolonization was highly effective in patients who completed the 10 decolonization treatment course.[11]

However, a new study determined the prevalence of high-level mupirocin and chlorhexidine resistance of *S. aureus* isolates (75% MRSA, 25% Methicillin sensitive *S. aureus*) recovered from patients with skin and soft tissue infections before and after mupirocin and chlorhexidine administration. Fritz and colleagues showed that mupirocin resistance occurred in 2.1% of all tested isolates, 0.9% of the isolates were resistant against chlorhexidine. Moreover they showed that *S. aureus* eradication of patients, who initially were colonized with a mupirocin or chlorhexidine resistant strain, was ineffective.[12] Another study showed that significant enhancement of using an alcohol/chlorhexidine solution in hospital hand hygiene before and after clinical staff intervention did not decrease MRSA infections in patients.[13]

Reports of increasing bacterial resistance to antibiotics and antimicrobial agents suggest that pathogens like MRSA will acquire more resistance in the near future. Beside the investigation and development of new drugs, it is absolutely necessary to focus also on other antibacterial approaches like antimicrobial PDT.

2. Photodynamic Mechanisms of Action

The mechanism of action of photodynamic inactivation of bacteria (PIB) is based on a non-toxic dye, termed photosensitizer (PS), molecular oxygen and visible light (400–700 nm). Upon irradiation, the PS can be excited from its energy ground state S_0 to higher energy states. The energy of these excited states is then available for chemical activation with molecules in the surrounding to produce reactive oxygen species (ROS) such as free radicals (type I reaction) or singlet oxygen (type II reaction) (Fig. 1).

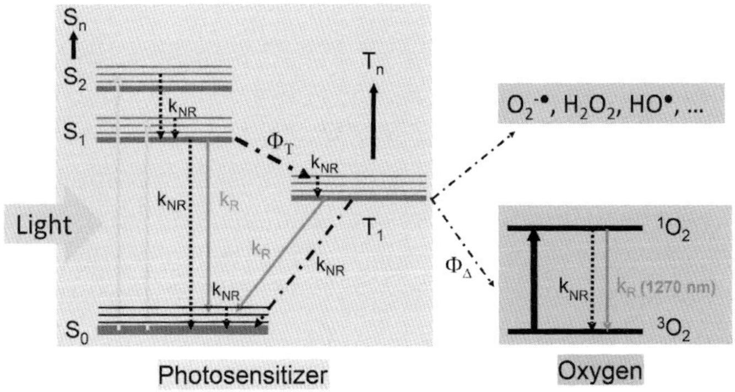

Fig. 1. Mechanism of photodynamic action.

After Absorption of a photon of appropriate wavelengths (400–700 nm) by the PS singlet state S_0 energy is stored in excited and short-lived singlet states S_1–S_n. Higher singlet states S_n relax via internal conversion (IC) to S_1. Firstly, the singlet state S_1 can return to the singlet state S_0 via IC (non-radiative rate K_{NR}) or secondly via the emission of fluorescence (radiative rate k_R).The third pathway is intersystem crossing from S_1 state to a metastable and long-lived triplet state T_1. That triplet state can turn back to S_0 ground state, or transfer energy or charge during collision with other molecules. Photodynamic reaction via type I transfer an electron or proton from the triplet state to a substrate generating radicals like hydrogen peroxide (H_2O_2), hydroxyl radical ($HO^\bullet$), and superoxide anion ($O_2^{-\bullet}$). Type II reaction includes the transfer of energy from T_1 to molecular oxygen in its triplet ground state ($^3O_2$) that leads to the generation of the highly reactive singlet oxygen ($^1O_2$) with the quantum yield Φ_Δ. Almost all $^1O_2$ molecules turn back to $^3O_2$ ground state via non-radiative processes (non-radiative rate K_{NR}). Only a very small amount can be directly detected by its luminescence at 1270 nm (radiative rate k_R).

In type I reactions protons or electrons can be transferred from the excited PS to oxygen or other substrates resulting in generation of radicals (e.g., super oxide anions, hydroxyl radicals). In type II reactions the excited PS transfers energy to molecular oxygen[14] and generates highly reactive singlet oxygen which photo-oxidizes biomolecules like lipids and proteins leading to lysis of cellular membranes and cell wall structures. Both type I and type II reactions can occur simultaneously, and the ratio between these processes depends on the type of PS used. ROS like singlet oxygen have a short lifetime ($<10^{-6}$ seconds) and consequently a limited radius of action ($<0.02\,\mu$m). Therefore, the location of cell damage correlates with the location of the PS molecules.

Thus, photo-inactivation of Gram-positive and Gram-negative bacteria is based on the concept that certain PS can accumulate in significant amounts inside the cell or attached to the cell wall (Fig. 2). The cell wall of bacteria is considered as a critical target to induce an irreversible damage of bacteria. The positive charge of the PS can promote a tight electrostatic interaction with negatively charged sites at the surface of the bacterial cell.[15]

It is known that Gram-positive bacteria are generally more susceptible to PDT as compared to Gram-negative species.[16] This difference is explained by the structural differences in the respective cell wall structures (Fig. 2). The cell wall of Gram-positive bacteria is characterized by a 40–80 nm thick outer peptidoglycan layer with no significant amount of lipids or proteins. Although the murein sacculus contains up to 100 peptidoglycan layers PS are able to cross the cell wall because of the more or less porosity.[17] Measurements of

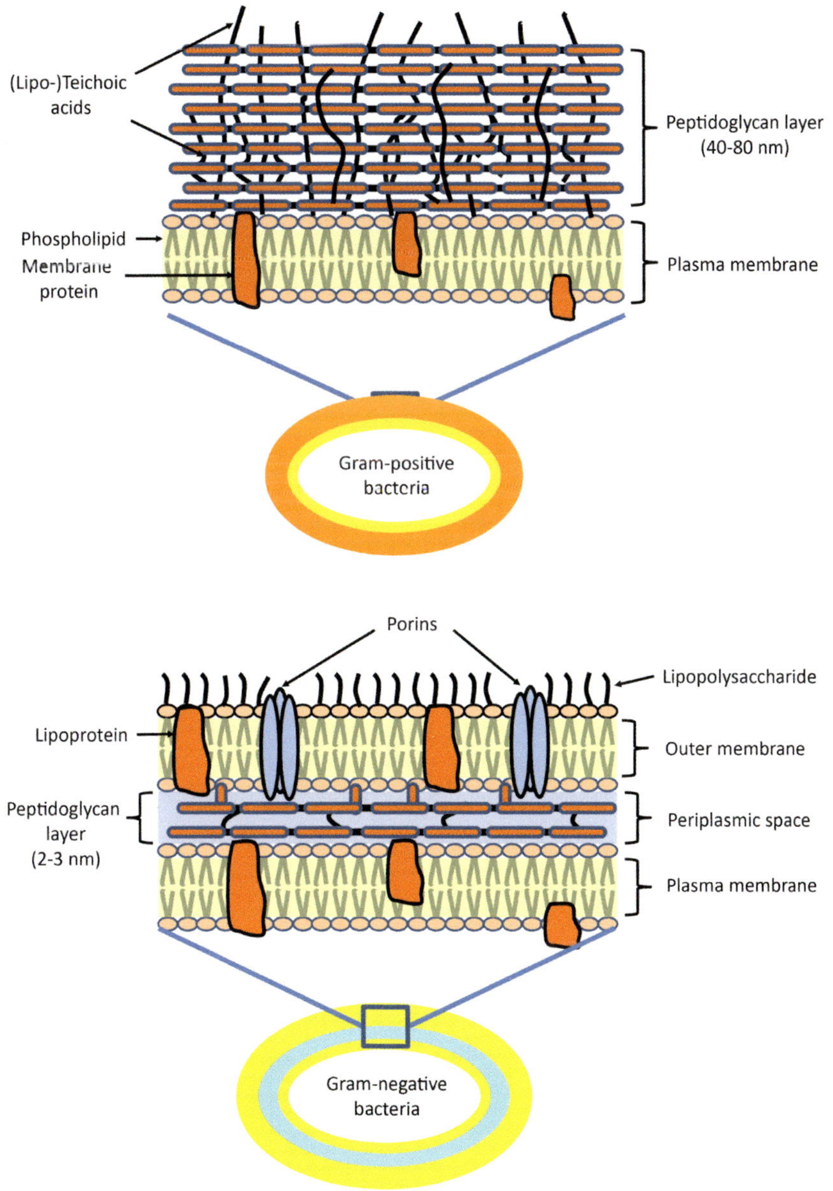

Fig. 2. Cell wall structure of Gram-positive and Gram-negative bacteria.

the penetration of polysaccharides, antimicrobial peptides and glycopeptides have shown that molecules of molecular weight in the range of 30–57 kDa can diffuse through this murein sacculus.[18,19]

In contrast, the cell wall of Gram-negative bacteria consist of three layers: (I) an inner cytoplasmatic membrane (II) a small peptidoglycan layer in the periplasmatic space and (III) an outer membrane with negatively charged lipopolysaccharides (LPS). This outer membrane effectively inhibits the penetration of PS molecules into the cell resulting in a less sensitivity against the PIB treatment in comparison to Gram-positive bacteria.[20]

Nutrients as well as antibiotics must first penetrate through these membranes before they reach their final destination or target. Fractionation and *in vitro* reconstitution of active principles, as well as the use of mutants deficient in the presumptive active component, have led us to conclude that most nutrients and most antibiotics penetrate the outer membrane of *Escherichia coli (E. coli)* and related organisms through water-filled channels produced by a class of proteins called porins.[20] The outer membrane of a single Gram-negative bacterium, such as *E. coli*, contains up to 10^5 water-filled porins.[21]

Positively charged PS are generally more efficient and can act at lower concentrations than neutral and anionic PS molecules.[15] In Gram-negative bacteria, the positive charge of the photosensitizer additionally enhances its uptake through the porins of the outer cell wall into the periplasmic layer.[20] Nevertheless, Gram-negative bacteria show a remarkable resistance to antimicrobial PDT.[22–24] Growth inhibition of *E. coli* by porphyrin-photosensitization was only possible in the presence of the nona-peptide polymyxin or Tris-EDTA which are membrane disorganizing substances.[25] This is due to the different cell wall structures of Gram-positive and Gram-negative bacteria (Fig. 2).

The schematic diagrams show the arrangement of the cell walls of Gram-positive (top) bacteria and Gram-negative (bottom) bacteria. Cell wall of Gram-positive bacteria shows a thickness in the range of 40–80 nm that consists up to 100 interconnecting layers of peptidoglycan. Teichoic acids are interwoven in the peptidoglycan layers. Some have a lipid attached (lipoteichoic acid). The cell wall of Gram-negative bacteria consists of a thin, inner wall composed of 2–3 layers of peptidoglycan (2–3 nm thick), a periplasmic space and an outer membrane (7 nm).[15] The outer membrane contains phospholipids, lipoproteins, LPS and proteins like porins. These porins are protein structures with water filled channels that allow transport of water soluble molecules of usually less than 600 Daltons.

3. PS for Direct Application on Microorganisms

In general, a PS used for antimicrobial PDT should exhibit special features that enable pronounced antimicrobial efficacy and low toxicity toward mammalian cells at the same time, the so-called '*therapeutic window*'. The PS should effectively absorb light that preferentially leads to the population of the triplet

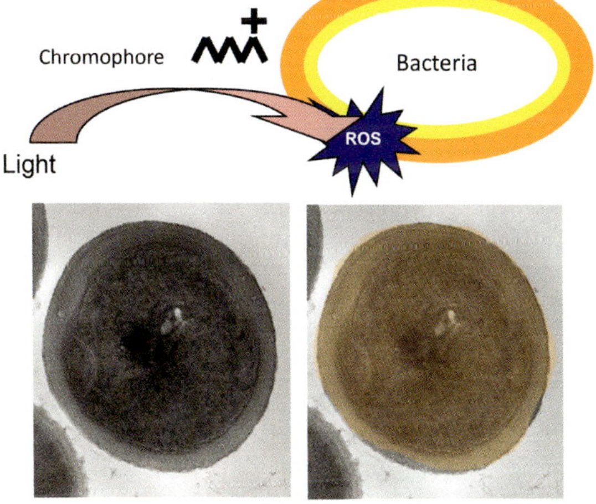

Fig. 3. Top: A schematic PS molecule with a positive charge on a alkyl chain attaches to the negatively charged surface of a microorganism cell (e.g. bacteria). The photosensitizer produces high amounts of ROS that can oxidatively damage the cell wall once the light is switched on. Bottom: In a real setting, the cationic PS will densely covers the entire cell surface leading to sufficient damage of the entire cell. The image pair (transmission electronic microscopy of *Bacillus atrophaeus* endospores) shows for example a bacterial endospore without photosensitizer (left) and coated with the PS (right, the color is artificially added).

T_1 state via intersystem crossing, whereas the conversion of the absorbed light to fluorescence or heat should be low. In contrast, the conversion of light energy via triplet T_1 state to generate ROS should be high. The ROS show a very short range of diffusion in the order of 10 nm, in particular in living cells.[26] Thus, the photosensitizer molecules should eagerly attach to the cell wall of microorganisms before the light is switched on for ROS production. Thus, photosensitizer molecules used in antimicrobial PDT exhibit different numbers of positive charges that enables the attachment to negatively charged cell surface (Fig. 3).

Depending on the time interval for incubation, the photosensitizer may be either attach to the cell surface only or may be taken up by the cells. The longer the incubation time, the higher should be the possibility of PS uptake. The photosensitizer molecules themselves should show only a small or even no toxicity and no mutagenicity to allow a safe application in humans.

Different chemical classes of positively charged PS, including phenothiazines (methylene blue and toluidine blue), phthalocyanines, porphyrins, chlorins, and fullerenes were extensively tested in the past years to kill Gram-positive bacteria and Gram-negative bacteria.[20,27–35] In general, these reports have shown that PS with an overall cationic charge can efficiently kill Gram-negative bacteria by photosensitization. *In vitro*, the use of such PS leads to a decrease of viable bacteria of more than 6 log10 steps depending on the PS concentration and the light dose. An explanation for the killing of Gram- negative bacteria by light activated porphyrins may be the impairment of cellular functions due to the localization of the molecules because of their positive charges, since meso-substituted, but negatively charged porphyrins have not shown toxicity against Gram-negative bacteria, despite the generation of singlet oxygen by both porphyrins.[36,37] Recently two new classes of PS were investigated regarding its use in antimicrobial PDT.

Firstly, Perinaphthenone molecules are known to produce very effectively singlet oxygen with a quantum yield close to unity.[38] However, these compounds usually show no positive charge with little or no effect in antimicrobial PDT. Therefore, Perinaphthenone molecules were chemically changed by adding positive charges at different sites of the chromophore.[39] Some of the new molecules showed an impressive cell killing effect against different bacteria species *in vitro* and even in biofilms.[40,41] The reduction of viable bacteria was in the range of 3–6 log10 orders (measured as colony forming units, CFU) that meets or even exceeds the standard of a successful antimicrobial procedure. Due to the high quantum yield of singlet oxygen production, the cell killing is solely mediated by the generation of singlet oxygen.

Secondly, many vitamin molecules are able to produce ROS such as singlet oxygen upon irradiation.[42,43] Thus, nature offers many dye molecules like riboflavin that show a high ability to generate singlet oxygen when exposed to light.[44] This effect is known e.g. as disadvantageous reaction in food chemistry.[45] Flavin molecules are abundant redox cofactors and of key importance for many biological processes and should have a safe potential to be used in humans, food, and other applications.[46] Therefore, new flavin derivatives were synthesized to create safe and effective PS against pathogens such as MRSA, multiresistant *Pseudomonas aeruginosa, Enterohaemorrhagic E. coli (EHEC)*, or *Acinetobacter baumannii*. The results showed that these novel vitamin derivatives allow an effective killing of Gram-positive bacteria and Gram-negative bacteria with an efficacy of up to 6 log10 steps and regardless the respective resistance patterns.[47]

Conclusion: Antimicrobial PDT can effectively kill various microorganisms by the generation of ROS. In light of short incubation times applied, it seems to be sufficient that the cationic PS attach to the negatively charged cell wall of the microorganisms. The oxidative damage mediated by the generated ROS upon light irradiation kill bacteria regardless their resistance pattern and their cell wall structure. Novel PS such as the vitamin derivatives (flavins) offer an effective and safe treatment of bacterial infections in animals and humans and beyond that.

4. PS for Immobilization on Surfaces — Antimicrobial Surfaces

It is well known that microorganisms can be distributed via contaminated surfaces, which are in alternating contact with different individuals.[48] Pathogenic microorganisms can survive on dry inanimate surfaces for months and years.[49] The skin of any other individual, who gets into contact with such a contaminated surface, can be colonized with such microorganisms that possibly lead to an infection. Therefore, it would be a great advantage to interrupt this transmission pathway by a self-disinfecting effect of surfaces. So far, different approaches and biomaterials were investigated that were recently summarized.[50,51] These technologies comprise bacteria repelling and anti-adhesive surfaces, intrinsically bioactive materials with antibacterial properties (e.g. Chitosan derivatives), bioactive antibacterial coatings (e.g. chlorhexidine grafting on modified titanium), and biomaterials delivering antimicrobials such as silver, copper or any other nanoparticles.[51] Silver nanoparticles were used as monolayer coating, which were firmly anchored on a surface avoiding the release towards microorganisms.[52]

Also photo-activation of surface coatings like antimicrobial PDT can be used to achieve pathogen killing on surfaces. Thin polymer layers containing a PS coat a surface that generates ROS upon light irradiation. After being generated, ROS should diffuse out of the thin coating, reach the microorganisms present on the surface, and kill them via oxidation processes (Fig. 4).

In contrast to standard antimicrobial PDT, the PS molecules are not attached to the microorganisms but are located in the coating. Thus, the diffusion of ROS is required to accomplish the killing of microorganisms on such self-disinfecting surfaces. Among all ROS species, singlet oxygen should play a major role due to its chemical and physical features. Being a neutral molecule, singlet oxygen ($^1O_2$) is — contrarily to other ROS — assumed to show normal diffusion in air, and the lifetime of $^1O_2$ is, besides its spontaneous decay, limited by collisions and reactions with air molecules.

In 1987 Dahl *et al.* already described a method to investigate the toxicity of diffusing $^1O_2$ using spatial separation of PS and bacteria.[53] The half-life of $^1O_2$ was determined by changing the distance between the site of singlet oxygen generation and the position of the bacteria by measuring the bacteria killing of singlet oxygen. These experiments yielded a half-life for singlet oxygen of about

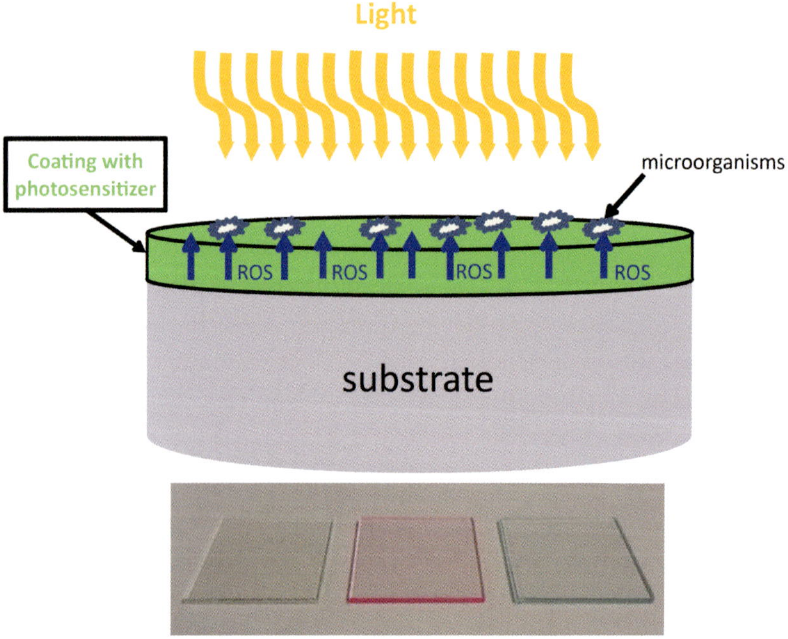

Fig. 4. Top: The scheme shows a substrate (e.g. plastic, glass, metal) that is coated with a thin film containing a PS. Upon light irradiation, ROS like singlet oxygen are produced in the coating. In particular, singlet oxygen can escape the coating via diffusion and reach the microorganisms on the surface. The microorganisms are killed by oxidative damage. Bottom: The image shows three plastic coupons, which were coated with different PS. The thickness of the coating is about 30 μm and the PS concentration in the coating in the range of 10–100 μMol/L.

24 milliseconds (ms) and confirmed findings of the singlet oxygen decay time in the range of milliseconds (35–86 ms) detected in water-saturated air at 5°C.[54] The generated singlet oxygen molecules reached a pathogen by diffusion, and the diffusion length of biologically relevant action was limited by the lifetime of singlet oxygen. These studies provided insight into an interesting field of antimicrobial PDT, where the PS and the targeted pathogen were separated during the PDT process.

It was already shown that polymerisation of a PS prevents oligomerisation and thus self-quenching of singlet oxygen that improved both, singlet oxygen formation and photostability of the PS.[55] Recently, Felgentrager *et al.* presented polyurethane (PU) surface coating that contained the meso-tetraphenylporphyrin (TPP) linked in PU polymer via OH-groups[56] Upon light irradiation the TPP in the surface coating produced singlet oxygen that was directly confirmed by the detection of its luminescence at 1270 nm. The diffusion of singlet oxygen out of the surface was confirmed by using an indirect method like the formation of tri-iodide.[57] After that, *S. aureus* were placed on such a surface and were subsequently irradiated. The PU-TPP surface coating showed a clear antimicrobial effect because the number of bacteria was reduced by up to 3 log10 orders after 30 min irradiation time.

Such a PDT surface coating can be applied as thin varnish to various surfaces consisting of plastic, glass or metal and are of great interest due to a preventive character of infections.[58–60] These surfaces may help to reduce the transmission of pathogens like multiresistant microorganisms, which is a great issue in hospital hygiene.[18] Since the photodynamic process of such surfaces does not necessarily lead to PS consumption, self-disinfecting coatings could offer a long-term and constant prevention of microorganism settlement and growth on any surface.[61]

An antimicrobial surface material was produced that is based on cellulose with incorporated phloxine B using a layer-by-layer deposition method.[62] Antimicrobial properties of the materials were tested in bacterial model systems as well as for decontamination of food and food contact surfaces. The developed materials showed significant self-decontaminating ability when under illumination. However, decontamination of food surfaces in contact with the developed materials was not achieved.

A recently published paper describes the synthesis and characterization of cellulose nanocrystals that were surface-modified with a cationic porphyrin.[63] The porphyrin was appended onto the cellulose surface via the Cu(I)-catalyzed Huisgen-Meldal-Sharpless 1,3- dipolar cycloaddition having occurred between azide groups on the cellulosic surface and porphyrinic alkynes. The resulting, generally insoluble, crystalline material showed excellent efficacy toward the photodynamic inactivation of *Mycobacterium smegmatis* and *S. aureus*, albeit only slight activity against *Escherichia coli*. The authors recommend such a surface for its use against a range of bacteria, with potential utilization in the health care and food preparation industries.

In another publication, antibacterial surfaces were prepared using a base polyethylene sheet topped with a layer containing a mixed powder of poly (vinylidene fluoride) and three different PS, rose bengal, toluidine blue O or methylene blue.[64] Photodynamic inactivation assays performed in nutrient broth demonstrated more than 4 log10 reduction of the attached *E. coli* after illumination for 24 h. More than 4 log10 reduction of *S. aureus* occurred even when the cultures were illuminated for only 6 h.

The use of high density poly(ethylene) (HDPE) was studied regarding its efficacy as antimicrobial surface. A two-layer system with one layer comprising PS-incorporated HDPE, and one layer comprising HDPE alone was investigated to demonstrate the versatility of that approach. The PS-incorporated materials were capable of reducing the adherence of viable MRSA by up to 3.6 log10 orders per centimeter of material surface.[65]

Conclusion: The major prerequisites for any surface coating are photostability of the PS, durability of the coating, and light irradiation. In case of high a light intensity, the antimicrobial effect on the surface should be achieved within seconds or minutes. In case of a low light intensity (e.g. ambient light), the time of irradiation should be as long as possible in the range of several hours.

5. Antimicrobial PDT in Animal Models

Animal models have become standard tools for the study of a wide array of antimicrobial therapies, in particular of wound infections, including antimicrobial PDT, and mice are by far the most frequently used species for wound infection models.[66] In order to facilitate the non-invasive monitoring of animal models of infection, a procedure was developed that uses genetically-engineered bioluminescent bacteria and a light sensitive imaging system to allow real-time visualization of infections.[67]

A mouse model of skin abrasion wound infected with MRSA was developed by Dai *et al.*[68] Bioluminescent MRSA was used to allow the real-time monitoring of the extent of infection in mouse wounds. PDT was performed with the combination of a polyethylenimine (PEI)-ce6PS and non-coherent red light. Antimicrobial PDT induced on average 2.7 log10 orders of inactivation of MRSA as judged by loss of bioluminescence in mouse skin abrasion wounds and accelerated the wound healing on average by 8.6 days in comparison to the untreated infected wounds.

Also highly water-soluble fullerene mono-adducts were applied to determine its antimicrobial effects against human pathogens like *S. aureus*, *E. coli*, and *A. baumannii* upon irradiation with UVA or white light. Two compounds out of the fullerenes were effective in a mouse model against Gram-negative third-degree burn infections determined by bioluminescence imaging.[35]

Another study investigated whether antimicrobial PDT combined with antibiotics used for MRSA could provide an additional benefit in controlling the infection.[69] The authors compared the effect of methylene blue alone, systemic administration of either LZD alone or VCM alone or the combination of antimicrobial PDT with either LZD or VCM. PDT alone was the most effective, whereas LZD alone was ineffective and VCM alone showed some benefit. Surprisingly the addition of LZD or VCM reduced the therapeutic effect of PDT alone. Considering that PDT in this mouse model stimulates neutrophils to be antibacterial rather than actively killing the bacteria, the authors propose that LZD and VCM might inhibit the activation of inflammatory cytokines without eradicating the bacteria, and thereby reduce the therapeutic effect of PDT.

Candida albicans pathogenicity was studied with a mouse model of systemic infection.[70] Sublethal dosis of methylene blue mediated antimicrobial PDT reduced the growth rate and the ability of *C. albicans* to form germ tubes compared to untreated cells. Survival of mice systemically infected with *C. albicans* pretreated with antimicrobial PDT was significantly increased compared to mice infected with untreated fungal cells. Antimicrobial PDT even increased *C. albicans* sensitivity to sodium dodecyl sulfate, caffeine, and hydrogen peroxide. The authors suggested that antimicrobial PDT may inhibit virulence factors and reduce *in vivo* pathogenicity of *C. albicans.*

PDT experiments in rats were performed to investigate the treatment of experimentally induced periodontitis.[71] The application of toluidine blue mediated antimicrobial PDT in reduced bone loss compared with scaling and root planning treatment. Additionally, rats treated with antimicrobial PDT exhibited reduced numbers of tartrate-resistant acid-phosphatase-positive cells and more proliferating nuclear antigen-positive cells in all treatment groups regardless of estrogen status. The authors concluded that antimicrobial PDT is an effective adjunctive therapy for the treatment of periodontitis in rats.

Antimicrobial PDT was studied using highly pure chlorine(6), which has been widely used as a second-generation PS.[72] The authors report effective inhibition of *in vitro* growth of several bacterial strains. To clarify a possibility for its clinical application, they examined *in vitro* and *in vivo* antimicrobial effects of chlorine(6) mediated PDT in mice models of skin infection of *S. aureus* Xen29. Inhibition zone analysis and CFU counts revealed that antimicrobial PDT effectively inhibited the *in vitro* bacterial growth and biofilm formation ability of *S. aureus* Xen29. *In vivo* experiments showed that mice receiving chlorine(6)- mediated PDT exhibited less intensity of bioluminescent signals, showing a significant inhibition of bacterial growth.

Antimicrobial photodynamic treatment was investigated with poly-lysine-conjugated chlorin p6 (pl-cp6) on pro-inflammatory cytokine expression and wound healing in a murine excisional wound model infected with *P. aeruginosaa.*[73] Treatment of infected wounds with a chlorine derivative and light reduced the bacterial load after 24 h up to 2.0 log10 steps. The treated wounds healed ~5 days earlier as compared to untreated controls and wound closure was not dependent on the light dose. Interestingly, 96 h post-treatment, the drug-treated wounds irradiated at 60 J/cm^2 showed considerable reduction of pro-inflammatory cytokines IL-6 (approximately five times) and TNF-α (approximately four times) compared to untreated controls. These results imply that chlorin-mediated PDT can reduce bacteria load and hyper-inflammatory response of infected wounds, leading to acceleration of wound healing.

Antimicrobial PDT was tested even in cutaneous leishmaniasis using carbaporphyrin ketals that are porphyrinoid compounds in which a pyrrole ring of a typical porphyrin macrocycle has been replaced by a ketal-substituted indene ring.[74] Its effectiveness *in vitro* against various pathogenic Leishmania species was determined to be in a micromolar range. Additionally, the effect of encapsulating the carbaporphyrin ketals in liposome formulations was tested. When testing the PS formulations *in vivo* using a hamster (*Mesocricetus auratus*) model of cutaneous leishmaniasis, the PS was active even in the dark, suggesting that the compound, once metabolized in the animal tissue, produces an active ingredient that does not seem to be photosensitive.

Conclusion: The results of the published animal studies clearly show that the models are very helpful to study the antimicrobial PDT under *in vivo* conditions. Different PS were tested against various microorganisms showing the good efficacy of antimicrobial PDT against sensitive and multiresistant bacteria, *Candida*, and *Leishmania*.

6. Antimicrobial PDT in Clinical Applications

It is major goal to apply antimicrobial PDT in humans to destroy pathogenic microorganisms. Due to a limited penetration depth of light into the tissue, the use of antimicrobial PDT is an optimal procedure to decolonize intact skin or mucosa and to kill bacteria in wounds or soft tissue infections.

One of the first clinical investigations were performed with 5-aminolevulinic acid (5-ALA), the precursor of the PS protoporphyrin IX, which has been widely applied in the killing of cells in cancerous and pre-cancerous lesions.[75] 5-ALA seems suitable for the treatment of human mycosis because *in vitro* studies have demonstrated that fungi can effectively be photosensitized by 5-ALA and killed at dose rates much lower than those that kill keratinocytes.[76] Its clinical efficacy has been investigated with an open pilot study enrolling nine patients with interdigital mycosis of the feet caused by *C. albicans*, *Trichophyton mentagrophytes* and *Trichophyton rubrum*.[77] Clinical and microbiological recovery was shown in six out of nine patients after one (four cases) or four (two cases) treatments. However, two patients (one with

C. albicans and one with *Trichophyton mentagrophytes* infection) had a persistent remission at a 4 weeks follow up visit. Overall tolerability was good in all patients. Another recent clinical open study conducted by Lee and co-workers investigated the efficacy of methyl 5-ALA (MAL)-PDT in six patients with recalcitrant *Malassezia* folliculitis.[78] After three sessions of MAL-PDT at 2-week intervals, inflammatory lesions decreased in four patients, slightly improved in one patient; the treatment was not effective in one patient.

In dentistry, a study compared the effect of subgingival ultrasonic scaling followed by repeated (three times) antimicrobial photodynamic therapy using methylene blue, ultrasonic scaling alone (U.S.), and scaling and root planing (SRP) with hand instruments for initial periodontal treatment.[79] The authors enrolled 27 systemically healthy patients with chronic periodontitis in the study. PDT reduced the proportion of positive sites after six months for *Treponema denticola* significantly more effective than US or SRP alone. Additionally, PDT resulted in a greater reduction of *Aggregatibacter actinomycetemcomitans* and *Tannerella forsythia* compared to mechanical debridement alone.

Antimicrobial PDT can be applied even in the root channel of teeth.[80] The authors applied the PS 5,10,15,20-tetra(m-hydroxyphenyl)chlorin (mTHPC) incorporated in liposomes and highly flexible invasomes on the endodontopathogenic species to kill *E. faecalis* in infected dental root canals. The results were compared to standard disinfection using 1 % chlorohexidine gel. Both mTHPC formulations showed a significant antimicrobial effect. A bacterial reduction by up to 3.6 log10 steps was ascertained. In contrast, chlorohexidine caused a decrease of bacteria number of only 1.2 log10 steps. It was found that both liposomal mTHPC formulations were capable to suppress the growth of *E. faecalis* inside the dentinal tubules up to 300 μm. These results are of great clinical importance because *E. faecalis* is resistant to several conventional antibacterial treatment measures.

First results of clinical trials were report on the efficacy of antimicrobial PDT in onychomycosis.[81] 80 patients with toenail onychomycosis were enrolled in a study, in which the effect of topical treatment with methylene blue light was compared to a treatment with 300 mg oral fluconazole per week. When assessing the efficacy 12 months after treatment, the authors reported a significant better response of the lesions to the photodynamic treatment.

Antimicrobial PDT was also studied in treating infected foot ulcers (62 patients) with a randomized, double-blind, parallel series, and placebo-controlled phase IIa trial.[82] The authors investigated the antimicrobial effect and tolerability of a single photodynamic treatment using novel cationic zinc phthalocyanine derivatives. No safety issues emerged from the analysis of adverse events. The authors stated that the photodynamic treatment was well tolerated and produced a significant reduction in germ load. They will perform further clinical trials to verify the efficacy of this approach as add-on to systemic antibiotic treatment.

Another study also investigated whether antimicrobial PDT can effectively be applied in bacterially colonized chronic leg ulcers and chronic diabetic foot ulcers to reduce bacterial load, and potentially lead to accelerated wound healing.[83] The authors used 3,7-bis(N,N- dibutylamino) phenothiazin-5-ium bromide as PS. Sixteen patients with diabetic foot ulcers were enrolled in a blinded, randomized, placebo-controlled, single-treatment, Phase IIa trial. Treatment was well tolerated with no reports of pain or other safety issues. In contrast to placebo, patients on active treatment showed a reduction in bacterial load immediately post-treatment. After 3 months, 50% of the patients with actively treated chronic leg ulcer showed complete healing, compared with 12% of patients on placebo. Here, the antimicrobial PDT treatment demonstrated a significant reduction in bacterial load. The authors also reported an apparent trend towards wound healing that will be studied with larger patient numbers.

Conclusion: The use of antimicrobial PDT in humans requires the clinical approval of the PS as medical drug. This in turn requires PS, which are non-toxic, non-mutagenic, effective against different species of bacteria and possibly other microorganisms, and patentable. So far, most of the clinical trials were performed in dermatology and dentistry. The studies in dentistry particularly comprised the treatments of periodontitis, root channel infection, and periimplantitis.

7. Conclusion and Advantages of Antimicrobial PDT

Microorganisms such as bacteria and viruses are champions of evolution and a few microbes have adapted to a point where they pose serious clinical challenges for humans. At the same time, new antibiotic and antiviral drugs are missing while the efficacy of common antibiotic drugs is dramatically decreasing.

Photodynamic inactivation of microorganisms proves to be an additional method to kill pathogenic microorganisms. It requires PS molecules that effectively generate ROS like singlet oxygen when exposed to visible light. The PS should show a high absorption coefficient for light in the visible part of the spectrum and a high quantum yield of ROS generation. To allow a broad and sustainable application, PS should be safe when applied in environment, food or humans. The technology optimally works in all fields, which can be reached by visible light and PS. Thus, the major application is decontamination of surfaces and also inside tissue down to a few millimeters.

Two different approaches are currently developed in antimicrobial PDT:

(i) The application of the PS molecules in a solution or suspension that enables attachment of the cationic ps molecules to the microorganism. Upon irradiation with visible light, ROS are then produced directly at the microorganisms leading to oxidative cell death. Appropriate formulations must be developed allowing an effective application of the PS on the area that needs disinfection.

(ii) The immobilization of the PS in a thin varnish that can coat different materials. Upon irradiation with visible light, ROS like singlet oxygen are then produced in that coating, diffuses to the surface and hence kill the microorganisms, which contaminate the surface. The major prerequisites for any surface coating are photostability of the PS, durability of the coating, and light irradiation. The technology should work with high light intensity of light sources (e.g., LED) or with low light intensity when the surface is exposed to ambient light.

With either procedure bacteria are killed regardless their resistance pattern and their cell wall structure. Novel PS such as the vitamin derivatives (flavins) offer an effective and safe treatment of bacterial infections in food, animals and humans and beyond that.

The results of the published animal studies show encouraging results and can assist in determining the optimal parameters for antimicrobial PDT under *in vivo* conditions. Different PS were tested against various microorganisms showing the good efficacy of antimicrobial PDT against normal and multiresistant bacteria, *Candida*, and *Leishmania*. The use of antimicrobial PDT in humans requires the clinical approval of the PS as medical drug. This in turn requires safe PS, which are non-toxic, non-mutagenic, effective against different species of bacteria and other microorganisms, and patentable. Most of the clinical trials were performed in dermatology and dentistry. The studies in dentistry particularly comprised the treatments of periodontitis, root channel infection, and periimplantitis.

References

1. C. A. Arias and B. E. Murray, *N. Engl. J. Med.* **360**(5), 439–443 (2009).
2. C. Daniel, *Drug-resistant bacteria and lack of new antibiotics pose 'catastrophic threat'. Nature News Blog* (2013). http://blogs.nature.com/news/2013/03/drug-resistant-bacteria-and-lack-of-new-antibiotics-pose-catastrophic-threat.html.
3. H. White, *National strategy for combating antibiotic resistant bacteria*, (2014) https://www.whitehouse.gov/the-press-office/2014/09/18/executive-order-combating-antibiotic-resistant-bacteria.
4. E. Nannini, B. E. Murray and C. A. Arias,*Curr. Opin. Pharmacol.* **10**(5), 516–521 (2010).

5. F. Layer, C. C. B. Strommenger, G. Werner and W. Witte,*Bundesgesundheitsblatt* 2012, **55**, 1377–1386 (2012).
6. R. M. Klevens *et al.*, *Clin Infect. Dis.* **42**(3), 389–391 (2006).
7. J. Melo-Cristino *et al.*, *Lancet* **382**, 9888 p. 205, 20 July (2013).
8. R. N. Master *et al.*, *Ann. N. Y. Acad. Sci.*, **1277**, 1–7 (2013).
9. B. Spellberg, *Lancet Infect. Dis.* **8**(4), 211–212; author reply 212–214 (2008).
10. B. Spellberg *et al.*, *Clin. Infect. Dis.* **38**(9), 1279–1286 (2004).
11. M. Buehlmann *et al.*, *Infect. Control Hosp. Epidemiol.* **29**(6), 510–516 (2008).
12. S. A. Fritz *et al.*, *Antimicrob. Agents Chemother.* **57**(1), 559–568 (2013).
13. P. D. Johnson *et al.*, *Med.J. Aust.* **183**(10), 509–514 (2005).
14. R. Kilger, M. Maier, R. Szeimies and W. Bäumler, *Chem. Phys. Lett.* **343**(5–6), 543–548 (2001).
15. T. Maisch *et al.*, *Photochem. Photobiol. Sci.* **3**(10), 907–917 (2004).
16. T. N. Demidova and M. R. Hamblin, *Antimicrob. Agents Chemother.* **49**(6), 2329–2335 (2005).
17. Z. Malik, H. Ladan and Y. Nitzan, *J. Photochem. Photobiol. B* **14**(3), 262–266 (1992).
18. C. L. Friedrich *et al.*, *Antimicrob. Agents Chemother.* **44**(8), 2086–2092 (2000).
19. R. Scherrer and P. Gerhardt, *J. Bacteriol.* **107**(3), 718–735 (1971).
20. H. Nikaido, E. Y. Rosenberg and J. Foulds,*J. Bacteriol.* **153**(1), 232–240 (1983).
21. H. Nikaido and E. Y. Rosenberg, *J. Bacteriol.* **153**(1), 241-52 (1983).
22. N. Kashef, G. Ravaei Sharif Abadi and G. E. Djavid, *Photodiagnosis Photodyn. Ther.* **9**(1), 11–15 (2012).
23. T. Maisch *et al.*, *J. Ind. Microbiol. Biotechnol.* **39**(7), 1013–1021 (2012).
24. G. P. Tegos and M. R. *Antimicrob. Agents Chemother.* **50**(1), 196–203 (2006).
25. J. Walther *et al.*, *FEMS Microbiol. Lett.* **290**(2), 156–163 (2009).
26. S. Hatz, L. Poulsen and P. R. Ogilby, *Photochem. Photobiol.* **84**(5), 1284–1290 (2003).
27. N. Komerik *et al.*, *Antimicrob. Agents Chemother.* **47**(3), 932–940 (2003).
28. T. Maisch *et al.*, *J. Appl. Microbiol.* **107**(5), 1569–1578 (2009).
29. A. S. Garcez *et al.*, *J. Endod.* **34**(2), 138–142 (2008).
30. X. Ragas, *Lasers Surg. Med.* **42**(5), 384–390 (2011).
31. R. A. Prates *et al.*, *J. Antimicrob. Chemother.* **66**(7), 1525–1532 (2011).
32. S. K. Sharma *et al.*, *Curr. Pharm. Des.* **17**(13), 1303–1319 (2011).
33. M. Wainwright *et al.*, *Lett. Appl. Microbiol.* **50**(5), 486–492 (2010).
34. Y. T. Yang *et al.*, *Lasers Surg. Med.* **45**(3), 175–185 (2013).
35. L. Y. Huang *et al.*, *Nanomedicine* **9**(2), 253–266 (2014).
36. M. Merchat *et al.*, *J. Photochem. Photobiol. B* **35**(3), 149–157 (1996).
37. F. Allegra and M. Jori, *Soc. Ital. Dermatol. Sifilogr. Sezioni. Interprov. Soc. Ital. Dermatol. Sifilogr.* **1955**(2), 106.
38. R. Schmidt *et al.*, *J. Photochem. Photobiol. A: Chem.* **79**(1–2), 11–17 (1994).
39. A. Spath *et al.*, *J. Med. Chem.* **57**(12), 5157–5168 (2014).
40. F. Cieplik *et al.*, *Free Radic. Biol. Med.* **65**, 477–487 (2013).
41. F. Cieplik *et al.*, *Int. J. Med. Microbiol.* **303**, 41–41 (2013).
42. A. Knak *et al.*, *Photochem. Photobiol. Sci.* **13**(5), 820–829 (2014).
43. W. Baumler *et al.*, *Photochem. Photobiol. Sci.* **11**(1), 107–117 (2012).
44. J. Baier *et al.*, *Biophys. J.* **91**(4), 1452–1459 (2006).
45. D. G. Bradley, H. J. Kim and D. B. Min, *J. Agric. Food Chem.*, **54**(16), 6016–6020 (2006).

46. S. O. Mansoorabadi, C. J. Thibodeaux and H. W. Liu, *J. Org. Chem.*, **72**(17), 6329–6342 (2007).
47. T. Maisch *et al.*, *PLoS One*, 2014 Dec 3; 9(12):e111792 (2014).
48. V. Decraene *et al.*, *J. Gen. Appl. Microbiol.* **54**(4), 195–203 (2008).
49. A. Kramer, I. Schwebke and G. Kampf, *BMC Infect. Dis.* **6**, 130 (2006).
50. K. Page, M. Wilson and I. P. Parkin, *J. Mater. Chem.* **19**(23), 3819–3831 (2009).
51. D. Campoccia, L. Montanaro and C. R. Arciola, *Biomaterials* **34**(34), 8533–8554 (2013).
52. A. Taglietti *et al.*, *Biomaterials* **35**(6), 1779–1788 (2014).
53. T. A. Dahl, W. R. Midden and P. E. Hartman, *Photochem. Photobiol.* **46**(3), 345–352 (1987).
54. W. R. Midden and S. Y. Wang, *J. Am. Chem. Soc.* **105**(13), 4129–4135 (1983).
55. J. Wahlen, *et al.*, *Adv. Synth. Catal.* **346**(2–3), 152–164 (2004).
56. A. Felgentrager *et al.*, *Phys. Chem. Chem. Phys.* **16**(38), 20598–20607 (2014).
57. J. Mosinger and B. Mosinger, *Experientia* **51**(2), 106–109 (1995).
58. V. Decraene, J. Pratten and M. Wilson, *Curr. Microbiol.* **57**(4), 269–273 (2008).
59. V. Decraene, J. Pratten and M. Wilson, *Infect. Control Hosp. Epidemiol.* **29**(12), 1181–1184 (2008).
60. S. Ismail *et al.*, *Infect. Control Hosp. Epidemiol.* **32**(11), 1130–1132 (2011).
61. M. Wilson, *Infect. Control Hosp. Epidemiol.* **24**(10), 782–784 (2003).
62. L. Brovko *et al.*, *J. Appl. Microbiol.* **117**(5), 1260–1266 (2014).
63. E. Feese *et al.*, *Biomacromolecules* **12**(10), 3528–3539 (2011).
64. R. Cahan *et al.*, *Photochem. Photobiol.* **87**(6), 1379–1386 (2011).
65. C. P. McCoy *et al.*, *PLoS One* **9**(9), e108500 (2014).
66. F. F. Sperandio, Y. Y. Huang and M. R. Hamblin, *Recent Pat. Antiinfect. Drug Discov.* **28**(2), 108–120 (2013).
67. T. N. Demidova *et al.*, *J. Photochem. Photobiol. B-Biol.* **81**(1), 15–25 (2005).
68. T. H. Dai *et al.*, *Lasers Surg. Med.* **42**(1), 38–44 (2010).
69. M. Tanaka *et al.*, *Photochem. Photobiol.* **89**(3), 679–682 (2013).
70. I. T. Kato *et al.*, *Antimicrob. Agents Chemother.* **57**(1), 445–451 (2013).
71. V. G. Garcia *et al.*, *J. Periodontol.* **84**(4), 556–565 (2013).
72. J. H. Park *et al.*, *Biol. Pharm. Bull* **35**(4), 509–514 (2012).
73. K. Sahu *et al.*, *Lasers Med. Sci.* **28**(2), 465–471 (2013).
74. V. M. Taylor *et al.*, *Antimicrob. Agents Chemother.* **55**(10), 4755–4764 (2011).
75. P. K. Lee and A. Kloser, *J. Drugs Dermatol.* **12**(8), 925–930 (2013).
76. G. Monfrecola *et al.*, *Photochem. Photobiol. Sci.* **3**(5), 419–422 (2004).
77. P. G. Calzavara-Pinton, *Photodermatol. Photoimmunol. Photomed.* **20**(3), 144–147 (2004).
78. J. W. Lee, B. J. Kim and M. N. Kim, *Lasers Surg. Med.* **42**(2), 192–196 (2010).
79. M. Petelin *et al.*, *Lasers Med. Sci.* **30**(6), 1647–56 Aug; (2015).
80. A. Ossmann *et al.*, *Clin. Oral. Investig.* 19(2), 373–84 Mar; (2015).
81. L. W. Figueiredo Souza, S. V. Souza and A. C. Botelho, *Dermatol. Ther.* **27**(1), 43–47 (2014).
82. E. Mannucci *et al.*, *Acta Diabetol.* **51**(3), 435–440 (2014).
83. S. Morley *et al.*, *Br. J. Dermatol.* **168**(3), 617–624 (2013).

Chapter 18

Antimicrobial Effects of Photosensitized Silicon Phthalocyanine (Pc) 4

Minh Lam[†], *Matthew L. Dimaano*[†], *Pranab K. Mukherjee*[‡], *Mahmoud A. Ghannoum*[‡], *Nancy L. Oleinick*[§] *and Elma D. Baron*[†,¶,*]

[†]*Department of Dermatology, Case Western Reserve University, University Hospitals Case Medical Center, Cleveland, OH 44106*

[‡]*Center for Medical Mycology, University Hospitals Case Medical Center, 11100 Euclid Avenue, Case Western*

[§]*Department of Radiation Oncology, Case Western Reserve University, Cleveland, OH 44106*

[¶]*Dermatology Service, Louis-Stokes VA Medical Center, Cleveland, OH 44106*

elma.baron@case.edu

Abstract

The incidence and prevalence of bacterial and fungal infections will continue to increase because of (i) more frequent use of broad-spectrum antibiotics and antifungals as well as immunosuppressive medications (ii) increased number of invasive medical procedures and (iii) higher incidence of neutropenia and HIV infections. In addition to the adverse side effects, the growing resistance against both antibiotic and antifungal agents has driven the continuing search for alternative treatment modalities. One promising option is photodynamic therapy (PDT). Numerous *in vitro* studies have clearly demonstrated activity of PDT against viruses, bacteria, yeast-like fungi and dermatophytes, using photosensitizers that include phenothiaziniums, porhyrins and phthalocyanines. Our lab has demonstrated that PDT sensitized by silicon phthalocyanine (Pc) 4 can efficiently induce cell death of *Candida albicans, Trichophyton rubrum* and *Staphylococcus aureus.* Because of the mechanism of action of PDT, which involves

*Corresponding author.

oxidative damage to cell membranes, microbial killing was observed in all species, regardless of their susceptibility or resistance to standard antimicrobial agents.

1. Introduction

The exploration for new antimicrobial agents against infections caused by opportunistic bacterial and fungal pathogens is accelerating at a rapid pace due to a rise of drug resistance. Aside from the issue of resistance, currently available drugs exhibit significant toxicity to various organ systems. Therefore, the search for more optimal treatments is warranted. Currently, one of the more encouraging candidates is photodynamic therapy (PDT). Initially developed to treat malignant tumors, PDT has been shown to be a promising treatment modality for dermatologic indications, both inflammatory and neoplastic, including cutaneous T-cell lymphoma, basal and squamous cell carcinoma.[1,2] In addition, PDT has been employed in non-malignant diseases, such as macular degeneration, recalcitrant viral warts, and acne as well as psoriasis.[3] In recent years, there is well-documented evidence to suggest that PDT is effective against viruses, bacteria and yeast-like fungi, including dermatophytes.[4] PDT involves administering a photosensitizing drug followed by delivering light of an appropriate wavelength, fluence and fluence rate, in the presence of molecular oxygen in tissue. The interaction of photosensitizer, light and oxygen, results in a cascade of events leading to the generation of singlet oxygen and other forms of reactive oxygen species (ROS) to exert cytotoxic effects on subcellular constituents, such as mitochondria, and eventually leads to the induction of apoptosis and tumor cell destruction.[2,5,6] The findings of apoptotic responses in pathogenic fungi, such as *Aspergillus fumigatus* and *C. albicans*,[7–13] for instance, have fueled the possibility of developing novel antifungal agents that would clear infections by activating the fungal suicide pathway. To successfully design and implement an ideal novel antifungal agent, however, one must be able to fully understand the fungal suicide pathway and find the known components that are unique or highly divergent from their mammalian counterparts. Targeting these "unique" components could provide insight and aid in designing novel antifungal agents that could clear infections, thus minimizing harm to the host.

2. PDT with Silicon Phthalocyanine (Pc) 4

Pc 4 is a second-generation photosensitizer developed at Case Western Reserve University (Fig. 1). Upon exposure to 675-nm red light, Pc 4 activated to an excited singlet state, which undergoes intersystem crossing to a triplet state.[14] The excited triplet Pc 4 interacts with molecular oxygen forming ROS,

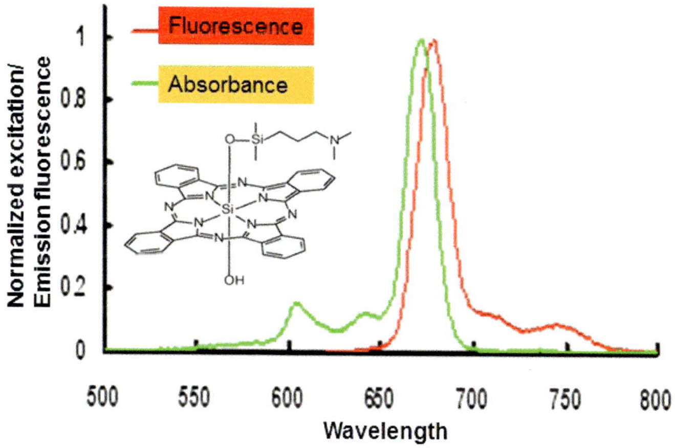

Fig. 1. Structure and absorbance and fluorescence spectra of Pc 4 — A silicon phthalocyanine with a dimethylamino-propylsiloxy ligand that links axially to the silicon in the center, Pc 4 is an effective, second-generation chemically pure photosensitizer. In culture medium, the absorbance and fluorescence maximum occurred at 672 nm and 675 nm, respectively. The absorbance spectrum may be ideal for tissue penetration and minimal cutaneous photosensitivity.

which cause oxidative damage to proteins and lipids within their immediate vicinity.[15,16] Pharmacokinetic data in mice[17] indicate that Pc 4 is much more rapidly cleared than the FDA-approved Photofrin® when delivered systemically, thereby minimizing the possibility of prolonged generalized photosensitivity.[18] Pc 4-PDT has demonstrated an excellent safety profile in a Phase 1 trial on cutaneous neoplasms.[19]

We are currently conducting two Phase 1 trials of topically applied Pc 4-PDT, for cutaneous T-cell lymphoma and psoriasis, in our institution. To demonstrate its versatility, we recently examined the cytotoxic effects of Pc 4-PDT in *C. albicans*,[10] *T. rubrum*[20] and *S. aureus* (*Dimaano, unpublished data*) *in vitro*. We concluded that the cytotoxic effects of Pc 4-PDT not only applied to all the common microbes, but also to their respective resistant strains. These preclinical studies are significant for they will help in developing an innovative strategy that utilizes inherent molecular cell death mechanisms without involving the genomic blueprint and thus potentially prevent resistant strains from emerging. Furthermore, the lack of mutagenicity and genotoxicity to both microbial and human DNA also favor PDT's long-term safety.[21]

3. Pc 4-PDT Affects Overall Cell Killing as Demonstrated by CFU Assay

To demonstrate the cytotoxicity of PDT sensitized by Pc 4, CFU assay was performed as previously described[10] with minor modifications. Briefly, following photoirradiation, *C. albicans, non-albicans, T. rubrum* and *S. aureus* cell suspensions that received a single exposure to Pc 4-PDT (1.0 μM Pc 4 and 2.0 J/cm^2) were serially diluted in 1×PBS then plated on agar (Becton Dickinson and Company, Sparks, MD, USA) plates and incubated at 37°C to allow colony formation. Aliquots of the cell suspension were plated onto 100-mm Petri dishes in amounts sufficient to yield at least 50–100 colonies per dish. After 24 h, colonies were counted by eye. For example, we tested and found that one dose of Pc 4-PDT has a similar cytotoxic effect, as demonstrated by CFU assay, in amphotericin B- and fluconazole-resistant strains of both *C. albicans* and *C. glabrata*. These strains were *C. albicans* strains 15428 (amphotericin B and fluconazole sensitive) and 9652 (amphtoericin B and fluconazole resistant) and *C. glabrata* strains 9193 (amphotericin B and fluconazole sensitive) and 1911 (amphotericin B and fluconazole resistant). All *C. albicans/non-albicans* strains were available from the culture collection in the Center for Medical Mycology of University Hospitals Case Medical Center. Most antifungal resistant strains, e.g., *C. albicans* 9652 and *C. glabrata* 1911, are clinical isolates obtained from patients who had failed amphotericin B and fluconazole therapy. Antifungal susceptibility testing was performed according to the M27-A methodology based upon the *National Committee for Clinical Laboratory Standards.* Table 1 shows the remaining organisms and their

Table 1. Pc 4 phototoxicity in different strains of *C. albicans/non-albicans, T. rubrum* and *S. aureus* — Pc 4-PDT induces ROS generation, affects metabolic activity (XTT) and reduces cell survival (CFU) equally regardless of antibiotic or antifungal sensitivity; ND, not determined.

Organisms	CFU	XTT	ROS
Candida albicans (SC-5134)	√	√	√
Candida albicans (GNDN-2346)	√	ND	ND
Candida albicans (15428)	√	√	ND
Candida albicans (9652)	√	√	ND
Candida glabrata (A239)	√	ND	ND
Candida glabrata (9193)	√	√	ND
Candida glabrata (1911)	√	√	ND
Candida kruzei (MRL-214)	√	ND	ND
Candida tropicalis (MRL-15020)	√	ND	ND
Candida dubliniensis (MRL-14973)	√	ND	ND
Candida parasilosis (CP-514)	√	ND	ND
Candida orthopsilosis (CO-907)	√	ND	ND
Candida metapsilosis (CM-694)	√	ND	ND
Candida keyfer (ATCC-46764)	√	ND	ND
Trichophyton rubrum (24602)	√	√	√
Trichophyton rubrum (MRL-666)	√	√	√
MSSA (ATCC 25923)	√	ND	√
MRSA-HA (PFGE type 300)	√	ND	√
MRSA-CA (ATCC 43300)	√	ND	√

respective drug-resistant strains which were tested and found equally affected by Pc 4-PDT.

4. Pc 4 Uptake into *C. albicans, T. rubrum* and *S. aureus*

We have previously shown that Pc 4 preferentially, but not exclusively, binds to mitochondria of a variety of mammalian cells.[22–24] In *C. albicans*, the majority of Pc 4 fluorescence was detected in the cytoplasmic compartment but not the cell wall, plasma membrane or nucleus, similar to its distribution in mammalian cells.[23] In *C. albicans*, Pc 4 displayed a punctate pattern of fluorescence primarily localized in a cytoplasmic region distinct from the plasma membrane (Fig. 2) and resembling mitochondria.[10] In addition, as previously found in mammalian cells, Pc 4 also binds to other intracellular organelles, presumably ER membranes, as well as other vesicular membranes. Similarly, Pc 4 uptake by *T. rubrum* has also been reported[20] and more recently uptake has been demonstrated in Methicillin-sensitive *Staphylococcus aureus* (MSSA), Hospital Acquired (HA)-Methicillin-resistant *Staphylococcus aureus* (MRSA) and Community Associated (CA)-MRSA (*Dimaano, unpublished data*).

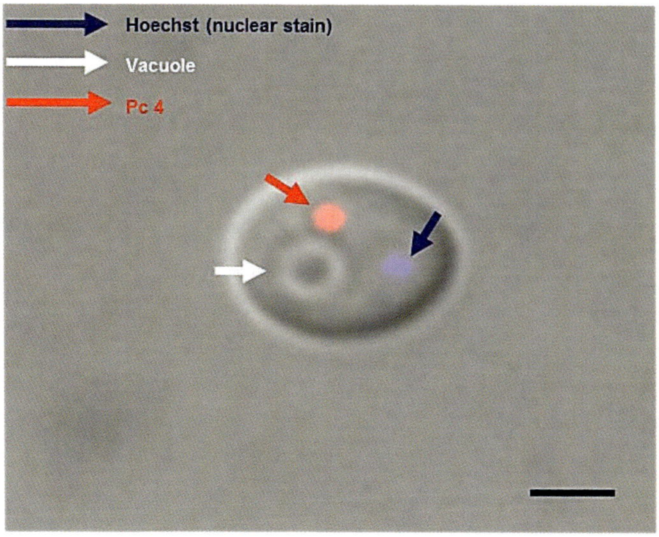

Fig. 2. Pc 4 uptake by *C. albicans*. Pc 4 fluorescence (*pseudo red*) was detected by confocal microscopy, suggesting that it penetrated the cell wall and diffused into the cytoplasmic compartment in *C. albicans* within an hour following 0.2 μM Pc 4 and 10 μg/mL Hoechst 33342 incubation. However, Pc 4 did not co-localize with the nucleus (*pseudo blue*); Scale bar, 5 μm.

5. Pc 4-PDT Affects Metabolic Activity and Induces Intracellular ROS, as Demonstrated XTT Assay and Confocal Microscopy, Respectively

Metabolic activity assay was performed using XTT (2,3-bis[2-methoxy-4-nitro-5-sulfophenyl]-2H-tetrazolium-5-carboxyanilide), (Sigma-Aldrich, St. Louis, MO, USA), which is converted by mitochondrial dehydrogenases of viable cells into a water-soluble orange formazan derivative through the reduction of the tetrazolium ring. A decrease in absorbance is indicative of a loss of metabolic activity. One h following photoirradiation of Pc 4-loaded cells, the metabolic activity of the *C. albicans, T. rubrum* and *S. aureus* cultures was assayed using XTT. The absorbance of the resulting orange solution was measured spectrophotometrically (Spectronic Genesys 5, Analytical Instrument, Golden Valley, MN, USA) at 492 nm. As an independent indicator of yeast metabolic activity and viability, we have also used the fluorescent probe FUN-1 (2-chloro-4-[2,3-dihydro-3-methyl-(benzo-1,3-thiazol-2-yl)-methylidene]-1-phenylquinolinium iodide) (Invitrogen, Carlsbad, CA, USA). This fluorescent probe typically enters vacuoles of viable yeast cells and forms cylindrical intravacuolar structures that fluoresce orange–green.[25,26] Since this process requires ATP, non-viable or metabolically inactive cells cannot incorporate FUN-1 and will exhibit minimal or no fluorescence. We showed that metabolic activity can in fact be reduced by almost 50% within 1 h following photoirradiation of Pc 4-loaded cells, as demonstrated by confocal microscopy in *C. albicans*[10] and *T. rubrum*.[20]

Confocal Microscopy — The formation of ROS was monitored by the conversion of nonfluorescent 2′,7′-dichlorofluorescin diacetate (H_2DCFDA) (Invitrogen, Carlsbad, CA) to fluorescent 2′,7′-dichloro-fluescin (DCF), as described previously.[22,27,28] Briefly, Pc 4-treated *C. albcians, T. rubrum* and *S. aureus* were loaded with 5 μM,2′,7′-dischlorofluorescin diacetate in complete growth medium for 30 min at 30°C. After loading, cells were washed twice with 1 × PBS. Images of DCF fluorescence were collected using a solid-state diode 488-nm laser and a 500–550-nm band-pass barrier filter on a confocal microscope.

We demonstrated the formation of ROS with *C. albicans* (SC5134) following Pc 4-PDT. Confocal images showed DCF fluorescence increased within 15 min after Pc 4-PDT (Fig. 3). With the same fluorescence probe, we have quantitatively measured the amount of intracellular ROS generation by performing a dose response (Pc 4 at concentration of 1.0 μM and light dose ranging from 0.3, 0.9, 1.5 to 2.0 J/cm^2) by flow cytometry with two strains of *T. rubrum*.[20] Our consistently showed the higher the Pc 4-PDT dose the

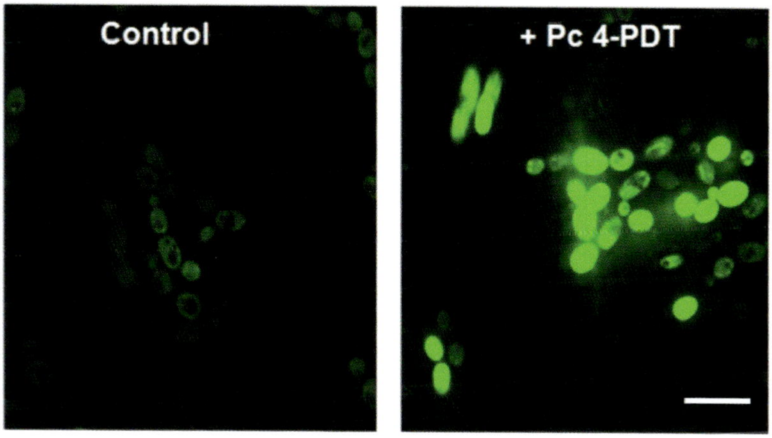

Fig. 3. Pc 4-PDT generates intracellular ROS — Pc 4-PDT induces intracellular ROS in *C. albicans*. To monitor ROS generation by confocal microscopy, Pc 4-treated *C. albicans* were loaded with 2′,7′-dichlorofluorescein, as described previously.[20] Shown are confocal microscopy images collected before and 15 min after Pc 4-PDT; Scale bar, 30 μm.

greater DCF fluorescence or ROS level in both 24602 (terbinafine sensitive) and MRL-666 (terbinafine resistant) cells.

Summary

Although PDT was initially developed to treat different forms of cancer, this minimally invasive therapy can also potentially be employed for the treatment of other conditions, such as opportunistic, multidrug-resistant pathogenic yeast-like fungi, dermatophytes and bacteria. In fact, given its cytotoxic effect on microorganisms, PDT is considered by many as a promising antimicrobial agent.[29–31] However, the mechanism of cell death by PDT in these microbes is not well defined. We have recently demonstrated that PDT with Pc 4 is effective against *C. albicans/non-albicans, T. rubrum* and *S. aureus.*

Pc 4 is shown to be readily taken up by these microbes within an hour of incubation. Immediately following photoirradiation, metabolic activities were attenuated, as measured by XTT and FUN-1.[10,20] Intracellular ROS formation could also be monitored by both confocal microscopy and flow cytometry.[20] Overall cytotoxicity was quantified by CFU assay for numerous strains of *C. albicans/non-albicans, T. rubrum* and *S. aureus* (Table 1).

Infections caused by opportunistic fungal pathogens, such as *C. albicans,* remain a huge healthcare problem, from limited skin and mucosal involvement to systemic life-threatening disease. In addition, infections with the more aggressive candidal species, such as *C. glabrata*, are now prevalent and largely responsible for the increased mortality associated with candidiasis in hopitalized patients. Invasive candidiasis is usually associated with indwelling medical devices, including dental implants catheters, and artificial joints, which can act as substrates for biofilm growth. Resistance to widely-utilized conventional antifungal therapy further increases the mortality rate, particularly in immunocompromised individuals. Therefore, alternative remedies are currently explored and warranted. PDT, which involves administering a photosensitizing drug followed by delivering light, may be employed as a novel therapeutic intervention or in combination with the current antifungal therapy to create a synergistic cytotoxicity effect. Moreover, our photosensitizing drug, Pc 4, has been shown (i) to induce programmed cell death, or apoptosis, following photoirradiation and (ii) to have no direct contact with DNA; thus, gene mutations that give rise to resistance are less likely to occur. Opportunistic yeast-like fungi, such as *C. albicans*, and bacteria, including *Streptococcus pneumonia*, have been found to possess the cellular makeup for apoptosis similar to mammalian cells[32,33]; however, the mechanism of initiation of PDT-induced apoptosis in pathogenic candida and bacterial organisms is not well characterized. Most importantly, we must also identify the endogenous molecular switches that trigger fungal and bacterial apoptotic pathways and reveal the known components that are unique or highly divergent from the mammalian counterparts. Essentially, the knowledge will fuel the possibility of developing ideal novel antifungal and antibacterial agents that would clear infections by activating the suicide pathway and causing minimal harm to the hosts.

At the same time, we are moving forward to characterize the fundamental mechanism of cell death induced by Pc 4-PDT in dermatophytes and *S. aureus* in clinical isolates. Furthermore, the effect of Pc 4-PDT on biofilms formed by *C. albicans* and *S. aureus* are currently being investigated in the laboratory.

Acknowledgments

This study was supported in part by the National Institutes of Health Grant (5P30AR039750) *via* the Skin Diseases Research Center (SDRC) and the Ohio Department of Development — Center for Innovative Immunosuppressive Therapeutics (TECH 09-023). We thank Dr. Malcolm E. Kenney (CWRU, Department of Chemistry) for providing Pc 4; Dr. Curtis J. Donskey (VA, Geriatric Research Education and Clinical Center) for MSSA, HA-MRSA and CA-MRSA. We also would like to thank Dr. Scott J. Howell for his comments on the manuscript and Mr. Yuntao Li and Ms. Sara Lin for their technical assistance.

References

1. P. Babilas, S. Schreml, M. Landthaler, and R. M. Szeimies, *Photodermatol. Photoimmunol. Photomed.* **26**, 118–132 (2010).
2. S. H. Ibbotson *Br. J. Dermatol.* **146**, 176–188 (2002).
3. F. Schmitt and L. Juillerat-Jeanneret, *Anticancer Agents Med. Chem.* **12**, 500–525 (2012).
4. M. Wainwright, Photodynamic antimicrobial chemotheapy (PACT). *J. ANtimicrob. Chemother.* **42**, 13–28 (1998).
5. T. J. Dougherty, *J. Laser Med. Surg.* **20**, 3–7 (2002).
6. N. L. Oleinick, R. L. Morris, and I. Belichenko, *Photochem. Photobiol. Sci.* **1**, 1–21 (2022).
7. Y. U. Baek, Y. R. Kim, H. S. Yim, and S. O. Kang, *FEBS Lett.* **556**, 47–52 (2004).
8. A. Fraser and C. James, *Trends Cell Biol.* **8**, 219–221 (1998).
9. B. Hao, S. Cheng, C. J. Clancy, and M. H. Nguyen, *Antimicrob. Agents Chemother.* **57**, 326–332 (2013).
10. M. Lam *et al.*, *Photochem. Photobiol* **87**(4), 904–909 (2011).
11. F. Madeo *et al.*, *Curr. Opin. Microbiol.* **7**, 655–660 (2004).
12. S. A. Mousavi and G. D. Robson, *Fungal Genet. Biol.* **39**, 221–229 (2003).
13. D. L. Richie, *et al.*, *Mol. Microbiol.* **63**, 591–604 (2007).
14. He, J. *et al.*, *Photochem. Photobiol.* **65**, 581–586 (1997).
15. T. J. Dougherty, *et al.*, *J. Natl. Cancer Inst.* **90**, 889–905 (1998).
16. N. L. Oleinick and H. H. Evans, *Radiat. Res.* **150**, S146–156 (1998).
17. M. J. Egorin, *et al.*, *Cancer Chemother. Pharmacol.* **44**, 283–294 (1999).
18. N. L. Oleinick, A. R. Antunez, M. E. Clay, B. D. Rihter, and M. E. Kenney, *Photochem. Photobiol.* **57**, 242–247 (1993).
19. E. D. Baron, *et al.*, *Lasers Surg. Med.* **42**, 728–735 (2010).
20. Lam, M. *et al.*, *Antimicrob. Agents Chemother.* **58**, 3029–3034 (2014).
21. F. Pereira Gonzales and T. Maisch, *Fungal Biol.* **116**, 1–10 (2012).
22. M. Lam, N. L. Oleinick, and A. L. Nieminen, *J. Biol. Chem.* **276**, 47379–47386 (2001).
23. N. S. Trivedi, H. W. Wang, A. L. Nieminen, N. L. Oleinick, and J. A. Izatt, *Photochem. Photobiol.* **71**, 634–639 (2000).
24. J. Usuda, *et al.*, *J. Biol. Chem.* **278**, 2021–2029 (2003).
25. J. Chandra, P. K. Mukerjee, and M. A. Ghannoum,, *Nat. Protoc.* **3**, 1909–1924 (2008).
26. P. J. Millard, B. L. Roth, H. P. Thi, S. T. Yue, and R. P. Haugland, *Appl. Environ. Microbiol.* **63**, 2897–2905 (1997).
27. T. L. Dawson, G. J. Gores, A. L. Nieminen, B. Herman, and J. J. Lemasters, *Am. J. Physiol.* **264**, C961–C967 (1993).
28. A. L. Nieminen, A. M. Byrne, B. Herman, and J. Lemasters, *J. Am. J. Physiol.* **272**, C1286–C1294 (1997).
29. J. P. Lyon, L. M. Moreira, P. C., de Moraes, F. V. dos Santos, and M. A. de Resende, *Mycoses* **5**, e265–275 (2011).
30. P. Calzavara-Pinton, M. T. Rossi, R. Sala, and M. Venturini, *Photochem. Photobiol.* **88**, 512–522 (2012).

31. F. Javed, L. P. Samaranayake, and G. E. Romanos, *Photochem. Photobiol. Sci.* **13**, 726–734 (2014).
32. M. Ramsdale, *Biochim. Biophys. Acta.* **1783**(7), 1369–1380 (2008).
33. K. Lewis, *Microbiol. Mol. boil. Rev.* **64**(3), 503–514 (2000).

Chapter 19

Photofrin-PDT from Bench to Bedside: Some Lessons Learned

Julia Levy and Ed Levy

Executive Chair of Scientific Advisory Board (retired), Quadra Logic Technologies (QLT Inc.), Vancouver, BC, Canada

Abstract

There are numerous clinical and scientific individuals carrying out Photodynamic Therapy (PDT) on patients throughout the world, but most have no idea how to go about obtaining permission to use new drugs or treatments for these patients. In this paper one example of this process by a new, small company is described. You will find that approval for patient treatment is daunting even when you think you are doing this correctly.

1. Introduction

Thomas Dougherty, a chemist working in a biological research group at the Roswell Park Cancer Institute in Buffalo, launched photodynamic therapy (PDT) as a treatment modality in the 1970s using the drug hematoporphyrin derivative (Hpd) — see Fig. 1 for a timeline. The Hpd he used was created from material originally made by Sam Schwartz, a physician at the University of Minnesota, who was using PDT at the Mayo Clinic for tumor detection. Dougherty's pioneering initiative and its sequlae deviated from the normal trajectory of drug development as practiced by large pharmaceutical companies. The product that was eventually approved as the first PDT drug was the active component of Hpd and was isolated by Dougherty in 1981 and named Photofrin (porfimer sodium). Photofrin was not chemically defined, consisting of a myriad of oligomers of hematoporphyrin and the mixture was unstable at temperatures above 2°C. Even more unusual was the fact that initial human treatments were undertaken without the blessing of the U.S. Food and Drug

1974	First patient ever treated with PDT. Drug used was hematoporhyrin derivative (HpD). Treatment conducted by Tom Dougherty at Roswell Park.
1979	After treating 30 patients, Dougherty is challenged by FDA for not having filed an IND.
1980	Dougherty and colleagues form the company, Oncology Research and Development (ORD).
1981	ORD develops Photofrin, the active component contained in HpD.
1982	Photofrin patent filed.
1984	J & J purchases rights to Photofrin by acquiring ORD. The PDT program is housed in a J&J subsidiary first called Photofrin Medical and later, Photomedica.
1985	First patent on Photofrin issues.
1986	J & J announces its intention to divest Photomedica.
1987	J & J sells Photomedica to QLT; in parallel QLT partners with Lederle laboratories of American Cyanamid Company.
1993	Approval in Canada for certain forms of bladder cancer (superficial).
1994	Approval in Japan for early stage non-small cell lung cancer.
	April — NDA submitted to FDA.
	September — QLT Reacquires all Photofrin Rights from American Cyanamid Oncologic Drug Advisory Committee meeting.
1995	Approval in U.S. for palliation of obstructive esophageal cancer.
1998	Approval in U.S. for early stage non-small cell lung cancer.
2000	QLT sells Photofrin rights to Axcan Pharma.
2003	Approval in U.S. for Barrett's Esophagus.

Fig. 1. Photofrin timeline.

Administration (FDA) because Dougherty and colleagues believed all aspects of the drug and treatment were in the state of New York and thus not within federal jurisdiction. The history of those early years of PDT with Photofrin have been documented elsewhere, but the development story that resulted in regulatory approvals in multiple jurisdictions has received less attention.[a] We are not attempting to produce a manual for drug development or to provide a systematic account of the development of Photofrin. Instead we discuss some aspects of the development process and highlight seven lessons learned by a very small start-up biotech company allied with a major pharmaceutical company. Although the story is centered on PDT with Photofrin, we think many features are similar to experiences by any small company taking a drug candidate from the laboratory bench to the bedside.

[a]The early stages of the Photofrin story are presented in T. J. Dougherty 2007. A personal history of photodynamictherapy. In *Regional Cancer Therapy*, ed. P. Schlag, 117–131. Humana Press, and in T. J. Dougherty, The Circuitous, Serendipitous Route to a New Cancer Therapy Memoirs of an Accidental Pioneer (in press). ("Personal" and "Memoir", respectively.)

In spite of the unusual process in introducing PDT to the medical community, the reported results of cancer treatments with the procedure were promising enough to spark the interest of Johnson and Johnson (J & J), a large and profitable pharmaceutical company. In 1984, J & J licensed Photofrin from Oncology Research and Development (ORD), the company Dougherty and his colleagues had formed. After some rocky starts with manufacturing, J & J continued clinical testing of the drug, taking the program through open label Phase II clinical trials.[b] J & J had no cancer franchise at that time although they had engineered the development of methoxsalen, a drug developed for the treatment of cutaneous T-cell lymphoma that involved extracorporeal photopheresis. And J&J had very little experience in dealing with the use of medical lasers, fiber optics and endoscopes required for PDT.

By 1986 J & J decided that they would discontinue development of Photofrin and were in the process of both searching for potential buyers of Photomedica and preparing to shut down the development program, including closing ongoing clinical trials. It was at this point of the story that the company the authors of this chapter were associated with came into the picture. The company was then called Quadra Logic Technologies, but subsequently renamed QLT Inc., the name we will use throughout this chapter.

We will briefly outline how QLT, and its big pharma partner, Lederle Laboratories, became involved and then devote the rest of the chapter to presenting some of the major lessons that a small biotech learned in the development process.[c] We will pay special attention to the differences between academic research and the development of products for the purpose of gaining approval of FDA and other regulatory bodies before they can be marketed.[d] Although it is not surprising to anyone that there are differences between doing research in the academic milieu, and doing the research and development required for commercialization, we believe that some of the devils in the details

---

[b]Normally developing a drug involves conducting three phases of clinical trials, that is trials with human subjects: Phase I trials are relatively small safety studies often conducted on healthy volunteers; phase II trials are dose-ranging; and then phase III are large-scale, comparative trials that often include hundreds or even thousands of patients. (See Lesson 4 below.)

[c]Lederle Laboratories, was a division of American Cyanamid Company (ACY). Due to a series of mergers and acquisitions in the the 1990s, neither American Cyanamid Company nor Lederle are part of today's biopharmaceutical scene.

[d]We will use FDA as a placeholder both for the U.S. Food & Drug Administration and for regulatory bodies in general. As can be seen from the Timeline in Fig. 1, FDA was not the first regulatory body to approve Photofrin, but FDA and the US market are the most important from the perspective of a biopharmaceutical company.

will be less than obvious even to those undertaking development outside of Photodynamic Therapy.

2. QLT Takes a Major Role in PDT[e]

In 1981 four academics from the University of British Columbia (UBC), and a businessperson, incorporated QLT in Vancouver, Canada. QLT was a company in search of a technology that could be turned into a product. Several lines of research and many business opportunities were investigated, but the most promising work was in the laboratory of one of the founding academics and one of the authors of this chapter, JL. JL had a number of monoclonal antibodies directed against a variety of tumor cell lines. Her research included the study of immunotoxins for targeting and killing tumors *in vivo*. She and her colleagues initiated studies using hematoporphyrin conjugated to antibodies directed to tumor-associated antigens. This was QLT's first step into the new field of PDT. She subsequently collaborated with a colleague at UBC, an expert in the field of porphyrin chemistry, Dr. David Dolphin. This collaboration resulted in the discovery of second-generation photosensitizers, including Visudyne (verteporfin).[f]

In 1987 JL gave a talk at the Hamilton Regional Cancer Centre in Ontario, and in informal discussions afterwards, a clinician explained that she was an investigator in a clinical trial using Photofrin-PDT to treat chest wall metastases of breast cancer. A number of her patients were benefiting significantly and she was very concerned that the trial would be terminated because the sponsor, J & J, had halted the development program. On the flight home JL decided to suggest to QLT's CEO, Jim Miller, that QLT look into manufacturing Photofrin and supplying it to Canadian investigators. That way QLT would build relationships with clinicians that would be useful when QLT's own photosenstizer began clinical trials some years hence. Miller, another of the academic founders, was nothing if not an aggressive CEO and his response was that QLT would purchase Photomedica from J & J. There were several hurdles in the way, not the least of which was that QLT didn't have the funds to purchase Photomedica, nor did QLT have the enormous

[e]A much more extensive account of the QLT story and JL's role can be found as part of the Life Sciences Oral History Program. See Jones M. *The life sciences foundation oral history program: Julia levy with Ed levy* (interview February 24, 2012). http://www.biotechhistory.org/wp-content/uploads/2014/07/LSF_OH_Julia_Levy.pdf.

[f]In 2000 Visudyne was approved by the U.S. FDA for the treatment of macular degeneration, a leading cause of blindness among the elderly. See http://www.accessdata.fda.gov/drugsatfda_docs/nda/2000/21-119_Visudyne.cfm.

financial and other resources necessary to mount a full-scale development program. Nevertheless, on December 21, 1987 a three-way deal involving J & J, Lederle Laboratories of American Cyanamid Company, and QLT was concluded.[g] Not only was this deal struck in what was lightening speed by the standards of pharmaceutical deal making, but also the terms were extremely creative. QLT acquired Photomedica. Lederle obtained marketing rights for Photofrin and purchased stock in QLT that underwrote both the acquisition and the development program.[h] In addition QLT engaged Lederle to lead the development, but QLT was to take part in every aspect of development — it was expected to be a true partnership.

3. Lessons

3.1. *Lesson 1: What Did We Get Ourselves Into?*

When QLT acquired the Photofrin development program it was ready to enter Phase III clinical trials, the last step before filing a New Drug Application (NDA), which is U.S. regulatory terminology for an application for marketing approval. Thus, many years of preclinical work and several years of Phase I and Phase II clinical trials had been successfully completed. Lederle followed its standard procedure of creating a project team with members from every discipline that would make a contribution to documents filed with FDA.[i] In effect the project team mirrored the structure of FDA, plus some additions such as marketing and sales.[j] At the outset QLT had few employees and none with drug development experience. As Tom Dougherty observed, "Since QLT had no experience in this area, it was not really clear what QLT was bringing to the party."[k] Nevertheless, the project team included QLTers. (Spoiler alert:

[g]According to the terms of J & J's original acquisition, Dougherty's approval of the purchaser was required. He was pleased to learn that Lederle was an interested party, but for reasons unspecified they terminated discussions. When the three-way deal was nearing completion, J&J asked for Tom's approval, inexplicably without revealing that Lederle was involved. See Tom Dougherty's account in "Personal" and "Memoir".

[h]At least that was the plan. It turned out that the development program cost way more than originally estimated.

[i]We are using FDA as a placeholder both for the U.S. Food & Drug Administration and for regulatory bodies in general. As can be seen from the Timeline, FDA was not the first regulatory body to approve Photofrin, but FDA and the U.S. market are the most important from the perspective of a biopharmaceutical company.

[j]It is important to note that FDA does not take economic considerations into account in its approval process. Neither does it regulate the practice of medicine. FDA's role is to pass judgment on the safety and efficacy of drugs and devices relative to the indications for which approval is sought.

[k]Dougherty, "Personal", p.125.

Although Lederle played a major role in preparing the written NDA by the time the application appeared before FDA's Oncolgic Drug Advisory Committee in 1995, Lederle was no longer in the picture and the defense of the NDA was under the supervision of Alexandra Mancini, QLT's experienced Director of Regulatory Affairs. Overall, the QLT team numbered a couple of hundred, representing almost every area of drug development.).

Early on QLT asked Lederle for an outline of their drug development process. Eventually they sent over a flow diagram that had to be projected onto a very large wall just so the writing was decipherable. The interaction among areas such as toxicology, pharmacology, chemistry and manufacturing, clinical, regulatory, etc. were so complex that we despaired that we would ever be able to make a contribution. (But we did — see Lesson 4 below.)

However, one factor was missing from their diagram: devices. The treatment regimen for PDT is a two-step process requiring the administration of a photosensitizer ("administration" in our case was by intravenous injection) and then using devices to illuminate the tumour with light of a wavelength appropriate for activating the photosensitizer. In the case of Photofrin, the optimal wavelength is red light at about 630 nm. In the FDA scheme of things this meant that PDT is a combination product and would require submissions to different administrative offices (divisions?) of the FDA: the Oncology Division of the Center for Drug Evaluation and Research (CDER) and the Center for Devices and Radiological Health (CDRH).[l] (More about the regulatory implications of combination products in Lesson 3 below.)

Those unfamiliar with the regulatory system may wonder, what's the big deal that PDT needed both a drug and a laser plus fibre optic conductors to deliver the light? It is a big deal. Lederle had no experience with devices. This is the case with many big pharma operations: Working with devices required whole categories of expertise that were not on board, so given a choice between PDT and other opportunities involving stand alone drugs, the familiar is more attractive.[m] In fact QLT had recognized at the outset that device expertise

[l]The third Center at FDA dealing with prospective human medications was the Center for Biological Evaluation and Research (CBER). CBER handles products derived from living material, especially biotechnology products and others related to blood. Since the starting material for Photofrin is animal blood, one might think it would be treated as a biologic, but its application to cancer meant it went to the Oncology Division of CDER.

[m]In fact, the device aspect of PDT was a factor that "turned off" many prospective drug companies in the cancer field for direct and indirect reasons. Directly, they had no expertise in devices, and indirectly the "customers" for PDT were usually not oncologists, whose treatments tended to be chemotherapy. Instead PDT's customers were surgeons or other interventional physicians that worked with machines.

was lacking and could become a barrier. The only person QLT hired from J & J's Photomedica team was their device technician.[n] By default QLT took responsibility for the device part of the project.

So the joint project team began the process of gearing up the development program. From the beginning some Lederle team members considered QLT personnel as interlopers and pointed out that their Lederle job description did not require them to instruct QLT personnel how to do drug development. But there were several others who regarded the QLT role as possibly benefitting the project and assuming some of the workload.[o] (More about QLT's contribution in Lesson 4 below.)

By far the biggest initial task the team faced was to decide which type of cancer indication would be the target of our Phase 3 trials. One of the main things we learned from that process was the diversity of expertise and factors that had to be considered when choosing a target and designing trials. Of course it was necessary to consult medical practitioners, and we did so at various stages, but it was also vital to take account of the basic science and the regulatory requirements. We will return to some of the latter below.

3.2. *Lesson 2: The 3 G's of Drug Development versus the Norms of Academic Research*

Until QLT acquired Photofrin, our "audiences" were primarily other academics, the medical and scientific communities, and investors,[p] with some slight, nodding acquaintance with regulatory systems. Developing Photofrin meant that regulatory requirements, processes, standards, and documentation took center stage. What did that mean on the ground or in the laboratory?

Regulatory systems revolve around 3Gs: Good Manufacturing Practices (GMP), Good Laboratory Practices (GLP), and Good Clinical Practices (GCP). Of course we cannot attempt to outline these extensively here, but we can convey the flavour of GMP and GLP.

---

[n] J & J's Photomedica team put in a bid to complete Photofrin development. We thought their price tag of $10M was much too high. Of course, as it turned out, it would have been a bargain at twice the price.

[o] Most notable among Lederle personnel who welcomed and mentored QLT members of the project were Lou Gura (Project Leader), Dr. Stuart Marcus (Medical Lead), and Dennis Foley (Regulatory Affairs).

[p] QLT had become a public company in 1986 when it was listed on the less than prestigious Vancouver Stock Exchange. By 1989 QLT was listed on the Toronto and NASDAQ Exchanges and delisted from VSE. See R. Voyer and P. Ryan, *The New Innovators: How Canadians Are Shaping The Knowledge-Based Economy* (James Lorimer & Company Ltd., Toronto, 1994), pp. 32–44.

Above we mentioned that JL proposed that QLT manufacture Photofrin and provide it to Canadian physicians whose supply had been cut off by J & J's decision to divest. That was a pipe dream. We had no idea what FDA requirements would have to be met manufacturing under either GMP or GLP. It wasn't simply a matter of chemistry; we could probably have produced the sort of material that other academics were using in their research. But the problem was that this was to be used in human clinical trials conducted under the auspices of FDA and Health Canada. The drug had to be produced under something close to GMP conditions, which meant that we would have to be able to prove and document that the material was within the specifications of the product that had been used in the trials and preclinical studies previously. To show that would have required us to bring on board experts in a wide variety of fields, not only in the usual areas of manufacturing such as formulations, quality control, etc., but also in Photofrin specific fields.

As for GLP, in general activities conducted under the rules of Good Laboratory Practices compare to academic research roughly what NASA activities around a space launch are to preparations for a family picnic. Of course we are not demeaning the quality of academic research; we are underscoring the level of documentation that goes into GLP research. For example, in the latter, the provenance of every reagent, vessel, and instrument has to be documented, and all of these are subject to audit. All data generated under GLP have to be recorded in specific workbooks that provide a copy of each page. Each page has to be signed off. Standard operating procedures (SOPs) for even the simplest process, the making of a buffer for example, have to be painstakingly documented and followed precisely.

3.3. *Lesson 3: How Significant is it that PDT is a Drug-device Combination and not just a Drug? Some Good News and Some Remaining Challenges*

As discussed above in Lesson 1, PDT being a drug-device combination presented challenges in many areas including prospective partners, clinicians, the logistics of clinical trials, and the regulatory process. However, we inherited something extremely fortunate about the regulatory challenges the drug-device feature that PDT might face: namely, FDA had established coordination and cooperation between the two Centers that would handle the drug and device filings.

The problem faced by many companies developing combination products, especially if the components fell under the jurisdictions of different Centers, was that there was no mechanism in place at FDA to coordinate the two (or more) parts of the application. So, for example, a product consisting of a syringe loaded with a drug or a treatment involving ultrasound enhanced by a drug, would in effect be dealing with two applications that were all but independent from FDA's perspective — neither timing nor analysis was coordinated.

But we had been lucky with PDT. The original interaction with FDA was with the Oncology Division of CDER and when they saw that the treatment involved devices and a drug, they consulted the device division, CDRH. That was the first stroke of good luck; the second was even more significant. It turned out that somewhat earlier CDRH actually conducted research of its own on PDT. So the device specialists could be of major assistance in helping the drug experts come to grips with this new therapy. In effect then the Oncology Division was the primary reviewer and CDRH, especially Richard Felten, acted as consultant to CDER and made recommendations about the device components.

A sequel: After eight years of congressional review of the groundbreaking Medical Device Amendments of 1976, the Safe Medical Devices Amendments of 1990 (SMDA) was enacted.[q] One of the many issues addressed in the Congressional investigations was the matter of combination products. In the lead-up to drafting the bill, the FDA convened a hearing about combination products. Lederle and QLT made a submission at that session. We were very much the exception. Almost every other presentation was a complaint about FDA's lack of coordination in handling combination products. In contrast, the gist of our submission was that whatever changes were made, they should allow for the arrangement that we had and that was working just fine: a combination product would be submitted to a lead office of FDA and other offices that had jurisdiction would act as consultants. We were very pleased to see that the approach we were using was embodied in the regulations in SMDA.[r] SMDA even did something fairly rare in bureaucratic circles by removing a silo. The FDA office leading the review of the combination product was empowered to pass judgment on the entire combination product, i.e. to approve or disapprove, no matter how many other offices would have had jurisdiction had the components of the combination been submitted separately.

[q] F. E. Samuel, *Health Affairs* **10**(1), 192–195 (1991).

[r] D. C. Schillinger, The office of combination products: Its roots, its creation, and its role. (2005). http://nrs.harvard.edu/urn-3:HUL.InstRepos:8852096.

3.4. *Some Remaining Challenges: Forging Alliances with Devices Companies*

One of the major challenges the PDT program faced was the acquisition and maintenance of partnerships with device companies. The underlying reason was straightforward: The drug development process is far, far longer, difficult and more uncertain than the process of getting devices on the market. By 1990 PDT had already been in the clinic for over 15 years and even then the size of the expected market was not clear. A large pharmaceutical company could in principle finance device development, but unease with the device world made that problematic for Lederle. We were able to create alliances with two laser companies, Coherent and Laserscope as well as with fiberoptic companies that could produce the catheters with tips that were specially designed for PDT and were further customized for particular treatment sites.[s]

3.5. *Lesson 4: What Could a Very Small, Start-up Company Contribute to a Partnership with a Very Large Company?*

We give two examples of contributions, and more importantly, why they are significant. The first is that the small start-up, being new to the field, can bring fresh approaches by thinking outside the boxes, which was easy for us since we had little idea where the boxes were. One of the main candidates for Phase III trials was esophageal cancer. We had some good Phase II data and there was a clear medical need. The exact requirements in terms of trial design of Phase III pivotal trials were not specified in the regulations, but the gold standard was that they be multi-centered, double-blinded, and comparative. The comparator was either standard-of-care, if there were one, or a placebo, if there were no approved standard of care. The regulations did state that there be Phase III "trials", so it was clear that there must be two. The clinical regulatory universe is divided into "indications", where an oncology indication, for example, is extremely specific with respect cancer type and often sub-type, as well as stage of cancer and order of treatment (first-, second-, third-line).[t] When the joint

---

[s]The fact that we had alliances with two laser companies was significant. The alliances did not apply in Japan, where most everything was unique to that country. Our partner there, Lederle Japan, formed an alliance with one device company. Even though one of our early approvals was in Japan — see timeline — Photofrin barely got off the ground in Japan because the required laser was super-expensive and cumbersome. It is still a mystery to us why Lederle Japan insisted on maintaining that alliance.

[t]Cancer treatments are approved at most for the indication(s) for which phase three trials have been conducted and their labels are so restricted. However, FDA approves drugs but does not regulate the practice of medicine, so many drugs, especially oncology ones, are used "off-label". Companies can only market or promote drugs for their label indications.

project team was preparing to approach FDA with a proposal for phase III trials, a QLTer piped up with this question: could we propose to FDA that we run a non-comparative Phase III on patients with very advanced esophageal cancer for whom there is no treatment option, and a placebo would be unethical, even if technically possible? After a prolonged pause, a senior Lederle person responded that the question was "touchingly naïve" and that a QLT person could pose it to FDA, but no one from Lederle would. At the meeting with FDA, a QLT person did so, and the most senior FDA person there (Robert Temple) stood up, walked around the room and said, "Bring me a trial with 8 such patients and if they can enjoy a glass of beer after your treatment, we will count that as one trial." As a result, we conducted one of the smallest pivotal trials ever, even though by enrolling 12 patients we exceeded the requirement. Conducting such a small trial probably saved the development program at least a year or two.

The second example illustrates that the smaller company may well know how to do some things better than the bigger company, especially early on, because the smaller company has been conducting the scientific research far longer. QLT had in depth understanding of photochemistry and photobiology. Lederle did not. In his memoir Tom Dougherty points out that when J & J took over the manufacture of Photofrin there were serious setbacks due to sterility and other problems. Of course, J & J either had the manufacturing expertise in-house or could access it, but large companies are heavily compartmentalized and there often is not an easy mechanism to cross boundaries to get help.

We encountered a similar situation to J & J's manufacturing issues when Lederle was planning a toxicology study that involved performing surgery on dogs that had received an injection of Photofrin. QLT was concerned about the possibility that surgical lighting could activate the photosensitizer and cause unwanted photodynamic effects. Lederle reminded us, emphatically, that they had vast experience in running toxicological experiments and QLT did not. The results were not as bad as we thought; they were much worse. Not only were the surgical lights bright, but also the operating theatre had skylights and it was a sunny day. Our canine subjects were badly injured. And as a result, we had to conduct additional studies to show that when proper precautions are taken, there are no cardiac side effects of PDT. The cost of these studies, like almost all those conducted under GLP, are more likely to run into the hundreds of thousands of dollars than in tens of thousands. (And of course this is in addition to the suffering of the canine subjects.)

To conclude this section, we want to emphasize that we are not lionizing our little company at Lederle's expense. The examples we provide here were the exceptions. Without our partner's expertise, mentorship and dedication, we could not have come close to succeeding.

3.6. *Lesson 5: Why Wouldn't a Company Make its Drug Widely Available to Scientific Researchers?*

Almost all of the progress that Photofrin had made toward becoming a treatment in medical practice had been due to the efforts, insights and discoveries of academic researchers. And most of the early members of QLT's PDT team were academics, so one of the first controversies we had with Lederle was centered on their efforts to curb the distribution of Photofrin to academic researchers. It took us a while to realize the rationale for their view, but in the end all agreed on a compromise. It is instructive to examine their reasoning.

At first we thought that it was only a matter of control for the sake of control, but eventually we began to appreciate their point. No doubt the desire to control was part of the story, but the overriding consideration is much more significant: The most ethical thing one can do in drug development is get the product approved and available to patients, of course only if the product stands up to the required standards.[u] Put another way, that means that any delay in the development process is denying medication to patients who need it.

We QLTers were focused on the fact that new research on Photofrin could provide information that could contribute to the success of the development program. But what is the link between delays and providing research material to academic investigators? Part of the answer goes back to control. Researchers request a product in development and usually have to sign a Material Transfer Agreement (MTA) that specifies, among other things, how the material will be used. However, there is no guarantee that the material will actually be used in the way specified.

The compromise was fair in our view, but not without its negative outcomes. For example, contrary to the MTA, a researcher to whom we sent Photofrin used it as a control in developing a new test. The result raised the possibility that Photofrin had a characteristic that had not been seen in any

[u]Clearly in this case the ethical consideration is also aligned with monetary ones, since the sooner the product is on the market, the sooner there are financial returns. However, we believe the ethical considerations play a significant role and are often neglected in discussion of drug development.

of the numerous previous studies. Of course, if this were correct, then it was something that had to be taken into consideration. If the study had been under our control, we would have first validated the new test and the results before publishing the outcome. Neither the test nor the result stood up in the long run, but every research result has to be reported and considerable effort had to be expended in conducting studies and submitting reports to convince regulatory authorities that there was no reason for concern.

In spite of such examples the compromise reached with Lederle resulted in Photofrin being made much more widely available to researchers than is usually the case with big pharmaceutical companies. The development effort was helped immensely by the work of a multitude of academic researchers who elucidated every aspect of Photofrin ranging from chemical features to mechanisms of action and beyond, all of which provided more information about Photofrin for designing trials and for the all-important interactions with FDA.

3.7. *Lesson 6: How do you Deal with the Academic Refrain that you Need a Better Molecule?*

It is a truism that almost every academic paper concludes that more research is needed. In the PDT world, the analog of that conclusion repeated at every PDT conference was, "We need a better photosensitizer." Of course in an important sense that conclusion is absolutely true. Photofrin had its shortcomings, as did the photosensitizer we developed as Visudyne. But this conclusion can also be deeply misleading, and counterproductive. We recall a meeting of mainly academic PDT researchers in Budapest when a respected chemist who had worked in both academia and industry stood up in exasperation and gave our colleagues a vigorous lecture explaining what was being misunderstood. The basic point is that if you are going to actually develop a drug, you have to "freeze" work on features of the candidate in order to achieve its development. Otherwise you will find yourself on a constant merry-go round and never get to approval! This is a tricky point. First we will note what "freezing" does not mean and then say what it does.

"Freezing" features of the program does not mean that research stops when development starts. In fact research continues all the way through development to approval and beyond. Certainly, much of the experimentation undertaken in development, especially when it is taking place in a highly regulated environment, is strictly the application of standard tests to new potential products. But in addition there can be, and in our case was, a plethora of research that can be called "use-inspired basic research."[V] That is, research undertaken to elucidate the behavior, operation and effects of the candidate drug in the context of its use, namely as a medicinal agent. It was such research that showed Photofrin's lipophilic nature, that it worked by activating singlet oxygen, and many other properties. In fact we would argue that Photofrin succeeded where several other photosensitizers failed precisely because of the quality of the research from the beginning and throughout development.

But "freezing" aspects of the program does mean something important. Basically it is a reflection of the fact that the development of a drug takes on average more than a decade, and changing course during the process can return the program to ground zero. We already saw a hint of that when we discussed manufacturing processes and noted that for regulatory purposes even what some would regard as an ever so slightly modified product is regarded by regulators as being a different entity requiring that toxicological and other work be redone on the "new" material. The point then is that if you continually switch to the next best molecule, then you will never, ever get anything approved.

A paradigm example of the importance of "freezing" was experienced by QLT when the company was preparing the NDA filing documents for Visudyne. Visudyne is a benzoporphyrin derivative composed of two enantiomers in equivalent concentration. On High Pressure Liquid Chromatography (HPLC) the two peaks did not resolve completely. One of the chemists in the company's manufacturing group decided to try a slightly different buffer to see if he could resolve the two peaks completely (this was after the process had been frozen so the scientist was really doing this out of curiosity). Unfortunately, when the HPLC was run, not only were the two enantiomers fully resolved, but two unique small peaks also appeared as impurities.

This was not a disaster because it was assumed that all batches of the drug used clinically also contained the two small impurities and the drug had been shown to have no intrinsic toxicities. But what the company was obliged to do at this point was to go back over every batch of Visudyne that had ever been used in the clinic or in toxicology experiments and demonstrate that the impurities were indeed present in all of them. In addition, the two impurities had to be fully characterized. They turned out to be two very minor molecules that were by-products of the manufacturing process. This extra work actually delayed

[V] In Pasteur's Quadrant, Donald Stokes identifies "use-inspired science" as a category of investigation that falls between pure applied and pure basic research. Having divided the universe of scientific investigation into quadrants, Stokes claims that use-inspired science is exemplified by Pasteur and thus its location is in Pasteur's quadrant.

the company filing their NDA with the FDA. But the experience underlines the importance of good record keeping and maintenance of samples of all batches.

In the case of Photofrin, late in the development program, the formulation was changed from a liquid to a freeze-dried solid, mainly for ease of handling and stability considerations. For a while it appeared that there were differences in the two formulations and that there were more toxicities associated with the freeze-dried one. There likely would have been an issue even if the alleged difference would have gone the other way, i.e. if there were fewer toxicities with the new formulation. That's because, as with the Visudyne example, the question is whether all the earlier work was actually carried out on the "same" product. At any rate, eventually all were convinced that there was no significant difference.

3.8. *Lesson 7: Riding the Scientific and Emotional Roller-coaster Called the Oncologic Drug Advisory Committee (ODAC)*

When the Lederle-QLT PDT team first met, our colleagues at Lederle appeared to be both exultant and exhausted from just having successfully brought another product through the ODAC, which is a committee of external experts convened to advise FDA on the merits of an NDA package. The QLT contingent thought our colleagues were exaggerating their efforts and over-dramatizing the process. We were wrong. The magnitude of effort required to file and defend an NDA in the Oncology field does not begin to be conveyed by citing such cold facts as these: the actual application consisted of 160 large binders of information. As for understanding the emotional side, Tom Dougherty summed it up, "It was one of the most excruciating experiences of my life."[W] Immediately after the Advisory Committee voted 12 to 1 to recommend to FDA that Photofrin PDT be approved for the palliation of obstructive esophageal cancer, Tom reflected that his... emotions were still bouncing around , and it took me several hours before I realized that the first PDT drug (which came out of our lab and clinic at Roswell Park) had actually been approved by the FDA — 23 years... after we carried out the first PDT experiments in a tiny lab in the basement of the hospital.[X]

[W] Dougherty, "Personal", 126.

[X] Ibid., p.127 Incidentally, the ODAC vote was somewhat confusing in our case because one of the questions posed was in the negative and it was difficult to tell what a positive vote meant.

We'll return to comments about the implications and importance of ODAC and the entire regulatory process, but first we will outline briefly what takes place at an ODAC session. The company, which in regulatory speak is called the "sponsor", submitted its NDA to FDA in April 1994. As is its practice, FDA had torn the binders apart and sent sections to the appropriate internal FDA reviewers: clinical, manufacturing, statistics, toxicology, pharmacokinetics, devices, formulations, etc. Then in September 1994 FDA convened an ODAC meeting to examine the submission. The committee consisted of 13 outside experts who heard QLT's oral defense of its submission — the only time during the session that the sponsor is scheduled to speak. This was followed by FDA's assessment and its questions posed to ODAC. These questions ranged from the quite narrow ones, to those that essentially asked whether the drug, or in this case the combination product, should be approved. After an extended discussion, the Committee votes on its responses to questions, including whether to recommend approval of the package. The non-binding recommendation was for approval and then in December 1994, FDA made its decision and issued an approval letter.

As Sponsors we sat through what seemed to us to be an interminable rambling discussion on the strengths and weaknesses of the project. Many ODAC members are clinical researchers with little knowledge of the process of drug development. The Sponsor is prohibited from interjecting or explaining any aspect of the discussion, so has to sit through sometimes patently incorrect assertions. Dougherty describes the experience as excruciating. He was not exaggerating.

4. Conclusion: Macro Reflections on the Drug Development Process

We conclude by reflecting on the drug development process as a whole. In the last decade there have been many trenchant criticisms of the biopharmaceutical industry.[y] All aspects of the industry and all categories of participants, including companies, regulatory bodies, and physicians, have been targeted. We emphasize that we believe many of these criticisms are well deserved. For example, in our view the magnitude of funds expended on sales and marketing is obscene, as are many of the ways in which those funds are used. However, we also believe

[y] Among the best known are ones by Angell and by Guzman. M. Angell, *The truth about the drug companies: how they deceive us and what to do about it* (1st edn.). (New York, Random House, 2004). And M. Goozner, *The $800 Million Pill: The Truth Behind the Cost of New Drugs.* (University of California, Berkeley, CA, Press, 2004).

that one class of criticisms of the drug development process is unfounded, namely charges that the process is in effect rigged because data are supplied by companies and are thus tainted and entirely suspect or worse. To be clear we certainly agree that every aspect of drug development could be improved: Choice of comparative treatments in clinical trials, the way "indications" are manipulated so as to permit an excess of me-too drugs, the focus on profit rather than need, etc. Nevertheless, at least in the case of new molecular entities aimed at the treatment of life-threatening conditions such as cancer, the studies, trials, data, and experiments submitted in support of an application are on balance genuine basic and applied science and so too are judgments about their scientific adequacy. Some of that is conveyed in our description of our ODAC meeting. To be sure, we see shortcomings there too: Some of the experts are not well briefed and seem not to be aware of the implications of their suggestions such as proposing additional clinical trials.

One way to convey that the drug development process is more rigorous than rigged is to pose what might appear to be a tangential question: Why do so many drugs fail even after they begin human clinical trials? Years of work and millions of dollars go into the development of a drug before it is actually tested in humans in clinical trials. And yet the overwhelming majority of drug candidates that begin clinical trials will fail to make it all the way through to approval. According to FDA statistics, roughly one in four or five drugs that begin clinical testing actually end up on the market.[z] Why do more than 75% of the drugs fail even at the clinical stage of drug development?

Sometimes trials are poorly designed or poorly administered and sometimes there are economic or market reasons for terminating a program, but the main factor is that not enough is known about the drug or the disease, or how the drug, its breakdown products and metabolites behave within the enormously complex system that is the human body.[aa] That is, drugs are largely defeated by Nature, or rather our lack of knowledge about Nature.

If the current drug development and clinical trial systems were as subject to corporate manipulation as some critics suggest, then companies would be guilty of being part of a vast conspiracy of dunces. There are of course enormous financial stakes involved in drug development and such stakes produce

---

[z]Of 100 drugs for which Investigational New Drug applications are submitted to FDA, about 70% will successfully complete phase 1 and go on to phase 2; about 33% of the original 100 will complete phase 2 and go to phase 3; 25 to 30 of the original 100 will clear phase 3; and about 20 of the original 100 will ultimately be approved for marketing. See K. Flieger, *Testing Drugs in People* (2000). URL: http://www.fda.gov/cder/about/whatwedo/testtube-3a.pdf.
[aa]I. Kola and J. Landis, *Nature Rev. Drug Discov.* **3**, 711–715 (2004). 2. **3**(2), 711–715.

distortions.[ab] But there is a multitude of scientists, technicians, officials, and executives in regulatory bodies, in companies, and in the universities where most of the initial discoveries are made, working honestly and diligently to evaluate the merits of drug candidates.

We are proud to have been a part of the drug development process. We are grateful to those who prepared the way before our company came on the scene, and we thank the hundreds of colleagues, partners, regulatory reviewers, physicians, nurses, patients and investors, and all others who contributed to bringing Photofrin from the laboratory bench to the bedside.

Acknowledgments

We would like to thank Tom Dougherty, Rashid Aziz, and Emily Marden for many helpful comments.

About the Authors

During the period that is the focus of this article – 1987–1995 – Julia Levy was initially Chief Scientific Officer of QLT and then, acting CEO and eventually, CEO. Initially, Ed Levy was the leader of QLT's PDT Project team, and then VP of Corporate Development. Julia Levy's academic career was in the field of immunology and continued throughout the period. Ed Levy taught philosophy of science until joining QLT in 1988.

[ab]The sizes of costs and returns arecontested, but no one argues they are small. One thing confounding estimates of costs of drug development is the scope of what is taken into account. In our experience the cost of developing a successful drug — which can be estimated by looking at the balance sheets of a biotech that has developed one drug — is in the range of $100–$200 M. But if we factor in the costs of failed drugs, the amount is doubled. And if we consider opportunity costs, the amount is doubled again. The last two estimates are usually based on less than full disclosure by large companies.

Index